T0325619

Handbook of Research on Developments and Trends in Wireless Sensor Networks:
From Principle to Practice

Hai Jin
Huazhong University of Science and Technology, China

Wenbin Jiang
Huazhong University of Science and Technology, China

INFORMATION SCIENCE REFERENCE

Hershey · New York

Director of Editorial Content: Kristin Klinger
Director of Book Publications: Julia Mosemann
Acquisitions Editor: Mike Killian
Development Editor: Beth Ardner
Publishing Assistant: Kurt Smith
Typesetter: Devvin Earnest
Quality control: Jamie Snavely
Cover Design: Lisa Tosheff
Printed at: Yurchak Printing Inc.

Published in the United States of America by
    Information Science Reference (an imprint of IGI Global)
    701 E. Chocolate Avenue
    Hershey PA 17033
    Tel: 717-533-8845
    Fax: 717-533-8661
    E-mail: cust@igi-global.com
    Web site: http://www.igi-global.com/reference

Library of Congress Cataloging-in-Publication Data

Handbook of research on developments and trends in wireless sensor networks : from principle to practice / Hai Jin and Wenbin Jiang, editors.
    p. cm.
  Includes bibliographical references and index.
  Summary: "This book showcases the work many devoted wireless sensor network researchers all over world, and exhibits the up-to-date developments of WSNs from various perspectives"--Provided by publisher.
  ISBN 978-1-61520-701-5 (hardcover) -- ISBN 978-1-61520-702-2 (ebook) 1.
Wireless sensor networks--Research. I. Jin, Hai. II. Jiang, Wenbin.
  TK7872.D48H357 2010
  681'.2--dc22
                    2009052441

British Cataloguing in Publication Data
A Cataloguing in Publication record for this book is available from the British Library.

List of Contributors

Table of Contents

Section 1
Introduction and General Issues

Wenbin Jiang, Huazhong University of Science and Technology, China
Hai Jin, Huazhong University of Science and Technology, China
Chen Yu, Huazhong University of Science and Technology, China
Chao Liu, Huazhong University of Science and Technology, China

Ruay-Shiung Chang, National Dong Hwa University, Taiwan
Shuo-Hung Wang, National Dong Hwa University, Taiwan

Yuan He, Hong Kong University of Science and Technology, China
Yunhao Liu, Hong Kong University of Science and Technology, China

Ricardo H. González G., Simon Bolivar University, Venezuela
Antonio A.F. Loureiro, Federal University of Minas Gerais, Brazil
Raquel A.F. Mini, Pontifical Catholic University of Minas Gerais, Brazil

Section 2
Protocols and Middlewares

Section 3
Security and Privacy

Section 4
Practices and Applications

Detailed Table of Contents

Section 1
Introduction and General Issues

Chapter 1

    Wenbin Jiang, Huazhong University of Science and Technology, China
    Hai Jin, Huazhong University of Science and Technology, China
    Chen Yu, Huazhong University of Science and Technology, China
    Chao Liu, Huazhong University of Science and Technology, China

This chapter gives a brief overview of the WSNs, including the basic concepts, architectures and protocols, current developments and trends, diversified applications, etc. It provides readers a basic view of what WSNs looks like wholly and also elementary background knowledge for reading the following chapters.

Chapter 2

    Ruay-Shiung Chang, National Dong Hwa University, Taiwan
    Shuo-Hung Wang, National Dong Hwa University, Taiwan

This chapter studies the deployment problem in wireless sensor networks. The deployment affects the efficiency and the effectiveness of sensor networks. It discusses different types of deployment strategies including deterministic or random deployment, and centralized or distributed deployment. It also reviews the measures of deployment while considering the connectivity and coverage in detail. The best coverage and the worst coverage are also discussed. An important issue is the energy efficiency in wireless sensor networks. It classifies the power conservation issue into 3 types in sensor networks. They are duty-cycle dependent, role dependent, and topology dependent methods. Finally, future trends in sensor deployment are proposed.

Chapter 3

Sensor networks have been widely deployed in all kinds of applications. In the near future, this chapter expects to witness the proliferation of sensor networks with a variety of functions that requires a comprehensive collaboration mechanism among them. Specific designs are necessary to manipulate a fabric of multiple sensor networks, facilitate the collaboration among them, and support efficient data sensing, aggregation, storage, transmission, and query processing. In this chapter, through comparison and analysis of existing works, it presents a new taxonomy of collaborative sensor networks based on the architectures and the methodologies of organizing different modules of a system. It finds that a cross-layer design simultaneously guaranteeing the energy efficiency of underlying sensor networks and the quality of services for users is a challenging issue which has not been well studied. This chapter also discusses the design space in this field and point out potential research directions.

Chapter 4

This work presents a survey of the main aspects of QoS that are being used on the Wireless Sensor Networks technology world. It describes approaches based on traditional networks, as well as new approaches that face with resources limitation, which is one of the main characteristics of WSN nodes. This work also describes how QoS management creates a series of challenges to deal with, and how some QoS features could be used to enable users to make a better profit of their resource limitations. In its opinion, an exposition of these topics will improve the set of techniques and strategies that designers and programmers could use to develop and implement satisfactory services level to their applications in WSNs.

Chapter 5

Maintaining reliable networks of low cost, low energy wireless sensor network (WSN) nodes is a major concern. One way to maintain a reliable network is to perform in-field testing on nodes throughout their lifetimes, identifying failing nodes so that they can be repaired or replaced. This chapter explores the requirements for a wireless test access mechanism, and introduces a method for remote execution of software-based self-test (SBST) programs. In an effort to minimize overall test energy consumption, an SBST method is derived that takes the least amount of microcontroller cycles, and is compatible with system-level optimizations such as concurrent test execution. To further reduce test energy, compression algorithms compatible with WSN nodes are explored for use on test programs. The efficacy of all proposed methods is evaluated experimentally, using current measurement circuitry applied to a WSN node.

Baha' Alsaify, University of Arkansas, USA
Haiying Shen, University of Arkansas, USA

This book chapter discusses power conservation problems in Wireless Sensor Networks (WSNs). The problem arises from the fact that WSNs have limited energy since sensors are powered with small batteries (due to the sensor size constraints). Currently, some energy-efficiency methods focus on reducing energy consumption by designing sensor hardware. Other methods enable sensors communicate with each other in an efficient manner by developing new communication protocols. Some communication protocols need to have extra information such as sensor locations for determining the best possible relays to deliver data to a Base Station. This chapter will first provide a survey on current power conservation methods. It will then discuss the efficiency and effectiveness of these methods, and propose possible solutions. Finally, it concludes the chapter with concluding remarks and open issues.

Section 2
Protocols and Middlewares

Junaid Ansari, RWTH Aachen University, Germany
Xi Zhang, RWTH Aachen University, Germany
Petri Mähönen, RWTH Aachen University, Germany

In this chapter, the practical and experimental aspects of medium access control protocol design for wireless sensor networks are discussed. It outlines the basic design principles of medium access control procedures and the general development trends from the perspective of real world implementations. it takes into account especially energy consumption and latency requirements. it share the authors' experiences gained in practical research and pre-commercial projects, wherein it has been tackling issues such as traffic awareness, co-existence of different protocols, spectrum agility for wireless sensor networks and porting of MAC-layers to different platforms. The contemporary focuses on medium access control design such as cross-layer approaches, multi-radio based protocols and radio wake-up based medium access solutions are also covered in this chapter.

Yunlu Liu, Beihang University, China
Weiwei Fang, Beihang University, China
Zhang Xiong, Beihang University, China

Transmission control is an important issue for supporting Quality of Services (QoS) in Wireless Sensor Networks (WSNs). This chapter gives an introduction to the transmission control relevant protocols and algorithms, often developed for particular applications to reflect the application-specific requirements. According to the functionality, the existing works can be categorized into three main types: congestion control, reliability guarantee and fairness guarantee. Due to the unique constraints of WSNs, these transmission control protocols are not "cleanly" placed on top of the network layer, but call for careful cross-layer design. In the end, some exciting open issues are presented in order to stimulate more research interest in this largely unexplored area.

Chapter 9

Qiang-Sheng Hua, The University of Hong Kong, China
Francis C.M. Lau, The University of Hong Kong, China

This chapter studies the joint link scheduling and topology control problems in wireless sensor networks. Given arbitrarily located sensor nodes on a plane, the task is to schedule all the wireless links (each representing a wireless transmission) between adjacent sensors using a minimum number of timeslots. There are two requirements for these problems: first, all the links must satisfy a certain property, such as that the wireless links form a data gathering tree towards the sink node; second, all the links simultaneously scheduled in the same timeslot must satisfy the SINR constraints. This chapter focuses on various scheduling algorithms for both arbitrarily constructed link topologies and the data gathering tree topology. It also discusses possible research directions.

Chapter 10

    Habib M. Ammari, Hofstra University, USA

This chapter studies duty-cycling to achieve both k-coverage and connectivity in highly dense deployed wireless sensor networks, where each location in a convex sensor field (or simply field) is covered by at least k active sensors while maintaining connectivity between all active sensors. Indeed, the limited battery power of the sensors and the difficulty of replacing and/or recharging batteries on the sensors in hostile environments require that the sensors be deployed with high density in order to extend the network lifetime. Also, the sensed data originated from source sensors (or simply sources) should be able to reach a central gathering node, called the sink, for further analysis and processing. Thus, network connectivity should be guaranteed so sources can be connected to the sink via multiple communication paths. Finally, wireless sensor networks suffer from scarce energy resources. A more practical deployment strategy requires that all the sensors be duty-cycled to save energy. With duty-cycling, sensors can be turned on or off according to some scheduling protocol, thus reducing the number of active sensors required for k-coverage and helping all sensors deplete their energy as slowly and uniformly as possible. It also extends the discussion to connected k-coverage with mobile sensors as well as connected k-coverage in a three-dimensional deployment area. Furthermore, It discusses the applicability of the protocols to heterogeneous wireless sensor networks.

    Tales Heimfarth, Federal University of Rio Grande do Sul, Brazil
    Edison Pignaton de Freitas, Federal University of Rio Grande do Sul, Brazil
    Flávio Rech Wagner, Federal University of Rio Grande do Sul, Brazil
    Tony Larsson, Halmstad University, Sweden

Wireless sensor networks (WSNs) are gaining visibility due to several sophisticated applications in which they play a key role, such as in pervasive computing and context-aware systems. However, the evolution of WSN capabilities, especially regarding their ability to provide information, brings complexity to their development, in particular for those application developers that are not familiar with the technology underlying and needed to support WSNs. In order to address this issue and allow the use of the full potential of the sensor network capabilities, the use of a middleware that raises the abstraction level and hides much of the WSN complexity is a promising proposal. However, the development of a middleware for WSNs is not an easy task. Systems based on WSNs have several issues that make them quite different from conventional networked computer systems, thus requiring specific approaches that largely differ from the current solutions. The proposal of this chapter is to address the complexity of middleware made for sensor networks, presenting a taxonomy that characterizes the main issues in the field. An overview of the state-of-the-art is also provided, as well as a critical assessment of existing approaches.

Section 3
Security and Privacy

    Jianmin Chen, Florida Atlantic University, USA
    Jie Wu, Florida Atlantic University, USA

Many secure mobile ad hoc networks (MANETs) and wireless sensor networks (WSNs) use techniques of applied cryptography. Numerous security routing protocols and key management schemes have been designed based on public key infrastructure (PKI) and identity-based cryptography. Some of these security protocols are fully adapted to fit the limited power, storage, and CPUs of these networks. For example, one-way hash functions have been used to construct disposable secret keys instead of creating public/private keys for the public key infrastructure. This survey of MANET and WSN applications presents many network security schemes using cryptographic techniques and give three case studies of popular designs.

Chapter 13
Riaz Ahmed Shaikh, Kyung Hee University, Korea
Brian J. d'Auriol, Kyung Hee University, Korea
Heejo Lee, Korea University, Korea
Sungyoung Lee, Kyung Hee University, Korea

Until recently, researchers have focused on the cryptographic-based security issues more intensively than the privacy and trust issues. However, without the incorporation of trust and privacy features, cryptographic-based security mechanisms are not capable of singlehandedly providing robustness, reliability and completeness in a security solution. Tthis chapter presents generic and flexible taxonomies of privacy and trust. It also gives detailed critical analyses of the state-of-the-art research, in the field of privacy and trust that is currently not available in the literature. This chapter also highlights the challenging issues and problems.

Chapter 14
Juan Hernández-Serrano, Universitat Politècnica de Catalunya, Spain
Juan Vera-del-Campo, Universitat Politècnica de Catalunya, Spain
Josep Pegueroles, Universitat Politècnica de Catalunya, Spain
Miguel Soriano, Universitat Politècnica de Catalunya, Spain

Wireless sensor networks (WSNs) are made up of large groups of sensor nodes that usually perform distributed monitoring services. These services are often cooperative and interchange sensitive data, so communications within the group of sensor nodes must be secured. Group key management (GKM) protocols appeared and were broadly studied, in order to ensure the privacy and authentication throughout the group life. However, GKM for WSNs is already challenging due to the exposed nature of wireless media, the constrained resources of sensor nodes and the need of ad-hoc self-organization in many scenarios. This book chapter presents the basis of GKM and its state-of-the art for WSNs. It analyzes the current non-resolved topics and presents a GKM proposal that solves some of these topics: it minimizes both the rekeying costs when the group membership changes and the routing costs within the group.

Chapter 15
Yan-qiang Sun, National University of Defense Technology, China
Xiao-dong Wang, National University of Defense Technology, China

Guaranteeing security of the sensor network is a challenging job due to the open wireless medium and energy constrained hardware. Jamming style Denial-of-Service attack is the transmission of radio signals that disrupt communications by decreasing the signal to noise ratio. These attacks can easily be launched by jammer through either bypassing MAC-layer protocols or emitting a radio signal targeted at blocking a particular channel. This chapter surveys different jamming attack models and metrics, and figure out the difficulty of detecting and defending such attacks. It also illustrates the existed detecting strategies involving signal strength and packet delivery ratio and defending mechanisms such as channel surfing, mapping jammed region and timing channel. After that, It explores methods to localize a jammer, and proposes an algorithm Geometric-Covering based Localization. Later, It discusses the future research issues in jamming sensor networks and corresponding countermeasures.

Section 4
Practices and Applications

Chapter 16

Brian J. d'Auriol, Kyung Hee University, Korea
Sungyoung Lee, Kyung Hee University, Korea
Young-Koo Lee, Kyung Hee University, Korea

Wireless sensor networks can provide large amounts of data that when combined with pre-processing and data analysis processes can generate large amounts of data that may be difficult to present in visual forms. Often, understanding of the data and how it spatially and temporally changes as well as the patterns suggested by the data are of interest to human viewers. This chapter considers the issues involved in the visual presentations of such data and includes an analysis of data set sizes generated by wireless sensor networks and a survey of existing wireless sensor network visualization systems. A novel model is presented that can include not only the raw data but also derived data indicating certain patterns that the raw data may indicate. The model is informally presented and a simulation-based example illustrates its use and potential.

Chapter 17

Natalija Vlajic, York University, Canada
Dusan Stevanovic, York University, Canada
George Spanogiannopoulos, York University, Canada

The use of sink mobility in wireless sensor networks (WSN) is commonly recognized as one of the most effective means of load balancing, ultimately leading to fewer failed nodes and longer network lifetime. The aim of this book chapter is to provide a comprehensive overview and evaluation of various WSN deployment strategies involving sink mobility as discussed in the literature to date. The evaluation of the surveyed techniques is based not only on the traditional performance metrics (energy consumption, network lifetime, packet delay); but, more importantly, on their practical feasibility in real-world WSN applications. The chapter also includes sample results of a detailed OPNET-based simulation study. The results outline a few key challenges associated with the use of mobile sinks in ZigBee sensor networks. By combining analytical and real-world perspective on a wide range of issues concerning sink mobility, the content of this book chapter is intended for both theoreticians and practitioners working in the field of wireless sensor networks.

This chapter first provides a brief survey on previous research works on network-wide broadcast services. It then revisits the network-wide broadcast problem by remodeling it with active/dormant cycles and showing the practical lower bounds for the time and message costs, respectively. It also proposes an adaptive algorithm named RBS (Reliable Broadcast Service) for dynamic message forwarding scheduling in this context, which enables a reliable and efficient broadcast service with low delay. The performance of the proposed solution is evaluated under diverse network configurations. The results suggest that the proposed solution is close to the lower bounds of both time and forwarding costs, and it well resists to the network size and wireless loss increases.

Chapter 19

Along this paper, an indoor positioning system that uses wireless ZigBee technology is presented an evaluated. In this system, mobile wireless devices measure the level of the received signal from reference nodes, which are placed in well-known positions. With this information, the position is estimated and presented to the user in a graphical way. A precision study is presented, being this study done in function of positions and numbers of reference nodes. It is also analyzed the presence of obstacles in the system, and a study of maximum distance inter nodes that allow positioning with a minimum of quality in the results is done.

Chapter 20

As sensor network deployments grow and mature there emerge a common set of operations and transformations. These can be grouped into a conceptual framework called Sensor Web. Sensor Web combines cyber infrastructure with a Service Oriented Architecture (SOA) and sensor networks to provide access to heterogeneous sensor resources in a deployment independent manner. This chapter presents the Open Sensor Web Architecture (OSWA), a platform independent middleware for developing sensor applications. OSWA is built upon a uniform set of operations and standard data representations as defined in the Sensor Web Enablement Method (SWE) by the Open Geospatial Consortium (OGC). OSWA uses open source and grid technologies to meet the challenging needs of collecting and analyzing observational data and making it accessible for aggregation, archiving and decision making.

Preface

Over the past half a century, our real world has been digitalized more and more deeply. Computer, communication, and control technologies have enormously changed our life styles. We can let computers play chess with world-class players by artificial intelligence. We can forecast coming weather of any city in several minutes by super computing. We also can contact any people whenever we want, and wherever we are, by wireless communication. When we immerse in this wonderful digital world happily, some flaws of this life mode have been ignored to some extent. Which are the ways of how to make the digital world touch the real world freely, effectively, and roundly. At present, most data that the digital world requires is collected and preprocessed manually by some special ways before it can accept them. How to make the digital world understand the real world is seamlessly an attractive research area. The appearance of the Wireless Sensor Network (WSN) has provided a good way for eliminating the gap between the digital world and the real one.

The WSN, as one of the top ten emerging technologies for the 21st century (MIT's Technology Review stated in 2003), has been developing at an accelerated pace in the past ten years. Much research work has been done to push it forward, referring to various aspects, including architecture, operating systems of nodes, routing protocols, data gathering, fusion, location mechanism, time synchronization, and so forth. Moreover, large numbers of promising applications have emerged and been deployed over various geographical areas such as infrastructure protection, scientific exploration, military surveillance, traffic monitoring and controlling, mining and shipping security, environment protection, animal tracking, military affairs, and so forth. With the conveniences brought by WSNs, our lives have been affected and changed largely in many ways.

However, there are still many challenges existing for WSNs. Typical ones include the unreliable wireless communication systems, large scale deployments, limited power availability, failures of nodes, and so forth. Attempting to conquer these difficulties, various routing shemes, power management policies, and data dissemination methods have been designed and implemented for WSNs. To view the latest developments, and to provide a seedbed for new breakthroughs for these challenges, this book brings the latest achievements and excellent studies of WSNs together.

General, this book is organized as four sections as follows:

Firstly, Section 1 introduces WSNs. Briefly including some general issues such as deployment strategies, taxonomy and design space for collaborative sensor networks, design features and challenges for QoS, infrastructure for testing nodes, power conservation techniques, and so forth.

Secondly, main attentions are paid to protocols and middlewares of WSNs in section 2. Topics include practical experiences and design considerations on medium access control protocols, transmission control protocols, joint link scheduling and topology control, connected k-coverage protocols for densely deployed WSNs, middleware support for WSNs, and so forth.

Then, security and privacy are discussed in section 3. Compared with traditional networks, security and privacy are more sensitive topics and harder jobs in WSNs, due to limited resources in them. Prominently, some important subjects are discussed in this book, referring to applied cryptography in secure mobile Ad Hoc networks and WSNs, privacy and trust management schemes, distributed group security, and jamming attacks and countermeasures, and so forth.

Finally, section 4 shows some representative practices and applications. Which are data visualizations, sink mobility, network-wide broadcast service, an indoor positioning system using wireless Zig-Bee technology, an open sensor Web architecture by integrating sensor networks with Web and cyber infrastructure, and so forth.

This book is one in the latest collection of wonderful works from many devoted WSN researchers all over the world, which exhibits the up-to-date developments of WSNs in various aspects. The primary target audience of this book are those who are interested in WSNs and related issues, including scholars, researchers, developers, postgraduate students, et al. In particular, the book will be a valuable companion and comprehensive reference source for students who are taking a course in WSNs. To provide the greatest reading flexibility, this book is organized in self-contained chapters.

Hai Jin
Wenbin Jiang

Acknowledgment

We would like to thank all the students and colleagues who helped us collect and arrange all materials, search various references from the internet and libraries, read our draft chapters, and send us perfect suggestions and ideas. They are: Zhen Tang, Chunqiang Zeng, Qing Long, Hao Li, Lei Zhang, Shuguang Wang, Wei Tong, et al.

We also want to give many thanks to all reviewers who have given professional comments for all chapters. Their excellent work guarantees the quality of this book.

Finally, we also express our thanks to the National Basic Research Program of China (973 Program, No.2006CB303000) which gave great support to this book.

Hai Jin
Wenbin Jiang

Section 1
Introduction and General Issues

Chapter 1
Introduction and Overview of Wireless Sensor Networks

Wenbin Jiang
Huazhong University of Science and Technology, China

Hai Jin
Huazhong University of Science and Technology, China

Chen Yu
Huazhong University of Science and Technology, China

Chao Liu
Huazhong University of Science and Technology, China

ABSTRACT

Wireless sensor networks (WSNs) are becoming increasingly popular, which is changing the way people perceive the world largely, as well as the living styles of human beings. To give readers a basic, wide view of WSNs and make them understood more deeply, this chapter introduces their various aspects briefly, including basic concepts, architectures and protocols, etc. Moreover, it discusses their recent developments, challenges and new trends, based on analysis of many meaningful references. Some classic applications are also shown to approve the popularity of WSNs.

INTRODUCTION

The computer is one of the most important inventions in the 20th century. It has extended the capacities of the brain of human beings hugely during the past 50 years. Many problems that are hard or even impossible for a man's brain have been addressed by computers in seconds. However, how the computer to get data from the real world to process is usually a key problem for scientists and engineers even today. At the early stage, programmers perforate some holes in the paper strips to let computer know what they want it to do (as the input to the computers). Later, keyboard, mouse, even touch-screen were invented to let the computer know the world. However, these tools are still passive and people-centered. People need to type or write char by char, word by word to let the computer understand the real world that they see and touch. It is still boring for most people.

DOI: 10.4018/978-1-61520-701-5.ch001

Is it possible to shorten the distance between the computer and the real world and let computer see, touch and understand the real world directly and freely? This is really an attractive topic and now becoming a hot research area. WSN is one of these good answers today. It can get large numbers of data referring to temperature, moisture, smoking, movement of some objects, etc., through perceiving the real world by various sensors. With the aid of wireless networks, these data are collected and sent to some computers to be processed. So we can regard WSNs as eyes, hands, noses and nerve systems of computer world. By applying WSNs, the computer world can touch and perceive the real world everywhere, even many places where human beings have never reached because of dangers, bad weather, complexity of environments, and other reasons. They are changing the ways we explore the world greatly. More and more researchers and organizations pay their attentions to WSNs.

WHAT ARE WIRELESS SENSOR NETWORKS?

The concept of WSNs was proposed by the U.S. military as early as 1970's. From then on, many research projects, applications and theories about WSNs have come forth. Recalling the history of the development of sensor networks, we can roughly divide it into several stages as follows (Chong, & Kumar, 2003).

In the 1970s, the emergence of some prototypes of sensor networks with traditional point-to-point transmission could be referred to as the first generation of sensor networks. In 1979, the Defense Advanced Research Projects Agency (DARPA) of USA launched the Distributed Sensor Networks Program (DSN). It was one of the representatives.

From 1980s to 1990s, processing and communications capabilities of sensor nodes were improved obviously, which made these nodes can work together by networking with each other.

However, during this period, research work still mainly focused on the military field such as, SensIT plan (Sensor Information Technology), WINS (Wireless integrated network sensors) launched by DARPA. It is regarded as the second generation sensor networks.

With the continuous development of related technologies such as microelectronics, wireless communications, network transmission, from the end of the 20th century, WSNs have attracted wide attentions from academia, military and industry, which really set off a high wave of the development of WSNs technologies. Many projects have been launched and many applications have been deployed in various fields, which include military, environment monitoring, health care, intelligent home, urban transportation, space exploration, public safety monitoring, etc. Some representative projects are C4KISR Plan of DARPA, ALERT of U.S National Weather Service, SSIM program of Wayne State University, etc.

Although WSNs have affected various aspects of our work and life, different people still have different understandings about WSNs.

Concepts

Generally, WSN is regarded as an emerging technology that combines the concept of wireless network with sensors. Significant advances in microelectronics technology, computing and wireless communications reduce the energy consumption, improve the ability of communication and extend functions of WSNs continuously, which also reflect the basic characteristics of WSNs. From the following five representative statements, we can see main features of WSNs.

Karl (2003) said that WSNs "combine simple wireless communication, minimal computation facilities, and some sort of sensing of the physical environment into a new form of network that can be deeply embedded in our physical environment, fueled by the low cost and the wireless communication facilities" (p.1).

Akyildiz, Su, Sankasubramaniam, & Cayirci (2002a) described WSN as the one "which has been made viable by the convergence of micro-electro-mechanical systems technology, wireless communications and digital electronics" (P. 393).

Lewis (2004) stated that WSNs "generally consist of a data acquisition network and a data distribution network, monitored and controlled by a management center. The plethora of available technologies makes even the selection of components difficult, let alone the design of a consistent, reliable, robust overall system." (p.11)

Sohrabi, Gao, Ailawadhi, & Pottie (2000) said that "Wireless sensor networks are part of a growing collection of information technology constructs which are moving away from the traditional desktop wired network architecture toward a more ubiquitous and universal mode of information connectivity" and "A wireless sensor network refers to a group of sensors, or nodes, linked by a wireless medium to perform distributed sensing tasks. Connections between nodes may be formed using such media as infrared devices or radios." (p.16)

Rentala, Musunui, & Gandlham (2002) regarded nodes in WSNs as wireless integrated network sensors that "combine micro-sensor technology and low power computing and wireless networking in a compact system. Sensor nodes are randomly dispersed over the area of interest and are capable of RF communication, and contain signal processing engines to manage the communication protocols and for data processing before transmission. The individual nodes have a limited capability, but are capable of achieving a big task through coordinated effort in a network that typically consists of hundreds to thousands of nodes." (p.1)

Moreover, Yick, Mukherjee and Ghosal (2008) stressed that WSNs should focus on the deployment and communication mechanisms.

Although, different researchers have different opinions about what WSNs are in some details, the basic elements, essential characteristics, the

fundamental architectures of WSNs have been uniformed gradually. The following will discuss these common aspects of WSNs further.

Basic Elements of Wireless Sensor Networks

Briefly, a WSN is made up of three elements: sensors, observers and sensing objects. Wireless network is the way of communication among sensors and observers. The basic functions of WSNs include sensing, information collecting, data processing and distributing, etc. One important characteristic of the WSN is that a group of sensors with limited resources and functions can achieve some large sensing task and cooperation. Moreover, some or all of nodes in one WSN can move. Nodes communicate with each other by Ad-hoc mode. Each node can act as a router's role, and each node has the ability of dynamic searching, locating and restoring connections.

- **Sensors:** A sensor mainly consists of sensing, transmission, storage and power units. Its tasks contain collecting the information from some object in the real world, storing sensing data, making simple calculation and transmitting them to some observer.
- **Observers:** Observers are the wireless sensor network users, who inquiry, collect and utilize the sensing information initiatively or passively. Observers can be men, computers or some other equipments. They deal with the sensing information by analyzing, mining before making some decision.
- **Sensing Objects:** Sensing objects are some targets such as tanks, soldiers, animals, and harmful gases, etc. that observers are interested. Generally, sensing objects are shown as some digital characterizations of some physical phenomena, chemical phenomena and others, including movement of objects, temperature, humidity, concentration of smoke, and so on. A WSN can sense a

number of objects within the distribution region of the network.

Features of Wireless Sensor Network

Usually, a WSN is regarded as a multi-hop self-organizing network system consists of diversified nodes that communicate with each other through wireless communication. Generally, wireless communication networks include mobile cell networks, wireless LANs, Bluetooth networks, Ad-hoc networks, etc. As a special kind of wireless communication network, the WSN consists of a large number of low-cost micro-sensor nodes deployed in the monitoring region. These sensor nodes cooperate with each other to sensing, gathering and processing the information of sensing objects in the network coverage area, and send it to observers. WSNs not only have the commonness of the wireless communications network, but also have their own distinctive features as follows:

1. **Application-Related:** WSNs obtain information from outside world through sensing the physical quantities of objects. In different applications, the requirements for network systems are different. It causes that the hardware platforms, software systems and communication protocols also vary greatly. This makes WSNs differ from the Internet that has unified communication protocols for the platform. To achieve efficient and reliable system goals, the design work should be related to each special application, which is the prominent feature of the WSNs, which is significant different from the traditional networks.

2. **Data-Centered:** A WSN is a task network. People usually only care certain value of the observed criterion in a certain area, and do not care about the observed data from some single node specifically. Its main task is to collect sensing data from a large number of nodes, merge and extract them, and convey the useful information to users effectively. When users want to inquire about some incident, they notify concerned events to the sensor network, not to certain single node.

3. **Large-Scale Distribution:** WSNs are usually deployed in some large stretch of areas intensively. The quantity and density of nodes will increase enormously when the areas expanding further. The areas monitored are often complex and hard to reach, so it is very difficult to make maintenance usually. Therefore, the software and hardware of the sensor network should have high robustness and strong fault-tolerance.

4. **Dynamic Topology:** Many reasons can change the topologies of WSNs dynamically. For example, there are often new nodes joining or leaving the networks; in some cases, sensor nodes might be mobile; some nodes are set to switch between work and sleep status discretionarily for power saving; some nodes may break down at any time due to various unpredictable reasons. With the dynamic changes of topological structures of the networks, WSNs should have the abilities of self-adjusting and reconstructing.

5. **High Reliability:** WSNs are often deployed in some abominable environments or unmanned areas over large scales. This makes the network online maintenance very difficult or even impossible. Therefore, it requires that sensor nodes should be very stable, difficult-to-damaged, and adaptable to various extreme environments. Additionally, to prevent the monitoring data from being stolen and observers from obtaining counterfeit information in some crucial WSNs, it is very important to apply some mechanisms of privacy and security for wireless sensor communication. All above require that the WSNs should have well fault-tolerance and robustness, in another word, high reliability.

Figure 1. Typical sensor node

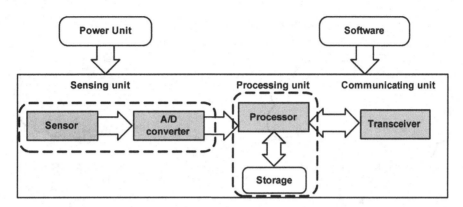

6. **Self-organization:** There are many unpredictable factors in the physical environments of networks. For example, the locations of the nodes can not be established in advance precisely; some nodes die due to energy exhausted or other reason; wireless communications quality subjected to environmental impacts can not be forecasted accurately; some network environment emergencies are uncontrollable. All above make the nodes should have the ability of self-organization. Without human intervention and any other pre-network facilities, the nodes can make their self-configuration and self-management automatically and quickly.

ARCHITECTURE AND PROTOCOLS OF WSNS

Hardware Structure of Sensor Node

In WSNs, the structures of sensor nodes mainly depend on the demands of the specific applications. However, we can still get the abstract common structure from diversified nodes. A typical hardware structure of a sensor node is shown as Figure 1. It mainly consists of the sensing unit, the wireless communication unit, the processing unit and the power unit.

The basic work steps of one wireless sensor node are: Firstly, the sensor node obtains some analog signal from the environment by the sensor. The signal got from environment maybe light, heat, movement, etc. Then, the analog signal is converted into digital signal by A/D converter. Processing unit receives the digital signal from A/D converter and deal with it locally, usually cooperating with a small-scale storage unit. Finally, the transceiver is responsible for transferring the data required to some control center through the wireless network, at the same time, receiving commands and data from it.

The power unit is one of the most important components of WSNs. Compare with other communication networks, the WSN pays more attention to the power unit since it's always limited and hard to replace, which usually makes the node unavailable directly. The power unit offers the energy to other parts of the node usually by batteries. However, the power of the batteries is always exhausted rapidly. So it is meaningful to make use of energy from the environment directly if possible, such as light energy, wave energy, etc.

Architecture Overview

The typical architecture of WSN can be seen in Figure 2. Generally, a classic WSN system contains: sensor nodes, sink nodes, manager nodes, etc. Firstly, a number of sensor nodes are deployed randomly in a sensor field that observ-

Figure 2. Architecture of WSN

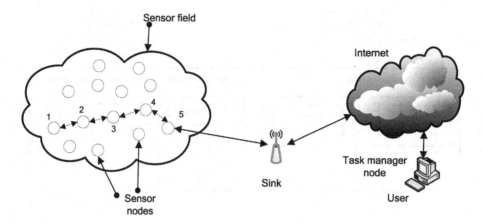

ers are interested in. These nodes communicate with each other through wireless channels to exchange information. Secondly, a sink node is arranged in or near the sensor field to collect the data got from the sensor nodes. Finally, the sink sends the data collected to the task manager node through Internet.

Due to various reasons such as the power supply, the capacity of transceiver, equipment price and weight, the communication distances between sensor nodes are usually limited, compared with wired sensor network. One node could just communicate with its neighboring nodes through wireless channels. Multi-channel routing mechanism and multi-hop router should be used when the nodes are out of its communication scope.

Like other communication networks, WSN includes a great number of nodes, which form the logic connections. The overall connections among nodes form the network topology. The basic architecture of network topology includes several categories such as tree connected network, ring connected network, mesh connected network, star connected network, bus connected network, hierarchical network, and so on (Akyildiz, Su, Sankasubramaniam, & Cayirci, 2002a; Akyildiz, Su, Sankasubramaniam, & Cayirci, 2002b). For example, in the hierarchical structure, nodes are divided into several levels according to their abili-

ties, which usually include base stations, the head nodes of clusters and ordinary sensor nodes. Base station has an obvious higher power than other nodes. Under ordinary circumstances, base station is considered as the gateway and data center to connect other networks. It's also regarded as the sink node. The cluster head is a leader of a subset of ordinary sensor nodes. It is responsible for collecting data from the subset and sending some commands and information to them.

Another thing associated with WSN's topology is the topology control. To maintain the logical connections of WSNs, the approaches and methods for topology control should consider some special WSN-relevant factors such as limited bandwidths, scarce energy, unstable infrastructures, etc. Some of the control methods should consider routing technologies simultaneously. Moreover, the WSN will change its logical topology, along with the movements or failures of some nodes.

Protocols of Wireless Sensor Networks

For other Ad-hoc networks, the design of network protocols mainly concentrates on network congestion, throughput, flow control, reliable transmission, etc. However, the WSN cares more about the power saving, fault-tolerance, cost per unit, and so on. There are several important protocols for wireless sensor networks such as medium access control protocols, network protocols and time synchronization and localization protocols, etc. They are to be introduced briefly as follows.

Medium Access Control Protocols

MAC protocol is mainly responsible for the coordination of shared network channels nodes access. Many researchers have presented a variety of different MAC protocol. For example, Ye, Heidemann and Estrin (2002) put forward S-MAC protocol (Sensor Medium Access Control) by revising IEEE 802.11 protocol, on the basis of the energy effectiveness of WSNs. Its main goal was to reduce the energy consumption as well as provide scalability. In the MAC protocol designing, the whole overhead of WSN must be taken into account firstly. Many researches on the MAC protocols tried various methods to increase the idle rates of the sensor nodes. For example, Zheng, Radhakrishnan & Sarangan (2005) proposed P-MAC protocol to further reduce the energy consumption of free listening.

Network Protocols

The following section briefly describes the network protocols from different points of perspectives which lead to different design schemas.

1. **Resource-Aware Routing:** (Schurgers, & Srivastava, 2001) Totally different from other networks, WSNs have limited energy and other resources, in many cases. So some resource-aware algorithms optimizing routing paths are desired, considering the large distributed system. Resource-aware routing methods pay more attention to the whole resource consumption and/or the utilization balance of all nodes, instead of taking care of one single node separately.

2. **Data-Centric Routing:** Usually, users are more concerned about the information got from WSN instead of the sensors of the WSN itself. So many data-centric routing protocols have been developed (Krishnamachari, Estrin, & Wicker, 2002). Generally, there are two roles in the hierarchies of the routing protocols. One is the data center which collects and saves information from WSN, and the other is a group of wireless sensors that are connected by data-centric routing protocols.

3. **Cluster-Based Routing:** (Ibriq, & Mahgoub, 2004) Cluster-based routing protocol is multi-level hierarchical. It groups wireless sensor nodes to relay the sensed data to the sink. Each group of sensors has a cluster head or gateway that is usually a more powerful node, compared with other common sensors. A cluster head performs some aggregation function on data received and sends it to some sink. There are also some special nodes between clusters, named gateways, which perform network communication issues.

4. **Geographic Routing:** (Yu, Govindan, & Estrin, 2001) In some cases, WSNs are more interested in the information of some geographic areas, which means the logical topology like above cluster-based infrastructure is subordinate against the physical distribution of the wireless sensors. They pay more attention to declining the overhead of message sending from one special physical space to another.

Time Synchronization and Localization Protocols

Time-synchronization is an important topic of WSNs. The famous Internet protocol, NTP (Network Time Protocol) is designed for a stable network infrastructure. However, it's not suitable for WSNs due to the distinction of their requirements for synchronization and the requirements of energy factors. There are some metrics of WSN time-synchronization to be especially important: precision, lifetime, scope, availability, efficiency, cost, etc. (Elson, & Estrin, 2001).

Localization protocols are designed to track the locations of diversified targets and WSN nodes themselves. There also exists another possible solution, GPS (Global Positioning System). However, in many cases, it's not a feasible solution for many cases due to the high consumption of cost, energy, computation power and space. Moreover, it is not suitable for some special environments such as indoor, pipe and coal mine. So the WSN-based approaches for localization are necessary for low cost and extreme environments. There have been a mount of researches issued (Boukerche, Oliveira, Nakamura, & Loureiro, 2007; Stoleru, Stankovic, & Son, 2008).

DEVELOPMENT AND APPLICATIONS OF WSNS

Current Development

There has been a lot of work done in developing WSNs in various aspects, including hardware devices, software and programming modes, operating systems, MAC and routing protocols, and many other issues like middleware, geographical location and coverage. Summarization of the current main developments of WSNs could help researchers better understand and identify the technical and theoretical issues which should be paid more efforts.

Development of Software

At present, there are still many WSN software products that are not flexible enough to meet the requirements of applications, in particular, the dynamic needs for services of multi-users. For example, most of them should be pre-installed. Once being done, they cannot be changed, which

is unable to adapt to the alteration of the environment. However, in some occasions, applications are variable. Recently, the development of software and programming model in WSNs simplified the development process of applications. For example, Agile (Fok, Roman, & Lu, 2005) increased the flexibility of the sensor networks. Another example is Mate (Levis, & Culler, 2002) developed by Berkeley, which supported a simple high-level programming representation. It has been mainly used in civil engineering, allowing network programming in an energy-efficient way. Furthermore, Sensorware (Boulis, Han, & Srivastava, 2003) has been developed by ULCA, which provided a powerful language model. It could share resources among nodes, and dynamically develop distributed algorithms according to the demands of users and systems, which has been mainly used for family activities. All technologies mentioned above realized some flexible approaches to make WSNs programmable.

Development of Operating System

Representative operating systems for WSNs' nodes include TinyOS (Levis, Madden, Polastre, Szewczyk, & et al., 2005), SOS (Han, Rengaswamy, Shea, Kohler, & Srivastava, 2005), Contiki (Dunkels, Grönvall, & Voigt, 2004), MANTIS (Bhatti, S., Carlson, J., Dai, H., &et al., 2005), etc. TinyOS was developed by UC Berkeley based on the event-driven. It achieved concurrence-intensive operation with minimal hardware support. It also provided a software framework based on components for application-oriented fields, with high energy-efficiency and flexibility to adapt to the external environment. However, TinyOS tasks used FIFO scheduling, which led to a weak real-time performance. Trumpler, & Han (2006) proposed an improved approach, named TinyMOS, which reduced the response time of system calls.

Developed by the Networked and Embedded Systems Lab (NESL) at UCLA, SOS was an operating system for mote-class sensors in WSNs. A common kernel was applied to implement messaging, dynamic memory, module loading and unloading, and other services. Dynamically loading software module was an important feature of SOS, which made the system support dynamic addition, modification, and removal of network services.

Contiki was developed by Adam Dunkels at the Networked Embedded Systems Group at the Swedish Institute of Computer Science. It was small, highly portable, multitasking, open-source, and usually used on a number of memory-constrained networked systems ranging from 8-bit computers to embedded systems. Complete with a graphical user interface, Contiki only required a few kilobytes of code and a few hundred bytes of RAM.

MANTIS was a multithreaded cross-platform embedded operating system for WSNs. It could support more complex tasks such as compression/aggregation and signal processing. However, it still achieved good memory efficiency, by implementing in a lightweight RAM footprint that fitted in less than 500 bytes of memory, including kernel, scheduler, and network stack.

Development of Protocols

Due to the limitation of WSNs resources and high relevance to applications, researchers have developed a variety of protocols for WSNs. MAC protocols and routing protocols are some key ones that affect the efficiency of the WSNs.

For MAC protocols, S-MAC protocol and PMAC protocol (Zheng, Radhakrishnan, & Sarangan, 2005; Demirkol, Ersoy, & Alagoz, 2006) were based on adaptive competitions. They adapted to the maintenance of local information and dynamic change of business and topology. GeRaF (Zorzi, & Rao, 2003) was based on the location information of sensor nodes, which simplified overhead by selecting relay nodes between receiving sensor nodes and sending nodes.

So far as routing protocols are concerned, they are responsible for the data packets transmission from source sensor nodes to destination nodes through WSNs, which have characteristics of energy-priority, data-centric, local-topology information and application-related.

Typical routing protocols include Flooding, SPIN, Directed Diffusion (Alazzawi, Elkateeb, & Ramesh, 2008; Intanagonwiwat, Govindan, Estrin, Heidemann, & Silva, 2003), etc. Flooding protocol did not require any maintenance expenses, and had a better fault-tolerance, but the disadvantages were information overlap and waste of resources. SPIN protocol adopted metadata description, taking into account data negotiation, eliminating data redundancy, saving energy. Directed Diffusion protocol made use of name-scheme diffuse data between sensor nodes to provide energy efficiency and to aggregate and cache data for sensor nodes.

Development of Middleware

Different WSNs have different design objectives and requirements. The fusion of these differences could be achieved through a flexible software architecture called middleware above physical sensor nodes. Without middleware, the development of WSNs would face many difficult issues such as the diversity of operating systems, complex and ever-changeable network environments, problems caused by inconsistency, performance, efficiency and security resulted from dispersive data processing. Middleware can be applied to simplify the development of software and improve the reusability, reliability, scalability of the software in practical applications. Generally, Middleware of WSN can be categorized into three classifications (Masri, & Mammeri, 2007; Henricksen, & Robinson, 2006).

1. **Database-Oriented Approaches:** They regard the whole WSN as a database and can be accessed by applications through SQL-

like methods. SINA (Shen, Srisathapornphat, & Jaikaeo, 2001) and TinyDB (Madden, Franklin, Hellerstein, & Hong, 2005) are two representative prototype systems. SINA adopted naming-based and location-aware mechanism to support the hierarchical structure and facilitate the expansion of the network. TinyDB established a tree and broadcasted queries to leaf nodes, then transmitted data back to the parent node, which could reduce the communication overhead. However, both SINA and TinyDB have weak reliability and mobility.

2. **Event-Driven Approaches:** They are also referred to as message-oriented approaches, which provide the information control paradigm, and suitable for large-scale WSNs. MiLAN (Heinzelman, Murphy, Carvalho, & Perillo, 2004) and DSWare (Bannach, Lukowicz, & Amft, 2008) are two representatives.

3. **TupleSpace-Oriented Middleware:** It utilizes tuple space communication paradigm, without locations of sensor nodes and signs of information. TinyLime (Curino, Giani, Giorgetta, & Giusti, 2005) was one representative. The core abstraction of TinyLime was the transiently shared tuple space. It expressed data as a basic data structure named tuple, and achieved process synchronization through reading, writing and movement of tuples.

Development of Localization and Coverage Algorithms

Localization in WSNs is designed to determine the locations of certain sensor nodes according to some algorithms with a small number of other known nodes. Localization algorithms can be divided into two categories based on the types of information used: distance-based localization algorithms and connectivity-based localization algorithms (Mao, Fidan, & Anderson, 2007).

For distance-based localization algorithms, inter-sensor distance measurements among sensors are the key factors that directly affect the precision of locations. Since many individual inter-sensor distance data are required, these algorithms are always time-consuming and power-consuming. Connectivity-based Localization algorithms take the advantages of low power consumption and low cost, but they can only achieve rough accuracy of the estimated locations.

The coverage of WSNs can be regarded as the allocation optimization of resources and the improvement of quality of service through nodes placement and routing paths selections for resource-constrained sensor networks. The coverage problems of WSNs can also be divided into two categories according to the different configurations of nodes, as the definite coverage and the random coverage (Li, Wan, & Frieder, 2003; Meguerdichian, Koushanfar, Potkonjak, & Srivastava, 2001).

The definite coverage is referred to as one that knows the locations of sensor nodes. On the contrary, random coverage considers the random distribution of sensor nodes and does not know the locations of nodes in advance.

The development of WSNs research in other issues such as time synchronization and models, could be found in some literatures (Krohn, Beigl, Decker, & Riedel, 2007; Tilak, Abu-Ghazaleh, & Heinzelman, 2002).

Applications

A number of useful applications of WSNs have emerged, including battlefield surveillance, agriculture utilities, environmental monitoring, medical insurance, industrial automation, and so on. Many literatures (Winkler, Tuchs, Hughes, & Barclay, 2008; Burrell, Brooke, & Bechwith, 2004; Lu, Qian, Rodriguez, Rivera, & Rodriguez, 2007; Dermibas, 2005; Baker, Armijo, Belka, Benhabib, & etc. 2007; Yamaji, Ishii, Shimamura, & Yamamoto, 2008; Carlos, Hernandez, & Pablo

2007; Puccinelli, & Haenggi, 2005, etc.) have mentioned diversified applications of WSNs in recent years.

Military Applications

WSNs can be utilized into several military fields such as equipment monitoring, real-time battlefield surveillance, object targeting, battlefield assessment, monitoring and searching for attack sources (Winkler, Tuchs, Hughes, & Barclay, 2008). A number of international institutions have started the wartime military researches.

Smart Sensor Web (Paul, 2000) is a Web-centric sensor network based on fusion of sensor information. It can provide situational awareness enhanced to fighters conveniently. It emphasized the multi-sensor fusion of large arrays of local sensors to provide real-time information to soldiers including weather, imagery, object information, mission planning, etc.

PinPtr (Maroti, Simon, Ledeczi, & Sztipano-vits, 2004; Puccinelli, & Haenggi, 2005) is an Ad-hoc sensor network developed by Vanderbilt University, which was used for the sniper localization. It could confirm the origination of gunfire sound by detecting the muzzle blast and the acoustic shock wave. Then, by comparing the arrival times of the sound at different sensor nodes, the position of the sniper could be obtained.

Another one is base protection (Winkler, Tuchs, Hughes, & Barclay, 2008). A certain number micro-magnetometer-sensor nodes were deployed around the headquarters, continuously monitoring the possible attack. It has gained much interest in military fields.

Agriculture Fields

A typical agriculture monitoring system usually consists of environmental monitoring nodes, base stations, communication systems, etc. Sensors with different functions can be deployed in the target region and used to monitor the subtle

changes of the climate for long time, including temperature, humidity, wind, atmosphere, rainfall, etc. The monitoring result can be used for scientific forecasts to help farmers alleviate disaster on time, make scientific cultivation and increase crop productivity.

In 2002, Intel established the world's first wireless vineyards in Oregon (Burrell, Brooke, & Bechwith, 2004). Sensor nodes were deployed in every corner of the vineyard to detect the soil temperature, humidity in the region every minute, which ensured the healthy growth of grapes. The researchers found that subtle changes of the climate in the vineyard could greatly affect the quality of wine. Through many years of data records as well as the correlated analysis, researchers have get a clear idea about the relationship between the quality of wine and grape growing in the process of sunshine, temperature and humidity. This is a typical example of precise agriculture and smart farming.

Environment Monitoring

WSNs have been widely used in ecological environment monitoring, biological population studies, meteorological and geographical research, floods and fire detection (Lu, Qian, Rodriguez, Rivera, & Rodriguez, 2007; Arampatzis, Lygeros, Manesis, 2005; Carlos, Hernandez, & Pablo 2007), etc. Some representative research targets of environmental monitoring applications are shown as follows:

By tracking the habitat, feeding habits of rare birds, animals and insects using WSNs, the study about endangered species can be carried; By deploying sensors nodes in rivers, the water level and the situation of contaminated water resources and related information can be monitored at any time; By deploying nodes in mountains of mud-rock flows, landslides and other natural disaster-prone areas, corresponding measures to prevent further accidents or decrease the loss can be taken in time once monitoring a warning in advance; By setting a large number of nodes in some forest protected areas to monitor the situation of potential fire at any time, alarms could be sent out immediately once the forest fires occur.

The University of Hawaii has done much research work on environment monitoring (Arampatzis, Lygeros, Manesis, 2005). Some sensor networks have been set in the Volcanoes National Park, trying to discover the influence of tiny climate change to living environment of the species at the brink of extinction.

Intelligent building monitoring can be regarded as a subcategory of environment monitoring. For example, in a bridge monitoring scenario (Burrell, Brooke, & n, 2004), the monitoring sensor nodes could record the automotive weight information of the cars passing the toll-booths, which was uploaded to the base-station for the bridge safe monitoring.

Medical Insurance

WSNs also play remarkable roles in the aspects about the health status detection, the hospital and drugs management, etc. Once sensors are attached to some patient, his body temperature, respiration, blood press and heart rate could be automatically recorded. Doctors can remotely monitor the patient conditions. Moreover, the employment of the sensor networks to collect people's physiological data plays an important role in the process of the development of some new drugs.

(Baker, Armijo, Belka, Benhabib, & et al., 2007) discussed five prototype applications, including infant monitoring, alerting the deaf, blood-pressure monitoring and tracking, and monitoring vital signs.

Industrial Automation

Industry automation is a meaningful application field for WSNs (Yamaji, Ishii, Shimamura, & Yamamoto, 2008; Carlos, Hernandez, & Pablo 2007; Puccinelli, & Haenggi, 2005).

1. **Industrial Safety:** Many WSNs have been used in many hazardous industrial environments such as coal mines, oil drilling, nuclear power plants, etc. These sensor networks can tell us who the employees are, what they are working, as well as their security situations and other important information. We also could deploy some wireless sensor nodes in the outfalls of some related plants, which could monitor the discharge of sewage.

2. **Advanced Manufacturing:** With the development of manufacturing technology, many kinds of equipments become increasingly complex including production lines and complex machines. To master the statuses of these equipments and find problems by early detections, some sensors are usually attached, which can effectively reduce accident rates.

3. **Warehouse Logistics Management:** With integration of multi-sensors, appropriate deployments and flexible networks, the WSNs can be used to control the temperature and humidity of food, vegetables, fruit, etc. in warehouses, by controlling the central air-conditioning systems. It provides efficient solutions for ensuring the quality and safety of inventory and reducing energy consumption.

CHALLENGES AND TRENDS OF WIRELESS SENSOR NETWORKS

Although, huge human and material resources have been devoted to the research work of WSNs, there are still many unaddressed problems and obstacles. Apperceiving these problems and obstacles will help researchers understand the main trend of the development of WSNs in the future. Here, we want to discuss these in several aspects respectively (Aboelaze, & Aloul 2005; Du, & Chen, 2008; Fang, & Zhang, 2008; Yoneki, & Bacon, 2005; Lionel, & Liu, 2008).

Sensor Nodes and Topology Control

There are some challenges in sensor nodes, whose abilities are subjected to the following restrictions (Yick, Mukherjee, & Ghosal, 2008):

1. **Communication Capability:** The communication bandwidths of sensors in WSNs always change and the ranges of communication coverage vary usually from a few dozen to several hundred meters. Moreover, frequent disconnections of the wireless channels often leads to failures of communications among sensors. So it's better for many sensors to be equipped with the functionality of offline working.

2. **Energy Supply:** In WSNs, many sensors are abandoned due to the exhaustion of their batteries. The problems of energy supply constraints have hindered the popularization of many WSNs applications. The commercialized wireless sender/receiver is far from enough to meet the needs of the sensor networks. In WSNs, most applications should be based on the premise of energy conservation.

3. **Computing Capability:** Most sensors in WSNs have embedded processors and memories, which can complete some simple information-processing tasks. However, as the capacities of the embedded processors and memories are usually limited, WSNs must know how to use multiple sensors with limited computing capacities to process distributed information jointly.

4. **Large Scale Deployment and Complexity:** In practice, environments are usually unknown or not fully controllable, so the deployments of WSNs are also challenges. The policies of the deployments of WSNs are different with different applications. For example, spreading sensors through the aircraft is an available way in remote harsh environments. However, it's not suitable for coal mine. Moreover, in the future, for the deployment of the large number of the sensor nodes, more attention will be paid to support the scalability of the node deployment with more effectiveness, fast deployment of large-scale nodes, and to guarantee better maintainability and fault tolerance.

5. **Topology Control and Heterogeneity:** The key point about topology control is the consideration on the autonomy and the adaptability. The autonomy means sensors can be deployed in unattended or inaccessible regions. Therefore, the sensors should interact as little as possible with base stations or administrators. For the adaptability, it means that sensors can adjust themselves to adapt to the changes of the environments. Generally, the topology challenge is how to control the power and the structure of topology, as well as minimize energy consumption to meet the requirements of the network coverage and connectivity. Moreover, it is important and meaningful for heterogeneous sensor nodes such as fixed ones and mobile ones to work together with some effective collaboration mechanisms. Large-scale collaborations among nodes are required by many applications of WSNs. The heterogeneity in software and hardware of sensor nodes is one of obstacles in the development of future WSNs, which also increases the difficulty for topology control.

6. **Mobility and Flexibility:** WSNs are networks with high-dynamic changes. The sizes, topologic structures and node densities of WSNs are all changeable. Some sensor nodes may fail due to some accidents or exhaustions of power. Some new nodes may join networks randomly. Some nodes may move away from current works. So the adaptive dynamic reconfiguration is an important research direction of WSNs. In the early years, the traditional studies about WSNs were based on the assumptions that

the locations of the nodes and receivers were fixed. Later, mobile nodes and receivers appeared. Recently, mobile ones and fixed ones have been deployed in some WSNs simultaneously. More and more WSN systems can adaptively adjust their topologies according to the changes of sensor nodes to enhance the flexibility of the networks, at the same time, greatly enhance the capacities of the networks.

Robustness and Fault-Tolerant

Wireless sensor nodes are often deployed in unattended environments, which make them vulnerable and easily become failure. The study on the reliability and fault-tolerant in WSNs is also a challenge (Akyildiz, & Erich 2006; Aboelaze, & Aloul 2005). On one hand, sensor nodes have always limited power supplies, which lead the nodes to fail easily. On the other hand, the wide distribution of nodes is arbitrary, resulting in difficulties in maintenance and replacement of the fault nodes. Therefore, the software and hardware of sensor nodes should have strong fault tolerance.

As far as different layers in WSNs, there are various fault-tolerant technologies. Physical layer requires sensor fault-tolerant technologies about hardware resources, including large-scale integrated circuits and related redundancy design techniques of hardware detection. Link layer mainly focuses on fault-tolerant coverage, fault-tolerant topology control, etc. Network layer pays main attention to fault-tolerant routing algorithms. Transport layer concerns on nodes fault detection, isolation mechanism to ensure data received correctly, etc. Application layer needs fault-tolerant models about data acquisition and data fusion. Studying on new approaches by using cross layer fault-tolerant structures are also novel challenges recently.

Low Cost

Cost plays a fundamental role in applications of WSNs, especially for applications with a large number of sensor nodes (Yick, Mukherjee, & Ghosal, 2008; Aboelaze, & Aloul 2005).

The administration and the maintenance are some important issues for cost. Specially, WSNs should be capable of self-configuration and self-maintenance. Self-configuration is the ability of network nodes to find other nodes and form into a structured, functioning network without human intervention. Self-maintenance is the ability of the network to detect and recover from faults appearing in either sensor nodes or network communication links again automatically.

Another considerable cost is the energy consumption of sensor nodes supplied by batteries. The requirement of reducing node energy consumption raises the demand for green energy solutions, especially for protocols with applicability to support efficient energy management. The technologies of multi-channel and dynamic channel allocations for energy research can be used to adapt to different business requirements and the changes of the network topology to improving energy efficiency and network performance. Since many traditional layered protocol stacks cannot meet requirements for the efficient utilization of resources such energies, memories of nodes, etc., some optimized cross-layer designs attract more and more attentions to reduce expenses and get better network performance, through combining the MAC layer, physical layer and network layer.

Generality

Despite a substantial amount of research works have been focused on various protocols in different layers of the WSN architecture, the lack of unified protocol stacks is still a big problem for WSNs (Akkaya, & Younis, 2005; Ibriq, & Mahgoub, 2004). So the generality is an obvious obstacle and challenge for WSN approaches (Tirkawi, & Fischer, 2009).

How to provide standards and general paradigms for WSNs is now a hot topic. However, unlike Internet, WSNs are applications-specific. The reasons listed below make it difficult to find general solutions for WSNs.

1. **The limitation of resources of the sensor nodes:** Many wireless sensors have limited resources such as small memory, weak computation capacity, limited energy, narrow bandwidth, etc.
2. **The lack of unified architecture:** Due to big differences among different applications, many different operating systems, routing protocols and wireless communication approaches have been developed, which have bricked a big obstacle for generality.
3. **Diverse physical hardware structures:** Unlike traditional computers that have unified architecture, different sensor nodes have very different physical structures. For example, there are various detectors built in sensors for detecting gas, temperature, humidity, vibration, etc.

To achieve generality, Tirkawi, & Fischer (2009) have discussed some schemes related to middleware, dynamic and mobile agents, semantic methods, standards, and COTS (commercial off the shelf). However, these schemes are still far from practice.

Security

Since WSNs' communication is based on wireless channels and lack of resources to make complex security algorithms, WSNs are easier to be attacked by outside hackers, compared with other traditional networks. It makes security a true challenge (Puccinelli, & Haenggi, 2005). Typical attacks to WSNs include (Du, & Chen, 2008) jamming, tampering, collision, exhaustion, selective forwarding attack, Sybil attack, flooding, clone attack, etc. How to avoid these attacks is a very big problem for many

WSNs. Du, & Chen (2008) also have discussed some solutions such as key management with low computational and low storage requirements, trusted server schemes taking base station as the key distribution center, key pre-distribution schemes by distributing key information among all sensor nodes before deployment, secure time synchronization by utilizing the long transmission range and other features of high-end sensors, secure routing by combining powerful high-end sensors and low-end sensors, etc.

Although, many approaches have been presented for secure issues of WSNs, there are still big gaps between the reality and the expectation. The cost-effectiveness and energy efficiency still pose great research challenge in the future.

CONCLUSION

The research on WSNs has become a new emerging direction of computer science. It has attracted more and more attentions of researchers for its up-and-coming essentiality for various aspects of human's life. This chapter gives a brief overview of the WSNs, including the basic concepts, architectures and protocols, current developments and trends, diversified applications, etc. It provides readers with a basic view of what WSNs look like and elementary background knowledge for reading the following chapters. More detailed and more technical contents about WSNs will come with following chapters.

ACKNOWLEDGMENT

This work is supported by the National Basic Research Program of China (973 Program) under grant No.2006CB303000.

REFERENCES

Aboelaze, M., & Aloul, F. (2005, March). *Current and Future Trends in Sensor Networks: A Survey.* Paper presented at the Second IFIP International Conference on Wireless and Optical Communications Networks (WOCN), Dubai, UAE.

Akkaya, K., & Younis, M. (2005). A survey on routing protocols for wireless sensor network. *Journal on Ad Hoc Networks (Elsevier), 3,* 325–349. doi:10.1016/j.adhoc.2003.09.010

Akyildiz, I. F., & Erich, P. (2006). Stuntebeck. Wireless underground sensor networks: Research challenges. *Journal on Ad Hoc Networks (Elsevier), 4,* 669–686. doi:10.1016/j.adhoc.2006.04.003

Akyildiz, I. F., Su, W., Sankarasubramaniam, Y., & Cayirci, E. (2002b). A survey on sensor networks. *IEEE Communications Magazine, 40,* 102–116. doi:10.1109/MCOM.2002.1024422

Akyildiz, I. F., Su, W., Sankasubramaniam, Y., & Cayirci, E. (2002a). Wireless Sensor Networks: A Survey. *Journal of Computer Networks, 38,* 393–422. doi:10.1016/S1389-1286(01)00302-4

Alazzawi, L. K., Elkateeb, A. M., & Ramesh, A. (2008, March). *Scalability Analysis for wireless sensor networks routing protocols.* Paper presented at 22nd International Conference on Advanced Information Networking and Applications, GinoWan, Okinawa, Japan

Arampatzis, T., Lygeros, J., & Manesis, S. (2005, June). *A Survey of Applications of Wireless Sensors and Wireless Sensor Networks.* Paper presented at the 13th IEEE Mediterranean Conference on Control and Automation, Limassol, Cyprus.

Baker, C. R., Armijo, K., Belka, S., Benhabib, M., Bhargava, V., Burkhart, N., et al. (2007, May). *Wireless sensor networks for home health care.* Paper presented at the International Conference on Advanced Information Networking and Applications Workshops (AINAW), Ontario, Canada.

Bannach, D., Lukowicz, P., & Amft, O. (2008). Rapid Prototyping of Activity Recognition Applications. *IEEE Pervasive Computing / IEEE Computer Society [and] IEEE Communications Society, 7*(2), 22–31. doi:10.1109/MPRV.2008.36

Bhatti, S., Carlson, J., Dai, H., et al. (2005). MANTIS OS: an embedded multithreaded operating system for wireless micro sensor platforms. *Mobile Networks and Applications archive, 10*(4), 563-579

Boukerche, A., Oliveira, H. A. B., Nakamura, E. F., & Loureiro, A. A. F. (2007). Localization Systems for Wireless Sensor Networks. *IEEE Wireless Communications, 14*(6), 6–12. doi:10.1109/MWC.2007.4407221

Boulis, A., Han, C. C., & Srivastava, M. B. (2003, May). *Design and Implementation of a Framework for Efficient and Programmable Sensor Networks.* Paper presented at the International Conference on MobiSys, San Francisco, CA.

Burrell, J., Brooke, T., & Bechwith, R. (2004). Vineyard computing: sensor networks in agricultural production. *IEEE Pervasive Computing / IEEE Computer Society [and] IEEE Communications Society, 3*(1), 38–45. doi:10.1109/MPRV.2004.1269130

Carlos, F., Hernandez, G., & Pablo, H. (2007). Ibarguengoytia-Gonzalez and etc. Wireless Sensor Networks and Applications: a Survey. *IJCSNS International Journal of Computer Science and Network Security, 7*(3), 264–273.

Chong, C. Y., & Kumar, S. P. (2003). Sensor Networks: Evolution, Opportunities, and Challenges. *Proceedings of the IEEE, 91*(8), 1247–1256. doi:10.1109/JPROC.2003.814918

Curino, C., Giani, M., Giorgetta, M., & Giusti, A. (2005). *TinyLIME: bridging mobile and sensor networks through middleware.* Paper presented at the International Conference on Pervasive Computing and Communications, Kauai Island, HI.

Demirkol, I., Ersoy, C., & Alagoz, F. (2006). MAC Protocols for Wireless Sensor Networks: a Survey. *IEEE Communications Magazine, 44*(4), 115–121. doi:10.1109/MCOM.2006.1632658

Dermibas, M. (2005). *Wireless sensor networks for monitoring of large public buildings (Technical Report)*. Buffalo, NY: University at Buffalo, Department of Computer Science and Engineering.

Du, X. J., & Chen, H. H. (2008). Security in Wireless sensor networks. *IEEE Wireless Communications, 15*(4), 60–66. doi:10.1109/MWC.2008.4599222

Dunkels, A., Grönvall, B., & Voigt, T. (2004, November). *Contiki - a Lightweight and Flexible Operating System for Tiny Networked Sensors.* Paper presented at Proceedings of the 29th Annual IEEE International Conference on Local Computer Networks, Tampa, FL.

Elson, J., & Estrin, D. (2001, April). *Time Synchronization for Wireless Sensor Networks.* Paper presented at the IEEE International Parallel and Distributed Processing Symposium (IPDPS), San Francisco.

Fok, C. L., Roman, G. C., & Lu, C. Y. (2005, June). *Rapid Development and Flexible Deployment of Adaptive Wireless Sensor Network Applications.* Paper presented at the 25th IEEE International Conference on Distributed Computing Systems (ICDCS), Columbus, OH.

Han, C., Rengaswamy, R. K., Shea, R., Kohler, E., & Srivastava, M. (2005, June). *SOS: A dynamic operating system for sensor networks.* Paper presented at the Third International Conference on Mobile Systems, Applications, And Services (Mobisys), Seattle, WA.

Heinzelman, W. B., Murphy, A. L., Carvalho, H. S., & Perillo, M. A. (2004). Middleware to support sensor network applications. *IEEE Network, 18*(1), 6–14. doi:10.1109/MNET.2004.1265828

Henricksen, K., & Robinson, R. (2006, December). *A Survey of Middleware for Sensor Networks: State-of-the-Art and Future Directions.* Paper presented at the ACM MidSens, Melbourne, Australia.

Ibriq, J., & Mahgoub, I. (2004, July). *Cluster-based routing in wireless sensor networks: Issues and challenges.* Paper presented at International Symposium on Performance Evaluation of Computer and Telecommunication Systems, San Jose, CA.

Intanagonwiwat, C., Govindan, R., Estrin, D., Heidemann, J., & Silva, F. (2003). Directed diffusion for wireless sensor networking. *IEEE/ACM Transactions on Networking, 11*(1), 2-16

Karl, H. (2003). *A short survey of wireless sensor networks* (Technical Report TKN-03-018). Technical University Berlin, Telecommunication Networks Group.

Krishnamachari, B., Estrin, D., & Wicker, S. (2002, June). *Modeling Data-Centric Routing in Wireless Sensor Networks.* Paper presented at the IEEE INFOCOM, New York.

Krohn, A., Beigl, M., Decker, C., & Riedel, T. S. (2007, June). *Collaborative Time Synchronization in Wireless Sensor Networks.* Paper presented at the International Conference on Networked Sensing Systems, Braunschweig, Germany.

Levis, P., & Culler, D. E. (2002, October). *Mate: a Tiny Virtual Machine for Sensor Networks.* Paper presented at the International Conference on Architectural Support for Programming Languages and Operating Systems (ASPLOS), San Jose, CA.

Levis, P., Madden, S., Polastre, J., Szewczyk, R., Whitehouse, K., Woo, A., et al. (2005). In W. Weber, J. M. Rabaey, and E. Aarts (Eds.) *TinyOS: An operating system for wireless sensor networks* (pp.115-148), Ambient Intelligence. New York: Springer-Verlag.

Lewis, F. L. (2004). Wireless Sensor Networks. In Cook, D.J. and Das, S.K. (Eds.), *Smart Environments: Technologies, Protocols, and Applications* (pp. 11-46). New York.John Wiley.

Li, X. Y., Wan, P. J., & Frieder, O. (2003). Coverage in wireless ad hoc sensor networks. *IEEE Transactions on Computers, 52*(6), 753–763. doi:10.1109/TC.2003.1204831

Lionel, M. N., & Liu, Y. H. (2008, June). *China's National Research Project on Wireless Sensor Networks*. Paper presented at the International Conference on Ubiquitous and Trustworthy Computing (SUTC), Taichung, Taiwan.

Lu, K. J., Qian, Y., Rodriguez, D., Rivera, W., & Rodriguez, M. (2007, November). *Wireless Sensor Networks for Environmental Monitoring Applications: A Design Framework*. Paper presented at the IEEE Global Telecommunications Conference, Washington, DC.

Madden, S. R., Franklin, M. J., Hellerstein, J. M., & Hong, W. (2005). TinyDB: An Acquisitional Query Processing System for Sensor Networks. *ACM Transactions on Database Systems, 30*(1), 122–173. doi:10.1145/1061318.1061322

Mao, G., Fidan, B., & Anderson, B. D. O. (2007). Wireless sensor network localization techniques. *Journal of Computer and Telecommunications Networking, 51*(10), 2529–2553.

Maroti, M., Simon, G., Ledeczi, A., & Sztipanovits, J. (2004). Shooter localization in urban terrain. *IEEE Computer, 37*, 60–61.

Masri, W., & Mammeri, Z. (2007, September). *Middleware for Wireless Sensor Networks: A Comparative Analysis*. Paper presented at the IFIP International Conference on Network and Parallel Computing Workshops, Dalian, China.

Meguerdichian, S., Koushanfar, F., Potkonjak, M., & Srivastava, M. B. (2001, April). *Coverage Problems in Wireless Ad-oc Sensor Network*. Paper presented at the IEEE INFOCOM, Anchorage, AK.

Paul, J. L. (2000, July). *Smart Sensor Web: Web-based exploitation of sensor fusion for visualization of the tactical battlefield*. Paper presented at the Third International Conference on Information Fusion, Paris, France.

Puccinelli, D., & Haenggi, M. (2005). Wireless sensor networks: applications and challenges of ubiquitous sensing. *IEEE Circuits and Systems Magazine, 5*(3), 19–31. doi:10.1109/MCAS.2005.1507522

Rentala, P., Musunui, R., & Gandlham, S. (2002). Survey on Sensor Networks (Technical Report UTDCS-33-02). Dallas, TX:University of Texas at Dallas.

Schurgers, C., & Srivastava, M. B. (2001, October). *Energy Efficient Routing In Wireless Sensor Networks*. Paper presented at the IEEE Military Communications Conference, Washington, D.C.

Shen, C., Srisathapornphat, C., & Jaikaeo, C. (2001). Sensor information networking architecture and applications. *IEEE Wireless Communications, 8*(4), 52–59.

Sohrabi, K., Gao, J., Ailawadhi, V., & Pottie, G. J. (2000). Protocols for self-organization of a wireless sensor network. *IEEE Personal Communications Magazine, 7*(5), 16–27. doi:10.1109/98.878532

Stoleru, R., Stankovic, J., & Son, S. (2008). On Composability of Localization Protocols for Wireless Sensor Networks. *IEEE Network, 22*(4), 21–25. doi:10.1109/MNET.2008.4579767

Tilak, S., Abu-Ghazaleh, N. B., & Heinzelman, W. (2002). A taxonomy of wireless micro-sensor network models. *ACM SIGMOBILE Mobile Computing and Communications Review, 6*(2), 28–36. doi:10.1145/565702.565708

Tirkawi, F., & Fischer, S. (2009). Generality Challenges and Approaches in WSNs. *Journal of Communications* [IJCNS]. *Network and System Sciences, 1*, 58–63.

Trumpler, E., & Han, R. (2006, May). *A systematic framework for evolving TinyOS*. Paper presented at the IEEE Workshop on Embedded Networked Sensors, Cambridge, MA.

Winkler, M., Tuchs, K. D., Hughes, K., & Barclay, G. (2008). Theoretical and practical aspects of military wireless sensor networks. *Journal of Telecommunications and Information Technology*, *2*, 37–45.

Yamaji, M., Ishii, Y., Shimamura, T., & Yamamoto, S. (2008, June). *Wireless sensor network for industrial automation*. Paper presented at the International Conference on Networked Sensing Systems (INSS), Kanazawa, Japan.

Ye, W., Heidemann, J., & Estrin, D. (2002, June). *An energy-efficient MAC protocol for wireless sensor networks*. Paper presented at the INFOCOM, New York.

Yick, J., Mukherjee, B., & Ghosal, D. (2008). Wireless sensor network survey. *Journal of Computer Networks*, *52*, 2292–2330. doi:10.1016/j.comnet.2008.04.002

Yoneki, E., & Bacon, J. (2005). *A survey of Wireless Sensor Network technologies: research trends and middleware's role (Technical Report UCAM-CL-TR646)*. University of Cambridge, Cambridge, UK.

Yu, Y., Govindan, R., & Estrin, D. (2001). *Geographical and Energy Aware Routing: a recursive data dissemination protocol for wireless sensor networks (Technical Report)*. UCLA Computer Science Department, Los Angelas.

Zheng, T., Radhakrishnan, S., & Sarangan, V. (2005, April). *PMAC: an adaptive energy-efficient MAC protocol for wireless sensor networks*. Paper presented at the IEEE International Parallel and Distributed Processing Symposium (IPDPS), Denver, CO.

Zhou, Y., Fang, Y. G., & Zhang, Y. C. (2008). Securing Wireless Sensor Networks: A survey. *IEEE Communication Surveys & Tutorials*, *10*(3), 6–28. doi:10.1109/COMST.2008.4625802

Zorzi, M., & Rao, R. R. (2003). Geographic random forwarding(GeRaF) for ad hoc and sensor networks: energy and latency performance. *IEEE Transactions on Mobile Computing*, *2*(4), 349–365. doi:10.1109/TMC.2003.1255650

KEY TERMS AND DEFINITIONS

Data-Centric: Refers to approaches, methods that focus on the data instead of processing nodes.

Generality: The generality can be regarded as the integration of the compatibility, interchangeability, and simplicity, etc., which make the system more extendable and standardized.

Localization: It aims at locating the places where various events take places exactly.

Middleware: A flexible software component that exists between the above application layer and the below network, link layers, which can simplify the development of software and improve its reusability, reliability and scalability.

Resource-Aware: Refers to methods that pay main attention to resources such as power, bandwidth, memory, etc.

Time Synchronization: It aims at synchronizing the time of all sensor nodes in a wireless sensor network, or making different events take place in some specified order.

Wireless Sensor Network: A network consisting of some wireless sensor nodes that are distributed spatially for cooperatively monitoring physical or environmental conditions.

Chapter 2
Deployment Strategies for Wireless Sensor Networks

Ruay-Shiung Chang
National Dong Hwa University, Taiwan

Shuo-Hung Wang
National Dong Hwa University, Taiwan

ABSTRACT

In this chapter, we study the deployment problem in wireless sensor networks. The deployment affects the efficiency and the effectiveness of sensor networks. We discuss different types of deployment strategies including deterministic or random deployment, and centralized or distributed deployment. We also review the measures of deployment while considering the connectivity and coverage in detail. The best coverage and the worst coverage are also discussed. An important issue is the energy efficiency in wireless sensor networks. We classify the power conservation issue into 3 types of sensor networks. They are duty-cycle dependent, role dependent, and topology dependent methods. Finally, future trends in sensor deployment are proposed.

INTRODUCTION

Applications of wireless sensor networks (WSNs) exist in many scenarios such as battlefield surveillance, reconnaissance of opposing forces and terrain, forest fire detection, drug administration in hospitals, and so on (Akyildiz, Su, Sankarasubramaniam, & Cayirci 2002). We can generally classify those applications into target tracking and area monitoring. In the target tracking scenarios, we concern if we can trace the moving object accurately. The number of sensors and the position of sensors affect the performance of tracking. In the area monitoring scenarios, we need to have enough sensors to avoid blind angle. It seems that a denser infrastructure cause a more effective WSNs. However, if not deployed well, a denser network will lead to a larger number of packet collisions and traffic congestions. The cost of larger sensors is another reason to devise good deploying strategy. Some factors to explain the importance of deployment strategy in WSNs are discussed below:

DOI: 10.4018/978-1-61520-701-5.ch002

- **Limited Energy:** A well-known characteristic is that wireless sensor nodes have limited energy and have difficult in recharging. According to the energy consumption model (Heinzelman, Chandrakasan, & Balakrishnan 2002), the longer the transmitting range is, the more energy the WSNs consume. For energy saving, proper distance among sensors is important for WSNs. Transmitting by multiple hops path is usually better than by directly (Heinzelman et al., 2002). The topology of WSNs affects the network lifetime considerably.

- **Transmission Jobs:** To prolong the lifetime of WSNs, we usually regularly schedule sleep intervals for sensors. Usually, multiple sensors sensing similar data will need to aggregate them to the source. Transmission jobs will cost a lot, if the WSNs don't have uniform sensing coverage.

- **Unprotected Area:** The monitoring area is not usually protected, especially for military applications. To prevent being invaded, deployment information is a good option to key management schema for WSNs (Wenliang, Jing, Han, & Shigang, 2004), that is, deployment strategy affects the key schema.

How to deploy a lot of sensor nodes into an area of interest is an important issue for wireless sensor networks. Usually, it affects whether the network operations are energy-efficient, robust, and reliable. Different deployment methods result in distinct routing methods. Varying routing methods consume sensor energy at a varied rate. That is, the deployment will affect the energy depletion and the lifetime of WSNs. Furthermore, various deployments lead to different sensor network topologies. Sensor connectivity will also vary. The connectivity of sensor nodes affects the robustness of WSNs. Consequently, how to deploy the senor nodes efficiently and effectively affects if the WSNs work smoothly and as desired.

Since a deployment strategy is important for WSNs, how do we evaluate the strategy? Coverage (Gage, 1992) and energy-efficiency are two major concerns for deployment. Owing to the characteristic of WSNs with nodes easy to be damaged, using only a node to monitor a fixed area is unreliable. Usually, even a small area needs to be covered by several sensors. So, we can consider coverage the base of executing fault tolerance in WSNs. So, how do we deploy sensors in an interested area with finite nodes to achieve a better coverage? Because the sensing range and the communicate range of a sensor are not the same, coverage does not equal connectivity. To make WSNs work smoothly, we should consider connectivity while discussing coverage. We explain the coverage problem and related issues in the section of the coverage problem. We also study how to evaluate the coverage of an area. The benchmarks include the best and the worst case coverage. They help sensors to be deployed with better coverage. Another important issue about WSNs is its lifetime. If sensors were deployed unbalanced, nodes in sparse area will deplete energy soon due to more frequent data flow and more distant transmission range. Therefore, suitable arrangement of sensors will make WSNs more durable. Other strategies such as duty cycle based and role based are also used to help energy-efficiency. These methods are all mentioned in this chapter.

The rest of the chapter is organized as follows. We discuss the classification of deployment strategy first. Then, we present a survey for the coverage problem. Next we show the energy-efficiency issue. The last section is the conclusions and future trends.

CLASSIFICATION OF DEPLOYMENT STRATEGY

In this section, we will first classify the deployment strategies. A simple taxonomy is shown in Figure 1. One branch is deployment with all

Figure 1. The taxonomy of the deployment

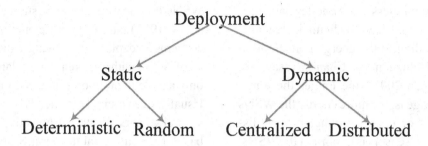

static sensor nodes and another is deployment with at least one mobile node. According to Younis, and Akkaya (2008), the former is called static device deployment and the latter dynamic device deployment. In static device deployment, it can be further classified into two groups based on whether or not the placement points for all sensor nodes are planned accurately in advance. The deterministic static deployment is the most elaborate because the sensor topology is usually designed beforehand for best performance. Some of deterministic protocols are grid-based deployment whose nodes are placed on the crossing points of the grids. However, in the situation with a lot of sensor nodes and the field unfriendly and uncontrollable, the deterministic static deployment may not work or can be difficult to deploy each sensor node as expected exactly. The random strategy is another choice. In random strategy, sensor nodes can be deployed according to uniform, Poisson, Gaussian, or other distribution model. As for the dynamic device deployment, it can be categorized into centralized and distributed methods. In the centralized method, some of sensor nodes such as cluster heads or base stations decide the position itself. Other nodes are deployed by cluster heads or base stations. On the other hand, the distributed protocols let every sensor node decide its own location.

DETERMINISTIC DEPLOYMENT

A deterministic strategy is a plan for deployment while considering the coverage, connectivity, and other factors in advance. There are many related literatures. In the literature (Wang, Hu, & Tseng, 2008), the authors proposed a placement solution while discussing the relation between sense range (r_s) and communication range (r_c). Given an interested area A to be deployed with sensor nodes, partition A into two types of regions. One is the single-row region whose maximum width is smaller than $\sqrt{3}\,r_{\min}$ and the other is the multi-row region. $r_{\min}$ is the smaller one between r_s and r_c. For a single-row region, find its bisector and place a sequence of sensor nodes along the bisector to satisfy connectivity and coverage. For a multi-row region, place sensor nodes row by row. Based on the relation between r_s and r_c, there are two cases to discuss. 1) $r_c < \sqrt{3}\,r_s$: Figure 2 (a) – (c) show three possible subcases. Sensor nodes are spaced with a distance of r_c on each row. Rows are separated by a distance of $r_s + \delta$, where $\delta = \sqrt{r_s^2 - r_c^2/4}$. When a new row is deployed, the first node in the row is shifted by a distance of $r_c/2$. For example, in Figure 2 (a), all nodes in the second row are shifted by a distance of $r_c/2$ from those in the first row horizontally. In the meantime, each row of sensor nodes can cover a belt-like region with a width of 2δ as in Figure 2. The coverage is guaranteed and the connectivity for each row is also guaranteed in case 1 under this arrangement. To connect the whole network with $r_c < \sqrt{3} r_s$, it needs an extra column of sensor nodes separated by a distance no larger than r_c to connect each row. 2) $r_c \geq \sqrt{3} r_s$: In this case, let $\sqrt{3} r_s$ replace r_c to be the distance between two adjacent sensor nodes. That placement is like a regular triangle lattice as shown in Figure 2(d). Furthermore, since $\sqrt{3} r_s$ is the distance between two adjacent nodes and r_c is larger than $\sqrt{3} r_s$, both coverage and connectivity are guaranteed.

RANDOM DEPLOYMENT

Random deployment methods decide the density of a network rather than calculate each node's position like those in deterministic strategies. Using uniform distribution, we decide the interested area and the number of sensor nodes first. Then deploy them as uniformly as possible by air-drop or other methods. Suppose there is a 2-D square area Ω, and we randomly place N sensors inside the area. If the distance between any two sensors is shorter than R, a wireless link exists between them. Define the size of the square area as L^2 and the deployment is uniformly random, the probability that there is a link between any two sensors is

$$p = \frac{\pi R^2}{L^2} \tag{1}$$

We assume that the average sensor density is constant and the communication range R is preset. The average degree of each sensor is also a constant (Zhipu, & Murray, 2007).

$$E[d] = Np = N\frac{\pi R^2}{L^2} \tag{2}$$

Another random strategy is choosing Poisson distribution as the basis for deployment. Sensor

nodes form a Poisson point process with density d and each node is active independently with probability p. Here, the eigenvalue λ of the Poisson distribution equals dp. The number of deployed sensor nodes is determined by the density and Poisson distribution instead of only by the density as in the uniform distribution (Santosh, Ten, Lai, & József, 2008; Zhang, & Hou, 2006). In the reference (Mhatre, Rosenberg, Kofman, Mazumdar, & Shroff, 2005), the authors find a minimum cost for Poisson distribution deployment with a lifetime constraint. Tow types of nodes, type 0 and type 1, in their assumption. Type 0 nodes are those with intensity λ_0 and battery energy E_0; type 1 nodes with intensity λ_1 and battery energy E_1. Type 0 nodes do the sensing while type 1 nodes act as the cluster heads. They develop a set of formulas to determine the optimum node intensities and node energies that guarantee a lifetime of at least T units, while considering connectivity and coverage of the surveillance area. The details of deduction can refer to Mhatre et al. (2005).

An interesting issue is which one is better, deterministic deployment or random deployment? The answer proposed by Santosh et al. (2008) is that deterministic deployment requires more sensor nodes than random deployment does to achieve the same level of coverage degree. However, the answer proposed by Zhang and Hou (2006) is completely different. Those two literature's seemingly contradictory results stem from their different conditions. The assumption proposed by Santosh et al. (2008) is that $p \to 0$. Under most conditions the probability for a node to be active does not converge to 1 or 0. Therefore, grid deployment has less density than uniform or Poisson distribution deployment (Zhang, & Hou, 2006).

CENTRALIZED DEPLOYMENT

In this classification, the locations of sensor nodes are decided by a few nodes such as base stations or

Figure 2. Examples for deterministic deployment: (a) $r_c < r_s$ (b) $r_c = r_s$ (c) $r_s < r_c < \sqrt{3}\ r_s$ (d) $r_c \geq \sqrt{3}\ r_s$

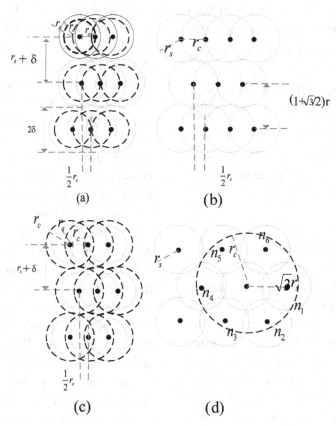

cluster heads. For a cluster-based sensor network, Virtual Force Algorithm (VFA) was suggested by Yi, and Krishnendu (2004). VFA is a famous protocol and is becoming a main solution to area coverage for homogeneous wireless sensor networks. In this method, the cluster head calculates its local area coverage and sends instructions to its members to move to the requested position. The main idea of VFA is to use potential field to avoid obstacle and decide the moving direction. A potential field is a commonly used method in mobile robotics, where they are typically applied to tasks such as local navigation and obstacle avoidance. The forces between nodes are classified into attractive and repulsive forces. When two nodes' locations get close, the repulsive force splits them farther. In the case of two distant nodes, the attrac-

tive force draws them closer. The force exerted on node i by node j is denoted by $\overrightarrow{F_{ij}}$.

$$\overrightarrow{F_{ij}} = \begin{cases} W_a(d_{ij} - D_{th}), \alpha_{ij} & if \quad d_{ij} > D_{th} \\ 0 & if \quad d_{ij} = D_{th} \\ W_r d_{ij}^{-1}, \alpha_{ij} + \pi & if \quad d_{ij} < D_{th} \end{cases} \quad (3)$$

where W_a, W_r are the virtual force coefficients, d_{ij} is the Euclidean distance between node i and node j, α_{ij} is the orientation of the line segment from node i to node j, and the threshold D_{th} is used to decide the status of repulsive or attractive force which is determined by users. The total force exerted on node i can be calculated by adding all forces contributed by other nodes in the network. Once the total force and orientation is determined, the new location of this node can also be decided. The algorithm proposed by Yi and Krishnendu (2004) needs the knowledge of obstacles implying a certain degree of a priori knowledge of the terrain. All cluster-heads use the information of obstacles and the initial deployment of sensor nodes to calculate the positions for redeployment. These cluster-heads need to have more powerful computational capabilities than other nodes for the centralized decision processes.

Chellappan et al. (2007) proposed deployment schemes under limited mobility constraint for movable sensors. Given an area divided into several clusters which have different number of nodes, the distributed self-deployment protocol uses the minimum moving hops of all nodes to make the whole area balanced with the number of nodes. The goal of method proposed by Chellappan et al. (2007) is to minimize the sensor movements while balancing the numbers of nodes in different regions. The main idea is to transfer the non-linear variance minimization problem into a linear optimization problem by assigning a proper weight to a region. The weight corresponds to the number of sensor nodes that the region needs. During the movement process, a region with a high weight has high priority to ask nodes to move in. Their method can be executed by a centralized node and is summarized as follows. 1) Collect the information on the number of nodes in each region. 2) Construct a graph with vertices and edges depending on the above information. 3) Determine the minimum cost maximum weighted flow in the graph. 4) According to the flow found in step 3, move those available mobile nodes to those regions which need them.

DISTRIBUTED DEPLOYMENT

In this kind of deployment, the new location of a mobile sensor node is decided by itself. An incremental and greedy deployment algorithm for a mobile sensor network was proposed by Andrew, Maja, and Gaurav (2002a). Nodes are deployed one at a time, with each node making use of data gathered from previously deployed nodes. The advantage of this approach is simplicity and clarity. Due to this sequential step, it usually takes more time to deploy than other concurrent methods.

One major bracket of distributed methods is virtual-force based deployment which is like the idea mentioned in the sub-section of centralized deployment. There are several examples here. Nojeong and Pramod (2003) proposed a "Distributed Self-Spreading Algorithm". This distributed algorithm defined a partial force calculation to decide the movement of nodes during the deployment process. Andrew, Maja, and Gaurav (2002b) presented a potential-field-based approach to handle the deployment problem. The idea of potential fields is also used to drive every moving node to an available destination (Zavlanos, & Pappas, 2008). Nearest neighbor coordination protocols are developed to ensure that distinct moving nodes are assigned to distinct destinations.

Chang and Wang (2008) use density control by each node to deploy sensor nodes concurrently. They make nodes to form clusters to achieve the target area density balance. Three phases are in this proposed algorithm. They are initialization, goal selection, and goal resolution. In initialization phase, every sensor node is active. When the sensor detects the surrounding, it can find obstacles and other sensors. Then it knows how large it is in its detection area. These sensors detect how many numbers of neighbor nodes they already have. After detecting, they send messages to their neighbor nodes and form the cluster. In goal selection phase, every sensor has the information about its local density, its neighbor node's density, the angles and the distances from neighbor nodes. They use these messages to decide the next position where they will move. The sensor moves to the location with low density. In goal resolution phase, to avoid forming a fragmented sensor network, the node's future position should be checked before moving. The node broadcasts its possible next position to its neighbors and waits for one "ok" or "suggested" message. Sometimes, the node will receive multiple "suggested" messages meaning that it can move. At that time, the node will choose the largest suggestion distance to move. The goal of the method proposed by Chang and Wang (2008) is to let sensor nodes deploy themselves like quickly spreading small molecules.

For deploying mobile sensors, Wang, Cao, Berman, La, and Thomas (2007) calculate how large the uncovered holes are and let the dynamic nodes to bid and move to cover them. They proposed proxy-based bidding protocol to execute hole-exchange for reducing the distance to which the mobile nodes move. Static sensor nodes detect coverage holes locally and estimate their sizes as bids. Mobiles nodes choose the highest bids and move in the largest coverage hole. This idea is to let mobile nodes move to locations where it is strongly requested.

THE COVERAGE PROBLEM

The coverage paradigm was formulated by Gage (1992) and is divided into three groups of useful behaviors: blanket (or field) coverage, where the object is to achieve a static arrangement of sensors to maximize the detection rate of the targets in an area; barrier coverage, whose objective is to achieve a static arrangement of sensors with the task of minimizing the probability of undetected target penetration through the barrier; and sweep coverage that essentially represents moving barrier coverage or can be achieved using random uncoordinated motion of sensors.

The coverage can be measured by both the quantity, such as the measure of the covered area, and the quality, such as the degree of coverage. No matter which deployment method is used, the covered area is an important metric of the deployment. Sufficient area coverage means a better chance for detecting the event or for executing the

surveillance. An important issue for the coverage is its degree. A k-coverage satisfies that every point of the interested area can be covered by at least k sensor nodes at any time. The k-coverage usually concerns the robustness of the WSNs. That is, a k-covered network provides multi-path for routing, loading balance for transmitting, and thorough surveillance. Satisfying the k-covered condition, a network also has the ability to achieve fault-tolerance. We can find at least k nodes to cover an area. If any one or some of the k nodes are damaged, we can be easy to find one be an active node for surveillance. Gupta and Younis (2003) provide an example to cluster nodes for fault tolerance. However, how can we know whether an interested area is k-covered or not, especially in a distributed way. This problem will be introduced in the sub-section of judging K-coverage. We also need to consider the connectivity problem while talking about the coverage for a smoothly running WSNs. When the coverage area is wide and there are many communicating holes, an event detected or a query request cannot reach its destination. We will discuss the coverage problem under the condition of connectedness. The range of sensing or communicating is also an important factor affecting the coverage. The relation between coverage and connectivity is discussed in the following. Finally, we explain the worst case and the best case coverage and introduce how to dynamically maintain the two cases in the sub-section of the worst and best case coverage.

JUDGING *K*-COVERAGE

One important problem is how we judge an interested area is k-covered or not. A primitive solution is to find out all regions divided by their degree of coverage, and check if each region is k-covered or not. Some papers like (Huang, & Tseng, 2003; Maxim, & Gaurav, 2002) also discuss the problem of determining whether every point in a given area of the sensor network is covered by at least

k sensor nodes. Huang and Tseng (2003) examine whether the perimeter of each sensor's sensing range is covered, instead of determining the coverage of each location. Because they don't need to scan each region to examine the k-coverage, the algorithm complexity is $O(d \log d)$ where d is the maximum number of sensors that are neighboring to a sensor. We describe the idea used by Huang and Tseng (2003) as follows. First, each sensor determines all angles of arch which intersect other sensor nodes' coverage. Those angles are denoted by $[\alpha_{j,L}, \alpha_{j,R}]$ as shown in Figure 3. Place all values, $\alpha_{j,L}, \alpha_{j,R}$, in an ascending order on a line segment $[0, 2\pi]$. Finally, traverse the line segment from left to right to determine the coverage of one sensor. For example, the range between $\alpha_{j5,L}$ and $\alpha_{j4,R}$ is at least 3-covered for its three overlapped segments. The coverage level of the area covered by the sensor node s_i is increased by 1, which is 4-covered.

COVERAGE AND CONNECTIVITY

A k-covered area doesn't equal a k-connected area in wireless sensor networks. The converse is also true that a connected area doesn't guarantee a covered area. In fact, the relationship between connectivity and coverage depends on the ratio of the communication radius and sensing radius. We summarize some relations between coverage and connectivity from related papers (Wang et al. 2003; Zhang, & Hou, 2005).

Theorem 1 (Wang et al. 2003; Zhang, & Hou, 2005): For a set of sensors that at least 1-cover a convex region A, condition $r_c \geq 2r_s$ is both necessary and sufficient to ensure that coverage implies connectivity. Here, r_c is the communicate range of a sensor node and r_s is the sensing range of a sensor node.

We can use Figure 4 to demonstrate the scenario of Theorem 1. To prove the "necessary" part, we devise a scenario in which coverage does not imply

connectivity if $r_c < 2r_s$. In Figure 4 (a), a sensor s with communicate range r_c and sensing range $2r_s$ is circled by 6 nodes such that they together cover the whole disk centered at s with radius r_c. Since the communicate range is smaller than twice of the sensing range, node s cannot connect the other nodes circling it. The network is covered by sensors but it is not connected.

We use Figure 4 (b) to explain the "sufficient" condition. For any two nodes u and v, $\overline{uv}$ denotes the line segment joining them. Each point p on $\overline{uv}$ is at least 1-covered. Let nodes x and y are two disconnected but consecutive nodes which cover the line segment. Then, with $|xy| \geq 2r_s$ and $|px| \leq r_s$, the distance between y and p is at least r_s according to the triangle inequality. Node y cannot cover the line segment and that contradicts the assumption. Therefore, the distance between every pair of two consecutive sensors is less than $2r_s$, and is thus less than r_c. Under the condition that $r_c \geq 2r_s$, a sensor network only needs to be configured to guarantee coverage in order to satisfy both coverage and connectivity.

Theorem 2 (Zhang, & Hou, 2005): For a set of sensors that K_s-cover a convex region A, if $r_c \geq 2r_s$ the interior connectivity is $2K_s$.

Figure 3. An example of judging k-coverage

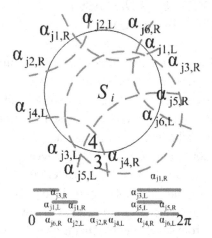

Figure 4. A necessary and sufficient condition for coverage and connectivity

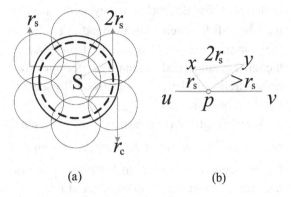

(a)   (b)

Here, we define boundary sensor as a sensor whose sensing range intersect with the boundary. All other sensor nodes in a region are interior sensor nodes except those boundary nodes. Usually, the interior sensor nodes have heavier traffic because of their locations. So, Theorem 2 highlights the property of interior connectivity. We define interior connectivity as the number of sensor nodes that must be removed to disconnect any two interior sensors in an interested area. From the above two theorems, we know that the boundary nodes that are located within r_s to the boundary are K_s connected; to the rest of the area, the interior connectivity is $2 K_s$.

THE WORST AND BEST CASE COVERAGE

Usually, the sensor distribution isn't uniform. The *k*-coverage is hard for all sensor nodes in an interested area. We consider another indicator for the quality of service of a sensor network. Meguerdichian, Koushanfar, Potkonjak, & Srivastava (2001) define two types of coverage: worst case coverage and best case coverage. In worst case coverage, attempts are made to quantify the quality of service by finding areas of lower observation from sensor nodes. On the contrary, in best

case coverage, finding areas of high observation from sensors are primary concern. From these issues, we may answer how well the network can observe a given area and what the chances are that an event occurring in a specific location will be detected. We can also find out the weak points in a sensor field to strengthen the deployment. Meguerdichian et al. (2001) proposed an algorithm to calculate the maximal breach paths (worst case coverage) and maximal support paths (best case coverage) in a sensor network. The maximal breach path is a path where its closest distance to any of the sensors is maximized. The maximal support path is a path where its farthest distance from the closest sensors is minimized. The authors use ideas from the Voronoi diagram (Aurenhammer, 1991) which is composed of line segments that are equidistant from neighboring sensors, and Delaunay triangulation (Mulmuley, 1994) which is formed by connecting nodes that share a common edge in the Voronoi diagram. They transform the continuous geometric problem into a graph problem and use the skills of Binary-Search and Breadth-First-Search to solve the problems. To stand to its definition, the maximal breach path must lie on the lines of Voronoi diagram, since any point deviating from the Voronoi line segments must be closer to at least one sensor. For the same reason, the maximal support path lies on the lines of the Delaunay triangulation of the sensors. The related example can be referenced by Meguerdichian et al. (2001). It must be noted that the maximal support path or the maximal breach path is not unique. In fact, there are many paths that can qualify as the maximal support path or as the maximal breach path.

After finding the maximal breath path and the maximal support path, the algorithm will calculate the breach weight and the support weight using edges weight. If new sensors can be deployed such that this breach weight is decreased, then the worst case coverage is improved. Similar to the worst case coverage formulation, if additional sensors can be deployed such that support weight

Figure 5. The best and worst case coverage radii

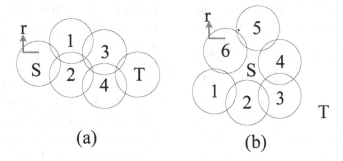

(a)   (b)

is decreased, then the best case coverage is improved. Li, Wan, and Frieder (2003) also proposed algorithms to find the best-coverage-path with the least energy consumption. Huang, Andréa, and Michael (2005) propose solution to dynamically maintain the two measures, the best case coverage and worst case coverage distance. They translate the best and worst coverage problems into graph connectivity problems. Define $U(r)$ to be the region on the plane composed of the union of all of the coverage disks when the coverage radius is r. $\overline{U(r)}$ is the complement of $U(r)$. For any two points S and T as in Figure 5, the best coverage radius is the minimum r such that S and T are in the same connected region of $U(r)$ while the worst coverage radius is the minimum r such that S and T are in two disconnected regions in $\overline{U(r)}$. They sort all sensors according to their distance to the point S and find a distance approximate to the best coverage radius. Since the algorithm use binary heap on the distance order, the update complexity is in $O(log\, n)$ time. A similar argument on the sequence of $\overline{U(r)}$ also gives an approximation of the worst coverage radius.

THE ENERGY-EFFICIENCY ISSUE

One of the major concerns of a deployment strategy is how to utilize it to make the WSNs energy-efficiency. We classify the energy effi-

ciency mechanisms into three categories which are duty cycle dependent, role dependent, and topology dependent. In the duty cycle dependent mechanism, they usually let several nodes do monitoring or communicating in turn. The goal of this mechanism is to make full use of every sensor node. In the role dependent mechanism, the assumption is that there are several types of nodes in the WSNs. Different types of nodes execute different functions. Since the energy consumption is related to the transmission range exponentially. Network topology will certainly affect the energy consumption. We survey important literatures for extending lifetime of the WSNs in this section.

DUTY CYCLE DEPENDENT

Duty cycle is the fraction of time when a sensor is on. Redundant sensors can reduce the duty cycle. A simple method is the design of a random sleep schedule (Huang, & Lin 2004). In this model, each sensor node enters the sleep state randomly and independently of each other. The network is essentially a collection of active/sleep process under random sleep schedule. This schedule does not guarantee coverage. However, we can infer the relation between deployment and duty cycle. Assume that static sensor nodes are deployed in a two-dimensional field as a stationary Poisson point process with density λ and the long term average sleep ratio of a sensor is p. Given an area A where A= $\lambda \pi r^2$ is the expected number of sensors deployed around the point of event with radius r, the probability that there are m sensor nodes in the area is $A^m e^{-A} \big/ m!$. The probability that a given point event is not covered by any active sensor in a given time slot is $P_u = \sum_{m=0}^{\infty} p^m \frac{A^m e^{-A}}{m!}$ $= e^{-A(1-p)}$. The associated conditional probability of uncoverage is $P_{u|c} = \frac{1}{1-e^{-A}} \sum_{m=0}^{\infty} p^m \frac{A^m e^{-A}}{m!}$ $= e^{-A(1-p)} (1-e^{-A})^{-1} (1-e^{-Ap})$. The relation between the deployment parameter (λ) and the duty cycle parameter (p) can be explained by the above equation.

Another kind of duty cycle is the design of coordinated sleep schedules. In this mode, sensor nodes coordinate with each other to decide when to enter sleep mode. Using this schedule leads to a more robust network, because it considers the coverage issue. The price for this kind method is the communication overhead and consumed energy. The idea of coordinated sleep is to let every node enter sleep mode in turn while maintaining coverage. An example is illustrated in Figure 6. Nodes 1, 2, 3, and 4 have its own sensing circle and node 1's sensing circle is covered by the other 3 nodes. Since even without node 1, there is no loss for the coverage. Node 1 can send request for its neighbors to ask to enter sleep mode. If node 1 gets a confirmation, it will wake up until the sleep time expires. Otherwise, if node 1 doesn't get the confirmation message, it will stay active for a period and resend the request later.

Since sensor nodes are deployed redundantly, in the coordinated sleep schedule, we can classify all sensor nodes into groups or clusters to achieve sleeping in turn to balance the energy consumption and prolong the lifetime of WSNs. Many papers discuss the related details. We need to consider the coverage and connectivity while processing clustering. Carle and Simplot (2004) select a connected dominating set of sensor nodes to monitor the network. A dominating set is a subset of network nodes in which each node is either in this subset or is a neighbor of a node in this subset. Let *S* be a subset of one node's all neighboring nodes. A node *u* is "fully covered" by a subset *S* if and only if (1). The set *S* is connected. (2). Any neighbor of *u* is a neighbor of at least one node from *S*, and (3). All nodes in *S* have higher priority than *u*. The priority of a node is decided by its energy level. A node belongs to the dominating set if and only if no subset fully covers it. The nodes belonging to a dominating set are gateway nodes; others are non-gateway nodes. These gateway nodes monitor all nodes in the sensor network. Figure 7 shows an example that uses the node identifier for priority. For node 1, *S*= {2, 3, 6} and *S* satisfies the three conditions, so nodes 2, 3, and 6 fully cover node 1. Consequently, node 1 is a non-gateway node. Nodes 2, 3 and 4's conditions are the same as node 1. For node 5, *S*= {6, 7} and node 5's additional neighbor, node 4, is not a neighbor node of *S*, so node 5 is not "fully covered" and it is a gateway node. For nodes 6 and 7, the corresponding set *S* is empty. Therefore, they are gateway nodes because their neighbors have no node in *S* to satisfy the second condition.

The dominating set only finds a covered and connected set of sensor nodes. How can we achieve the sensor scheduling for *k*-coverage? That is, given a sensor network with *n* sensors that can provide *k*-coverage for the monitored region, schedule the activities of the sensors such that at

Figure 6. An example of coordinated sleep schedules

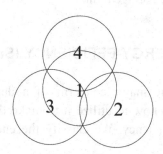

Figure 7. An example for dominating set (gateway nodes)

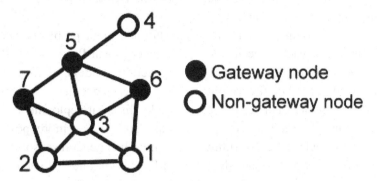

any time, the whole region can be k-covered and the network lifetime is maximized (Yingshu, & Shan, 2008). The algorithm proposed by Yingshu and Shan (2008) finds a collection of subsets, and each subset can k-cover the whole monitored region. As a result, we can use those subsets to monitor the interest area in turn while achieving maximum network lifetime. The idea of k-perimeter-covered was proposed by Huang and Tseng (2003) using a greedy algorithm, *PCL-Greedy-Selection* (GS). A monitored region is k-covered if and only if each sensor in the area is k-perimeter-covered. The *Perimeter Coverage Level* (*PCL*) of a sensor is the number of other sensor nodes which cover any point on the sensor' perimeter of sensing area. A smaller *PCL* means that the sensor node density is smaller. GS iteratively constructs subsets and chooses a node from the interested area one by one depending on a sensor' *PCL* for each subset. The algorithm GS picks up a sensor with lower density first to join a subset. Those nodes with lower density mean that they are distributed as widely as possible. In this way, more sensor nodes are left to join other subsequent subsets. We may get redundant sensor nodes in GS. So, we still have to check each sensor node in each subset to see if a sensor node' removing will affect the coverage of its own subset. According to the steps in GS, the time complexity is O(n^2 dlog(d)) where n is the number of the sensor nodes and d is the maximum number of neighbors of n sensor nodes.

ROLE DEPENDENT

The common three types of nodes in wireless sensor network are 1) microsensor nodes, 2) relay nodes (aggregation and forwarding nodes), and 3) base stations. We focus on the deployment of the general sensor nodes in above section. However, the other two kinds of nodes, relay nodes and base stations, are also critical to the lifetime of WSNs. We discuss both of them in this section. Here, we assume that relay nodes and base stations are powerful than general sensor nodes in transmitting range. All other sensor nodes have the same transmission radius r.

Relay Nodes

We can consider the relay nodes deployment problem in two aspects. 1) What is the minimum number of relay nodes we should deploy to make sure that there is a path between any pair of sensor nodes. 2) How to deploy a set of relay nodes to improve the connectivity of a given network. In this sub-section, we consider that sensor nodes are vertices (V), connections between two nodes are edges (E), and the distances between two connectable nodes are weights (W) in a graph (G).

The first aspect is exactly the Steiner minimum tree with minimum number of Steiner points and bounded edge length problem (SMT-MSPBEL). For this NP-hard problem (Garey, & Johnson, 1979; Lin, & Xue, 1999), Errol and

Guoliang (2007) proposed a polynomial time 7-approximation algorithm. Based on the topology of sensor nodes, the algorithm computes a minimum spanning tree. For each edge in the minimum spanning tree, they decide how many nodes should be deployed depend on the weight of edge. If the weight is smaller than the twice sensor communication range ($2r$) and larger than the sensor communication range, the algorithm places a relay node at the midpoint of the edge. If the weight is larger than $2r$, it places a relay node at the point of the edge every relay node's communication range.

Ibrahim, Seddik, and Liu (2007) also discuss how to deploy relay nodes to aim at improving the network connectivity of a given network. Assume $|V| = n$, $|E| = m$, and n_c is equal the number of candidate relay nodes. For an edge e_l, $1 \leq l \leq m$, connecting nodes v_i, and v_j, $\{ v_i, v_j \} \in V$, define the edge vector $a_l \in R^n$, where $a_{e,i} = 1$, $a_{e,j} = -1$, and the rest is zero. The matrix $A \in R^{n \times m}$ of the graph G is the matrix with l-th column given by a_l. The $n \times n$ Laplacian matrix L is defined as $L = AA^T = \sum_{l=1}^{m} a_l a_l^T$. The diagonal entry $L_{i,i}$ is the degree of node i. If $(v_i, v_j) \in E$, $L_{i,j} = -1$. Otherwise, $L_{i,j} = 0$. The second smallest eigenvalue of L, $\lambda_2 (L)$, called *Fiedler value*, represents the algebraic connectivity of the graph G (Fiedler, 1973). The authors use it to measure how connected a graph is. They first divide the network area into n_c equal square regions. Secondly, they find a value K, $K < n_c$, to get the maximum $\lambda_2 (L)$. After deciding the number of relay nodes to deploy, they find where to deploy a relay node in each region. They then divide the region into n_c sub-regions, and find a sub-region to make *Fiedler value* maximum. The algorithm repeats the division until no improvement is possible in the *Fiedler value*.

Base Stations

The placement problems can be formulated as Integer Linear Programs (ILPs) by probabilistic detection models. Different constraints with guaranteed coverage, connectivity, or robustness make different deployment strategies. Since the base station has more powerful computation ability than general sensor nodes do, the way of ILPs is more suitable for solution of placement of base station. In the initial deployment situation, we have the joint problem of determining sensor nodes, relay and base station placements. From different aspects such as minimum sensor nodes, costs, or energy, we have different benchmark based on bandwidth, distance, energy and so on. Another most common situation for base station is that they are fixed in some position. Usually, a sensor network has only a single sink or a base station. However, in the case of several base stations, the data traffic generated by any sensor node may split and be sent to multiple different BSs. Another issue is how to optimally match a sensor node to a proper BS to maximize the network lifetime. Hou, Shi, and Sherali (2006) also use linear programming constraining with traffic flow and energy consumed to find the upper bound of network lifetime. Each base station has its own traffic flow coming from different relay nodes or sensor nodes under the lifetime restriction. If the flow from one relay/sensor node is larger than a specified threshold, the relay/sensor node selects the base station as its destination. In the case of not being larger than a threshold, the node chooses the largest one among those to all base stations. To summarize the method, they use a sequential fixing procedure to find the destinations for all nodes.

TOPOLOGY DEPENDENT

With the same reason for static deterministic deployment, the topology dependent deployment wants to achieve better performance by considering the whole interested area. In this sub-section we focus on the energy efficiency issue with network topology. Minimizing the number of active sensor nodes is correlated with maximizing the lifetime of a sensor network. Here, we discuss three types of common topology for sensor networks and show the required number of active nodes for each topology while considering coverage and connectivity. Three patterns illustrated in Figure 8 are square-grid based, hexagonal-grid based, and strip based topology (Iyengar, Kar, & Banerjee, 2005). We assume that sensing range is equal to the communicating range of a sensor node and the circles mean sensing range of sensor nodes. In the square-grid based topology, the sensor nodes are located at the grid-point as in Figure 8(a). Each node is in the transmission range of four other neighbor nodes. For a $R \times R$ square region, there are (R/r) grid lines in both horizontal and vertical directions. So, R^2/r^2 nodes are in the square region, and the spatial density of the square-grid based topology is $1/r^2$. That means we need R^2/r^2 active nodes for a square-grid based topology. In the hexagonal-grid based topology, the sensor nodes are located at the alternate vertices of the hexagon, and one at the center of each hexagon as in Figure 8(b). Each node is in the transmission range of three other neighbor nodes. For a $R \times R$ square region, there are $(2R/(r+r/2))$ horizontal lines to calculate the active nodes. The gap between two adjacent active nodes in each horizontal line is $\sqrt{3}\,r$. So, $4R^2/3\sqrt{3}\,r^2$ nodes are in the square region, and the spatial density of the hexagonal-grid based topology is approximately $0.769/r^2$. That means we need $4R^2/3\sqrt{3}\,r^2$ active nodes for a hexagonal-grid based topology. In the strip based topology, a strip is the collection of sensors horizontally spaced by r as in Figure 8 (a) and the active sensor nodes are located as in Figure 8(b). Only one set of sensor nodes is used to connect those horizontal strips for communication as in 4 shaded circle shown in the right side

of Figure 8(d). The gaps between two adjacent sensor nodes are r in horizontal and $(1+\sqrt{3}/2)$ r in vertical. For a $R \times R$ square region, there are $(R/(1+\sqrt{3}/2)\,r)$ strips to calculate the active nodes. The gap between two adjacent horizontal active nodes in each strip is r. So, there are $R^2/(1+\sqrt{3}/2)r^2$ nodes in the square region, and the vertically string of active nodes can be negligible when R is larger. Therefore, the spatial density of the strip based topology is approximately $0.536/r^2$. That means we need $R^2/(1+\sqrt{3}/2)r^2$ active nodes for a strip based topology. The redundancy of strip-based is the smallest among those three pattern mentioned above. It means the situation is theoretically the best choice for topology based deployment while considering connectivity and coverage. However, only one vertically strip to connect all communication of sensor nodes can not ensure the quality of network.

Another method is topology adjustment deployment to decide the proper location one by one while considering the minimum power consumption. An incremental self-deployment algorithm was proposed by Wei and Cassandras (2005). The addition of a sensor node is a 3-step process. 1) Determining the bottleneck node: The bottleneck node is a sensor node that a newly adding node connects to make the most improvement to the power conservation of the whole network. The authors design an inner force field for each sensor node to find a bottleneck node. The inner force field, which is the same idea as virtual force mentioned in the section of classification of deployment strategy, is the force applied to a node and generated by one of its link to other node. We can get the composition of forces field applied on the node. If the composition force is zero, the node is in the equilibrium state and corresponds to the minimum-power deployment policy. In step 1, we find a node with the largest inner force. 2) Enumerating topology adjustment options: The bottleneck node indicates the area which needs a new node mostly. When a new sensor node is

Figure 8. Examples for a square-grid based topology(a) and the hexagonal-grid based topology(b), a strip (c) and the strip based topology (d)

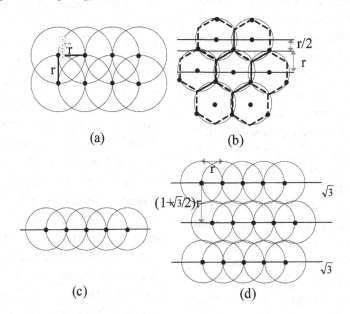

(a) (b)

(c) (d)

inserted in a network, the network topology will change. In this step, enumerate probable topologies while considering the relatives (like parents and children in a tree data structure) between a bottleneck node and a newly inserted node. 3) Obtaining a new deployment: With the set of possible new topologies, find the one with the least power consumption.

CONCLUSION AND FUTURE TRENDS

In this chapter, we introduce the deployment strategies and taxonomy to compare the related literatures. Static deployment strategies focus on those fixed sensors after the initial deployment. Dynamic deployment strategies provide moving direction for mobile nodes. Both of those strategies consider connectivity and coverage. We also use different views from coverage and energy efficiency separately to consider deployment strategies. The coverage requirement determines the quality of sensor network. The quality means

the robustness which is high probability to detect an event and the efficiency which is how fast a sensor network can reply to an event. The energy efficiency means how long a sensor network can function. In fact, these two requirements are conflicting optimization goals for a sensor network due to its limited energy. A k-coverage deployment with a large k incurs the cost of exhausting network lifetime rapidly. A sensor network with stingy energy consumption will reduce event detection probability or cause detection delay. Energy efficiency and coverage is a tough trade-off problem. Shansi, Qun, Haining, Xin, and Xiaodong (2007) proposed an analytical model for the trade-off between detection quality and energy consumption. The detailed mathematical analysis can be seen in the reference (Shansi et al., 2007). In summary, the solution depends on the goal of sensor network application. That is, based on different scenarios, we should choose appropriate sensing schedule such as coverage first or energy efficiency first.

In the future, we have the vision that sensor networks will surround our environment and we can even query the WSNs by connecting to the Internet. We need the sensor nodes to reply the messages fast enough. We also hope the sensor network can get all information back even for those places that originally have no sensor nodes by moving some mobile sensors. That means two trends of deployment. 1.) The nodes' deployment should adjust their locations by the demand of people instead of just coverage. 2.) Due to the integrated environment, the energy utilization must consider the related communication equipments such as cell phones, notebooks, or even vehicles to achieve a longer lifetime and better utilization.

REFERENCES

Akyildiz, I., Su, W., Sankarasubramaniam, Y., & Cayirci, E. (2002). Wireless sensor networks: A survey. *Computer Networks*, *38*, 393–422. doi:10.1016/S1389-1286(01)00302-4

Andrew, H., Maja, J. M., & Gaurav, S. S. (2002a). An Incremental Self-Deployment Algorithm for Mobile Sensor Networks. *Autonomous Robots Special Issue on Intelligent Embedded Systems*, *13*, 113–126.

Andrew, H., Maja, J. M., & Gaurav, S. S. (2002b). Mobile Sensor Network Deployment using Potential Fields: A Distributed, Scalable Solution to the Area Coverage Problem. *Proceedings of the 6th International Symposium on Distributed Autonomous Robotics Systems* (pp. 299-308). Fukuoka, Japan.

Aurenhammer, F. (1991). Voronoi Diagrams – A Survey Of A Fundamental Geometric Data Structure. *ACM Computing Surveys*, *23*, 345–405. doi:10.1145/116873.116880

Carle, J., & Simplot, D. (2004). Energy Efficient Area Monitoring by Sensor Networks. *IEEE Computer*, *37*(2), 40–46.

Chang, R. S., & Wang, S. H. (2008). Self-Deployment by Density Control in Sensor Networks. *IEEE Transactions on Vehicular Technology*, *57*(3), 1745–1755. doi:10.1109/TVT.2007.907279

Chellappan, S., Wenjun, G., Xiaole, B., Dong, X., Bin, M., & Kaizhong, Z. (2007). Deploying Wireless Sensor Networks under Limited Mobility Constraints. *IEEE Transactions on Mobile Computing*, *6*(10), 1142–1157. doi:10.1109/TMC.2007.1032

Errol, L. L., & Guoliang, X. (2007). Relay Node Placement in Wireless Sensor Networks. *IEEE Transactions on Computers*, *56*(1), 134–138. doi:10.1109/TC.2007.250629

Fiedler, M. (1973). Algebraic connectivity of graphs. *Czechoslovak Mathematics Journal*, 298-305.

Gage, D. W. (1992). Command control for many-robot systems. *The Nineteenth Annual AUVS Technical Symposium* (pp. 22-24). Huntsville, Alabama, USA, Garey, M.R., & Johnson, D.S. (1979). *Computers and Intractability: A Guide to the Theory of NP-Completeness*. New York: W.H Freeman and Co.

Gupta, G., & Younis, M. (2003). Fault-tolerant clustering of wireless sensor networks. *2003 IEEE Wireless Communications and Networking*, *3*, 1579-1584.

Heinzelman, W., Chandrakasan, A., & Balakrishnan, H. (2002). An application-specific protocol architecture for wireless microsensor networks. *IEEE Transactions on Wireless Communications*, *1*(4), 660–670. doi:10.1109/TWC.2002.804190

Hou, Y. T., Shi, Y., & Sherali, H. D. (2006). Optimal base station selection for anycast routing in wireless sensor networks. *IEEE Transactions on Vehicular Technology*, *55*(3), 813–821. doi:10.1109/TVT.2006.873822

Huang, C. F., & Lin, M. (2004). Network coverage using low duty-cycled sensors: random & coordinated sleep algorithms. *Proceedings of the 3rd international symposium on Information processing in sensor networks* (pp. 433-442). Berkeley, California, USA.

Huang, C. F., & Tseng, Y. C. (2003). The coverage problem in a wireless sensor network. []. San Diego, California, USA.]. *ACM WSNA, 03,* 115–121. doi:10.1145/941350.941367

Huang, H., Andréa, W. R., & Michael, S. (2005). Dynamic Coverage in Ad-Hoc Sensor Networks. *Mobile Networks and Applications, 10*(1-2), 9–17. doi:10.1023/B:MONE.0000048542.38105.99

Ibrahim, A. S., Seddik, K. G., & Liu, K. J. R. (2007). Improving Connectivity via Relays Deployment in Wireless Sensor Networks. *IEEE Global Telecommunications Conference,* (pp. 1159–1163).

Iyengar, R., Kar, K., & Banerjee, S. (2005). Low-coordination Topologies for Redundancy in Sensor Networks. *Proceedings of the 6th ACM International Symposium on Mobile Ad Hoc Networking and Computing* (pp. 332-342), Urbana-Champaign, IL, USA.

Li, X. Y., Wan, P. J., & Frieder, O. (2003). Coverage in Wireless Ad Hoc Sensor Networks. *IEEE Transactions on Computers, 52*(6), 752–763.

Lin, G., & Xue, G. (1999). Steiner tree problem with minimum number of Steiner points and bounded edge-length. *Information Processing Letters, 69,* 53–57. doi:10.1016/S0020-0190(98)00201-4

Maxim, A. B., & Gaurav, S. S. (2002). *Multi-robot Dynamic Coverage of a Planar Bounded Environment* (Technical report). U.S.A., University of Southern California, Robotic Embedded Systems Laboratory.

Meguerdichian, S., Koushanfar, F., Potkonjak, M., & Srivastava, M. (2001). Coverage Problems in Wireless Ad-Hoc Sensor Networks. *Proceedings of IEEE Infocom,* (pp. 1380-1387).

Mhatre, V. P., Rosenberg, C., Kofman, D., Mazumdar, R., & Shroff, N. (2005). A minimum cost heterogeneous sensor network with a lifetime constraint. *IEEE Transactions on Mobile Computing, 4*(1), 4–15. doi:10.1109/TMC.2005.2

Mulmuley, K. (1994). *Computational Geometry: An Introduction Through Randomized Algorithms.* New Jersey, USA: Prentice-Hall.

Nojeong, H., & Pramod, K. V. (2003). An Intelligent Deployment and Clustering Algorithm for a Distributed Mobile Sensor Network. *Proceedings of the 2003 IEEE International Conference on Systems, Man & Cybernetics,* vol.5 (pp. 4576-4581).

Santosh, K., & Ten, H., Lai, & József, B. (2008). On k–coverage in a mostly sleeping sensor network. *Wireless Networks, 14*(3), 277–294. doi:10.1007/s11276-006-9958-8

Shansi, R., Qun, L., Haining, W., Xin, C., & Xiaodong, Z. (2007). Design and Analysis of Sensing Scheduling Algorithms under Partial Coverage for Object Detection in Sensor Networks. *Transactions on Parallel and Distributed Systems, 18*(3), 334–350. doi:10.1109/TPDS.2007.41

Wang, G., Cao, G., Berman, P., La, P., & Thomas, F. (2007). Bidding Protocols for Deploying Mobile Sensors. *IEEE Transactions on Mobile Computing, 6*(5), 515–528. doi:10.1109/TMC.2007.1022

Wang, X., Xing, G., Zhang, Y., Lu, C., Pless, R., & Gill, C. D. (2003). Integrated Coverage and Connectivity Configuration in Wireless Sensor Networks. *Proceedings of the First ACM Conference on Embedded Networked Sensor Systems* (pp. 28-39). Los Angeles, CA.

Wang, Y. C., Hu, C. C., & Tseng, Y. C. (2008). Efficient Placement and Dispatch of Sensors in a Wireless Sensor Network. *IEEE Transactions on Mobile Computing, 7*(2), 262–274. doi:10.1109/TMC.2007.70708

Wei, L., & Cassandras, C. G. (2005). A minimum-power wireless sensor network self-deployment scheme. *IEEE Wireless Communications and Networking Conference*, Volume 3, (pp. 1897–1902).

Wenliang, D., Jing, D., Han, Y. S., & Shigang, C. (2004). A key management scheme for wireless sensor networks using deployment knowledge. *Twenty-third Annual Joint Conference of the IEEE Computer and Communications Societies Publication, Vol. 1.* (pp. 586-597).

Yi, Z., & Krishnendu, C. (2004). Sensor Deployment and Target Localization in Distributed Sensor Networks. *ACM Transactions on Embedded Computing Systems, 3*(1), 61–91. doi:10.1145/972627.972631

Yingshu, L., & Shan, G. (2008). Designing *k*-coverage schedules in wireless sensor networks. *Journal of Combinatorial Optimization, 15*(2), 127–146. doi:10.1007/s10878-007-9072-6

Younis, M., & Akkaya, K. (2008). (in press). Strategies and Techniques for Node Placement in Wireless Sensor Networks: A Survey. *Ad Hoc Networks, 6*(4), 621–655. doi:10.1016/j.adhoc.2007.05.003

Zavlanos, M. M., & Pappas, G. J. (2008). Dynamic Assignment in Distributed Motion Planning With Local Coordination. *IEEE Transactions on Robotics, 24*(1), 232–242. doi:10.1109/TRO.2007.913992

Zhang, H., & Hou, J. C. (2005). Maintaining Sensing Coverage and Connectivity in Large Sensor Networks. *Ad Hoc and Sensor Wireless Networks, 1*(2), 89–123.

Zhang, H., & Hou, J. C. (2006). Is Deterministic Deployment Worse than Random Deployment for Wireless Sensor Networks? *IEEE International Conference on Computer Communications* (pp. 1-13). Barcelona, Spain.

Zhipu, J., & Murray, R. M. (2007). Random consensus protocol in large-scale networks. *46th IEEE Conference on Decision and Control* (pp. 4227-4232).

KEY TERMS AND DEFINITIONS

Connectivity: The connectivity is an important measure of its robustness as a network. A network has good connectivity, which means any node in this network can connect another directly or indirectly.

Coverage Problem: The problem about how large is an area covered by sensors.

Deployment Strategy: The method about how to place sensors in an interested area.

Dynamic Network: In a WSN, it includes at least one mobile node.

Energy Efficiency: In a wireless sensor network, we need sensors to work in a longer lifetime. So, we need sensor with low energy consumption when they work.

Sensor Network: A wireless sensor network (WSN) is a wireless network consisting of spatially distributed autonomous devices using sensors to cooperatively monitor physical or environmental conditions, such as temperature, sound, vibration, pressure, motion or pollutants, at different locations.

Static Network: In a WSN, sensor nodes are stationary and will not move after the initial deployment.

Chapter 3
Collaborative Sensor Networks:
Taxonomy and Design Space

Yuan He
Hong Kong University of Science and Technology, China

Yunhao Liu
Hong Kong University of Science and Technology, China

ABSTRACT

Sensor networks have been widely deployed in all kinds of applications. In the near future, we expect to witness the proliferation of sensor networks with a variety of functions that requires a comprehensive collaboration mechanism among them. Specific designs are necessary to manipulate a fabric of multiple sensor networks, facilitate the collaboration among them, and support efficient data sensing, aggregation, storage, transmission, and query processing. In this chapter, through comparison and analysis of existing works, we present a new taxonomy of collaborative sensor networks based on the architectures and the methodologies of organizing different modules of a system. We find that a cross-layer design simultaneously guaranteeing the energy efficiency of underlying sensor networks and the quality of services for users is a challenging issue which has not been well studied. We also discuss the design space in this field and point out potential research directions.

1. INTRODUCTION

Sensor networks have attracted a lot of attention during the past decade. Due to the recent advances in wireless communication and micro-electronic technologies, both the price and size of sensors decrease quickly. Today's applications of sensor networks range from personal to mission critical systems including scientific observation, digital life,

home automation, environment surveillance, traffic monitoring, and so on (Gao, Massey, Selavo, Welsh, & Sarrafzadeh, 2007; Li & Liu, 2007b; Szewczyk, Mainwaring, Polastre, Anderson, & Culler, 2004; Yang, Li, & Liu, 2007; Zhang, Sadler, Lyon, & Martonosi, 2004). Many of them are developed and promoted by governments, enterprises and public organizations, offering continuous collection of real-time information, satisfying the requirement of people's daily life.

DOI: 10.4018/978-1-61520-701-5.ch003

In the foreseeable future, we expect to witness the proliferation of sensor networks with a variety of functions that requires a comprehensive collaboration mechanism among them. Specific designs are necessary to manipulate a fabric of multiple sensor networks, facilitate the collaboration among them, and support efficient data sensing, aggregation, storage, transmission, and query processing. Previous studies in sensor networks, however, mainly focus on the performance and efficiency inside a single sensor network.

In this chapter, we broaden the view into the scope of multiple sensor networks. The term "collaborative sensor networks" refers to the sensing systems with the following characteristics: first, it is an information system providing live sensing and query processing services; second, multiple heterogeneous sensor networks are integrated in; third, sensor networks in the system collaborate with each other in accomplishing the querying tasks.

We present a new taxonomy for describing and classifying infrastructures and designs of collaborative sensor networks. The taxonomy categorizes the collaborative sensor networks based on their systematical architectures and functionalities. This contributes a deep and comprehensive view to the current studies. We also discuss and summarize the challenging problems and potential issues in the community of collaborative sensor networks.

The remainder of the chapter is organized as follows. Section 2 briefly discusses the state-of-art in sensor networks. An overview of collaborative sensor networks is presented, covering all the elements from a systematical point of view. Section 3 elaborates the taxonomy of collaborative sensor networks. Challenging issues and potential research directions in this field are discussed in Section 4. We conclude this chapter in Section 5.

2. OVERVIEW

2.1 Background and Motivation

Sensor networks pose a number of unique technical challenges due to the following factors, such as Ad hoc deployment, unattended operation, untethered power supply, and dynamic changes in both connectivity and environments. Designs of sensor networks are thus emphasized to be energy-efficient and highly adaptive, which generally can be categorized into two classes: software and hardware. Research works in both classes have achieved remarkable successes in the past decade. Software focuses on algorithms, protocols, operating systems, middleware, and programming languages designed for networks of embedded sensors in various application scenarios (He, Stankovic, Lu, & Abdelzaher, 2005; Li & Liu, 2007a; Lian, Naik, Chen, Liu, & Agnew, 2007; Xiao, Chen, & Zhou, 2008). Hardware includes the work on sensor platforms, tools, implementations, measurements, and sensing devices, etc. Specifically, algorithms in sensor networks are mainly interested in energy-efficient sensing, data delivery, localization, and routing techniques. We have also seen a number of successfully applied operating systems (e.g. T-kernel (Gu & Stankovic, 2006)), sensor motes (e.g. Mica (Hill & Culler, 2002)), programming languages (e.g. nesC (*nesC: A Programming Language for Deeply Networked Systems*)), and real deployments of sensor networks (e.g. GreenOrbs (*GreenOrbs*), OceanSense (*OceanSense*, ; Yang et al., 2007)).

Most of the existing works, however, only concentrate on the issues in a single sensor network. Networked sensors are regarded to be geographically close to each other and with homogeneous functionalities, while different sensor networks are seldom considered related. For example, sensor networks in a sea-depth monitoring sensor networks are deployed in a constrained area of offing. They are uniformly equipped with acoustic sensing devices and generate homogeneous data of sea depth. Many similar cases may be found in other scenarios, such as traffic navigation, environmental surveillance, and home automation. Even if the data from multiple sensor networks are naturally related in a common application, such correlations are particularly studied. Up to now people still prefer to process heterogeneous sensory data statically and locally.

On the other hand, sensor networks gradually become a class of facilities for more and more citizen users, which produce increasing needs of live sensory data from real world. For example, a scheduling system of a harbor requires information of weather, seawater, harbor business, and the roadway traffic from the city center to the harbor. Another example is geospatial observation. Subaerial sensor networks such as thermometers, dogvanes, and seismoscope, as well as the aero observers such as satellites must collaborate to achieve comprehensive knowledge of the global environments. And it must be an inevitable trend that collaboration of multiple heterogeneous sensor networks become an essential feature in the future networked sensing systems.

Obviously, the state-of-arts in traditional sensor networks, which take homogeneity and isolation of sensor networks for granted, cannot satisfy the above requirements. Researchers have begun to study the collaborative mechanisms for multiple sensor networks, providing integrated sensing services for various user applications (Pepe, Borgman, Mayernik, & Wallis, 2007; Reddy et al., 2007). Due to the heterogeneity and energy-constrained features, we may meet the following challenges to enable the approach to collaborative sensor networks. The first obstacle is the heterogeneities of data, locations, communication protocols, and functionalities. Automatic data collection, aggregation / fusion, and query processing is essential in such architectures. A uniform interface and formalization is required to interpret, exchange, and share data. It becomes much more complicated and difficult to achieve these functionalities in the context of multiple sen-

sor networks. Moreover, to integrate the multiple independent sensor networks into a collaborative sensing and querying process is also hard, especially when the sensor networks are geographically distributed, heterogeneous, and managed by different authorities. When considering the requirements from the application level, query diversity and efficiency of bandwidth utilization are two other important factors concerned. Most of the existing works, however, fail to address these challenging problems.

2.2 Building Blocks and Techniques

In this subsection, we briefly introduce the building blocks and techniques in collaborative sensor networks (He et al., 2008). Generally a system of collaborative sensor networks can be divided into several main modules, including sensing entities (i.e. sensor networks or distributed sensor motes), data and query processor, query input, and data output. Query processor consists of three parts, i.e. sensing control, query scheduling, and data aggregation. In practical systems, these three parts are not necessarily deployed together or separately. For example, data aggregation might run on the sinks of sensor networks, while sensing control and query scheduling might run on a central management server. Figure 1 illustrates a typical flow chart. Queries are accepted from the external users. After interpretation and scheduling, queries are translated into concrete sensing actions. The module of sensing control globally coordinates the sensing actions and interacts with all the sensing entities (i.e. sensors / sensor networks). It issues sensing requests and retrieve live sensory data from those sensing entities, which are then aggregated and stored by the module of data aggregation. The remainder of this subsection briefly introduces the above building blocks.

Figure 1. The flow chart of typical collaborative sensor networks

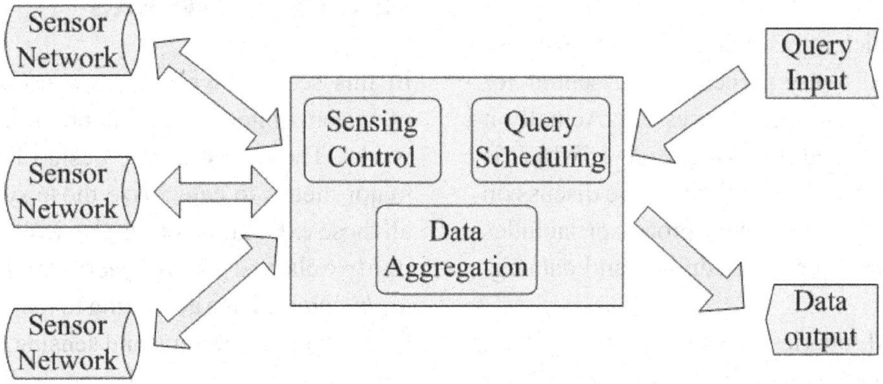

2.2.1 Infrastructure Design

The infrastructure design refers to the methodology (or mechanism) to organize different modules. Because collaborative sensor networks are deployed for different purposes, the ways for such an integrated system to control sensing, schedule queries, and share data might differ much from each other.

The major difference lies in the system architecture. As we elaborate in Section 3, some existing works adopt flat architecture, where sensor networks and entities are autonomous and self-organize into a fully decentralized system. There is no central server or centralized data storage. Among them there are some systems, where sensor networks altruistically collaborate with each other so that they can act as a whole for a common purpose. On the other hand, there are hierarchical systems, where sensor networks are controlled and scheduled by centralized servers or management services.

2.2.2 Query Input and Data Output

There are mainly two types of methods, with respect to the manners of query input and data output: *web-based* and *web-free*. *Web-based* method is mostly adopted by hierarchical systems, which provides a web portal for accepting external queries and exhibiting the sensory data and queried results. Certain interfaces are required for the users to input queries in a computer-comprehensible way. Moreover, the sensory data and queried results are usually visualized into graphical figures, charts, and so on. *Web-free* method is mostly adopted in flat architectures, where there is not necessarily any website present. The sensor networks, especially represented by their sinks, can be the query issuers and data consumers.

When we consider the orientation of data flows inside the system, there are also two types of forms: *horizontal* and *vertical*. *Horizontal* data flows correspond with the flat architecture. Since sensor networks are organized in a flat layer, queries and data both flow horizontally inside the layer. The queries may be issued from and sent to any point in the network, such as a sensor network or simply a sensing device. Similarly, the sensory data are shared among *au pair* sensor networks. On the contrary, in most hierarchical systems, the main orientation of data flows is *vertical*. Queries are accepted from the upper web portals, interpreted and scheduled by the intermediate processing modules, and then delivered to the lower sensor networks. After sensing the required data, the results are returned in the reverse direction. Note that there are also partial horizontal data flows in the hierarchical systems because sensor networks need to share data for collaborations.

2.2.3 Query Processor

No matter which infrastructure the system owns, the module of query processor is essential for collaborations and query processing. Actually it can be decomposed into parts, each of which is deployed on different positions, as we discuss on Fig. 1. The mission of query processor includes sensing control, query scheduling, and data aggregation.

First of all, queries injected into the system, no matter issued by sensor networks themselves or by external users, may be substantive. Therefore they should be reasonably scheduled before delivered to the underlying sensor networks for the efficiency of sensing and data reusing.

Scheduled queries are translated into a series of concrete sensing actions, which are coordinated by the module of sensing control. When multiple sensor networks are concerned in a querying task, a well-scheduled sensing process can save a lot of energy than a poorly-scheduled one. This difference becomes more apparent when different areas covered by the sensor networks have different popularities. It is a challenging issue to design energy-efficient algorithms for query scheduling and sensing control in the context of multiple sensor networks.

Data aggregation is the process to fuse sensory data, store them, and combine them into queried results. The aggregated data is not only useful for live queries, but also reusable for the subsequent queries demanding the same data, as long as the stored data do not expire. Since sensor networks can continuously generate huge data volumes, it is obviously unacceptable to store all of them. Some existing aggregation methods work only for a single network, but can not be directly migrated to multiple sensor networks. How to aggregate the sensory data during sensing and combine the heterogeneous data into usable queried results is another challenging issue.

3. TAXONOMY OF COLLABORATIVE SENSOR NETWORKS

In this section, we present a novel taxonomy of systems integrating collaborative sensor networks. The infrastructure design is used as the major metric to categorize those systems. First, all those existing works are basically categorized into two classes: *flat architecture* and *hierarchical architecture*. Then according to the differences in the designs of querying and sensing, each class is sorted into several sub-classes, as detailed in the remainder of this section.

3.1 Flat Architecture

The first class of sensor networks is called flat architecture. In this class, sensor networks, as well as some independent sensing devices, are connected equally in a common network. There is not any centralized controlling unit or server deployed in the whole system. Sensor networks and devices collaborate with each other for a common purpose on the whole, or representing the users controlling them as query issuers and data consumers. According to the roles of sensor networks in a whole system, flat architecture systems can be classified with *unified model* and *autonomous model*.

3.1.1 Unified Model

The first collaborative mechanism for sensor networks is proposed by Delin et al. in (Delin, 2002). In their proposal named Sensor Web (Note that it has same name with the web-based sensing systems introduced in Section 3.2.2, but has totally different meanings), different types of sensing devices owned by a common authority, are geographically distributed and connected as a unified networking system.

The basic concept of a network of sensors has been proposed since last century. The feature of the unified model lies in the ability of the indi-

Figure 2. The sketch of Sensor Web

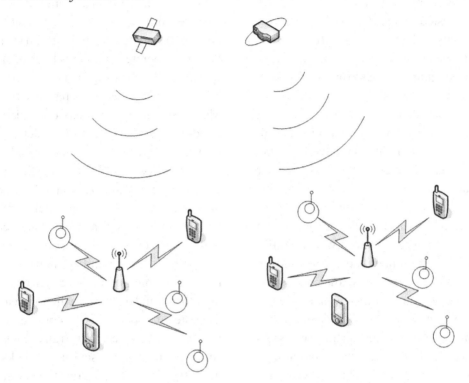

vidual sensing entities in a Sensor Web to act and coordinate as a whole. Heterogeneous sensors are included in the system and interact with each other through wireless (sometimes wired) communications. It thus allows the whole system to be synchronized and with synthetical actions, which differs from many other networks. In addition, the individual entities of a Sensor Web are all equal with one another. In the Sensor Web architecture, it does not require special gateways or routing mechanisms to enable each of the individual pieces communicate with one another or an end user. Data sensed by a sensor are delivered to and utilized by some other sensors such that all the sensors can act a whole for their common functionalities. By definition, a Sensor Web is an autonomous stand-alone sensing system, which does not need to assemble resources among sensor networks through the presence of the World Wide Web to function.

Figure 2 sketches a Sensor Web system applied for observation on geospace. Computational devices and sensors, including stationary sensor motes, mobiles, cameras, and satellites, are connected through wireless communications and organized into an integrated system. It differs much from the conventional sensor networks in two points: first, it has heterogeneous sensing devices with different functionalities and features; second, the sensors are geographically distributed, even very far away from each other. Communication ranges are no longer assumed to be limited especially for those aerospace monitors. There are also some wireless base stations for transmission relays deployed on the ground.

As we introduced, the above described system is deployed for geospace observation. We may notice that there is neither any external user nor centralized controlling unit. Basing on their common purpose of observation, all the sensing entities act collaboratively to accomplish their tasks. For example, a tornado is detected by a sensor in some area such that the sensors in the adjacent areas react to increasing their sensing frequency. Nephogram generated by the satellites might be requested by the subaerial stations for synthetical analysis. Increased rainfall detected by the subaerial sensors will trigger another sensors deployed in mountainous areas to predict disasters like debris flow and landslide.

Basically a Sensor Web pod (i.e. a sensor) is a physical platform that can be orbital or terrestrial, fixed or mobile, and might even have real-time accessibility via the Internet. Sensor-to-sensor communication is both omni-directional and bi-directional where each sensor sends out sensory data to every other sensor in the network. Two important characteristics of Sensor Web that differ it from ordinary sensor networks are as follows: first, a Sensor Web usually involves heterogeneous sensing devices; second, on-the-fly data processing and interoperation can occur within the Sensor Web itself and the system subsequently reacts as a coordinated collective whole to the incoming data stream. For example, instead of having uncoordinated smoke detectors, a Sensor Web can react as a single, spatially-dispersed, fire locator.

The scheme of Sensor Web is suitable for and already applied in environmental monitoring and control. Today, there have been a variety of Sensor Web field deployments with systems spanning as many as 6 miles and running continuously for over 3 years. Sensor Webs have been field tested in many environments including the gardens at the Huntington Library for botanical conditions including soil moisture and temperature, Antarctica to monitor microclimate conditions for extreme life detection, and, in cooperation with the University of Arizona, in the Central Avra Valley Storage and Recovery Project for flooding detection. Sensor Webs have also proved valuable in urban search and rescue, as well as infrastructure protection.

We may meet the following issues to enable a unified architecture like Sensor Web: 1) Data representation and query interpretation. Since heterogeneous sensors need to collaborate, a uniformly recognized interface for representing the sensory data and interpreting various queries from sensors themselves is a crucial part in the design. 2) Data fusion, sharing and transmission. Sensors generate continuous huge data volumes, while not all the sensory data on a sensor are useful and required by the other participating sensors. Meanwhile, there is not any centralized or dedicated data storage point in this model, which means data should be processed and shared within the sensors themselves. Considering the fairness and energy efficiency of sensing and data transmission in such systems, designing efficient protocols is still important in the energy-constrained context. 3) Real-time interoperation. As collaborative sensor networks of the unified model are deployed to realize a common purpose such environmental monitoring and disaster prediction, real-time and accurate interoperability is required to guarantee the quality of services. When an event is detected, some corresponding sensor networks must be instantly informed and react accurately on receiving the notification. Reliable, accurate, and efficient data transmission is thus important such that topology construction and routing protocols should be appropriately designed.

3.1.2 Autonomous Model

The second sub-class of flat architecture is autonomous model. Deployed and managed by different authorities, sensor networks in this model are treated as autonomous entity. Moreover, there are usually some computational resources/units other than sensor networks integrated in the system. By far there are not many existing works adopting autonomous model. We just briefly introduce a typical example Sensor-Grid in (Tham & Buyya, 2005).

Grid computing is essentially the union of heterogeneous computational servers connected with high-speed network connections. It enables convenient sharing of resources of CPU, memory, storage, content and databases by users and applications. The combination of sensor networks

Figure 3. The Sensor Map architecture

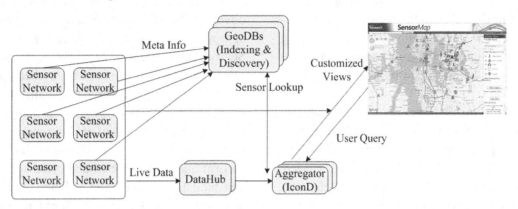

and grid computing, called Sensor-Grid, enables the complementary strengths and characteristics of sensor networks and grid computing to be realized in a single networking platform. While sensor networks generate continuous real-time data of physical worlds, the computation-intensive processing such as decision-making, analytics, data mining, optimization, and prediction can be carried out on the grid nodes. In words, sensor networks give "ears" and "eyes" to grid computing, while grid computing strengthen sensor networks with powerful computing capabilities. Therefore, the combination of them enables real-time and accurate surveillance, analyzing, and decision making, which are all critical parts in the applications of environmental monitoring, disasters prediction, and first-aid systems.

Considering the features of Sensor-Grid, both sensor networks and grid nodes should be autonomous entities. Therefore, besides the challenges listed in Section 3.1.1, some other issues should also be critical in designing collaborative mechanisms with the autonomous model. First, load balance is important to fully utilize the grid computing resources and provide efficient services. Second, pricing and marketing of both sensory data and computational resources is an interesting issue when Sensor-Grid becomes an open online market of sensing and computational tasks. Third, data privacy and authority should be

concerned if the owners of sensor networks and grid nodes are competitive with each other.

3.2 Hierarchical Architecture

Most of the existing works of collaborative sensor networks adopt hierarchical architectures, where centralized units or services are deployed to manage their sensing components, collect and store the sensory data, and handle the queries from external users. Web portals are usually considered as an essential part in those systems. Generally there are two major sub-classes: *centralized model* and *service-oriented model*.

3.2.1 Centralized Model

Centralized model refers to the methodology that centralized servers or proxies are used to manage the sensing, query processing, data input, and result output with collaborative sensor networks.

S. Nath et al propose SensorMap in (Nath, Liu, & Zhao, 2007). SensorMap represents a new class of applications that rely on real-time sensor data and its mash-up with the geocentric web to provide instantaneous accesses to environmental data and timely decision support. The platform also transparently provides mechanisms to store and index data, to process queries, to aggregate and present results on the web interface based on

Windows Live Local. As a convenient interface connecting the web portal and heterogeneous sensing devices, it plays the function of encouraging the community to publish more live data on the Web and enable people to build useful services on top of it.

Figure 3 illustrates the architecture of Sensor-Map. Current SensorMap prototype is a centralized Web portal consisting of four components: the *GeoDB* database, the *DataHub* web service, the *Aggregator* for creating icons, and the *SensorMap GUI*. Several types of sensor networks, such as thermometers, video cameras, traffic monitors, and parking slots sensors are included in the system.

GeoDB is a database storing sensor metadata. The meta-info of sensors includes information such as publisher name, sensor location, sensor name, sensor type, data type, unit, sensor data access methods, and free text descriptions. The users are supported to base their queries on sensor types, descriptive keywords, and geographic locations, such as the list of cameras along a route or the average temperature that thermometers report inside a geographic region. The indexing is implemented as table-valued functions in a SQL server. *GeoDB* uses geo-indexing techniques to efficiently support these types of queries.

DataHub is in charge of collecting raw data from sensors. It provides two ways to make real-time data accessible on SensorMap. Sensors with public web interfaces can register their URL directly to *GeoDB*. The SensorMap client then uses these URLs to fetch real-time data. For sensors which have Internet connections but no URL (such as those in mobile phones or behind firewalls), DataHub provides a simple interface to cache sensory data. The sensors are clients for DataHub and can send data in real time using standard Web service calls. The *Aggregator* or the *SensorMap GUI* directly retrieves those cached data from DataHub rather than try to contact the sensors.

Because there might be hundreds (even thousands) of sensors deployed in the map viewed

by a user, directly exhibiting each sensor will obviously clog the whole SensorMap. Therefore appropriate aggregation is necessary to make SensorMap usable. The *Aggregator* creates icons representing sensors that can be mashed up with maps. It accepts queries from the client, interprets them, and then sends the geographic components of the queries to the *GeoDB*. After querying the metadata of sensors, a set of sensors that satisfy a client query are selected and contacted. Their real-time data are retrieved and then aggregated accordingly. By doing so, SensorMap provides useful summarization of data to the client; for example, when a user is browsing at a city level, instead of showing him hundreds of temperature sensors in the city, SensorMap shows only one icon with aggregated results. What aggregation is performed by the *Aggregator* depends on sensor types. For example, for data collected from thermometers, average and standard deviation of temperatures reported by thermometers in a neighborhood are displayed.

The *GUI* of Web portal is the most distinctive feature of SensorMap. It is based on Windows Live Local and therefore supports features such as zooming, panning, street maps, satellite images, and 3D views. In addition, it enables end users to inject queries to available sensors. SensorMap currently supports three types of queries: geographic queries, type queries, and free-text queries. Geographic queries draw geometric shapes directly on the map specifies (for example, within a region, near a route); Type queries specify sensor types within the view-port; Free-text queries with keywords describe sensors specify. The *GUI* displays the overlays of the results that the Aggregator returns on Windows Live Local maps. *GeoDB* and *Aggregator* are transparent to both the data publishers and users. The *GUI* further enables users to save views (geographical region or sensor-type filters) on the client machine as cookies for quick retrieval.

From the above introduction, we may find that SensorMap adopts a fully centralized architecture.

While the whole system are divided into several modules according to their different functionalities, *GeoDB*, *DataHub*, *Aggregator*, and the *GUI* interface all reside in a central server. Data collection, aggregation, visualization, sensor indexing, and query processing are all executed by the server locally. SensorMap works reasonably well currently when only a few sensor networks are included and user queries are not too frequent. But it will probably be argued that such architecture lacks scalability.

The following challenges should be addressed in a centralized model like SensorMap. First, data publishing becomes complicated when there are different types of sensors. Data sources may have very different interfaces such as proprietary communication protocols, data presentation, and accessibility. And networked sensors, even the Internet ready ones, are typically behind firewalls due to management boundaries and security concerns. Therefore interoperability and extensibility is a critical challenge to collect and publish heterogeneous sensory data. To meaningfully aggregate data from multiple sources, we need common representations or formalizations of sensor types and units.

Second, scalable data management is important with respect to energy efficiency. As queries from users might be frequent and interested in overlapping subsets of sensors, it is obviously unacceptable to resolve all the queries on demand with real-time sensings. Data cached should be fully reused as long as they do not expire. Moreover, when approximate results are tolerated, sensors should be appropriately selected to construct a subset, which as a whole is able to provide approximate information with satisfying precision for a user. Sampling and inference among sensors should be suitably designed in these scenarios.

Although it is out of discussion of this chapter, sensor discovery should be a practical challenge. From the web portal of SensorMap (*SensorMap*), we have noticed some unqualified data being visualized. SensorMap provides an open portal to publish sensory data. But there is not rigorous verification mechanism to ensure that the data registered are legal and not adversary. Accordingly, automatic verification and testing techniques for sensors are necessary in the future design of SensorMap.

3.2.2 Service-Oriented Model

As sensor network deployments begin to grow there emerges an increasing need to overcome the obstacles of connecting and sharing heterogeneous sensor resources. Common data operations and transformations exist in deployment scenarios and can be encapsulated into a layer of software services that hide the complexity of the underlying infrastructure from the application developer. Services are defined for common operations including data aggregation, scheduling, resource allocation, and sensor discovery. Combing sensors and sensor networks with a Service-Oriented Architecture (SOA) (Foster, 2005) is an important step forward to present sensors as important resources which can be discovered, accessed, controlled via the World Wide Web. It envisions the opportunity for connecting geographically distributed sensors and computational resources into an integrated system.

X. Chu et al propose an open sensor web architecture named NOSA (Chu, Kobialka, Durnota, & Buyya, 2006). It is built upon the Sensor Web Enablement (SWE) standard defined by the Open Geospatial Consortium (OGC) (Percivall & Reed, 2006), which consists of a set of specifications, including SensorML, Observation & Measurement, Sensor Collection Service, Sensor Planning Service and Web Notification Service.

Figure 4 illustrates a typical collaboration between services and data encodings of SWE. In general, SWE is the standard developed by OGC that includes specifications for interfaces, protocols, and encodings that enable discovery, access, obtaining sensory data, as well as data-processing services. The following are the five primary specifications for SWE:

Figure 4. A typical collaboration within sensor web enablement framework

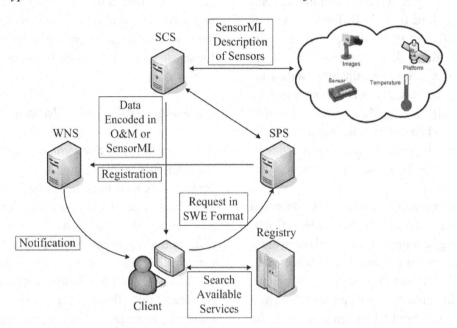

1. Sensor Model Language (SensorML) is an information model and XML encoding language that describes and specifies a single sensor with regard to discovery, query, and control of sensors.

2. Observation and Measurement (O&M) is an information model and XML encodings for observations and measurement.

3. Sensor Collection Service (SCS) is a service to fetch observations from a single sensor or a collection of sensors, which conforms to the information model of O&M. It is also used to describe the sensors by utilizing SensorML.

4. Sensor Planning Service (SPS) is a service to help users build a feasible sensor collection plan and to schedule requests (or queries) to sensors and sensor networks.

5. Web Notification Service (WNS) is a service to manage web client sessions and notify the client about the output of the requested services.

SWE only provides the basic standards of how an integrated system of sensors and sensor net-

works is constructed, but does not have any reference implementation or working system available to the community. Hence, there are many design issues to consider, including all of the common issues faced by other distributed systems such as security, multithreading, transactions, maintainability, performance, scalability, and reusability. NOSA mainly focuses on providing an interactive development framework, an open and standards-compliant Sensor Web services middleware, and a coordination language to support the development of various sensor applications.

NOSA consists of four basic layers. From bottom to top, they are Sensor Fabric Layer, Application Services Layer, Sensor Application Development Layer, and Application Layer.

Sensor Fabric Layer is the collection of numerous sensors and sensor networks. Besides the above core services of SWE, there are several other important services in the service layer of NOSA. The Sensor Directory Service provides the capability of indexing, searching for, and registering remote sensing services and resources. The Sensor Coordination Service enables the interaction and interoperation between sensing entities (i.e. sensors and sensor networks), which monitors, detects, and responds different kinds of events. The Sensor Data Grid Service publishes and maintains replicas of collected sensory data. The Sensor Grid Processing Service collects the sensory data and processes it on top of the Sensor Data Grid Service. The development layer focuses on providing useful tools in order to ease and accelerate the developments of sensor applications.

The main contribution of NOSA is to enable a SOA for multiple sensor networks. It facilitates the collaboration of multiple sensor networks by hiding the complexity of operations in underlying sensor networks. Concrete operations such as data aggregation, scheduling, resource allocation, query planning, and resource discovery are made transparent to the application layer users. For the application developers who want to develop complex applications based on geographically distributed sensors and sensor networks, such as researchers, NOSA provides a convenient platform as well as powerful toolkits.

Another instance of service-oriented model is IrisNet (Gibbons, Karp, Ke, Nath, & Seshan, 2003). IrisNet appears similarly with SensorMap in the sense that they both adopt tools like XML to describe sensory data and provide visualized results for queries on a web portal. But IrisNet differs much from SensorMap because it adopts a decentralized architecture. A web portal in IrisNet is not as indispensable as that in SensorMap.

Despite differences between sensor types, developers need a generic data acquisition interface to access sensors. In IrisNet, the nodes that implement such interfaces are called sensing agents (SAs). SAs collect raw sensory data from multiple sensors (possibly of different types). The sensor types can range from webcams and microphones to thermometers and pressure gauges. A SA host receives one or more raw sensor feeds from directly attached sensors and stores them in circular shared memory buffers. Multiple services can share an SA and access those buffers without causing conflicts.

On the other hand, services must store the service-specific data the SAs produce in a distributed database. In IrisNet, the nodes that implement such distributed databases are called organizing agents (OAs). Web-based sensing services are developed upon a group of orchestrated OAs. Consequently, each OA participates in only one sensing service (a single physical machine can run multiple OAs). The OAs involved in a single service should collect and organize sensor data to answer a particular set of queries relevant to the service.

In general, SAs take charge of sensing controls and data aggregation from sensor networks, while OAs are in charge of storing data and resolving queries. They two, together with the underlying sensor networks and a web portal, construct the main body of IrisNet.

Figure 5(a) shows a small portion of an XML document representing an example of the parking-space-finder service's schema in IrisNet. The schema describes the static (for example, <handicapped>) and dynamic (for example, <available>) data as well as the hierarchy that the service uses. The database schema provided by the service owner identifies the database's hierarchical portion using special ID tags. Figure 5(b) shows the hierarchy's tree representation corresponding to Figure 5(a), formed by the ID-tagged nodes.

IrisNet adopts a distributed manner of data caching. The XML-described databases are partitioned into pieces and stored on different OAs, probably different machines over Internet. How to resolve a query through multiple databases is a traditional issue in the research field of distributed databases. When the data required by a query is stored on multiple OAs, the path to resolve the query has much impact on the response latency and data transmission overhead. Basing on the hierarchical naming system, IrisNet designs a distributed algorithm to for OAs to select an efficient querying path. Moreover, sensory data stored in OAs frequently change and must be dynamically updated, it brings further challenges for IrisNet to keep consistency among OAs and

Figure 5. XML description and the hierarchy

```
<Region id="CHN">
 <state id="HKSAR">
  <city id="HK">
   <neighborhood id="Kowloon">
    <block id="block1">
     <parkingSpace>
      <handicapped>yes</handicapped>
      <available>yes</available>
     </parkingSpace>
    </block>
     <block id="block2">
      <parkingSpace>
       <available>yes</available>
      </parkingSpace>
     </block>
     <block id="block3">
      <parkingSpace>
       <available>yes</available>
      </parkingSpace>
     </block>
    </neighborhood>
    <neighborhood id="New Territories">
     <block id="block1">
      <parkingSpace>
       <available>no</available>
      </parkingSpace>
     </block>
    </neighborhood>
   </city>
  </state>
 </Region>
```

(a)

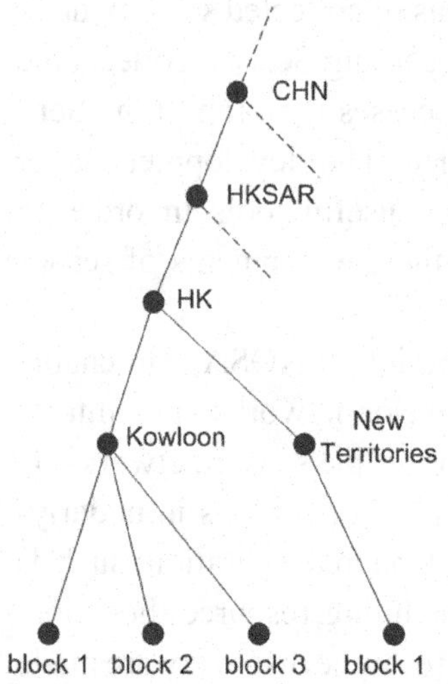

(b)

provide real-time queried results for users, which preserve data consistency.

4. DESIGN SPACE AND FUTURE DIRECTIONS

Research on collaborative sensor networks is an emerging field attracting more and more attentions. In this section, we discuss the design space of collaborative sensor networks, in the order from underlying sensor networks to application level services. We also point out some critical issues as future research directions.

4.1 Formalization and Description of Sensor Networks

Interoperability is a feature that distinguishes collaborative sensor networks from traditional sensor networks. Heterogeneous sensors are encompassed in an integrated system. Thus a uniform framework describing sensor networks is necessary to facilitate their collaborations. Essential properties, such as sensor type, data attributes, energy, and communication protocols, should be formalized so that we can support efficient information sharing among sensor networks, as well as standardized querying interfaces for the application users.

Moreover, to develop an open sensing system that allows dynamic plug-and-play operations of sensing resources, intelligent sensor discovery and registry is also important. A uniform formalization of sensing resources is also essential for such functionalities.

4.2 Interconnecting Sensor Networks

We have seen different schemes to interconnecting sensor networks in Section 3. For example, in order to support real-time sensings and to act as a whole for a common purpose, sensor networks in the unified model are connected in an ad-hoc network, where no central server is present. Efficient data sharing and reactions can thus be realized. On the contrary, in the centralized model of SensorMap, the powerful server efficiently resolves queries and visualizes the sensory data for the web users. IrisNet adopts a partially centralized scheme. Sensor networks are organized by SAs in each service. But data storage and query processing is distributed among a group of OAs. Based on the above investigations, we believe interconnection methodology is a critical point in the design and the choice of it depends on the manners that sensor networks collaborative in.

4.3 Sensing Control and Query Scheduling

As sensor networks become proliferated, to integrate heterogeneous sensor networks and support multi-user data sharing in a single sensing system is turn into truth. This brings further challenges both in the underlying sensor networks and in the upper querying mechanisms.

Sensor networks are well-known to be energy constrained. How to support substantive queries with limited energy resources is a big problem. Fortunately, there have been a lot of energy-efficient techniques proposed for sampling, data aggregation, and data transmission in conventional sensor networks. But they can not be directly migrated into the context of collaborative sensor networks. A more reasonable scheme is synthetical consideration of sensings and queries. Specifically, sampling techniques are proposed to save energy and get approximate information of a sensor network. Now it becomes more challenging because the output samples should not only

well brief the sensor networks, but also satisfying the requirements of precisions and scopes of individual users.

Another potential direction is to aggregate queries and to batch them in periodical sensing tasks. Considering the diversity of queries, however, it is non-trivial to do so. Queries have different priorities and deadlines, while different areas covered by the sensor networks might receive different amount of interest. All these diverse factors must be considered in designing an energy-efficient mechanism of sensing control, especially when the energy or bandwidth resources are limited (Li, Yan, Ganesan, Lyons, & Shenoy, 2007).

Considering the query scheduling at a higher level, we may find more interesting issues. In the decentralized architectures, a complex query (either ad-hoc query or stream query) concerning multiple sensor networks is resolved by a sequence of the sensor networks (He et al., 2008).

If the query is resolved with data cached in distributed databases of sensory data, optimizations should be made to minimize the response latencies and transmission overhead. This is similar with the issue of query processing in distributed databases (Ozsu & Valduriez, 1999). The differences between the two issues are summarized as follows: A distributed database is distributed into partitions/fragments that may be replicated, while each sensor network in the collaborative system is unique. Live sensory data reside only in their original sensor networks, except minor information exchanges among them. Besides, due to the nature of sensor networks, the total in-network energy consumption is the first-place metric when we optimize query processing. Oppositely, the goal of query optimization in distributed database is to minimize the cost of bandwidth, which is incurred when data is forwarded from one node to another. Meanwhile, when considering the heterogeneity (diverse periods of validity, diverse sizes, etc) of the cached sensory data, the scheduling of query processing should be carefully addressed to ensure the correctness of query results while minimizing the energy consumption of query processing.

Another case is that the query is on-demand resolved by real-time sensings in the underlying sensor networks. A school of schemes have proposed to adopt correlation-based query scheduling in a single sensor network (Babu, Motwani, Munagala, Nishizawa, & Widom, 2004; Deshpande, Guestrin, Madden, Hellerstein, & Hong, 2004, 2005). Similarly, when it comes to the context of multiple sensor networks, more or less there are natural correlations among the data of different sensor networks. Data in the upstream sensor networks can help to infer partial information of the downstream ones. Due to the heterogeneity of sensory data in different sensor networks, mining and quantifying the correlations among them is an open issue which has not been well studied before. Moreover, given the diversity of correlations, it is a very interesting but challenging issue to design an energy efficient algorithm to schedule real-time query processing in collaborative sensor networks.

4.4 Scalable and Reliable Data Management

In the context of multiple sensor networks, huge amount of sensory data are continuously generated and expire relatively fast. When the system scales up, centralized storage of sensory data is obviously not efficient. Therefore distributed storage is adopted. But it brings additional problems. First, to ensure data completeness, redundant data caching should be adopted. The sensory data, however, has different periods of validity and importance. To make it even worse, the cache data expire fast and should be updated frequently. As a result, the data consistency is hard to maintain among multiple distributed storages. Moreover, when the whole data set is partitioned into pieces, it is non-trivial to design efficient addressing mechanisms and guarantee load balance among storages.

On the other hand, some security issues related to data management are also important. First, sensor networks are usually deployed and managed by different owners. The collaborations run well if we assume all the owners are altruistic. This is not always true, however. It is probably the real case that data from a sensor network are only available to certain customers. Such selective trading relations are all encompassed and incarnated in a common framework of collaborative sensor networks. Rigorous authentication and auditing mechanisms are thus crucial to ensure data privacy and avoid profit loss of the owners. Furthermore, data integrity is also important for the users. The natural dynamics of sensory data make it even more difficult to check data integrity when data are forward through multiple sensor networks.

4.5 Other Issues

Besides the issues discussed above, there are some other issues in the design space, which are also important with respect to the applicability of collaborative sensor networks in a wide variety of applications.

Queries can be posed anywhere. The motivation of collaborative sensor networks is to meet the requirements of citizen users for live data from the real world. A possible direction is to develop ubiquitous services for users all around the world. Then it would be very beneficial if we can exploit any locality between the querier and the queried data. Because the sensed data are inherently coupled to a physical location, many queries will be posed within tens of miles of the data. Location-aware query interpretation and processing will greatly reduce the processing overhead.

Marketing and pricing. The proposal of Sensor-Grid actually illuminates a potential direction of sensor applications. Since sensor networks are owned by different authorities, it is probably feasible to introduce marketing and pricing mechanisms into the service-oriented systems. Although these have been extensively studied (Lalis & Karipidis, 2002; MacKie-Mason, Osepayshvili, Reeves, & Wellman, 2004; Shetty, Padala, & Frank, 2003; Walsh, Wellman, Wurman, & MacKie-Mason, 1998), but the context of sensor networks brings new challenges. For example, given the prices of sensory data, it will be an essential task for the service providers to estimate the cost of processing a complex query. Whether the estimated cost is minimized and accurate directly determines the profit of service owners. Therefore how to minimize the processing cost with the give prices becomes an optimization problem.

QoS-aware design. Being an information service, the systems of collaborative sensor networks share some characteristics with other web-based services, such as search engine. QoS related factors, such as the response latency and the accuracy of results, should be optimized in a QoS-aware design. Considering the diversity of applications and users that pose queries, different querying tasks have priorities and overhead, while those tasks will probably share some critical resources during query processing, such as intermediate data forwarders, sensors, and cached data. It is definitely challenging to coordinate conflicting querying tasks and meet the QoS requirements of all the users with limited resources.

5. CONCLUSION

Research on collaborative sensor networks is becoming a hot spot in the community. In this chapter, we introduce the building blocks and techniques of collaborative sensor networks. Taxonomy of collaborative sensor networks based on the architectures and the methodologies of organizing different modules of a system is presented. We discuss the design space in this field and point out potential research directions.

REFERENCES

Babu, S., Motwani, R., Munagala, K., Nishizawa, I., & Widom, J. (2004). *Adaptive Ordering of Pipelined Stream Filters.* Paper presented at the ACM SIGMOD International Conference on Management of Data, Paris.

Chu, X., Kobialka, T., Durnota, B., & Buyya, R. (2006). *Open Sensor Web Architecture: Core Services.* Paper presented at the Intelligent Sensing and Information Processing.

Delin, K. A. (2002). The Sensor Web: A Macro-Instrument for Coordinated Sensing. *Sensors, 2,* 270–285. doi:10.3390/s20700270

Deshpande, A., Guestrin, C., Madden, S., Hellerstein, J. M., & Hong, W. (2004). *Model-Driven Data Aquisition in Sensor Networks.* Paper presented at the International Conference on Very Large Data Bases.

Deshpande, A., Guestrin, C., Madden, S., Hellerstein, J. M., & Hong, W. (2005). *Exploiting Correlated Attributes in Acuqisitional Query Processing.* Paper presented at the IEEE International Conference on Data Engineering.

Foster, I. (2005, May 6). Service-Oriented Science. *Science, 308,* 814–817. doi:10.1126/science.1110411

Gao, T., Massey, T., Selavo, L., Welsh, M., & Sarrafzadeh, M. (2007). *Participatory User Centered Design Techniques for a Large Scale Ad-Hoc Health Information System.* Paper presented at the HealthNet.

Gibbons, P. B., Karp, B., Ke, Y., Nath, S., & Seshan, S. (2003). IrisNet: An Architecture for a World-Wide Sensor Web. *IEEE Pervasive Computing / IEEE Computer Society [and] IEEE Communications Society, 2*(4), 22–33. doi:10.1109/MPRV.2003.1251166

GreenOrbs. Retrieved, from http://www.greenorbs.org/

Gu, L., & Stankovic, J. A. (2006). *t-kernel: Providing Reliable OS Support to Wireless Sensor Networks.* Paper presented at the ACM Conference on Embedded Networked Sensor Systems, Boulder, CO.

He, T., Stankovic, J. A., Lu, C., & Abdelzaher, T. F. (2005). A Spatiotemporal Protocol for Wireless Sensor Network. *IEEE Transactions on Parallel and Distributed Systems, 16*(10), 995–1006. doi:10.1109/TPDS.2005.116

He, Y., Li, M., Liu, Y., Zhao, J., Huang, W., & Ma, J. (2008, May 27). *Collaborative Query Processing among Heterogeneous Sensor Networks.* Paper presented at the HeterSANET co-located with the ACM International Symposium on Mobile Ad Hoc Networking and Computing, Hong Kong, China.

Hill, J., & Culler, D. (2002). Mica: A Wireless Platform For Deeply Embedded Networks. *IEEE Micro, 22,* 12–24. doi:10.1109/MM.2002.1134340

Lalis, S., & Karipidis, A. (2002). *An Open Market-Based Architecture for Distributed Computing.* Paper presented at the IEEE International Parallel & Distributed Processing Symposium.

Li, M., & Liu, Y. (2007a). *Rendered Path: Range-Free Localization in Anisotropic Sensor Networks with Holes.* Paper presented at the ACM International Conference on Mobile Computing and Networking.

Li, M., & Liu, Y. (2007b). *Underground Structure Monitoring with Wireless Sensor Networks.* Paper presented at the IEEE International Conference on Information Processing in Sensor Networks.

Li, M., Yan, T., Ganesan, D., Lyons, E., & Shenoy, P. (2007). *Multi-user Data Sharing in Radar Sensor Networks.* Paper presented at the ACM Conference on Embedded Networked Sensor Systems, Sydney, Australia.

Lian, J., Naik, K., Chen, L., Liu, Y., & Agnew, G. (2007). Gradient Boundary Detection for Time Series Snapshot Construction in Sensor Networks. *IEEE Transactions on Parallel and Distributed Systems.*

MacKie-Mason, J. K., Osepayshvili, A., Reeves, D. M., & Wellman, M. P. (2004, June 3-7). *Price Prediction Strategies for Market-Based Scheduling.* Paper presented at the 14th International Conference on Automated Planning & Scheduling, Whistler, British Columbia, Canada.

Nath, S., Liu, J., & Zhao, F. (2007). SensorMap for Wide-Area Sensor Webs. *IEEE Computer Magazine, 40*(7), 90–93.

nesC: A Programming Language for Deeply Networked Systems. Retrieved, from the World Wide Web: http://nescc.sourceforge.net/

OceanSense. Retrieved, from the World Wide Web: http://www.cse.ust.hk/~liu/Ocean/

Ozsu, M. T., & Valduriez, P. (1999). *Principles of Distributed Database Systems* (second ed.)New York: Prentice Hall.

Pepe, A., Borgman, C. L., Mayernik, M., & Wallis, J. C. (2007). *Knitting a Fabric of Sensor Data Resources.* Paper presented at the the ACM Workshop on Data Sharing and Interoperability on the World-wide Sensor Web, Cambridge, MA.

Percivall, G., & Reed, C. (2006). OGC Sensor Web Enablement Standards. *Sensors & Transducers Journal, 71*(9), 698–706.

Reddy, S., Chen, G., Fulkerson, B., Kim, S. J., Park, U., Yau, N., et al. (2007). *Sensor-Internet Share and Search - Enabling Collaboration of Citizen Scientists.* Paper presented at the ACM Workshop on Data Sharing and Interoperability on the World-wide Sensor Web, Cambridge, MA.

SensorMap. Retrieved, from http://atom.research. microsoft.com/sensormap

Shetty, S., Padala, P., & Frank, M. (2003). *A Survey of Market Based Approaches in Distributed Computing* (Technical Report TR03-13). Florida, USA: University of Florida Computer & Information Science & Engineering

Szewczyk, R., Mainwaring, A. M., Polastre, J., Anderson, J., & Culler, D. E. (2004). *An analysis of a large scale habitat monitoring application.* Paper presented at the ACM Conference on Embedded Networked Sensor Systems.

Tham, C. K., & Buyya, R. (2005). SensorGrid: Integrating Sensor Networks and Grid Computing. *CSI communication* (Grid Computing).

Walsh, W. E., Wellman, M. P., Wurman, P. R., & MacKie-Mason, J. K. (1998). *Some Economics of Market-Based Distributed Scheduling.* Paper presented at the the 18th international Conference on Distributed Computing Systems.

Xiao, B., Chen, H., & Zhou, S. (2008). Distributed Localization Using a Moving Beacon in Wireless Sensor Networks. *IEEE Transactions on Parallel and Distributed Systems, 19*(5), 587–600. doi:10.1109/TPDS.2007.70773

Yang, Z., Li, M., & Liu, Y. (2007). *Sea Depth Measurement with Restricted Floating Sensors.* Paper presented at the IEEE Real-Time Systems Symposium.

Zhang, P., Sadler, C. M., Lyon, S. A., & Martonosi, M. (2004). *Hardware design experiences in ZebraNet.* Paper presented at the ACM Conference on Embedded Networked Sensor Systems.

KEY TERMS AND DEFINITIONS

Collaborative Sensor Networks: Collaborative sensor networks refer to the sensing systems with the following characteristics: first, it is an information system providing live sensing and query processing services; second, multiple heterogeneous sensor networks are integrated in; third, sensor networks in the system collaborate with each other in accomplishing the querying tasks.

Data Aggregation: Data Aggregation is an operation that aggregates values from multiple (two or more) contributing data characteristics, such as taking the sum, average, max, or min of the data.

Query: A query is defined as a precise request for information retrieval from a database or an information system such as a sensor network.

Query Processing: Query processing refers to the process to answer a query to a database or an information system, which usually involves interpreting the query, searching through the space storing data, and retrieving the results satisfying the query.

Sensing Control: Sensing control refers to the mechanisms, either implemented on the sensor motes or running on the base station(s) of a sensor network, to control, coordinate, and schedule the process of sensing activities of the sensor motes in the network.

Sensor Mote: A sensor mote, also known as a sensor node, is a sensing device that is capable of performing physical sensing, gathering sensory information, and computing. Usually it is also able to communicate with other connected sensor motes.

Sensor Network: A sensor network is a network consisting of spatially distributed sensor motes to cooperatively monitor physical conditions and environments, such as temperature, humidity, vibration, pressure, illumination, sound, images, etc.

Chapter 4
QoS:
Requirements, Design Features, and Challenges on Wireless Sensor Networks

Ricardo H. González G.
Simon Bolivar University, Venezuela

Antonio A.F. Loureiro
Federal University of Minas Gerais, Brazil

Raquel A.F. Mini
Pontifical Catholic University of Minas Gerais, Brazil

ABSTRACT

This chapter presents a survey of the main aspects of QoS that are being used on the Wireless Sensor Networks technology world. We describe approaches based on traditional networks, as well as new approaches that face with resources limitation, which is one of the main characteristics of WSN nodes. This chapter also describes how QoS management creates a series of challenges to deal with, and how some QoS features could be used to enable users to make a better profit of their resource limitations. In our opinion, an exposition of these topics will improve the set of techniques and strategies that designers and programmers could use to develop and implement satisfactory services level to their applications in WSNs.

INTRODUCTION

WSNs are a class of wireless ad hoc networks that are different from conventional networks (Heinzelman et al., 1999) by several reasons: its small footprint, its low cost design, and its nodes scarce resources such as: limited supply of energy, limited storing and computation power, low bandwidth and energy-conserving communication. These limitations prevent them from running sophisticated network protocols, and limit inter-nodes communication. However WSNs technology also offers a series of features described as advantages by Krishnamurthy et al. (2005), Shen et al. (2004), and Paavola (2007), where can be highlighted: the ability to capture online information directly from field devices, the reduction of deployment and maintenance costs, compared to wired connection infrastructure on heavy duty or isolated environments, the improvement on area visibility due to

DOI: 10.4018/978-1-61520-701-5.ch004

the lack of wires, enabling a better visualization of devices, and the decentralization of monitoring and automation functions. All these features can be used to improve the collection of information from the real world

In WSNs there are different types of roles that each node could play: the Sensor Node (SN) role, which covers activities related to the capture of information from the real physical world (sensed data), the Routing Node (RN) role, that deals with the relay of sensed and management data from SNs to the Sink, and finally the Sink role that gathers any collected information from the whole network or the one that comes from only a subset of network nodes. The Sink behaves as a portal or link between the user of the applications and the wireless sensor nodes. Some nodes in the network could play more than a role at a time and, in fact, it is very common that many nodes play both the SN and the RN roles.

Due to the fact that WSNs have to interact directly with their environment to collect data, some of their characteristics are expected to be very different from other more conventional data networks. This differentiation implies a series of challenges that WSNs have to face, and that was described by Chen and Varshney (2004), Younis et al. (2004) and Wang et al. (2006). These topics include features such as: severe resource constraints, unbalanced traffic, data redundancy, network dynamics, energy balance, scalability, multiple sinks, multiple traffic types, information about packet relevance, energy and delay trade-off, etc.

Meanwhile, some of the main QoS goals in conventional wireless networks are to provide high throughput, low delay or high bandwidth efficiency. For a sensor network, one of the main concerns is to enlarge the network lifetime. To reach this QoS goal, network designers must save energy, probably giving up performance in other aspects of the operation such as delay and bandwidth utilization. This occurs because each node depends on limited power energy sources,

such as batteries, that could not be enough to cover node energy expenses for long periods of time, especially when they are operating in hostile or remote regions, where they cannot expect replacement (Bhatnagar et al., 2001).

This chapter comprises an introduction to general characteristics of QoS on WSNs. Its first Section presents a description of the QoS requirements that could be defined on WSNs. The second Section describes some strategies and techniques that could be used to reach some requirements in WSNs, including a survey of some projects that manage some QoS features. The third Section presents some challenges and open research topics about dealing with QoS on WSNs. Finally, the last Section describes some conclusions about present and future of QoS on WSNs.

QOS REQUIREMENTS IN WSNS

QoS can be defined as the collective effect of service performance which determines the degree of satisfaction of a user of the service (ITU-T, 1994). Different applications require different QoS and if they do not meet the requirements, the resulting behavior will be unsatisfactory (Chen & Varshney, 2004).

When a wireless system has to work with QoS, it can offer two different alternatives: Soft or Hard QoS. Soft QoS means that there could exist time periods when the required QoS is not guaranteed due to path breaking or network partition (Chen & Nahrstedt, 1999). This situation is also called a best-effort service. On the other hand, if a Hard QoS (Lutfiyya et al., 2001) is offered, then there is a guarantee that resources will be available when they are required. This Hard QoS should be guaranteed by statically allocating resources based on worst-case scenario. Hard QoS is important in systems that must meet their timing constraints to avoid disastrous consequences, but it is very difficult to reach in wireless systems where it is hard to manage a reliable delivery mechanism on the air interfaces.

Many of the current researches on QoS on WSNs are focused on end-to-end delay. Others are focused on protocols that are energy aware to maximize the lifetime of the network, dealing with scalability features for a large number of sensor nodes, or tolerant nodes damage and battery exhaustion situations (Akkaya & Younis, 2003). In order to get a better understanding of QoS features we are going to expose some of the most relevant QoS requirements that could be used in WSNs applications.

REAL TIME REQUIREMENTS

The sensed data should reflect the physical status of the sensing environment (Felemban et al., 2005). Because sensor data could be valid only for limited time duration, it needs to be delivered within a time bound. Different sensed data could have different deadlines depending on the dynamics and nature of the sensed environment. In order to reach a better understanding of these requirements, we describe two of the most common metrics used to define real time requirement:

Latency: Is defined as the time length between the occurrence of an event and its detection at Sink level. In surveillance and control applications, event detection needs to be reported in real time so that appropriate action can be taken as soon as possible (Lu et al., 2004).

Throughput: Is a measure of how much work is being processed by unit of time. The throughput requirement also varies within different applications. There are some applications that need to measure certain variables with a fine temporal resolution. In such applications, the more data the Sink receives is better. In other applications, such as fire detection, a single but timely report should be enough (Lu et al., 2004).

ENERGY-AWARE REQUIREMENTS

In some applications, the conservation of energy that is directly related to network lifetime is considered relatively more important than the quality of sent data (Al-Karaki & Kamal, 2004). When the energy level is low, the network may be required to reduce the quality of results in order to reduce energy dissipation. Hence, energy-aware routing protocols are needed to manage these situations.

Many WSNs protocols consider energy efficiency as its main objective and assume data traffic with unconstrained delivery requirements. However, there is a growing interest in applications with demands on end-to-end latency or reliability, which have created additional challenges to WSNs world. Transmission of data in such cases requires an adequate awareness of both factors: energy and QoS, in order to ensure the efficient use of the sensor resources and the effective access to the gathered measurements (Younis et al., 2004). Therefore, designing energy-aware algorithms is becoming a challenging task for extending the lifetime of WSNs.

RELIABILITY REQUIREMENTS

Depending on their contents, sensed data could have different reliability requirements. For example, in a forest monitoring applications, one temperature measure that is in a normal range can be delivered to the control center, tolerating a certain percentage of message loss. On the other hand, if a sensor detects an abnormally high temperature, data should be delivered to the control center with a very high probability, since it can be a sign of fire (Felemban et al., 2005). Some other aspects related to reliability on WSNs are:

- **Robustness:** (Hadim & Mohamed, 2006) The network should remain operational even if certain well defined failures occur

Figure 1. A summary of WSN strategies that produce improvement in QoS metrics

Network Layer	QoS Strategy	Produce Improvement in QoS Metric
Application Layer	Compression	Latency, Available Bandwidth
	Adjust Sensing Rate	Network Lifetime, QoD
	Data Reduction Techniques	Latency, Available Bandwidth
Network Layer	Routing Protocols	Reliability, Latency, Network Lifetime
	Packets Priority	Reliability, Latency, Network Lifetime
Link Layer	Energy Aware MACs	Network Lifetime
	Selection of Low Interference Channels	Reliability, Network Lifetime
	MAC that Avoid Collisions	Reliability, Network Lifetime
Physical Layer	Channel Surfing	Reliability, Network Lifetime
	Modifying Signal Power	Network Lifetime
	Using Low Interference Channel	Reliability, Network Lifetime

- **Fairness:** In many applications, particularly when bandwidth is scarce, it is important to ensure that the Sink receives information from all sources in a fair manner, trying to share wireless media between every participant (Lu et al., 2004).

HETEROGENEOUS TRAFFIC REQUIREMENTS

Some sensed data is created aperiodically by detection of critical events at unpredictable points in time (Felemban et al., 2005). In addition, there are other types of sensed data generated from periodic monitoring of environmental status. In that way, sensor network applications could have a mixture of these traffic types with different QoS requirements for each one.

In many applications, there exists information with different importance levels that can be classified as critical and non critical. Each kind of traffic could have different QoS requirements inside the same network to reach application goals. In those cases, a service differentiation mechanism is needed in order to guarantee the appropriate management of each traffic type.

All these previously mentioned requirements may impact other dimensions of the WSNs design space, such as coverage and resources usage.

DESIGN STRATEGIES AND TECHNIQUES TO DEAL WITH QOS IN WSNS

In WSNs, the wireless medium is employed to coordinate nodes activities and to transmit data collected to a Base Station. Thus, QoS management should implement strategies for prioritization of specific flow of messages. Next, we briefly mention some QoS techniques that can be used in each network layer of the WSNs protocol stack (Correia et al., 2005), in order to deal with prioritization strategies. A summary of some of these strategies can be appreciated in Figure 1.

One of the functions of **the physical layer** is to avoid the interference with other networks or natural sources of radiation. In case of the quality of different wireless channels could be measured, the channel with lower interference level should be selected in order to reduce collisions and interferences, because the occurrence of these events conduce to message retransmission that consume battery power and contribute to drain nodes batteries earlier. In some WSNs implementations, there is the possibility to use different wireless channels, this characteristic could be seized by a Channel Surfing strategy (Xu et al., 2007), where as soon as some interference that could disrupt the communication would be detected, then the affected nodes or the whole WSN can change its RF transmissions to other channel in order to avoid interference and improve the signal reception. Another strategy at the physical layer consists of a reduction of the mote signal strength to a minimum acceptable value; in that way, communication could be maintained, but using as low energy power as possible in order to make a better use of its battery power and reducing the possibility of interferences with other further nodes

In **the link layer** the scheduling of medium access and the sequence of packets to send, could be changed to meet some QoS requirements. These changes may be achieved by packet reordering, by priority control and by admission policies. In this layer, it is also possible to adjust the amount of control packets sent, in order to increase the QoS of any important data packet.

In **the network layer**, similar to the link layer, it is possible to use packet prioritization. In wired networks, packet priority is assigned by flow or individually, using packet queuing policies such as Token Bucket and Weighted Fair Queuing (Peterson & Davie, 2003), among others. Using priorities, some packets should be routed through different paths, in order to improve some features, such as latency or reliability.

At **the application layer**, as well as in wired networks, audio and video applications can employ adaptive compression techniques, which adjust to network conditions in order to reduce the bandwidth used. This approach can be applied to sensor networks, adjusting the frequency in which data is sent (Sankarasubramaniam et al., 2003). In the same way, some data reduction techniques could be used to decrease the amount of bandwidth used, but without losing much relevant information (Aquino et al., 2008).

Traditionally, QoS strategies have to deal with prioritization of different flows, giving a better opportunity to more relevant packets, which means to define and manage different services level.

SERVICE DIFFERENTIATION

This is one of the mechanisms that are used to manage QoS traffic and warrant the reliable delivery of network packets (Akkaya & Younis, 2003). The importance of service differentiation can be appreciated when different data packets might have different transmitting relevance, as is described by Bhatnagar et al. (2001), in a forestal fire detection system. In this environment, if a sensor detects a 60F measure on a normal spring day, it is considered a normal value, but a temperature of 1000F in the same situation is unexpected and it could be an indication of a forest fire. From a monitor's point of view, the packet containing the 1000F temperature is much more important, and the network should make a better effort in delivering it.

Service Differentiation models, such as the **Integrated Services** (IntServ) model (Braden et al., 1997), give a special treatment to packets from a given flow. Each router could store per-flow states and maintains a token bucket to control the flow of forwarding packets.

On the other hand, the **Differentiated Services** (DiffServ) model (Blake et al., 1998), does not provide any guarantee about flows. Instead, the edge routers mark a Per-Hop-Behavior (PHB) in the packet header. The core routers use the PHB as information in their packet scheduling decision. Packets from different flows that have the same PHB should be treated identically.

The communication medium in WSNs is the air, which is more error prone than wired links. Because of this fact, wireless links may be highly unpredictable and could show undesired error rates. Thus, in wireless sensor networks two of the primary service differentiation metrics that should be used are:

- **Reaching Probability:** In WSNs the higher priority packets should have a better probability to be received than lower ones. Typically, the reaching probability to the

Sink would be based on the wireless link error rates along its path. In traditional networks, the main way of providing reliability is through the use of acknowledgments, which usually work on an end-to-end basis. For sensor networks, this end-to-end acknowledgment scheme may not work efficiently, because the reaching probability to the Sink would be reduced significantly with their hop distance. If a hop-by-hop acknowledgment mechanism is deployed, the number of retransmissions would depend on local channel error. Further, an acknowledgment based scheme requires that a packet is kept at a node, until its acknowledgment is received, although memory limitation may make it difficult to keep every packet at a sensor node.

- **Latency:** A high priority packet should reach the Sink node with a lower latency than a low priority one. This would mean that a low priority packet could take longer routes to make way for higher priority packets on short routes.

A SURVEY OF PROJECTS THAT MANAGE SOME QOS FEATURES

Many of the Wireless Sensor Networks protocols with QoS support are surveyed by Stankovic et al. (2003), Chen and Varshney (2004), Younis et al. (2004), Wang et al. (2006), Wang et al. (2008), and Xia and Zhao(2008). In these works, some basic protocols such as SAR (Sohrabi et al., 2000), SPEED (He et al., 2003), Energy-Aware protocol (Akkaya & Younis, 2003), and RAP (Lu et al., 2002) are described. These protocols require some energy and consider only one QoS factor, i.e. either fault tolerance or real-time. Felemban et al. (2005), considered these two QoS factors together, but data-centric style of communication and energy-awareness in its communication protocol

are not supported. At middleware and application layers, some works had been reported by Sharifi et al. (2006), Heinzelman et al. (2004), Delicato et al. (2005), Alex et al.(2004), and Hadim and Mohamed (2006). Also, Ruiz et al. (2004) uses services and functions for fault detection without recovery. And Lim (2002) deals only with fault recovery. None of these two later works consider energy consumption in their solutions and additionally introduce lots of overhead in time and resource usage.

Chen and Varshney (2004) show that some of the existing research efforts related to the QoS in WSNs can be classified into three large categories: traditional end-to-end QoS, reliability assurance as QoS, and application-specific QoS. We extend this classification to include categories that concerns exclusively to resource limitation. Because this is one of the main characteristics on WSNs. A brief review of QoS aware protocols in these categories is presented in the following subsections.

End-to-End Delay and Latency QoS

Delay and latency are one of the main end-to-end common parameters of QoS that is managed in WSNs and other Networks. A lot of efforts are being done at the research community to manage these parameters. Some of the more relevant works in this subject are:

SPEED (He et al., 2003) is a QoS routing protocol for WSNs that provides soft real-time with some end-to-end guarantees. This protocol requires that each node maintains information about its neighbors and uses geographic forwarding to find a good path. In addition, SPEED strives to ensure a certain speed for each packet in the network, so that each application can estimate the end-to-end delay by considering the distance to the Sink and the speed of the packet before making the admission decision (Younis et al., 2004). Moreover, SPEED can provide congestion avoidance when the network is overloaded. The delay estimation at each node is basically made by calculating the elapsed time when an ACK is received from a neighbor as a response to a transmitted data packet. By looking at the delay values, SPEED selects the next path node that could meet the speed requirement. A backpressure-rerouting module is used to prevent voids when a node fails to find a next hop node, and to eliminate congestion by sending messages back to the source nodes, so that they will pursue new routes. SPEED does not consider any energy metric in its routing protocol.

The Energy-Aware QoS routing protocol (EAQoS) (Akkaya & Younis, 2003) uses a cost function based on energy consumption and error rate, to find an optimal path for data from a sensor node to the Sink, while meeting end-to-end delay requirements. In order to support both, best effort (non real-time) and real-time traffic at the same time, EAQoS uses a class-based queuing model to send its packets. It also uses a K-least cost path algorithm to find a set of candidate routes, where each route is checked against the end-to-end constraints and the one that provides maximum throughput is picked. EAQoS not only should find paths that meet the requirements for real-time traffic, but also needs to maximize the throughput for non-real time traffic as well. A bandwidth ratio r represents the amount of bandwidth to be dedicated to the real-time and non-real-time traffic on a particular outgoing link in case of congestion. Simulation results show that the proposed protocol consistently performs well regarding QoS and energy metrics. However, one initial drawback is that the same r-value is set for all nodes, which does not provide flexible adjusting of bandwidth sharing for different links. The protocol is extended by Younis et al.(2004) by assigning a different r-value for each node in order to achieve a better link utilization. Akkaya and Younis (2004), in a later work, describes an approach that pursues a multi-hop packet relaying technique in which, to minimize transmission, employs a Weighted Fair Queuing (WFQ) packet scheduling methodology along with leaky bucket constrained data sources,

in order to provide soft real-time guarantees for data delivery. Such employment of WFQ at each node provides a service differentiation between best effort and delay-constrained traffic.

In the work Middleware Layer Mechanism for QoS Support presented by Sharifi et al. (2006), three different types of nodes are considered: Sink, Cluster Heads and Common Nodes. In a Sink node, an Analyzer can differentiate between WSNs services and application QoS requirements (e.g. service response time); the Analyzer utilizes the information stored in the Sink node (i.e. min and max delays and region of service) to assign criticality to the application interest, and select appropriate Cluster Heads. If the interest analyzer detects that the WSN cannot service the interest with the specified QoS requirements, it will send an alarm to the user application. The Selector component that runs in the Cluster Head, assigns critical interests to Common Nodes with minimum transmission delay and desirable residual energy. The Monitoring component, in each Cluster Head, monitors Common Nodes' operations and network residual energy to reprogram Common Nodes parameters so that the mechanism can meet message deadlines; other common nodes may service other interests or may be idle. The Scheduler in Cluster Heads and Sink nodes balances energy consumption and common nodes' operations to schedule high-priority messages first. If there are multiple Sink nodes with different QoS requirements, the Interest Analyzer assigns a Sink identifier to the interest. When the Interest Analyzer sends an interest to the network, the Cluster Heads know which Sink node is interested in this service, therefore the interest is sent to the specified Sink. A Cluster based organization is used to guarantee on-time delivery of messages to the Sink nodes and user applications, based on: energy, interest criticality, and sensor nodes' default information such as delay and data rate.

The Tai et al. (2007) work studies the effects of data relaying in WSNs. A three-tier hierarchy is proposed, which generalizes to a two-tier WSNet with multiple Sinks, to address the conflicting issues between energy efficiency and QoS, expressed as the delay in delivering data to the base station. Energy requirements are translated into capacities assigned to the nodes in the network. The objective is to minimize the sum of the total energy spent on communication and the penalties incurred from the violation of delay constraints. The architecture could use two different strategies. In the first strategy, data packets originating from the same source are sent to the base station, possibly through several different paths (splitting). In the second strategy, exactly one path is used for this purpose (not splitting relaying traffic). Tai et al (2007) perform an empirical analysis that quantifies the performance gains and losses of the splittable and unsplittable traffic allocation strategies with delay-constrained traffic. Their experiments show that splitting traffic does not provide a significant advantage in energy consumption, but can afford strategies for relaying data with a lower delay penalty when using a model based on soft-delay constraints.

Schurgers et al. (2001) explores the energy-latency tradeoffs in wireless communication using techniques such as modulation scaling. An important observation of Prabhakar et al. (2001) is that in many channel coding schemes, the transmission energy can be significantly reduced by lowering power and increasing the duration of transmission. Using these techniques, Yu et al (2004) consider a real time scenario where the raw data gathered from the source nodes must be aggregated and transmitted to the Sink within a specified latency constraint, they also present algorithms to minimize the overall energy dissipation of the sensor nodes in the aggregation tree, subject to the latency constraint. They propose two off-line problem solutions using a numerical algorithm for the optimal solution, a pseudo-polynomial time approximation algorithm based on dynamic programming and an on-line distributed protocol that relies only on the local information available at each sensor node within the aggregation tree.

Multi-Speed Routing Protocol (MMSPEED) is a packet delivery mechanism described by Felemban et al. (2005), that is spread through network and Medium Access Control (MAC) layer offering some probabilistic QoS guarantee in WSNs. MMSPEED provides QoS in two quality domains: timeliness and reliability. In this way, packets can choose the most proper combination of service options depending on their timeliness and reliability requirements. For the service differentiation in the timeliness domain, the proposed mechanism provides multiple network-wide speed options extending the idea of single network-wide speed guarantee presented by He et al. (2003). For the service differentiation in the reliability domain, they exploit the inherent redundancy of dense sensor networks by realizing probabilistic multipath forwarding that depends on the packet's reliability requirements. All previously described QoS provisioning are used in a localized way without global network information, by employing localized geographic packet forwarding information and augmented with a dynamic balance strategy, which compensates the local decision inaccuracy as a packet travels towards its destination. In this way, MMSPEED can guarantee end-to-end requirements in a localized way. Simulation results (Felemban et al., 2005), show that MMSPEED provides QoS differentiation, improving the effective capacity of a sensor network in terms of number of flows that meet reliability and timeliness requirements.

While some energy-aware MAC protocols have been proposed specifically for sensor networks (Xu & Saadawi, 2001; Woo & Culler, 2001), there are not too many researches that combine real time scheduling techniques and energy-awareness. Some of the researches that do are:

The Power-aware reservation based MAC (PARMAC) (Adamou et al., 2001), which is an energy-aware protocol primarily designed for mobile ad hoc networks but that can be applied to sensor networks; however, it should be considered that nodes do not move. In PARMAC, the network is divided into grids, where every node is assumed to reach all the other nodes within the same grid area. Time is divided into fixed length frames, which are composed of Reservation Period (RP), and Contention Free Period (CFP). In each RP, nodes within a grid cell exchange 3 messages to reserve the slots for data transmission and reception. Data is then sent in CFP. If the reservation can be done before the deadline for real-time packets, then delay bounds can be provided. The protocol saves energy by minimizing the idle time of the nodes and allowing the nodes to sleep during CFP. Moreover, the control packet overhead and the packet retransmissions are minimal, achieving significant energy savings. PARMAC assumes that all stations are synchronized using a global synchronization scheme.

Sensor MAC or S-MAC (Ye et al., 2002) was derived from PARMAC and consists of three major components: periodic listen and sleep, collision and overhearing avoidance, and message passing. It achieves good scalability and collision avoidance by utilizing a combined scheduling and contention scheme. They synchronize nodes and let them periodically sleep, if otherwise, they would be in the idle listening mode. In the sleep mode, a node will turn off its radio, saving battery power. However, the latency is increased, since a sender must wait for the receiver to wake up before it can send out its data. S-MAC also adopts the RTS/CTS and ACK mechanisms for unicast messages to address collision avoidance and the hidden terminal problem. In case that a collision occurs, a Backoff Delay strategy is used to avoid its repetitions. S-MAC also breaks the long message into many small fragments, and transmits them in a burst. Using RTS/CTS packets, they reserve the medium for transmitting all the fragments. Every time a data fragment is transmitted, the sender waits for an ACK from the receiver. If it fails to receive the ACK, it will extend the reserved transmission time for one more fragment, and re-transmit the current fragment immediately.

Caccamo et al. (2002) have proposed an implicit prioritized access protocol which utilizes the Earliest Deadline First (EDF) scheduling algorithm (Liu & Layland, 1973), in order to ensure timeliness for real-time traffic. The idea is to take advantage of the periodic nature of the sensor data traffic to create a schedule rather than use control packets for channel reservation. A sensor network architecture composed of several hexagonal cells is considered. In order to avoid channel interferences, seven (7) different frequency channels are used. Within each cell, all the nodes are assumed to be fully connected, so that there will be no hidden channel problem. Intra-cell messages are exchanged, inside each cell, in multi-cast using EDF with implicit contention. On the other hand, inter-cell messages are exchanged among neighboring cells using more capable router nodes. Enabling multicast within each cell provides elimination of redundant data since only one message is transmitted out of the cell after intra-cell message exchanges. Simulation results show that the protocol performs better in terms of throughput and average delay in heavy load conditions when compared to CSMA/CA with the RTS/CTS option disabled.

The RAP protocol proposes a communication architecture (Lu et al, 2002), where a priority is assigned to each packet based on its requested velocity, in order to be queued at the network layer. The work also proposes an approach using different queues, where each priority corresponds to a range of requested velocities, i.e. the deadline and closeness to the gateway. Assuming that packets that miss their deadlines are useless, priority queues actively drop packets that have missed their deadlines to avoid wasting bandwidth. This strategy ensures a prioritization in the MAC layer to deal with packet delay. RAP does not consider any energy metric.

The DMAC protocol (Lu et al., 2004) is designed to solve the interruption problem due to activation/sleep duty cycles of other WSN MAC protocols. DMAC allows continuous packet forwarding by giving to the sleep schedule of a node an offset that depends on its depth on the tree. Activation/sleep duty cycles of other proposed

MAC protocols suffer from a data forwarding interruption problem, whereby not all nodes on the multihop path to the Sink, are notified of data delivery in progress, resulting in significant sleep delay. DMAC adjusts the duty cycles adaptively according to the traffic load in the network. A data prediction mechanism is also proposed in DMAC, which uses a More-To-Send (MTS) packet in order to alleviate problems pertaining to channel contention and collisions. Data prediction is employed to solve the problem when each single source has low traffic rate but the aggregated rate at an intermediate node is larger than what the basic duty cycle can handle. The interference between nodes with different parents could cause a traffic flow to be interrupted because the nodes on the multihop path may not be aware of the interference. MTS packet is also used to remain active when a node fails to send a packet to its parent due to interference. Some simulation results (Lu, 2004) show that by exploiting the application specific structure of data gathering trees, DMAC provides significant energy savings and latency reduction while ensuring high data reliability.

Woo and Culler (2001) propose an Adaptive Rate Control (ARC) mechanism that enables the media access control protocol to assist in achieving fair bandwidth delivery to the base station, for nodes in a multihop network. Their adaptive rate control uses loss as collision signal to adjust the transmission rate in a similar manner to the congestion control used by TCP. While TCP's congestion control is end-to-end over a network with many independent flows, Woo and Culler propose an adaptive rate control that works collectively at every node in the network, since each node is both a router and a sender, and where routing is done at the application level. They propose an implicit mechanism which passively adapts the rate of transmission of both, original and route-thru traffic, without the use of any MAC control packets. Their techniques reduce the amount of packets in the network, which increases the available power for data packets. Similar efforts to deal with congestion, such as RTS-CTS and

DATA-ACK handshake series, can constitute up to 40% of the overhead in their platform. They use Implicit Acknowledgement (IACKs), an advantage of wireless multihop networks, in which the acknowledgments role could be freely obtained when the receiving node re-routes the packet to its parent. This feature enables the elimination of the explicit ACK control packet.

As well as those protocols, an end-to-end metric such as latency, defined as end-to-end delay from sensor to Sink, is being used in Dousse et al. (2004), where a result shows that is positive to establish analytical bounds on the latency of a sensor network, that has random active and sleeping periods for their motes, in order to save its battery power.

EQoSA (Baroudi, 2007) is a general framework for adaptive Energy and QoS Aware that uses a scheduling based MAC protocol, where the time is divided into rounds and each round starts with a setup phase followed by a number of sessions. Each session stalls with a contention window followed by a number of slots reserved for those users who expressed their wishes to transmit packets and the cluster head approved their assignment. The proposed protocol has two unique features; the dynamic size of each session and the ability to accommodate bursty traffic sensors by allocating more than one data slot in each session. Discrete event driven simulation were extensively ran to evaluate the EQoSA against Bit-Map MAC protocol (a modified LEACH) and it shows an outstanding improvement both in the end-to-end packet delay as well as in the total consumed energy. These results make EQoSA a viable candidate MAC protocol for real time applications such as video and imaging applications.

But delay and latency are not the only QoS parameters that are being used in WSNs.

Reliability Assurance

Another common metric used on QoS in WSNs is Reliability. Some end-to-end reliability issues for WSNs are applied in:

Bhatnagar et al. (2001) and Deb et al. (2003a; 2003b) propose an Adaptive Forwarding Scheme (AFS) that uses priority levels to support service differentiation. Based on the criticality of data inside a packet, different priority levels are assigned: a single path forwarding scheme is used for low priority level, and a multipath forwarding scheme for high priority level. Each priority level maps to a desired reliability for data delivery. This strategy provides reliability by introducing redundancy in the number of paths between a source and the monitor (Sink). The number of paths is controlled by the packet priority and the network conditions. Each packet is forwarded along its default path. In the multipath forwarding scheme, nodes that receive a copy of a packet with a priority level l_i, forward it with a probability F_i to provide a desired level of reliability. Since pure local knowledge may not be adequate for a node to decide whether or not to forward a packet, the monitor node reinforces the forwarding probability periodically. This means that the number of copies of a packet transmitted in the network would be proportional to its priority. In this approach, the concept of reliability is still based on end-to-end service. The novelty of their work is that they consider the need for information-awareness and adaptability to channel errors as they happened with differentiated allocation of network resources for each situation.

Sankarasubramaniam et al. (2003) propose a new reliable transport scheme (ESRT) for WSNs. This is developed to achieve reliable event detection in WSNs with minimum energy expenditure. More importantly, their solution is based on a non-end-to-end concept. The solution includes a congestion control component that serves the dual purpose of achieving reliability and conserving energy. The reliability of event detection is controlled by the Sink, which has more power than sensors. It is worth noting that this chapter brings up the concept of non-end-to-end service. However, their solution only resides in an individual transport layer. Furthermore, it does not consider other important QoS factors.

Tilak et al. (2002) show the advantages of modifying the sensor mode reporting rate, so that an efficient point on the reliability curve is met. They study the effect of the infrastructure configurations on phenomenon driven and continuous driven network delivery models. They work with the case where the amount of collected data is more than what is necessary and study performance in terms of network efficiency, as well as meeting the application accuracy and latency demands.

EEAC (Energy-Efficient Asynchronous Clustering) (Luo et al., 2008) is proposed as a QoS-oriented events-driven asynchronous clustering protocol, which can deliver traffic in a timely and reliable manner. The EEAC clustering algorithm starts asynchronously according to a probability, determined by cluster-heads' data transmission rate and residual energy. Nowadays, most clustering proposals were based on periodic synchronous approaches, which require time synchronization and are not suitable for events-driven applications. EEAC avoids time synchronization and adopts composite formula to elect cluster-heads. Simulation results show that EEAC ensures the real-time transmission of sensitive data, reduces the packet loss rate, and evenly distributes nodal energy consumption, thus prolonging network lifetime.

Su and Almaharmeh (2008) describe an integration module for the wireless sensor networks and the Internet, in order to facilitate the flow of packets between the Internet and the WSNs. It uses a self-adapting QoS function to match changes on the traffic pattern. They consider that each WSN is able to sense a number of different attributes or topics, where each available topic should be registered within a registration authority and it could have two main arguments: the priority level and the reliability level. The priority level tells, how urgent is the data within the network, while the reliability argument tells how reliable the collected data is. The priority information associated with a topic during a communication session will be used to determine the link resources, such as bandwidth, that should be used. The reliability information associated with each topic is crucial to the applications on the Internet to filter out the data that does not satisfy the reliability requirements of that application.

On WSNs, other less traditional QoS features should be considered, that are inherent to their scarce power resources.

Extending Network Lifetime

The main goal of some QoS efforts is to extend the Network Lifetime to get a reasonable coverage of events for a large period of time.

Perillo and Heinzelman (2003) provide application QoS through the joint optimization of sensor scheduling and data routing. To save battery power, sensors reporting rate or data resolution can be adjusted, or the sensors can be scheduled to be turned off for a large period of time. This optimization can also extend the lifetime of a network considerably compared with approaches that do not use intelligent scheduling, even when combined with power-aware routing algorithms. In this work proposal, the routes should be chosen so that the nodes, those are more critical for used as sensors, should be avoided by the other in their routes. On the other hand, when determining the length of time for which a sensor set should be used, it is important to consider that the affected sensors are not only those that are active in the set, but also those being used in the chosen path(s) to the data sink. Actually, Perillo and Heinzelman's goal is to balance the application reliability with efficient energy consumption, providing in this way a type of application Quality of Service (QoS). Bhardwaj et al. (2002) solve a similar sensor management optimization problem, but they are more focused on efficiently sending event-triggered sensor data to the Sink, and the model is aimed more toward tracking applications.

Hadim and Mohamed (2006), work with a middleware called MILAN (Heinzelman, 2004) (Middleware Linking Applications and Networks) that lets sensor network applications specify their quality needs and adjusts the network characteristics to increase its application lifetime while still meeting those needs. To accomplish these goals, applications represent their requirements through specialized graphs that incorporate state-based changes in application needs. Based on this information, MILAN makes decisions about how to control the network, as well as the sensors themselves, to balance application QoS and energy efficiency, extending the network lifetime. Unlike traditional middleware that sits between the application and the operating system, it has an architecture that extends into the network protocol stack. As MILAN is intended to sit on top of multiple physical networks, an abstraction layer that is provided, allows network specific plug-ins to convert MILAN commands into protocol-specific commands that are passed through the usual network protocol stack.

Finally, the Simple Hierarchical Routing Protocol (SHRP) (Abbas et al., 2007) is a protocol whose primary objective is to extend the network lifetime, by routing messages through good quality path. Where the path quality could be expressed as a combination of: the remaining battery power on its nodes, the links with low interferences levels, and the short paths to arrive to the Sink. In order to contribute to save energy power, some redundant information about sensed value will not be sent. To provide certain reliability features, SHRP uses also other parameters such as LQI (Link Quality Indicator), and the number of hops to reach the Sink node. These metrics are managed in routing tables. Each node sends periodically to its neighbors a message called Network Information Message (NMI), which includes some updates to these metrics values. As soon as this message is received, a routing node can update its routing table. In an intuitive form, we can say that a good neighbor is a node that has enough battery to guarantee the shipment of a message and that offers a reliable way, in terms of radio signal, to reach the Sink node. In addition, this protocol is also able to inform to a central point about any possible disconnection caused by a reduction of battery power or by a long interference period.

Network Lifetime as a Requirement

In some research development, instead of reducing communication delay or extending the network lifetime, they set as its main goal to reach a specified network lifetime, because in that way a system could operate enough time to reach its original design goals. Some different works are being developed in this topic, including:

Mini et al. (2004) propose the use of a finite energy budget for each network activity, specifically for the construction of an energy map. This kind of map is a representation that shows information about the node's remaining energy in each part of the network, and that can be used to apply some routing strategies that aid in an efficient energy management. In the construction of this energy map, the term finite energy budget means that each node can spend a certain amount of energy in the process of map construction. In this context, energy can be represented as the number of packets that each node can send to the monitoring node. The goal is to construct the best energy map under the constraint that each node cannot send more than a certain number of packets. Is also required that sensed errors rate should be almost constant the whole time, firing some warning messages otherwise, and managing the energy budget so that it should be finished exactly at the end of a desired network lifetime. To reach these goals in each instant, each node should decide if it will send another energy information packet or not. This work proposes that the send decision should be taken according to a certain probability p. The value of p could be adaptable and could depend on: the number of packets that a node could send with its remaining energy power, the error between the predicted and the correct energy value, and its network estimated lifetime. Some simulation results (Mini et al, 2004) show that network lifetime could be

reached using the proposed finite energy budget model. Showing that for different sizes of budget, the measurement error is almost constant during all the simulation time, and the budget finishes at the end of the simulation. This feature is a clear indication that the adaptive process achieved a good performance in these scenarios.

Amirijoo et al. (2007) propose to provide lifetime predictability by automatically adjusting the Quality of the Data (QoD), which means to modify the sampling period of sensor nodes. Their approach is focused on enabling that a Communication Server (CS) or Sink reaches a specified lifetime. This device is battery operated, and uses a satellite system to achieve communication of the WSNs with their users. All network components must cooperate to manage the lifetime of the network, and specifically of the Sink, to successfully complete the designated mission. Although, it is desirable that all nodes meet the network lifetime, if the Sink fails to meet the lifetime, it could have a greater impact compared to a sensor failing. Amirijoo et al (2007) made an evaluation of their approach, which provides an actual lifetime within 2% of the specified one, even when some variations in workload and communication link quality occur.

Qu et al. (2007) describe an improvement to the precision of collected data by minimizing the total error bound, under the constraint that a predefined network lifetime should be achieved. They propose an optimization problem to deal with this situation by combining the changing pattern of sensor readings, the residual energy of sensor nodes, and the communication cost from the sensor node to the base station (Sink). Their proposals were evaluated by simulation experiments. If the energy of each sensor node was unlimited, they can send every information update to the base station in order to collect all the information that is being generated. However, as sensor nodes are energy-constrained, it could be difficult for the sensor nodes to send all their collected data to the Sink. Because of this fact, only when a new sensed value X_i exceeds a previous one in a certain threshold, then X_i will be sent to the base station. Otherwise, there is no need to transmit the X_i value immediately and some aggregation techniques could be used to gather as much information as possible in a later data send. In that way, communication cost and energy consumption could be reduced.

Other Application-Specific QoS Features

There are some additional QoS aware protocols and approaches that are based on other less traditional QoS metrics, some of them are:

The Sequential Assignment Routing protocol (SAR) (Sohrabi et al., 2000) is the first protocol for sensor networks that includes explicitly a notion of QoS in its routing decisions (Akyildiz et al., 2002; Şohrabi et al., 2000), to minimize an average weighted QoS metric throughout the lifetime of the network. SAR applies a table-driven multi-path approach that strives to achieve energy efficiency and fault tolerance to gather WSNs Information. SAR creates trees, rooted at one-hop neighbors of the Sink, using available energy resources on each path and the priority level of each packet as QoS metrics. On the process to create those trees, multiple paths from Sink to sensors are formed, but only one of them is actually used, according to the energy resources and additive QoS metric; other paths are kept as backup. Failure recovery is done by enforcing the routing table consistency between upstream and downstream nodes on each path. Any local failure triggers a procedure for path restoration. Simulation results show that SAR offers less power consumption than the minimum-energy metric algorithm, which focuses only on the energy consumption of each packet without considering its priority. SAR maintains multiple paths from nodes to Sink. Although, this allows fault-tolerance and easy recovery, the protocol suffers from the overhead of maintaining the tables and states at each sensor node, especially when the number of nodes is large. SAR does not use redundant routes to split the load and effectively boost the usable bandwidth.

Iyer and Kleinrock (2003) define QoS as the optimum number of sensors that should be sending information at any given time, because of a sensor's replenishments or deaths (e.g., more sensors being dropped from an airplane or being recharged with additional battery power). Determining the number of active sensors is a very important issue, because in any sensor network we want to accomplish two things: (1) maximize the lifetime of the network by having a sensors periodically powered-down to conserve their battery power, and (2) having enough sensors powered-up and sending packets toward the information sinks so that enough data is being collected to gather the desired information (a live sensor is defined as a sensor that has not run out of battery power). They utilize the Sink to communicate QoS information to each of the sensors using a broadcast channel, exploiting the mathematical paradigm of the Gur Game to dynamically adjust the resolution of the QoS of the data that is receiving from the sensors, in order to define the optimum number of sensors that should be active in the WSN.

Kay and Frolik (2004) also use a spatial resolution to define WSNs QoS in terms of how many of the deployed sensors are active. This measure is intuitively applicable where the expectation is that the number of sensors deployed exceeds the minimum number of nodes required for system functionality. Over-deployment enables the network to maintain a desired QoS throughout the network's lifetime, even when some sensor nodes fail or run out of energy. Herein, network lifetime is defined as the duration of time for which the desired QoS (i.e., the desired number of active sensors in the network) is maintained. The method utilizes a strategy of rewards and punishments that control the operating state of sensors using Gur Game and the ACK strategies in cluster networks, in order to control the number of active sensors while minimizing energy usage. Examples of the application of this kind of usage include: target tracking systems, structural health monitoring systems, and remote sensing systems, where the number of active sensors may need to be increased or decreased based upon events (e.g. target acquisition, detection of an exceeding structural load, and rapid temperature increase).

In other papers, such as in Meguerdichian et al. (2001a; 2001b), QoS is also defined as coverage or exposure. The basic idea is how to cover the desired area of interest without leaving sensing holes so that unexpected events can be detected as quickly and reliably as possible. The deployment of sensors can be pre-defined or random.

Chen and Heinzelman (2005) consider the bandwidth constraint when studying QoS-aware routing for supporting real-time video or audio transmission. Much of this work is targeted at finding a feasible route from a source to a destination without considering current network traffic or application requirements. Therefore, the network may easily become overloaded with too much traffic and the application has no way to improve its performance under a given network traffic condition. While this may be acceptable for data transfer, many real-time applications require QoS support from the network. They believe that such support can be achieved by either finding a route to satisfy the application requirements or offering network feedback to the application when the requirements cannot be met. They propose a QoS-aware routing protocol that incorporates an admission control scheme and a feedback scheme to meet the QoS requirements of real-time applications. The novel part of this routing protocol is the use of the approximate bandwidth estimation to react to network traffic. Chen and Heinzelman's approach implements these schemes by using two bandwidth estimation methods, one based on a listening estimation of the node ratio free and busy times, and the other one based on dissemination of information about the bandwidth using a Hello message. Both approaches are used to find the residual bandwidth available at each node to support new streams. Chen and Heinzelman (2005) show simulation results of its QoS-aware routing protocol that show that the packet delivery ratio increases greatly, and packet delay and energy dissipation decrease significantly; meanwhile the overall end-to-end throughput is not impacted, compared with routing protocols that do not provide QoS support.

Alex et al. (2004) describe MidFusion a middleware that uses the Bayesian theory paradigm to support sensor network applications performing information fusion. MidFusion discovers and selects the best set of sensors on behalf of applications, depending on the QoS requirements and the QoS that can be provided by the sensor networks in a transparent way. Unlike MILAN it does not need a priori knowledge of the type of sensors. MidFusion and MILAN both have no standard way to represent application requirements.

Delicato et al. (2005) describe other middleware that provides an abstraction layer between applications and the underlying network infrastructure, and it also keeps the balance between application QoS requirements and the network lifetime. It monitors both network and application execution states, performing a network adaptation whenever it is needed. Such middleware is in charge of decisions about communication protocols, network topologic organization, sensor operation modes, and other infrastructure functions typical of WSNs. It uses the greedy algorithm to choose nodes that best satisfy the application requirements. Delicato et al. (2005) also use QoS parameters such as data accuracy and energy-awareness in its evaluation of network resource use.

Xia and Zhao (2007) describe the deadline miss ratio as a metric for network quality of service. In a control loop, a deadline miss occurs when the actuator does not receive a control command by its deadline, which is by default equal to the sampling period. An actuator is a special sensor node with the capacity to introduce some modification in the operational parameters of the physical system.

There is a lack of flexibility in time-triggered sampling when the system operates in resource constrained environments with variable workload. As a solution for this problem they propose to adapt the sampling periods at runtime. Xia et al (2007) design a Fuzzy Logic Control based on QoS Management (FLC-QM) to dynamically adjust the sampling period of relevant sensors in a way that the deadline miss ratio is kept at a desired level. This solution is generic, scalable, and easy to implement. It can simultaneously address multiple QoS problems such as delay, packet loss, and network utilization. Simulation results (Xia et al, 2007) demonstrate the effectiveness of the proposed FLC-QM approach.

QOS CHALLENGES AND OPEN RESEARCH

In the context of WSNs, efficient resource usage not only means efficient bandwidth utilization, but also a minimal usage of energy. Even during network overload periods, the most important traffic should still have its QoS requirements satisfied in the presence of different network conditions. Based on previously described challenges, Chen and Varshney (2004) identifies a series of questions and open research issues for QoS support from 1 to 9, that we have increased in this work from 10 to 13, based on others work made by Younis et al. (2004) and Xia (2008):

1. **Simpler QoS models:** Traditional DiffServ and InterServ models could be useful to apply in WSNs. Maybe novel and simple QoS models are required to identify the architecture for QoS support in WSNs. Cross layer instead of traditional layered design may be helpful to work out with a simpler model.

2. **QoS-aware data dissemination protocols:** Routing protocols with support to QoS-constrained should manage traffic packets, while minimizing energy consumption. These protocols should support some priority schemes, which enable the sending of at least the high-priority traffic even on overloaded traffic situation, or under a highly dynamic network.

3. **Services:** What kind of non-end-to-end services can WSNs provide? Are traditional best efforts, warranted, and differentiated services still feasible in this new paradigm?

4. **QoS support based on collective QoS parameters:** It is very interesting to explore the support mechanisms for data delivery models using collective QoS parameters, such as collective latency, collective packet loss, collective bandwidth, etc. Furthermore, how should these mechanisms differ from those in traditional networks?

5. **Redundancy trade-offs:** Data redundancy can be exploited to improve information reliability. However, this model increases the energy expenses because of the processing of the redundant data. If we introduce data fusion, expenses can be reduced in order to save energy, but it also introduces delays into the network. The optimum trade-off among reliability and energy expenses is still an unsolved problem.

6. **Adaptive QoS assurance algorithms:** It is desirable to maintain QoS throughout the network life time, instead of having a gradual decay of quality as time progresses. This prevents gaps in data sets received by the Sink originated by network dynamics.

7. **Service differentiation:** What should the criteria of differentiation be among services? Should it be based on traffic types, data delivery models, sensor types, application types, or the content of packets? Considering the memory and processing capability limitations, we cannot afford to maintain too many flow states in a node. Thus, it is desirable to control network resource allocation into a few differentiated traffic classes.

8. **QoS support via a middleware layer:** If QoS requirements from an application are not feasible in the network; the middleware may negotiate a new QoS with both the application and the network.

9. **QoS control mechanisms:** Sensors sometimes may send excessive data, wasting precious energy, while they may also send non-useful or inadequate data at other times, so that the expected quality of the application cannot be met. It is desired that QoS control how the algorithms manage these situation.

10. **Node mobility:** Another interesting issue is the consideration of node mobility. Most of the current protocols assume that the sensor nodes and the Sink are stationary. However, there might be situations where the Sink, and possibly the sensors, need to be mobile. In such cases, the frequent update of the position of nodes and the propagation of that information through the network may excessively drain the node's energy. Clever QoS routing and MAC protocols are needed in order to handle the overhead of mobility and topology changes in energy-constrained environment (Younis et al., 2004).

11. **Other possible future research includes the integration of sensor networks with IP-based networks (e.g. Internet):** Most of the applications on security and environmental monitoring require that the data collected from the sensor nodes be transmitted to a server so that further analysis can be done. On the other hand, the requests from the user could be made to the sink through the Internet. Since the routing requirements of each environment (wsn and Internet) are different, further research is necessary for handling QoS requirements under these kinds of situations (Younis et al., 2004).

12. **Dealing with not independent factor:** Overcoming bandwidth limitation, effective energy and delay trade-off, handling buffer size limitation, reliability, supporting multiple traffic types and the removal of redundancy have been identified as the main technical challenges for supporting QoS requirements in WSNs. Few research projects have started to tackle only a subset of these issues, leaving lots of room for future research (Younis, 2004). Some of them can affect other parameter behavior, i.e. reducing energy consumption through a decrease of radio uptime will increase battery node lifetime, but also would increase the end-to-end node communication delay. How to deal with more than a parameter at a time could be a challenging task.

13. **Mixed Traffic:** Diverse applications may need to share the same WSN, introducing periodic and aperiodic data as much as different message characteristics (Xia, 2008). Even more interesting things can occur when nodes from different network in the same physical space could cooperate with each other. Different networks could decide to cooperate if some services advantages could be reached that justify this cooperation, getting a better network performance than if these networks worked isolated from each other (Vaz et al., 2008).

CONCLUSION

This chapter presents the state of the art of QoS on WSNs. Starting with a series of concepts, requirements, and strategies and following with a survey of tools and projects that are dealing with some QoS aspects on WSNs, giving a broad view of what is being done and what can be done. This survey allows us to identify that, even though there are many works that use traditional approaches to QoS, considering metrics such as end-to-end delay, throughput and bandwidth limitation, there are a lot of new approaches to QoS to deal with the limited resources that each sensor node could have. In these approaches, two main statements could be noticed. The first one is that many interesting tools and strategies, which are being used in traditional networks, could not be applied in WSNs, because there is not enough memory, processing power or storage capacity on a sensor node to manage these implementations, which were originally designed for computer equipment with better performances. The second aspect to deal with is that nodes have to work with a limited power source, such as batteries, due to this fact, some energy-aware considerations have to be taken in order to work with an energy budget in such a way that the used approaches could warrant that the network would be alive, available, and supporting measurement and control process, for a reasonable period of time.

As a complement to this survey, a list of open researches or hot topics is also presented in order to illustrate which could be the challenges that could be faced on QoS for WSNs to satisfy current and future user information requirements.

REFERENCES

Abbas, C. B., Gonzalez, R., Cardenas, N., & Villalba, L. J. (2007). A proposal of a Wireless Sensor Network Routing Protocol. *IFIP International Federation for Information Processing, Volume 245, Personal Wireless Communications*. 410–422. Boston: Springer.

Adamou, M., Lee, I., & Shin, I. (2001, December). *An energy efficient real-time medium access control protocol for wireless ad-hoc networks*. Paper presented at WIP session of IEEE Real-time systems symposium, London.

Akkaya, K., & Younis, M. (2003, May 19-22). An energy-aware QoS routing protocol for wireless sensor networks. In *Proceedings.of the 23rd International Conference on Distributed Computing Systems Workshops, 2003*. (pp 710-715).

Akkaya, K., & Younis, M. (2004). Energy-aware delay-constrained routing in wireless sensor networks. *International Journal of Communication Systems, 17*, 663–687. doi:10.1002/dac.673

Akyildiz, I. F., Su, W., Sankarasubramaniam, Y., & Cayirci, E. (2002, March). Wireless sensor networks: a survey. *Computer Networks, 38*(4), 393–422. doi:10.1016/S1389-1286(01)00302-4

Al-Karaki, J. N., & Kamal, A. E. (2004, December). Routing Techniques in Wireless Sensor Network: a Survey. *IEEE Wireless Communications., 11*(6), 6–28. doi:10.1109/MWC.2004.1368893

Alex, H., Kumar, M., & Shirazi, B. (2004). MidFusion: middleware for information fusion in sensor network applications. In *Proceedings of International Conference on Intelligent Sensors, Sensor Networks and Information (ISSNIP)*. Australia.

Amirijoo, M., & Son, S. H., & Hansson, J. (2007. June). *QoD adaptation for achieving lifetime predictability of WSN nodes communicating over satellite links*. Paper presented in Fourth International Conference on Networked Sensing Systems (INSS).

Aquino, A. L., Figueiredo, C. M., Nakamura, E. F., Frery, A. C., Loureiro, A. A., & Fernandes, A. O. (2008, March). *Sensor Stream Reduction For Clustered Wireless Sensor Networks*. Paper presented at ACM Symposium on Applied Computing 2008 (SAC 2008), Fortaleza, Brazil.

Baroudi, U. (2007). EQoSA: Energy And QoS Aware MAC For Wireless Sensor Network. In *Proceeding of the 9th International Symposium on Signal Processing and its Applications, IEEE, 1*(3), 1306-1309

Bhardwaj, M., & Chandrakasan, A. (2002). Bounding the lifetime of sensor networks via optimal role assignments. In *Proceedings of INFOCOM 2002*.

Bhatnagar, S., Deb, B., & Nath, B. (2001, September). *Service Differentiation in Sensor Networks*. Paper presented at Fourth International Symposium on Wireless Personal Multimedia Communications.

Blake, S., Black, D., Carlson, M., Davies, E., Wang, Z., & Weiss, W. (1998, December). *An Architecture for Differentiated Services*. RFC 2475, IETF.

Braden, R., Zhang, L., Berson, S., Herzog, S., & Jamin, S. (1997, September). *Resource reSerVation Protocol (RSVP) version 1, Functional Specification*, RFC 2205, IETF.

Caccamo, M., Zhang, L. Y., Sha, L., & Buttazzo, G. (2002, December). An Implicit Prioritized Access Protocol for Wireless Sensor Network. In *Proceedings of the IEEE Real-Time Systems Symposium*.

Chen, D., & Varshney, P. K. (2004). QoS Support in Wireless Sensor Networks: A Survey. In *Proceedings of the International Conference on Wireless Networks (ICWN 2004), Las Vegas*.

Chen, L., & Heinzelman, W. B. (2005, March). QoS-aware Routing Based on Bandwidth Estimation for Mobile Ad Hoc Networks, *IEEE Journal on Selected Areas of Communication, Special Issue on Wireless Ad Hoc Networks, 23*(3).

Chen, S., & Nahrstedt, K. (1999, August). Distributed Quality-of-Service Routing in ad-hoc Networks. *IEEE Journal on Selected Areas in Communications, 17*(8).

Correia, L. H., Macedo, D. F., dos Santos, A. L., & Nogueira, J. M., (2005, April). *Issues on QoS Schemes in Wireless Sensor Networks*. (Tech. Rep. RT.DCC.004/2005). Universidade Federal de Minas Gerais DCC: Brasil, Belo Horizonte.

Deb, B., Bhatnagar, S., & Nath, B. (2003a, September). *Information Assurance in Sensor Networks*. Paper presented at 2nd International ACM Workshop on Wireless Sensor Networks (WSNA), San Diego CA.

Deb, B., Bhatnagar, S., & Nath, B. (2003b, October). *ReInForM: Reliable Information Forwarding using Multiple Paths in Sensor Networks*. Paper presented at 28th Annual IEEE conference on Local Computer Networks (LCN 2003), Bonn, Germany.

Delicato, F. C., Pires, P. F., Rust, L., Pirmez, L., & Ferreira, J. (2005). *Reflective middleware for wireless sensor networks*. Paper presented at 20th Annual ACM Symposium on Applied Computing (ACM SAC), USA.

Dousse, O., Mannersalo, P., & Thiran, P. (2004). Latency of wireless sensor networks with uncoordinated power saving mechanisms. In *Proceedings of the ACM MobiHoc 2004*.

Felemban, E., Lee, C., Ekici, E., Boder, R., & Vural, S. (2005, March). Probabilistic QoS guarantee in Reliability and Timeliness Domains in Wireless Sensor Network. In [Miami, FL.]. *Proceedings of the IEEE INFOCOM, 4*, 2646–2657.

Hadim, S., & Mohamed, N. (2006). Middleware Challenges and Approaches for Wireless Sensor Networks. *IEEE Distributed Systems Online, 7(3)*, March 2006.

He, T., Stankovic, J. A., Lu, C., & Abdelzaher, T. (2003, May). SPEED: A stateless protocol for realtime communication in sensor networks. In *Proceedings of the International Conference on Distributed Computing Systems*, Providence, RI.

Heinzelman, W., Kulik, J., & Balakrishnan, H. (1999, August). Adaptive Protocols for Information Dissemination in Wireless Sensor Networks. In *Proceedings of the 5th ACM/IEEE Mobicom Conference*, Seattle, WA.

Heinzelman, W. B., Murphy, A. L., Carvalho, H. S., & Perillo, M. A. (2004). Middleware to support sensor network applications. *IEEE Network Magazine Special Issue, 18*(1), 6–14. doi:10.1109/MNET.2004.1265828

ITU-T. Telecommunication Standardization Sector of International Telecommunication Union. ITU-T Recommendation E.800. (1994, August). *Telephone Network and ISDN Quality of Service, Network Management and Traffic Engineering: Terms and definitions related to quality of service and network performance including dependability. Retrieved August 20, 2008,* from http://www.itu.int/rec/T-REC-E.800-199408-I/en

Iyer, R., & Kleinrock, L. (2003). *QoS Control for Sensor Networks.* Paper presented at the IEEE International Communications Conference (ICC 2003), May 11-15. Anchorage, AK.

Kay, J., & Frolik, J. (2004). *Quality of Service Analysis and Control for Wireless Sensor Networks.* Paper presented at the 1st IEEE International Conference on Mobile Ad-hoc and Sensor Systems. October 24-27. Fort Lauderdale, FL.

Krishnamurthy, L., Adler, R., Buonadonna, P., Chhabra, J., Flanigan, M., Kushalnagar, N., et al. (2005). Design and deployment of industrial sensor networks: experiences from a semiconductor plant and the north sea. In *Proceedings of the 3rd international conference on Embedded networked sensor systems SenSys '05,* (pp 64-75). ACM Press.

Lim, A. (2002). Support for reliability in self-organizing sensor networks. In *Proceedings of the 5th International Conference on Information Fusion* (pp 973-980) IEEE Press, vol. 2, 2002.

Liu, C. L., & Layland, J. W. (1973). Scheduling algorithms for multiprogramming in hard real time environment. *Journal of the ACM, 20*(1), 46–61. doi:10.1145/321738.321743

Lu, C., Blum, B. M., Abdelzaher, T. F., Stankovic, J. A., & He, T. (2002). RAP: A Real-Time Communication Architecture for Large-Scale Wireless Sensor Networks. In *Proceedings of the IEEE Real-Time and Embedded Technology and Applications Symposium (RTAS 2002),* San Jose, CA, September 2002.

Lu, G., Krishnamachari, B., & Raghavendra, C. S. (2004, April). An Adaptive Energy-Efficient and Low-Latency MAC for Data Gathering in Wireless Sensor Networks. In *Proceedings of the 18th International Parallel and Distributed Processing Symposium (IPDPS'04),* 26-30 April 2004. (pp 224)

Luo, J., Liu, H. Z., Li, R. F., & Bao, L. (2008) QoS-Oriented Asynchronous Clustering Protocol in Wireless Sensor Networks. In *Proceeding of the IEEE Intl Conference on Wireless Communication, Networking and Mobile Networking (WICOM),* Dalian, China, Oct. 12-14, 2008.

Lutfiyya, H., Molenkamp, G., Katchabaw, M., & Bauer, M. (2001). *Issues in managing soft QoS requirements in distributed systems using a policy-based framework.* Paper presented at 2nd International Workshop on Policies for Distributed Systems and Networks.

Meguerdichian, S., Koushanfar, F., Potkonjak, M., & Srivastava, M. B. (2001a). Coverage Problems in Wireless Ad-hoc Sensor Networks. In *. Proceedings of IEEE Infocom, 2001,* 1380–1387.

Meguerdichian, S., Koushanfar, F., Qu, G., & Potkonjak, M. (2001b). *Exposure in Wireless Ad-hoc Sensor Networks. Paper* presented at Mobile Computing and Networking, (pp. 139-150).

Mini, R., Loureiro, A., & Nath, B. (2006). Energy Map Construction for Wireless Sensor Network under a Finite Energy Budget. In *Proceedings of the Seventh ACM/IEEE MSWiM'04,* October 4-6, 2004, Venezia, Italy.

Paavola, M., (2007, December). *Wireless Technologies in Process Automation – Review and an Application Example.* (Report A No 33), Hawai, USA: University of Oulu. Control Engineering Laboratory.

Perillo, M., & Heinzelman, W. (2003, May). Providing Application QoS Through Intelligent Sensor Management. In *Proceedings of the 1st IEEE International Workshop on Sensor Network Protocols and Applications (SNPA '03)*.

Peterson, L. L., & Davie, B. S. (2003). *Computer Networks: A Systems Approach*, 3rd Edition. San Francisco: Morgan Kaufmann Publishers Inc.

Prabhakar, B., Uysal-Biyikoglu, E., & Gamal, A. E. (2001). Energy-efficient transmission over a wireless link via lazy packet scheduling. In *Proceedings of the IEEE InfoCom, 2001*.

Qu, W., Li, K., Masaru, K., & Takashi, N. (2007). *An Efficient Method for Improving Data Collection Precision in Lifetime-adaptive Wireless Sensor Networks*. Paper presented at IEEE International Conference on Communications ICC '07. 24-28 June. (pp 3161-3166) Glasgow, UK.

Ruiz, L. B., Siqueira, I. G., & Oliverira, L. B. (2004). *Fault management in event-driven wireless sensor networks*. Paper presented at 7th ACM/IEEE Int. Symposium on Modeling, Analysis and Simulation of Wireless and Mobile Systems, Italy.

Sankarasubramaniam, Y., Akan, O. B., & Akyildiz, I. F. (2003). *ESRT: Event-to-Sink Reliable Transport in Wireless Sensor networks*. Paper presented at MobiHoc 2003, Annapolis, Maryland, June 2003.

Schurgers, C., Aberhorne, O., & Srivastava, M. B. (2001) *Modulation scaling for energy-aware communication systems*. Paper presented at IS-LPED 2001, (pp. 96–99).

Sharifi, M., Taleghan, M. E., & Taherkordi, A. (2006).). A Middleware Layer Mechanism for QoS Support in Wireless Sensor Networks. Paper presented at *the 4th IEEE International Conference on Networking* (ICN´06), Mauritius, April 23-29.

Shen, X., Wang, Z., & Sun, Y. (2004). Wireless Sensor Networks for Industrial Applications. In *Proceedings of the 5Ih World Congress on Intelligent Control and Automation*. (pp 3636—3640) June 15-19, Hangzhou, P.R. China.

Sohrabi, K., Gao, J., Ailawadhi, V., & Pottie, G. J. (2000, October). Protocols for self-organization of a wireless sensor network. *IEEE Personal Comm. Magazine*, 7(5), 16–27.

Stankovic, J. A., Abdelzaher, T. F., Lu, C., Sha, L., & Hou, J. C. (2003). Real-time communication and coordination in embedded sensor networks. In *Proceedings of the IEEE, vol. 91, no. 7*, July 2003.

Su, W. & Almaharmeh, B. (2008, April) QoS Integration of the Internet and Wireless Sensor Networks. *WSEAS Transactions on Computers. 4, (7)*.

Tai, S., Benkoczi, R. R., Hassanein, H., & Akl, S. G. (2007, June). QoS and data relaying for wireless sensor networks. *Journal of Parallel and Distributed Computing, 67*(6), 715–726. doi:10.1016/j.jpdc.2007.01.009

Tilak, S., Abu-Ghazaleh, N. B., & Heinzelman, W. (2002). Infrastructure tradeoffs for sensor networks. In *Proceedings of the ACM International Workshop on Wireless Sensor Networks and Applications Workshop, 2002*.

Vaz, P. O., da Cunha, F. D., Almeida, J. M., Loureiro, A. A., & Mini, R. A. (2008). The problem of cooperation among different wireless sensor networks. In *Proceedings of the 11th international symposium on Modeling, analysis and simulation of wireless and mobile*. Vancouver, Canada (pp 86-91) ISBN:978-1-60558-235-1

Wang, C., Sohraby, K., Li, B., Daneshmand, M., & Hu, Y. (2006) A Survey of Transport Protocols for Wireless Sensor Networks. *IEEE Network Magazine*. May/June 2006 (pp 34-40)

Wang, M. M., Cao, J. N., Li, J., & Das, S. K. (2008, May). Middleware for wireless sensor networks: A survey. *Journal of Computer Science and Technology, 23*(3), 305–326. doi:10.1007/s11390-008-9135-x

Woo, A., & Culler, D. (2001, July). A Transmission Control Scheme for Media Access in Sensor Networks. In *Proceedings of the ACM/IEEE Conf. on Mobile Computing and Networks (MobiCOM 2001),* Rome.

Xia, F. (2008). QoS Challenges and Opportunities in Wireless Sensor/Actuator Networks. *Sensors, 8*(2), 1099–1110. doi:10.3390/s8021099

Xia, F., Zhao, W., Sun, Y., & Tian, T. (2007, December). Fuzzy Logic Control Based QoS Management in Wireless Sensor/Actuator Networks. *Sensors, 7*(12), 3179–3191. doi:10.3390/s7123179

Xia, F., & Zhao, W. H. (2007). Flexible Time-Triggered Sampling in Smart Sensor-Based Wireless Control Systems. *Sensors, 7*(11), 2548–2564. doi:10.3390/s7112548

Xu, S., & Saadawi, T. (2001, June). Does the IEEE 802.11 MAC Protocol Work Well in Multihop Wireless Ad Hoc Networks? *IEEE Comm. Magazine,* June 2001.

Xu, W., Trappe, W., & Zhang, Y. (2007) Channel surfing: defending wireless sensor networks from jamming and interference. *IPSN'07,* April 25-27 (pp 499-508) Cambridge MA.

Ye, W., Heidemann, J., & Estrin, D. (2002, June). An Energy-Efficient MAC Protocol for Wireless Sensor Networks. In *Proceedings of the IEEE INFOCOM 2002,* New York.

Younis, M., Akayya, K., Eltowiessy, M., & Wadaa, A. (2004, January). On Handling QoS Traffic in Wireless Sensor Networks. In *Proceedings of the International Conference HAWAII International Conference on System Sciences (HICSS-37) vol. 40, no. 8,* (pp. 102-116). Hawaii, USA.

Yu, Y., Krishnamachari, B., & Prasanna, V. K. (2004). Energy-latency tradeoffs for data gathering in wireless sensor networks. In *Proceedings of IEEE Infocom 2004.*

KEY TERMS AND DEFINITIONS

Differentiated Services (DiffServ) Model: It is a model that classifies information packets flows into different service levels. The idea is to give the same treatment to the flows associated to an specific service level, but offering special treatments to some services depending on the system objectives.

Energy Aware Routing Protocol: It is a routing protocol that uses some information about power energy characteristics to take its routing decisions. Some of the more commons energy power features used by these protocols are: the amount of battery power spent by sending a message to a node, the amount of battery power spent by sending a message through a specific path, or the remaining battery power of a node.

Latency: It is the amount of time since an event occurs, in the physical world, and its detection in the WSNs Sink node, in order to be registered and processed by system users.

Network Lifetime: It is the amount of time that a Wireless Sensor Network would be fully operative. One of the most used definitions of network lifetime is the time at which the first network node runs out of energy to send a packet, because to lose a node could mean that the network could lose some functionalities. But, is also possible to use a different definition, in which some nodes could die or run out of battery power, whenever other network nodes could be used to capture desired information or to route information messages to their destination.

QoS: The establishment of some goals or requirements to application performance metrics, in order to reach some user's degree of satisfaction, about the services provided by a system. Different applications and users could require different QoS metrics and levels, and if they do not meet the requirements, the system behavior will be unsatisfactory.

Reliability: It is a measure of how much everything is working properly on the system. A fully reliable network would be that where: there is no one packet lost, each network component is working without appreciable faults, and each system component would have enough time and resources to do its work.

Throughput: It is a measure of how much work is being done by a system in each unit of time.

Wireless Sensor Network: It is a set of related devices with small size and some resource limitations that are used to collect data from the physical world, and gather them through a wireless interface to a place were this information would be used. In some cases a wireless sensor node could be used in automation closed loops to make some control activities, very close to the place where the physical process is located.

Chapter 5
Infrastructure for Testing Nodes of a Wireless Sensor Network

Bojan Mihajlović
McGill University, Canada

Željko Žilić
McGill University, Canada

Katarzyna Radecka
McGill University, Canada

ABSTRACT

Maintaining reliable networks of low cost, low energy wireless sensor network (WSN) nodes is a major concern. One way to maintain a reliable network is to perform in-field testing on nodes throughout their lifetimes, identifying failing nodes so that they can be repaired or replaced. This chapter explores the requirements for a wireless test access mechanism, and introduces a method for remote execution of software-based self-test (SBST) programs. In an effort to minimize overall test energy consumption, an SBST method is derived that takes the least amount of microcontroller cycles, and is compatible with system-level optimizations such as concurrent test execution. To further reduce test energy, compression algorithms compatible with WSN nodes are explored for use on test programs. The efficacy of all proposed methods is evaluated experimentally, using current measurement circuitry applied to a WSN node.

INTRODUCTION

Wireless sensor networks (WSN) have become available for use in various industrial control, environmental monitoring, and military applications. Typical WSN applications require that the network be reliable and maintainable in order to be useful. Such constraints necessitate a unique approach to the design and testing of WSN nodes. It may seem counter-intuitive that reliable networks can be built with what are often inexpensive nodes that are individually unreliable. In fact, one way to maintain a reliable network is to test nodes throughout their lifetimes in order to identify failing nodes so that they can be repaired or replaced. This is especially vital when nodes are deployed in inhospitable environments that accelerate their failure rate. This work is aimed at addressing the aggregation of

DOI: 10.4018/978-1-61520-701-5.ch005

quality issues in the operation of WSNs: correct operation, reliability, availability, and operation under strict energy constraints.

While much work has been dedicated to the manufacturer testing of embedded systems such as WSN nodes, this chapter addresses the in-field testing of nodes in a deployed WSN. There are several obstacles in realizing this type of testing scenario, starting with the lack of a testing infrastructure. Such as infrastructure needs to define how testing is carried out at a network-level and how test programs and test responses are stored and communicated. There is also a need for test programs that achieve an adequate test quality, or fault coverage, of individual node components and of the node as a whole. All the while, nodes must operate while using as little energy as possible. This poses significant challenges in maintaining wireless signal quality, and consequently, reliable network operation. It means that the overhead of performing testing must be kept to a minimum, so as not to severely impact the operational life of the node. These obstacles are addressed in this chapter, and the approach used is briefly introduced in the following subsections.

TESTING INFRASTRUCTURE

To achieve a reliable WSN, one requires an infrastructure for performing remote node tests. Until recently, the most efficient and often only in-field self-testing involved using built-in self-test (BIST) hardware within wireless nodes. Since this is often not possible due to the performance, area, and energy overheads (Krstic, Lai, Cheng, Chen, & Dey, 2002), a promising type of software-based testing is introduced as an effective alternative. These software-based self-test (SBST) programs work by using an existing microcontroller unit (MCU) instruction set to perform self-testing of all digital and mixed-signal components on a WSN node. The SBST programs allow an equivalent test quality to be achieved while minimizing energy consumption.

An infrastructure is presented that allows a basestation to distribute SBST programs to nodes in the network, remotely execute the SBST programs, and return test results back to the basestation. A method is also presented that harnesses regular network nodes as *helper nodes* in characterizing the wireless links that they can establish with their neighbours. The resulting testing infrastructure allows detection of failed node and even predication of failing nodes to be achieved.

To ensure that nodes are properly tested, any test that is employed must meet an acceptable coverage of modeled faults (fault coverage). The SBST programs that are used must then be built to test the entire node, and must cover every testable component on-board the node. Testable components include the MCU, memory, RF module, and sensors, as these are the primary components of any WSN node. A known-good test result for each test must also be calculated and stored separately. The result can then be compared with actual results remitted by nodes.

ENERGY EFFICIENCY

Since nodes are power-sensitive devices whose power sources are often on-board batteries, network quality can suffer if some or all nodes exhaust their energy reserves prematurely. Any overhead energy consumption must be minimized, such as the running of self-test programs. To do this, several energy-saving techniques are introduced which can reduce test energy consumption and test time:

- **Test optimization:** Test time is decreased by selecting the most efficient set of instructions to achieve the same test quality
- **Test combination:** There is an inherent overlap in testing separate systems on the same node. The coverage of each test is analyzed and redundancy eliminated

Figure 1. Generic node architecture

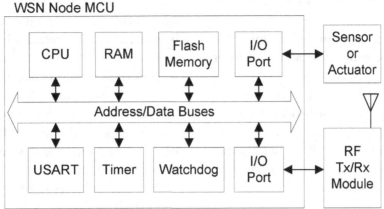

- **Test concurrency:** By reordering and re-scheduling tests, test energy and test time can be reduced
- **Test program compression:** Compressing test programs reduces communication and, in turn, the energy required to perform testing

By taking this approach, we address WSN quality issues that are currently impediments to correctly operating, reliable, available, and energy-efficient networks.

BACKGROUND

The introduction of ultra-low power network protocols such as IEEE 802.15.4 and its overlay protocol, Zigbee, has allowed wider use of a new class of devices. The new protocols enable devices to save energy otherwise expended in frequent communication, and are well-suited to wireless sensor networks (WSNs). A typical WSN is composed of sensor nodes and a basestation. The nodes' main function is to collect data through environmental sensors, as well as to control actuators to physical processes. At minimum, each sensor node contains an embedded MCU, wireless transceiver, antenna, environmental sen-

sors/actuators, and a battery, as seen in Figure 1. To minimize cost, components can be COTS parts assembled on a printed circuit board (PCB) containing a printed dipole antenna. The MCU is used to collect and process sensor data, which is stored on-board its embedded memory. The MCU also performs wireless network communication through the attached RF module, in infrequent and low-power packet transmissions.

Data harvested from the environment is relayed by nodes to the basestation through the network. The basestation serves as a focal point of the network, where sensor readings are aggregated and actuator control is focused. A sensor node itself may be required to refine data before announcing its results, or communicate with neighbouring sensor nodes. The following are general network-level requirements that a robust WSN should meet:

- **Availability:** The network should be online and available to the applications running on-board both the basestation and nodes.
- **Scalability:** Networks with thousands of nodes may become commonplace in environmental monitoring or military applications. The system must scale to accommodate such uses.
- **Self-Organization:** Harsh operational environments leads to a high rate of node

failures. A robust network must be able to circumvent failed nodes and organize to a new functional state.

- **Timing:** Real-time operation must be available to support applications where it is necessary.

- **Data Aggregation:** Sending data from individual nodes in large networks to a single basestation may overwhelm available network bandwidth. The ability of some nodes to poll data from a cluster of their peers can alleviate this pressure.

Node-specific requirements include the following, but are not limited in scope to a node, as failure to meet them can have a network-level effect:

- **Low Power Consumption:** Nodes are often battery operated and must remain functional for months or years. Since manual replacement of batteries is often not possible, minimizing power consumption is critical in achieving a robust network.

- **Low Cost:** Since wireless sensor networks are potential replacements for present wired sensor networks, they must remain low-cost in order to make them a viable alternative.

WSN AVAILABILITY

To ensure that a WSN can operate for a long time, a reliability assessment must be performed on the failure rates of its individual nodes. Even if redundancy techniques or highly reliable hardware are utilized, nodes can be deployed in harsh or even hostile environments that accelerate their failure. Of particular concern are failures that can sever the network and render a portion inaccessible. These failures affect network availability, which can be defined as the probability that the entire network is available for use. The primary challenge is to maintain a working application in the face of node failures, while detecting and even predicting the failure of sensor nodes and their components. The probability that a network will be restored to full operation in a given amount of time after a failure has occurred is defined as network serviceability (Chiang, Zilic, Radecka, & Chenard, 2004) or maintainability (Lala, 2001). In reality the calculation of such a metric is a complex task (Callaway, 2004), although some models exist, such as the Exponential Failure Law (EFL) defined in (Lala, 2001).

WSNs are subject to time-varying component connectivities due to factors impacting their wireless links (Chiang et al., 2004). Phenomena such as multipath fading, the "hidden terminal" problem (Callaway, 2004), and electromagnetic interference (EMI) contribute to the complexity of calculating an important metric such as availability. For the purpose of this paper, the concern is on overall system-level availability of a WSN application. So *perceived availability* can be defined as the probability that a WSN application is functioning correctly over a period of time (Chiang et al., 2004). Since WSNs are by nature distributed systems, the definition of reliability can be divided into component-level (local) and process-level (global) (Hariri & Huitlu, 1995; Raghavendra, Kumar, & Hariri, 1988). While component-level reliability deals with the reliability of individual elements which comprise the distributed system, such as nodes, process-level reliability includes all processes, hardware components, and communication channels which constitute the WSN.

Component-level failures can affect process-level reliability, which in-turn affects perceived availability. For example, a component-level failure of a network gateway node can cause the node to fail, and the network to split, which will affect perceived network availability. If node failures could be detected in such a scenario, then network down-time can be minimized. Periodic node testing is one method to ensure that nodes are continually operating correctly, while a survey of other methods can be seen in (Koushanfar,

Potkonjak, & Sangiovanni-Vincentelli, 2004). Since nodes can be repaired or replaced after they are found to be defective, testing can play an essential role in maintaining a high degree of network availability. The following section will explore various system-level testing options available for use on WSN nodes.

SYSTEM-LEVEL TESTING

Since WSN nodes are made to be deployed in inhospitable or even hostile environments, it is often impossible to obtain physical access to the devices in order to perform functional testing on them. Accelerated failures can also be attributed to the inhospitable environments in which WSNs might be used. Environmental conditions can compromise the reliability of node operation and by extension, the availability of the WSN itself. To achieve a high reliability and availability in a sensor network, node failures must be detected and/or averted during network operation. Thus, the in-field testing of individual nodes throughout their lifetimes is essential in predicting and detecting node failures and, in turn, improving network availability. Higher-priced systems, such as those in military applications, achieve enhanced reliability by using redundancy and highly reliable parts. However, such configurations have been shown to be unsuitable to power sensitive devices (Hariri & Huitlu, 1995). The cost and power consumption of such methods can quickly become prohibitively large on WSN nodes. It is clear that an alternative way of achieving the needed reliability and availability is needed that allows devices to retain a small power footprint and cost.

There are several methods of functional testing which can be used for system-level testing of nodes, including:

- Boundary Scan
- Hardware Built-in Self-Test (BIST)
- Software-based Self-Test (SBST)

Boundary scan methods require dedicated pins and larger, modified flip-flops to be physically incorporated onto a chip. Devices known as automatic testing equipment (ATE) are traditionally interfaced physically with the device-under-test (DUT). The ATE executes automatic test-pattern generation (ATPG) algorithms which create test vectors to stimulate potential faults based upon some fault model. The test vectors are shifted into the chip through the dedicated pins, the tests executed, and test responses shifted out. To overcome the need for physical connectivity, some research has been done in creating overlays which allow remote testing capability, as in the boundary scan that wirelessly interacts with the DUT in (Chiang et al., 2004). The process of shifting test vectors and responses can also consume a disproportionate amount of power for the testing of a sensor node. Using BIST resolves this issue by using dedicated hardware on-board the DUT to generate test vectors to output a single resulting test signature. However, hardware BIST is often not possible due to the performance, area, and energy overheads of dedicated test circuitry (Krstic et al., 2002). Since BIST is most often not included on sensor node components, creating a custom chip would also be relatively expensive when there are cheaper common off-the-shelf (COTS) parts available.

Recent work in software-based self-test (SBST) programs (Zhang, Zilic, & Radecka, 2006; Kranitis, Paschalis, Gizopoulos, & Xenoulis, 2005; Paschalis & Gizopoulos, 2005) offers a way to achieve high quality testing with a small performance overhead and no area penalty. The SBST programs utilize an existing MCU's instruction set to perform a self-test of all digital and mixed-signal components on a WSN node. Various test sequences are run through the system bus, peripherals, transceiver, and MCU itself in order to uncover faults based on some fault model. A summary of the combined test results, also known as a *test signature*, can then be generated. The signature can either be directly passed onto a basestation, or compared to a known-good result stored within the SBST program itself. The recent work in SBST can be attributed to its numerous advantages in certain systems over traditional methods, including:

- Testing while chip is running at full functional speed (at-speed testing) (Chen & Dey, 2001)
- Generation of deterministic test vectors from SBST program code
- Unique signature responses contained within SBST program code and compared by MCU (Chen & Dey, 2001)
- No need for expensive automatic test equipment (ATE)
- In-field testability
- Energy efficiency

This self-testing approach is considered an inexpensive way of achieving reliability, availability, and serviceability in a WSN. There are other compelling reasons for the use of SBST, such as certain tests that can determine if the hardware or software of a node has been tampered with by an intruder. The same interface is compatible with providing manufacturer testing, where tests are broadcast to multiple nodes simultaneously. This reduces the need for individual node testing using expensive automated testing equipment (ATE) or alternative on-board testing interfaces. Recent work in developing SBST programs for embedded processors includes the work of (Kranitis et al., 2005), which was used as the basis for an overall WSN node testing methodology proposed in (Zhang, Zilic, & Radecka, 2006; Zhang, 2005).

The SBST approaches used thus far on microprocessors can be classified into two categories. The first includes (Shen & Abraham, 1998; Batcher & Papachristou, 1999) that have a high level of abstraction and are functionally oriented. The second category includes (Kranitis et al., 2005, Chen & Dey, 2001; Kranitis, Paschalis, Gizopoulos, & Zorian, 2002), which are structurally oriented and

require structural fault-driven test development. From the first category, (Shen & Abraham, 1998) requires a lengthy test set, whereas the approach of (Batcher & Papachristou, 1999) is not purely a software-based method and requires some extra hardware. In the second category, a gate level netlist is necessary in (Chen & Dey, 2001), which is difficult information to obtain for COTS chips because of intellectual property (IP) constraints. The SBST methodology in (Kranitis et al., 2002) is based upon the application of deterministic test patterns targeting structural faults of individual processor components. The work of (Kranitis et al., 2005) builds upon (Kranitis et al., 2002) by defining test priorities for different processor components. The SBST proposed in (Kranitis et al., 2005) is also used as the basis for the methodology proposed in (Zhang, Zilic, & Radecka, 2006; Zhang, 2005), and shares desirable characteristics with functional testing, like high-level test development using the instruction set. The work takes a lower-level approach in its use of RTL information and uses a divide-and-conquer method to target individual components with the stuck-at fault model. This allows a high fault coverage of more than 95%, and is thus used as a model in the development of the SBST programs seen here.

In the early 1980's, an s-graph model at the register transfer level (RTL) was proposed (Thatte & Abraham, 1980; Brahme & Abraham, 1984) to represent a microprocessor, and used functional-level fault models for instruction-level test generation. Many further graph-based functional testing methods were later proposed, such as (van de Goor & Verhallen, 1992; Joshi & Hosseini, 1998), but suffered from the need for large amounts of manual effort in order to produce a relatively low fault coverage. The application of such test sets is also a lengthy process on microprocessors with large numbers of registers and instructions, since most of these methods relied on external ATE to deliver input test patterns and compare test responses.

REDUCTION OF TEST DATA VOLUME

As the speed of modern chips has increased, the speed of the testing interfaces to these devices has not followed suit. This, in addition to increasing design complexities have contributed to longer test times of devices seen as of late, as noted by the Semiconductor Industry Association (2005). Compression of test data aims to reduce the volume of information transferred between a DUT and its tester. This is often done to reduce the overall time required to test a device, since the majority of test time is spent in the sending of test data and the receiving of a response. As a summary, the work thus far performed in test data compression has largely fallen into the following groups (Khoche, Volkerink, Rivoir, & Mitra, 2002):

1. Compression of fully-specified test vectors
2. Compression of incompletely-specified test vectors
3. Hardware Built-In Self-Test (BIST)
4. Compression/compaction of test responses

Methods 1, 2, and 4 require ATE to be connected to a DUT, where a remote interface could serve for in-field testing. For each test executed, a compressed test vector known as a *test cube* is generated by the ATE and transferred to the node where it is decompressed on-the-fly. The test is then executed and a response sent back to the ATE. In method 1, the test cube is a directly compressed test vector generated by the ATPG algorithm, while method 2 can employ many different schemes to further compress test vectors. In this method, the fact that most test vectors effectively contain many don't-care (X) bits is used to allow even greater compression. A type of *lossy compression* (Salomon, 2004) is applied, which is effective when only a few select bits of a test vector need to be exactly reproduced on the DUT. In Method 4, test response vectors can similarly be reduced

in size. Compression can be performed on-board the DUT by the same algorithm as Method 1, or test responses can be compacted to signatures. These signatures are generated in much the same way as a simple hash, often by linear-feedback shift-register (LFSR) hardware.

All test communication for the aforementioned methods is done through a boundary scan interface. However, it has already been shown that the secret keys of cryptographic chips can be compromised using boundary scan attacks, and it is expected that boundary scan chains will become increasingly inaccessible on future production chips (Hely, Bancel, Flottes, & Rouzeyre, 2006). The test methods are also limited by the speed of the boundary scan chain, which can be significantly slower than a typical integrated circuit's operational frequency while incurring an area overhead. The exception is BIST, which can run at-speed and is improved through deterministic BIST methods (Liang, Hellebrand, & Wunderlicht, 2001), but is expensive in terms of physical device area and disallows the use of many COTS parts.

TESTING INFRASTRUCTURE

The testing of nodes is accomplished through the use of SBST programs within a testing infrastructure. Such an infrastructure for distributing and executing tests was found to be lacking from the literature. As part of the test protocol, SBST programs are dynamically loaded as needed onto nodes by the basestation, whose job it is to also collect back test responses signaling a pass or fail condition. A node-level view of the protocol functionality can be seen in Figure 2.

The test protocol also defines network-level tests, such as the scenario shown in Figure 3, where shaded arrows represent test channels. It can be seen that the Node Under Test (NUT) performs most of its testing through self-test but is also evaluated by neighbouring nodes. Testing is initiated by the basestation, which transfers a

Figure 2. General description of WSN node testing protocol

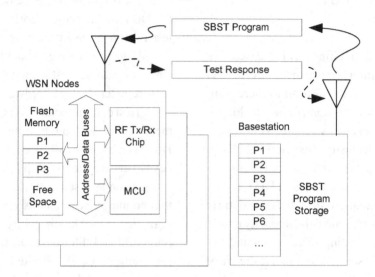

compressed SBST program to the NUT. The on-board MCU decompresses the SBST program and stores it in embedded flash memory, enabling the basestation to remotely activate NUT self-testing. Once activated, the NUT executes the SBST program, which begins by executing MCU core self-testing, and continues to include embedded RAM, embedded flash memory, on-board data/address buses, and peripherals such as sensors. A resulting signature is calculated and sent back to the basestation, where it is compared to a node's known-good signature. If a signature is deemed correct, the RF module and antenna have also effectively been tested as operating correctly on a pass/fail level. For a more comprehensive evaluation of RF module and antenna performance, several nodes neighbouring the NUT can be used to track a decline in communication efficiency. An additional test program can be distributed to instruct the NUT and its neighbours to communicate at predetermined power levels. The results are sent back to the basestation, which tracks and analyzes changes over time to anticipate failing RF modules, and hence, failing nodes.

RF MODULE TEST SCHEME

Traditional test schemes that can be applied to WSN node RF transceivers work by looping transmitter output back into the receiver, and capture a test response in the baseband of the receiver (Dabrowski, 2003; Halder, Bhattacharya, Srinivasan, & Chatterjee, 2005; Ozev, Orailoglu, & Olgaard, 2002). A digital signal processor (DSP) is commonly used to analyze the test response at the receiver end. The advantage of this method is that receiver and transmitter subsystems are decoupled, but the testing method requires the addition of hardware to provide the loopback path.

An SBST method is introduced here to test RF modules, consisting of an RF transceiver and on-board printed antenna, with the aid of other nodes in the network. Several RF operating metrics are characterized and compared with some specifications of the IEEE 802.15.4 (2006) communication standard, parts of which are seen in Table 1. Instead of specifying tests for each component of an RF module, the module is considered in its entirety and functionally verified based upon published specifications.

Figure 3. Node testing architecture

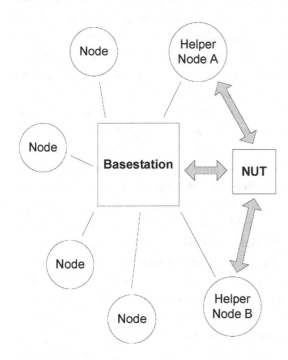

TRANSMISSION POWER AND RECEIVER SENSITIVITY

To ascertain RF module transmission power and receiver sensitivity, a simple SBST test is introduced here. Test packets are sent in long bursts from the basestation to the NUT, where their reception is tallied. After completion, a test response is sent from the node to the basestation containing the packet error rate (PER) of the transmission. The receiver sensitivity of the node can be calculated with the Friis transmission equation (Stutzman & Thiele, 1997). Using the same method, the

roles of basestation and node can be swapped to find NUT transmission power. The transmission power that achieves the required PER can be determined by varying the power in steps, as this is a software-programmable value on many RF transceivers.

ADJACENT CHANNEL AND ALTERNATE CHANNEL REJECTION

Many COTS RF transceivers use radio bands that many other devices are licensed to operate on. Since there is a high likelihood of interference between devices, IEEE 802.15.4 specifications require RF modules to reject the interference generated by other devices on neighbouring channels. A test is thus proposed for the ability of the RF module to perform this task. Here, *adjacent channel* refers to the channels closest in frequency to the channel being used, while *alternate channels* are those in-turn closest to the adjacent channels. For example, if channel 13 is being used, channels 12 and 14 are adjacent, while channels 11 and 15 are alternate.

An adjacent/alternate channel rejection test can be constructed as seen in Figure 4. In this scenario, the NUT is sent a long burst of packets from the basestation on a particular channel, while the helper node generates interference on either the adjacent or alternate channel. Using data from the receiver sensitivity test, transmission power on channel n_2 is set so that received signal strength on that channel is slightly higher

Table 1. Select RF specifications of the IEEE 802.15.4 protocol

Specification	Requirement	
Transmission Power	Minimum output power	-3 dBm
Receiver Sensitivity	Minimum input signal power yielding a packet error rate (PER) of < 1%	-85 dBm
Adjacent Channel Rejection	Min. ratio of the adjacent channel signal level to desired signal level with PER of < 1% (interference with adjacent channel)	0 dB
Alternate Channel Rejection	Min. ratio of the alternate channel signal level to desired signal level with PER of < 1% (interference with alternate channel)	30 dB

Figure 4. Test framework of adjacent channel and alternate channel rejection

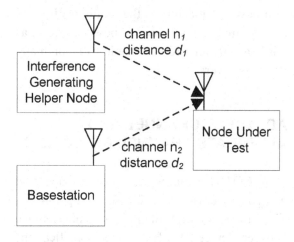

than the minimum specification. This signal on channel n_2 is made purposely weak so that it will be sensitive to adjacent/alternate channel interference. Setting distances $d_1 = d_2$, the transmission power on channel n_1 is found which causes a PER of 1% in channel n_2. Again using the Friis equation, the adjacent/alternate channel rejection ratio can be found.

ENERGY EFFICIENCY OF MCU TESTING

In this section, the dynamic programming idea of (Aho & Johnson, 1976) is used to construct the SBST programs proposed in (Kranitis et al., 2005) for testing an MCU. Since SBST programs are created using the MCU instruction set, a typical instruction uses one of many addressing modes and contains parameters such as a source register, destination register, and operand value. Energy optimizations can thus be performed based upon *program length* (number of cycles), *instruction type*, and *instruction parameters*.

Software energy consumption can be shown to be proportional to program length, but some published works including (Nikolaidis & Laopoulos, 2001; Chang, Kim, & Lee, 2000) indicate

that there is a dependency between instruction energy consumption and the values of instruction parameters. These instruction parameters are collectively referred to as *energy-sensitive factors*. The list of energy optimization criteria are addressed in our discussion by the use of *instruction combination, instruction selection,* and *operand selection*, respectively.

We experimentally verify that an instruction's energy consumption is proportional to the Hamming distance between the previous and current values of its energy sensitive factors. The average current drawn by the MCU can be measured as it repeatedly executes certain instructions or a short instruction sequence with different configurations of energy sensitive factors. From this, it can be determined that executing the same instructions with different addressing modes has a differing energy cost. Such measurements offer the opportunity to employ instruction-level energy reduction methods like instruction selection/combination which uses the least number of MCU cycles, as well as operand selection with least Hamming distance and weight.

INSTRUCTION SELECTION AND COMBINATION

The goal of instruction selection is to reduce program length by replace existing instructions with ones that require the least amount of cycles, while maintaining equivalent fault coverage. Instruction combination involves finding overlapping fault coverage between separate component tests. By combining component tests into a test superset, redundant testing can be eliminated.

Consider the testing of register files using the March X algorithm (van de Goor, 1993), where the operation set performed on each register is O_{REG} = *{Write 0, Read 0, Write 1, Read 1}*. Here, the most intuitive implementation of the *WriteX* sub-operation could be the set of instructions *I(Reg, WriteX) = mov X, R_N*. In Table 2a, we compare the set of instructions used to implement the O_{REG} = *{Write 0}* operation before and after instruction selection. As is typical in low-power MCUs, there is a differing execution time for instructions depending on the addressing mode used. In I_1, immediate-mode addressing of the *mov* instruction would, for example, take two clock cycles. The register-mode addressing used in all but the first instruction of I_2 would then take one clock cycle. Using instruction selection would save *n-1* clock cycles in the execution of the I_2 test over that of I_1, while properly propagating faults and conserving fault coverage.

It is also common that the testing of one MCU component results in collateral coverage of other, non-targeted components. For example, arithmetic component testing can cause collateral coverage of the data bus. Instruction combination is useful in eliminating overlapping fault coverage in order to reduce total SBST program length, while also conserving fault coverage. Two or more component tests should be combined into an SBST program superset, where redundant instructions can be identified and eliminated.

OPERAND SELECTION

Operand selection aims to reduce test energy consumption by minimizing the Hamming weight of all instruction operands and the Hamming distance of successive operands. Setting *don't–care* bits in

the operand to logic 0 serves to minimize Hamming weight, while a minimum bitwise change between the operands of successive instructions minimizes Hamming distance. Minimizing Hamming weight works on the principle that less energy is required for the transfer of logic 0 than of logic 1 in the binary coding used on MCU data buses. Minimizing Hamming distance reduces transistor switching, and thus CMOS switching power, which makes up the majority of power consumption in integrated circuits like MCUs.

Consider ALU testing, where a set of instructions performs a single O_{ALU} = *{add with carry}* operation. In Table 2b the set of instructions I_4 implements the operation *add with carry* using two 0x8000 operands, which have a combined Hamming weight of 2. This is the lowest Hamming weight of any two operands that satisfy the test, and considerably lower than the combined Hamming weight of 32 in the operands of I_3. Additionally, instruction selection is performed on the second instruction of I_4 to reduce program length. Considering all of the instruction-level energy reduction methods covered in this section, and the SBST method proposed in (Kranitis et al., 2005), the pseudo-code for our energy-saving SBST method can be seen in Figure 5.

Table 2. (a) The operation {Write 0} before and after instruction selection (b) the operation {add with carry} before and after operand selection

a)	before	after		b)	before	after
	I_1:	I_2:			I_3:	I_4:
	mov 0, R_1	mov 0, R_1			mov 0xFFFF, R_N	mov 0x8000, R_N
	mov 0, R_2	mov R_1, R_2			mov 0xFFFF, R_M	mov R_N, R_M
	...	...			add R_N, R_M	add R_N, R_M
	mov 0, R_N	mov R_{N-1}, R_N				

Figure 5. Energy-reduced SBST algorithm

```
Extract the set of all microcontroller components C = C₁, C₂, ..., Cₙ;
Extract the set of all operations each component performs O = O₁, O₂, ..., Oₘ;
Extract the set of instruction sequences
    I(C,O) = I₁, I₂, ..., Iₚ, during execution, which cause component C to perform operation O;
Reorder C with the test priority from high to low;
for ( i = 1, I < n+1, i++ )
    Choose Cᵢ from set C;
    for ( j = 1, j < m+1, j++ )
        Choose Oⱼ from set O;
        Choose test sequence Iₖ from I(Cᵢ,Oⱼ) that has the smallest program length;
        Set operand values of Iₖ to those with the least Hamming distance and weight;
        Apply Iₖ with chosen operands to the input of Cᵢ and propagate the result to primary
        outputs;
    endfor
endfor
```

ENERGY EFFICIENCY OF EMBEDDED MEMORY TESTING

There is a need for SBST programs that test the flash memory often found on WSN nodes. Flash memory is prone to failure over time and is a prime candidate for the implementation of the SBST concept. Unlike RAM, once flash memory has once been written to, it cannot be overwritten without first performing a block-erase. This makes conventional memory testing methods such as (van de Goor, 1991) unsuitable. Instead, a March-type algorithm called *March FT* is proposed in (Yeh, Wu, Cheng, Chou, Huang, & Wu, 2002), which can be seen in Equation 1. It has the highest fault coverage of all published approaches for most conventional faults and for all disturb faults. Here we aim to reduce the software energy required to perform the March FT algorithm. The extending of the single-bit-convert (SBC) memory addressing to flash memory is explored, as well as the interleaving of flash memory testing with other component tests.

$$(f); \downarrow (r1,w0, r0); \updownarrow (r0); (f); \uparrow (r1,w0, r0); \updownarrow (r0) \quad (1)$$

where:

$\uparrow$ and $\downarrow$ are increasing and decreasing address orders, respectively.
(ex.: $\uparrow$ means from address 0 to address $n-1$).
$\updownarrow$ means address order is irrelevant.
n is the total size of memory to be tested.
$w0/1$ is writing 0 or 1 into a cell, respectively.
$r0/1$ is reading a cell, expecting a value of 0 or 1, respectively.
f is the erasing of a block of flash memory.

SINGLE-BIT-CONVERT (SBC) ADDRESSING FOR MEMORY TESTING

The SBC memory addressing scheme was introduced in (Cheung & Gupta, 1996) as an energy-optimized way of addressing RAM for testing. It is an alternative to the way incremental addressing counts up or down through all testable memory addresses. SBC works by minimizing the Hamming distance between consecutive addresses, which reduces the switching activity on the address bus. In (Cheung & Gupta, 1996), address bus transitions are reduced by 50%, which accounts for a total energy consumption reduction of 18% to 77% depending on the size of RAM. As a candidate for use in March-type testing algorithms, SBC addressing can be used for any ↕ portions where incremental addressing is not explicitly enforced.

INTERLEAVING FLASH MEMORY TESTING WITH OTHER TESTS

When a block-erase or chip-erase is performed on flash memory, the segment is generally filled with a sequence of logic 1 bits. Once a flash-write operation resets any bit to logic 0, the segment containing the bit must be block-erased to reset it back to logic 1. This erase operation is much slower than a read or write operation on flash memory. A block of embedded flash memory on-board an MCU, for example, can be erased by executing a series of instructions located in either flash memory or RAM. When flash memory is subjected to an erase operation, all timing control is transferred to an embedded flash controller and MCU execution is stalled until the erase operation completes. The MCU resumes execution after completion, provided a different segment than the one containing program code was erased. However, if the erase operation is initiated from RAM, the MCU will execute code from RAM while the flash erase is taking place. This allows the otherwise wasted MCU idle time to be used to execute other tests from RAM, and is the premise of the test interleaving shown in Figures 6a and 6b.

The efficiency of interleaving flash memory testing with other component tests depends upon several factors. The size of the flash block being erased, and hence its timing, must approximately match the time required to execute an unrelated component test. Figure 6a shows test routines requiring differing amounts of time and average power. In Figure 6b, the *other component test* is rescheduled into the same time slot as the flash-erase operation. This potentially causes a reduc-

tion in software energy consumption based upon the principle that executing simultaneous tests requires less energy than the sum of the energies of executing individual tests. This is because an MCU has an overhead power consumption, known as a baseline, simply being powered-on and ready to execute instructions. Equations for test time and test energy reductions can be seen in Equations 2 and 3, respectively.

$$Time\ Reduction\ (\%) = \frac{Min(T_{FE}, T_O)}{T_{TOTAL}} \times 100 \quad (2)$$

$$Energy\ Reduction\ (\%) = \frac{E_B - E_A}{E_B} \times 100 \quad (3)$$

Where $T_{TOTAL} = T_{CPU} + T_{FE} + T_{FP} + T_{FR} + T_O$ (T_{CPU}, T_{FE}, T_{FP}, T_{FR}, T_O are defined in Figure 6a). E_B and E_A are the total software energy consumed before and after time interleaving, respectively.

ENERGY EFFICIENCY THROUGH REDUCTION OF TEST DATA VOLUME

It can be seen that none of the methods mentioned in the previous sections can be applied to compressing the volume of test data contained in SBST programs. Since the programs either self-generate test vectors or contain them within the program code, a way to reduce test data volume is to compress the SBST programs themselves. In the following section we investigate the viability of compressing SBST programs by characterizing their structure.

CHARACTERIZATION OF SBST PROGRAM CODE

Data compression is the process of removing redundancy from a data set in order to represent

Figure 6.

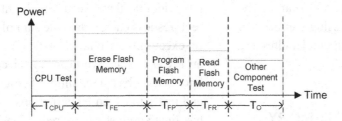

(a) Concept of Time Interleaving Individual Node Tests - Before Time Interleaving

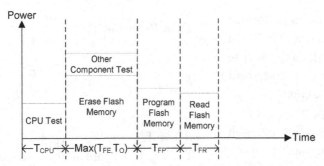

(b) Concept of Time Interleaving Individual Node Tests - After Time Interleaving

the original with a smaller set. The less random a data set, the more redundancy that can be extracted to yield better compression. If the compression of SBST programs is to be considered, the programs themselves should be characterized as being of low entropy. An analysis is therefore performed on three real SBST programs from (Zhang, Zilic, & Radecka, 2006) used in testing WSN nodes, denoted P_n. As a useful measure of entropy, a histogram of symbol frequencies is produced by inputting SBST program machine code into MATLAB 7.0, with a symbol size of 1 byte. Code compilation is targeted at the Texas Instruments MSP430 MCU that is part of our WSN research platform.

Figure 7 shows that the frequency of some symbol occurrences is reasonably high in the measured SBST program. The many peaks and valleys visually depict that the program composition is somewhat redundant and thus a good candidate for compression. A similar analysis of two additional SBST programs found similarities in all of their

frequently occurring bytes, which indicates that the programs share similar characteristics. These similarities can be attributed to:

1. **Instruction set structure:** The target MCU instruction set is composed of instructions for single-operand arithmetic, two-operand arithmetic, and conditional branching. Single-operand instructions share a 6-bit prefix such that the first byte of each instruction is *0x10–0x13* (bytes 16–19 in decimal). Likewise, the first byte of conditional branch instructions falls into the range *0x20–0x23* (bytes 32–35 in decimal).

2. **Commonly used instructions and operands:** Certain instructions are used more often in testing than others. For example, an efficient way to perform certain flash memory March testing is to write to an array of data elements (*mov* instruction), then read the array using an exclusive-or (*xor* instruction) between consecutive data

elements. Certain operands are also more prevalent than others, such as passing *0x00* in the immediate addressing mode to test stuck-at-one faults in buses.

3. **Shared code between programs:** There is inherent overlap between SBST programs if they are to be executed independently of each other. Segments of code holding constants and performing hardware initialization must often be replicated.

In the following section, several general-purpose compression algorithms are explored for use on SBST programs due to the lack of appropriate SBST-specific techniques in the literature. Algorithm attributes are compared with each other and against node-specific requirements.

GENERAL-PURPOSE COMPRESSION ALGORITHMS

There has been extensive work performed on entropy coding algorithms over the years, a summary of which can be seen in (Lelewer & Hirschberg, 1987). To uncover an effective method of compressing SBST programs, in this section three groups of prominent algorithms are evaluated: Lempel-Ziv-Welch (LZW), Dynamic Huffman, and Bentley-Sleator-Tarjan-Wei (BSTW). The algorithms are compared against the following node-specific requirements:

1. **Low energy consumption:** By transferring compressed SBST programs instead of uncompressed ones, the power-hungry RF transceiver can have a smaller duty cycle and energy can be saved on-board a node.
2. **Low cost:** Of the components in a typical node, a disproportionate amount of cost is attributed to the MCU. A selection should therefore be carefully made in choosing an MCU meeting the minimum requirements of the WSN application. Since a

typical application such requires modest resources, an MCU with small amounts of processing power, RAM, and flash memory is often enough. It is typical for an MCU to contain 1kB of RAM, which must be taken into account when selecting a compression algorithm. This is important since the price difference between a common MCU with 5kB of RAM compared to 1kB can be almost double according to Texas Instruments (n.d.).

LZW Compression

The Lempel-Ziv-Welch (LZW) algorithm (Welch, 1984) was innovative in its approach to codebook construction, since new symbols are created out of combinations of symbols already in the codebook (Lelewer & Hirschberg, 1987). This means that the algorithm takes $O(n^2)$ time (where n is the set of source symbols), since both source symbol and codebook matching can vary in length. To take advantage of the full benefits of the LZW algorithm, the codebook size should be allowed to grow to several times that of the symbol set size. This allows more complex symbol combinations to be constructed for greater compression efficacy. Since the MCU and transceiver chip on-board a WSN node typically have register boundaries set at one or two-byte increments, making efficient use of registers requires a minimum codebook size of 256 symbols. Using even this small symbol size, the LZW algorithm would quickly occupy the limited memory resources of the node. Even if the codebook is limited to $4S$ entries (where S denotes the number of unique symbols of the codebook), in a worst-case scenario the remaining $3S$ entries would require about 300kB of memory. Nevertheless, this algorithm is considered because of its ubiquity and as a measure of the compression performance that can be achieved with larger amounts of memory.

Figure 7. Data byte frequency of SBST program P₁

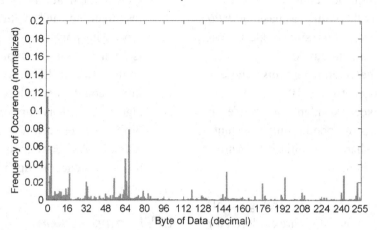

Dynamic Huffman Coding

Dynamic Huffman coding is an adaptive compression scheme that, unlike static Huffman coding (Huffman, 1952), dynamically generates its codebook as compression is taking place. One of the schemes known as Dynamic Huffman coding was made by Faller (1973), Gallager (1978), and Knuth (1985), after which it came to be known as Algorithm FGK. Different improvements were later made by Vitter (1987). As source symbols are mapped to codes, they are placed into a codebook with a hierarchical tree data structure. The observed frequency of source symbols is used to swap branches of the tree in order to continually give frequently occurring symbols smaller codes. As with all adaptive methods, compression performance improves as experience is built. The advantage these methods hold over original Huffman compression is that they only require $O(n)$ time, and that the codebook need not be transmitted since the receiver may reconstruct it by adapting his codebook lockstep with the sender.

While dynamic Huffman techniques offer a complexity less than that of LZW compression, the implications of maintaining a large hierarchical tree codebook are that it too requires significant memory resources. Each codebook node consists of the symbol frequency and the symbol itself,

while the Huffman code of each symbol can be inferred from its position in the tree. Since nodes and branches of the tree are often swapped, in a practical implementation they must also be addressed with pointers which themselves use memory. The result of tracking symbol frequency is that memory usage grows logarithmically as n increases. To limit memory usage, an upper limit to symbol frequency can be declared, in which case the algorithm will become static beyond that limit. Alternatively, the frequency tables can be deleted and statistics gathering restarted, but this negates the experience that the codebook has built.

BSTW Algorithm

This algorithm by Bentley, Sleator, Tarjan and Wei (BSTW) (1986) is an adaptive compression method simpler than the dynamic Huffman coding that can outperform static Huffman in some cases (Bentley et al., 1986). As source symbols are input, the most frequent ones are given the shortest codes. In coding a source symbol, the algorithm outputs a code based upon its location in the codebook, then uses a Move-to-Front (MTF) scheme to place that symbol at the front of the codebook. This scheme ensures that symbols occurring frequently are efficiently encoded with small codes, especially if they appear in bursts.

Figure 8. Data byte frequency of BSTW-reordered SBST Program P_1 vs. geometric distribution $y = P(1 - P)^x$ (where $P = 0.08$)

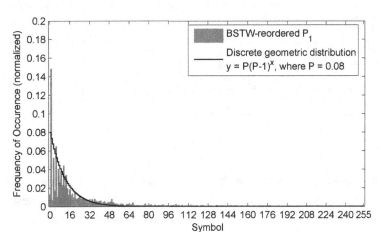

More importantly, symbol frequency is not explicitly tracked, as this is implicit in a symbol's position in the codebook and does not require additional memory.

The algorithm itself requires only $O(n)$ time, and transmission of the codebook is also unnecessary for the same reasons as with dynamic Huffman techniques. In BSTW the outputted codes are actually indices to a changing codebook. The source data set is remapped from its original distribution into a data set resembling a *geometric distribution*, which can be imagined as source symbols being sorted into an order of descending frequency. An example of a reordered version of SBST program P_1 can be seen in Figure 8, whose original representation is in Figure 7. Also seen in Figure 8 is a superimposed geometric distribution which approximates the BSTW-reordered P_1.

Several implementation options for BSTW are possible, where the complexities of accessing the codebook data structure differ. The pseudo-code for one version of the algorithm is depicted in Figure 9, which in addition to a codebook uses an equally-sized structure to cross-reference codebook data. A linear search of $O(n)$ is then saved since $2S$ memory is used instead of only S in the smallest implementation. After applying the BSTW algorithm, the resulting dataset must be re-encoded with a scheme such as universal codes or Golomb-Rice codes in order to achieve a smaller representation (compression). Such coding schemes give optimal or near-optimal results to data sets following geometric distributions. In the following section the three groups of compression algorithms that have been presented will be compared to each other to find a good candidate for use on SBST programs.

COMPRESSION ALGORITHM COMPARISON

In evaluating a compression algorithm for its potential suitability, the metrics of algorithmic complexity and memory footprint can be used as direct measures of meeting relevant node-specific requirements. These metrics are therefore compared in Table 3 between the three aforementioned algorithms, where n represents the number of symbols processed, and S denotes the number of unique symbols of the codebook. The following section explores coding schemes which need to be applied to some compression methods, including BSTW, for unambiguous decompression to be possible.

Figure 9. BSTW algorithm pseudo-code

```
// codebook stores symbols, indices stores indices of codebook, inverse-indexed by symbol.
for ( i = 0, I < size_of_source, i++ )
    index = indices [current_symbol]
    destination = index                                 // write index to destination
    if ( index ≠0 )
        Remove current symbol from codebook and shift codebook to fill hole
        Fill indices with new index to codebook
        Place current_symbol at front of codebook and fill indices with its index
    endif
    Increment source and destination by a symbol
endfor
```

The coding schemes are evaluated based upon the same requirements as the algorithm comparison: complexity and memory requirements.

Static Coding

Applying the BSTW algorithm described above on a set of source data coded in binary would generate a reordered dataset still in binary coding. In order to achieve compression, the dataset must be transformed into a smaller representation. Re-encoding the dataset using a static code can achieve this, since static codes are effectively a simplified form of compression that makes little or no use of probability estimation. Instead, they are generated in such a way as to be optimally suited to one particular probability density function (PDF). For example, the output data of the BSTW algorithm has been shown in Figure 8 to approximately conform to geometric distribution, so a static code suited to a geometric distribution could be used to express it. A code containing intrinsic delimiter bits can also be generated so that boundaries between codewords can be detected, which

allows the resulting dataset to be unambiguously decoded. Codes such as these are said to contain the *prefix property*, such that no code is a prefix of any other code in the dataset. In the following discussion several types of coding are evaluated for complexity and memory requirements, as per the requirements already discussed.

Universal Codes

Universal codes are often used in conjunction with adaptive schemes because of the benefits of their prefix property. These codes map positive binary-coded integers into variable-length binary codewords. If the source symbol set is remapped in descending order of frequency (*monotonically decreasing*), such as BSTW-remapped Figure 8 is close to being, the advantage of using them includes the property that the resulting codes will be within a constant factor of the optimal code (Lelewer & Hirschberg, 1987). Asymptotically optimal universal codes include Elias-δ and Fibonacci codes, while Elias-γ codes are not (Lelewer & Hirschberg, 1987). All three codes

Table 3. Compression algorithm comparison – complexity and memory usage

Compression Algorithm	Complexity over Source Data	Codebook Memory Usage (worst case)	
Lempel-Ziv-Welch (LZW)	$O(n^2)$	$(S-1) +$	$\dfrac{n(n+1)}{2}$
Dynamic Huffman (Algorithm FGK)	$O(n)$	$(S-1) \log_2(n) + S + Pointers$	
Algorithm BSTW	$O(n)$	$2S$	

include intrinsic inter-symbol delimiters, and a comparison between the mapping of their symbols can be seen in Table 4. Each code is catered to a different probability distribution, but notable differences includes the fact that errors in an Elias-γ code are often not recoverable, while in Fibonacci codes a 1-bit error can at most cause the loss of 3 symbols before the decoder is resynchronized. Each of the codes is also generated according to algorithms of differing complexity, but Fibonacci codes are especially difficult to develop as the most popular method is recursive. This property would likely preclude Fibonacci codes from being generated as needed for *on-the-fly* decoding, and an implementation would need to store the codes in system memory for lookup purposes.

Rice Coding

Rice coding (Rice, 1979), a special form of Golomb coding (Golomb, 1966), is a form of static coding that allows for a degree of probability estimation to accommodate different source symbol PDFs. Codes are generated by dividing a source symbol by a *divisor*, where the quotient of the division operation is represented in unary coding and the remainder in binary coding. Unary and Rice codes for some symbols can be seen in Table 4, both of which can be seen to also include inter-symbol delimiters. Rice codes differ from Golomb codes in that their divisor setting is restricted to powers of 2. This restriction ensures that the division operation can be performed on an MCU by simple bit-shift instructions, while the remainder is a logical bit-mask with the original symbol. Such a simple implementation requires a minimum of memory and allows codes to be generated deterministically on-the-fly as source symbols are processed. Compared to the generation algorithms of universal codes already reviewed, Rice coding is potentially of equal or less complexity. Other promising advantages of using Rice coding include the ability to vary the divisor setting between datasets without penalty, making the coding scheme more flexible than universal coding.

EXPERIMENTAL RESULTS

WSN Research Platform

Energy measurement and RF module characterization experiments are performed with a real WSN node, known as the *WSN research platform*, consisting of a Texas Instruments MSP430F149 MCU and an RF module, assembled on a custom printed circuit board (PCB). The RF module is made up of a Chipcon CC2420 wireless transceiver, employing the IEEE 802.15.4/ZigBee protocol at 2.4GHz, and a printed dipole antenna. The transceiver has 8 programmable transmission power levels which range from 0dBm to -25dBm. It also has an integrated received signal strength indicator (RSSI) which gives a discrete value for received signal power that can be read from an internal register. The transmit and receive gains of the on-board antenna, respectively G_T and G_R, have both been experimentally found to be -8 dB from prior work in (Chenard, Zilic, Chu, & Popovic, 2005).

Energy consumption measurements are made by connecting the WSN research platform to a custom current-measurement circuit depicted in Figure 10. Both the research platform and current-measurement circuit are developed in-house (Zhang, Zilic, & Radecka, 2006; Zhang, 2005). The output of the current-measurement circuit includes a voltage that is measured across a known resistance, giving an instantaneous current reading. Small currents are thus amplified to easily-measured voltages. This allows instantaneous power to be found, which when plotted over time, can be integrated to find total node energy consumption. All current-measurement circuit outputs are analyzed in this way using an Agilent Infiniium 54830D 600MHz mixed-signal oscilloscope. A portion of the node data bus (D_0–D_3) is forwarded to the oscilloscope to differentiate between test routines. The energy consumed by those test routines can then be calculated with Equation 4.

Table 4. Comparison of symbol coding for various static codes

Symbol	Static Code				
	Elias-δ	Elias-γ	Fibonacci	Unary	Rice*
0	0	0	11	1	1 00
1	1000	100	0 11	01	1 01
2	1001	101	00 11	001	1 10
3	10100	11000	10 11	0001	1 11
4	10101	11001	000 11	00001	01 00
5	10110	11010	100 11	000001	01 01
6	10111	11011	010 11	0000001	01 10
7	11000000	1110000	0000 11	00000001	01 11
					* with divisor setting of 4.

$$E = \frac{V_{dd}}{R} \int_{t_N}^{t_M} V \, dt \qquad (4)$$

where:

E is energy, in joules.

V_{dd} is the supply voltage to the node.

t_N and t_M are the measurement start and finish times, respectively.

$V = f(I)$ is the voltage output of the current measurement circuit.

I is the instantaneous current drawn by the node.

R is the resistance across which V is measured.

RF MODULE CHARACTERIZATION

The previously described RF module test scheme is tested here against select IEEE 802.15.4 specifications shown in Table 1. Experimental results for this RF module characterization can be seen in Table 5, which can be seen to all exceed the required specification.

The low-power IEEE 802.15.4 communication protocol has been developed to favor the use of battery-powered nodes. Such nodes are able to save energy by duty-cycling their MCU and RF transceiver, wherein they would spend most of their operational lives in a sleep mode. The protocol allows nodes to listen for periodic beacon transmissions in order to determine if a message is pending for them. Since beacon frequency can be varied, this mechanism allows the application designer to decide on a balance between node energy consumption and message latency.

Table 6 shows energy consumption measurements of various node operating modes. It can be seen that mode 4 costs much less energy than mode 5. A periodic beacon network therefore offers a sizable node power savings, and its frequency can be catered to the timing requirements of the application in order to maximize energy savings. This can be done to periodically transfer and activate SBST programs on the node. To save additional energy, the node transmission power level should be set as low as possible while maintaining the required PER on the link.

Table 5. RF module test results

Specification	Requirement	Measured Result
Transmit Power	Minimum: -3 dBm	-1 dBm
Receiver sensitivity	Maximum: -85 dBm	-88 dBm
Adjacent channel rejection	Minimum: 0 dB	7 dB
Alternate channel rejection	Minimum: 30 dB	43 dB

Figure 10. Current measurement method

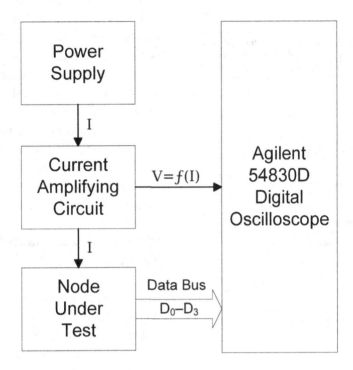

MCU TESTING EFFICIENCY

The energy efficiency ideas for MCU testing we have previously discussed are experimentally tested here on our WSN research platform. Table 7 shows the energy consumption of an MCU SBST program before and after optimizations. The measurements show that operand and instruction selection achieved a 21.2% energy reduction in the running of the test, while also achieving a 20.1% test time reduction.

EMBEDDED MEMORY TESTING EFFICIENCY

Two techniques were previously proposed to improve the energy efficiency of embedded memory SBST programs: SBC addressing, and the interleaving of flash memory testing with other tests. Here we compare the energy consumption of SBC addressing against binary addressing, and

determine the energy effect of interleaving flash memory testing with either RAM testing or RF module testing. The testing modes, themselves portions of component tests, can be seen in Table 8. Modes 5–7 are the first three elements of March FT algorithm for flash memory testing while modes 8 and 9 are the main elements of the March X algorithm used for RAM testing. Modes 8 and 9 are made to iterate on a single RAM memory block 10 times in order to approximate the time required for mode 5. Since our WSN research platform contains only 2kB of RAM, this is used

Table 6. RF module operating modes

Mode	Mode Description	Current (mA)
1	Send 5 packets with maximum power level	20.0
2	Send 5 packets with medium power level	15.0
3	Send 5 packets with minimum power level	12.0
4	Sleep between periodic beacon packets	0.39
5	Continuous listen on RF channel	20.7

Table 7. Node energy consumption during mcu testing

Metric	Unmodified Test	After Operand Selection	After Operand and Instruction Selection
Energy (μJ)	2.558	2.430	2.014
CPU cycles	940	940	751

Table 8. Test mode descriptions

Mode	Mode Description
5	Flash memory block erase (512 bytes)
6	Flash memory block test (r1, w0, r0) (512 bytes)
7	Flash memory block test (r0) (512 bytes)
8	RAM block test (w0) (10 blocks = 5 kB)
9	RAM block test (r0) (10 blocks = 5 kB)
A	RF Module Initialization
B	RF packet transmission (4 packets)

as a strategy to show the maximum energy savings effect of fully interleaving a flash memory block erase.

SBC Addressing of Flash Memory

The current consumption of SBC addressing compared to binary addressing can be seen in Figure 11. As expected, the average current of the SBC addressing can be seen to be slightly lower than that of binary addressing. However, SBC addressing requires greater test time, which means its energy consumption *(power × time)* is greater than that of binary addressing. Interestingly, this result is contrary to the one found in (Cheung & Gupta, 1996). The increased test time can be attributed to the need for extra instructions such as *shift, xor,* and *mov*, which convert addresses from binary to SBC. The additional current consumption of these extra instructions then exceeds the energy savings of using SBC addressing.

Interleaving of Flash Memory Testing with Other Tests

The concept we introduced of interleaving flash memory testing with other tests is experimentally tested here by running RAM testing and RF module testing during a flash memory block erase operation. Energy measurements for the interleaving of flash memory erase and RAM testing can be seen in Table 9, which show a 13.4% energy reduction and a 20.7% time reduction with the use of interleaving. Similarly, results for the interleaving of flash memory erase with RF module testing seen in Table 10 yield a 10.7% energy reduction and 14.2% time reduction with the use of interleaving. In the case of interleaving RAM testing, it is apparent

Figure 11. Energy comparison of binary and SBC addressing

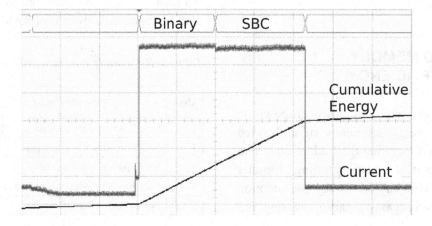

that modes 8 and 9 are slightly synthetic in their use of a 10-iteration block operation. However, a meaningful energy savings would result in the case of an MCU with at least 5 kB of RAM to be tested. It should also be noted that the increased energy consumption seen in Table 10 over results of Table 9 for the same test modes is a result of the contribution of the RF module to total node energy consumption.

SBST PROGRAM COMPRESSION

The first experiment compares the compression ratios of the three general-purpose algorithms previously presented on three real SBST programs.

Results include an estimate of the minimum memory requirements for using each algorithm, based upon knowledge of their functioning and required data structures. The second experiment measures the energy required to receive compressed and uncompressed SBST programs on a real WSN node.

Experimental Setup

In the experiment comparing compression ratios, three real SBST programs from (Zhang, Zilic, & Radecka, 2006; Zhang, 2005) (denoted P_n) are used with algorithm families LZW, Dynamic Huffman, and BSTW. The SBST programs are compiled into machine-code for the MCU used

Table 9. Interleaving flash memory testing with RAM testing

Mode	Test Item	Energy Before Interleaving (μJ)	Energy After Interleaving (μJ)
5	Flash memory erase	143.10	
8	RAM ($w0$)	73.06	189.94
9	RAM ($r0$)	78.54	
6	Flash ($r1, w0, r0$)	413.16	421.77
7	Flash ($r0$)	7.92	7.86
	Total Energy (μJ)	715.78	619.57
	Total Time (ms)	91.24	72.3

Table 10. Interleaving flash memory testing with RF module testing

Mode	Test Item	Energy Before Interleaving (μJ)	Energy After Interleaving (μJ)
5	Flash memory erase	231	
A	RF Initialization	280.83	1276
B	RF packet transmission	983	
6	Flash ($r1, w0, r0$)	723.03	707.46
7	Flash ($r0$)	13.2	12.87
	Total Energy (μJ)	2236	1996
	Total Time (ms)	105.2	90.24

Table 11. Achieved compression ratio vs. memory usage for three SBST programs (denoted P_n)

Algorithm		Compression Ratio (%)			RAM (Bytes)
		P_1	P_2	P_3	
Uncompressed		0	0	0	0
Lempel-Ziv-Welch (LZW)		38.8	32.2	55.4	7936†
Dynamic Huffman (Algorithm FGK)		15.8	19.4	27.5	1728‡
BSTW (MTF)	Rice Codes (divisor = 8)	7.7	16.7	18.4	512
	Rice Codes (divisor = 16)	18.8	22.5	23.9	
	Rice Codes (divisor = 32)	17.3	19.7	19.5	
	Rice Codes (divisor = 64)	9.9	10.3	10.7	
	Elias-γ Codes	8.6	24.2	19.2	
	Fibonacci Codes	15.6	23.9	22.2	
BSTW (swap)	Rice Codes (divisor = 8)	-31.9	-24.7	-29.6	512
	Rice Codes (divisor = 16)	-1.4	3.1	-0.6	
	Rice Codes (divisor = 32)	7.4	10.3	7.6	
	Rice Codes (divisor = 64)	5.3	7.8	5.3	
	Elias-γ Codes	-2.8	-14.7	5.1	
	Fibonacci Codes	8.3	0	13.4	

† codebook of 4096 strings, where the first 256 symbols are 1 byte, and the rest at least 2 bytes (very conservative estimate). ‡ assuming 256 symbols, 16 kB blocks, and 2 byte pointers.

in our *WSN research platform*. Compression ratio results for algorithm LZW are collected using the UNIX utility *compress* v.4.2.4, while for the dynamic Huffman family of algorithms, an implementation of algorithm FGK created by Toub (n.d.) is used. To evaluate the performance of several static coding methods, two variations of algorithm BSTW are actually implemented in MATLAB 7.0. Functions for determining the length of various static codes are also implemented in MATLAB 7.0, which are applied to the BSTW-remapped data. As an example, the equation for determining the length of a Rice-encoded symbol is given in Equation 5.

The energy measurement experiment is performed with the WSN research platform. The energy contribution of only the transceiver is found by measuring the node energy consumption while the MCU is in a low-power sleep state and the transceiver is operating in *listen mode*. Since the current draw of the *listen mode* approximates

the published current draw of *receive mode* of the CC2420 chip according to Chipcon AS (2004), it is used to isolate the transceiver component of total WSN node power consumption. Such an approximation is necessary since the transceiver would otherwise never be in *receive mode* with the MCU in a low-power sleep state.

$$\left\lfloor \frac{symbol}{divisor} \right\rfloor + 1 + \log_2 (divisor) \qquad (5)$$

Compression Ratio Comparison

While algorithm BSTW has been introduced as solely using the MTF heuristic, in this experiment an alternate heuristic denoted *swap* is also explored. In the MTF heuristic, when a new symbol is encountered it is moved to the front of the list, while using the swap heuristic it is exchanged with the element at the front of the list. A comparison of compression ratios and memory

Table 12. Contribution of current-draw for components when transceiver is receiving data

Component	Average Current (mA)
MCU operation	4.327
Transceiver operation	20.970

Table 13. Energy usage for wireless reception of three SBST programs (denoted P_n)

Wireless Reception of SBST Program Type	Energy Usage (mJ)		
	P_1	P_2	P_3
Uncompressed	14.727	0.779	9.420
Compressed	11.955	0.603	7.167

usage for the aforementioned algorithms can be seen in Table 11.

Algorithm LZW gives both the best compression ratios across all three SBST programs, as well as the greatest memory usage. Since results are collected using a compiled utility, the minimum memory requirements are estimated to be 8kB, although the value is likely much higher. Even this very conservative estimate is enough to disqualify algorithm LZW from use on WSN nodes for requiring excessive memory resources, although its results are a useful benchmark for gauging the relative compression performance of the other algorithms. Programs P_1 and P_2 are compressed by LZW in the 30% range while P_3 sees a 55% reduction.

The memory requirements for algorithm FGK are estimated from the analysis in Table 3 for the same reasons as algorithm LZW, although it is found to require approximately 1.7kB of memory. Achieved compression ratios exceed those of BSTW (MTF) with Rice codes for divisor settings of 8 and 64, for all three SBST programs. Compared to BSTW (MTF) with universal codes, performance is better for P_1 and P_3 by a small margin.

Algorithm BSTW with static codes has the lowest memory requirements of those compared (see Table 3), at 0.5kB. When heuristic MTF is combined with Rice coding with a divisor of 32, better performance over algorithm FGK is seen for P_1 and P_2 by a slim margin. Using a divisor of 16, the performance of algorithm FGK is exceeded by a slightly larger margin for the same programs. Algorithm FGK gives better results for P_3 in all cases at a cost of 3x the memory, although

by a slim 3.6% compared to Rice coding with a divisor of 16.

Energy Expenditure Comparison

Reducing the volume of data received by the node directly leads to a reduction in node energy consumption, since the transceiver presents the highest power consumption of the system. In this experiment the goal is to quantify the energy savings of transferring compressed SBST programs to a WSN node from a basestation. Algorithm BSTW and Rice coding (with a divisor of 16) is used for its compatibility with the memory limitations of nodes. Both compressed and uncompressed versions of the same three SBST programs from Table 11 are transmitted, and the total energy consumption of the node is measured in receiving the programs.

When the transceiver is receiving data, the ratio of current draw of the MCU to the transceiver is found to be approximately 1:4.5, seen in Table 12. In isolating the transceiver current-draw, the energy usage for receiving both compressed and uncompressed SBST programs can be seen in Table 13. The result is that the reception of compressed SBST programs over uncompressed ones yields an energy savings of 18.8%, 22.5%, and 23.9% for programs P_1, P_2, and P_3, respectively. These energy savings are directly proportional to the achieved compression ratios seen in Table 11.

CONCLUSION

In this chapter we present an infrastructure for the distribution and remote execution of SBST tests, as well as for the remitting of test responses. As part of the testing scenario, both self-testing and testing through *helper nodes* is used to verify the correct operation of node components. Once uncovered, failed and failing nodes can be repaired or replaced before they affect network availability.

SBST programs are constructed with instructions requiring the least amount of cycles, while their operands are selected to contain the least Hamming distance and weight. This reduces test energy consumption, verified through experimental results collected from a current measurement circuit connected a WSN research platform. Further, a March-family algorithm is used to test embedded flash memory, where it is shown that the energy optimizations traditionally yielded by SBC addressing are not efficient in this case. The time interleaving of embedded flash memory tests with other components tests is used to reduce both test time and energy, while the use of *helper nodes* also enables the verification of RF module specifications.

Finally, SBST program compression is explored through general-purpose algorithms in an effort of further reduce the energy consumption associated with the distribution of test programs. While most algorithms require more memory than is available on a typical node, the BSTW algorithm is found to suit such devices. A compression scheme involving the BSTW algorithm combined with the memory-less Rice coding is found to give the best overall compression ratio for devices with small amounts of memory.

REFERENCES

Aho, A., & Johnson, S. (1976). Optimal Code Generation for Expression Trees. *Journal of the ACM*, *23*(3), 488–501. doi:10.1145/321958.321970

Batcher, K., & Papachristou, C. (1999). Instruction randomization self test for processor cores. In *Proceedings of the VLSI Test Symposium* (pp. 34-40).

Bentley, J. L., Sleator, D. D., Tarjan, R. E., & Wei, V. K. (1986). A locally adaptive data compression scheme. *Communications of the ACM*, *24*(4), 320–330. doi:10.1145/5684.5688

Brahme, D., & Abraham, J. A. (1984). Functional testing of microprocessors. *IEEE Transactions on Computers*, (C-33), 475–485. doi:10.1109/TC.1984.1676471

Callaway, E. H. (2004). *Wireless sensor networks: architectures and protocols*. Boca Raton, FL: Auerbach Publications.

Chang, N., Kim, K., & Lee, H. G. (2000). Cycle-accurate energy consumption measurement and analysis: Case study of ARM7TDMI. In *Proceedings of the International Symposium on Low Power Electronics and Design* (pp. 185-190).

Chen, L., & Dey, S. (2001). Software-based self-testing methodology for processor cores. *IEEE Transactions on Computer-Aided Design of Integrated Circuits and Systems*, *20*, 369–280. doi:10.1109/43.913755

Chenard, J.-S., Zilic, Z., Chu, C. Y., & Popovic, M. (2005). Design methodology for wireless nodes with printed antennas. In *Proceedings of the Design Automation Conference* (pp. 291-296).

Cheung, H., & Gupta, S. K. (1996). A BIST methodology for comprehensive testing of RAM with reduced heat dissipation. In *Proceedings of the International Test Conference* (pp. 386-395).

Chiang, M. W., Zilic, Z., Radecka, K., & Chenard, J.-S. (2004). Architectures of increased availability wireless sensor network nodes. In *Proceedings of the International Test Conference* (pp. 1232-1241).

Chipcon, A. S. (2004). *CC2420 data sheet* (revision 1.4). Oslo, Norway: Chipcon AS.

Dabrowski, J. (2003). BiST model for IC RF-transceiver front-end. In *Proceedings of the International Symposium on Defect and Fault Tolerance in VLSI Systems* (pp. 295-302).

Faller, N. (1973). An adaptive system for data compression. *Record of the 7th Asilomar Conference on Circuits, Systems, and Computers* (pp. 593-597).

Gallager, R. G. (1978). Variations on a theme by huffman. *IEEE Transactions on Information Theory*, *24*(6), 668–674. doi:10.1109/TIT.1978.1055959

Golomb, S. W. (1966). Run-length encodings. *IEEE Transactions on Information Theory*, *12*(3), 399–401. doi:10.1109/TIT.1966.1053907

Halder, A., Bhattacharya, S., Srinivasan, G., & Chatterjee, A. (2005). A system-level alternate test approach for specification test of RF transceivers in loopback mode. In *Proceedings of the International Conference on VLSI Design* (pp. 289-294).

Hariri, S., & Huitlu, H. (1995). Hierarchical modeling of availability in distributed systems. *IEEE Transactions on Software Engineering*, (21): 50–56. doi:10.1109/32.341847

Hely, D., Bancel, F., Flottes, M. L., & Rouzeyre, B. (2006). Secure scan techniques: A comparison. In *Proceedings of the International On-Line Testing Symposium* (pp. 119-124).

Huffman, D. A. (1952). A method for the construction of minimum-redundancy codes. In . *Proceedings of the IRE*, *40*(9), 1098–1101. doi:10.1109/JRPROC.1952.273898

Institute of Electrical and Electronics Engineers. (2006). *Wireless medium access control (MAC) and physical layer (PHY) specifications for low rate wireless personal area networks (LRWPANs)* (IEEE standard 802.15.4).

Joshi, B. S., & Hosseini, S. H. (1998). Efficient algorithms for microprocessor testing. In *Proceedings of the Annual Reliability and Maintainability Symposium* (pp. 100-104).

Khoche, A., Volkerink, E., Rivoir, J., & Mitra, S. (2002). Test vector Compression Using EDA-ATE synergies. In *Proceedings of the VLSI Test Symposium* (pp. 97-102).

Knuth, D. E. (1985). Dynamic huffman coding. *Journal of Algorithms*, *6*(2), 163–180. doi:10.1016/0196-6774(85)90036-7

Koushanfar, F., Potkonjak, M., & Sangiovanni-Vincentelli, A. (2004). Fault tolerance in wireless sensor networks. In I. Mahgoub and M. Ilyas (Eds.), *Handbook of Sensor Networks* (section VIII), Boca Raton, FL: CRC Press.

Kranitis, N., Paschalis, A., Gizopoulos, D., & Xenoulis, G. (2005). Software-based self-testing of embedded processors. *IEEE Transactions on Computers*, *54*, 461–475. doi:10.1109/TC.2005.68

Kranitis, N., Paschalis, A., Gizopoulos, D., & Zorian, Y. (2002). Effective software self-test methodology for processor cores. In *Proceedings of Design, Automation and Test in Europe* (pp. 592-597).

Krstic, A., Lai, W.-C., Cheng, K.-T., Chen, L., & Dey, S. (2002). Embedded software-based self-test for programmable core-based designs. *IEEE Design & Test of Computers*, *19*(4), 18–27. doi:10.1109/MDT.2002.1018130

Lala, P. K. (2001). *Self-checking and fault-tolerant digital design* (1st ed.). San Francisco: Morgan Kaufmann.

Lelewer, D. A., & Hirschberg, D. S. (1987). Data compression. *ACM Computing Surveys*, *19*(3), 261–296. doi:10.1145/45072.45074

Liang, H.-G., Hellebrand, S., & Wunderlicht, H.-J. (2001). Two-dimensional test data compression for scan-based deterministic BIST. In *Proceedings of the International Test Conference* (pp. 894-902).

Nikolaidis, S., & Laopoulos, T. (2001). Instruction-level power consumption estimation embedded processors low-power applications. *International Workshop on Intelligent Data Acquisition and Advanced Computing Systems: Technology and Applications* (pp. 139-142).

Ozev, S., Orailoglu, A., & Olgaard, C. V. (2002). Multi-level testability analysis and solutions for integrated bluetooth transceivers. *IEEE Design & Test of Computers, 19*(5), 82–91. doi:10.1109/MDT.2002.1033796

Paschalis, A., & Gizopoulos, D. (2005). Effective software-based self-test strategies for on-line periodic testing of embedded processors. *IEEE Transactions on Computer-Aided Design of Integrated Circuits and Systems, 24*, 88–99. doi:10.1109/TCAD.2004.839486

Raghavendra, C. S., Kumar, V. K. P., & Hariri, S. (1988). Reliability analysis in distributed systems. *IEEE Transactions on Computers, 37*, 352–358. doi:10.1109/12.2173

Rice, R. F. (1979). *Some practical universal noiseless coding techniques* (JPL Technical Report No. 79-22). Pasedena, CA: Jet Propulsion Laboratory.

Salomon, D. (2004). *Data compression: The complete reference (3rd ed.)*. New York: Springer.

Semiconductor Industry Association. (2005). *International technology roadmap for semiconductors*: Test and test equipment. *Computer, 37*(1), 47–56. doi:10.1109/MC.2004.1260725

Shen, J., & Abraham, J. A. (1998). Native mode functional test generation for processors with applications to self test and design validation. In *Proceedings of the International Test Conference* (pp. 990-999).

Stutzman, W. L., & Thiele, G. A. (1997). *Antenna Theory and Design (2nd ed.)*. New York: John Wiley & Sons.

Texas Instruments Corporation. (n.d.). *MSP430 ultra-low power microcontrollers, MSP430x1xx – flash ROM no LCD – price list per 1000 units*. Retrieved February, 2007, from http://focus.ti.com/paramsearch/docs/parametricsearch.tsp?familyId=911&sectionId=95&tabId=1527&family=mcu

Thatte, S. M., & Abraham, J. A. (1980). Test generation for microprocessors. *IEEE Transactions on Computers, C-29*, 429–441. doi:10.1109/TC.1980.1675602

Toub, S. (n.d.). *Adaptive Huffman compression*. Retrieved July, 2002, from http://www.gotdotnet.com van de Goor, A. (1991). *Testing semiconductor memories: theory and practice*. New York: John Wiley & Sons.

van de Goor, A. (1993). Using march tests to test SRAMs. *IEEE Design & Test of Computers, 10*(1), 8–14. doi:10.1109/54.199799

van de Goor, A., & Verhallen, T. (1992). Functional testing of current microprocessors (applied to the Intel i860). In *Proceedings of International Test Conference* (pp. 684-695).

Vitter, J. S. (1987). Design and analysis of dynamic huffman codes. *Journal of the ACM, 34*(4), 825–845. doi:10.1145/31846.42227

Welch, T. A. (1984). A technique for high-performance data compression. *IEEE Computer, 17*(6), 8–19.

Yeh, J.-C., Wu, C.-F., Cheng, K.-L., Chou, Y.-F., Huang, C.-T., & Wu, C.-W. (2002). Flash memory built-in self-test using march-like algorithms. *IEEE International Workshop on Electronic Design, Test and Applications* (pp. 137-141).

Zhang, R. (2005). *Energy reduced software-based self-testing for wireless sensor network nodes*. Master of Engineering thesis, McGill University, Canada.

Zhang, R., Zilic, Z., & Radecka, K. (2006). Energy efficient software-based self-test for wireless sensor network nodes. In *Proceedings of the VLSI Test Symposium* (pp. 186-191).

KEY TERMS AND DEFINITIONS

Automatic Test Pattern Generation (ATPG): An algorithm by which test vectors are automatically generated from a design specification.

Built-In Self-Test (BIST): A hardware module that generates test vectors, applies them to testable components, and constructs a test signature from the aggregated results.

Energy-Sensitive Factors: The parameters of instructions that a microcontroller executes which affect its total energy consumption.

Packet Error Rate (PER): The proportion of packets that have been received over a wireless link containing one of more errors.

Perceived Availability (of a Network): The probability that a network is functioning correctly over a period of time, as seen by a user of that network.

Rice Coding: A form of static coding that allows for a degree of probability estimation to accommodate different source symbol probability densities.

Software-Based Self-Test (SBST) Program: Software created to test the components of the microcontroller is it executing on, and/or attached peripherals.

Chapter 6
Power Conservation Techniques in Wireless Sensor Networks

Baha Alsaify
University of Arkansas, USA

Haiying Shen
University of Arkansas, USA

ABSTRACT

This chapter discusses power conservation problems in Wireless Sensor Networks (WSNs). The problem arises from the fact that WSNs have limited energy since sensors are powered with small batteries (due to the sensor size constraints). Currently, some energy-efficiency methods focus on reducing energy consumption by designing sensor hardware. Other methods enable sensors to communicate with each other in an efficient manner by developing new communication protocols. Some communication protocols need to have extra information such as sensor locations for determining the best possible relays to deliver data to a Base Station. In this chapter we will first provide a survey on current power conservation methods. We will then discuss the efficiency and effectiveness of these methods, and propose possible solutions. Finally, we conclude the chapter with concluding remarks and open issues.

INTRODUCTION

Due to the recent advancements in the wireless networks, and the manufacturing of small, costless and energy-efficient devices, WSNs have been developing in a rapid speed. WSNs consist of a large number of energy-limited sensors that are used to monitor some area. Since WSNs are comprised of a large number of collaborating sensors (usually in the order of thousands), the cost of these sensors needs to be minimized. These sensors usually operate in areas where it is difficult for humans to work in. For example, WSNs are used to monitor a battle field, monitor the temperature fluctuations in the south and north poles, and work for extended periods in high temperature deserts. One of the future applications for WSNs is to be deployed in space or on the surface of planets.

Due to the size and cost constraints, the sensors are powered with small batteries. These batteries do not hold much energy in them. Also, since the sensors operate in usually inaccessible territories,

DOI: 10.4018/978-1-61520-701-5.ch006

Figure 1. A typical WSN

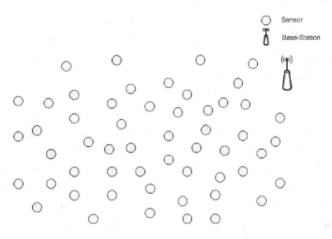

it is not possible to change the sensors' batteries. Therefore, power conservation is important to reduce energy consumption in the WSN environment. The main goal for energy conservation is to enable sensors work for the longest possible time (i.e. network life time) while maintaining the quality of service (QoS). The QoS in the WSN environment refers to the network ability to cover 100% of the area that's being monitored.

Figure 1 shows a typical WSN. Usually WSNs have several hundred sensors and a Base Station. The function of the sensors is to collect data and send it to the Base Station. Another function for the sensors is to serve as a bridge between other sensors and the Base Station. The reason for the need of relays is because the communication range for a sensor is very limited due to energy constraints. In addition, the required energy to transmit data grows with the increase of the transmission range. In a WSN, all of the sensors' data will be sent to a Base Station. The Base Station collects the information, organizes them, deletes any redundant information, and put the information in a readable format for users to read. Usually the Base Station has higher processing capability than the scattered sensors. It also has larger size and more stored energy, and may have a power link that provides with constant power supply.

Figure 2. Sensors with their transmission ranges

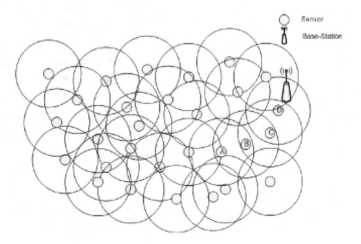

Figure 2 illustrates how the sensors collaborate to deliver their data to the Base Station. In the figure, sensor A wants to deliver its data to the Base Station, it uses sensors B, C, D as relays.

There are a number of techniques that can be used to reduce the energy used in the data collection and access operation. Some of these techniques deal with the sensors hardware (Zhong, 2007). They try to reduce the operational voltage for the sensors. Moreover, they looked into how to schedule the packets in a way that consumes the least amount of energy (Zhong, 2007). Another group of methods to reduce energy consumption is to employ sleep/awake intervals. They allow sensors not be active 100% of the time. There are a number of definitions for "sleeping" in the context of WSNs. It could mean that the only inactive part of a sensor is the radio since it is the most energy consuming part of sensor hardware. The radio is either the transmitter or the receiver (AKA transceiver). Figure 3 shows a simple sensor construction with the radio components.

Figure 3. Sensor's internal design

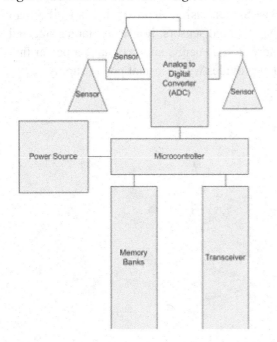

In another possible sleeping mode, the sensor hardware except the preprocessor is completely shutdown. The function of the preprocessor is to awake the sensor when it is time for the sensor to be active.

Another way to reduce energy consumption by the sensors is arranging the sensors monitor a fraction of the area (e.g. 80%, 75%). This method uses the idea of sleep/awake scheduling to activate a group of sensors for a specific period of time. These activated sensors will monitor part of the area, which will increase the system's lifetime. However, it will also affect the QoS of the network since some events may occur at the uncovered area. Energy can also be saved by controlling where data is stored. There are three methods that can be used to store and exchange sensed data.

- **Storing data locally:** In this method, the sensors store sensed data in its local memory banks. However, the sensor memory is limited and it cannot hold much information. Thus, this method is not suitable for applications that require information to be gathered very frequently (from the Base Station point of view). On the other hand, if sensed data is collected in the Base Station, the Base Station needs to initiate queries to the sensors for queried data. The queries in this case will be broadcasted throughout the network, which is energy draining if the queries are issued in a very intensive manner.

- **Storing data at a designated location within the network:** This method requires a function that processes the keywords of gathered data, and the location of the node that should hold the data is decided based on the resultant hash value. If the Base Station wants to recover some data, it will hash the keywords for the desired data and determine the location that holds the data and then send the query. In this case, the number of queries roaming among sensors is smaller than the "Storing data locally" approach since the location of the data is known. In other words, only one query is needed to retrieve the information. In this method, the sensors should be able to determine their locations. A location can be either a real geographical location or a certain point or node in the WSN.

- **Sending data to the Base Station:** In this method, each node sends data directly to the Base Station immediately after sensing the data. This, no node is responsible of storing data. This method will not be effective if events frequently happen. In this case, it will drain the energy in both the sensor detecting the event and in the relay sensors. It will also create congestions in the network due to a large number of exchanged messages. The congestions lead to losing of a high number of messages, and hence a high retransmission rate. Therefore, this method is suitable for an environment with high events occurring rate.

A sensor may receive data from multiple sources to be delivered to the Base Station. Functioning as relay node, it gathers the data from all the sources and forwards it as one message. However, if the sensor just appends the data, it will end up with one giant message. To deal with this problem, the sensor can apply aggregation functions to the data including: SUM, AVERAGE, MIN, MAX and etc. As a result, the amount of data sent in one message will be substantially reduced. The amount of required energy to send a message increases exponentially with the increase of data size.

The rest of this chapter is organized in the following manner. The "Power Conservation Techniques" section will present a survey of the works for energy conservation in WSNs. Then, some problems in the area of energy conservation will be identified and possible solutions will be presented in the "Why Energy Conservation" section. After that, some insight about the future of WSNs will be presented on the "Future Trends" section. Finally a conclusion that summarizes the chapter will be the end.

POWER CONSERVATION TECHNIQUES

Reducing energy consumption has received increasing attention in the recent years. Current techniques to reduce the operational energy consumption can be divided into two main groups: (1) Hardware and (2) Software. The first group manipulates a sensor's hardware such as IC design. In this chapter, we focus on the second group.

In order to effectively design an energy-efficient routing protocol in a WSN, a number of issues need to be taken into consideration (Al-karaki, 2004):

- **Node[1] Deployment:** Node deployment is critical to designing a routing protocol. There are two possible node deployment methods:
  - **Deterministic Deployment:** in this method, nodes are manually placed in the monitored area. Because the nodes' locations are known in advance, there is no need to start a localization process, which is used to determine the location of the nodes at any time of the network life time. Because WSNs are usually used in areas that are difficult for humans to enter, this method is not commonly used. In deterministic deployment, the nodes are usually placed in a way that supports fast event reporting where data routing path is short.
  - **Random Deployment:** In this method nodes are randomly scattered in the monitored area. One of the possible ways to deploy nodes in a random

manner is to throw them out from a plane flying over the desired area. Once the nodes reach the ground, they will start probing their neighbors and form bonds with them. In other words, they will form an ad-hoc network. Nodes usually form clusters to provide the outmost network connectivity relying on the nodes with the most residual energy as cluster heads (CHs). The nodes with the most connections can also be chosen as CHs. The responsibility of a CH includes collecting the information from the nodes in its cluster and forwarding them to the Base Station.

- **Quality of Service:** This criterion is application specific. It means that the application the nodes serve will specify the degree of QoS. QoS can be described as how fast the sensed events will be reported to the Base Station. In applications that are designed for battles, fast information delivery is crucial for the security and safety of a whole country. QoS can also be measured by the percentage of coverage or connectivity between the sensors in the field.

- **Data Reporting Model:** The choice of the data reporting model depends primarily on two factors:
  ◦ The application
  ◦ The time criticality of the data

There are four models that can be used to report the data from the individual sensors back to the Base Station:

  ◦ **Time-driven:** A time-driven model is suitable for applications that require periodic status delivery. Sensors in this model will periodically switch on their sensors, sense the surrounding environment, and send its data to the Base Station.

  ◦ **Event-driven:** An event-driven model is used when reaction to certain events is crucial. Any sudden change in the sensed values will trigger the sensors, and the change will immediately be reported to the Base Station.

  ◦ **Query-driven:** This model is very similar to the event-driven model, but instead of a change in the sensed value, the arrival of a query will trigger the delivery of data. The queries will be sent from a Base Station to the sensors. If the Base Station knows the location of the data, it will send it immediately to the sensor having the data. Otherwise, it simply broadcasts the query to all the nodes in the network, and only the one(s) with the intended data will respond.

  ◦ **Hybrid:** This model is a combination of the previous models. It entirely depends on the application the sensors serve.

- **Node/Link Heterogeneity:** Heterogeneity means that nodes in a WSN environment differ in power levels, communication range, computational power, and etc. The routing mechanism must be able to compensate for the differences between nodes. If the routing protocol uses clusters to organize the nodes, the node with the highest power level can be chosen to be a CH.

- **Fault Tolerance:** The routing protocol needs to be able to self repair. Self repair means that the protocol is able to go around failed nodes or links in delivering data. This is due to possible hardware failures and interference, which may break the links between nodes. Scalability: usually a WSN contains hundreds, even thousands of nodes. Scalability means that the routing protocol will be able to cope with that huge number of nodes and provide a routing mechanism that will provide energy savings and fast data delivery.

- **Networks Dynamics:** Dynamics is not that much of an issue in WSNs since most sensors are stationary. Nodes sometimes move from their original places. The reason for the movement may relate to some environmental issues, or in some rare cases, the nodes themselves have the ability to change location. The protocols need to dynamically reconnect links between nodes to provide network connectivity and area coverage.

- **Transmission Media:** WSNs as their name suggest are formed by placing several sensors on an area. These nodes connect with each other by wireless channels. The same problems that affect the wireless medium will affect the performance of WSNs. The main problem of the wireless medium is its sensitivity for interference. This sensitivity will result in a high rate of retransmissions, which in turn will be very energy draining.

- **Connectivity:** It is one of the most important properties the WSNs should have. The higher density (i.e., nodes/area unit), the higher connectivity the network has. By having a dense connected network, the WSN will be fault tolerant. This means that the network will be able to be operational even if a large number of nodes fail.

- **Coverage:** Coverage is the reason for the existence of WSNs. The goal for WSN is to cover and monitor an area and report the events that occur in that area. Each sensor will be responsible to cover a certain area based on its transmission range. Any event that happens in that area will be reported to the Base Station having that sensor as the origin of the data.

- **Data Aggregation:** Data aggregation means that some basic mathematical operations will be performed on the sensed data at relay sensors. The reason for having aggregation functions is to reduce the size

of the message to be delivered. That reduction will provide fast data delivery and also cheap data delivery (energy wise).

Taking all of the previous characteristics into consideration, numerous protocols have been proposed for energy conservation. A survey of most of the protocols is presented in the following:

Sensor Protocols for Information via Negotiation (SPIN)

Sensor Protocols for Information via Negotiation (SPIN) (Kulik, 2002) is proposed to reduce energy consumption. This protocol assumes that each node in the system is a Base Station, and it sends its data to every one of these presumably base stations. The advantage of having such data dissemination is that the user (or the Base Station) can query any one of the nodes and receive the desired data. SPIN has two key innovations for energy efficiency.

- **Negotiation:** Nodes negotiate with each other before sending the data. If the node that data are intended to be sent to already has that particular data, then no data transmission will occur. Otherwise, the sensor that has the data will send it. Negotiation helps to insure that only useful information is exchanged between the sensors and no overlap information is exchanged.

- **Resource Adaptation:** Each one of the sensors has its own resource manager. Before a transmission, a node will poll its resource information before the transmission. The application will be responsible for contacting the resource manager and for the sensor's status and power levels.

The negotiation in SPIN performs using a three way handshaking technique. A negotiation starts by sending an advertisement message called ADV. The ADV message contains metadata about the data the original sensor wants to send. The receiver will check its memory banks to see if it already has the data. If it has the data, it does nothing. Otherwise, it replies back using a special message called request for data (REQ). When the REQ message is received by the sensor advertising for the data, it will immediately send the data (i.e., the whole package) to the sensor that sent the REQ message.

Figure 4 shows a graphical representation for the handshaking procedure. In the figure, the middle node is the node that sensed the event. It will try to decimate its data to its neighbors. The middle node assumes the dark nodes do not have the data. The middle node will send ADV messages to all of its neighbors. The ones that do not already have the data will respond back with a REQ message. The middle node will then send the information to those nodes. There are a number of families of the SPIN protocol: SPIN-1 and SPIN-2. The described SPIN protocol is called SPIN-1. SPIN-2 is a threshold-based resource aware protocol. This SPIN version is energy efficient and make use of the resource manager. It uses the same handshake procedure as SPIN-1, but only starts a negotiation procedure if it knows that it can complete the transaction without its power level dropping below a threshold value.

Low Energy Adaptive Clustering Hierarchy (LEACH)

Another energy-efficiency protocol is called Low Energy Adaptive Clustering Hierarchy (LEACH) (Heinzelman, 2000). LEACH is a cluster based routing protocol. The authors of this protocol specify that only 5% of all the nodes in the system have the opportunity to act as CHs in any given round. They assume that the Base Station has a fixed location and that all the sensors are homogeneous in capability and power levels. Because of the second assumption, the selection of the CHs is done randomly. In LEACH, the CHs have the responsibility to gather data from nodes in their cluster, combine them using some aggregation technique, and send the final aggregated data to the Base Station. Data aggregation can be helpful in avoiding information overload, and to provide reliable data from several unreliable sources. The lifetime of a network using LEACH protocol is divided into rounds. Each one of these rounds has two operational phases:

Figure 4. Handshaking in SPIN

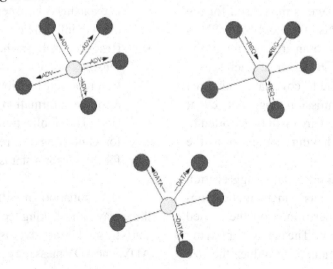

- **Advertisement Phase:** Each node in the system will decide whether to become a CH or not. The decision will be based on how many times the node has already become a CH in previous rounds, and the percentage of CHs in the network (e.g., 5% in the paper). Specifically, the node punches the information into a pre-determined formula and if the result is less than a threshold value, the node will become a CH. Once becoming a CH, the node sends an advertisement message to the rest of the nodes. After receiving the advertisement messages from the CHs, each one of the nodes decides which CH is the most appropriate for itself.

- **Cluster Set-Up Phase:** After each node decides which cluster it belongs to, it informs that CH about being a part of its cluster. During the setup time period, the CHs keep their receivers on.

The reason behind dividing the network lifetime into rounds is to spread the energy usage among the nodes in the system. If there was only one round, then the CH is fixed, which means that the CH will be the same for the whole system lifetime. Consequently, the CH has more responsibility than other nodes in the cluster. It uses more energy than the rest of the nodes, and drains its energy faster than the others. Having a dead CH means that, the cluster will not be able to perform its duties, this will lead to poor network performance.

Threshold-Sensitive Energy Efficient Protocols (TEEN & APTEEN)

Two other protocols that are energy-efficiency are called Threshold-sensitive Energy Efficient Protocols (TEEN and APTEEN) (Manjeshwar, 2001; Manjeshwar, 2002). These protocols are designed to be well suited for time critical applications in addition to energy aware performance.

TEEN and APTEEN are cluster based routing. In TEEN, the CH will send its members a value called hard threshold. The sensors continuously sense the environment and only relay information to the Base Station if the sensed value exceeds the specified hard threshold. Another value given to the cluster members is called the soft threshold. Soft threshold indicates the required change in the sensed value to trigger transmitter activation. Therefore, the hard threshold specifies the point that if reached, the transmitter will activate and it will start reporting data. This in turn will reduce the number of transmissions by restricting the transmitter to work if and only if the sensed values are in the hard threshold range. The soft threshold will further reduce the number of transmissions by specifying the rate of change that needs to occur while the sensed values are in the hard threshold range to activate the transmitter. In other words, if there is no soft threshold, the nodes will report the events as long as those events are in the hard threshold range. But with soft threshold, the events will not be reported unless they are changing in a rate faster than what is specified in the soft threshold and they are in the hard threshold range. If the application wants an accurate view of the environment, the soft threshold value can be small. This means that the rate of transmission will go up. Energy savings using the TEEN model can be achieved by the reduction in the number of transmissions. Therefore, there is a trade-off between accuracy and energy savings in TEEN. APTEEN, on the other hand, provides even more energy savings. The CH sends the following four values to the nodes in its cluster:

- **Attributes:** The set of parameters that users are interested in. In other words, they are the types of events the sensors will provide data for.
- **Thresholds:** It contains the values for both the hard and soft thresholds.
- **Schedule:** A Time Division Multiple Access (TDMA) slot scheduling. This

scheduling will be assigned to nodes in the cluster to allow simultaneous communication between the nodes in the cluster and the CH. Each one of the members will be assigned a time slot to communicate within.

- **Count Timer:** It specifies the shortest time period between two successive reports from one node. It can be specified as a multiple of a TDMA slot.

APTEEN gives the user the flexibility to choose the count timer and the threshold values. Giving the application or the user the ability to control those values gives the application with the ability to control the power consumption ratios of the system. One enhancement to the TEEN and APTEEN protocols is that if two or more nodes are in close proximity, one node of this group can send the data to the CH and the other nodes can go to sleep. The reason that justifies this enhancement is that close sensors usually sense the same events. Therefore, if they all were to report their data to the Base Station, data overlap will occur. Arranging one node to report data and the rest to sleep not only prevents data overlapping, but also provides energy savings. It also provides energy balancing if the role of data reporting can be rotated between close-by sensors. This enhancement can only be implemented when nodes are stationary.

Minimum Cost Forwarding Algorithm (MCFA)

Another way to reduce energy consumption is to implement a method for the sensors in the system to forward their data using paths that consume the least possible amount of energy. An example of this method is the Minimum Cost Forwarding Algorithm (MCFA) proposed by Fan (2001). Every sensor in the system using MCFA knows the cheapest (i.e. energy wise) path to the stationary Base Station. There are two main operations tackled by the algorithm:

- How does each node determine its minimum cost? Initially all the nodes in the system have a cost equal to infinity. The sink[2] broadcasts a special message called the advertise message (ADV) to its neighbors. This message initially has a value of zero as its cost. Each one of the neighbors will calculate the cost for sending this message to the broadcast origin (the node that sent the ADV message), add that cost to the message value, and broadcast it to its neighbors. Then, each one of the new neighbors will calculate the cost to broadcast the message to the broadcast origin (the sensor the message had been received from), and add that cost to the one in the message. After calculating the new cost value, the cost field in the ADV message will be updated with the new value and the ADV message will broadcast again. If a node receives an ADV message with less cost, it will update its cost value and rebroadcast the ADV message. Since the nodes depend entirely on the cost information they have, the need for node ID is eliminated. The process used to route data without the need for ID's will be outlined in the next bullet.

- How to send messages along the minimum cost path? If a node senses some event(s), it is required to report it back to the Base Station. The node that wants to report some data will build a message that contains the event's data and its minimum path cost. To discover such a path, it will first broadcast a message. The nodes that receive that message will then decrement the cost between itself and the message source from the cost information in the message. If the resultant value is equal to its minimum cost value, then the path is on the minimum cost path. Only the nodes that have their minimum cost information equal to the result of the previous operation will accept the message and broadcast it in the same manner. Let's use an example to further illustrate this concept. Figure 5 shows a graphical representation of the example. Assume that node A wants to deliver a message to the sink. Nodes B and C are A's neighbors. The minimum cost value for nodes A, B, and C are 70, 40, and 50 respectively. The broadcast cost is equal to 30. The operation 70-30 = 40 will be executed on both B and C. Then both B and C will compare the result with their minimum cost values. After conducting the comparison, only B will accept and rebroadcast the message with the value 40 in the message cost.

Fractional Coverage Schemes (FCS)

Instead of developing new protocols to govern the behavior of the WSNs, sleep/awake methods only need minor modifications to the WSNs. Fractional Coverage Schemes (FCS) proposed by Ye (2006) is an example of such methods used to conserve energy. FCS also has some relaxation of the QoS constraint. The network life time is divided into rounds. In each round a subset of nodes is selected based on node power levels to perform the sensing functionality. Different subsets are chosen for different rounds to insure energy load balancing. After the deployment phase, the Base Station broadcasts a hello message. Sensors that receive the hello message rebroadcast it. Each node in the system calculates node density in its surrounding, based on the hello message rate of arrival, which is based on the hello messages received from its neighbors. After calculating the node density, the network selects the subset of nodes to be active. Selecting the active nodes has two phases:

- **Advocation Phase:** In this phase, several nodes called pioneer nodes are selected. Each node in the system has the probability of becoming a pioneer node with

probability T_i. The probability depends entirely on the surrounding node's density, and the ideal node's density. The ideal node density is determined by the application requirements.

- **Competition Phase:** The residual energy in the sensor's battery is used as a competing factor. The competition takes place in three stages:
  - **Sending and receiving competition messages:** Each one of the nodes will send a special message that contains its residual energy value. The node will then wait until it receives competition messages from all of its surrounding competing nodes. If the node has the highest energy residual, it will then send a win message.
  - **Receiving win message:** If a node receives a win message from one of its neighboring competing nodes, it will conclude that it does not have the highest stored energy. The node will then send a give up message.
  - **Receiving a giving up message:** If a node receives a give up message, and it has already sent a win message, then it is the winner of the competition and it will stay active until the end of the round.

The feature of this method that distinguishes it from other methods is that it does not require that a 100% of the area be monitored. Based on the application, a percentage of the area (e.g. 70%, 80%, etc.) will be monitored. This directly affects the system's QoS, since monitoring 100% of the area is a way to measure the QoS. An enhanced version of FCS is to add another condition on the selection of nodes at any round. The selection of nodes needs to be such that it ensures the coverage of previously uncovered areas.

Figure 5. MCFA forwarding

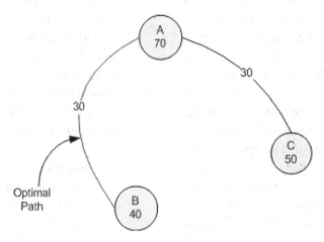

Directed Diffusion

Another way to conserve energy in WSNs is Directed Diffusion presented by Intanagonwiwat (2000). Directed Diffusion has three main goals:

- **Achieve significant energy savings**
- **Adapt to a small subset of network paths:** This means that if the network experiences some interference, then there is a way to successfully function (delivering data) without affecting the QoS.
- **Achieve robust multipath data delivery:** This means that data and events can be delivered by the use of many alternatives rather than just one single path. That is necessary because it ensures the system's ability to function in case that one of the nodes on the data path breaks.

The Directed Diffusion method is an example of a method that reduces energy consumption while dealing with a multi-sink environment. This means, there is more than one sink in the system. Those sinks are responsible for initiating an interest (queries) in some events. Directed Diffusion works as following:

- **A sink transmits its interest in some events:** The event the sink wants information about is specified by a set of attributes and values. For example, if a Base Station wants to track the movement of a four legged animal, it sends an interest as follows:
  - **Type:** Four legged animal
  - **Interval:** 10 ms; report back every 10 ms
  - **Rect:** [(0,100),(100,200)]; define the area to look at
- **When a sensor receives an interest, it starts sensing its area looking for that specific event:** A sensor may receive more than one interest, either from the same sink or from other sinks. Once the sensor receives the interest, it will store it in its data banks and broadcast the interest to its neighbors to alert them to look for that event.
- **If a sensor receives two interests that overlap with each other, it will send a single interest to its neighbors:** The new interest contains information about the two interests merged into one message. This way, instead of sending two messages, a single message will achieve the same purpose with less amount of energy.

• **When a node detects an event, it will broadcast the event to its neighbors:** When a neighbor receives an event notification from its neighbor, it will check its database to see if any interest went through it that asks for that specific event. If no interest was found in the node database, then it will simply drop the message. If an interest was found, it will broadcast it to its neighbors and only nodes with that interest will pick up the message and broadcast it again.

The energy saving technique this method uses is simply to aggregate and combine interests (queries) into one message. This aggregation will reduce the number of messages in the system which will reduce the broadcasting burden on the nodes that results in energy consumption reduction. The responses from the nodes to the Base Station can also be combined, which provides less number of exchanged messages and helps in reducing the energy consumption of the system.

Rumor Routing

A variation of the Directed Diffusion method presented in the previous section is called Rumor Routing (Raginsky, 2002). Rumor Routing is used in the case the events do not occur very frequently, but the Base Station sends too many queries about those events. The Rumor Routing protocol sends a message called agent that contain within it metadata about that event. The agent will move within the network leaving a trail in each one of the nodes it visits. The way this routing mechanism reduces energy consumption is by avoiding flooding the network with queries or events.

The model this work proposes assumes a set of densely distributed wireless sensor nodes with a short but symmetric transmission range. Symmetric transmission ranges in this model will exist between two neighboring nodes if these nodes can communicate with each other. Each node maintains a table that contains the set of events it witnessed. Once a node witnesses an event, it will add that event to its table with a parameter zero distance from that event. It will also send a special message called an agent. This agent is a long lived packet that travels between the nodes in the network propagating some information about the event. Once it is encounters a node in its path, the agent leaves information about the event and how to get to it (it will specify the node that will lead to the event. In other words, the one it came from). If an agent encounters a path that leads to the event with less cost, it will update its information that leads to that event based on the least cost path. But how to retrieve an event? The Base Station will send a query to the network. This query will roam the network from one node to another. It will keep on roaming the network in a random manner until it reaches a node that has already seen an agent for that event or until it's "time to live" value expires. If the time to live value expired, the Base Station will simply generate another query and wait for response. If the query encounters a node that has seen an agent that holds information about the event in question, it will start tracing back the nodes that lead to the node that witnessed the event by using the information the agent left behind. Once the node that witnessed the event reads the query, it will send the data to the Base Station using the information in the query (the query records the path it took to reach the source node). This method takes advantage of the fact that the probability of two lines in a plain meet is approximately 96%.

The main theme for most of the introduced methods is to reduce the number of messages exchanged in the system, the size of the message itself, or reduce both. An assumption in most of the papers is that data sensing and processing at the nodes cost far less than forwarding this data to other nodes. In some rare cases (Alippi, 2007), this assumption does not hold. In Alippi (2007), the authors propose an adaptive sampling algorithm to dynamically estimate the optimal sampling frequency of the signal (event) to be monitored. Obtaining the optimal sensing rate means minimizing activities in both the sensors and the transceiver. This minimization will reduce energy consumption while maintaining acceptable data accuracy.

DISCUSSION OF THE TECHNIQUES

Issues, Controversies, and Problems

As mentioned in the previous section, energy conservation is an issue that needs to be dealt with in order to effectively implement WSN technology. All of the above techniques are valid ones, but each one of them has its own disadvantages. For example, Heinzelman (2000) presents a way to reduce energy consumption in WSNs, but it assumes that the nodes in the system are homogeneous. This assumption is not a valid assumption since even if the nodes were homogeneous the moment they were deployed, they will become heterogeneous. The reason for such change is because the nodes responsibility and experience differ based on their location. That means, if a node is located by the highway, it will report more events than a node located at an inner small road (in case that the event is reporting car movements). Another drawback

in the previous sections is in Fan (2001). In this paper, they proposed a system that sets up the cost from each node to the Base Station in advance. This system will be ineffective and useless if it has been used in a place that suffers from high interference occurring. The reason for being ineffective in the presence of interference is because interference cut the links between nodes in a WSN. Having a path that use a broken link will simply disable the whole path. The only solution for this problem is to broadcast the data, hoping that it will reach the sink. Broadcasting the data will not conserve energy. It will drain any energy left in the system leaving it virtually inactive. In Mao (2006), the authors propose an energy conservation scheme called FCS. In this system the whole area will not be monitored. Therefore, it does not deal with the situation that a crucial event happens in the uncovered area. For example, if this system is deployed as a part of an early defense alarm, a missile moving through the uncovered area will not be detected. These weaknesses are just the tip of the iceberg.

Solutions and Recommendations

A solution for the previous weaknesses can be found through combining some of the works together. For example, if we combine (Zhong,

Table 1. Techniques outlined in this chapter[3].

Technique	Hierarchy	How To Conserve Energy	Disadvantages
SPIN	Clustering	By the use of Negotiation.	Consumes a lot of energy if the events rate of occurrence is high.
LEACH	Clustering	Use Clustering.	Assumes that nodes are homogeneous.
TEEN & APTEEN	Flat	Control the transmission recurrences.	Needs special attention to the values of Hard & Soft threshold to effectively conserve energy.
MCFA	Flat	Setup Link Costs.	A problem will occur if a link gets broken.
FCS	Flat	Select part of the nodes to be active at any given time.	Does not cover 100% of the total area to cover.
Directed Diffusion	Flat	By using data aggregation.	Nodes need some location awareness.
Rumor Routing	Flat	Send agent from the source node through the network.	Uses random agent and query deployment.

2007) with (Kulik, 2002), we can achieve energy consumption reduction in both proposals. Another solution to some of the problems is by increasing the number of sinks in the system. By doing so, each node can send its data to its nearest sink. Determining the closest sink can be done by measuring the signal strength of that sinks. The sink with the strongest signal is the closest. By sending the data to the nearest sink, the path length will be reduced, which will substantially reduce the energy consumption in the overall network. Also using one sink in the system will lead to congestions in the vicinity of the sink, which will cause energy crop at the nodes located in the sink's vicinity. Messages will also be dropped at a high rate in that congested area. Hence, having more than one sink will help balance the load among the network. It will also produce a smaller number of dropped messages, which will reduce the amount of retransmissions, leading to less consumed energy and prolonging the network lifetime.

FUTURE TRENDS

WSNs are an emerging technology that is rapidly taking its place among the most useful technologies in the world. From our point of view, there are two major routes the development of WSNs will take:

- **Developing more powerful hardware for the nodes:** Currently WSNs are developed to monitor a specific area on the surface of the earth. This area can be a battle field, the poles or the hot burning deserts. For example, in battle fields, WSNs help the soldiers on the ground to identify sources of attacks. They can use a technology called Acoustic Source Location (Svaizer, 1997) to locate the sources of intrusions. WSNs can pinpoint the source of intrusions and relay live information to the headquarters about the current status of the battle. The

live information helps the commander make accurate decisions during a battle. In the future, WSNs can be used to monitor the bottom of the deep seas and oceans. They can also be deployed in outer space to act as an early alarm system against asteroids that are in a collision course with earth. Mounting wireless sensors in space is different than mounting them on earth. First of all, the size of the nodes is not that much of a problem. The reason for not being a problem is that on earth having thousands of big nodes in a relatively small area may affect the quality of the collected samples due to possible interference from other nodes. While in space, no matter how big the nodes are, they are still tiny compared to the monitored space area. Of course, having bigger nodes will cause delivery problem, but this is another issue we have no interest to tackle here. Secondly, the communication power needed in space environment exceed that on earth. This is because of the difference in communication in an area on the earth surface than that in a region of space. In space, we are dealing with three dimensions instead of two dimensions on earth. The distance between two nodes in space is in kilometers instead of meters. In addition, in space, the communications between the sensors themselves and the Base Station will be affected by interference caused by the sun's radiation, which is not present on earth. Therefore, a new kind of sensors' transmitter needs to be developed to deal with these interferences. WSNs can also be used in the future to monitor the deep dark parts of the oceans and seas. Scientists are discovering new creatures living in the deepest part of the seas every day. Most of these discoveries happen by coincidence due to washed up creatures on the ocean's shore. To monitor these parts of the seas, new hardware needs to be developed. The hardware needs to have the ability to work under water for a long time, and to be able to withstand the high pressure. A new transmission technology needs to be added to sensors to facilitate their communication. The reason for the need of a new transmitter is because usually sensors use high frequency signals to communicate in air medium, which provides a fast communication between nodes. In water, high-frequency signals cannot penetrate the medium. To solve this problem, we need a transmitter using a low frequency signals. Although this solution will facilitate the communication between sensors, it will create some other issues and problems. One of these possible problems is that lower-frequency signals provide slow and possibly unreliable communication. This is because low frequencies are more affected by noise than high-frequency signals. Another problem is that systems using lower frequencies need substantially longer antenna than systems using high-frequency signals. Therefore, size is an issue that the new underwater technology needs to deal with. As a trivia, some of the submarines use an antenna that has a length ranging between 610 meters and 730 meters (Lockheed Martin Corporation, 2006), and an in-line amplifier is used to boost the signals inside the antenna.

- **Combining the current technology of WSNs with the technology in other areas:** In the future, it is highly possible that the WSNs will not exist on their own. Actually, there has been a lot of work on integrating WSNs with Radio Frequency Identifier (RFID) technology. These two technologies are very similar to each other. WSNs are used to monitor an area, while RFID is used to track objects, and they are unable of monitor the changes in an area. Hence, by combining both technologies,

we can monitor and track objects that are moving in an area. At the same time, we can monitor what changes are taking place in that area. In other words, if we use the hybrid technology, we can effectively see how subjects will be affected by changes in the environment.

Another use of RFID technology is to design small devices called implants. One of the uses of these implants is to put them in pets to track their whereabouts and for the pet centers to identify them. There is another version of these RFID's that designed to be used under human skin. They are being used with the elderly to alarm emergency agencies in case an accident happens. They also are used with soldiers so their bodies can be identified if they got misshaped beyond recognition.

In (Li, 2008), the authors of the paper proposed a system that uses RFID implants on animals and sensors equipped with RFID reader. The function of the sensors is to monitor the environment changes. In addition to monitoring the environment, if one of the animals gets close to a sensor, the sensor will be able to read the RFID implants and determine the identity of that animal. Using of the proposed system, we can effectively monitor animals over a long period of time.

Another research area that can be very helpful in the development of WSNs is the advances that can be made in the area of energy storage. Giving the sensors a small and rich power source will be extremely helpful in designing an effective WSN. One type of batteries that are widely used in sensor networks is Lithium Polymer batteries. The reason why Lithium batteries are preferred over other forms of batteries is because of its thin size compared to other batteries. The battery size plays a major role in deciding which type of batteries to use, simply because the sensor nodes have volume. By having a small battery, the saved space can be used to accommodate other necessary hardware. Another way to look at the energy storage is to design a way for the sensors to harvest surrounding renewable energy (especially the sun energy). Renewable energy a big research area these days, Germany is starting to move toward harvesting the wind energy. They are currently the biggest exporter for wind renewable equipments in the world. The usage of renewable energy in Germany has jumped to 14% of the total energy generation power in 2007 (Wikipedia, 2009).

The previous examples are only the tip of the iceberg on what the future holds regarding the use of the WSNs. There is still much to be investigated in WSNs. Research can be on the hardware used inside the node (e.g., processing unit of a node, a sensor, or a transmitter). It can also be on the operating system used inside a node. Moreover, it can be on the protocols used for exchanging information between nodes.

CONCLUSION

WSNs comprise of two essential devices, a Base Station and a set of sensors with wireless communication capability. The function of the sensors is to collect data about events that occurres, and relay the data to the Base Station. The function of the Base Station is to take orders from a user who deploys the network, converts these orders to queries, and propagates them into the network. It also has the responsibility of collecting the sensor responses to quires. One of the major obstacles that limit the use of a WSN is their limited energy. In this chapter, we showed the numerous techniques that can be used to reduce the consumed energy. Some of these techniques force some nodes to go to sleep while keeping the other portion active to perform the network's function. Some techniques aggregate data at the relay sensors to reduce the number of exchange message, while other techniques discover in advance the least-cost path to the Base Station.

The future holds many advances in the area of WSNs. WSNs are extremely helpful in the battlefield. WSNs can be deployed in space or in the deepest parts of the oceans.

WSNs are one of the technologies that can change the way we do things. One of the spinoffs of WSN technology is the Radio Frequency Identification (RFID) technology. The public knows that technology by the name of implants. These implants are already in use by the public who put them in their pets. A new implant has been designed to be used by human to assist the elderly monitor their health conditions.

ACKNOWLEDGMENT

This research was supported in part by U.S. NSF grants CNS-0834592 and CNS-083210.

REFERENCES

Al-Karaki, J. N., & Kamal, A. E. (2004). Routing techniques in wireless sensor networks: a survey. *IEEE Wireless Communications, 11*(6), 6–28. doi:10.1109/MWC.2004.1368893

Alippi, C., Anastasi, G., Galperti, C., Mancini, F., & Roveri, M. (2007). Adaptive Sampling for Energy Conservation in Wireless Sensor Networks for Snow Monitoring Applications. *IEEE International Conference on Mobile Adhoc and Sensor Systems.* (pp.1-6).

Fan, Y., Chen, A., Songwu, L., & Zhang, L. (2001). A scalable solution to minimum cost forwarding in large sensor networks. *Tenth International Conference on Computer Communications and Networks.* (pp.304-309).

Heinzelman, W. R., Chandrakasan, A., & Balakrishnan, H. (2000). Energy-Efficient Communication Protocol for Wireless Microsensor Networks. *Hawaii international Conference on System Sciences: Vol. 8.*

Intanagonwiwat, C., Govindan, R., & Estrin, D. (2000). Directed diffusion: a scalable and robust communication paradigm for sensor networks. *6th Annual international Conference on Mobile Computing and Networking.* (pp. 56-67).

Kulik, J., Heinzelman, W., & Balakrishnan, H. (2002). Negotiation-based protocols for disseminating information in wireless sensor networks. *Wireless Networks Journal, 8*(2/3), 169–185. doi:10.1023/A:1013715909417

Li, Z., Shen, H., & Alsaify, B. (2008). Integrating RFID with Wireless Sensor Networks for Inhabitant, Environment and Health Monitoring. *14th IEEE International Conference on Parallel and Distributed Systems.* (pp.639-646).

Lockheed Martin Corporation. (2006). *Buoyant Wire Antenna System, Submarine Communications.* Marion, MA: Lockheed Martin Corporation.

Manjeshwar, A., & Agrawal, D. P. (2001). TEEN: a routing protocol for enhanced efficiency in wireless sensor networks. *Parallel and Distributed Processing Symposium.* (pp. 2009-2015).

Manjeshwar, A., & Agrawal, D. P. (2002). AP-TEEN: A Hybrid Protocol for Efficient Routing and Comprehensive Information Retrieval in Wireless Sensor Networks. *Parallel and Distributed Processing Symposium.* (pp. 195-202).

Mao, Y., Chan, E., Chen, G., & Wu, J. (2006). Energy Efficient Fractional Coverage Schemes for Low Cost Wireless Sensor Networks. *26th IEEE International Conference on Distributed Computing Systems Workshops.* (pp. 79-79).

Raginsky, D., & Estrin, D. (2002). Rumor routing algorithm for sensor networks. *1st ACM international Workshop on Wireless Sensor Networks and Applications.* (pp. 22-31).

Svaizer, P., Matassoni, M., & Omologo, M. (1997, April). Acoustic Source Location in a Three-Dimensional Space Using Crosspower Spectrum Phase. *IEEE international Conference on Acoustics, Speech, and Signal Processing: Vol. 1.* (pp. 231-234).

Wikipedia, The free Encyclopedia (2009). *Renewable energy in Germany.* Retrieved January 27, 2009, from http://en.wikipedia.org/wiki/Renewable_energy_in_Germany

Zhong, X., & Xu, C. Z. (2007). Energy-Aware Modeling and Scheduling for Dynamic Voltage Scaling with Statistical Real-Time Guarantee. *IEEE Transaction on Computers Journal, 56*(3), 358–372. doi:10.1109/TC.2007.48

Zhong, X., & Xu, C. Z. (2007). Energy-Efficient Wireless Packet Scheduling with Quality of Service Control. *IEEE Transaction on Mobile Computing journal, 6*(10), 1158-1170.

ADDITIONAL READING

Akyildiz, I. F., Melodia, T., & Chowdury, K. R. (2007). Wireless multimedia sensor networks: A survey. *IEEE Wireless Communication Journal, 14*(6), 32–39. doi:10.1109/MWC.2007.4407225

Akyildz, I. F., Su, W., Sankarasubramaniam, Y., & Cayirci, E. (2002). Wireless sensor networks: a survey. *Computer Networks Journal, 38*(4), 393–422. doi:10.1016/S1389-1286(01)00302-4

Boukerche, A., Oliveira, H. A. B., Nakamura, E., F., & Loureiro, A. A. F. (2007). Localization systems for wireless sensor networks. *IEEE Wireless Communication Journal, 14*(6), 6-12.

Gnanapandithan, N., & Natarajan, B. (2007). Decentralised sensor network performance with correlated observations. *International Journal of Sensor Networks, 2*(3/4), 179–187. doi:10.1504/IJSNET.2007.013198

Hai, L., Bolic, M., Nayak, A., & Stojmenovic, I. (2008). Taxonomy and Challenges of the Integration of RFID and Wireless Sensor Networks. *IEEE Network Journal, 22*(6), 26–35. doi:10.1109/MNET.2008.4694171

Hong, L., Yonghe, L., & Das, S. K. (2007). Routing Correlated Data in Wireless Sensor Networks: A Survey. *IEEE Networks Journal, 21*(6), 40–47. doi:10.1109/MNET.2007.4395109

Hsin-Mu, T., Tonguz, O. K., Saraydar, C., Talty, T., Ames, M., & Macdonald, A. (2007). Zigbee-based intra-car wireless sensor networks: a case study. *IEEE Wireless Communication Journal, 14*(6), 67–77. doi:10.1109/MWC.2007.4407229

Jaekyu, C., Yoonbo, S., Taekyoung, K., & Yanghee, C. (2007). SARIF: A novel framework for integrating wireless sensor and RFID networks. *IEEE Wireless Communication Journal, 24*(6), 50–56.

Li, J., & Yu, M. (2007). Sensor coverage in wireless ad hoc sensor networks. *International Journal of Sensor Networks, 2*(3/4), 179–187. doi:10.1504/IJSNET.2007.013202

Mainwaring, A., Culler, D., Polastre, J., Szewczyk, R., & Anderson, J. (2002). Wireless sensor networks for habitat monitoring. *1st ACM international Workshop on Wireless Sensor Networks and Applications*, (pp. 88-97).

Mengjie, Y., Mokhtar, H., & Merabti, M. (2007). Fault management in wireless sensor networks. *IEEE Wireless Communication Journal, 14*(6), 13–19. doi:10.1109/MWC.2007.4407222

Min, C., Gonzalez, S., & Leung, V. C. M. (2007). Applications and design issues for mobile agents in wireless sensor networks. *IEEE Wireless Communication Journal, 14*(6), 20–26. doi:10.1109/MWC.2007.4407223

Negri, L., Zanetti, D., Montemanni, R., & Giordano, S. (2008). Power-Optimized Topology Formation and Configuration in Bluetooth Sensor Networks: An Experimental Approach. *Ad Hoc & Sensor Wireless Networks Journal, 6*(1-2), 145–175.

Onur, E., Ersoy, C., Delic, H., & Akarun, L. (2007). Surveillance Wireless Sensor Networks: Deployment Quality Analysis. *IEEE Networks Journal, 21*(6), 48–53. doi:10.1109/MNET.2007.4395110

Sohrabi, K., Gao, J., Ailawadhi, V., & Pottie, G. J. (2000). Protocols for self-organization of a wireless sensor network. *Personal Communications Journal, 7*(5), 16–27. doi:10.1109/98.878532

Stojmenovic, I. (2006). Localized network layer protocols in wireless sensor networks based on optimizing cost over progress ratio. *IEEE Network Journal, 20*(3), 21–27. doi:10.1109/MNET.2006.1580915

Wang, C., Sohraby, K., Li, B., Daneshmand, M., & Hu, Y. (2006). A survey of transport protocols for wireless sensor networks. *IEEE Network Journal, 20*(3), 34–40. doi:10.1109/MNET.2006.1637930

Wang, L., & Xiao, Y. (2006). A survey of energy-efficient scheduling mechanisms in sensor networks. *Mobile Networks and Applications Journal, 11*(5), 723–740. doi:10.1007/s11036-006-7798-5

Wang, M., Cao, J., Liu, M., Chen, B., Xu, Y., & Li, J. (2009). Design and implementation of distributed algorithms for WSN-based structural health monitoring. *International Journal of Sensor Networks, 5*(1), 11–21. doi:10.1504/IJSNET.2009.023312

Wang, Q., Zhu, Y., & Cheng, L. (2006). Reprogramming wireless sensor networks: challenges and approaches. *IEEE Network Journal, 20*(3), 48–55. doi:10.1109/MNET.2006.1637932

Wang, Y., Wu, H., & Tzeng, N. (2007). Energy-Efficient Communication in Wireless Sensor Networks: An Integrated Approach on Power Control and Load Balancing. *Ad Hoc & Sensor Wireless Networks Journal, 3*(4), 287–310.

Wikipedia (2009). *Radio-frequency identification.* Retrieved May 11, 2009, from http://en.wikipedia.org/wiki/RFID.

Wikipedia (2009). *Wireless Sensor Networks,* Retrieved May 11, 2009, from http://en.wikipedia.org/wiki/Wireless_Sensor_Networks.

Xu, W., Ma, K., Trappe, W., & Zhang, Y. (2006). Jamming sensor networks: attack and defense strategies. *IEEE Network Journal, 20*(3), 41–47. doi:10.1109/MNET.2006.1637931

Younis, O., Krunz, M., & Ramasubramanian, S. (2006). Node clustering in wireless sensor networks: recent developments and deployment challenges. *IEEE Network Journal, 20*(3), 20–25. doi:10.1109/MNET.2006.1637928

KEY TERMS AND DEFINITIONS

Cluster Head: A node in a cluster that is responsible for collecting data from the sensors in its cluster and relay these data to the Base Station. The role of Cluster Head usually rotates between nodes in the cluster.

Clusters: A way nodes use to organize themselves. Nodes that are relatively close to each other usually belong to the same cluster.

Data Aggregation: Techniques that are used in relay nodes to combines data from different sources together.

Heterogeneous Nodes: Nodes that differ from each other.

Homogeneous Nodes: Nodes that have the same properties, capabilities, and resources.

Node: A small limited energy device that is responsible for detecting events in its vicinity and report them back to the Base Station.

Sink: A device that is placed at the edge of the Wireless Sensor Network that has the responsibility of sending requests to the sensors and receiving responses for these requests to present them to the user or organization that deployed the network.

Transmission Range: It is defined by the maximum distance a node can send its data to.

ENDNOTES

[1] In this chapter, Node and Sensor are used interchangeably.

[2] In this chapter, Sink and BaseStation (BS) are used interchangeably.

[3] Refer to the method description for information about the table used notation.

Section 2
Protocols and Middlewares

Chapter 7
Practical Experiences and Design Considerations on Medium Access Control Protocols for Wireless Sensor Networks

Junaid Ansari
RWTH Aachen University, Germany

Xi Zhang
RWTH Aachen University, Germany

Petri Mähönen
RWTH Aachen University, Germany

ABSTRACT

In this chapter, the practical and experimental aspects of medium access control protocol design for wireless sensor networks are discussed. We outline the basic design principles of medium access control procedures and the general development trends from the perspective of real world implementations. We take into account especially energy consumption and latency requirements. We share our experiences gained in practical research and pre-commercial projects, wherein we have been tackling issues such as traffic awareness, coexistence of different protocols, spectrum agility for wireless sensor networks and porting of MAC-layers to different platforms. The contemporary focuses on medium access control design such as cross-layer approaches, multi-radio based protocols and radio wake-up based medium access solutions are also covered in this chapter.

INTRODUCTION

Medium Access Control (MAC) protocols and Link Layer Control protocols are responsible for

DOI: 10.4018/978-1-61520-701-5.ch007

reliable and efficient transfer of information across physical links. MAC protocols in wireless sensor networks control radio activities and coordinate nodes to access the shared communication medium. Since wireless medium is inherently broadcast in nature, the MAC-layer is a very important part of wireless systems. If the nodes do not coordinate while transmitting, data from different nodes transmitted at the same time will be superimposed and consequently the packet transmission may be lost due to collisions.

MAC protocols are designed based on different network characteristics and application requirements. Wireless Sensor Networks (WSNs) have a broad range of applications. These include the military domain, such as battlefield surveillance (Alshrabi *et al.*, 2008), environmental and habitat monitoring (Szewczyk *et al.*, 2004), healthcare applications (Kim *et al.*, 2007), home automation (Mozer, 2004), traffic control (Chen *et al.*, 2005), etc. In most of the application scenarios, sensor nodes are battery powered and deployed at remote locations or in massive scale. This makes battery service or replacement difficult. Sensor networks are expected to be in operation for years to make a WSN solution economically viable. The classical performance metrics for network services include throughput, latency, fairness, delivery rate, etc. However, due to the limited energy supply to sensor nodes and the requirement for long lifetime, reducing power consumption in WSNs has become one of the primary objectives for designers (Demirkol *et al.*, 2006).

Sources of energy consumption in sensor nodes include radio communication, computation and sensing activities. The energy consumption of sensing activities is independent of the communication protocol stack and thus out of scope for this chapter. Radio communication bears a much heavier energy consumption than computation. MAC protocols, which directly control radio activities, play an important role in determining the power consumption in data communication for sensor networks. Therefore, WSN MAC protocol designs are focused more towards energy conservation than towards classical performance metrics. Although WSN applications still impose different requirements on latency, throughput and reliability, energy consumption remains as the most important optimization goal. This makes the design of MAC protocols for WSNs unique and

challenging. One of the challenges is that power consumption is often difficult to be estimated theoretically since it depends on implementation specific knowledge.

Vast literature exists on fundamentals of MAC protocols, especially on their theoretical underpinnings. In this chapter, we focus on some of the practical issues of sensor network MACs that are of high interest to engineers and researchers wishing to implement and deploy MAC solutions on a real hardware platform in real-world applications. Thus, we bias our discussion towards providing specific lessons learned from implementation projects in order to bridge the gap between theory and practice. Furthermore, we extend our discussion to include comments on the implementation complexity, memory usage, fault tolerance and the availability of compatible hardware for realizing the design. This chapter begins with the background information on the trends of MAC protocol development and their analysis from a practical perspective. We have implemented a number of MAC solutions based on various suggested methods in order to meet the power consumption limits in different projects. A detailed discussion on the practical aspects of design, implementation, performance evaluation and the integration of these MAC protocols is presented in this chapter. These designs consider traffic awareness in MAC protocols, multi-radio platforms and wake-up radio based solutions. Empirical studies on the coexistence of a sensor network with other networks, and spectrum agility aspects are also highlighted. The cross-layer design aspects are covered with two experimental designs and portable MAC architectures are introduced afterwards. Finally, we conclude the chapter.

MEDIUM ACCESS CONTROL TRENDS AND SOLUTIONS

A number of WSN MAC protocols have been developed with different design goals and techniques for various application scenarios. These protocols are fundamentally different from the ones designed for ad hoc networks because of the special characteristics exhibited by WSNs. Before going into the discussion on the practical experiences on MAC protocol design and implementation, we will have a look at the characteristics of WSNs and the development trends of MAC protocols.

Characteristics of Wireless Sensor Networks

In the following, we list some of the main characteristics of WSNs:

- Sensor networks are composed of thousands of battery powered nodes geographically dispersed in an ad hoc fashion. Owing to the battery constraints, sensor nodes use low transmit power levels. This results in a short communication range and a high density of deployed nodes in a certain geographical area.
- The data generation rate in sensor networks is low and therefore results in long idle periods. Also, some sensor networks (e.g. for monitoring applications) are event based; when the desired event of interest occurs, high data rates are needed. Overall these application characteristics result in intermittent traffic.
- Sensor networks exhibit dynamism because of mobility in the network, existing nodes dying out or new nodes joining the network.
- Most sensor networks are dedicated to perform a single application task, which means that sensor nodes in a particular network cooperate in servicing a single application. Therefore, application level fairness is more important than node level fairness.

Low-transmit power in sensor networks results in short-ranged sensor nodes exercising

multihop communication. The more hops a packet needs to go through before reaching the destination, the higher the probability of packet loss is. Therefore, guarantees for packet delivery from MAC protocol are desired. Long idle periods in sensor networks give opportunities for energy conservation schemes by turning off the sensor nodes when they are idle. Since the nodes in a sensor network collaboratively work for a single dedicated application, the overall network lifetime becomes the prime consideration instead of the battery life-time of individual nodes. Therefore, it is important to analyze MAC-layer from a network point of view.

Main Sources of Energy Wastage

The main sources of energy wastage in MAC protocols have been identified to be idle listening, packet collisions, overhearing, and control packet overhead (Ye *et al.*, 2002).

Idle Listening

Idle listening is the most significant source of energy wastage. It refers to the situation when a receiver node listens to an idle channel in anticipation of upcoming packets. The most effective approach to reduce energy wastage in idle listening is to turn off the radio during the idle period. Three main approaches are used to achieve this goal: Time Division Multiple Access (TDMA), scheduled contention and Low Power Listening (LPL). TDMA avoids idle listening by assigning different time slots to different nodes. Nodes are only active in the assigned time slots and are put into sleep during the irrelevant slots. Scheduled contention works based on a similar principle as TDMA. Nodes are scheduled to wake up at the same time to listen and/or contend for the channel. Idle listening is reduced since nodes know when other nodes transmit. LPL, on the other hand, refers to the practice where nodes wake up asynchronously for a short period of time and sense the channel for activity. If the channel is found to be idle, nodes are put back to sleep.

Packet Collisions

Packet collisions occur when two or more nodes transmit simultaneously. This can result in the interference of the two packets on the receiving node so that none of the packets can be decoded (received) correctly. Packet collisions require the retransmission of the collided packets. Retransmissions not only waste energy but also add extra latency. Collision avoidance is usually one of the basic requirements for random access based MAC protocols. TDMA can guarantee a collision free network since all nodes are assigned different time slots for packet transmission and reception. However, TDMA has its own practical constraints such as the scalability problem and the need for a control mechanism to schedule assignments. Even in a collision free network, retransmissions are inevitable due to interference and fading which are characteristics of wireless channel. Forward error correction (FEC) based channel coding schemes can be used to lower the residual bit error rate. Selecting a right FEC scheme, however, requires careful consideration since decoding can require a lot of computing power. Also, channel coding requires transmitting extra bits and therefore a price has to be paid for the additional overhead both in transmission and reception.

Overhearing

Overhearing refers to a node receiving packets that are not destined to it. Since wireless medium is inherently broadcast, nodes may overhear packets which are not destined to them. Receiving data packets and decoding them consume a large amount of current and causes energy wastage for the overhearing nodes. A node overhears when it cannot decide if the data packet is addressed to it. The destination address is usually transmitted before the data bytes begin, so that a node should be able to quickly decide if it makes sense to receive the rest of the bytes.

Control Packet Overhead

Control packet overhead refers to the control information that is needed to be transmitted in addition to the real payload data. Apart from classical header overhead, there are potentially many other overheads in the case of WSNs, e.g. information regarding the coordination of nodes for data exchange, synchronization messages for slot assignments, node wake-up schedules, etc. Although control packet overhead is inevitable, there are ways to minimize it. Designing compact protocol data structures, piggy-backing the control information along with the data packets or packing more data packets together with the same control overhead thereby increasing the ratio of data size to control size are a few ways to minimize the overhead.

Energy Efficient Techniques

One of the characteristics of sensor network MAC protocols is that the radio is not kept on all the time. Since the traffic rate in sensor networks is typically low, nodes can avoid or minimize idle listening and save significant amounts of energy. Sensor nodes periodically turn on and off the radio in order to conserve the energy. The periodic wake-up and sleep cycles form duty cycle behaviour. The duty cycle can be expressed as

$$D = \frac{\tau}{T},$$

where D is the duty cycle, and τ is the time duration that radio is in the listening mode (known also as the on time or the active time). The symbol T denotes period which the listen/sleep cycle is repeated. It is also known as the check interval or sampling period.

Figure 1. Concept of duty cycling in wireless sensor networks

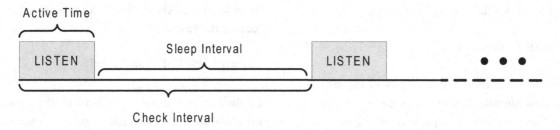

Duty cycling can result in remarkable energy savings with low duty cycle values. However, a low duty cycle means increased latency for data communication. This implies a trade off between the energy consumption and the latency in duty cycling based MAC protocols. Duty cycling requires that nodes have their active periods aligned to each other for data communication. Different controlling strategies have been designed to coordinate the active periods of the nodes. Different types of duty cycling schemes have different amount of associated overhead for controlling and signaling. In the following, we describe three common techniques used in practice.

Non-Contended Strict Time Slot Assignments

TDMA MAC protocols inherently perform duty cycling as nodes exchange information in their assigned slots and sleep during the rest of the slots (Demirkol *et al.*, 2006). Slot assignment needs to be coordinated by one of the nodes. Owing to the dynamic nature of the network and clock drifts, the slot schedule must be updated periodically. The update frequency depends on the accuracy of the crystal oscillators and the network dynamics. The interval between two slots assigned to a particular node depends upon the number of nodes in the neighbourhood and determines the duty cycle of the TDMA based MAC protocol. The slot assignment and maintenance has an associated control traffic overhead that must be taken into account.

Common Active/Sleep Schedules

One of the ways to achieve common active periods for data communication is to synchronize the network so that all the nodes in the neighbourhood wake-up at the same time (Demirkol *et al.*, 2006). This is achieved either through explicit synchronization frames or by indication for example in RTS/CTS control frames. One of the nodes in the neighbourhood initiates synchronization by transmitting a synchronization frame and dictates its own sleep schedule which is followed by other nodes. A synchronization update is performed periodically to counter the effect of the jitter developed over time and to let new nodes join the synchronized network.

Preamble Sampling

Preamble sampling technique allows nodes to wake-up periodically in an asynchronous fashion (Polastre *et al.*, 2004). The transmitter needs to send a long preamble of duration equal to or greater than the check interval to ensure that all the potential receivers detect the preamble sequence. The data packet follows the preamble sequence. A receiving node wakes up, listens to the preamble and then waits for the upcoming data.

There is no need for network wide time synchronization such as periodic broadcasting of synchronization frames in preamble sampling protocols. However, there is an associated overhead per data packet and the price is paid in terms of the transmission of a preamble sequence

Figure 2. Preamble sampling technique

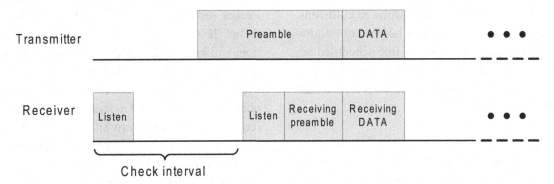

every time when data needs to be transmitted. In a typical WSN, where the traffic load is low, preamble sampling protocols outperform schedule based schemes. They are also comparatively more practical to use especially in the case of highly dynamic networks because they allow node asynchronicity. However, when the nodes exercise a very low duty cycle, the transmitter needs to transmit an extremely long preamble which can be very costly in terms of used energy. The minimization of the preamble length without compromising the savings from low duty cycle activities remains a challenging design goal for preamble sampling protocols.

Classification of MAC Protocols

Next we provide a short chronological literature review of some of the representative MAC protocols used in practice. We briefly describe the pros and cons of the selected MAC protocols, and discuss how they influenced the future protocol designs. There are two major streams of MAC protocols for WSNs, namely the TDMA based protocols and the contention based MAC protocols. In contention based protocols, there are two main sub-branches: IEEE 802.11 inspired MAC protocols which use RTS/CTS/DATA/ACK and the slotted access principle. These protocols also enforce common active/sleep schedules among nodes to allow low duty cycle operations. The second subclass consists of the preamble sampling

based protocols, which include basic CSMA/CA features and operate at a low duty cycle.

TDMA-Based MAC Protocols

TDMA based protocols divide a channel into time slots. They explicitly assign transmission and reception slots to nodes and allow them to sleep for the rest of the slots. The number of slots is usually fixed as adjusting is typically a complicated task. TDMA based protocols are suitable for WSNs, if there is a base station or cluster head which communicates with all nodes and thus provides the slot assignments. One of the major advantages of TDMA based MAC protocols is the possibility of implementing collision free systems, where nodes only transmit in their assigned slots. One can achieve a very low duty cycle since nodes sleep in the unassigned slots.

TRAMA

The TRAffic adaptive Medium Access protocol (TRAMA) (Rajendran *et al.*, 2003) is a TDMA based energy efficient collision free MAC protocol. It creates schedules which allow nodes to access a single channel in a collision free manner. Time is divided into random access and scheduled access periods. The random access period is used by the nodes to regularly broadcast their neighbour node identities and traffic information which can be used to build up two hop neighbour information and compute a collision free schedule. However, the channel access for the random access period is contention based, so it is prone to collisions. Therefore, data packets and schedule information is propagated using collision free scheduled access periods. Traffic load information is used to give priority to busy nodes in order to transmit packets in slot assignment.

LMAC

The Lightweight Medium Access protocol (LMAC) (van Hoesel *et al.,* 2004) is one of the TDMA based MAC protocol with reduced design and implementation complexity. Each node controls one time slot and the time slot is reused at a non-interfering distance. Unlike most of the traditional TDMA protocols, time slots are not assigned by a central controller. Nodes select their time slots based on the two-hop neighbourhood slot occupancy information. On one hand, LMAC reduces the need to have a central manager for slot assignment while on the other, every node needs to maintain neighbourhood information. During the initial network setup phase, nodes suffer from the possibility of packet collisions since they are unaware of their neighbours and need to contend for the control of time slots. When the setup is complete, nodes can communicate in a collision-free fashion. Each message transmitted by a node contains a control information part and a data payload part. All nodes in the neighbourhood listen to the control information part and decide based on the destination address whether to continue receiving the data portion or not.

BitMAC

BitMAC (Ringwald *et al.,* 2005) is a deterministic, collision free and robust MAC protocol tailored to dense sensor networks. It is designed for radios supporting On-Off-Keying modulation, multi-channel communication and synchronized bit transmission. It also assumes a static spanning tree topology for data collection networks. BitMAC eliminates the preamble of reply messages from a synchronized node. This technique can only be used on radios which provide a bit-level interface. Packet radios such as CC2420 (Texas Instruments, 2008a) are not suitable for this optimization. Unlike LMAC where every node controls a dedicated time slot, BitMAC allocates slots only to those nodes which need to send data. The sink node sends out beacon message and collects the requests to transmit from all nodes and assigns time slots based on the requests. This optimizes channel utilization but implies significant control overhead. In the case where the receiver node is interested in the aggregation of sets of data from the transmitters, the transmitting nodes access the channel concurrently and transmit in a bit synchronized way without collisions.

TDMA based MAC protocols result in better collision avoidance performance than contention based-protocols. However, TRAMA and LMAC both need to keep two-hop neighbourhood information and compute the schedule and priorities which add to the computational and communications workload of the node. Since all nodes need to be awake during the random access period for the exchange of control information, it is difficult to achieve a low duty cycle state and therefore the energy spent on the control overhead is significant. BitMAC is restricted to only one type of sensor network topology and application. It also imposes restrictions on the hardware platform, which makes it less suitable for general usage.

TDMA based MAC protocols usually require centralized or cluster-based control, which is not guaranteed in all WSNs. It is also not easy to change the slot assignments dynamically, thereby leading to scalability problems. Furthermore, in the TDMA frame, there are some unused slots in which none of the nodes transmit. As a result, time slots remain underutilized. Some schemes exercise un-utilized slot stealing from other nodes to increase the throughput and channel utilization. However, scalability and dynamism handling remain big practical constraints for TDMA based protocols.

Contention-Based Protocols

In contention-based protocols, two or more nodes may simultaneously attempt to use the medium for communication. Therefore, the nodes need to contend for the channel (Tannenbaum, 2003). This may lead to packet collisions in contrast to TDMA based protocols. Therefore, the protocols are designed to eliminate packet collisions. Unlike TDMA type protocols, contention based protocols are not handicapped by the scalability of the network.

S-MAC

The basic idea of Sensor-MAC (Ye *et al.*, 2002), a contention based protocol, is in part similar to IEEE 802.11 MAC protocol. In order to reduce the overhearing problem, it uses in-channel signaling to put each node to sleep when its neighbour is transmitting to another node. S-MAC uses RTS/CTS/DATA/ACK handshake for unicast packets in order to reduce collisions and overhearing by putting irrelevant nodes after receiving RTS/CTS packets to sleep. Scheduling the active periods reduces the energy consumption but on the other hand explicit SYNC frames, which are used for node synchronization, increase control packet overhead. In general, S-MAC trades off per-hop fairness and latency for energy reduction, but the protocol designers claim that end-to-end fairness and latency are not affected. Application level performance is a more meaningful metric for a WSN rather than an individual node performance.

T-MAC

T-MAC (van Dam *et al.*, 2003) is also an IEEE 802.11 inspired MAC protocol. The major breakthrough of this protocol is that it introduced the idea of adaptive duty cycling. It automatically adapts the duty cycle to the current traffic. Instead of using a fixed length active period, T-MAC uses a time out mechanism to dynamically determine the length of the active period. If a node does not detect any activity within a certain predefined time

out interval, it goes to sleep. This scheme reduces the amount of energy wasted in idle listening in which the receiving nodes wait for potential incoming packets. However, exercising this time-out mechanism might lead to early sleeping problem. This means that a receiving node goes to sleep while another node still has a message intended for it. This phenomenon occurs when the nodes are not synchronized.

Both S-MAC and T-MAC are contention based protocols, but have a certain degree of resemblance to TDMA based protocols since the nodes operate based on wake-up schedules. These protocols inherit the problem from TDMA based protocols where the overhead of maintaining synchronization is high. However, they perform better for broadcasting data packets since all nodes follow the same schedule, unlike in TDMA based protocols where nodes wake up in different time slots to avoid collision. Although TDMA based protocols typically have a broadcast time slot where all nodes wake up for synchronization, this slot is mainly used for control purposes instead of data transmission. Contention based scheduling protocols are also more realistic and practical to implement as they do not require nodes to form communication clusters, thus eliminating the need for inter-cluster communication and interference management.

Another category of contention based MAC protocols are the preamble sampling MACs, which share the advantages of easy support for broadcast and avoid the need for scheduling and RTS/CTS handshake. They are based on an adaptive preamble sampling method to reduce the duty cycle and minimize idle listening.

B-MAC

B-MAC is a CSMA/CA based preamble sampling protocol (Polastre *et al.*, 2004), which allows the nodes to follow their own sleep and wake-up cycles in an asynchronous manner. This protocol contains four main functions: Clear Channel Assessment (CCA) for determining the occupancy status of the channel, packet back-offs, link layer ACKs and low power listening. These functions can be turned on and off based on application needs and are provided through the flexible Application Programming Interfaces (API). B-MAC uses a specialized CCA mechanism, which takes the background noise into account when making decisions for channel assessment. The signal strength is sampled and added to a FIFO queue. Only the median sample from the queue is taken and the noise floor estimation is updated using an exponentially weighted moving average with variable coefficient (Polastre, 2003). In B-MAC, a receiving node periodically wakes up for a short time and listens to the channel for activity. If there is no activity, it goes to sleep and wakes up on the next cycle. When a transmitting node wants to send a packet, it transmits a long enough preamble to ensure that this channel activity is detected by the destined receiving node. After detecting channel activity the receiving node extends its active period to receive the data.

B-MAC works well in both static and mobile scenarios since all nodes exercise the same duty cycle in an asynchronous fashion and there is no need to maintain neighbourhood information. The long preamble makes sure that all nodes in the neighbourhood receive the data that follows the preamble. However, retransmissions in B-MAC are very costly because both the data and long preamble need to be retransmitted. This causes large energy wastage and imparts high latency. Preamble sampling MACs in general do not use RTS/CTS control frames. This reduces control overhead. On the other hand it causes MACs to potentially suffer from hidden terminal problem (Ansari *et al.,* 2007) and reduced data reliability.

WiseMAC

The length of the preamble in B-MAC depends directly on the duty cycle of the receiver. For a low duty cycle network, the length is very long and thus energy consuming for the transmitter. WiseMAC (El-Hoiydi *et al.,* 2003) reduces the preamble length for unicast packets by keeping a neighbourhood sleep schedule at all nodes. All nodes have independent channel sampling schedules. When the destination node receives a complete data frame, it sends back an acknowledgement packet to the source node which includes its channel sampling timing information. In this scheme, the source node tries to implicitly synchronize with the destination node by calculating the sampling offset. The preamble length cannot be reduced to zero because of the timing impreciseness coming from the crystal oscillators' inaccuracies. WiseMAC estimates the length of the preamble based on the crystal jitters accumulated over two subsequent packet transmissions and the offsets between transmitter and receiver sleep schedules. When one node transmits to another node frequently, the length of the preamble is significantly reduced, providing high energy efficiency. WiseMAC optimizes the preamble length but adds extra complexity to the network. The optimization applies only when the packets are unicast and when traffic load is high. If the packets need to be broadcasted, the transmitter needs to transmit a complete preamble. Since the reduction of the preamble length is based on the knowledge of the offset between two nodes, the advantage of WiseMAC can only be shown when there is a high enough traffic for timing information to be updated frequently. In a dynamic network where the neighbours of a node change every minute, the optimization gains do not apply either. In WiseMAC, the preamble length is ideally shortened so that only the destination nodes listen to the preamble. However, due to timing inaccuracies, there exists a possibility that other nodes in the neighbourhood overhear the preamble.

Hybrid Protocols

Since both TDMA based and contention based protocols have their own distinct characteristics and pros and cons in different application scenarios, it is natural to combine these two approaches to utilize their advantages and offset their weaknesses. In the following, we describe Z-MAC (Rhee *et al*., 2005) which is hybrid protocol based on TDMA and CSMA.

Z-MAC

Z-MAC adapts itself to the level of contention in the network and combines the features of both CSMA and TDMA. Under low contention, it behaves like CSMA and under high contention, it acts like TDMA based MAC protocol. The switching between two behaviours is dynamic. Time slot assignment is performed at the deployment time. Unlike in classical TDMA, a node may transmit in Z-MAC during any time slot and therefore performs carrier-sensing before transmission. The owner of the slot is given higher priority over other contending nodes by having a smaller initial contention window size. When the network is overloaded with traffic, which means there is high contention for slots, nodes always transmit in their own slots as is the normal behaviour of TDMA protocols. Under low contention, the slots can be used by all the nodes and this constitutes a CSMA like behaviour.

The main drawback of hybrid protocols is their fairly complex signaling overhead. The implementation of the protocols themselves is also slightly more complex, and these drawbacks may limit the applicability of the hybrid protocols.

Multi-Channel MAC Protocols

Multi-channel MAC protocols are well investigated for wireless ad hoc networks since they offer better spectral utilization and can provide better performance in terms of throughput and latency. Since the main objectives are different, a naïve migration of classical ad hoc protocols to sensor networks is not wise. In WSNs, power savings are achieved by using multi-channel MAC protocols. Using a signaling channel independent from the data channel used for packet transmissions enables a receiving node to transmit a 'busy tone'

over the control channel and to let other nodes become aware of the on going transmission. This separate signaling channel also enables nodes to determine when and for how long they can power themselves off. Real implementations of multi-channel MAC protocols are not very common because of the hardware limitations on sensor node and a higher economic price of multi-channel radio transceivers.

DCMA/AP

Dual Channel Multiple Access with Adaptive Preamble (DCMA/AP) protocol is designed especially for WSNs and uses two separate channels simultaneously (Ruzzelli, 2006). The data channel is used for preamble and data packet transmissions while the control channel may indicate reception-in-progress to avoid hidden and exposed terminal problems. When control channel is free, i.e. all the neighbouring nodes are idle, a node may start to transmit the preamble and data. The length of the preamble is greater than the inactive period of the node so that it can be detected by the receiving node. When the receiving node detects the preamble, it sends a Reception-In-Process (RIP) message on the control channel and contains the information of the duration of the data reception. When the neighbouring nodes wake up and detect the message, they go to sleep for the entire duration as indicated in the RIP. The major advantage of using dual channel enabled transceiver is that while receiving data the node can send RIP that prevents the neighbouring nodes from wasting energy on overhearing.

Table *1* shows the comparison of a few selected MAC protocols in terms of scalability, mobility support and synchronization requirements.

In this section we discussed the various approaches and trends in the WSN MAC research focusing on the pros and cons of each technique from the perspective of real world implementation and deployment. We also presented some representative protocols for each of the technique in order to show the evolutionary trends of MAC

Table 1. Comparison of MAC protocols

	Channel access mechanism	Scalability	Mobility support	Synchronizes the network
TRAMA	TDMA	Low	Low	Yes
LMAC	TDMA	Low	Low	Yes
BitMAC	TDMA	Medium	Low	Yes
S-MAC	CSMA/CA schedule based	Medium	Low	Yes
T-MAC	CSMA/CA schedule based	Medium	Low	Yes
B-MAC	CSMA/CA preamble sampling based	High	High	No
Wise-MAC	CSMA/CA preamble sampling based	Medium	Medium	Yes
Z-MAC	CSMA/CA + TDMA	High	Medium	Yes
DCMA/AP	CSMA/CA preamble sampling based	High	High	No

design in WSNs. We have seen that the preamble sampling technique better suits to the requirements of many WSN applications, especially ones with dynamic characteristics, asynchronized nodes and low traffic conditions. The following sections describe various protocol designs in detail, addressing the practical issues normally encountered during the real world implementation, deployment and desired performance metrics obtainment. The previously described MAC approaches will help the readers better understand the design considerations explained in the next sections.

TRAFFIC AWARENESS IN MEDIUM ACCESS CONTROL PROTOCOLS

Wireless sensor networks have a wide range of applications and these are characterized by different traffic load conditions and traffic types. The traffic generated by sensor network applications often fluctuates in time (Langendoen, 2007). The peak loads may drive the network into congestion, leading to high latency, whereas low traffic loads cause energy wastage in idle listening. For example, frequent waking up of nodes in a network with scarce traffic load results in high energy wastage. Conversely, an infrequent wake-up schedule can make a network incapable of handling high traffic loads and give a poor quality of service. Therefore,

load adaptation is an important feature for MAC protocol designs in order to have lower power consumption and to achieve higher lifetime of a sensor network. There has been a number of MAC protocols designed with traffic load adaptability such as the TDMA based TRAMA and the duty cycle based TA-MAC (Gong *et al.*, 2005). TDMA protocols allow a natural way to adapt the duty cycle as per traffic loads, since the nodes listen to the slots only on a need basis. If there is data to be received, the nodes wake up in the slot of the transmitting node to receive data. Furthermore, time slots can be stolen from the nodes not needing them in order to allow bursty transmission. For instance, in the scheduled duty cycle based approach (Gong *et al.*, 2005), the S-MAC contention window is expanded when there is congestion in the medium and contracted when there is less or no congestion. This allows contention resolution and helps in minimizing energy consumption. Furthermore, it helps avoiding the added latency associated with retransmissions.

Besides the traffic loads, traffic types, i.e. the link level addressing, also have a direct effect on energy consumption. A traffic type is categorized into unicast, multicast or broadcast. Although wireless medium is broadcast in nature, energy consumption in the network still heavily relies on the different types of traffic mainly due to significantly different control overheads. As mentioned earlier, in a TDMA based network, unicast means transmitting the data packets at the receiver's time slot while broadcast may mean transmitting the same data multiple times at different time slots (Ringwald *et al.*, 2008). In preamble sampling MAC protocols, the preamble length can be optimized based on the traffic load and traffic type information. In WiseMAC, for example, the preamble length for unicast transmission can be shortened by exchanging the neighbourhood sleep schedule information inside the ACK packets as described before.

In MFP-MAC (Bachir *et al.*, 2006) protocol, preamble is divided into smaller tiny preamble frames instead of a long monolithic preamble sequence. Preamble frames contain information about the destination address and the time-offset for the data frame following the preamble. An asynchronous node can receive one of the tiny framelets and thus know if it is the destination or not. After framelet reception, it goes to sleep. If the node is the destination, as indicated in the micro-frame, it wakes up again at the exact time of the transmission of the data packet. This way MFP-MAC saves energy in receiving the uninformative preamble sequence. The nodes may also estimate the channel occupancy time and extend their sleeping period. If the data size to be transmitted is small, the control overhead of the micro-frame is becomes relatively significant. Furthermore, waking the radio again for data reception after receiving an MFP becomes more energy expensive than piggy-backing the data to the repeating micro-frames. In B-MAC+ (Avvenuti *et al.*, 2006), the data packet is repeated in the preamble sequence and a receiving node simply receives one data packet after waking up and later goes to sleep. Our experiments using the CC2420 radio transceiver and the CC1000 (Texas Instruments, 2008) radio transceiver have shown that repeating the packet in the preamble sequence or piggybacking the data into the micro-frames is more energy conserving for the receiving nodes below a certain data size. For larger data size, the

non-addressed nodes have to pay a high packet reception price, which is wasted. Another optimization technique on the preamble for unicast packet transmission is the preamble-strobing technique as used in X-MAC (Buettner *et al.*, 2006). Data packet is transmitted repetitively as the preamble. After transmitting each packet, a node probes for the packet acknowledgement. If the acknowledgement is not received within a certain time interval, the node transmits the packet again. If the acknowledgement is received, the transmitter stops further packet transmission. The amount of energy saved depends on the wake-up offset between the transmitter and the receiver. In the worst case, the transmission duration of preamble is a node's sleep interval. Traffic Aware MAC (TrawMAC) (Ansari *et al.*, 2008b) combines the various optimizations on the preamble length as described above. For smaller data size, the data is piggybacked into the preamble frames. For larger data-size, micro-frames are repeatedly transmitted followed by the data-packet. TrawMAC has the ability to handle a large data packet from the application and transmit it in the data-frame-train format after the channel is reserved using a single preamble. Data size dependent optimizations include the use of piggybacking data into the preamble and transmitting frame-trains. These give TrawMAC the ability to work in an energy efficient manner and adapt to variable traffic loads. In unicast transmission, the data is immediately transmitted after the destination node strobes the micro-frame.

In unicast transmission, TrawMAC uses preamble strobing and also makes use of the destination node's sleep schedule to schedule the transmission of the preamble frame. Like WiseMAC, TrawMAC incorporates the clock jitter between the transmitter and the receiver. Figure 3 shows the TrawMAC's behavior for a unicast transmission. In Figure 3a we show the case for a large data size where the data frames follow the micro-frame preambles (MFPs). The reader should note that although the data transmission

Figure 3. Operational cycle for unicast transmission. (a) Shows micro-frame preambles (MFPs) transmission followed by data frames when data size is large. (b) Shows data-frame preambles (DFPs) transmission where data-payload is piggybacked to preamble- frames when data size is small

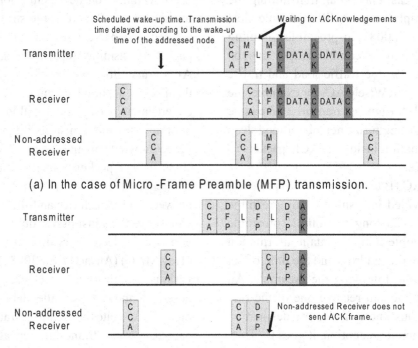

(a) In the case of Micro -Frame Preamble (MFP) transmission.

(b) In the case of Data -Frame Preamble (DFP) transmission.

is scheduled earlier, the transmitter delays the transmission of MFP incorporating the wake-up schedule of the receiving node plus margin for the clock drift accumulation. The receiving node performs Clear Channel Assessment (CCA) and keeps on listening (L) to the medium for the arrival of the MFP. The transmitter sends the data packet(s) after the receiver acknowledges an MFP. A non-addressed node sleeps after receiving the MFP. Figure 3b illustrates the case for a small data size, where the data is piggy-backed in the preamble frames in the form of data frame preambles (DFPs). The transmitter is shown incorporating the clock-drift between the transmitter and receiver. However, a reduction in the length of the preamble size is achieved.

For broadcast transmissions (cf. Figure 4), TrawMAC divides the monolithic preamble into micro-frames. If there are many broadcast transmissions, the transmitting node may deplete the battery earlier because of the need to transmit a long preamble sequence corresponding to the sleep interval. For a static topology with a fixed neighbourhood size, it may be more energy efficient to transmit multiple unicast packets instead of a single broadcast packet. TrawMAC has the intelligence to realize a broadcast transmission as multiple unicast transmissions when energy conservation can be achieved. TrawMAC's ability to combine neighbourhood sleep scheduled based preamble optimization with preamble strobing has practical advantages in mobile environments.

SPECTRUM AGILITY

With the ever growing popularity of wireless devices and networks, the wireless spectrum is getting congested. This requires an efficient use of the spectral resources and the need for coexistence

Figure 4. Operation cycle for broadcast transmission

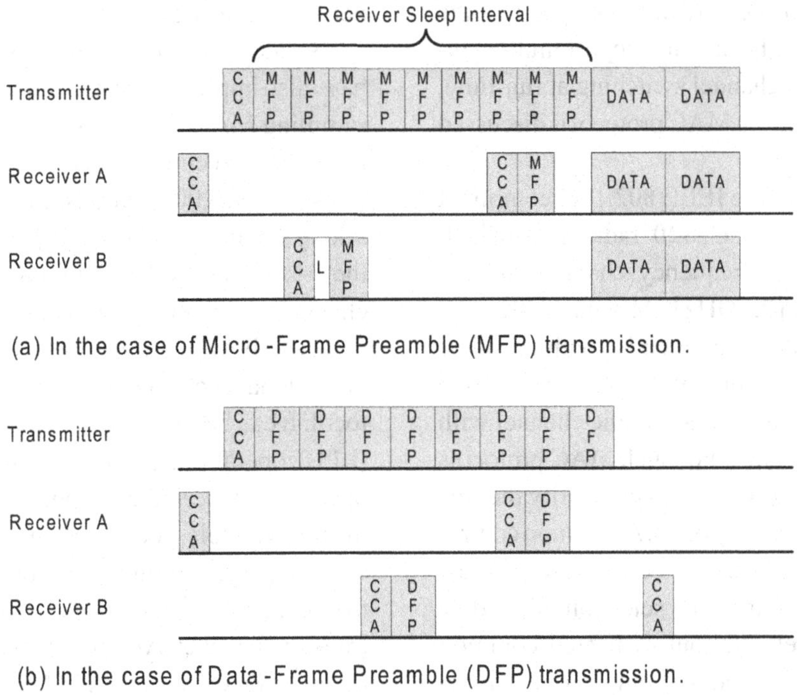

(a) In the case of Micro-Frame Preamble (MFP) transmission.

(b) In the case of Data-Frame Preamble (DFP) transmission.

support. Coexistence in WSNs can be categorized into two forms based on the transmit power levels and the degree of spectral occupancy: coexistence of a WSN with other networks as well as the coexistence of multiple WSNs. The coexistence issue in the context of WSNs has not yet been extensively researched. However, this issue has been well investigated in classical wireless networks, for instance, in the case of IEEE 802.11b/g and IEEE 802.15.3, which both operate in 2.4 GHz ISM frequency band.

In sensor networks, traffic load is generally low and the wireless channels are usually under-utilized. Dedicating a wireless channel for each network is not only a waste of resources but also difficult to implement due to the increasing scarcity of the wireless spectrum. Most of the commercially available sensor nodes such as Crossbow Inc.'s TelosB, MICAz and iMote2 use IEEE 802.15.4 compliant radio transceivers operating in the unregulated 2.4 GHz ISM frequency band because of the availability of high bandwidth and

reasonable transmission range. At the same time, other networks based on IEEE 802.11b/g and IEEE 802.15.1 standards have become very popular and can cause severe interference. Furthermore, microwave ovens can generate strong interference in the same band. Since they are high power electrical appliances, the interference has detrimental impact on the sensor node communication, even for those using spread spectrum techniques (Zhou *et al.*, 2006). Since sensor nodes are usually power constrained and use low transmit power levels, they always remain handicapped when competing for the same wireless band against other more powerful devices. It has been shown by empirical studies (Petrova *et al.*, 2007) that under moderate and high traffic conditions, IEEE 802.15.4 based low power sensor node radios suffer huge packet loss due to interference from different WLAN standards in 2.4 GHz band. Therefore, in order to avoid unnecessary collisions and retransmissions, spectrum agility becomes an important feature for MAC protocols. In order to realize spectrum agility, support from both hardware and software is required. From the hardware perspective, the transceiver should be able to support multichannel operation and channel switching at run-time. Software support from MAC protocol is discussed in this section.

Let us consider the IEEE 802.15.4 compliant Texas Instrument's CC2420 radio transceiver, which supports 16 frequency channels with 5 MHz separation in 2.4 GHz ISM band. It offers the possibility for data communication with any of the 16 channels. MAC protocols must be designed to support data communication in the channel with minimum interference. For such MAC protocols spectrum scanning is required to identify the free channel in the frequency band. A common control channel with little interference can be used to establish an agreement on the selection of the data exchange channel. Without dedicated common control channel the nodes should run a channel assignment algorithm to agree on selecting the data channel. In the design of Multi-Radio MAC protocol (MR-MAC) (Ansari *et al.*, 2007a), a low bandwidth control channel in the low frequency band with low interference is used as the common control channel to coordinate the operation and exchange control messages which contain information of the selected channel for transmission of data packets. Ideally, spectrum scanning for the free data channel should be performed every single time before initiating data communication. However, it is an energy consuming activity, especially if data transmission is frequent. In our experiments, we reduced the number of channel scannings by defining a minimum time interval between two new subsequent scans. If the interval between successive data transmissions is less than the minimum scan interval, the channel is used. If the interval between subsequent data transmissions is larger than defined minimum, i.e. scarce traffic load conditions, spectrum is scanned again before data transmission. This minimum interval is determined based on the knowledge of the congestion levels and the level of spectrum

occupancy in the environment where the sensor networks are deployed.

Besides using a dedicated control channel, we have also implemented a spectrum agile preamble sampling MAC that scans all the available channels while executing a low-power-listening cycle (Ansari *et al.*, 2009). A transmitting node transmits a long preamble followed by the data in one of the free channels. By scanning all the available channels sequentially, the receiving node is able to detect activity in the channel. Figure 5 shows the operational cycle of the spectrum agile MAC protocol. It can be seen that in Low-Power-Listening (LPL) operation, Node A exercises duty cycle operation by periodically polling all the channels in the available pool. Node B performs carrier sensing operations in order to determine which of the channels is unoccupied before attempting to transmit a packet. An asynchronously waking up node, Node C, detects the MFP in the data communication channel. The destination address and the timing information in the MFP let the receiver know when to wake up again to receive the data packets. All the optimizations techniques on the preamble in TrawMAC as discussed in the section on Traffic Awareness in Medium Access Control are applied in this spectral agile medium access control protocol as well.

Scanning all the channels is quite exhaustive for the nodes and leads to high energy consumption in idle listening. When a channel is found free and data communication is established, both the receiver and the transmitter increase the weight associated with that channel, which makes it more likely to be used next time. Once the receiving node finds activity in the channel, it determines if it is caused by an external interferer or another transmitting node. If the receiver is unable to sniff an understandable data frame within a certain time-out interval, it decides that the activity in the medium is caused by an external interferer. If understandable data frames are received, the receiver listens to the packet being transmitted and stops scanning of other channels. A potential transmitter

Figure 5. Operational cycle of the spectrum agile MAC protocol. Node A performs the Low-Power-Listening (LPL) operation by scanning sequentially all of the available channels upon waking up. Node B performs carrier sensing in all of the available channels before attempting to transmit a packet. Node C, upon its asynchronous wake-up, scans the channels and detects the data transmission from Node B. Upon receiving a complete MFP, Node C is able to know the timings of the data frames. It wakes up later to receive the data frames

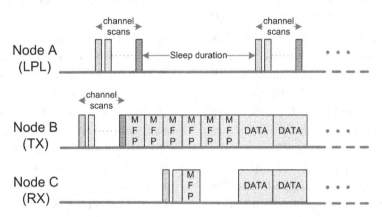

makes sure that no other node is transmitting at the same time. If there is data being transmitted, the potential transmitter refrains from sending data and tries to receive data. If a certain channel is found occupied multiple times by an external interferer, its associated weight is reduced so much that it is dropped from the pool of the available channels. If the channel pool is reduced and the quality of the channel weights is decreased, the sensor nodes expand the pool size. In this way the channel pool size is dynamic. The channel selection and weight assignment is governed through a light-weight exponential weighting algorithm. A similar analytical work is done by Fischer *et al.* (2007). Classical cognitive MAC protocols also learn the temporal spectral occupancy behaviours from other networks and try to adapt themselves accordingly. Due to energy constraints, radios in sensor networks do not provide access to modulation hardware components and do not have feature modulation detection capabilities in order to identify the types of the interferers. Little can be inferred about the type of the interferer based on just energy detection, which is the only metric provided by sensor network radios. Furthermore,

more advanced cognitive techniques based on learning algorithms remain incommodious to sensor networks because of the computational and battery resource limitations.

COEXISTENCE OF MEDIUM ACCESS CONTROL PROTOCOLS

Different MAC protocol designs suit different application scenarios. It is quite likely that there are a large number of sensor networks that will be operating at the same frequency. Due to different application domains and deployments it is also probable that many WSNs will be using different MAC protocols. Hence, there is a need for the protocols to share the medium and coexist in the same wireless spectrum in a symbiotic manner. The interference caused by different devices operating in the same frequency band can drastically affect the performance. In studying the behaviour of devices operating in the same wireless band, besides considering the physical layer parameters of the radio transceiver like transmit power, modulation scheme, receiver sensitivity, operating frequency, it is also important to consider the distances between the devices, operating duty-cycles, frame sizes, traffic loads and the noise floor.

We have conducted extensive studies on the co-existence behaviour of MAC protocols for sensor networks in (Ansari *et al.*, 2007). Our experimental results have shown that the behaviour of MAC protocols is very different when they share the same wireless channel as compared to their performance metrics in isolated environments. This is indeed a practical issue which drastically affects the performance. Most of the MAC protocols are designed and evaluated in isolated environments and without interference considerations.

We implemented nanoMAC (Ansari *et al.*, 2007), which is an IEEE 802.11 inspired scheduled contention based WSN MAC protocol on TelosB platform. We conducted experiments on the coexistence of nanoMAC and B-MAC implemented on the TelosB platform operating in the same IEEE 802.15.4 channel with the same transmit power and the clear channel assessment thresholds. In order to have a fair comparison, we placed all the sensor nodes equidistantly in a perfect wireless range of each other. We also made sure that there is no external interference in the channel. Enough traffic was generated from both MAC protocols to saturate the channel. It was found that nanoMAC was unable to transmit virtually anything, while B-MAC captured the channel all the time. This was because of the much stronger clear channel assessment algorithm of nanoMAC, which requires making the decision for the presence of carrier in the channel on a greater number of CCA samples. Also, nanoMAC is a non-persistent CSMA/CA based protocol while B-MAC transmits as soon as it senses the channel free. Once the CCA algorithms for the two protocols were made the same and nanoMAC's persistency was set to be 100%, we observed packet transmissions from nanoMAC as well. This suggests that in the future when many MAC solutions exist in the same wireless channel, features for symbiotic coexistence like persistency, back-offs, transmit powers and maximum radio channel occupancy rules need to be introduced.

MULTI-RADIO MEDIUM ACCESS CONTROL PROTOCOLS

Traditionally wireless sensor nodes have only one radio interface. It limits the capability of a sensor node by restricting the transmission ranges, operating frequency, data rates, power consumption etc. Chen *et al.* (2006) discuss the issues of heterogeneity and mobility in sensor networks when using multiple radios on a single sensor node. Multiple radios on a single sensor node platform also enable simultaneous data transmission and reception, thereby leading to high data rates and low latencies as realized in (Kohvakka *et al.*, 2006). This high performance multi-radio WSN platform uses four identical radio chips, which can be switched on and off according to the changing requirements of the application. The sensor node can also be configured with different levels of capabilities depending on the role in the network such as source, sink, etc. Apart from using identical radios, sensor node platforms can be realized by using radios operating in different frequency bands, thereby exploiting the characteristics of different radios and the advantages of different frequency bands. MAC protocol designs using multiple radio frequency bands or channels are not novel in WSN research community. Both the PicoRadio project (Otis, 2005) and STEM (Schurgers *et al.*, 2002) advocate the idea of using a second channel/frequency band for control messages. Oppermann *et al.* (2004) show that per bit energy required for high data rate radios is lower than low data rate radio. For instance, the power consumption per bit of the IEEE 802.15.4 compliant radios providing 250 kbps data rates are much lower than RF Monolithics, Inc.'s TR1000 radio chips, operating in 916.50 MHz, providing data rates of 115.2 kbps/19.2 kbps. Based on this idea, we have built a prototype (cf. Figure 6) us-

Figure 6. Prototype of hardware platform

ing TelosB sensor node platform with an external CC1000 radio chip interfaced as the second radio transceiver (Ansari *et al.*, 2008a).

CC1000 radio operates in the 433 MHz frequency band, which is used only by a few common devices such as garage door openers and weather stations operate with a low radio usage. This frequency band is used to handle traffic on the control channel and is able to support low

Table 2. Operation break-down of energy consumption

Operation Break-down	Energy Cost [µJ]
CCA	200
Turning on CC1000 to Tx Mode	109
Turning on CC1000 to Rx Mode	65.2
Truning off CC1000	6.22
Turning on CC2420 to Tx Mode	15.6
Turning on CC2420 to Rx Mode	15.7
Turning off CC2420	0.313
Operation Break-down	Power Cost [mW]
CC1000 in Transmission Mode	31
CC1000 in Reception Mode	25
CC2420 in Transmission Mode	52.4
CC2420 in Reception Mode	55.4

power channel polling. The CC2420 radio on TelosB operates in the crowded 2.4 GHz band and offers a higher data rate as compared to CC1000 radio. It is used only for burst data packets exchange. Since the CC1000 radio draws much less current in the radio communication than the CC2420 radio, the dual radio platform can offer an energy saving solution by lowering down the energy consumption on preamble sampling and control overhead. At the same time, it is able to provide bursty data transmission with low latency. Furthermore, spectrum agility is achieved through CC1000 serving as the out-of-band channel for establishing an agreement on the CC2420 data channel. Table 2 shows the energy and power consumption break down for the prototypic board operating with a supply voltage of 3V. It can easily be observed that CC1000 radio is better for performing low-power-listening while CC2420 with very fast data rates is effectively better for bursty data transmission/reception.

We carried out comparative studies of the advantages of multi-radio platforms and a suitable preamble sampling Multi-Radio Medium Access Control (MR-MAC) protocol running on it against the B-MAC protocol (Ansari *et al.* 2008). B-MAC

Figure 7. Power consumption performance of transmitters and receivers running MR-MAC, B-MAC on MICA2 and TelosB at different duty cycle (adapted from Ansari et al.,2008a; © 2008 IEEE)

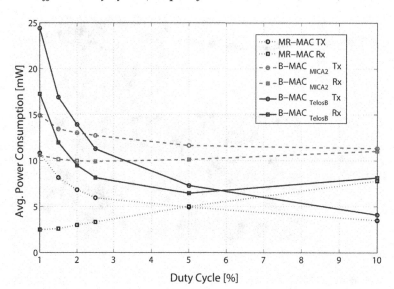

WAKE-UP RADIO BASED SOLUTIONS

was run on two platforms where TelosB platform uses CC2420 radio chip and MICA2 platform uses CC1000 radio chip. We run a sample application transmitting 100 bytes every second in one transmitter and one receiver configuration. The power consumption performance of transmitters and receivers running MR-MAC, B-MAC on MICA2 and TelosB at different duty cycles was measured as shown in the Figure 7. The transmission rate is chosen to be 1 Hz and data size to be 100 bytes. It can be seen that MR-MAC achieves higher energy efficiency than B-MAC running on both the platforms at lower duty cycles.

One of the challenging issues with multiple radios is to set the transmit power levels appropriately especially when the radios are operating in different frequency bands. Ideally, the transmit power levels of the two radios should be such that for the desired packet error rates given by the application requirement, the two radios achieve the same transmission range. Loosely speaking the radio range of the data-channel must not be shorter than that of the control channel.

Sensor network applications are typically characterized by long idle durations and intermittent traffic patterns. The traffic loads are so low that overall idle duration energy consumption starts to dominate. Low duty cycle MAC protocols are used to reduce the idle duration energy consumption. However, lowering down the duty cycle in favour of energy consumption leads to higher latency, which makes it unsuitable for many applications. Wake-up solutions are very effective to address this problem. Gu and Stankovic (2005) proposed to use radio triggered wake-ups for WSNs in order to avoid idle listening during the duty cycling operation and to provide instant wake-ups in response to radio transmission. Similar to the approach of RFID technology, their idea is to use only passive components in order to collect energy from ongoing radio transmissions. When the induced power at the receiving antenna is large enough, the wake-up circuit interrupts the microcontroller attached to the circuit. Upon a trigger, the microcontroller turns on the on-board radio chip and the sensor node starts receiving a

Figure 8. Wake-up receiver block diagram

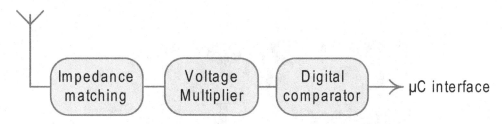

packet over the radio. Since this solution does not employ any kind of node addressing scheme, radio triggers lead to turning-on all the nodes within the transmission range and therefore, energy is wasted for the non-addressed nodes. Additionally, any interfering signal of high enough strength in the operating frequency band can cause undesired node wake-ups. Sinusoidal tone detector circuits consume very low energy and only cause wake-ups upon detecting sinusoid of a particular frequency. Sinusoidal tones of different frequencies can also be used for the addressing purposes. However, this scheme complicates the transmitter and receiver circuitry. We have designed a low cost Radio Triggered Wake-up circuit with Addressing Capabilities (RTWAC) (Ansari *et al.*, 2008) attached to a sensor node that allows the on-board radio module to be switched off unless required for data communication. An out-of-band modulated signal is transmitted and then received by the radio triggered wake-up circuit to interrupt the microcontroller from sleep mode to active mode. The microcontroller interprets the destination address and command messages which are transmitted over the modulated wake-up signal. This allows the non-addressed sensor node to switch back to sleep instantly. The wake-up addresses are shared with the addresses used by the MAC protocol running on the on-board radio of the sensor node. The command message field in the wake-up packet is used for specifying different types of tasks to be executed like turning on the radio for data communication, starting sensor data collection, logging the data to the EEPROM etc. In many applications, indicating the execution of

different tasks using the wake-up circuit is more energy saving than sending the same short command message over the communication radio. For the purpose of data communication between the nodes, a sophisticated CSMA/CA based MAC protocol is used, which avoids collision of packets and effectively handles congestion problems. The wake-up circuit essentially consists of three main components as shown in Figure 8: antenna front end, Voltage Multiplier (VM) and a threshold detector. The amount of power gathered at the output of the antenna is very small. The induced voltage which alternates at the radio frequency is not high enough to interrupt the digital logic of the microcontroller. Our prototype circuit board, working at 868 MHz, uses a five staged VM for 10 times amplification. VM is also, is used to detect the slowly varying envelope signal from the modulated high frequency carrier. The diodes in VM operate at high frequencies and can be turned on at very low forward voltages. In our prototype, we used low threshold RF Schottky diodes HSMS-2852 (Avago Technologies, 2008).

We attached this circuit to the MSP430 series microcontroller on the TelosB sensor node platform (c.f. Figure 9) which operates in Low Power Mode 3 (LPM3). Upon receiving an RF signal, the microcontroller switches from LPM3 to LPM1 and quickly interprets the wake-up signal. One of the major practical issues regarding radio triggered wake-ups is the operating range, which depends upon the induced power at the antenna. The voltage level at the output of the VM circuit is compared against a certain pre-defined noise level threshold. If the level is

Figure 9. A hybrid node consisting of an external wake-up circuit interfaced to a TelosB sensor node platform

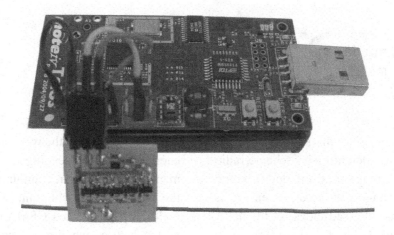

higher than the threshold, the microcontroller is triggered to interpret the data. If the threshold is chosen to be very high, the range is reduced and vice versa. False triggers are more likely to be invoked if the threshold is low. We noticed that in 868 MHz, the noise level threshold value has to be high because of the out-of-band leakages from the adjacent GSM band. With empirical adjustments of the threshold, we were able to obtain a maximum range of approximately 10m. Another issue that we encountered was the polarization effect of the half-wave dipole antennae that we used in our design. This can be solved if one of the antennae (preferably of the radio triggering wake-up transmitter) is circularly polarized.

Table 3. Criteria for the choice of operating frequency

Frequency range	Communication distance	Antenna size	Hardware circuit complexity and size
100-135 kHz	-	-	++
13.56 MHz	-	-	+
433 MHz	+	-	+
868 MHz	+	+-	+
2.4 GHz	+-	+	+

We also investigated the possible frequency ranges that best suit radio triggered wake-ups. Our main requirements that strongly influenced the choice of frequency were:

- Communication range
- Size of antenna and hardware circuit
- Availability of the appropriate electronic components
- License free Industrial, Scientific and Medical (ISM) frequency bands

For comparison, we chose the following ISM frequency ranges - 13.56 MHz, 433 MHz, 868MHz, 2.4 GHz and one RFID specific range of 100–135 kHz. Table 3 summarizes the important parameters based on the above mentioned criteria:

We conducted a comparative study of RTWAC with a classical low-power CSMA/CA (B-MAC) protocol operating at different duty cycles. Figure 10 and Figure 11 suggest that RTWAC has latency comparable to 1% duty cycle and energy consumption of 0.01% duty cycle B-MAC protocol implemented on TelosB. We designed an extremely low power asset tracking system (Ansari *et al.*, 2008c), where we combined the advantages of radio triggered wake-ups with CSMA/CA based duty cycle MAC. RTWAC helps in reducing the idle listening power consumption and minimizing the communication latency, while CSMA/CA MAC is used for reliable and efficient data communication that is not possible on the wake-up radio channel.

CROSS-LAYER DESIGN ASPECTS

Sensor network protocols are generally designed in isolation of each other and in distinctly layered manner. This decoupling of protocols and low degree of interaction ensue to sub-optimal performance. Violating the strict layering among the protocols by combining common functionalities, exchanging network related information by param-

Figure 10. The average power consumption comparison of RTWAC against a low-power listening MAC protocol at different duty cycles implemented on CC2420 radio (adapted from Ansari et al., 2008; © 2008 IEEE)

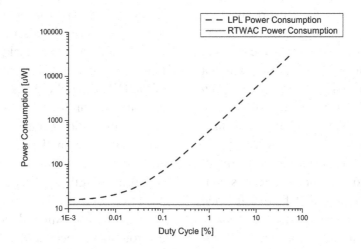

Figure 11. The average latency comparison of RTWAC against a low-power listening MAC protocol implemented on CC2420 radio at different duty cycles radio (adapted from Ansari et al., 2008; © 2008 IEEE)

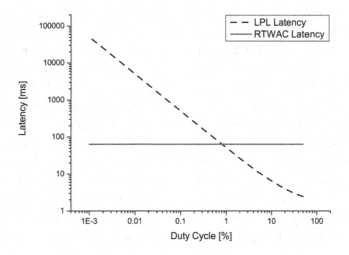

eter passing and suppressing redundancies among protocols can lead to improved performance. Since sensor networks have constraints in energy budget, computational power and communication capabilities, cross-layer design is important to achieve a better performance.

In strictly layered architectures, protocols of different layers may be modified independently which greatly simplifies their implementation. However, due to large overhead caused by maintaining a strict API for message exchange among the layers, layered architecture does not provide good support for flexibility and energy efficiency. Energy consumption can be reduced by allowing some degree of flexibility at the network layer and allowing information sharing across the layers (Su & Lim, 2006). In order to reduce the layering overhead, new schemes and protocol design that closely operate in cross-layer fashion have been proposed (Jurdak, 2007). Available resources are allocated over different layers of the protocol stack at the expense of independence between layers.

Most WSN MAC protocols expose a number of flexible application level parameters as controllable knobs. Knobs consist of MAC internal parameters like duty-cycle values, frame sizes, back-off values, persistency values, clear channel assessment durations, acknowledgements etc., as well as radio parameters like transmit power, receiver sensitivity thresholds, etc. These parameters can be set appropriately in accordance with the application requirements in a flexible manner both at the pre-deployment time and at run-time to achieve better performance. The work (Oldewurtel *et al.,* 2008) presents cross-layer architecture of PHY/MAC and Distributed Source Coding (DSC) for a typical sensor data gathering application at a sink or gateway node. We exploit the spatio-temporal redundancy in the measurements at the sensors and adjust the MAC parameters to achieve low energy consumption. We first identify the most influential parameters of the MAC protocol for energy consumption and then tweak these parameters at the deployment

and run-time according to the application requirements. These parameters include duty cycle values, packet sizes, initial/congestion back-off values and packet acknowledgments. The duty cycle of the operation is selected according to the sensory data sampling rate determined by the application requirements. Scarce data gathering allows the use of lower duty cycle values, which in return saves energy by avoiding frequent periodic wake-ups. In distributed source coding, the code rate selection is based on the degree of correlation in the sensor readings. A higher correlation leads to more compressed data and smaller MAC payload sizes, which leads to better data delivery rates. Also the temporal correlation in the sensor readings can reduce packet retransmissions by estimating the sensor readings of lost packets locally. We have found that lowering the transmit power levels of the CC2420 radio gives only an incremental gain in terms of the overall energy consumption because the radio operates in transmit state for a very small duration. We have found that the overall power consumption performance is improved while satisfying the application requirements using this cross-layer design approach.

We have also explored the cross-layer design of a routing protocol, a MAC protocol and some specifics of the radio layer. Three perspectives are covered in our cross-layer design, 1) Information exchange includes traffic patterns, traffic loads and battery status of the nodes. We also 2) combine commonalities like sleep schedule and neighbourhood information between routing and MAC layers. This eliminates redundant data structures and avoids the need for extra traffic in explicitly gathering the control information. 3) Piggybacking additional battery status information inside the CC2420 radio's auto-acknowledgement packets helps in load balancing across exercising different links in the network. The routing protocol chooses different routes to the sink to prevent overloading and early battery depletion of a particular node. One aspect of our design is the modular approach with well-defined interfaces for parameter passing,

Figure 12. Cross layer design based on TrawMAC, N-Safelinks routing and CC2420 radio model

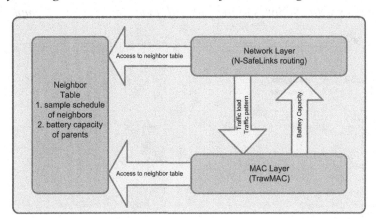

metadata exchange and a distinct modular block for common/combined functionalities. Combining the commonalities among protocols and sharing information across protocol layers elude the need for extra resources spent in gathering the same information explicitly by individual layers. Exposing network related metadata to the networking protocols enable them to make efficient use of the available energy and helps in avoiding extra control overhead. We validated our design using the features of TrawMAC protocol, N-SafeLinks routing protocol (Ikikardes *et al.*, 2007) and CC2420 radio model, as illustrated in Figure 12. N-SafeLinks routing protocol keeps *N* number of definite links to a certain node.

TrawMAC adapts its preamble length to the traffic type and traffic load generated by the multi-hop N-Safelink routing protocol. Also, the MAC protocol gets the battery status of nodes in the acknowledgement packets and helps the routing protocol to perform load balancing. The battery status indication is piggybacked into the five reserved bits of the zero-payload auto-ACK packets of the CC2420 radio. The neighbourhood sleep schedule information of the nodes is used by the MAC to optimize the preamble length. The neighbourhood information is updated on-the-fly as part of the MAC preamble frames and there is no need for sending explicit control frames in the N-safelink routing.

We conducted simulations for comparative lifetime studies of a small network with our cross-layered design approach to the strictly layered network stack. We considered a network with one child node, two/three parent nodes (representing the 2 and 3 safe links, respectively) and a sink node. Figure 13 shows that the percentage of the increased lifetime becomes higher when the traffic load increases. An increasing percentage is observed when more parent nodes are associated to a child node. This is due to the fact that the load balancing scheme fairly distributes the traffic from child to its parent nodes.

PORTABLE MEDIUM ACCESS CONTROL DESIGN

MAC protocols for WSNs are generally implemented in a monolithic fashion on a particular sensor node platform. Monolithic implementation restricts the flexibility of porting protocols from one platform to another and thus the usage of a particular MAC implementation on different platforms for different applications. A very generic implementation of the MAC protocol allows code reusability but limits the possibility of exploiting all the radio chip specific features supported in the silicon. Flexible MAC development with a uniform set of APIs is one of the

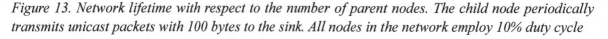

Figure 13. Network lifetime with respect to the number of parent nodes. The child node periodically transmits unicast packets with 100 bytes to the sink. All nodes in the network employ 10% duty cycle

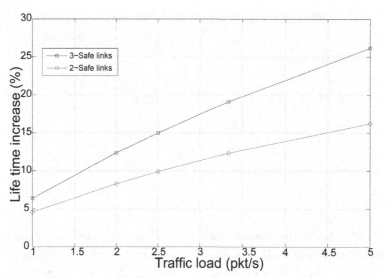

desirable features for easy development and fast porting. In principle, a MAC design can be independent of the underlying hardware (radio transceiver and the microcontroller). However, it is generally more energy efficient to exploit hardware supported features than implementing the code blocks in software on a general purpose processor. Therefore, the efficiency of a particular WSN MAC implementation relies to a good extent on the capabilities and features of the underline radio transceiver and microcontroller. The nature of MAC protocols also has a direct impact on the ease of implementation and portability onto other platforms. TDMA type of MAC schemes require more precise interaction with timers, and in the absence of appropriate primitives in the system software, protocol implementers may resort to platform specific solutions. Furthermore, in many MAC implementations, there is a high degree of dependency on the functionalities offered by the radio chip. Klues *et al.* (Klues *et al.* 2007) advocate a component oriented Medium access control Layer Architecture (MLA). They have ported different MAC protocols using MLA in TinyOS 2.0.1 and have found out significant code reusability and were able to obtain comparative performance metrics as those of monolithic implementations. In line to the same philosophy of code reusability and furthermore ease of portability, we describe a scheme which is applicable to virtually all the sensor node hardware platforms and WSN operating systems.

In the following, we categorize the dependencies from the hardware and software perspective. Hardware dependencies include the diverse nature of radio transceivers and the microcontrollers. The radio parameters include factors such as operating frequencies, supported channels, transmission powers, modulation schemes as well as the type of checksums, address recognition features, maximum allowable size of the radio packet etc. Sensor network class microcontrollers like Texas Instruments' MSP430, Atmel's ATMega128L, NXP Semiconductors' CoolFlux etc. have different timer engines, register lengths, interrupt nesting capabilities and interrupt prioritization support. Apart from the hardware inconsistencies, the operating systems have also different architecture and behaviour. Certain operating systems are single-threaded and the whole MAC protocol needs to be implemented in the same thread. In general, multi-threaded operating systems provide more control on the program execution, hardware usage and timings but they also add overhead in terms of memory and execution latency. In multi-threaded operating system environment, most medium access operations can be executed outside of interrupt services routines, in a micro-kernel task with the highest priority. This approach integrates the MAC into the system as a single component. Control over timings is also facilitated; processes can be initiated on timing deadlines. Hardware control or drivers play a key role in a MAC protocol implementation and these include features like:

- **Memory management for packet handling:** Buffers are needed by the MAC for its internal operations exclusively as well as by data passing to the upper layers.
- **Interrupt control for MAC events:** MAC protocols use interrupts for timing as well as for detecting events such as the reception of a packet. These interrupts trigger code execution e.g. time synchronization algorithms based on the time stamps made on packet reception at the MAC layer. These codes are usually small and fast to be executed. MAC protocols return to a state where new events can be detected as soon as possible. MAC events are used for controlling MAC code execution.
- **Inter-process communication:** MAC processes may need to interact, e.g. one process starts to operate when informed by another process that the node is synchronized. Such interaction can be made either via global variables which have different values according to the current operation status, or via semaphores, which block code execution until another process releases the blocking semaphore.
- **Radio chip control:** MAC protocols need access to radio chip functionalities for control on MAC functions and physical attributes. Switching radio status, setting radio

transmission power and enabling acknowledgments are examples of such control. The whole chain comprises the radio driver software necessary to control radio attributes as well as the handling of the physical interface to transfer bytes between the microprocessor and the radio.

In the following, we define the top level interface of a MAC protocol presenting the MAC functionality to the networking layer and also to the application. It hides all the underlying MAC internals. For most of the MAC protocols, the top level APIs can essentially be categorized in the following functionalities:

- **Packet Sending Interface:** The network layer requires being able to send the assigned payload to the specified destination.
- **Packet Receiving Interface:** The network layer requires MAC to notify the reception of data and deliver the payload received.
- **Configuration Interface:** The configuration interface is meant to set and correspondingly obtain the values of the various parameters of the MAC protocols and the underlying radio layer. For instance, these may include the radio channel to be used, selection of the transmit power levels etc. It also has functionalities like switching on/off the radio stack, setting the duty cycle and sampling periods, enabling/disabling the acknowledgements, setting the values for the initial and congestion back-offs, setting the persistency value and getting metadata like the signal strength values, time stamp etc. of the received packet.

We highlighted the importance of portability aspects of MAC protocols, the issues associated with the portability and implementation owing to the diversified nature of the hardware platform and the operating system capabilities. We also categorically identified the key functionalities to be provided by the hardware in order to implement a MAC protocol in a portable manner. Furthermore, we defined the three basic interface abstractions for implementing a portable MAC protocol on much diversified hardware/software platforms without compromising the performance of platform specific implementations.

CONCLUSION

In this chapter we discussed the requirements and practical design issues associated with WSN MAC protocols. In particular, we focused on the issue of power conservation in MAC designs. Following a short overview of different approaches and research trends for MAC protocols, we gave a detailed description of our experiences on experimentations with various MAC designs and their real senor node testbed implementations. We described a traffic aware MAC protocol, which is capable of handling variable traffic loads in an energy efficient manner, including discussions on various optimizations on traffic type based preambles. The issue of spectral agility and cognition in sensor network protocols was highlighted since spectrum agility has pragmatic importance due to limited spectral resources. A spectral agile MAC solution was presented as an example to handle congested spectrum. We shortly described our empirical results on the coexistence of different MAC protocols in the same spectrum. Based on our experiments, a simple scheme for symbiotic coexistence of the protocols was outlined. The design of a multi-radio based MAC protocol was presented; it uses radios operating in high and low frequency bands in order to achieve remarkable energy savings as compared to single radio based MAC solutions. We provided the design details of our extremely low power radio triggered wake-up solution with addressing capabilities. We have combined the radio triggered wake-ups with a duty cycling MAC in an effort to achieve the best out

of the two approaches. Radio triggered wake-up gives high energy savings in idle listening and allows instant wake-ups for data communication, while the usual CSMA/CA based MAC has efficient channel access properties. We presented the cross-layer design aspects of MAC protocols in sensor networks for achieving power consumption optimizations. The chapter also discussed a cross-layer optimization experiment which set the operating parameters of distributed source coding and MAC protocol for power consumption gains. In addition, a MAC-routing cross-layer design for enhancing the lifetime of the sensor network was described. Finally, we outlined a portable MAC design architecture, which facilitates implementation and eases code migration from one platform to another.

REFERENCES

Ali, M., Saif, U., Dunkels, A., Voigt, T., Römer, K., & Langendoen, K. (2006). Medium access control issues in sensor networks. *SIGCOMM Computer Communication Review, 36*(2), 33–36. doi:10.1145/1129582.1129592

Alsharabi, N., Fa, L. R., Zing, F., & Ghurab, M. (2008). Wireless sensor networks of battlefields hotspot: Challenges and solutions In *Proceedings from WiOPT'08: The 6th International Symposium on Modeling and Optimization in Mobile, Ad Hoc, and Wireless Networks and Workshops*, Berlin, Germany.

Ansari, J., Ang, T., & Mähönen, P. (2009*)*. Spectrum Agile Medium Access Control Protocol for Wireless Sensor Networks. *Demonstrated at ACM SIGCOMM*. Barcelona, Spain.

Ansari, J., Pankin, D., & Mähönen, P. (2008). Radio-triggered wake-ups with addressing capabilities for extremely low power sensor network applications. In *Proceedings from PIMRC'08: The 19th Personal, Indoor and Mobile Radio Communications Symposium*, Cannes, France.

Ansari, J., Pankin, D., & Mähönen, P. (2008c). Demo abstract: Radio-triggered wake-ups with addressing capabilities for extremely low power sensor network applications. Presented at EWSN'08: *The 5th European Conference on Wireless Sensor Networks,* Bologna, Italy.

Ansari, J., Riihijärvi, J., Mähönen, P., & Haapola, J. (2007). Implementation and Performance Evaluation of nanoMAC: A Low-Power MAC Solution for High Density Wireless Sensor Networks. *Int. Journal of Sensor Networks, 2*(5), 341–349. doi:10.1504/IJSNET.2007.014361

Ansari, J., Zhang, X., & Mähönen, P. (2007a). Demo abstract: Multi-radio medium access control protocol for wireless sensor networks. In *Proceedings from SenSys'07: The 5th ACM Conference on Embedded Networked Sensor Systems*, Sydney, Austrailia.

Ansari, J., Zhang, X., & Mähönen, P. (2008a). Multi-radio Medium Access Control Protocol for Wireless Sensor Networks. In *Proceedings from WEWSN'08: The 1st International Workshop on Energy in Wireless Sensor Networks*, Santorini, Greece.

Ansari, J., Zhang, X., & Mähönen, P. (2008b*).* Poster Abstract: Traffic Aware Medium Access Control Protocol for Wireless Sensor Networks. *Presented at EuroSys'08.* Glasgow, Scotland.

Avago Technologies. *(n.d.). HSMS-285x-series surface mount zero bias schottky detector diodes data sheet.* Retrieved December 02, 2008, from, http://www.avagotech.com

Avvenuti, M., Corsini, P., Masci, P., & Vecchio, A. (2006). Increasing the Efficiency of Preamble Sampling Protocols for Wireless Sensor Networks. In *Proceedings from IWCMC'05: The First Mobile Computing and Wireless Communication International Conference*, Amman, Jordan.

Bachir, A., Barthel, D., Heusse, M., & Duda, A. (2006). Micro-frame preamble MAC for multihop wireless sensor networks. In *Proceedings from ICC'06: The IEEE International Conference on Communications 2006*, Istanbul, Turkey.

Buettner, M., Yee, G., Anderson, E., & Han, R. (2006). X-MAC: A short preamble MAC protocol for duty-cycled wireless networks. In *Proceedings from Sensys'06: The 4th ACM Conference on Embedded Networked Sensor Systems*, Boulder, CO.

Chen, C., & Ma, J. (2006). MEMOSEN: Multi-radio Enabled MObile Wireless SEnsor Network. In *Proceedings from AINA'06: The IEEE 20th International Conference on Advanced Information Networking and Applications*, Vienna, Austria.

Chen, W., Chen, L., Chen, Z., & Tu, S. (2005). A Realtime Dynamic Traffic Control System Based on Wireless Sensor Network. In *Proceedings from ICCP'05: The 34th International Conference on Parallel Processing Workshops*, Washington D.C.

Dam, T. V., & Langendoen, K. (2003). An adaptive energy-efficient MAC protocol for wireless sensor networks. In *Proceedings from Sensys'03: The 1st International Conference on Embedded networked sensor systems,* Los Angeles.

Demirkol, I., Ersoy, C., & Alagöz, F. (2006). MAC Protocols for Wireless Sensor Networks: A Survey. In *IEEE Communications Magazine*, April.

El-Hoiydi, A., Decotignie, J. D., Enz, C., & Roux, E. L. (2003). Poster abstract: WiseMAC, an ultra low power MAC protocol for the WiseNET wireless sensor network. In *Proceedings from Sensys'03: The 1st International Conference on Embedded networked sensor systems*, Los Angeles, CA, USA.

Fischer, S., Petrova, M., Mähönen, P., & Vöcking, B. (2007) Distributed Load Balancing Algorithm for Adaptive Channel Allocation for Cognitive Radios. In *Proceedings from CrownCom'07: The Second International Conference on Cognitive Radio Oriented Wireless Networks and Communications*, Orlando, USA.

Gong, H., Liu, M., Mao, Y., Chen, L., & Xie, L. (2005) Traffic adaptive MAC protocol for wireless sensor network. In *Proceedings from ICCNMC'05: The 2005 International Conference on Computer Networks and Mobile Computing*, Zhangjiajie, China.

Gu, L., & Stankovic, J. (2005). Radio-Triggered Wake-Up for Wireless Sensor Networks. *Real-Time Systems, 29*, 157–182. doi:10.1007/s11241-005-6883-z

Ikikardes, T., Hofbauer, M., Kaelin, A., & May, M. (2007). A robust, responsive, distributed tree-based routing algorithm guaranteeing n valid links per node in wireless ad-hoc networks. In *Proceedings from ISCC'07: The 2007 IEEE International Symposium on Computers and Communications*. Aveiro, Portugal.

Jurdak, R. (2007). *Wireless ad hoc and sensor networks: A cross-layer design perspective.* New York: *Springer Publishers.*

Kim, S., Pakzad, S., Culler, D., Demmel, J., Fenves, G., Glaser, S., & Turon, M. (2007). Health Monitoring of Civil Infrastructures Using Wireless Sensor Networks. In *Proceedings from IPSN'07: The 6th international conference on Information processing in sensor networks*, New York.

Klues, K., Hackmann, G., Chipara, O., & Lu, C. (2007). A component based architecture for power-efficient media access control in wireless sensor networks. In *Proceedings from Sensys'07: The 5th international conference on Embedded networked sensor systems,* Sydney, Austrailia.

Kohvakka, M., Arpinen, T., Hännikäinen, M., & Hämäläinen, T. D. (2006). High-performance multi-radio wireless sensor networks platform. In *Proceedings from REALMAN'06: The 2nd international workshop on Multi-hop ad hoc networks: from theory to reality,* New York.

Langendoen, K. (2007). Medium Access Control in Wireless Networks. *Volume II: Practice and Standards,* 1st ed. New York: Nova Science Publishers.

Mozer, M. C. (2004). *Lessons from an Adaptive House In Smart environments: Technologies, protocols, and applications.* Hoboken, NJ: J. Wiley & Sons.

Oldewurtel, F., Ansari, J., & Mähönen, P. (2008). Cross-layer design for distributed source coding in wireless sensor networks. In *Proceedings from SENSORCOMM'08 The Second International Conference on Sensor Technologies and Applications,* Cap Esterel, France.

Oppermann, I., Stoica, L., Rabbachin, A., Shelby, Z., & Haapola, J. (2004). UWB wireless sensor networks: UWEN-a practical example . *IEEE Communications Magazine, 42,* 527–532. doi:10.1109/MCOM.2004.1367555

Otis, B. (2005). *Ultra-low power wireless technologies for sensor networks.* Ph.D. dissertation, University of California, Berkeley, USA.

Petrova, M., Wu, L., Mähönen, P., & Riihijärvi, J. (2007). Interference measurements on performance degradation between colocated IEEE 802.11g/n and IEEE 802.15.4 networks. In *Proceedings from ICN'07: The 6th International Conference on Networks,* Sainte-Luce, Martinique.

Polastre, J. (2003). *Sensor Network Media Access Design,* (Technical Report*),* University of California, Berkeley, USA.

Polastre, J., Hill, J., & Culler, D. (2004). Versatile low power media access for wireless sensor networks. In *Proceedings from Sensys'04: The second international conference on Embedded networked sensor systems*, Baltimore, USA.

Rajendran, V., Obraczka, K., & Garcia-Luna-Aceves, J. (2003). Energy efficient, collision-free medium access control for wireless sensor networks. In *Proceedings from Sensys'03: The first international conference on Embedded networked sensor systems*, Los Angeles, CA.

Rhee, I., Warrier, A., Aia, M., & Min, J. (2005). Z-MAC: a hybrid MAC for wireless sensor networks, In *Proceedings from Sensys'05: The third international conference on Embedded networked sensor systems*, San Diego, USA.

Ringwald, M., & Römer, K. (2005). BitMAC: A Deterministic, Collision-Free, and Robust MAC Protocol for Sensor Networks. In *Proceedings from EWSN'05: The Second IEEE European Workshop on Wireless Sensor Networks,* Istanbul, Turkey.

Ringwald, M., & Römer, K. (2008). Poster abstract: BurstMAC low idle overhead and high throughput in one MAC protocol. Presented at EWSN'08: *The 5th European Confererence on Wireless Sensor Networks*, Bologna, Italy.

Ruzzelli, A. G., O'Hare, G., Jurdak, R., & Tynan, R. (2006). Advantages of Dual Channel {MAC} for Wireless Sensor Networks. In *Proceedings from COMSWARE'06: The First International Conference on COMmunication System softWAre and MiddlewaRE*, New Delhi, India.

Schurgers, C., Tsiatsis, V., Ganeriwal, S., & Srivastava, M. (2002). Optimizing sensor networks in the energy-latency-density design space. *IEEE Transactions on Mobile Computing*, 70–80. doi:10.1109/TMC.2002.1011060

Su, W., & Lim, T. L. (2006). Cross-layer design and optimization for wireless sensor networks. In *Proceedings from SNPD-SAWN'06: The 7th ACIS Intl. Conference on Software Engineering, Artificial Intelligence, Networking, and Parallel/Distributed Computing*, Washington, USA.

Szewczyk, R., Mainwaring, A., Polastre, J., Anderson, J., & Culler, D. (2004). An Analysis of a Large Scale Habitat Monitoring Application. In *Proceedings from Sensys'04: The 2nd international conference on Embedded networked sensor systems*, Baltimore, MD.

Tannenbaum, A. (2003). *Computer Networks PA*. Upper Saddle River, New Jersey: Prentice Hall.

Texas Instruments. (2008) *CC1000 data sheet*. Retrieved December 11, 2008, from http://focus.ti.com/lit/ds/symlink/cc1000.pdf van Hoesel, L., & Havinga, P. (2004). A lightweight medium access protocol (LMAC) for wireless sensor networks. In *Proceedings from INSS'04: The First International Workshop on Networked Sensing Systems*, Tokyo, Japan.

Texas Instruments. (2008a) *CC2420 data sheet.*, Retrieved December 11, 2008, from http://focus.ti.com/lit/ds/symlink/cc2420.pdf

Ye, W., Heidemann, J., & Estrin, D. (2002). An Energy-Efficient MAC protocol for Wireless Sensor Networks. In *Proceedings from Infocom'02: The 21st Conference on Computer Communications*, New York.

Zhou, G., Stankovic, J., & Son, S. H. (2006). Crowded Spectrum in Wireless Sensor Networks. In *Proceedings from EmNetS'06: The third Workshop on Embedded Networked Sensors*, Cambridge, MA.

KEY TERMS AND DEFINITIONS

Carrier Sense Multiple Access (CSMA): CSMA refers to the mechanism of a node to find the channel free from any other on going transmission before attempting to transmit a packet. Simultaneous transmission of packets by different nodes can lead to packet collisions, which makes it incomprehensible at the receiving node(s).

Clear Channel Assessment (CCA): Clear channel assessment refers to the method of finding the state of the channel, which is suitable for data transmission.

Contention based Channel Access: In contention based channel access, all the nodes compete with each other for accessing the channel in order to transmit a packet.

Cross-layer Optimization: Cross-layer optimization in networks refers to obtaining enhanced network performance by exploiting the information across layers of the networking stack. Strict layered approach does not allow exchanging certain parameters and metadata that favour system performance.

Low Power Listening (LPL): Low power listening refers to the operation in which nodes periodically turn on and off the radios.

Medium Access Control (MAC): Medium Access Control refers to the procedure for accessing the shared the medium (wired/wireless) for data communication.

Preamble Sampling/Channel Polling Protocols: Preamble sampling or channel polling protocols refer to a set of CSMA/CA based protocols where all the nodes wake-up asynchronously, sniff for the activity in the medium and go back to sleep. A long preamble is transmitted before the data so that all the waking-up nodes notice an upcoming data packet.

Time Division Multiple Access (TDMA): Time Division Multiple Access is a channel access scheme with time slices or time slots for each of node in the neighbourhood. Each node receives data in its assigned time slot.

Chapter 8
Transmission Control Protocols for Wireless Sensor Networks

Yunlu Liu
Beihang University, China

Weiwei Fang
Beihang University, China

Zhang Xiong
Beihang University, China

ABSTRACT

Transmission control is an important issue for supporting Quality of Services (QoS) in Wireless Sensor Networks (WSNs). This chapter gives an introduction to the transmission control relevant protocols and algorithms, often developed for particular applications to reflect the application-specific requirements. According to the functionality, the existing works can be categorized into three main types: congestion control, reliability guarantee and fairness guarantee. Due to the unique constraints of WSNs, these transmission control protocols are not "cleanly" placed on top of the network layer, but call for careful cross-layer design. In the end, some exciting open issues are presented in order to stimulate more research interest in this largely unexplored area.

INTRODUCTION

The transmission control protocol plays an important role in achieving transport reliability and guaranteeing information fidelity. Given that an interesting event has indeed been detected, the relevant information has to be reported reliably from the event spot to the aggregating sinks, which are often located several hops away. Meanwhile, users may often want to disseminate task settings or query commands through the sink node, which also requires reliable data delivery from sinks to individual or groups of sensor nodes. Failure to deliver important data in the network can hurt the applications' objective, and even defeat the very reason why the sensor network was deployed.

In the Internet, both the network layer and the transport layer play an important role in achieving data transport reliability. The network layer offers a best-effort service to provide available routing paths among network nodes, while the transport layer is responsible for achieving reliable delivery, conges-

DOI: 10.4018/978-1-61520-701-5.ch008

Figure 1. Transmission control protocols: A taxonomy

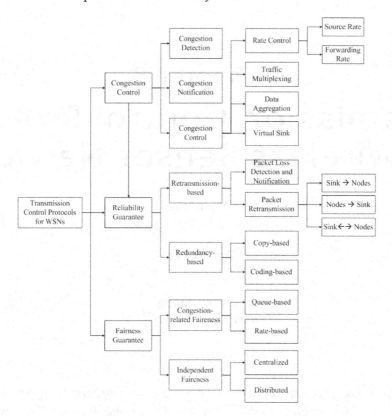

tion alleviation and fairness guarantee. Unlike in the Internet, transmission reliability in the sensor network can't be provided predominantly by transport layer mechanisms of end systems, such as TCP and UDP. Rather, the inherent tradeoffs between transmission reliability, QoS provision and resource constraints require the adoption of cross-layer solutions.

To address the transmission control issues in wireless sensor networks, several protocols have been proposed in the literature. According to the specific functionality, they can be categorized into the following three types:

1. **Congestion Control:** Congestion occurs when offered traffic load exceeds available network capacity. It can not only degrade information fidelity but also waste node energy. Therefore, congestion must be effectively and efficiently controlled, either to avoid it or alleviate it. Generally, there

are three mechanisms that deal with this problem: congestion detection, congestion notification and congestion control.

2. **Reliability Guarantee:** Reliability guarantee requires the ability to detect and repair packet loss in the network. The protocols for dealing with this issue in WSNs can be sorted into two main categories: retransmission-based and redundancy-based.

3. **Fairness Guarantee:** The constraint resources in WSNs are shared with all sensor nodes. The transmission control protocols need to provide fair resource allocation among all sensor nodes according to the specific fairness policy, so that the sinks can obtain a fair amount of observation data from the whole network. In general, the fairness can be implemented as an integrated part of the congestion control protocol or as an independent control protocol.

Currently, there is no single solution covering all the above design points. However, relevant solutions have been developed for a single point or small point sets. A detailed taxonomy of the transmission control paradigms in WSNs is shown in *Figure 1*.

BACKGROUND

Transmission control protocols in WSNs are confronting some unique environments and problems. There are some factors that negatively impact the data transmission in WSNs:

1. The wireless link is an open and lossy media, in which the transmission error rate is relatively higher than that in the wired network, due to path loss via distance attenuation and shadowing by obstacles;
2. The packet reception at the receiving node can be interfered by the other nodes that transmit simultaneously in the same area;
3. The routing paths may become unavailable due to death, movement or energy exhaustion of nodes;
4. The node buffer for the packets to be transmitted will overflow when the traffic load exceeds the channel capacity;
5. The transmission control has to comply with the stringent resource constraints in energy, memory and communication of sensor nodes.

An option would be to use a protocol within the traditional network directly as the approach providing reliable and efficient communication for WSNs. However, the traditional transmission control protocols that are designed for the Internet, i.e., UDP and TCP, can not be directly applied to WSNs. On the one hand, UDP is a best-effort service without any consideration on reliable transmission. One the other hand, there exist some mismatch points between TCP and most of sensor network paradigms:

1. TCP follows an end-to-end design philosophy. The end systems are responsible for the task of transmission control, while the intermediate nodes only receive and forward the packets. In WSNs, the intermediate nodes may perform in-network processing or data aggregation to multiple received packets. Such operations are deemed as packet loss by TCP, which will in turn start the retransmission immediately;
2. The handshake mechanism for setting up a TCP connection is too complex and time-consuming for WSNs, therefore is not suitable for the real-time applications. Besides, maintaining the connection state is not easy in WSNs due to dynamic topology changes;
3. TCP strives for perfect reliability and can't tolerate any packet losses. However, the WSN applications in general only require a certain number of packets collected by the sink to achieve application-specific information fidelity;
4. The retransmission is controlled by acknowledgement and timeout mechanisms in TCP. In WSNs, the length of data packet is commonly very short. The transmission of a large number of ACK packets will incur considerable energy consumption. If a packet is lost, all sensor nodes on its routing path will be invoked for the retransmission from source node to the sink;
5. The packet out-of-order due to packet loss or multipath transmission will induce the wrong response of TCP. The sender will frequently enter the phase of congestion control, which degrades the transmission efficiency;
6. TCP relies on the concept of individually addressable nodes with globally unique addresses or network-wide unique addresses. However, the WSNs are data-centric, in which the nodes collaborate to perform sensing tasks to provide data information to the end user. To save limited energy, sensor nodes can use addresses with shorter length, e.g., locally-unique addresses, geographic addresses or content-based addresses.

PROTOCOLS

Congestion Control

Congestion in WSNs has negative impacts on network performance and application objective, i.e., indiscriminate packet loss, increased packet delay, larger energy consumption and severe fidelity degradation. Therefore, it must be effectively and efficiently controlled, either to avoid it or mitigate it. Typically, there are three methods that can work together to handle this problem: congestion detection, congestion notification, and congestion control (*Figure 1*).

We present comparisons of protocols for congestion control as in Table 1. The notations of the table title are explained as follows: A: protocol; B: the direction of data transmission; C: detection

method; D: feedback method; E: transmission consumption; F: computation consumption; G: extra hardware support; H: reliability measurement; I: control mode.(Fang, Qian, & Liu, 2008)

Congestion Detection and Notification

Accurate and efficient congestion detection plays an important role in congestion control of WSNs. It is a basal proceeding of congestion control and always works as a part of control algorithms. The methods for congestion detection in WSN are listed as follows:

1. **Buffer Occupancy:** (Chen & Yang, 2006; Cheng Tien & Ruzena, 2004; Hull et al., 2004; Sumit et al., 2006; Yogesh et al., 2003) It is a common approach used in traditional

Table 1. Comparison of congestion control protocols

A	B	C	D	E	F	G	H	I
ESRT(Yogesh, zg, r, & Ian, 2003)	Sensor->Sink	Buffer	ICN	Moderate	Low	No	Event-based	Centralized
PORT(Zhou, Lyu, Liu, & Wang, 2005)	Sensor->Sink		ECN	Moderate	Moderate	No	Event-based	Both
IFRC(Sumit, Ramakrishna, Ramesh, & Konstantinos, 2006)	Sensor->Sink	Buffer	ICN	Low	Low	No	Event-based	Distributed
Fusion(Hull, Jamieson, & Balakrishnan, 2004)	Sensor->Sink	Buffer	ICN	Low	Low	No	Event-based	Distributed
CCF(Cheng Tien & Ruzena, 2004)	Sensor->Sink	Buffer	ICN	Low	Moderate	No	Event-based	Distributed
Buffer-based(Chen & Yang, 2006)	Sensor->Sink	Buffer	ICN	Low	Low	No	Event-based	Distributed
PCCP(Chonggang, Chonggang, Sohraby, Lawrence et al., 2006)	Sensor->Sink	Congestion degree	ICN	Moderate	Moderate	No	Event-based	Distributed
CODA(Chieh-Yih, Shane, & Andrew, 2003)	Sensor->Sink	Buffer & Channel	ECN	High	Moderate	No	Event-based	Both
CAR(Kumar R, 2006)	Sensor->Sink			Moderate	Low	No	Packet-based	Both
BGR(Popa, Raiciu, Stoica, & Rosenblum, 2006)	Sensor->Sink	Buffer & Channel	ICN	Low	Low	No	Packet-based	Distributed
CONCERT(Laura, Andrew, & Sergio, 2005)	Sensor->Sink			Low	High	No	Event-based	Distributed
PREI(Edith, Yangfan, Michael, & Jiangchuan, 2006)	Sensor->Sink			Low	High	No	Event-based	Distributed
Siphon(Chieh-Yih, Shane, Andrew, & Jon, 2005)	Sensor->Sink	Buffer & Channel	ECN	Low	Low	Yes	Packet-based	Centralized

networks. It includes several different rules, such as instantaneous buffer state, buffer state forecast and so on. Take the buffer state forecast as an example. As shown in *Figure 2*, where b_k, b_{k-1} are the buffer fullness levels at the time k and k-1, B is the buffer size, Δb is the buffer length increment observed between time k and k-1. Then at the time $k+1$, if $b_k + \Delta b > B$, the sensor node infers that it is going to experience congestion in the next time slot.

$$\Delta b = b_k - b_{k-1}$$

This method is easy to implement, while gaining relatively lower accuracy(Chieh-Yih et al., 2003; Iyer, Gandham, & Venkatesan, 2005).

2. **Channel Loading:** (Chieh-Yih et al., 2003; Chieh-Yih et al., 2005; Popa et al., 2006) It gives accurate information about how busy surrounding network is. However, it has limited effect such as in detecting large-scale congestion caused by data impulses from sparsely located sources that generate high-rate traffic. Besides, listening to the channel activeties all the time will incur excessive energy consumption in the node and thus is not practical for energy-constrained WSNs. Therefore, CODA(Chieh-Yih et al.,

2003) proposes to use a combination of the present and past channel loading conditions, obtained through a low-cost sampling technique, and the current buffer occupancy to infer congestion.

3. **Congestion Degree**(Chonggang,Chonggang, Sohraby, Lawrence et al., 2006) It is defined as the ratio of service time over inter-arrival time, which incorporates information about mean packet inter-arrival (t_a^i) and mean packet service times (t_s^i) at the MAC layer. The congestion degree ($d(i)$) is computed as follows.

$$d(i) = t_s^i / t_a^i$$

After detecting congestion, the congestion information should be propagated from the congested node to the upstream nodes or source nodes or the sink for further control. The methods for disseminating congestion information can be categorized into:

1. **Explicit Congestion Notification:** (ECN) (Chieh-Yih et al., 2003; Chieh-Yih et al., 2005; Zhou et al., 2005) The control messages containing congestion information will be propagated to the involved sensor nodes of congestion. The control message could be given prioritized access to the wireless medium for fast propagation (Hull et al.,

Figure 2. Buffer occupancy in sensor nodes

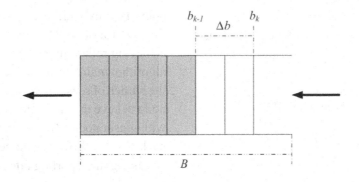

2004). This method inevitably introduces additional overhead and excessive energy consumption.

2. Implicit Congestion Notification (ICN) (Chen & Yang, 2006; Cheng Tien & Ruzena, 2004; Chonggang, Chonggang, Sohraby, Lawrence et al., 2006; Hull et al., 2004; Popa et al., 2006; Sumit et al., 2006; Yogesh et al., 2003): It piggybacks congestion information in the header of data packets. Taking advantage of the broadcast nature of wireless channel, sensor nodes listen to the neighbor node to get congestion information. In this manner, transmission of an additional control message is avoided.

Congestion Control Algorithms

1. Traffic Control

Traffic control is a popular method to control congestion in WSNs. It includes source rate control, forward rate control and the mixture of the two.

A.   Source Rate Control

Y.Sankarasubramaniam et al. propose an Event-to-Sink Reliable Transport algorithm (ESRT)(Yogesh et al., 2003), which is the first study of reliable transport in WSNs from the event-to-sink perspective. The algorithm is designed for event driven WSNs that relies on the collective effort of numerous sensor nodes.

Five characteristic regions are proposed based on the reporting rate of source nodes and good put at sink node as follows:

(NC,LR): $f<f_{max}$ and $\eta \prec 1-\varepsilon$ (No Congestion, Low Reliability)

(NC,HR): $f \leq f_{max}$ and $\leq \eta \succ 1+\varepsilon f_{max}$ and $\eta \succ 1+\varepsilon$ (No Congestion, High Reliability)

(C,HR): $f>f_{max}$ and $\eta \succ 1$ (Congestion, High Reliability)

(C,LR): $f> f_{max}$ and $\eta \leq 1$ (Congestion, Low Reliability)

OOR: $f<f_{max}$ and $1-\varepsilon \leq \eta \leq 1+\varepsilon$ (Optimal Operating Region)

Where N=No, C=Congestion, L=low, H=high, R=Reliability, OOR=Optimal Operating Region, f is reporting rate of source nodes, f_{max} is the upper limit of f beyond which the network reliability drops. η is defined as:

$$\eta = \frac{r}{R}$$

where r is the number of received data packets at the sink and R is the number of data packets required. And in a slot i, the network is in the state $S_i \in \{(NC,LR),(NC,HR),OOR,(C,HR),(C,LR)\}$. The states are dynamically transferred to another one for the goal of the state of OOR via adjusting the reporting rate of source nodes as follows:

$$f_{i+1} = \begin{cases} \dfrac{f_i}{\eta_i} & S_i = (NC,LR) \\ \dfrac{f_i}{2}(1+\dfrac{1}{\eta_i}) & S_i = (NC,HR) \\ f_i & S_i = (OOR) \\ \dfrac{f_i}{\eta_i} & S_i = (C,HR) \\ f_i^{(\eta^i/k)} & S_i = (C,LR) \end{cases}$$

Authors verify the performance of ESRT by both analytical and simulation results. The congestion control of ESRT serves the dual purpose of achieving reliability and conserving energy. The algorithm mainly runs on the sink, which release the burden of sensor nodes but also needs the sink node to have the ability to directly communicate with sensor nodes. However, the contributions of packets from different sources may be different. Adjusting the reporting rates of the sensor nodes in such an undifferentiated manner is not the most energy-efficient way to increase the knowledge of the physical phenomenon(Zhou et al., 2005).

Yangfan Zhou et al. propose a price-oriented reliable transport protocol (called PORT) to deal with the problems in ESRT, such as adjusting the report rates of sources in an undifferentiated manner.

PORT adjusts source-reporting rates based on an optimization approach as follows:

$$minimize \ \sum_{i=1}^{m} (t_i \times p_i)$$

$$s.t. \ u = f(t_1, t_2, ..., t_m) \geq u'$$

Where t_i is the incoming packet rate of source i, p_i is the communication cost of source i, m is the number of the sources, $u = f(t_1, t_2, ..., t_m)$ is the current fidelity obtained and u' is the minimum fidelity required on the phenomenon of interest. The sink should determine the reporting rates of sources so that the energy consumption of the network is minimized, subject to the constraint that the fidelity of the phenomenon knowledge cannot exceed a given tolerable minimum value. The control messages are transmitted to the source nodes hop by hop or via broadcast.

Sumit Rangwala et al. propose an Interference-Aware Fair Rate Control algorithm (called IFRC) (Sumit et al., 2006). IFRC is consisted of three inter-related components: one that measures the level of congestion at a node, another that shares congestion information with potential interferers, and a third that adapts the rate using an Additive Increase Multiplicative Decrease (AIMD) control law.

Each Node shares its congestion state with its potential interferers based on congestion detection via queue occupancy. And it should abide three rules:

1. If it tests congestion locally via a certain upper threshold U, it would inform all its neighbors its condition and also its congested children's

2. It couldn't be with a rate which exceeds that of any congested neighbor or any congested child of the neighbor

3. It couldn't be with a rate which exceeds that of its parent

If a node isn't constrained by any rules above, it adapts its rate additively. Every $1/r$ seconds, its rate is incremented by δ/r, where δ is a parameter set based on per-network condition (it is set to be $(r_{init})^2$)and r is the rate of the node. If the node is congested indicated by the congestion detection component, it halves its current r.

The performance of IFRC is tested on a 40-node real network test-bed. There is no packet loss due to queue overflow in the experiment, while well fairness guarantee is gained.

B. Forward Rate Control

Forward rate control could be implemented locally and adaptively, while being real time relative to source rate control. For the global coordination always used by source rate control requires additional control packets, while being less robust. These problems should be taken seriously in WSNs.

Bret Hull et al. propose a mitigating congestion algorithm (called Fusion)(Hull et al., 2004). Fusion combines hop-by-hop flow control, source rate limiting scheme and prioritized MAC layer together into one strategy.

In hop-by-hop flow control, each sensor sets a congestion bit in the header of every outgoing packet. The congestion detection is done by detecting the output queue falling below a high water mark α (α=0.25). Rate limiting meters traffic being admitted into the network to prevent unfairness toward sources far from a sink using a token bucket scheme. A prioritized MAC ensures that congested nodes have prioritized access right to the channel. If a sensor node is in congestion, its backoff window is one-fourth the size of an uncongested sensor's backoff window.

Authors evaluate the congestion control algorithm in an indoor wireless sensor network test-bed. One of the important contributions of the work is the elaborate implementation and analysis done on the real WSN test-bed.

Cheng Tien Ee and Ruzena Bajcsy propose a congestion control and fairness algorithm (Cheng Tien & Ruzena, 2004). There are two types of congestion concerned here, which are the congestion due to the simultaneous transmission and that due to high packet generation rate.

The basic concept of the algorithm is as follows:

1. Measure the average rate r at which packets can be sent from this mote
2. Divide the rate r among the number of children motes downstream n, to give the per-node data packet generation rate $r_{data} = r/n$, adjust the rate if queues are overflowing or about to overflow
3. Compare the rate r_{data} with the rate $r_{data, parent}$ sent from the parent, use and propagate the smaller rate downstream

The algorithm is simple and distributed which fits WSNs very much. However, the rate adjusting is only based on the condition of one layer without the whole flow consideration.

The forward rate control would finally transfer the congestion to the source nodes oppositely. The phenomenal of packets losses would occur without further source rate control.

C. Integrated Traffic Control

Based on the discussion above, both the two approaches have their respective pros and cons. Some researchers then propose integrated methods of the two, such as PCCP(Chonggang, Chonggang, Sohraby, Lawrence et al., 2006) and

CODA(Chieh-Yih et al., 2003) . That fits the feature of WSNs that the nodes would forward while generating data.

Priority-based Congestion Control (PCCP) is proposed by Chonggang Wang to avoid/reduce packet loss while guaranteeing weighted fairness and supporting multipath routing with lower control overhead. It designs the congestion control approach through a scheduler between network layer and MAC layer, which maintains two queues: one for source traffic and another for transit traffic. WFQ or WRR algorithm should be used to guarantee network fairness.

PCCP designs a priority-base rate adjustment algorithm (PRA) to give each sensor node a priority index. It sets three priority index: Source Traffic Priority ($SP(i)$), Transit Traffic Priority($TP(i)$) and Global Priority ($GP(i)$)($GP(i)= SP(i)+ TP(i)$).

$$r_{rvc}^i = \begin{cases} \min(\frac{r_{svc}^i}{d(P_{i,0})}, \frac{1}{t_s^i}) & O(P_{i,0}) \leq O'(P_{i,0}), d(P_{i,0}) \prec d'(P_{i,0}) \\ \min(\frac{1}{t_s^{P_{i,0}}} * \frac{GP(i)}{GP(P_{i,0})}, \frac{1}{t_s^i}) & O(P_{i,0}) \geq O'(P_{i,0}), d(P_{i,0}) \succ d'(P_{i,0}) \end{cases}$$

$$r_{src}^i = r_{svc}^i * \frac{TP(i)}{GP(i)}$$

where $P_{i,0}$ denotes the parent node of node i, $d(P_{i,0})$ denotes the congestion degree, $d'(P_{i,0})$ denotes the previous congestion degree, t_s^i denotes the packet inter-arrival time of node i, $O(P_{i,0})$ denotes the number of offspring nodes of $P_{i,0}$, $O'(P_{i,0})$ denotes the previous number of offspring nodes, r_{src}^i denotes the source rate and r_{svc}^i denotes the scheduling rate.

PCCP guarantees the approximate same reporting rate of the nodes with equal source traffic priority and the nodes with higher source traffic priority get higher rate and bandwidth. However it introduces more extra control consumption with setting priority in packets. It adjusts the reporting rate concerned with weighted fairness, which meanwhile provides fairness guarantee but no information fidelity guarantee.

Congestion Detection and Avoidance (CODA) is to alleviate congestion(Chieh-Yih et al., 2003). CODA comprises two mechanisms: open-loop hop-by-hop backpressure and closed-loop, multi-source regulation. It uses a combination of the channel loading conditions and the current buffer occupancy to infer detection of congestion. When a node detects congestion, it broadcasts backpressure messages, which are propagated upstream toward the source. In the condition of persistent congestion, closed-loop regulation operates over a slower time scale of asserting congestion control over multiple sources from a single sink.

Open-loop control is more appropriate for transient congestion, whereas closed-loop control is better for persistent congestion.

2. Traffic Multiplexing

Most of the routing protocols in WSNs adopt the single-path unicast by using metrics such as energy consumption and routing hops. Due to the many-to-one traffic pattern, congestion will be incurred when traffic flows merge at an intermediate node. If the intersecting traffic flows travel towards destinations at different locations, such congestion can be resolved by traffic multiplexing(Kumar R, 2006; Popa et al., 2006).

The congestion aware routing (CAR) (Kumar R, 2006)is based on resource control, and introduces packet differentiation in data transmission. It dynamically discovers the congestion zone (conzone) and enforces differentiated routing based on conzone and data priority. While high priority packets are routed inside the conzone, low priority packets are forced to use the routing paths outside the conzone.

The biased geographical routing (BGR)(Popa et al., 2006) is integrated with two congestion control algorithms: In-Network Packet Scatter (IPS) and End-to-End Packet Scatter (EPS). BGR introduces "bias", which is actually the deviation angle, in the geographical routing for splitting congested traffic. IPS alleviates transient congestion by splitting traffic immediately before the congestion areas, while EPS alleviates long term congestion by splitting the flow at the source, and performing rate control at the same time.

The traffic multiplexing approaches try to avoid the fidelity degradation which may occur in traffic control schemes. However, it requires more resources available for using. Besides, it is not suitable for the case of convergent traffic, in which all traffic flows move towards a single sink. The congestion will reappear somewhere nearer the sink.

3. Data Aggregation

To reduce energy consumption for data transmission, in-network processing operations, such as data aggregation, could be performed by the sensor node to multiple correlated data packets that it received in a short period of time. This technology can also be used to alleviate network congestion (Edith et al., 2006; Laura et al., 2005).

The CONCERT algorithm(Laura et al., 2005) is proposed to specifically use adaptive data-aggregation as a method to counteract network congestion. In scenarios where the monitored area is accessible, the data aggregator nodes can be placed in appropriate positions where congestion is expected to occur; instead, in scenarios where the areas is not accessible, mobile aggregator devices could be used.

In PREI algorithm(Edith et al., 2006), the whole area is divided into multiple grids with same size. In a grid, the aggregating node first filter out some data with significant deviation, then perform aggregation to the remaining data. The aggregation can be easily extended to multiple levels, where a reporting node is responsible for further collection and aggregation of the data from the aggregating nodes in surrounding grids.

4. Virtual Sink

WSNs exhibit a unique "funneling" effect (Chieh-Yih et al., 2005) which is a product of the distinctive many-to-many, hop-by-hop traffic pattern, and results in a significant increase in traffic intensity and even network congestion as events move closer toward the sink. To solve this problem, Siphon proposes to exploit the availability of a small number of all-wireless, multi-radio virtual sinks that can be randomly distributed or selectively placed across the sensor field. Virtual sinks are the sensor nodes equipped with a secondary long-range radio interface based on IEEE 802.11, in addition to their primary low power mote radio. A secondary ad hoc network that is rooted at the real physical sink is formed among all virtual sinks. When congestion occurs in the sensor network, the congested traffic can be redirected into the secondary network timely. When both networks are overloaded, the backpressure mechanism will be triggered to reduce the source traffic. Siphon is implemented on a real sensor network using TinyOS on Mica2 motes and the Stargate platform. However, the increase of extra hardware and the effective deployment of virtual sinks are two problems left unresolved in Siphon.

Reliability Guarantee

Reliability guarantee requires detecting and repairing packet loss in a WSN. The protocols to deal with this problem in WSNs could be sorted into two main categories as shown in Fig.1: retransmission-based protocols and redundancy-based protocols. We would discuss these principles in more detail.

We present comparisons of protocols for reliability guarantee as in Table 2. The notations of the table title are explained as follows: A: protocol; B: the direction of data transmission; C: detection method; D: feedback method; E: transmission consumption; F: computation consumption; G: extra hardware support; H: reliability measurement; I: control mode.(Fang et al., 2008)

Retransmission-Based Protocols

In retransmission-based protocols, sensor nodes have to hold the transmitted packet copy in the buffer until this packet is received by the next hop node correctly. Otherwise, the packet should be retransmitted.

1. Loss Detection and Notification

A typical loss detection mechanism writes a sequence number in each packet header. The

Table 2. Comparison of reliability guarantee protocols

A	B	C	D	E	F	G	H	I
PSFQ(Chieh-Yih, Andrew, & Lakshman, 2002)	Sink->Sensor	Packet id	NACK	Moderate	Low	No	Packet-based	Distributed
GARUDA(Seung-Jong, Ramanuja, Raghupathy, & Ian, 2004)	Sink->Sensor	Packet id	NACK	High	Low	Yes	Packet-based	Distributed
RMST(Stann & Heidemann, 2003)	Sensor->Sink	Packet id	NACK	Low	Low	No	Packet-based	Distributed
RBC(Hongwei, Anish, Young-ri, & Mohamed, 2007)	Sensor->Sink	Buffer id	IACK & ACK	Low	Moderate	No	Packet-based	Distributed
BRTM(Tezcan, Wenye, & Mo-Yuen, 2005)	Bidirectional	Packet id	NACK & ACK	Low	Low	No	Both	Centralized
AFS(Felemban, Lee, Ekici, Boder, & Vural, 2005)	Sensor->Sink			High	Low	No	Packet-based	Centralized
Reinform(Deb, Bhatnagar, & Nath, 2003)	Sensor->Sink			High	Moderate	No	Packet-based	Distributed
MMSPEED(Felemban et al., 2005)	Sensor->Sink			High	Moderate	No	Packet-based	Distributed
GRAB(Fan, Gary, Songwu, & Lixia, 2005)	Sensor->Sink			Moderate	Moderate	No	Packet-based	Distributed
ErasureCode(Sukun, Fonseca, & Culler, 2004)	Sensor->Sink			High	High	No	Packet-based	Distributed

continuity of sequence number can be used to detect packet loss. There are three ways to notify the sender about the loss, which are given as follows (Chonggang, Chonggang, Sohraby, Bo et al., 2006):

1. **ACK (Acknowledgment):** The sender starts a timer each time it transmits a packet, while the receiver sends back an acknowledgement message after receiving the packet. If the timer expires, the sender retransmits the packet and restarts the timer; otherwise, it cancels the timer and deletes the packet copy in the buffer. This approach consumes a lot of bandwidth and energy resources for frequent acknowledgments.

2. **NACK (Negative ACK):** The receiver keeps silence if it receives the packet in right sequence, and sends back a NACK message if it detects the packet loss. This method consumes less bandwidth and energy resources than ACK, while it needs to know the boundary of the transmission sequence and can't handle the single packet transmission.

3. **IACK (Implicit ACK):** The sender keeps detecting the channel state after it transmits a packet. If the packet is overheard being forwarded again, this implies that the packet has been successfully received and therefore acknowledged implicitly. This approach avoids control message overhead and is more energy efficient. However, it depends on whether the sensor nodes have the capability to overhear the physical channel. It would not be feasible in the case where the transmission is corrupt or the channel is not bidirectional or the sensor nodes access the physical channel using Time Division Multiple Access (TDMA)-based protocols.

2. Retransmission Mechanism

Retransmission mechanism may be either end-to-end or hop-by-hop. In the end-to-end approach, such as TCP, STCP(Iyer et al., 2005), the end-points (destination or source) are responsible for loss detection and notification. In the hop-by-hop method, intermediate nodes detect and notify packet loss. The hop-by-hop methods are more suitable for WSNs due to less resource consumption and more efficient operations than those in end-to-end approaches (Chieh-Yih et al., 2002; Stann & Heidemann, 2003). We introduce several relevant protocols as below.

A. Sink-to-Node Reliability

In WSNs, sink-to-node mode is usually used to send control messages such as queries to all or multi-nodes from sinks via the way of broadcast or multicast.

The Pump Slowly Fetch Quickly (PSFQ) algorithm (Chieh-Yih et al., 2002)addresses the case where an ordered block of packets is to be delivered from one sink node to a set of sensor nodes. The data source pumps the packets one after another into the network, using a large period and a broadcast mechanism. The node receiving those packets stores them into the internal buffer and forwards them to downstream nodes if they are received in-sequence. An intermediate node receiving an out-of-sequence segment does not forward it immediately, but quickly requests the missing segments from the upstream neighboring node, which is called a fetch operation. Between two pumps, multiple fetch trials can be made. Therefore, pumping is slow and fetching is quick. Proactive fetch is introduced into the mechanism to resolve the problem in NACK. That is a node sets a timer each time it receives a new packet. Upon expiration of this timer, the node enters the fetch mode and requests all the missing segments from its upstream neighbors. However, PSFQ uses lots of timer which complicate the control process.

A scalable approach for reliable downstream data delivery (FARUDA) described by Nurcan Tezcan et al. addresses a work focusing on reliable downstream data delivery(Seung-Jong et al.,

2004). When the sink sends the first packet, it stamps the packet with a bId. When a node receives the first packet successfully, it increments its bId by one, and sets the resulting value as its own bId, which is the approximate number of hops from the sink to it. A node is a core node if its bId is $3i$ ($i \in$ N). In the period of messages transmission, only core nodes keep the messages. It uses out-of-sequence forwarding method. A downstream core node initiates a request for a missing packet if it receives an A-map packet from an upstream core node with the corresponding bit set of the numbers of the packets they haven't received. Other nodes overhear the channel. If the packets are all available in the core node, they ask the core node for all the packets. FARUDA also uses a pulsing based solution for reliable short-message delivery. The Wait-for-First-Packet (WFP) pulse the solution used is a small finite series of short duration pulses. That needs to implement the equipments. FARUDA provides smart methods to deal with the reliability problem in WSNs.

B.   Node-to-Sink Reliability

Reliable Data Transport (RMST) proposed by Fred Stann and John Heidemann in (Stann & Heidemann, 2003) presents analysis and experiments resulting in specific recommendations for implementing reliable data transport in WSNs. It introduces selective ARQ: a combination of No ARQ and ARQ in MAC Layer Design. In this scheme packets sent to asingle neighbor employ a stop-and-wait ARQ mechanism, while packets sent to multiple neighbors have no ARQ. It adds back channel to single direction routing for loss notification in routing layer. It uses two transport layer paradigms: End-to-End Selective Request NACK and Hop-by-Hop Selective Request NACK. The operation is only done at the sinks and needs no cache at nodes in the former mechanism, while each caching node on the reinforced path from source to sink caches the fragments in the later one. End-to-End Positive ACK is adopted in

application layer. In this approach a sink requests to receive a large data entity, which is fragmented at the source.

RMST uses a serial of mechanisms from MAC layer to application layer. This fits the multifaceted influence in reliable guarantee by the physical, MAC, network and transport layer.

Reliable Bursty Convergecast (RBC) proposed by Hongwei Zhang et al. in(Hongwei et al., 2007) addresses three issues: To improve channel utilization, to reduce ack-loss and to alleviate retransmission-incurred channel contention. A window-less block acknowledgment scheme that guarantees continuous packet forwarding and replicates the acknowledgment for a packet is used to improve channel utilization and to reduce ack-loss. Differentiated contention control is used to alleviate retransmission-incurred channel contention. It also designs mechanisms to handle varying ack-delay and to reduce delay in timer-based retransmissions.

RBC resolves the problems in slide window that the sequent packets need to wait a long time for forwarding to improve the forward efficiency. However, the complicated queuing structure and the overhearing and control consumption shouldn't be ignored.

C.   Bidirectional Reliability

A bidirectional reliable transport mechanism (BRTM) (Tezcan et al., 2005) addresses both node-to-sink and sink-to-node reliable transport. Bidirectional reliability is achieved by transmitting ACK/NACK packets between the sink and essential nodes, which are selected using a weighted-greedy algorithm based on the residual energy of nodes and rotated in time. A lightweight ACK mechanism incorporating with congestion control is used for reliable event transfer from nodes to sink, while NACK solves the reliable query delivery from sink to nodes.

BRTM uses end-to-end approach to provide reliable guarantee. The efficiency of this method is lower than hop-by-hop manner in the most applications of WSNs.

Redundant Sending

1. Copy-Based Redundancy

When the deployment density is high, the nodes can create multiple copies for a forwarded packet, and transmit these copies to multiple neighboring nodes, so that the reliability can be guaranteed through data redundancy and multipath routing. The critical problem is how many neighboring nodes should the current node selects as the next hop forwarders to meet the reliability requirement.

The ReInForM(Deb et al., 2003)approach is based on the idea of sending multiple packet copies along multiple, randomly chosen routes. For node i, the required number of paths P_i can be obtained as:

$$P_i = \frac{\log(1 - r_i)}{\log(1 - (1 - e_i)^{h_i})} - (1 - e_i)$$

where h_i is the hop distance from node i to the sink, e_i is the error rate of node i, and r_i is the reliability requirement. If node i is the source node S, r_s can be set by itself. Otherwise, r_i is given as:

$$r_i = 1 - (1 - (1 - e_s)^{h_s - 1})^P$$

Then node i selects the next hop forwarders from the neighboring nodes with hop distance of $(h_i - 1)$, h_i and $(h_i + 1)$, with the number according to the proportion of $1:(1 - e_i): (1 - e_i)^2$. Node j is decided to be a forwarder when $P_j \geq 1$, otherwise it becomes a forwarder with probability P_j. This approach didn't take energy reserved and error rate of the selected node into consideration. Besides, routing loop can be caused by choosing nodes with larger hop distance to the sink.

In MMSPEED(Felemban et al., 2005), a sending node selects the next hop forwarding nodes according to the error rates between the sender-receiver pair. Each node i maintains the recent average of packet loss rate $e_{i,j}$ to each immediate neighboring node j. Using $e_{i,j}$, the end-to-end reachability of a packet from node i to the final destination d via a neighboring node j as follows:

$$RP_{i,j}^d = (1 - e_{i,j})(1 - e_{i,j})^{\left| dist_{j,d} / dist_{i,j} \right|}$$

where $[dist_{j,d} / dist_{i,j}]$ is hop count estimation from node j to the final destination d. Meanwhile, the total reaching probability TRP with initial value of zero can be updated as follows:

$$TRP = 1 - (1 - TRP)(1 - RP_{i,j}^d)$$

Node i keeps on adding the neighboring nodes into the forwarding set until $TRP > P^{req}$, where P^{req} is the end-to-end reachability requirement of the forwarded packet. The value of P^{req} is initially set by the source node according to the packet type or other metrics, then updated at each hop and sent to the corresponding node at the next hop. MMSPEED also employs reliability back-pressure mechanism to remedy the problem of local decision in a more global scope, i.e., the downstream node can ask for the upstream node to reduce the assigned P^{req} for itself, if it can't satisfy the requirement even with all possible forwarding nodes.

GRAdient Broadcast(GRAB)(Fan et al., 2005) integrates the gradient-based routing with limited broadcast. Some state information is carried within the data packet: a, the amount of credit assigned to the packet at the source; C_{source}, the cost of the source to send a packet to the sink; $P_{consumed}$, the amount of energy that has been consumed from the source to the current hop; C_{sender}, the cost of the current sender that broadcasts the packet. To collect data reports, the sink first builds a cost field by propagating advertisement packets in the whole network. The cost at a node is the minimum energy overhead to forward a packet from this node to the sink. The node delivers the data by broadcasting, and only receivers with a smaller routing cost may forward the packet at each hop. Then the packet is forwarded by successive nodes of decreasing costs and won't traverse back to the same node again. The receiving node with lower cost value first compares two values R_α and R_{thresh}:

$$R_\alpha = \frac{\alpha - (P_{consumed} + C_{receiver} - C_{source})}{\alpha},$$

$$R_{thresh} = \left(\frac{C_{receiver}}{C_{source}}\right)^2$$

If $R_\alpha > R_{thresh}$, the node is considered as have sufficient credit to use. It broadcasts at a power to reach multiple neighbors towards the sink. Otherwise, the node should forward the packet along its minimum cost path to minimize the total cost. If a node receives multiple packet copies with same data content, it only forwards the packet once. This approach limits the broadcast range in a narrow forwarding mesh so that the reliability can be achieved with less energy consumption. However, it assumes the error rates are the same for each node.

2. Coding-Based Redundancy

Erasure code is an error correcting mechanism widely used in network applications, with which we can reconstruct m original messages by receiving any m out of n code words ($n > m$). If n is sufficiently large compared to the loss rate, the transmission reliability can be achieved without retransmission. It can provide similar reliability guarantee with fewer requirements on bandwidth and storage than copy-based approaches. It has been integrated within BVR routing(Rodrigo et al., 2005) and Directed-Diffusion routing(Chalermek, Ramesh, Deborah, John, & Fabio, 2003) to improve transmission efficiency in WSNs(Djukic & Valaee, 2006; Sukun et al., 2004).

A particular erasure code algorithm called Reed-Solomon code is used in (Sukun et al., 2004)

and (Djukic & Valaee, 2006). We simply describe its main idea here. For a given data, it is broken down into m messages w_0, w_1, w_2..., w_{m-1}. Then the code word vector $P_{n\times 1}$ can be given as:

$$P_{n\times 1} = \begin{pmatrix} p(x_1) \\ p(x_2) \\ p(x_3) \\ \vdots \\ \vdots \\ p(x_{n-1}) \\ p(x_n) \end{pmatrix} = AW = \begin{pmatrix} 1 & x_1 & \cdots & x_1^{m-1} \\ 1 & x_2 & \cdots & x_2^{m-1} \\ 1 & x_3 & \cdots & x_3^{m-1} \\ \vdots & \vdots & & \vdots \\ \vdots & \vdots & & \vdots \\ 1 & x_{n-1} & \cdots & x_{n-1}^{m-1} \\ 1 & x_n & \cdots & x_n^{m-1} \end{pmatrix} \begin{pmatrix} w_0 \\ w_1 \\ \vdots \\ w_{m-1} \end{pmatrix}$$

A is called as Vandermonde matrix with element $A(i, j) = x_i^{j-1}$ where each x_i is nonzero and distinct from each other. If we have any m rows of A and their corresponding $P(x)$ values, vector W which contains coefficients of polynomial can be obtained, which can then be organized into original messages.

However, the coding-based approaches require the hardware equipped with high computing capability processing unit, which also brings more energy consumption for data encoding and decoding.

Fairness Guarantee

Fairness guarantee has been one of the most fundamental problems in networking research. Relatively less attention has been paid in the emerging area of WSNs. The fairness guarantee in wireless network is quite different from wired network due to the time varying nature of the wireless channel as well as channel contention. Furthermore, a WSN as a special wireless network also has its own features such as energy consumption consideration, data aggregation consideration and so on. These features introduce more challenges in the research of fairness guarantee.

Fairness is defined specifically for different goals in WSNs. In general, it includes fairness based on congestion control and "pure" fairness that focuses on flow control. Fairness is a hot research direction in wired and wireless networks. It includes max-min fairness, proportional fairness and so on, which are also important in WSNs. In this section, we would analyze these problems in detail with several fairness mechanisms. There is a tradeoff between the price and effect. Some of these mechanisms gain limited improvements with little resource consumption additions, while some others obtain relative much better results but with more extra consumptions.

We present comparisons of protocols for fairness mechanisms in Table 3. The notations of the table title are explained as follows: A: protocol; B: problem type; C: if based on congestion control; D: end to end or hop by hop; E: control method; F: energy consumption consideration; G: throughput consideration; H: data aggregation consideration; I: weight consideration; J: control mode.

Fairness Guarantee with Flow Control

The fairness guarantee with flow control in wireless networks is much different from that in wired networks. The main difference is that flows in a wireless network not only consume bandwidth usefully on the links they are active on but also wastefully on links that they interfere with. Moreover, there is heterogeneity in the amount of interference (i.e., bandwidth wastage) that each flow may cause. The fundamental differences between wired and wireless networks demands a fresh look on the problem of fair and efficient rate control algorithms for wireless networks in general and wireless sensor networks specifically(Sridharan & Krishnamachari, 2007).

There are also some differences between wireless sensor networks and traditional wireless networks due to some unique features of wireless sensor network itself, such as consideration on energy consumption, data aggregation and so on.

1. Max-Min Fairness

Avinash Sridharan and Bhaskar Krishnamachari propose a max-min fairness algorithm for WSNs (Sridharan & Krishnamachari, 2007).They use additive increase strategy to achieve fair rate allocation while trying to maximize network utilization. The work concerns with both fairness and efficiency.

The algorithm firstly formulates the problem of maximizing the network utilization subject to a max-min fair rate allocation constraint, which is in the form of two coupled linear programs.

P1:

max Y

$$s.t. \; R_{in} + N \times R_{noise} \le B$$

Table 3. Fairness

A	B	C	D	E	F	G	H	I	J
MNUMF(Sridharan & Krishnamachari, 2007)	Fairness	No	End to end	Rate	No	Yes	No	No	Centralized
LMFDC(Shigang, Shigang, Yuguang, & Ye, 2007)	Fairness	No	End to end	Rate	Yes	Yes	No	Yes	Centralized
AEFR(Krishnamachari & Ordonez, 2003)	Fairness	No	End to end	Rate	Yes	Yes	No	Yes	Centralized
FQ(Jangeun & Sichitiu, 2003)	Fairness	No	Hop by hop	Queue	No	No	No	Yes	Distributed
EPS(Cheng Tien & Ruzena, 2004)	Fairness	No	Hop by hop	Rate	No	No	No	No	Distributed
CFRC(Li et al., 2007)	Fairness	No	Hop by hop	Rate	No	No	Yes	Yes	Distributed
PCCP(Chonggang, Chonggang, Sohraby, Lawrence et al., 2006)	Fairness	No	Hop by hop	Queue	No	No	No	Yes	Distributed
AFA(Shigang & Zhan, 2006)	Fairness	No	End to end	Rate	No	No	No	Yes	Distributed
IFRC(Sumit et al., 2006)	Fairness	No	Hop by hop	Rate	No	No	No	Yes	Distributed

$$R_{in} = C \times R_{src}$$

$$R_{noise} = C \times R_{src} + R_{src}$$

$$r_{src}^{(i)} \geq Y_i \forall i \in T$$

P2:

$$\max \sum_{i \in T} r_{src}^{(i)}$$

$$s.t.\ R_{in} = C \times R_{src}$$

$$R_{noise} = C \times R_{src} + R_{src}$$

$$r_{src}^{(i)} \geq Y^* \forall i \in T$$

where Y is the minimum rate among all flows, R_{in} is the total input rate arriving at each node, R_{noise} is the total output rate exiting from a node, R_{src} is the rate allocated to each source, C representing the parent-child relationships on the data gathering tree. The first linear program identifies the max-min rate allocation, while the second maximizes the sum-rate subject to the constraint determined by the solution of the first problem.

Then a dual based approach is proposed, which is sub-optimal and partly distributed (the decision making is not completely local, it relies partially on information exchange with the root). Firstly, each node allocated at least the max-min available bandwidth. Then, the rest bandwidth is allocated to the node which would incur least interfere and contribute the largest throughput to the network with the same source flow assignment. The process would proceed, allocating the remaining bandwidth to the node with second optimality and so on.

This mechanism gives a good basis to guarantee fairness and efficiency in wireless sensor networks.

Shigang Chen et al. propose a max-min assignment and forwarding schedule based on two linear programs(Shigang et al., 2007), which is

called Lexicographic Max-min Fairness. It is the first work that solves the max-min rate assignment problem for congestion-free end to end flows without fixed routing paths in WSNs.

The mechanism firstly calculates the maximum common rate (MCR) which is the smallest max-min rate that is greater than rate r (initial r=0) by the first linear program. Then it calculates the maximum single rate (MSR) which is the maximum feasible rate for a certain node assumed that all other nodes keep their rates in the previous step. And it makes sure if some nodes get up to their max-min rates and goes to the first step. The process would proceed until all nodes have got up to their max-min rates.

It has been proved that the assignment is the one and only one max-min optimal rate assignment by the authors. The work also concerns with energy constraint, medium contention and weight consideration.

Bhaskar Krishnamachari and Fernando Ordonez propose formal mathematical models of the WSNs based on non-linear optimization and use these models to analyze the impact of fairness constraints on network performance(Krishnamachari & Ordonez, 2003). Fairness refers to absolutely even partition of the network throughput for each node here.

Two optimization based formulations are proposed(Krishnamachari & Ordonez, 2003). The first one which is shown below balances the competing minimum energy and maximum information objectives by limiting the total energy on each node and maximizing the total information based on fairness constraints. The objective to maximize the total information routed to the sink. The formula (a)-(f) are constraints, which separately ensure equal or more flow for output than input of nodes, proportional fairness throughput, pre-specified finite amount of energy for every node, wireless channel setup and non-negativity property to flow and transmission power variables. The second one balances the competing minimum energy and maximum information objectives by limiting the minimum information to be extracted to the sink and minimal the energy required based on fairness constraints. The only differences in the second one are the different objective function which is now $\min \sum_{i=1}^{n}(\sum_{j=1}^{n+1} P_{ij} + \sum_{j=1}^{n} f_{ji}C)$ to minimize the total energy usage and the constraint (d) which is now $\sum_{j=1}^{n} f_{jn+1} \geq f_{\min}$ to ensure total information obtained by the sink.

$$\max \sum_{j=1}^{n} f_{jn+1}$$

$$s.t. \sum_{j=1}^{n} f_{ji} \leq \sum_{j=1}^{n+1} f_{ij} \quad (a)$$

$$\sum_{j=1}^{n} f_{ji} + R_i \geq \sum_{j=1}^{n+1} f_{ij} \quad (b)$$

$$\sum_{j=1}^{n+1} f_{ij} - \sum_{j=1}^{n} f_{ji} \leq \alpha_i \sum_{j=1}^{n} f_{jn+1} \quad (c)$$

$$\sum_{j=1}^{n+1} P_{ij} + \sum_{j=1}^{n} f_{ji}C \leq E_i \quad (d)$$

$$f_{ij} \leq \log(1 + \frac{P_{ij}d_{ij}^{-2}}{\eta}) \quad (e)$$

$$f_{ij} \geq 0, P_{ij} \geq 0 \quad (f)$$

where f_{ij} represents flow from node i to j, R_i represents the source rate of node i, α_i represents the fraction of proportional fairness of node i, P_{ij} represents transmission power, E_i represents pre-specified energy for node i, and d_{ij} represents the distance from node i to j.

The authors have obtained preliminary results for the optimizations using an interior point algorithm and done some experiments to verify these solutions can output more information using less energy based on proportional fairness. More extensions would be taken such as network aggregation based on the two optimization models.

2. Fairness Based on Queue Management

Jangeun Jun and Mihail L.Sichitiu propose a serial scheduling approaches based on fair queuing(Jangeun & Sichitiu, 2003). Fair queuing is implemented locally to schedule the processing of local and relayed traffics.

The problem is incurred by the phenomena that the node closest to the gateway will gradually but completely starve the node further away from the gate way as the offered load at each node increases. The first fair queuing is proposed to solve this problem. Using the default queuing scheme, each node can be modeled as shown in *Figure 3*(f1 is the originating traffic flow and f2 and f3 are the relayed ones). It is clear that f1 will receive more bandwidth and starve others. There would also be some problems if the offered load is high enough to saturate the network. A significant amount of bandwidth is wasted due to over-injected packets. A per-flow queuing to gain per-flow fairness and a MAC priority scheme are proposed as shown in *Figure 4*. The differentiated service for per-flow is provided by giving different weights.

The authors present the simulation results of their fair queuing methods with four flows. The simulation results match closely the analytically estimated values. These schedules can be used within wireless sensor networks directly. They would be more perfect for the WSNs if concerning with energy consumption, data aggregation and so on.

Fairness Integrated with Congestion Control

Fairness and congestion are two relevant but distinct problems. Resolving congestion does not mean fairness guarantee. However, some control mechanisms, such as rate adjustment or queue schedule, can be adopted to solve both these problems.. We would introduce some fairness algorithms proposed based on congestion control as follows.

Figure 3. Single network-layer queuing scheme

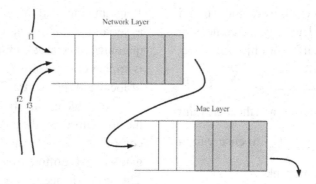

1. Fairness Based on Traffic Control

Shigang Chen et al. propose a distributed aggregate fairness model(Shigang & Zhan, 2006), which is based on the congestion avoidance mechanism, and a localized algorithm (AFA) that implements the model. The fairness here refers to weighted end to end fairness.

Firstly, the congestion avoidance mechanism ensures no packet drop due to queue overflow. The goal of the aggregate fairness guarantee is for every node to assign bandwidth to the upstream links in proportion to their aggregate flow weights, which include the proportion of source flow and flow weight. AFA realizes the aggregate fairness via rate limit. When sensor i is congested, it computes a rate limit for each upstream neighbor k as follows

$$l_{k,i} = \frac{F_{k,i}}{F_i} r_i$$

where $l_{k,i}$ is the rate limit of flow pass through link(k,i), $F_{k,i}$ and F_i are the aggregate flow weight of link(k,i) and node i. r_i is the rate of node i. The assignment of a local rate limit is as follows

$$l_i = \frac{w(i)}{F_i} r_i$$

where l_i is the rate limit of local rate of node i, $w(i)$ is the weight of flow generated by node i.

That means the rate limit is proportional to aggregate flow weights. After an upstream neighbor enforces a rate limit, it may become congested also. Then it will enforce rate limits in a similar way to its upstream neighbors. This process repeats towards the data sources. Eventually all affected data sources will adjust their rates to the fair bandwidth shares defined by the rate limits.

Authors give several serial simulations to test the fairness guarantee, energy efficiency, packet drops, weight fairness results and convergence speed of the algorithm. The results verify that AFA could gain good performance in these aspects.

Sumit Rangwala et al. propose an interference-aware fair rate control method (IFRC) based on dynamical congestion control(Sumit et al., 2006). Fairness here refers to nodes shouldn't get more bandwidth than their parents and congestion neighbors or neighbors of their parents. That also guarantees the network performance.

Each node adjusts its rate dynamically based on the test of its queue length to avoid congestion. And it should abide by three rules:

- If it tests congestion locally via a certain upper threshold U, it would inform all its neighbors the states of itself and its congested children (the reason for this information is to extend the fairness control to all the potential interferential nodes).

- The data rate could not exceed that of any congested neighbor or any congested child of the neighbor
- The data rate could not exceed that of its parent

The three rules above guarantee the interferential nodes of any congestion node have the equal fair rate.

Authors evaluate the performance of IFRC on a real indoor wireless sensor network testbed. As an aside, implementation on real testbed is a great challenge in research of wireless sensor networks. The results show approximately the same throughput of all nodes, no buffer overflow and good adaptable ability. The authors also introduced several IFRC extensions including weighted fairness, multiple base stations and network with partly transmission nodes.

2. Fairness Based on Queue Management

Cheng Tien Ee and Ruzena Bajcsy propose two different algorithms that implement fairness: Probabilistic Selection (PS) and Epoch-based Proportional Selection (EPS) based on per-child packet queue and maintenance of per-child tree size(Cheng Tien & Ruzena, 2004). The fairness here refers to the uniform quantity of packets received by the sink node from each node. The work was based on the congestion control algorithm described in the preceding part in (Cheng Tien & Ruzena, 2004).

The main work for PS is to fix on the probability of choosing queue Q(i) for a node A, where Q(i) is the queue storing packets from C(A)(the set of the children of A). Below is the choosing formula, where only queues that are backlogged, that is, queues that are not empty are concerned with.

Probability of choosing queue Q(i)

$$= \frac{subTreeSize(i)}{1 + \sum_{j \in C(A) and Q(j) backlogged} subTreeSize(j)}$$

The EPS introduces an epoch to ensure transmission from each queue exactly the same multiple times the number of nodes serviced by that queue. All queues work in FIFO model to keep that exactly the same quantity of packets from each child will be transmitted within one epoch.

The authors present correctness proof for both PS and EPS. They are also evaluated based on congestion control in the preceding part in (Cheng Tien & Ruzena, 2004) both in the simulator and on Mica nodes.

Chonggang Wang et al. propose a congestion control algorithm, where fairness is an important objective based on congestion control(Chonggang, Chonggang, Sohraby, Lawrence et al., 2006). Fairness here refers to weighted fairness which means the sink can get different (but in a weighted fair way) throughput from sensor nodes.

Figure 4. Weighted per-flow queues at the network layer with MAC-layer QoS Suppor0074

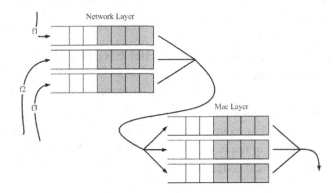

The weighted fairness is guaranteed by Priority-based Rate Adjustment (PRA), which is a scheduler with two sub-queues between the network layer and the MAC layer. Each sensor node here is given a priority index, based on which the data rate would be adjusted to guarantee the weighted fairness. The rate adjustment follows the rules that nodes with the same priority index get the same source rate and nodes with a larger source priority index receive higher source rate and higher bandwidth.

Simulations carried out by the authors show that PCCP effectively supports flexible weighted fairness through configuration of priority index of each sensor node.

FUTURE RESEARCH DIRECTIONS

We summarize some of these directions and give some pertinent references as follows:

1. The congestion can be avoided in advance by exploiting the data correlation to control traffic initiation within the event area. The redundant network transmission can be well reduced without a large degradation of data fidelity.
2. The existing schemes address either congestion control or reliability guarantee, while few of them investigate both problems systematically. A good transmission control protocol should consider both of them together with issues like energy-efficiency and other performance metrics.
3. The control protocol should try to comply with some new architecture and schemes emerged in WSNs, such as address-free network, in-network processing and clustering architecture.
4. Some light-weighted coding schemes can be used in reliability guarantee. For example, Tornado codes are appealing in many network applications in that Tornado codes are

easily and fast encoded and decoded, just requiring the simple exclusive-or (XOR) operation.
5. The existing protocols rarely consider the node mobility. Since the node mobility will affect the transmission negatively, it is necessary to provide support for mobile sensor networks in new protocols.
6. When designing protocols for providing fairness guarantee, some unique features of the WSNs should be taken into consideration, such as the energy constraint.

CONCLUSION

In this chapter, we presented a comprehensive survey of transmission control techniques in wireless sensor networks which have been presented in the literature. Overall, these works can be classified based on the functionality into three categories: congestion control, reliability guarantee and fairness guarantee. We also highlight the design tradeoffs between reliability and overhead in some of the control paradigm, as well as the advantages and the disadvantages of each control technique. Although many of these techniques look promising, there are still many challenges that need to be solved in the WSNs. We highlighted those challenges and pinpointed some future research directions in this regard. However, transmission control in sensor networks have so far attracted less attention in the research community than, e.g. MAC or routing protocols. There will be accordingly plenty of room for more interesting research.

REFERENCES

Chalermek, I., Ramesh, G., Deborah, E., John, H., & Fabio, S. (2003). *Directed diffusion for wireless sensor networking* (Vol. 11, pp. 2-16). Piscataway, NJ: IEEE Press.

Chen, S., & Yang, N. (2006). Congestion Avoidance Based on Lightweight Buffer Management in Sensor Networks. *Parallel and Distributed Systems . IEEE Transactions, 17*(9), 934–946.

Cheng Tien, E., & Ruzena, B. (2004). *Congestion control and fairness for many-to-one routing in sensor networks.* Paper presented at the Proceedings of the 2nd international conference on Embedded networked sensor systems, Baltimore, MD.

Chieh-Yih, W., Andrew, T. C., & Lakshman, K. (2002). *PSFQ: a reliable transport protocol for wireless sensor networks.* Paper presented at the Proceedings of the 1st ACM international workshop on Wireless sensor networks and applications. Atlanta, GA

Chieh-Yih, W., Shane, B. E., & Andrew, T. C. (2003). CODA: congestion detection and avoidance in sensor networks. Paper presented at the *Proceedings of the 1st international conference on Embedded networked sensor systems.* Los Angelas

Chieh-Yih, W., Shane, B. E., Andrew, T. C., & Jon, C. (2005). Siphon: overload traffic management using multi-radio virtual sinks in sensor networks. Paper presented at the *Proceedings of the 3rd international conference on Embedded networked sensor systems.*

Chonggang, W., Sohraby, K., Bo, L., Daneshmand, M. A. D. M., & Yueming Hu, A. Y. H. (2006). A survey of transport protocols for wireless sensor networks. *Network, IEEE, 20*(3), 34–40. doi:10.1109/MNET.2006.1637930

Chonggang, W., Sohraby, K., Lawrence, V., Bo Li, A. B. L., & Yueming Hu, A. Y. H. (2006). *Priority-based Congestion Control in Wireless Sensor Networks.* Paper presented at the Sensor Networks, Ubiquitous, and Trustworthy Computing, 2006. IEEE International Conference.

Deb, B., Bhatnagar, S., & Nath, B. (2003). *Re-InForM: reliable information forwarding using multiple paths in sensor networks.* Paper presented at the Local Computer Networks, 2003. LCN '03. In Proceedings of 28th Annual IEEE International Conference.

Djukic, P., & Valaee, S. (2006). *WLC12-4: Reliable and Energy Efficient Transport Layer for Sensor Networks.* Paper presented at the Global Telecommunications Conference, 2006. GLOBECOM '06. IEEE.

Edith, C. H. N., Yangfan, Z., Michael, R. L., & Jiangchuan, L. (2006). *Reliable Reporting of Delay-Sensitive Events in Wireless Sensor-Actuator Networks.* Paper presented at the Mobile Adhoc and Sensor Systems (MASS), 2006 IEEE International Conference.

Fan, Y., Gary, Z., Songwu, L., & Lixia, Z. (2005). Gradient broadcast: a robust data delivery protocol for large scale sensor networks (Vol. 11, pp. 285-298). The Netherlands: Kluwer Academic Publishers.

Fang, W.-W., Qian, D.-P., & Liu, Y. (2008). *Transmission control protocols for wireless sensor networks.* Ruan Jian Xue Bao/Journal of Software, *19*(6), 1439-1451.

Felemban, E., Lee, C. G., Ekici, E., Boder, R., & Vural, S. (2005). *Probabilistic QoS guarantee in reliability and timeliness domains in wireless sensor networks.* Paper presented at the INFOCOM 2005. 24th Annual Joint Conference of the IEEE Computer and Communications Societies, Hongwei, Z., Anish, A., Young-ri, C., & Mohamed, G. G. (2007). *Reliable bursty convergecast in wireless sensor networks* (Vol. 30, pp. 2560-2576). Oxford, UK: Butterworth-Heinemann.

Hull, B., Jamieson, K., & Balakrishnan, H. (2004). *Mitigating congestion in wireless sensor networks.* Cambridge MA: MIT Computer Science and Arti_cial Intelligence Laboratory

Iyer, Y. G., Gandham, S., & Venkatesan, S. (2005). STCP: a generic transport layer protocol for wireless sensor networks. Paper presented at the Computer Communications and Networks, 2005. ICCCN 2005. In *Proceedings of the 14th International Conference.*

Jangeun, J., & Sichitiu, M. L. (2003). *Fairness and QoS in multihop wireless networks.* Paper presented at the Vehicular Technology Conference, 2003. VTC 2003-Fall. 2003 IEEE 58th.

Krishnamachari, B., & Ordonez, F. (2003). *Analysis of energy-efficient, fair routing in wireless sensor networks through non-linear optimization.* Paper presented at the Vehicular Technology Conference, 2003. VTC 2003-Fall. 2003 IEEE 58th.

Kumar, R. R. H., Cao GH, etc. (2006). *Congestion aware routing in sensor networks.* Retreived (n.d.). from, http://nsrc.cse.psu. edu/tech_report/NAS-TR-0036-2006.pdf

Laura, G., Andrew, T. C., & Sergio, P. (2005). CONCERT: aggregation-based CONgestion Control for sEnsoR neTworks. Paper presented at the *Proceedings of the 3rd international conference on Embedded networked sensor systems.*

Li, S., Li, S., Liao, X., Peng, S., Zhu, P. A. Z. P., & Jiang, J. A. J. J. (2007). Credit based Fairness Control in Wireless Sensor Network. *Paper presented at the Software Engineering, Artificial Intelligence, Networking, and Parallel/Distributed Computing, 2007. SNPD 2007. Eighth ACIS International Conference.*

Popa, L., Raiciu, C., Stoica, I., & Rosenblum, D. S. (2006). Reducing Congestion Effects in Wireless Networks by Multipath Routing. Paper presented at the Network Protocols, 2006. ICNP '06. *Proceedings of the 2006 14th IEEE International Conference.*

Rodrigo, F., Sylvia, R., Jerry, Z., Cheng Tien, E., David, C., Scott, S., et al. (2005). Beacon vector routing: scalable point-to-point routing in wireless sensornets. Paper presented at the *Proceedings of the 2nd conference on Symposium on Networked Systems Design \& Implementation - Volume 2.*

Seung-Jong, P., Ramanuja, V., Raghupathy, S., & Ian, F. A. (2004). A scalable approach for reliable downstream data delivery in wireless sensor networks. Paper presented at the *Proceedings of the 5th ACM international symposium on Mobile ad hoc networking and computing.*

Shigang, C., Shigang, C., Yuguang, F., & Ye, X. (2007). Lexicographic Maxmin Fairness for Data Collection in Wireless Sensor Networks. *Mobile Computing . IEEE Transactions, 6(7),* 762–776.

Shigang, C., & Zhan, Z. (2006). Localized algorithm for aggregate fairness in wireless sensor networks. Paper presented at the *Proceedings of the 12th annual international conference on Mobile computing and networking.*

Sridharan, A., & Krishnamachari, B. (2007). Maximizing Network Utilization with Max-Min Fairness in Wireless Sensor Networks. *Paper presented at the Modeling and Optimization in Mobile, Ad Hoc and Wireless Networks and Workshops, 2007. WiOpt 2007. 5th International Symposium.*

Stann, F., & Heidemann, J. (2003). RMST: reliable data transport in sensor networks. Paper presented at the Sensor Network Protocols and Applications, 2003. *Proceedings of the First IEEE. 2003 IEEE International Workshop.*

Sukun, K., Fonseca, R., & Culler, D. (2004). Reliable transfer on wireless sensor networks. *Paper presented at the Sensor and Ad Hoc Communications and Networks, 2004. IEEE SECON 2004. 2004 1st Ann. IEEE Communications Society Conf.*

Sumit, R., Ramakrishna, G., Ramesh, G., & Konstantinos, P. (2006). *Interference-aware fair rate control in wireless sensor networks* (Vol. 36, pp. 63-74): ACM.

Tezcan, N., Wenye, W., & Mo-Yuen, C. (2005). A bidirectional reliable transport mechanism for wireless sensor networks. *Paper presented at the Military Communications Conference, 2005. MILCOM 2005. IEEE.*

Yogesh, S. zg, r, B. A., & Ian, F. A. (2003). ESRT: event-to-sink reliable transport in wireless sensor networks. Paper presented at the *Proceedings of the 4th ACM international symposium on Mobile ad hoc networking \& computing.*

Zhou, Y., Lyu, M. R., Liu, J., & Wang, H. (2005). *PORT: A Price-Oriented Reliable Transport protocol for wireless sensor networks*, Chicago, IL: IEEE.

ADDITIONAL READING

Chen, S., & Yang, N. (2006). Congestion Avoidance Based on Lightweight Buffer Management in Sensor Networks. *Parallel and Distributed Systems . IEEE Transactions, 17*(9), 934–946.

Cheng Tien, E., & Ruzena, B. (2004). Congestion control and fairness for many-to-one routing in sensor networks. Paper presented at the *Proceedings of the 2nd international conference on Embedded networked sensor systems*, Baltimore, MD

Chieh-Yih, W., Shane, B. E., & Andrew, T. C. (2003). CODA: congestion detection and avoidance in sensor networks. Paper presented at the *Proceedings of the 1st international conference on Embedded networked sensor systems.*

Chieh-Yih, W., Shane, B. E., Andrew, T. C., & Jon, C. (2005). Siphon: overload traffic management using multi-radio virtual sinks in sensor networks. Paper presented at the *Proceedings of the 3rd international conference on Embedded networked sensor systems.*

Chonggang, W., Sohraby, K., Bo, L., Daneshmand, M. A. D. M., & Yueming Hu, A. Y. H. (2006). A survey of transport protocols for wireless sensor networks. *Network, IEEE, 20*(3), 34–40. doi:10.1109/MNET.2006.1637930

Chonggang, W., Sohraby, K., Lawrence, V., Bo Li, A. B. L., & Yueming Hu, A. Y. H. (2006). Priority-based Congestion Control in Wireless Sensor Networks. Paper presented at the Sensor Networks, Ubiquitous, and Trustworthy Computing, 2006. *IEEE International Conference.*

Djukic, P., & Valaee, S. (2006). WLC12-4: Reliable and Energy Efficient Transport Layer for Sensor Networks. Paper presented at the *Global Telecommunications Conference, 2006. GLOBECOM '06. IEEE.*

Edith, C. H. N., Yangfan, Z., Michael, R. L., & Jiangchuan, L. (2006). Reliable Reporting of Delay-Sensitive Events in Wireless Sensor-Actuator Networks. Paper presented at the *Mobile Adhoc and Sensor Systems (MASS), 2006 IEEE International Conference.*

Fan, Y., Gary, Z., Songwu, L., & Lixia, Z. (2005). *Gradient broadcast: a robust data delivery protocol for large scale sensor networks* (Vol. 11, pp. 285-298): Kluwer Academic Publishers.

Fang, W.-W., Qian, D.-P., & Liu, Y. (2008). Transmission control protocols for wireless sensor networks. Ruan Jian Xue Bao/Journal of Software, *19*(6), 1439-1451.

Felemban, E., Lee, C. G., Ekici, E., Boder, R., & Vural, S. (2005). Probabilistic QoS guarantee in reliability and timeliness domains in wireless sensor networks. Paper presented at the *INFOCOM 2005. 24th Annual Joint Conference of the IEEE Computer and Communications Societies.*

Hongwei, Z., Anish, A., Young-ri, C., & Mohamed, G. G. (2007). *Reliable bursty convergecast in wireless sensor networks* (Vol. 30, pp. 2560-2576): Butterworth-Heinemann.

Hull, B., Jamieson, K., & Balakrishnan, H. (2004). *Mitigating congestion in wireless sensor networks*, Baltimore, MD.

Iyer, Y. G., Gandham, S., & Venkatesan, S. (2005). STCP: a generic transport layer protocol for wireless sensor networks. Paper presented at the *Computer Communications and Networks, 2005. ICCCN 2005.*

Kumar, R. R. H., Cao GH, etc. (2006). *Congestion aware routing in sensor networks*. Retreived (n.d.). from, http://nsrc.cse.psu. edu/tech_report/ NAS-TR-0036-2006.pdf

Laura, G., Andrew, T. C., & Sergio, P. (2005). CONCERT: aggregation-based CONgestion Control for sEnsoR neTworks. Paper presented at the *Proceedings of the 3rd international conference on Embedded networked sensor systems*.

Li, S., Li, S., Liao, X., Peng, S., Zhu, P. A. Z. P., & Jiang, J. A. J. J. (2007). Credit based Fairness Control in Wireless Sensor Network. Paper presented at the Software Engineering, Artificial Intelligence, Networking, and Parallel/Distributed Computing, 2007. SNPD 2007. In *Proceedings of the Eighth ACIS International Conference*.

Popa, L., Raiciu, C., Stoica, I., & Rosenblum, D. S. (2006). Reducing Congestion Effects in Wireless Networks by Multipath Routing. Paper presented at the *Network Protocols, 2006. ICNP '06. Proceedings of the 2006 14th IEEE International Conference*.

Rodrigo, F., Sylvia, R., Jerry, Z., Cheng Tien, E., David, C., Scott, S., et al. (2005). Beacon vector routing: scalable point-to-point routing in wireless sensornets. Paper presented at the *Proceedings of the 2nd conference on Symposium on Networked Systems Design \& Implementation - Volume 2*.

Shigang, C., Shigang, C., Yuguang, F., & Ye, X. (2007). Lexicographic Maxmin Fairness for Data Collection in Wireless Sensor Networks. Mobile Computing. *IEEE Transactions on, 6*(7), 762–776.

Shigang, C., & Zhan, Z. (2006). Localized algorithm for aggregate fairness in wireless sensor networks. Paper presented at the *Proceedings of the 12th annual international conference on Mobile computing and networking*.

Sridharan, A., & Krishnamachari, B. (2007). Maximizing Network Utilization with Max-Min Fairness in Wireless Sensor Networks. Paper presented at the *Modeling and Optimization in Mobile, Ad Hoc and Wireless Networks and Workshops, 2007. WiOpt 2007. 5th International Symposium*.

Sukun, K., Fonseca, R., & Culler, D. (2004). Reliable transfer on wireless sensor networks. Paper presented at the *Sensor and Ad Hoc Communications and Networks, 2004. IEEE SECON 2004. 2004 First Annual IEEE Communications Society Conference*.

Sumit, R., Ramakrishna, G., Ramesh, G., & Konstantinos, P. (2006). *Interference-aware fair rate control in wireless sensor networks* (Vol. 36, pp. 63-74): ACM.

Tezcan, N., Wenye, W., & Mo-Yuen, C. (2005). A bidirectional reliable transport mechanism for wireless sensor networks. Paper presented at the *Military Communications Conference, 2005. MILCOM 2005. IEEE*.

Yogesh, S. zg, r, B. A., & Ian, F. A. (2003). ESRT: event-to-sink reliable transport in wireless sensor networks. Paper presented at the *Proceedings of the 4th ACM international symposium on Mobile ad hoc networking & computing*.

Zhou, Y., Lyu, M. R., Liu, J., & Wang, H. (2005). *PORT: A Price-Oriented Reliable Transport protocol for wireless sensor networks*. Chicago, IL.

KEY TERMS AND DEFINITIONS

Congestion Control: Control mechanisms to detect, notify and avoid or mitigate the traffic congestion in the network.

Fairness Guarantee: Schemes to gain fair or weighted fair resource allocation among all sensor nodes, so that the sinks can obtain a fair or weighed fair amount of observation data from the whole network

Queuing Management: Schemes to adjust the packet transfer via queue schedule.

Reliability Guarantee: Schemes to detect and repair packet loss in the network.

Traffic Control: Methods to control data flow in the network, which includes source rate control, forward rate control and so on.

Transmission Control: Control mechanisms working on transport layer to gain more QoS of the network, including congestion control, reliability guarantee, fairness guarantee and so on.

Wireless Sensor Networks: Wireless networks consisting of spatially distributed autonomous devices using sensors to cooperatively monitor physical or environmental conditions, such as temperature, sound, vibration, pressure, motion or pollutants, at different locations.

Chapter 9
Joint Link Scheduling and Topology Control for Wireless Sensor Networks with SINR Constraints

Qiang-Sheng Hua
The University of Hong Kong, China

Francis C.M. Lau
The University of Hong Kong, China

ABSTRACT

This chapter studies the joint link scheduling and topology control problems in wireless sensor networks. Given arbitrarily located sensor nodes on a plane, the task is to schedule all the wireless links (each representing a wireless transmission) between adjacent sensors using a minimum number of timeslots. There are two requirements for these problems: first, all the links must satisfy a certain property, such as that the wireless links form a data gathering tree towards the sink node; second, all the links simultaneously scheduled in the same timeslot must satisfy the SINR constraints. This chapter focuses on various scheduling algorithms for both arbitrarily constructed link topologies and the data gathering tree topology. We also discuss possible research directions.

INTRODUCTION

Cross-layer design of wireless ad-hoc and sensor networks has received increasing attention in the past several years (Goldsmith & Wicker, 2002). Most of these work focused on the interplay among the physical, MAC and network layer, resulting in various joint designs of power control, modulation and coding, link scheduling and routing. Very few

of them, however, have considered joint design with topology control. Topology control (Gao et al., 2008; Santi, 2005) is the strategy to tune the sensors' transmitting powers so that the sensor nodes collectively can maintain a certain global topology such as connectivity. The goal in topology control is to minimize the sensors' power consumption while trying to provide sufficient network capacity. Topology control plays a very important role in wireless sensor networks: first, the packets are sent via radio

DOI: 10.4018/978-1-61520-701-5.ch009

transmissions by which there must be a connected topology (or other topologies, such as t-spanner) to guarantee that the information collected at each sensor can be forwarded to the other sensors; second, since all the sensor nodes are power limited, energy efficiency is a fundamental challenge in sensor networks; third, a high throughput capacity can ensure the collected information to be more quickly sent to the sink nodes, which is crucial in many critical sensor applications. The higher throughput capacity achieved by topology control in wireless sensor networks can be realized by reducing the network's interferences (Wattenhofer et al., 2001), the degree of which is generally considered to be directly related to the sensor network's maximum node degree (Wang & Li, 2003). Such interference degree, as well as the other graph-based interference models developed later by many other researchers (Schmid & Wattenhofer, 2006), however, can not accurately reflect the actual capacity gains of wireless sensor networks in reality. For example, it has been shown that low node degree does not necessarily mean low interference degree (Burkhart et al., 2004), and a higher graph-based interference degree does not necessarily mean lower network capacity (Hua & Lau, 2008). In this chapter, we study two related joint link scheduling and topology control problems, the goal of which is to minimize the number of timeslots used to schedule all the wireless links (transmissions) in any given topology or a specifically constructed topology. Here the number of timeslots used corresponds to the reciprocal of the network capacity.

SYSTEM MODEL AND PROBLEM DEFINITIONS

System Model

We have the following assumptions: (1) All the stationary wireless sensors are arbitrarily located on a plane, and each sensor is equipped with an omni-directional antenna; (2) we assume a single channel and half-duplex mode, which means each sensor can not send to or receive from more than one node, nor to receive and send at the same time; (3) the link capacity is fixed, which means increasing the transmission power only increases its transmitting range but not its capacity; (4) time is slotted with equal durations; (5) we assume the signal-to-interference-plus-noise ratio (SINR) model is applied, which is a popular model approximating the physical reality of signal transmission in a wireless network. The SINR model is more realistic than the graph-based interference models, which also makes our link scheduling problems much more challenging.

The SINR ratio at the receiver of a link i can be represented as (Gupta & Kumar, 2000):

$$SINR_i = \frac{g_{ii} \cdot p_i}{n_i + \sum_{j=1, j \neq i}^{Q} g_{ij} \cdot p_j} \geq \beta$$

(The SINR model)

where p_i denotes the transmission power of link i's transmitter i_s; n_i is the background noise at link i's receiver i_r; g_{ii} and g_{ij} are the link gain from i_s to i_r, and that from link j's transmitter j_s to i_r, respectively; Q denotes the number of simultaneous transmissions with link i; β is the SINR threshold which is larger than or equal to 1. Here the numerator $g_{ii} \cdot p_i$ means the received power at i_r. In the denominator, $g_{ij} \cdot p_j$ means the attenuated power of p_j at i_r and it is regarded as the interference power for link i, thus $\sum_{j=1, j \neq i}^{Q} g_{ij} \cdot p_j$ means the accumulated interferences caused by all the other simultaneous transmissions. Since we do not consider fading effects and possible obstacles in wireless transmissions, the link gain can be represented by an inverse power law model of the link length, i.e., $g_{ii} = 1 / d^{\alpha}(i_s, i_r)$ and $g_{ij} = 1 / d^{\alpha}(j_s, i_r)$. Here $d(,)$ is the Euclidean distance function, and α is the path loss exponent which is equal to 2 in free space, and varies between 2 and 6 in urban areas.

We define a non-negative $Q \times Q$ link gain matrix $H = (h_{ij})$ such that $h_{ij} = \beta \cdot g_{ij} / g_{ii}$, for $i \neq j$, and $h_{ij} = 0$, for $i = j$. We also define a noise vector $\eta = (\eta_i)$ such that $\eta_i = \beta \cdot n_i / g_{ii}$. With these definitions, we can rewrite the SINR inequality as $p_i \geq \sum_{j=1}^{Q} h_{ij} \cdot p_j + \eta_i$. By using a vector-matrix notation, the above inequality becomes $P \geq HP + \eta$, or $(I - H)P \geq \eta$. If there is only one transmitting link, i.e., no interferences from other links, the SINR model degenerates into the SNR (Signal to Noise Ratio) model, which is shown below:

$$p_i \geq \beta \cdot n_i \cdot d^{\alpha}(i_s, i_r) \text{ (The SNR model)}$$

Obviously, the SNR model defines the minimum power of link i's transmitter i_s to use such that the receiver i_r can successfully decode the packet. We now define the spectral radius $\rho(H)$ of the H matrix as $\rho(H) = \max_i | \lambda_i(H) |$ where $\lambda_i(H)$ stands for the ith eigenvalue of H. Let r_i and c_j represent the ith row sum and jth column sum of H, and we have: $r_i = \sum_j h_{ij}$ and $c_j = \sum_i h_{ij}$. Now according to (Andersin et al., 1996), we know the matrix H is a non-negative irreducible matrix. Also by compiling the propositions proposed in (Pillai et al., 2005, Zander, 1992b), we have the following useful properties of the H matrix:

Property 1: $\rho(H)$ increases when any entry of H increases.

Since $h_{ij} = \beta \cdot d^{\alpha}(i_s, i_r) / d^{\alpha}(j_s, i_r)$, we can see that $\rho(H)$ can be reduced by either reducing the threshold value β, the length of any links or by selecting the links which can result in larger $d(j_s, i_r)$ values.

Property 2: $\min_i(r_i) \leq \rho(H) \leq \max_i(r_i)$; $\min_j(c_j) \leq \rho(H) \leq \max_j(c_j)$.

Property 3: $(I - H)^{-1} > 0$ if and only if $\rho(H) < 1$.

Property 4: The power vector $P^* = (I - H)^{-1} \cdot \eta$ is Pareto-optimal in the sense

that $P^* \geq P$ component-wise for any other non-negative P satisfying $(I - H)P \geq \eta$.

Problem Definitions

In this chapter, we study two closely related joint link scheduling and topology control problems. The first (MLSAT) is for given arbitrary link topologies, and the second (MLSTT) is for forming a data gathering tree topology. Note that, from the following problem definitions, we can easily see that if the tree topology has been constructed, MLSTT becomes a special case of MLSAT. The MLSAT problem is a prominent open problem (Locher, Rickenbach & Wattenhofer, 2008). However, as we will see, how to construct the tree topology plays a very important role in the scheduling length. In addition, for the two problems, we assume each link has one packet to be transmitted. In this case, we can take the totally used timeslots T (the scheduling length) as the frame length, which means that the scheduling sequence will be repeated in the subsequent frames, i.e., $X_{i,t} = X_{i,t+k \cdot T}$ ($0 < t \leq T$; k is a positive integer; $X_{i,t}$ equals 1 if link i transmits in timeslot t and 0 otherwise).

Problem **MLSAT** (Minimum Frame Length Link Scheduling for Arbitrary Topologies):

Given n links which are arbitrarily constructed over arbitrarily located sensors on a plane, assign each link's transmitting sensor a power level and a timeslot, such that all the links scheduled in the same timeslot satisfy the SINR constraints and the number of timeslots used is minimized.

Problem **MLSTT** (Minimum Frame Length Link Scheduling for a Data Gathering Tree Topology):

Given n sensors arbitrarily located on a plane, connect these sensors to form a data gathering tree towards the sink such that the number of timeslots used to schedule all the constructed links under the SINR model is minimized.

Figure 1. Left, ten arbitrarily located nodes in a plane; right, nine arbitrarily constructed links over the ten nodes

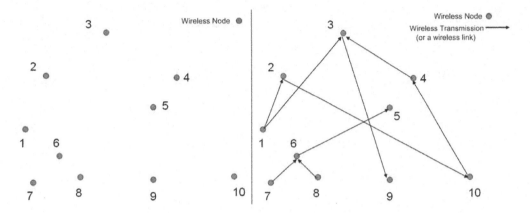

Examples

We give some examples to illustrate the MLSAT and the MLSTT problems. The example in Figure 1 has nine arbitrarily constructed links over ten arbitrarily located nodes on a plane. The MLSAT problem is to assign each transmitting node a power level and a timeslot such that the minimum number of timeslots to successfully schedule all these nine links is achieved, i.e., each transmitting node will be able to send a packet to its receiver. Due to the half-duplex constraint, only the set of links which do not share a common node can be potentially scheduled in the same timeslot. This set of links must form a matching. Then, due to

the SINR model, only those links that satisfy the SINR constraints can successfully send a packet to their receivers. In order to minimize the total timeslots, we may try to schedule as many links as possible in each timeslot. This, however, means more cumulative interferences which could make all the links fail to transmit in the worst scenario. In addition, the aggregated interferences $(\sum_{j=1,j\neq i}^{Q} g_{ij} \cdot p_j = \sum_{j=1,j\neq i}^{Q} (p_j / d^\alpha(j_s, i_r)))$ are directly related to the transmission powers and the geometric distribution of the senders and receivers.

For the MLSTT problem, the links are not already given. Figure 2 shows two different ways

Figure 2. (left) A data gathering tree by the nearest component connector algorithm; (right) A data gathering tree by the minimum spanning tree algorithm

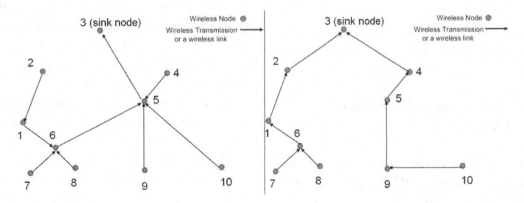

Figure 3. Two pair-wise transmission examples

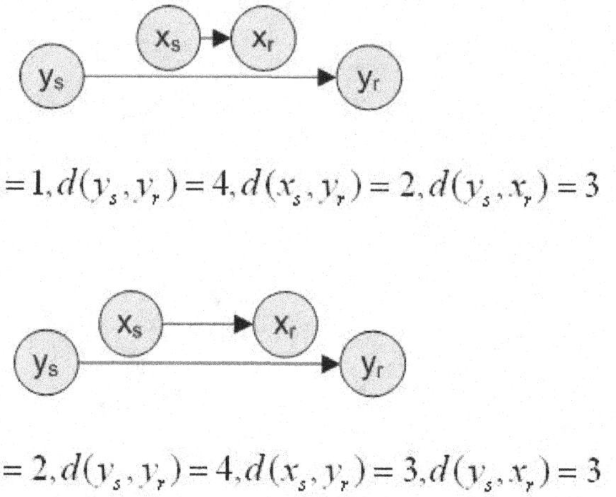

(a) $d(x_s, x_r) = 1, d(y_s, y_r) = 4, d(x_s, y_r) = 2, d(y_s, x_r) = 3$

(b) $d(x_s, x_r) = 2, d(y_s, y_r) = 4, d(x_s, y_r) = 3, d(y_s, x_r) = 3$

to connect the sensors as a tree towards the sink node. The left side is a tree constructed via the nearest component connector algorithm presented in (Fussen, 2004), and the right side is constructed by a minimum spanning tree algorithm. By our discussion about interferences in the last paragraph, different ways of connecting the nodes may result in different scheduling lengths. In tackling the MLSTT problem, the scheduling strategy must be jointly considered with the topology construction algorithm.

We will give further examples in the following to elucidate our link scheduling problems under the SINR model.

THE FOUR FACTORS THAT IMPACT THE SCHEDULING LENGTH

In this section, we discuss the four factors that have a significant influence on the scheduling length of our scheduling problems.

Power Assignment Strategies Make a Difference

We give a pair-wise transmission example in Figure 3(a). According to (Hua&Lau, 2006), we have the following two facts.

FACT 1: Based on the SINR model, if we employ constant power assignment, then in order to ensure simultaneous wireless transmissions (x_s, x_r) and (y_s, y_r), the following two inequalities must hold: $d(y_s, x_r) > \beta^{1/\alpha} \cdot d(x_s, x_r)$ and $d(x_s, y_r) > \beta^{1/\alpha} \cdot d(y_s, y_r)$.

FACT 2: Based on the SINR model, if we employ linear power assignment, then in order to ensure simultaneous wireless transmissions (x_s, x_r) and (y_s, y_r), the following two inequalities must hold: $d(y_s, x_r) > \beta^{1/\alpha} \cdot d(y_s, y_r)$ and $d(x_s, y_r) > \beta^{1/\alpha} \cdot d(x_s, x_r)$.

So for the two transmissions in Figure 3(a), if we employ constant power assignment, we must guarantee $d(x_s, y_r) > \beta^{1/\alpha} \cdot d(y_s, y_r)$ for simultaneous transmissions. Since $\beta \geq 1$, this inequality does not hold and the two transmissions can not be simultaneously scheduled. Similarly, if we employ linear power assignment, we must guarantee $d(y_s, x_r) > \beta^{1/\alpha} \cdot d(y_s, y_r)$, and this inequality does not hold either; so the two transmissions can not be simultaneously scheduled.

Although both constant and linear power assignment can not concurrently schedule the two links, there does exist a power assignment that can simultaneously schedule these two transmissions. For example, if we set $\alpha = 4, \beta = 2, n_i = n_j = 1, p_x = 80$ and

$p_y = 3150$, we can compute the SINR values for transmissions (x_s, x_r) and (y_s, y_r). Since

$$SINR_x = (80 / 1^4) / (1 + 3150 / 3^4) \doteq 2.001 > 2$$
and
$$SINR_y = (3150 / 4^4) / (1 + 80 / 2^4) \doteq 2.051 > 2$$

we can see that these two links can be scheduled in the same timeslot.

From this example, we can conclude that, in order to minimize the total scheduling length, picking the right power assignment strategy is of paramount importance.

Link Topologies Make a Difference

Link topology refers to the geometric distributions of all the senders and receivers of the wireless links. We take the two of the links in Figure 3(b) as an example. Also according to (Hua&Lau, 2006), we have the following fact.

FACT 3: Based on the SINR model, for any pair-wise wireless transmissions (x_s, x_r) and (y_s, y_r), if we have $d(x_s, y_r) \cdot d(y_s, x_r) \le \beta^{2/\alpha} \cdot d(x_s, x_r) \cdot d(y_s, y_r)$, then there does not exist any power assignment strategy that can simultaneously schedule the two links.

For example, for the two links in Figure 3(b), if we set $\alpha = 4$ and $\beta = 2$, then since

$$d(x_s, y_r) \cdot d(y_s, x_r) = 3 \cdot 3$$
$$= 9 < \beta^{2/\alpha} \cdot d(x_s, x_r) \cdot d(y_s, y_r)$$
$$= 2^{2/4} \cdot 2 \cdot 4 \Box 11.31$$

we can see that there are no power assignment strategies that can schedule the two links in the same timeslot.

From this example, we can see that, in a joint link scheduling and topology control algorithm, we must construct a topology that can avoid as many as possible of these pair-wise wireless links that cannot be simultaneously transmitted. In other words, we must find a topology that can take full

advantage of power control to schedule as many links as possible in every timeslot.

Length Diversities Make a Difference

The link topology shown in Figure 4 has a length diversity of 1. This link topology is called a parallel link array and is borrowed from (Baccelli et al. 2006). We now give a theorem which states that this link topology can be scheduled in a constant number of timeslots.

THEOREM 1: The parallel link array given in Figure 4 can be scheduled in m timeslots where m is a constant that satisfies $m \ge (2\alpha\beta / (\alpha - 1))^{1/\alpha}$.

PROOF: In each timeslot, as shown in Figure 4, we just pick all the links where each pair of nearby links has equal horizontal separation distance $d = mh$. If we can prove that all of these links can be successfully scheduled in one timeslot, we can then deduce that the total links can be scheduled in m timeslots. So we need to prove that all the links we pick in each timeslot do satisfy the SINR constraints.

Suppose we use constant power assignment, i.e., all the simultaneously scheduled links employ the same transmission power P. The SINR value for every link i scheduled in the same timeslots is:

$$SINR_i = \frac{P / h^\alpha}{n_i + \sum_k (2 \cdot P / ((k^2 m^2 + 1)^{\alpha/2} h^\alpha))}$$
$$\ge \frac{P / h^\alpha}{n_i + \sum_k (2 \cdot P / (k^\alpha m^\alpha h^\alpha))}$$
$$= \frac{m^\alpha}{n_i \cdot m^\alpha \cdot h^\alpha / P + 2 \cdot \sum_k (1 / k^\alpha)}$$

Suppose the transmission power $P \gg n_i \cdot m^\alpha \cdot h^\alpha$. Then due to a standard Riemann Zeta Function, the above SINR inequality becomes

$$SINR_i \doteq \frac{m^\alpha}{2 \cdot \sum_k (1 / k^\alpha)} \ge \frac{m^\alpha \cdot (\alpha - 1)}{2 \cdot \alpha}$$

Figure 4. Parallel link array with equal lengths and equal horizontal separation distances. (Solid circles mean the transmitters, the arrows mean the receivers)

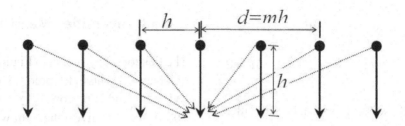

So as long as $m \geq (2\alpha\beta / (\alpha - 1))^{1/\alpha}$, we have $SINR_i \geq \beta$. This completes the proof.

Another link topology is given in Figure 5, whose length diversity equals n which is the number of the links. We call this link topology co-centric exponential node chain and it was first used in (Moscibroda et al. 2007). We set the path loss exponent $\alpha = 3$, the background noise $n_i = 0$ and the threshold $\beta = 2$. The best heuristic link scheduling algorithm so far employs a novel nonlinear power assignment strategy for this link topology, which is presented in (Moscibroda et al. 2007); it can schedule all of these n links in $O(\log n)$ timeslot. Here we need to point out that, considering arbitrary power assignment strategy, no better upper bound or any lower bound of this link topology's scheduling length is known.

From the above, we can see that, for a set of links, the length diversity of the link topology plays a very important role in the scheduling length. Since the links which have constant length diversity can be scheduled in constant scheduling length, we can conclude that the smaller the length diversity of the link topology, the smaller the scheduling length it tends to have. So for the joint link scheduling and topology control algorithm, we should try to construct a link topology which has smaller length diversity. In addition, for link topologies which have large length diversity, such as the co-centric exponential node chain, a good power assignment strategy together with a clever scheduling algorithm is necessary for minimizing the scheduling length.

The Scheduling Policies Make a Difference

We consider the co-centric link topology again. We set the path loss exponent $\alpha = 3$ and the threshold $\beta = 2$. Now according to Fact 3, for any two nearby links, i.e., link i and link $i+1$, there are no power assignment strategies that can schedule the two links in the same timeslot. Suppose now we employ a kind of link removal based scheduling algorithm: First, we try to schedule all the links in the same timeslot; if failure we then choose in each timeslot to remove either the link with the longest length or the link with the shortest length. We repeat these steps until all links have been scheduled. By using this kind of algorithm, we can see that only one link can be scheduled in each timeslot and thus the scheduling length is n. As we have mentioned earlier, there is a clever algorithm given in (Moscibroda et al. 2007) that can schedule all links in time $O(\log n)$. This algorithm works as follows: Let L_i denote the set of links whose lengths d_i satisfy $2^i \leq d_i < 2^{i+1}$, then the algorithm schedules all the links in the link set i ($\bigcup_{k=0}^{n/\log n - 1} L_{i+k\log n}$) where $0 \leq i < \log n - 1$ in one timeslot. By using their nonlinear power assignment, it can be shown that the algorithm can schedule all of these links in one timeslot while satisfying the SINR constraints. Thus all the links can be scheduled in $O(\log n)$ timeslots.

From the example, we can see that designing an efficient scheduling algorithm is the key to our

Figure 5. Co-centric exponential node chain (all the links' senders and receivers are located on the same line with link i's sender's coordinate as $(-2^{(i-1)}, 0)$ and link i's receiver's coordinate as $(2^{(i-1)}, 0)$ (i is from 1 to n)

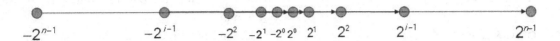

link scheduling problems. We also need to reiterate the important point that the link scheduling algorithm should be jointly designed with the power assignment strategies. For example, according to Theorem 2 in the following, we can see that, if we employ either constant or linear power assignment, there does not exist any scheduling algorithm that can schedule all the links in an efficient manner.

THEOREM 2: By using either constant or linear power assignment, no matter what kind of scheduling policies we propose, all the links in the link topology given in Figure 5 can only be scheduled in n timeslots, i.e., only one link can be scheduled in each timeslot.

PROOF: According to Fact 1 and Fact 2, there are no scheduling policies that can make any two links in Figure 5 schedulable in the same timeslot. This finishes the proof.

HEURISTIC ALGORITHMS FOR MLSAT

We begin this section with the hardness analysis of the MLSAT problem. If we do not allow power control, the MLSAT problem was first proven NP-hard in (Goussevskaia et al. 2007) by using a reduction from the partition problem. Very recently, by assuming that the maximum allowable power is bounded or the available powers are bounded, Völker et al. (2009) proved that the MLSAT problem is also NP-hard even with power control. For arbitrary power levels, the hardness of the MLSAT problem with power

control is still unknown. But for a special case, even with power control, Fu et al. (2009) proved that the minimum length link scheduling problem with arbitrary traffic demands with consecutive transmission constraints is NP-hard. Consecutive transmission constraints mean that each link must be consecutively scheduled. Thus we have to turn to heuristic algorithms to tackle the MLSAT problem. All the existing heuristics for MLSAT can be largely classified as either a top-down or a bottom-up approach. In a top-down approach, the heuristic would first try to pick the maximal number of links (a matching) which do not violate the half-duplex constraint, and then find a maximal link independent set which does not violate the SINR constraints by removing one link at a time. This process will continue until all links have been scheduled. In a bottom-up approach, the heuristic would pick each link incrementally to see if the union of the selected links satisfies the half-duplex and SINR constraints; if not, the link is discarded. This process continues until it finds a maximal link independent set, and until all the links have been scheduled. Since the top-down approach is based on removing one link at each step, it can also be called a link removal based scheduling approach; similarly, since the bottom-up approach is based on incrementing one link at each step, it can also be called a link incremental based scheduling approach. Based on Property 3, since all the heuristic link scheduling algorithms reduce the problem of finding whether there are positive power assignments that satisfy the SINR constraints to the spectral radius checking problem, the time complexities of these algorithms are dominated by the matrix eigenvalue computation. The time complexity for the $n \times n$ matrix eigenvalue computation and matrix inversion problem is $O(n^3)$ (Pan & Chen, 1999).

Top-Down Approach

The first link removal based scheduling algorithm called SRA (Step-wise Removal Algorithm) is proposed by Zander (1992a). For a set of non-adjacent links, this algorithm defers the link which has the maximum value $\max(r_i, c_i)$. The rationale behind this algorithm is based on Property 2, i.e., the spectral radius of the link gain matrix is bounded by the maximum value of the row sum r_i or the column sum c_i. So the SRA algorithm aims to minimize the upper bound of the spectral radius in each removal step. Note that the CSCS (Combined Sum Criterion Selection) algorithm presented in (Fu et al., 2008) is actually the same as SRA. Instead of minimizing the upper bound of the spectral radius, the Step-wise Optimal Removal Algorithm (SORA) proposed by Wu (1999) defers the link whose removal can minimize the spectral radius directly in each step. However, different from SRA which needs only $O(n)$ eigenvalue computations, SORA needs $O(n^2)$ eigenvalue computations. Zander (1992b) proposed another algorithm called LISRA (Limited Information Stepwise Removal Algorithm). In this algorithm, assuming all the links employ the same transmission powers, the link with the minimum SINR value is excluded in each step. SMIRA (Step-wise Maximum Interference Removal Algorithm) (Lee et al., 1995) would compute for each link the larger interference value between the received cumulative interferences from other links and the interferences it caused to all the other links, and then it postpones the link which has the largest interference value. For each link in the WCRP algorithm proposed by Wang et al. (2005), the algorithm first computes a so-called MIMSR (Maximum Interference to Minimum Signal Ratio) value, and then all the links whose MIMSR values exceed some pre-

determined threshold are removed in each step. Also for the set of non-adjacent links, the heuristic algorithm in (Das et al., 2005) discards the link with the maximum row sum value r_i in the link gain matrix.

Having covered the link removal algorithms for non-adjacent links, we now turn to the algorithms for the set of arbitrarily constructed links. To our current knowledge, the two-phase link scheduling algorithm in (Elbatt & Ephremides, 2002) is the first solution to the joint link scheduling and power control problem for ad-hoc networks. In the first phase, this algorithm uses a separation distance to find a "valid" link set, which is also a subset of some maximal matching of the original links. Here, the larger the separation distance, the fewer the number of links in the "valid" link set found. In the second phase, this algorithm tries to find an "admissible" link independent set satisfying the SINR constraints by using the LISRA algorithm in each link removal step. A variation of the two-phase link scheduling algorithm has been presented in (Li & Ephremides, 2007). This algorithm first defines a link metric which is a combination of the link's queue length and the number of blocked links (the number of links sharing either a transmitter node or a receiver node of the current link). Then it finds a maximal matching by greedily selecting a link with the longest queue length and the fewest blocked links (the lowest link metric value). There are two differences between the two-phase scheduling and its variation algorithm: the first is that the variation algorithm sets the separation distance value as zero, which means it tries to find a maximal matching but not a subset; the second difference is that, in order to find an admissible link independent set, the variation algorithm defers the link with the largest link metric, i.e., the link with the shortest queue length and the maximum number of blocked links.

The ISPA (Integrated link Scheduling and Power control Algorithm) algorithm in (Behzad & Rubin, 2007) first constructs a generalized power-based interference graph, which is very similar to the pair-wise link conflict (infeasible) graph proposed in (Hua & Lau, 2008 & 2006). The subtle difference between the two interference models is that the power-based interference graph takes maximum allowable power into account. Note that the links in this graph form a subset of some maximal matching of the original links. Then, by using the minimum degree greedy algorithm (MDGA), the ISPA algorithm finds a maximal number of links which satisfy the SINR constraints pair-wisely. Third, they use the SMIRA algorithm as the pruning method to find a maximal number of links that satisfy the SINR constraints. Fourthly, in a "maximality stage" they try to find more links to be added to the link independent set.

Different from all the previously mentioned link removal based scheduling algorithms, the Algorithm A in (Kozat et al., 2006) first defines each link's effective interference as the corresponding column sum (c_i) in the link gain matrix, and then it finds a maximum matching of the links directly instead of finding a maximal matching or even a subset of the maximal matching. If the maximum matching does not satisfy the SINR constraints, the link with the maximum effective interference is discarded in each link removal step. This process is repeated until all links have been scheduled.

Bottom-Up Approach

As mentioned earlier, the bottom-up approach is based on scheduling each link incrementally. The main difference between the top-down and bottom-up scheduling approaches is that, for a set of non-adjacent links, the top-down approach always consists of two phases, i.e., the link matching searching phase (either a maximum matching, a maximal matching or even just a matching) and the link removal based scheduling phase. The bottom-up approach, however, can directly schedule the links one by one without first finding a link matching. So we can largely classify the bottom-up approach into two categories: matching based scheduling and non-matching based

scheduling. We will first study the non-matching based algorithms since most state-of-the-art link incremental based scheduling algorithms directly schedule the links one by one without first finding a link matching.

Non-Matching Based Algorithms

The first polynomial time approximated link scheduling algorithm called GreedyPhysical is given in (Brar et al., 2006). This algorithm, however, is designed for random networks, which means that the approximation bound can not be generalized to arbitrarily constructed links. Moreover, the algorithm does not use packet-level power control, which means that all the links in the same timeslot employ the same transmission powers. Since this algorithm is designed for links with arbitrary link demands, which means different links may have a different number of packets to be transmitted, it can be easily applied to the unit link demand case; the algorithm first sorts all the links in the decreasing order of their interference numbers. The interference number of a link refers to the number of links which do not share a common node with the current link and can not be concurrently scheduled with it under the SINR model. The algorithm then greedily schedules these links, from the link with the largest interference number to the link with the fewest interference number.

Since the algorithm GreedyPhysical is only an approximation for random networks, Hua & Lau (2008) have given the first polynomial time approximate algorithm for arbitrary link topologies, i.e., solving the MLSAT problem. This algorithm is based on the exponential time exact scheduling algorithm for MLSAT. To the best of our knowledge, this is also the first nontrivial exponential time exact algorithm for MLSAT. By taking advantage of the inclusion-exclusion principle which has been successfully applied in exact graph coloring algorithms, the authors have devised an $O^*(3^n)$ time algorithm called ESA_MLSAT which is also a bottom-up based scheduling algorithm.

In addition, if exponential space is allowed, the time complexity can be reduced to $O^*(2^n)$. Here the $O^*(\cdot)$ notation is used to suppress the poly-logarithmic factor. With these exact scheduling results, the approximation algorithm first partitions all the links into $O(n / \log n)$ groups, and then uses the exact scheduling algorithm ESA_MLSAT in each group. It can thus achieve a polynomial time approximation with an approximation factor $O(n / \log n)$.

The Primal Algorithm proposed in (Borbash & Ephremides, 2006) is designed originally for some kind of "superincreasing" link demands, which means when we sort the link demands in a non-increasing order, each link with a higher demand is greater than or equal to the sum of all the links with lower demands. This algorithm first finds the link with the largest link demand, and then all the other links which can be pair-wisely scheduled with the current link under the SINR model. After that the algorithm schedules these two link sets with the duration of the link with a lower link demand. And then the algorithm checks how many packets have not been transmitted for the link with the largest link demand and schedules this single link packet by packet. The algorithm repeats these steps until all the packets have been transmitted. The authors of this paper have shown that this polynomial time greedy algorithm is optimal for these 'superincreasing' link demands. We can adapt the algorithm to arbitrary link demands by first sorting the links in a decreasing order of their traffic requirements, and then picking each link in order using the bottom-up approach. Obviously, this method can not guarantee the optimal scheduling length for cases with arbitrary link demands.

Also designed for arbitrary link demands, the IDGS (Increasing Demand Greedy Scheduling) algorithm presented in (Fu et al., 2008) first sorts the links in an increasing order of their link demands; and then in each timeslot it picks the link with the lowest link demand, and then it switches to pick the links in a reversed order, i.e., select-

ing the link with the highest link demand using a bottom-up approach.

We now introduce the two non-matching based scheduling algorithms proposed in (Li & Ephremides, 2007). The simplified scheduling algorithm first sorts the links in an increasing order of their link metrics, and then picks each link in order while giving it a power level which is the smaller value of its linear power assignment (a power assignment proportional to its link length to the power of the path loss exponent) and its maximum allowable power level. If any SINR constraints are violated then it defers it to the next timeslot. The second joint link scheduling and power control algorithm (JSPCA) behaves similarly to the simplified scheduling algorithm with the difference that the former one assigns the power levels with the values calculated from the Pareto-optimal power vector P^* (Property 4) rather than the pre-determined power assignments. Compared with the two-phase link removal algorithm and the simplified scheduling algorithm, the authors have shown that the JSPCA algorithm can greatly improve the network performance in terms of throughput and delay. The link scheduling and power control algorithm (LSPC) proposed in (Ramamurthi et al., 2008) first constructs a conflict graph based on the node-exclusive interference model (links sharing a common node can not be concurrently scheduled), and then sorts the links either in an increasing order or in a decreasing order of the node degrees. Finally it schedules the links in order using the bottom-up approach. Note that if we employ the increasing order and if we do not consider a backlogged system (without considering the links' queuing lengths), the LSPC algorithm becomes the same as the JSPCA algorithm introduced in (Li & Ephremides, 2007).

For the throughput maximization problem for single hop links, i.e., to compute the maximum number of packets transmitted on these links in a fixed frame length, Tang et al. (2006) first formulated it as a mixed integer linear programming (MILP) problem, and then they relaxed it as a linear programming problem. In order to generate a link's ordering for the proposed serial linear programming rounding algorithm (SLPR), the authors also relaxed the SINR requirement. By solving the linear programming problem, they sort the links in a decreasing order of the fractional values of the scheduling variables. Finally the greedy SLPR algorithm incrementally schedules these links using the bottom-up approach. The intuitive idea of this link ordering is that, the larger the fractional value of the scheduling variable calculated from the relaxed SINR model, the higher the probability of this link satisfying the original SINR requirement. Note that although this is a polynomial time algorithm, it suffers from an extremely high worst case computational complexity $O(n^8 \cdot M_{LP})$, where n is the number of the links and M_{LP} is the number of binary bits required to store the data.

We now introduce another class of non-matching based scheduling algorithms which feature a kind of nonlinear power assignment. This power assignment can overpower the short links, which means that on one hand, compared with constant power assignment, long links can use larger powers; on the other hand, short links can receive relatively larger power compared with linear power assignment. The nonlinear power assignment is first introduced in an algorithm for the MLSTT problem (Moscibroda & Wattenhofer, 2006) and has subsequently been used for the MLSAT problem. In (Moscibroda, Wattenhofer & Zollinger, 2006), by using the nonlinear power assignment, the authors study the relationship between the graph-based interference model which is called the in-interference degree and the SINR model. The in-interference degree of a node stands for the number of other transmitters whose transmission ranges cover this node. And the largest in-interference degree of a node is called the in-interference degree of the topology. This chapter concludes that the scheduling length of the MLSAT problem is upper bounded by the product of the in-interference degree of the topology and the square of the logarithmic function of

the number of the links. From this, we can see that a lower in-interference degree greatly shortens the scheduling length. In a later paper (Moscibroda, Oswald & Wattenhofer, 2007), the authors propose a low disturbance scheduling algorithm called LDS. This algorithm can generate a polylogarithmic scheduling length for a topology with low disturbances. Here low disturbance is characterized by a parameter called ρ - $disturbance$ which can also be regarded as the density of the links' distribution. For a link's ρ - $disturbance$, the algorithm first computes the number of other links' transmitters (receivers) located in the current link transmitter's (receiver's) range (the link's length divided by the value ρ which is greater than or equal to 1), and then the larger value is the link's ρ - $disturbance$. The maximum ρ - $disturbance$ of all the links becomes the ρ - $disturbance$ of the topology. With this parameter, the authors prove that the scheduling length of the MLSAT problem is upper bounded by the ρ - $disturbance$ of the topology multiplied by the product of the square of the logarithmic function of the number of the links and the square of the ρ value. From this, we know that a sparse link topology with a lower ρ - $disturbance$ can significantly reduce the scheduling length.

Matching Based Algorithms

In this section, we discuss some link incremental scheduling algorithms which are based on either a link matching or a superset of a link matching.

The Algorithm B proposed in (Kozat et al., 2006) is originally designed for minimizing the total power consumption, but it can be adapted for the minimum frame length link scheduling problem with a few modifications. Similar to Algorithm A given in the same paper which uses a top-down approach, the Algorithm B first finds a maximum matching of the unscheduled links; second, it sorts all the links in the maximum matching in a decreasing order of their effective interferences; third, the algorithm can then be adjusted to pick each link in order using the bottom-up approach. The authors have shown that Algorithm B can schedule more links in a timeslot than the top-down approach based Algorithm A.

Recently, Hua (2009) introduced a maximum directed cut based scheduling framework called MDCS. The fundamental differences between this framework and all the other state-of-the-art scheduling algorithms lie in two aspects: the MDCS framework uses a maximum directed cut which also contains a maximum matching as the building block for each phase's scheduling; and in each scheduling phase, the MDCS framework employs a link incremental based scheduling algorithm with novel scheduling metrics. We borrow an illustrative example (c.f. Figure 6) from Hua (2009) to briefly explain the rationale behind MDCS. First, we notice that finding a maxi*mum* matching in the bottom up based scheduling algorithms is preferred to finding a maxi*mal* matching or even just a matching. The reason is that, compared with the maximal matching or just a matching, the maximum matching can offer more potential links that can be scheduled in the same timeslot. Second, we can see that adding more links in the maximum matching can offer more potential links to be scheduled in the same timeslot. Since there may be more than one maximum matching, this step can be taken as diversifying the maximum matching found. For example, in the following example, there are $3n+1$ links and any maximum matching consists of $n+1$ links. Here we suppose

the found maximum matching is composed by link 1 and links from links $2n+2$ to $3n+1$. Now we can add links from links 2 to $n+1$ to this maximum matching. Thus if any link in the added links can be concurrently scheduled with the links in the found maximum matching, there will be fewer links in the subsequent scheduling phases which could lead to much fewer timeslots to schedule all the links. The problem then is how to add the non-matched links to the maximum matching. Examining the link gain matrix H, we can see that if adding a link to the maximum matching can make a link's transmitter (receiver) become another link's receiver (transmitter), the denominator of some element of the link gain matrix would become infinity which is very undesirable for any scheduling or removal metrics built upon the elements of the link gain matrix. So the problem boils down to finding the maximum directed cut upon a maximum matching. Also taking Figure 6 as an example, the found maximum directed cut comprises the maximum matching consisting of link 1 and links from links $2n+2$ to $3n+1$ and all the other links excluding link $n+1$. For more details of this maximum directed cut based scheduling framework and the various scheduling metrics, please refer to (Hua, 2009).

ALGORITHMS INEFFICIENCY ANALYSES

In this section, we give some inefficiency results for both top-down and bottom-up based link scheduling algorithms.

THEOREM 3: The following top-down based link scheduling algorithms have a worst case lower bound of $\Omega(n)$: the two phase scheduling algorithm (Elbatt & Ephremides, 2002), the variation of the two phase scheduling algorithm (Li & Ephremides, 2007), the ISPA algorithm (Behzad & Rubin, 2007), the Algorithm A (Kozat et al., 2006) and the heuristic link scheduling in (Das et al., 2005).

Figure 6. An example for the MDCS scheduling framework

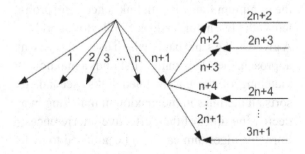

PROOF: Since the two phase scheduling algorithm and the ISPA algorithm use LISRA and SMIRA as their link removal algorithms respectively, the inefficiency results of the two link removal algorithms (Theorem 5.2 in (Moscibroda, Oswald & Wattenhofer, 2007)) can be directly applied here. For the other three scheduling algorithms, we can make use of the co-centric exponential node chain given in Figure 5. We can set the path loss exponent $\alpha = 3$, the background noise $n_i = 0$ and the threshold $\beta = 2$. For the variation of the two phase scheduling algorithm, since all the links have the same number of blocked links (zero), the links removed in each step are link 1 to link n-1, so only one link (link n) can be scheduled in the first timeslot. These removal steps will be repeated in the following n-1 timeslots. For the Algorithm A and the heuristic link scheduling, since they either use the link gain matrix column sum or row sum as their link removal metrics, the links removed in each step are either in an increasing order of their links' lengths or in a decreasing order of their links' lengths. However, both orders will result in $\Omega(n)$ scheduling lengths. This completes the proof.

THEOREM 4: The two bottom-up based link scheduling algorithms, i.e., the simplified scheduling algorithm in (Li & Ephremides, 2007) and the GreedyPhysical algorithm in (Brar, Blough & Santi, 2006), have a worst case lower bound $\Omega(n)$.

PROOF: We make use of the co-centric exponential node chain. Since all the links form a matching, the algorithm can schedule the links in a decreasing order of their lengths. So depending on the value of maximum allowable transmission power, the corresponding power assignments can be either linear power assignments, constant power assignments, or the long links employing constant power assignments while the remaining short links would employ linear power assignments. By using the inefficiency results of both constant and linear power assignments (Theorem 3.1 and 3.2 in (Moscibroda & Wattenhofer, 2006))

or Theorem 4.1 in (Hua & Lau, 2006), we can complete the proof for the simplified scheduling algorithm. Similarly since the GreedyPhysical algorithm does not employ packet-level power control, which means that all the links in the same timeslot use the same transmission powers (the links in different timeslots may use different powers), Theorem 4.1 in (Hua & Lau, 2006) can be directly applied here. This completes the proof for the GreedyPhysical algorithm.

PROPOSITION 5: Let's suppose there is a link topology whose pair-wise link conflict (infeasible) graph (Hua & Lau, 2008) is as shown in the following figure, then any link incremental scheduling algorithms in the order of $[1..n]$ will result in a scheduling length of $\Omega(n)$. However, a much fewer or even a constant number of timeslots is possible if we schedule the links in the upper and lower parts of this conflict graph respectively. This can be realized by the step-wise least discarded link incremental scheduling algorithm called SLDIA proposed in (Hua, 2009). This algorithm incrementally schedules the link whose addition in the current link independent set can discard the fewest number of links in the remaining links.

From this proposition, we have the following three corollaries.

COROLLARY 6: The link incremental scheduling algorithms which use the node degree in the pair-wise link conflict graph as the scheduling metric has a worst case lower bound of $\Omega(n)$.

Figure 7. A pair-wise link conflict (infeasible) graph

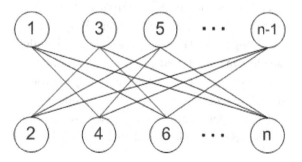

COROLLARY 7: Since all the links have unit link demand in MLSAT, the link incremental scheduling algorithms which use the link demands as a scheduling metric, such as the Primal Algorithm in (Borbash & Ephremides, 2006) and the IDGS algorithm in (Fu, Liew & Huang, 2008), have a worst case lower bound of $\Omega(n)$.

COROLLARY 8: Let's further suppose all the links in this link topology have the same number of blocked links, then the link incremental scheduling algorithms which use the number of blocked links as the scheduling metric, such as the JSPCA algorithm in (Li & Ephremides, 2007) and the LSPC Algorithm in (Ramamurthi et al., 2008), have a worst case lower bound of $\Omega(n)$.

JOINT TOPOLOGY CONSTRUCTION AND LINK SCHEDULING FOR MLSTT

In this section, we review some joint topology construction and link scheduling algorithms for the MLSTT problem. The first algorithm for this problem is given in (Moscibroda & Wattenhofer, 2006) in the context of fulfilling the connectivity property of all the arbitrarily located nodes on the plane. If we remove the last step of this algorithm, i.e., adding the links from the sink node to all the other nodes, this algorithm can be directly used for MLSTT. Since this algorithm employs the nonlinear power assignment and is targeted for narrow band networks, we call it NPAN. The NPAN proceeds in phases, where each phase comprises all the links in the nearest neighbor forest constructed over the sink nodes of the links in the previous phase. Here the sink node means the node with no outgoing links. This scheduling algorithm partitions the links in each phase into different groups based on the links' lengths, and then it incrementally schedules each link in the selected groups with the nonlinear power assignment. Since there are $O(\log n)$ groups in each scheduling phase (n is the number of the nodes) and there are $O(\log n)$ phases, by combining the scheduling length of each group which is

bounded by $O(\log^2 n)$, the total scheduling length is $O(\log^4 n)$. In a follow-up paper (Hua & Lau, 2006), the authors have studied how the wide-band networks would affect the poly-logarithmic scheduling length. They prove that, for a wide-band network with processing gain m, the scheduling length can be reduced to $O(\log(n / m) \cdot \log^3 n)$. This result shows that a higher processing gain can greatly shorten the scheduling length, especially when $m = \Theta(n)$. In addition, the paper also points out that the poly-logarithmic scheduling length is achieved at the expense of total power consumption which is an exponential function of the number of the nodes. Now if we do not schedule the links in each phase but rather to schedule the links when the tree topology has been constructed after $O(\log n)$ steps, the scheduling length can be reduced to $O(\log^3 n)$. This result is derived from the paper (Moscibroda, Wattenhofer & Zollinger, 2006) which proves that the scheduling length for arbitrary topologies is bounded by the in-interference degree of the topology times $O(\log^2 n)$, and the in-interference degree of the iteratively constructed tree topology is $O(\log n)$. By using a slightly different nonlinear power assignment in (Moscibroda, 2007), the scheduling length has been further reduced to $O(\log^2 n)$. In this chapter, the algorithm first iteratively constructs the tree through the nearest component connector algorithm (Fussen, 2004) which is almost the same as the nearest forest connection algorithm. Second, the algorithm partitions all the links in constant number of groups based on the links' lengths. The final result is reached since the scheduling length for each group is $O(\log^2 n)$. We call this scheduling algorithm NPAN-IPSN07. Here we take note that the results of (Hua & Lau, 2006) can be easily extended to all these follow-up nonlinear power assignment based scheduling algorithms.

Instead of iteratively connecting all the arbitrarily located nodes with either the nearest forest connection algorithm or the nearest component connector algorithm, Hua (2009) has recently proposed another joint topology construction and link scheduling algorithm based on first constructing a minimum spanning tree. The algorithm does not use nonlinear power assignment based scheduling algorithms but rather the proposed maximum directed cut based scheduling framework (MDCS).

ALGORITHM COMPARISONS

In this section, we compare the scheduling lengths generated by various link scheduling algorithms for both the MLSAT and MLSTT problems introduced in this chapter.

Comparisons of Algorithms for MLSAT

First we show how the arbitrary link topologies are generated. For any n arbitrarily located nodes in a 2000×2000 m² plane, we randomly select a link's transmitter and receiver subject to the constraint that they are different nodes on the plane. We then repeat this process until a number of n different links (either with different transmitters or receivers) have been constructed. So in this topology construction some nodes may not be used (Figure 8 is an example). In this simulation, we set the path loss exponent $\alpha = 4$ and the threshold $\beta = 2$. In fact we have also tested all the scheduling algorithms for the other (α, β) values. Since the arbitrarily generated link topology is a very dense link topology (c.f. Figure 8), if we choose a smaller α value or a larger β value, all of the scheduling algorithms can schedule at most one link in each timeslot which would make performance comparison impossible. However for some other (α, β) values which either have a larger α value or a smaller β value, all the algorithms behave similarly with the $\alpha = 4$ and $\beta = 2$ setting. So we only give the simulation results for the $(\alpha = 4, \beta = 2)$ case. We implemented seven bottom-up based scheduling algorithms: the MDCS scheduling

framework (Hua, 2009), the adjusted Algorithm B (Kozat, Koutsopoulos & Tassiulas, 2006), the GreedyPhysical algorithm in (Brar, Blough & Santi, 2006) with packet level power control, the JSPCA algorithm in (Li & Ephremides, 2007), the LSPC algorithm in (Ramamurthi et al., 2008), the LDS algorithm in (Moscibroda, Oswald & Wattenhofer, 2007) and the first fit based link increment scheduling algorithm. Here by first fit based link incremental scheduling algorithm, we mean that we just greedily schedule the links in its unsorted order with the bottom up approach. In addition, in order to differentiate from the JSPCA algorithm, the LSPC algorithm employs a decreasing order of the number of blocked links to incrementally schedule the links. Note that for the LDS algorithm, since its scheduling length relies on the parameter ρ, we have tested different ρ values and find that LDS can achieve the shortest scheduling length when $\rho = 1$, so we set $\rho = 1$ in our simulation. We also implement one top-down based scheduling algorithm which uses the link removal algorithm SORA. This algorithm first finds a maximum matching in each scheduling phase; then it employs SORA as the link removal algorithm. The reasons we use SORA as a representative for top down based link removal algorithms are: first, the simulation results in (Wu, 1999) have shown that, compared with SRA and SMIRA, SORA has the lowest outage probability and a better throughput capacity; second, for the co-centric exponential node chain topology, our own simulation result shows that the SORA algorithm can schedule it with the number of timeslots no more than that by the nonlinear power assignment based link scheduling algorithm given in (Moscibroda, Oswald & Wattenhofer, 2007); third, compared with all the other link removal based scheduling algorithms which have worst case lower bound $\Omega(n)$ where n is the number of the links, the scheduling length lower bound for the SORA algorithm is still unknown. Note that, we have tested these scheduling algorithms over ten sets of link topologies with the number

Figure 8. An arbitrary link topology example with 20 links constructed over 20 nodes

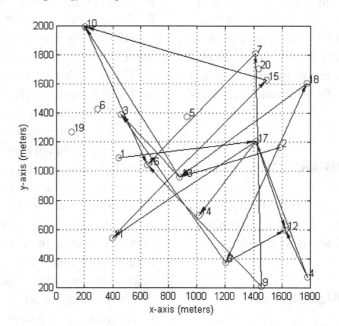

of links ranging from 20 to 110. And for each set of topology, we compute the average scheduling length over 10 different instances. In addition, for all the scheduling algorithms except LDS, we use the Pareto-optimal power assignment with no maximum allowable power limitations. This assumption, however, can be removed if we set the same maximum allowable power for all the scheduling algorithms.

The final scheduling results can be seen in Figure 9. Now we can sort these eight scheduling algorithms in an increasing order of their scheduling lengths: MDCS, the adjusted Algorithm B, LSPC, first fit, JSPCA, SORA, the adjusted GreedyPhysical with power control and LDS. We have the following observations from this ordering. (1) In matching based link scheduling algorithms, adding more links to the maximum matching in each scheduling phase can significantly reduce the scheduling length. This can be seen from the scheduling lengths of MDCS, the adjusted Algorithm B and the matching based link removal algorithm SORA. (2) Matching based link scheduling algorithms greatly outperform the non-matching based link scheduling algorithms

in terms of their scheduling lengths. This can be seen from the scheduling lengths of Algorithm B and the other four non-matching based scheduling algorithms (LSPC, first fit, JSPCA and Greedy-Physical). This observation is further strengthened by the result that even the matching based link removal algorithm SORA can generate fewer scheduling lengths than the non-matching based link incremental scheduling algorithms (the adjusted GreedyPhysical and LDS). (3) Compared with the top down and bottom up based scheduling algorithms, especially for all the matching based link scheduling algorithms, link incremental scheduling algorithms can greatly reduce the scheduling lengths compared with the link removal algorithms. This can be seen from the scheduling lengths of the algorithms MDCS, the Adjusted Algorithm B, LSPC, first fit, JSPCA and SORA. (4) The Fail First principle which corresponds to first selecting the link with the largest scheduling metric value outperforms the Succeed First principle which corresponds to first selecting the link with the smallest scheduling metric value. This is supported by the results from LSPC and JSPCA. (5) Since our generated arbitrary link topologies

Figure 9. Comparisons of scheduling lengths over arbitrary link topologies with different algorithms

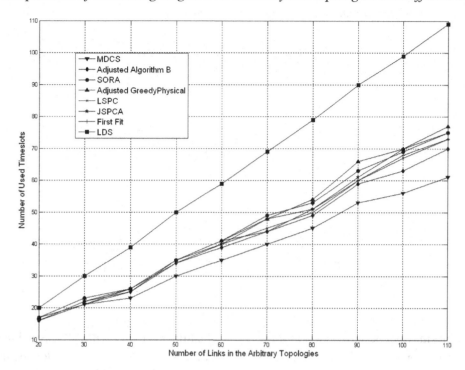

bear large $\rho_{-}disturbance$ values (Figure 8 is an example whose $\rho_{-}disturbance$ value could be as large as the number of links when $\rho = 1$), the low disturbance scheduling (LDS) generates the longest scheduling lengths (it almost schedules one link in each timeslot!). Since both the tree topologies shown in Figure 10 have much smaller $\rho_{-}disturbance$ values, we will see how LDS performs in these sparse link topologies in the next section.

Comparisons of Algorithms for MLSTT

All the nodes are also arbitrarily located in a $2000 \times 2000 \, m^2$ plane and we set the path loss exponent $\alpha = 4$ and the threshold $\beta = 20$. The reason for setting a much higher threshold value here is that the constructed tree topologies (c.f. Figure 10) are sparse link topologies, while the arbitrary link topologies (c.f. Figure 8) are dense link topologies. If we use larger α values or smaller β values, then

all the scheduling algorithms will generate almost the same very short scheduling lengths which would make the comparisons impossible. On the other hand, we have also tested the scheduling algorithms for other smaller α values or larger β values, and all the scheduling algorithms behave similarly with the setting $\alpha = 4$ and $\beta = 20$. So we omit these similar simulation results here. In Figure 10, the left side is a tree topology iteratively constructed by the nearest component connector (NCC) algorithm while the right side is a minimum spanning tree constructed over the same node set. Besides the MDCS scheduling framework and the LDS algorithm, we also implement the NPAN-IPSN07 algorithm which is currently the fastest nonlinear power assignment based link scheduling algorithm which can schedule the NCC-tree (tree constructed with NCC algorithm) in time $O(\log^2 n)$ (Mosciborda, 2007). And since the in-interference degree of a MST topology can be $O(n)$ we can not use the NPAN-IPSN07 algorithm to schedule the links in the MST topology since the

Figure 10. Different tree topologies over the same set of nodes (Left: iterative nearest component connector construction; Right: minimum spanning tree construction)

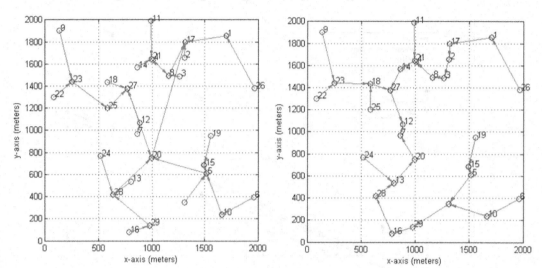

SINR constraints may not be satisfied. So for the MST topology, we apply the MDCS and the LDS scheduling algorithms, and for the NCC tree, we can also apply the NPAN-IPSN07 algorithm. But for the NPAN-IPSN07 algorithm, we must pay attention to the background noise value n_i since the scheduling length is also dependent on this parameter. Note that, in this algorithm, when the background noise $n_i < (\alpha - 2) / (2\beta \cdot (\alpha - 1))$ the SINR constraints can not be guaranteed by the proposed nonlinear power assignment (the reason is that the SNR model must be satisfied). So in this simulation, we set all the n_i to have the same value which is a little bit larger than $(\alpha - 2) / (2\beta \cdot (\alpha - 1))$ since we have found that a much larger n_i value can greatly increase the scheduling length.

The scheduling results are shown in Figure 11. From this figure we have the following observations: (1) the MST topology always yields much shorter scheduling lengths no matter which scheduling algorithm is used; (2) combined with Figure 4, for the MST and NCC tree topologies having much lower ρ - $disturbance$ values, LDS generates shorter scheduling lengths; although the reduction is not that significant, the reduction of

scheduling lengths with MDCS is huge; (3) for both MST and NCC tree topologies, the MDCS algorithm always achieves the shortest scheduling lengths; (4) for NCC tree, compared with the NPAN-IPSN07 algorithm, MDCS achieves a much shorter scheduling length.

CONCLUSION

This chapter reviews all the state-of-the-art polynomial time link scheduling algorithms under the SINR model. We have studied these algorithms through theoretical analyses as well as using simulation. We can draw some conclusions from the results. First, for both dense and sparse link topologies, the maximum directed cut based scheduling framework MDCS significantly outperforms all the other state-of-the-art link scheduling algorithms in terms of scheduling length. Second, our results show that connecting all the nodes (sensors) on a plane with the minimum spanning tree topology can greatly shorten the scheduling lengths, which means that the data gathering speed can be significantly increased. Third, matching based scheduling algorithms help reduce the schedul-

Figure 11. Comparison of scheduling lengths over different tree topologies with different algorithms

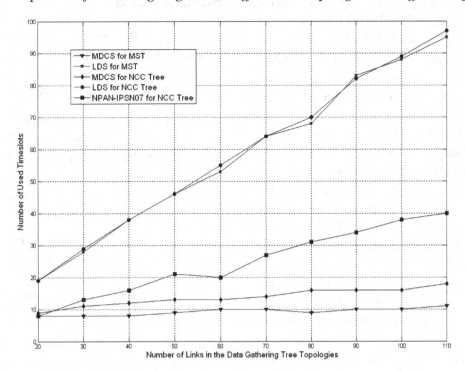

ing length compared with non-matching based scheduling algorithms. Fourth, link incremental based scheduling algorithms can greatly shorten the scheduling length compared with link removal based scheduling algorithms. Moreover, the time complexities of link incremental based scheduling algorithms are much lower than those of link removal based scheduling algorithms (Kozat, Koutsopoulos & Tassiulas, 2006).

There are many open problems in this research area that warrant further attention and investigation. Here we could only touch upon a small subset of them. For more open problems, please refer to (Hua & Lau, 2006, 2008, Hua, 2009 & Hua et al., 2009a,2009b).

First, instead of assuming each link having one packet to transmit, we can study the general minimum length link scheduling problem with arbitrary traffic demands over the links. Although the hardness of the MLSAT problem under power control has been proven to be NP-hard (Fu et al., 2009), the general minimum length link scheduling problem without consecutive transmission constraints is still open.

Second, although there are some approximated algorithms for either the MLSAT problem or the general minimum length link scheduling problem (Goussevskaia et al., 2009, Halldorsson & Wattenhofer, 2009), all their approximation ratios are obtained with the assumption of no power control. So it seems necessary to study approximation algorithms under power control. Furthermore, it would be interesting to study whether there are some inapproximability results for the minimum length link scheduling problems.

Third, although a polynomial time approximation algorithm for MLSAT has been proposed in (Hua & Lau, 2008), it is a centralized algorithm. A localized/distributed algorithm, where each sensor only has limited knowledge of the whole network, is necessary for sensor networks that may experience many changes dynamically. For example, we may want a sensor node to decide its transmission power locally while guaranteeing higher throughput capacity and lower power consumption.

Fourth, for the asymptotic upper bound of the scheduling length of the MLSTT problem, our simulation results and analyses have shown that the currently fastest $O(\log^2 n)$ bound (Moscibroda, 2007) can be further reduced, which needs a novel scheduling algorithm. Moreover, a non-trivial lower bound is also needed.

Fifth, it will be interesting to consider more layers of the sensor networks, such as the networking layer. For example, a joint link scheduling, topology control and routing solution with a much shorter provable scheduling length can be very challenging (Chafekar et al., 2007).

Sixth, it will also be interesting to consider other joint link scheduling and topology control problems. For example, we can consider the minimum frame length link scheduling problem for either a k-connected topology or a t-spanner topology.

Seventh, for a small number of links, it is possible to design some efficient exact algorithms for either the MLSAT or the general minimum length link scheduling problems. These problems can be formulated as a set covering problem (Hua & Lau, 2008) or as a set multi-covering problem (Hua et al., 2009a, 2009b).

Finally, it should be worthwhile to take the packets' arriving rates into account (i.e., stochastic network) when trying to solve the joint link scheduling and topology control problems (Joo, Lin & Shroff, 2008).

ACKNOWLEDGMENT

This research is supported in part by a Hong Kong RGC-GRF grant (7136/07E), the national 863 high-tech R&D program of the Ministry of Science and Technology of China under grant No. 2006AA10Z216, the National Science Foundation of China under grant No. 60604033, and the National Basic Research Program of China Grant 2007CB807900, 2007CB807901.

REFERENCES

Andersin, M., Rosberg, Z., & Zander, J. (1996). Gradual removals in cellular PCS with constrained power control and noise. *Wireless Networks*, *2*(1), 27–43. doi:10.1007/BF01201460

Baccelli, F., Bambos, N., & Chan, C. (2006). Optimal Power, Throughput and Routing for Wireless Link Arrays. In *25th IEEE International Conference on Computer Communications, Joint Conference of the IEEE Computer and Communications Societies (INFOCOM)*. Barcelona, Spain: IEEE.

Behzad, A., & Rubin, I. (2007). Optimum integrated link scheduling and power control for multihop wireless networks. *IEEE Transactions on Vehicular Technology*, *56*(1), 194–205. doi:10.1109/TVT.2006.883734

Borbash, S. A., & Ephremides, A. (2006). Wireless link scheduling with power control and SINR constraints. *IEEE Transactions on Information Theory*, *52*(11), 5106–5111. doi:10.1109/TIT.2006.883617

Brar, G., Blough, D., & Santi, P. (2006). Computationally efficient scheduling with the physical interference model for throughput improvement in wireless mesh networks. In M. Gerla, C. Petrioli, & R. Ramjee (Ed.), *Proceedings of the 12th annual international conference on Mobile computing and networking (MOBICOM)* (pp.2-13). Los Angeles: ACM.

Burkhart, M., Rickenbach, P., von Wattenhofer, R., & Zollinger, A. (2004). Does Topology Control Reduce Interference? In J. Murai, C. E. Perkins, & L. Tassiulas (Eds.), *Proceedings of the 5th ACM Interational Symposium on Mobile Ad Hoc Networking and Computing (MOBIHOC)* (pp.9-19). Tokyo: ACM.

Chafekar, D., Kumar, V. S., Marathe, M., Parthasarathy, S., & Srinivasan, A. (2007). Cross-layer latency minimization for wireless networks using SINR constraints. In E. Kranakis, E. M. Belding, & E. Modiano (Eds.), *Proceedings of the 8th ACM Interational Symposium on Mobile Ad Hoc Networking and Computing (MOBIHOC)* (pp.110-119). Montreal, Quebec: ACM.

Das, A. K., Marks, R. J., II, Arabshahi, P., & Gray, A. (2005). Power controlled minimum frame length scheduling in TDMA wireless networks with sectored antennas. In *24th IEEE International Conference on Computer Communications, Joint Conference of the IEEE Computer and Communications Societies (INFOCOM)*. Miami, FL: IEEE.

ElBatt, T., & Ephremides, A. (2002). Joint scheduling and power control for wireless ad-hoc networks. In *21st IEEE International Conference on Computer Communications, Joint Conference of the IEEE Computer and Communications Societies (INFOCOM)*. New York: IEEE.

Fu, L., Liew, S., & Huang, J. (2008). Joint power control and link scheduling in wireless networks for throughput optimization. In *Proceedings IEEE ICC*. Beijing, China: IEEE.

Fu, L., Liew, S., & Huang, J. (2009). Power controlled scheduling with consecutive transmission constraints: complexity analysis and algorithm design. In *28th IEEE International Conference on Computer Communications, Joint Conference of the IEEE Computer and Communications Societies (INFOCOM)*. Rio de Janeiro, Brazil: IEEE.

Fussen, M. (2004). *Sensor networks: interference reduction and possible applications*. Diploma Thesis, Distributed Computing Group, ETH Zurich, Switzerland.

Gao, Y., Hou, J. C., & Nguyen, H. (2008). Topology control for maintaining network connectivity and maximizing network capacity under the physical model. In *27th IEEE International Conference on Computer Communications, Joint Conference of the IEEE Computer and Communications Societies (INFOCOM)*. Phoenix, AZ: IEEE.

Goldsmith, A. J., & Wicker, S. B. (2002). Design challenges for energy-constrained ad hoc wireless networks. *IEEE Wireless Communications*, *9*(4), 8–27. doi:10.1109/MWC.2002.1028874

Goussevskaia, O., Halldorsson, M., Wattenhofer, R., & Welzl, E. (2009). Capacity of arbitrary wireless networks. In *28th IEEE International Conference on Computer Communications, Joint Conference of the IEEE Computer and Communications Societies (INFOCOM)*. Rio de Janeiro, Brazil: IEEE.

Goussevskaia, O., Oswald, Y. A., & Wattenhofer, R. (2007). Complexity in geometric SINR. In E. Kranakis, E. M. Belding, & E. Modiano (Eds.), *Proceedings of the 8th ACM Interational Symposium on Mobile Ad Hoc Networking and Computing (MOBIHOC)* (pp.110-119). Montreal, Quebec: ACM.

Gupta, P., & Kumar, P. R. (2000). The capacity of wireless networks. *IEEE Transactions on Information Theory*, *46*(2), 388–404. doi:10.1109/18.825799

Hajek, B. E., & Sasaki, G. H. (1988). Link scheduling in polynomial time. *IEEE Transactions on Information Theory*, *34*(5), 910–917. doi:10.1109/18.21215

Halldorsson, M., & Wattenhofer, R. (2009). Wireless communication is in APX. In *36th International Colloquium on Automata, Languages and Programming (ICALP)*, Rhodes, Greece: Springer-Verlag.

Hua, Q.-S. (2009). *Scheduling wireless links with SINR constraints*. PhD Thesis, The University of Hong Kong.

Hua, Q.-S., & Lau, F. C. M. (2006). The scheduling and energy complexity of strong connectivity in ultra-wideband networks. In E. Alba, C.-F. Chiasserini, N. B. Abu-Ghazaleh, & R. Lo Cigno (Eds.), *Proceedings of the 9th International Symposium on Modeling Analysis and Simulation of Wireless and Mobile Systems (MSWiM)* (pp.282-290). Terromolinos, Spain: ACM.

Hua, Q.-S., & Lau, F. C. M. (2008). Exact and approximate link scheduling algorithms under the physical interference model. In M. Segal, & A. Kesselman (Eds.), *Proceedings of the DIALM-POMC Joint Workshop on Foundations of Mobile Computing* (pp.45-54). Toronto, Canada: ACM.

Hua, Q.-S., Wang, Y., Yu, D., & Lau, F. C. M. (2009a). Set multi-covering via Inclusion-Exclusion. *Theoretical Computer Science, 410*(38-40), 3882–3892. doi:10.1016/j.tcs.2009.05.020

Hua, Q.-S., Yu, D., Lau, F. C. M., & Wang, Y. (2009b). Exact algorithms for set multicover and multiset multicover problems. In Y. Dong, D.-Z. Du, O. H. Ibarra (Eds.), *Proceedings of the 20th International Symposium on Algorithms and Computation (ISAAC)* (pp. 34-44). Hawaii: Springer.

Joo, C., Lin, X., & Shroff, N. B. (2008). Understanding the capacity region of the greedy maximal scheduling algorithm in multi-hop wireless networks. In *27th IEEE International Conference on Computer Communications, Joint Conference of the IEEE Computer and Communications Societies (INFOCOM)*. Phoenix, AZ: IEEE.

Kozat, U. C., Koutsopoulos, I., & Tassiulas, L. (2006). Cross-layer design for power efficiency and QoS provisioning in multi-hop wireless networks. *IEEE Transactions on Wireless Communications, 5*(11), 3306–3315. doi:10.1109/TWC.2006.05058

Lee, T. H., Lin, J. C., & Su, Y. T. (1995). Downlink power control algorithms for cellular radio systems. *IEEE Transactions on Vehicular Technology, 44*(1), 89–94. doi:10.1109/25.350273

Li, Y., & Ephremides, A. (2007). A joint scheduling, power control, and routing algorithm for ad hoc wireless networks. *Ad Hoc Networks, 5*(7), 959–973. doi:10.1016/j.adhoc.2006.04.005

Locher, T., von Rickenbach, P., & Wattenhofer, R. (2008). Sensor networks continue to puzzle: selected open problems. In S. Rao, M. Chatterjee, P. Jayanti, C. S. R. Murthy, & S. K. Saha (Eds.): *9th International Conference on Distributed Computing and Networking (ICDCN)* (pp.25-38). Kolkata, India: Springer.

Moscibroda, T. (2007). The worst-case capacity of wireless sensor networks. In T. F. Abdelzaher, L. J. Guibas, & M. Welsh (Eds.), *Proceedings of the 6th International Conference on Information Processing in Sensor Networks (IPSN)* (pp.1-10). Cambridge, MA: ACM.

Moscibroda, T., Oswald, Y. A., & Wattenhofer, R. (2007). How optimal are wireless scheduling protocols? In *26th IEEE International Conference on Computer Communications, Joint Conference of the IEEE Computer and Communications Societies (INFOCOM)* (pp. 1433-1441). Anchorage, AK: IEEE.

Moscibroda, T., & Wattenhofer, R. (2006). The complexity of connectivity in wireless networks. In *25th IEEE International Conference on Computer Communications, Joint Conference of the IEEE Computer and Communications Societies (INFOCOM)*. Barcelona, Spain: IEEE.

Moscibroda, T., Wattenhofer, R., & Weber, Y. (2006). Protocol design beyond graph-based models. In Proceedings 5th Workshop on Hot Topics in Networks (HotNets-V) (pp.25-30). Irvine, California: ACM.

Moscibroda, T., Wattenhofer, R., & Zollinger, A. (2006). Topology control meets SINR: the scheduling complexity of arbitrary topologies. In S. Palazzo, M. Conti, & R. Sivakumar (Eds.), *Proceedings of the 7th ACM International Symposium on Mobile Ad Hoc Networking and Computing (MOBIHOC)* (pp.310-321). Florence, Italy: ACM.

Pan, V. Y., & Chen, Z. Q. (1999). The complexity of the matrix eigenproblem. In *Proceedings of the 31st Annual ACM Symposium on Theory of Computing (STOC)* (pp. 507-516). Atlanta, GA: ACM.

Pillai, S. U., Suel, T., & Cha, S. (2005). The perron-frobenius theorem and some of its applications. *IEEE Signal Processing Magazine, 22*(2), 62–75. doi:10.1109/MSP.2005.1406483

Ramamurthi, V., Reaz, A. S., Dixit, S., & Mukherjee, B. (2008). Link scheduling and power control in wireless mesh networks with directional antennas. In *Proceedings IEEE ICC*. Beijing, China: IEEE.

Santi, P. (2005). *Topology control in wireless ad hoc and sensor networks*. Chichester, UK: John Wiley and Sons.

Schmid, S., & Wattenhofer, R. (2006). Algorithmic models for sensor networks. In *Proceedings 20th International Parallel and Distributed Processing Symposium (IPDPS)*. Rhodes Island, Greece: IEEE.

Tang, J., Xue, G., Chandler, C., & Zhang, W. (2006). Link scheduling with power control for throughput enhancement in multi-hop wireless networks. *IEEE Transactions on Vehicular Technology, 55*(3), 733–742. doi:10.1109/TVT.2006.873836

Völker, M., Katz, B., & Wagner, D. (2009). *On the complexity of scheduling with power control in geometric SINR*. Technical Report 15, ITI Wagner, Faculty of Informatics, Universität Karlsruhe.

Wang, K., Chiasserini, C.-F., Rao, R. R., & Proakis, J. G. (2005). A joint solution to scheduling and power control for multicasting in wireless ad hoc networks. *EURASIP Journal on Applied Signal Processing,* (1): 144–152. doi:10.1155/ASP.2005.144

Wang, Y., & Li, X.-Y. (2003). Localized construction of bounded degree planar spanner for wireless ad hoc networks. In *Proceedings of the DIALM-POMC Joint Workshop on Foundations of Mobile Computing* (pp.59-68). San Diego, CA: ACM.

Wattenhofer, R., Li, L., Bahl, P., & Wang, Y.-M. (2001). Distributed topology control for power efficient operation in multihop wireless ad hoc networks. In *20th IEEE International Conference on Computer Communications, Joint Conference of the IEEE Computer and Communications Societies (INFOCOM)* (pp. 1388-1397). Anchorage, AK: IEEE.

Wu, Q. (1999). Performance of optimum transmitter power control in CDMA cellular mobile systems. *IEEE Transactions on Vehicular Technology, 48*(2), 571–575. doi:10.1109/25.752582

Zander, J. (1992a). Performance of optimum transmitter power control in cellular radio systems. *IEEE Transactions on Vehicular Technology, 41*(1), 57–62. doi:10.1109/25.120145

Zander, J. (1992b). Distributed cochannel interference control in cellular radio systems. *IEEE Transactions on Vehicular Technology, 41*(3), 305–311. doi:10.1109/25.155977

KEY TERMS AND DEFINITIONS

Constant (Uniform) Power Assignment: If all the concurrently scheduled links employ the same transmission power, we call it a constant (uniform) power assignment.

Length Diversity: A notion to describe the number of magnitudes of lengths in a set of links $N = \{1, ..., n\}$. In particular, the length diversity of N is: $d(N) = |\{m \mid \exists i \in N : \lceil \log(d_{ii}) \rceil = m\}|$ is the length of link i)

Linear Power Assignment: If each link in the concurrently scheduled links employs the transmission power which is proportional to the corresponding link's length (the distance from the transmitter to the receiver) to the power of the path loss exponent, we call it a linear power assignment.

Link Independent Set: A set of links which can be concurrently scheduled under the SINR model.

Nonlinear Power Assignment: We use $d(\{i\})$ to denote the length diversity of all the links $\{i\}$ scheduled in the same timeslot. And we sort the links in a non-increasing order of their lengths. Then we assign the τ value (the power scaling exponent) to each link ($1 \leq \tau \leq d(\{i\})$), and the lower the length magnitude of the links, the higher the τ value. In particular, the links with the lowest length magnitude have the highest τ value of $d(\{i\})$, and the links with the highest length magnitude have the lowest τ value of 1. Then if the link i uses the transmission power $p_i = (f)^\tau \cdot (d_{ii})^\alpha$, we say it is a nonlinear power assignment. Here f is a function of the parameters α, β, n_i and the number of the links.

Pareto-Optimal Power Assignment: According to Property 4 of the link gain matrix H, if we set the transmission powers based on the power vector $P^* = (I - H)^{-1} \cdot \eta$, we call it a Pareto-optimal power assignment.

SINR Model: A specific interference model which is dependent on the so called signal-to-interference-plus-noise-ratio (SINR). In this model, we say that a link i has been successfully scheduled if and only if the power received by the link's receiver i_r from its corresponding transmitter i_s is at least a factor β higher than the sum of the received powers from the other concurrently scheduled links' transmitters plus the background noise n_i. Here the received power attenuates with distance, i.e., it equals to the transmitted power divided by the distance between the sender and receiver to the power of the path loss exponent α.

Topology Control: Adjustment of the links' transmission powers so that these links fulfill a network-wide property, such as connectivity, low interference and capacity improvement.

Wireless Link: A wireless transmission comprised by a source node (transmitter) and a destination node (receiver).

Chapter 10
Connected k–Coverage Protocols for Densely Deployed Wireless Sensor Networks

Habib M. Ammari
Hofstra University, USA

ABSTRACT

In this chapter, we study duty-cycling to achieve both k-coverage and connectivity in highly dense deployed wireless sensor networks, where each location in a convex sensor field (or simply field) is covered by at least k active sensors while maintaining connectivity between all active sensors. Indeed, the limited battery power of the sensors and the difficulty of replacing and/or recharging batteries on the sensors in hostile environments require that the sensors be deployed with high density in order to extend the network lifetime. Also, the sensed data originated from source sensors (or simply sources) should be able to reach a central gathering node, called the sink, for further analysis and processing. Thus, network connectivity should be guaranteed so sources can be connected to the sink via multiple communication paths. Finally, wireless sensor networks suffer from scarce energy resources. A more practical deployment strategy requires that all the sensors be duty-cycled to save energy. With duty-cycling, sensors can be turned on or off according to some scheduling protocol, thus reducing the number of active sensors required for k-coverage and helping all sensors deplete their energy as slowly and uniformly as possible. We also extend our discussion to connected k-coverage with mobile sensors as well as connected k-coverage in a three-dimensional deployment area. Furthermore, we discuss the applicability of our protocols to heterogeneous wireless sensor networks.

INTRODUCTION

Recent advances in miniaturization, low-cost and low-power circuit design, and wireless communi-cations have led to a new technology, called *wireless sensor networking*. A wireless sensor network consists of tremendous number of low-cost, low-power, and tiny communication devices, named *sensors*. Like nodes (or computers, laptops, etc) in traditional wireless networks, such as mobile ad hoc

DOI: 10.4018/978-1-61520-701-5.ch010

networks (MANETs), the sensors have data storage, processing, and communication capabilities. Unlike MANET nodes, the sensors have an extra functionality related to their sensing capability. Precisely, the sensors are densely deployed in a physical environment for monitoring specific phenomena. Wireless sensor networks can be used for a wide variety of applications dealing with monitoring (health environments monitoring, seism monitoring, etc), control, and surveillance (battlefields surveillance). Sensors sense some phenomenon in the environment and perform in-network processing on the sensed data at intermediate nodes before sending their results to a central gathering component, called the *sink*.

Sensing coverage is an essential functionality of wireless sensor networks. However, it is also well-known that coverage alone in wireless sensor networks is not sufficient because data originated from *source sensors* are not guaranteed to reach the *sink* for further analysis. Thus, *network connectivity* should also be considered for a network to function correctly. In wireless sensor networks, *coverage* and *connectivity* have been jointly addressed in an integrated framework. While coverage is a metric that measures the quality of surveillance provided by a network, connectivity provides a means to the *source sensors* to report their sensed data to the *sink*. Some real-world applications, such as intrusion detection, may require high degree of coverage, and hence large number of sensors to enable accurate tracking of intruders. For such highly dense deployed and energy-constrained sensors, it is necessary to duty-cycle them to save energy. Thus, the design of coverage configuration protocols for wireless sensor networks should minimize the number of *active sensors* to guarantee the degree of coverage of a field required by an application while maintaining connectivity between all active sensors. Hence, the first challenge is the determination of the number of sensors required to remain active to k-cover a sensor field. Given that sensors have limited battery power and wireless sensor networks

are generally randomly and hence highly dense deployed, the second challenge is the design of an efficient scheduling protocol that decides which sensors to turn *on* (active) or *off* (inactive) for k-coverage of a field.

The objective of this chapter is to study the problem of connected k-coverage in wireless sensor networks, which can be stated as follows:

Connected k-Coverage Problem: Given a field, a set S of sensors, and $k \geq 3$, select a minimum subset $S_{min} \boxtimes S$ of sensors such that each point in the field is k-covered, all sensors in S_{min} are connected and have the highest remaining energy.

Because the problem of selecting a minimum subset of sensors to k-cover a field is NP-hard (Gupta et al., 2006), we propose energy-efficient centralized, clustered, and distributed *approximation* algorithms to solve the *connected k-coverage* problem using as small number of sensors as possible.

The remainder of this chapter is organized as follows. First, we present the major challenges in the design of connected k-coverage protocols, specify the network model, and review related work. Second, we discuss connected k-coverage protocols for two-dimensional and three-dimensional wireless sensor networks. Third, we describe future and emerging trends. Then, we conclude the chapter.

BACKGROUND

In this section, we discuss the major challenges that face the design of connected k-coverage protocols for wireless sensor networks. Then, we specify the network model for the design of our own protocols for connected k-coverage configurations in densely deployed wireless sensor networks. Moreover, we present a sample of existing studies of k-coverage and connectivity in wireless sensor networks as well as some related work in connected k-coverage.

Challenging Problems

The design of connected *k*-coverage protocols for wireless sensor networks faces challenging problems due to several constraints. These constraints are imposed by the limited capabilities and resources of individual sensors, the nature of deployment fields, and the specific requirements of sensing applications.

Sensor Capabilities and Resources

The sensors suffer from several scarce capabilities and resources, such as storage, battery power (or *energy*), computation, communication, and sensing, with energy being the most critical resource. Indeed, the network lifetime depends on the lifetime of its individual sensors, which in turn depends on their energy. Precisely, when the energy of a sensor reaches a certain threshold, the sensor will become faulty and will not be able to function properly, which will have a major impact on the network performance. Also, those sensor capabilities are *stochastic* in nature, where the shape of the sensing and communication ranges of the sensors is not necessarily a disk and their energy cannot be determined with certainty. Considering those capabilities as *deterministic* may be interesting to only simplify enough their analysis and extract some theoretical results. Thus, algorithms designed for sensors should be as energy-efficient and realistic as possible to extend their lifetime and hence prolong the network lifetime.

Location Management

Sensor location management is another major challenge in the design of deployment strategies to achieve a certain degree of coverage. In most of the protocols designed for wireless sensor networks, the sensors are aware of their locations through either the use of *global positioning system* (GPS) receivers or some localization technique (Bulusu et al., 2000). On the one hand, a GPS receiver-based solution provides the sensors with highly accurate locations but is not cost-effective for densely deployed sensors given that each sensor should be equipped with a GPS receiver. On the other hand, the use of a localization technique does not require any additional cost but may not guarantee high sensor location accuracy.

Deployment Field

The energy of the sensors poses a big challenge for network designers especially in hostile environments, such as a battle-field, where it is impossible to access the sensors, and replace and/or recharge their batteries. Thus, sensor deployment could be a challenging problem when the sensors are to be deployed in a monitoring field, such as battlefields, using a *deterministic* strategy. Although this type of sensor deployment strategy enables covering the field with a minimum number of sensors to achieve any degree of coverage requested by the underlying sensing application, it may not be an always-feasible approach. Indeed, there are several applications that may require *multiple-level coverage*, where each location in the field is covered by more than one sensor at the same time. Such a deterministic deployment strategy may not also be practical in densely deployed sensor networks whose number may reach several thousands. Because of all of these reasons, the sensors are, in general, randomly deployed in the field. Unfortunately, a random deployment strategy may not guarantee the requested degree of coverage by an application. In particular, a random deployment strategy may yield a *coverage hole*, where some areas in the field are not well covered or not covered at all. Furthermore, this type of deployment strategy may result in a *connectivity hole*, where these randomly deployed sensors form at least two groups that cannot communicate with each other. This is why the sensors are, generally, deployed with uniform high density to achieve any requested degree of coverage while maintaining network connectivity.

Time-Varying Characteristics

Wireless sensor networks are mainly characterized by their several inherent time-varying properties, such as coverage and connectivity, due to their dynamic topology, which is defined by the sensors and communication links between the sensors. Indeed, the latter is susceptible to change because of several factors, such as low energy of the sensors, new sensor that have joined the network, and some sensors that have left the network, and sensor mobility. Hence, any network topology change will have an impact on the communication between the sensors. Therefore, the design of connected *k*-coverage configuration protocols for wireless sensor network has to account for the dynamics of the network topology to ensure full *k*-coverage of the deployment field while keeping the network connected constantly. Indeed, connectivity between the sensors is necessary so they can interact with each other and collaborate for the successful accomplishment of their mission, i.e., *k*-coverage of the target field.

Network Scalability, Heterogeneity, and Mobility

The design of connected *k*-coverage protocols should consider network scalability. That is, these protocols should scale with the number of sensors in the network, sensor mobility, and the size of the deployment field. For large-size networks, the number of sensors could be on the order of thousands. Also, the sensors may not necessarily be static given that sensor mobility helps achieve better quality of coverage (Liu et al., 2005). Furthermore, the sensors could be heterogeneous with regard to storage, processing, communication, sensing, and energy resources. In real-world applications, wireless sensor networks are composed of heterogeneous sensors that have a potential to increase the network lifetime and reliability without causing significant increase in its cost (Xing et al., 2005). Indeed, deploying

heterogeneous sensors reduces the probability of simultaneous failure of the entire neighbor set of a sensor (Ammari & Das, 2009).

Sensing Application Needs

The needs of a sensing application in terms of coverage and connectivity could be time-dependent. In particular, the degree of coverage required by an application, such as intruder detection and tracking, may not be the same based on the intruder activities. Indeed, in the absence of any intrusion activity, the coverage degree should be low. Then, as soon as an intruder (or external agent that accesses a monitored area) is detected, the degree of coverage should increase for a better intruder tracking, thus involving more sensors to keep track of the intruder movements. Also, a network may be designed to monitor one or several regions of interest in a field. That is, more than one mission could be accomplished, and each mission may have its specific demands in terms of coverage and connectivity. Thus, the sensors should be able to self-organize and adapt themselves so they guarantee the network coverage and connectivity that are needed for such situations. Specifically, the sensors should be autonomous so they can make their own decisions based on the information received from the rest of the network. Thus, the design of connected *k*-coverage protocols must account for the sensing applications needs.

Network Model

In this section, we present some assumptions about the characteristics of the sensors, and particularly, their sensing and communication ranges. Relaxation of some widely used assumptions for coverage and forwarding in wireless sensor networks will be discussed in the last two sections.

In the description of our proposed protocols for connected *k*-coverage in wireless sensor networks in the third Section, we assume that all the sensors are static and isotropic, i.e., their sensing ranges as well as their communication ranges are modeled by disks, called *sensing* and *communication disks*, whose radii are r and R, respectively. Moreover, each sensor has a unique *id* (an integer, for instance) and is aware of its own location through GPS (Global Positioning System) or a localization technique (Bulusu et al., 2000).

Literature Review

In this section, we review existing work relating to the problem of ensuring connected k-coverage in two-dimensional and three-dimensional wireless sensor networks.

Two-Dimensional Connected k-Coverage

The issue of determining the required number of sensors to achieve full coverage of a desired region was addressed in (Adlakha and Srivastava, 2003). Precisely, an exposure-based model was proposed to find the sensor density based on the physical characteristics of the sensors and the properties of the target. The minimum number of sensors needed to achieve k-coverage with high probability was showed in (Kumar et al., 2004) to be approximately the same regardless of whether the sensors are deployed deterministically or randomly, if the sensors fail or sleep independently with equal probability. Necessary and sufficient conditions for 1-covered, 1-connected wireless sensor grid network were given in (Shakkottai et al., 2005). Also, a variety of algorithms have been proposed to maintain connectivity and coverage in large wireless sensor networks (Shakkottai et al., 2005).

The first combined study on k-coverage and connectivity was proposed in (Xing et al., 2005). They proposed a coverage configuration protocol (CCP) based on the degree of coverage of the sensing application. The k-coverage set and the k-connected coverage set problems were formalized in terms of linear programming and

two non-global solutions were proposed for them (Yang et al., 2006). An optimal deployment pattern for achieving k-barrier coverage was established, efficient global algorithms for checking k-barrier coverage of a given region were developed, and it was showed the non-existence of localized algorithms for testing the existence of global barrier coverage (Kumar et al., 2005). To address this limitation, localized algorithms so sensors can locally determine the existence of local barrier coverage were proposed in (Chen et al., 2007). Moreover, optimal polynomial-time algorithms were proposed to solve the sleep-wakeup problem for the barrier coverage model using sensors with equal and unequal lifetimes (Kumar et al., 2007).

A directional sensors-based approach for network coverage was proposed in (Ai & Abouzeid, 2006), where the coverage region of a sensor depends on its location and orientation. The coverage problem in heterogeneous sensor networks was discussed in (Lazos & Poovendran, 2006a, 2006b). They formulated the coverage problem as a set intersection problem and derived analytical expressions, which quantify the coverage achieved by stochastic coverage. Efficient distributed algorithms to optimally solve the best-coverage problem with the least energy consumption were proposed in (Li et al., 2003). Optimal polynomial time worst and average case algorithms for coverage calculation based on the Voronoi diagram and graph search algorithms were proposed in (Megerian et al., 2005, Megerian et al., 2001b). In (Huang & Tseng, 2003), polynomial-time algorithms, in terms of the number of sensors, were presented for the coverage problem formulated as a decision problem. A distributed algorithm was proposed in (Abrams et al., 2004) to partition a wireless sensor network into k covers, each of which contains a subset of sensors that is activated in a round-robin fashion such that as many areas are monitored as frequently as possible. A survey of a variety of approaches on energy-efficient coverage problems is in (Cardei & Wu, 2006).

In (Bai et al., 2006), an optimal deployment strategy to achieve both full coverage and 2-connectivity regardless of the relationship between communication and sensing radii of the sensors was proposed. In (Huang et al., 2007), the relationship between coverage and connectivity of wireless sensor networks was studied and distributed protocols to guarantee both their coverage and connectivity were proposed. The problem of sensor selection to provide both sensing and connectivity was addressed in (Datta et al., 2007) and an approach for solving it based on the concept of connected dominating set was proposed. A joint scheduling scheme based on a randomized algorithm for providing statistical sensing coverage and guaranteed network connectivity was presented in (Liu et al., 2006). A distributed algorithm to keep a small number of active sensors in a network regardless of the relationship between sensing and communication ranges was proposed in (Zhang & Hou, 2005). It was also proved that if the original network is connected and the identified active nodes can cover the same region as all the original nodes, then the network formed by the active nodes is connected when the communication range is at least twice the sensing range (Tian & Georganas, 2005). A probabilistic Markov model was proposed to solve the problem of minimizing power consumption in each sensor while ensuring coverage and connectivity (Yener et al., 2007).

In (Gupta et al., 2006), centralized and distributed algorithms for connected sensor cover were proposed so the network can self-organize its topology in response to a query and activate the necessary sensors to process the query. Datta et al. (2005) proposed two self-stabilizing algorithms to the problem of minimal connected sensor cover (Gupta et al., 2006). In (Zhou et al., 2005), a distributed and localized algorithm using the concept of the kth-order Voronoi diagram was proposed to provide fault tolerance and extend the network lifetime, while maintaining a required degree of coverage. Control and coordination algorithms

were designed for a multi-vehicle network with limited sensing and communication capabilities (Cortes et al., 2004). Also, adaptive, distributed, and asynchronous coverage algorithms were proposed for mobile sensing networks. Indeed, it was proved that mobility can be used to improve coverage in wireless sensor networks (Liu et al., 2005).

Three-Dimensional Connected k-Coverage

The study of coverage and connectivity in three-dimensional wireless sensor networks, such as underwater sensor networks (Akyildiz et al., 2005), has gained relatively less attention in the literature compared to that of two-dimensional wireless sensor networks. Alam & Haas (2006) proposed a placement strategy based on Voronoi tessellation of a three-dimensional space, which creates truncated octahedral cells. Huang et al. (2004) proposed a polynomial-time algorithm to solve the α-Ball-Coverage (α-BC) problem whose goal is to check α-coverage of a three-dimensional region. Pompili et al. (2006) proposed a deployment strategy for three-dimensional communication architecture for underwater acoustic sensor networks, where sensors float at different depths of the ocean to cover the entire three-dimensional region. Poduri et al. (2006) discussed some difficulties encountered in the design of three-dimensional wireless sensor networks, such as ensuring network connectivity in the case of uniform random deployment and restrictions imposed by the environment structure on sensor deployment. Ravelomanana (2004) investigated fundamental properties of randomly deployed three-dimensional wireless sensor networks for connectivity and coverage, such as the required sensing range to guarantee certain degree of coverage of a region, the minimum and maximum network degrees for a given communication range, and the network hop-diameter.

CONNECTED *K*-COVERAGE CONFIGURATION PROTOCOLS

In this section, we discuss our proposed protocols for connected *k*-coverage configurations in two-dimensional and three-dimensional deployment fields.

Two-Dimensional Network Deployment

Unlike most previous work, we use a deterministic approach to compute the number of sensors needed to fully *k*-cover a field while maintaining network connectivity. The closest work to ours is the one proposed by Xing et al. (2005), and hence we will compare our protocols to theirs. First, we provide a tight characterization of *k*-coverage in wireless sensor networks. Second, based on this characterization, we propose centralized and distributed connected *k*-coverage protocol in two-dimensional deployment fields.

k-Coverage Characterization

We want to compute the *active sensor spatial density* required to *k*-cover a field. To do so, we should compute the maximum size of a convex area *A* of a field so that *A* is *k*-covered with exactly *k* sensors. Intuitively, the distance between any point in *A* and each of the *k* sensors should be at most equal to the radius of their sensing disks. Lemma 1 gives an upper bound on the *width* of such a convex area.

Lemma 1: Let r be the radius of the sensing disks of the sensors and $k \geq 3$. A convex area A is *k*-covered when exactly k homogeneous sensors are deployed in it, if the width of A does not exceed r.

Now, we present *Helly's Theorem* (Bollobás, 2006) (page 90), a fundamental result of convexity theory, which characterizes the intersection of convex sets.

Helly's Theorem (Bollobás, 2006): Let E be a family of convex sets in IR^n such that for $m \geq n + 1$ any m members of E have a non-empty intersection. Then, the intersection of all members of is non-empty.

Theorem 1, which is an instance of Helly's Theorem (Bollobás, 2006), will help us characterize the intersection of sensing disks of *k* sensors and compute the sensor spatial density required to fully *k*-cover a field. Indeed, Lemma 2 is an instance of Helly's Theorem, where $n = 2$ and $k = m$. This lemma will help us compute a tight bound on the active sensor spatial density required to *k*-cover a field.

Theorem 1: Let $k \geq 3$. The intersection of *k* sensing disks is not empty if and only if the intersection of any three of those *k* sensing disks is not empty.

Following Theorem 1, Lemma 2 states a *sufficient condition* for complete *k*-coverage of a field.

Lemma 2: Let r be the radius of the sensing disks of the sensors and $k \geq 3$. A field is *k*-covered if any Reuleaux triangle region of width r in the field contains at least k active sensors.

Proof: First, we compute the maximum area that is *k*-covered with exactly *k* sensors. Let a be the intersection area of the sensing disks of *k* sensors. From Lemma 1, the width of a should be upper-bounded by r so that any point in a is *k*-covered by these *k* sensors. Let us first consider the case of three sensors. Using the Venn diagram given in Figure 1a, the maximum size of the intersection area of the sensing disks of the sensors s_1, s_2, and s_3, so that the distance between any pair of sensors is at most equal to r, is obtained when s_1, s_2, and s_3 are symmetrically located from each other. This area, called *Reuleaux triangle* (Weisstein, 2009a), is denoted by $RT(r)$ and has a constant width equal to r. Given that the intersection area of *k* sensing disks is at most equal to that of three sensing disks such that the maximum distance between any pair of sensors is at most equal to

Figure 1. (a) Intersection of three disks and (b) their Reuleaux triangle

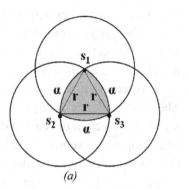
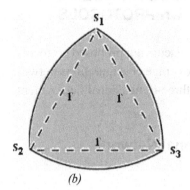

(a)   *(b)*

r, the maximum size of a is equal to the area of $RT(r)$, which we call *slice*. Thus, any point in a is k-covered with exactly k active sensors deployed in a. Since this applies to any $RT(r)$ region (or *slice*) in a field, it is guaranteed that the field is k-covered. This completes the proof.

As will be discussed later in subsequent sections, our sensor selection scheme exploits the overlap between adjacent slices in a field to select a minimum number of active sensors for full k-coverage of the field. Precisely, two adjacent slices intersect in a region shaped as a *lens* (also known as the *fish bladder*) so that the sides of their associated regular triangles fully coincide (Figure 2b). Note that k sensors located in the lens of two adjacent slices, say C_1 and C_2, k-cover the area associated with their union. Indeed, the distance between any of these k sensors located in the lens and any point in the area of the union

of both C_1 and C_2 is at most equal to r. Lemma 3 states this result.

Lemma 3: It is guaranteed that k active sensors located in the lens of two adjacent slices in a field k-cover both slices.

Theorem 2, which exploits the results of Lemma 3, refines the result of Lemma 2 by stating a *tighter sufficient condition* for complete k-coverage of a field.

Theorem 2: Let $k \geq 3$. A field is guaranteed to be k-covered if for any slice in the field, there is at least one adjacent slice such that their lens contains at least k active sensors.

Theorem 3, which exploits the result of Theorem 2, computes the minimum sensor spatial density required for complete k-coverage of a field.

Theorem 3: Let $k \geq 3$. The *minimum active sensor spatial density* required to guarantee k-coverage of a field is given by

Figure 2. (a) A slice and (b) intersection of two adjacent slices

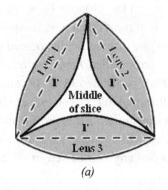
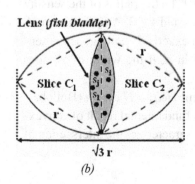

(a)   *(b)*

$$\lambda(r,k) = \frac{6\ k}{(4\ \pi - 3\ \sqrt{3})\ r^2}$$

where r is the radius of the sensing disks of the sensors.

Proof: It is easy to check that the area $\|\ Area(r)\ \|$ of the union of two adjacent slices is computed as

$$\|\ Area(r)\ \| = 2\ A_1 + 4\ A_2 = (4\ \pi - 3\ \sqrt{3})\ r^2\big/\ 6$$

where $A_1 = \sqrt{3}\ r^2\big/4$ is the area of the central equilateral triangle of side r and $A_2 = (\pi\big/6 - \sqrt{3}\big/4)\ r^2$ is the area of each of the three curved regions α (Figure 1a). By Theorem 2, k sensors should be deployed in the lens of two adjacent slices in order to k-cover both of them. Thus, the *minimum sensor spatial density* that guarantees k-coverage of a field is equal to

$$\lambda(r,k) = k\big/\ \|\ Area(r)\ \| = 6\ k\big/(4\ \pi - 3\ \sqrt{3})\ r^2$$

This completes the proof.

One may suggest that the *maximum area* that is guaranteed to be k-covered with exactly k sensors is a circle of radius $r\ /\ 2$. Fortunately, it is easy to check that our density $\lambda(r,k)$ is smaller than the one corresponding to the configuration where k active sensors are deployed in a circle of radius $r\ /\ 2$. In other words, $\lambda(r,k) = 6\ k\big/(4\ \pi - 3\ \sqrt{3})\ r^2 < 4\ k\big/\pi\ r^2$. This finishes the proof.

Centralized Clustered Protocol

In this section, we present our centralized, randomized connected k-coverage (CERACC$_k$) protocol for two-dimensional wireless sensor networks. First, we provide a sufficient condition for full k-coverage of a field. Our proposed protocol to achieve full k-coverage of a field consists of two phases, namely *sensor field slicing* and *sen-*

sor selection. Next, we describe both phases in details.

Sensor Field Slicing: In general, the sink is connected to an infinite source of energy, such as a wall outlet, and thus can be viewed as a *line-powered* node (Yarvis et al., 2005) that has no energy constraint. Hence, we assume that the sensor field slicing task is done by the sink in each *scheduling round* (or simply *round*). In order to exploit the result of Theorem 2, we propose a *slicing* scheme that divides a field into overlapping *slices*, such that two adjacent slices intersect in a *lens*. Intuitively, this implies that a field is sliced into regular triangles of side r. The result of this slicing operation is called *slicing grid*. Indeed, the regular triangle yields a *perfect tiling* of a two-dimensional Euclidean space.

Now, we address the question: *How can a slicing grid be randomly generated in each round so that all the sensors have the same chance to be selected for k-coverage?* First, the sink randomly generates one point p_1 in a field. To randomly determine a second point p_2, the sink generates a random angle $0 \leq \theta \leq 2\ \pi$ so that the segment $\overline{p_1 p_2}$ forms an angle θ with the x-axis centered at p_1 and the length of $\overline{p_1 p_2}$ is r. Then, the sink deterministically finds a third point p_3 to form the first regular triangle (p_1, p_2, p_3), called refere*nce triangle* (Figure 3a). All other regular triangles of the slicing grid are computed based on the reference triangle. Figure 3a shows a randomly generated slicing grid of a square field.

Sensor Field Clustering: In addition to slicing a field, we assume that the sink is also responsible for forming *clusters of slices* from the randomly obtained slicing grid. Precisely, each cluster consists of *at most* six adjacent slices forming a *disk*. Because of the random generation of slicing grids and the geometry of a field, some clusters consist of an entire disk, and hence called *interior clusters*, while others are formed by a portion of a disk, and hence called *boundary clusters*. Figure 3b shows a clustered sensor field. More-

Figure 3. (a) random slicing grid and (b) clustered sensor field

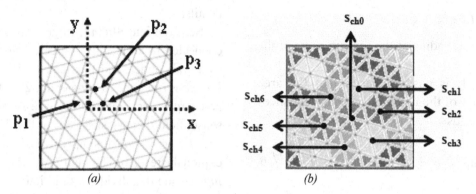

(a)	*(b)*

over, for each cluster, the sink selects a sensor, called *cluster-head*, which is located as near as possible to the center of its cluster. The random generation of slicing grid ensures that *all* sensors are *equally likely* to act as cluster-heads in each round. Each cluster is defined by one point, i.e., (*x,y*) coordinates, representing its center and at most six other points defining its slices (or slice portions for a non-complete cluster). These seven points define the *slicing information* of a cluster, which the sink would broadcast to its corresponding cluster-head. Next, we define the notions of *interior* and *boundary lenses*, which will be used in the description of our protocols.

Notice that each cluster overlaps with at most six others as shown in Figure 3b. By Lemma 3 and Theorem 2, sensors located in the boundary lenses of a given cluster should be selected first in order to minimize the necessary total number of active sensors to achieve full *k*-coverage of a field. However, this would require certain *coordination* between cluster-heads.

Cluster-Heads Coordination and Sensor Selection: Each cluster-head is in charge of selecting some of its sensing neighbors to *k*-cover its cluster based on its *slicing information*. Precisely, each cluster-head exploits the overlap between the slices of its cluster as well as the overlap between its slices and those of its adjacent cluster-heads to select a minimum number of its sensing neighbors to *k*-cover its cluster. We assume that

each sensor advertises its remaining energy to its sensing neighbors at the start of a round when it turns itself *on*. Each cluster-head s_{ch_i} maintains a list of remaining energy of its sensing neighbors, $\Pi(s_{ch_i}) = [\pi(s_j) : s_j \in SN(s_{ch_i})]$. It uses this list to select the ones with high remaining energy to stay active by sending a *SELECT* message including the cluster-head's *id* as well as the *id*'s of all selected sensors. This would avoid those ones with low remaining energy and help the sensors deplete their energy as slowly and uniformly as possible. We assume that at the beginning of each round, *all* the sensors are active. Those ones which are selected by their corresponding cluster-heads would remain active during the underlying round, while the others turn themselves *off* (or go to sleep). For the sensor selection, each cluster-head assigns *priorities* to sensors located in boundary lenses, interior lenses, and middle of slices in descending order. That is, sensors located in boundary lenses have high priority to be selected based on Theorem 2. Given that each cluster has at most six slices, each cluster-head manages at most six interior lenses and at most six boundary lenses. On the one hand, each cluster-head is responsible for selecting sensors from its interior lenses without any coordination with its adjacent cluster-heads. On the other hand, each cluster-head coordinates with at most six adjacent cluster-heads to select sensors from its boundary lenses to *k*-cover its cluster with a minimum number of sensors. For

Figure 4. Sleep-wakeup scheduling for k-coverage

ALGORITHM 1: Centralized-Randomized-Connected-k-COVERAGE (CERACC$_k$)
(* This code is run in each round *)
/* The following code is run by the sink */
1. Slice randomly a field into adjacent slices
2. Select a set *CH* of cluster-heads for all clusters in such a way they are located as close as possible to clusters' centers
3. Broadcast the selected set *CH* with all *slicing information*
/* The following code is run by each cluster-head */
4. Select sensors from its cluster (high prority to boundary lenses) and coordinate with all adjacent cluster-heads to k-cover a cluster based on the remaining energy of the sensors
5. Return

instance, in the case of a disk, its cluster-head, say s_{ch_0} (Figure 3b), would advertize the subsets S_1, ..., S_6 of sensors selected from its six boundary lenses to its adjacent cluster-heads s_{ch1}, ..., s_{ch6}, respectively.

Theorem 4 states a condition for *connectivity* based on the structure of the clusters of slices.

Theorem 4: *Let* $k \geq 3$. A *k-covered* wireless sensor network is *connected* if $R \geq r$, where r and R stand for the radii of the sensing and communication disks of the sensors, respectively.

Proof: First, each cluster is connected if $R \geq r$. Also, cluster-heads are connected to each other via active sensors. Thus, a *k-covered* wireless sensor network is guaranteed to be connected if $R \geq r$. This completes the proof.

Theorem 5 characterizes the performance of CERACC$_k$. in terms of the number of sensors to achieve *k*-coverage of a field.

Theorem 5: CERACC$_k$ selects a minimum number of active sensors in each round for complete *k*-coverage of a field.

Proof: Each cluster-head coordinates with its adjacent cluster-heads and ensures that its cluster is *k*-covered, where each of its slices is covered by exactly *k* sensors. Hence, by Theorem 2, each slice of a field is *k*-covered with a minimum number of active sensors. Thus, CERACC$_k$ guarantees that a field is *k*-covered with a total minimum number of active sensors. This finishes the proof.

The pseudo-code of our centralized, randomized connected *k*-coverage (CERACC$_k$) protocol for two-dimensional wireless sensor networks is given in Figure 4.

Distributed Protocol

Here, we discuss our distributed, randomized connected *k*-coverage (DIRACC$_k$) protocol for two-dimensional wireless sensor networks.

Overview: Each sensor s_i randomly slices its sensing range into six overlapping Reuleaux triangles of width equal to r_i, the radius of s_i's sensing range. Based on the result of Theorem 1, a sensor s_i randomly picks one of the *three-lens flowers* (as shown in Figure 5) and checks whether its sensing range is *k*-covered. A sensor starts first by choosing the sensors located in the three lenses of the selected three-lens flower to remain active and *k*-cover its sensing range based on their remaining energy and the radii of their sensing ranges. Specifically, a sensor s_i selects k sensors from each of the lenses of the three-lens flower whose radii of their sensing ranges is at least equal to r_i, the radius of the sensing range of s_i. Then, it activates them by sending AWAKE messages. When a sensor receives an AWAKE message, it becomes active and broadcasts a NO-TIFICATION message to inform all its neighbors. Every sensor keeps track of its active neighbors.

Figure 5. (a) First and (b) second three-lens flowers of s_i

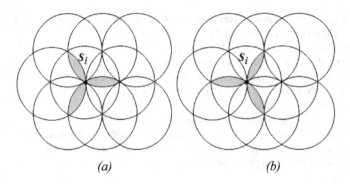

(a)   (b)

Figure 6. Distance for connectivity

Also, a sensor will turn itself active if its sensing range is not k-covered.

Lemma 4 states the condition under which a k-covered wireless sensor network is connected under the assumption that k-coverage is achieved by DIRACC$_k$.

Lemma 4: *Let* $k \geq 3$. *A k-covered wireless sensor network is guaranteed to be connected if the radii R and r of the communication and sensing disks of the sensors, respectively, satisfy* $R \geq \sqrt{3}\ r$.

Proof: Consider Figure 6. It is easy to check that the distance d between a pair of sensors located at the corners of two adjacent slices is equal to $d = \sqrt{3}\ r$. This completes the proof.

State Transition Diagram: The state transition diagram associated with DIRACC$_k$ is given in Figure 7. At any time, a sensor can be in one of the three states: READY WAITING, and RUNNING.

- **Ready:** A sensor is listening to AWAKE messages.
- **Waiting:** A sensor is neither communicating with other sensors nor sensing a field, and thus its radio is turned off. However, after some fixed time interval, it switches to the Ready state to receive AWAKE messages if its neighbors decide to do so.
- **Running:** A sensor can communicate with other sensors and sense the environment.

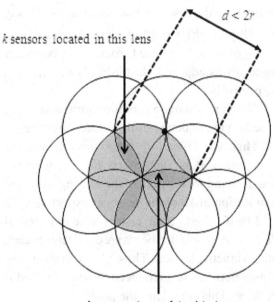

k sensors located in this lens

$d < 2r$

k sensors located in this lens

At the start of the monitoring task, all sensors are in the Ready state except one, say s_0, which is in the Running state. The single sensor s_0 in the Running state is one of the communication neighbors of the sink that is chosen randomly by the sink to activate some of its sensing neighbors to achieve k-coverage of its sensing disk. Each of these selected sensors should have at least a sensing range whose radius is at least equal to r_0, the radius of the sensing range of s_0. Likewise, each of these selected sensors will in turn run the same selection algorithm to k-cover their sensing

Figure 7. State transition diagram of DIRACC$_k$

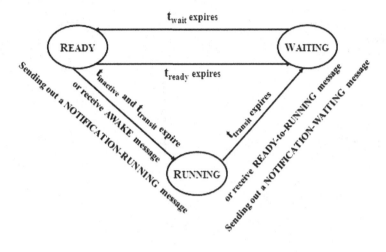

disks. This chain of sensor activations continues until the field is *k*-covered. As mentioned earlier, when a sensor is selected by any other sensor, it will send out a NOTIFICATION message to inform its neighbors.

While in the Ready state, a sensor keeps track of its sensing neighbors that are in the Running state. If it finds out that its sensing area is *k*-covered, it will switch to the Waiting state. It is not cost-effective to guarantee that a sensor is not selected more than once during one round. Indeed, guaranteeing disjoint subsets of selected sensors requires much coordination between the sensors, thus introducing unnecessary overhead.

For energy savings, a sensor may wish to switch from the Running state to the Waiting state. For this purpose, a sensor broadcasts a RUNNING-to-WAITING message and waits for some transit time $t_{transit}$. If $t_{transit}$ expires and it has not received any Running-to-Waiting message, it switches to the Waiting state and sends a NOTIFICATION-WAITING message, where it can stay there for t_{wait} time. When t_{wait} expires, it switches to the Ready state, where it can stay in this state for t_{ready} time.

If a sensor finds out that its sensing area is not *k*-covered, it will broadcast a READY-to-RUNNING message and wait for some transit time $t_{transit}$ before switching to the Running state. If after $t_{transit}$ it has

not received any other READY-to-RUNNING message, it will switch to the Running state and sends a NOTIFICATION-RUNNING message; otherwise, it will switch to the Waiting state.

Performance Evaluation

In this section, we evaluate the performance using a high-level simulator written in the C programming language. We consider a square sensor field of side length 1000m where 16000 sensors are randomly and uniformly deployed. We use the energy model given in (Ye et al., 2003), where energy consumption in transmission, reception, idle, and sleep modes are 60 mW, 12 mW, 12 mW, and 0.03 mW, respectively. Following (Zhang and J. Hou, 2005), we define *one unit of energy* as the energy required for a sensor to stay idle for 1 second. We assume that the initial energy of each sensor is 60 Joules enabling a sensor to operate about 5000 seconds in reception/idle modes (Ye et al., 2003). All simulations are repeated 200 times and the results are averaged. In all simulations, we consider $r = 25$ m and $k = 3$ unless specified otherwise. Also, the energy consumption is due to all activities of each sensor necessary to achieve *k*-coverage.

In this section, we compare our *k*-coverage protocol DIRACC$_k$ with the Coverage Configura-

Figure 8. DIRACC$_k$ compared to CCP (a) k vs. n_a and (b) total remaining energy vs. time

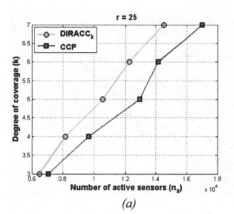

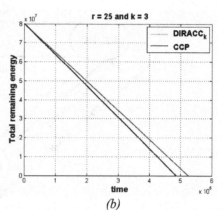

(a)   (b)

tion Protocol (CCP) (Xing et al., 2005) for a fair comparison. When the communication range of sensors is at least double their sensing range, Xing et al. (2005) showed that full coverage implies network connectivity, and hence no other mechanism to guarantee connectivity is necessary. Otherwise, Xing et al. (2005) integrated CCP with a topology maintenance protocol (SPAN) (Chen et al., 2002) to guarantee both coverage and connectivity.

Figure 8a plots the degree k of coverage versus the number n_a of active sensors for DIRACC$_k$ as compared to CCP. It shows that DIRACC$_k$ requires less active sensors than CCP to achieve the same coverage degree, thus yielding significant energy savings. This is due not only to a higher number of active sensors required by CCP, but also to the communication overhead caused by the exchange of messages between active sensors running CCP to coordinate among themselves and provide the requested k-coverage service. Thus, CCP consumes more energy than DIRACC$_k$ as shown in Figure 8b. While CCP requires SPAN to provide connectivity between active sensors when $R < 2\ r$, DIRACC$_k$ does not need such a topology maintenance protocol as all it requires is that $R \geq \sqrt{3}\ r$, thus ensuring connectivity provided that k-coverage is guaranteed.

Figure 9a plots n_a versus R while Figure 9b plots n_a versus r for different ratios $\alpha = R / r$.

Given the result reported in (Zhang & Hou, 2005) with respect to the relationship between r and R for real-world sensor platforms (i.e., $R \geq r$), we consider only the case ($R \geq r$). For our DIRACC$_k$ protocol, we consider $\alpha \geq \sqrt{3}$, and hence any increase in the communication range of sensors would not have any impact on the performance of DIRACC$_k$. It would, however, affect the performance of CCP. As can be observed, n_a decreases as R increases. Indeed, SPAN would require less number of sensors to maintain connectivity between active sensors as R increases. However, at some point, (surprisingly enough, this point corresponds to $R \geq 2\ r$), the number n_a of active sensors required for k-coverage does not decrease any further. Indeed, when $R \geq 2\ r$, SPAN is not needed at all as both k-coverage and $R \geq 2\ r$ guarantee connectivity. Similarly, the performance of CCP improves as the ratio α increases, i.e., R increases (Figure 9b). That is, less number of sensors is needed to provide k-coverage and connectivity.

Three-Dimensional Network Deployment

Our work is complementary to existing ones, especially those few works which dealt with three-dimensional wireless sensor networks. Indeed,

Figure 9. DIRACC$_k$ compared to CCP (a) n_a vs. R and (b) n_a vs. r

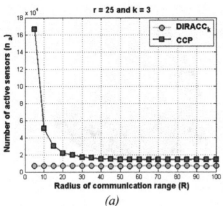

(a)

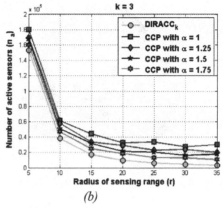

(b)

Alam & Haas (2006) considered only 1-coverage and proposed deterministic sensor placement strategies to achieve full coverage of a three-dimensional space. However, as we mentioned earlier, 1-coverage is not always enough given that sensors are not highly reliable and some applications, such intruder detection and tracking, require high coverage of a target field. First, we show the problem that we encounter when we apply the same reasoning in the previous section to a more general case of three-dimensional fields. Then, we present our energy-efficient distributed k-coverage protocol for three-dimensional wireless sensor networks and evaluate its performance.

The Curse of Dimensionality

Here, we define and analyze the k-coverage problem in three-dimensional wireless sensor networks. Then, we present our fully distributed k-coverage protocol.

Minimum connected k-coverage Problem: given a set S of sensors and an integer $k \geq 4$, select a minimum subset of sensors such that each point in a three-dimensional field is k-covered and all selected sensors are mutually connected.

In this section, we analyze the above problem from the perspective of the shape of a three-dimensional region C in a three-dimensional field

corresponding to full k-coverage. In other words, we want to determine the shape of C so that it is guaranteed to be k-covered when exactly k sensors are deployed in it. Clearly, the *breadth* of C should be less than or equal to the radius r of the sensing spheres of sensors so that each location in C is within the sensing spheres of these k sensors. Since our goal is to achieve k-coverage of a three-dimensional field with a minimum number of sensors, the volume of C should be maximum, and hence the breadth of C must be equal to r. Therefore, our problem reduces to the problem of finding the shape of this three-dimensional region C that has a constant breadth equal to r. In order to solve this problem, we first consider *Helly's Theorem* (Bollobás, 2006), which is stated above. The sensing range of a sensor can be viewed as a *point set*. Furthermore, by our assumption, the sensing range of a sensor is modeled as a sphere, which is a convex set. Thus, from Helly's Theorem, we infer that a three-dimensional $(n = 3)$ convex region C is k-covered by exactly k sensors ($\| E \| = k$) if and only if C is 4-covered by any four $(m = 4)$ of those k sensors, where $k \geq 4$. Now, let us identify the shape of a three-dimensional convex region C whose breadth is constant and equal to the radius r of the sensing spheres of sensors, and hence has a maximum volume. More importantly, this region C should

Figure 10. (a) Intersection of four symmetric spheres and (b) their Reuleaux tetrahedron

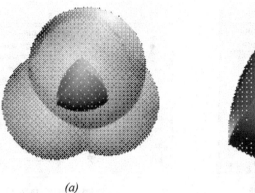
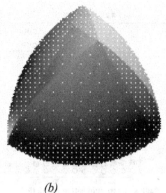

(a)   *(b)*

be guaranteed to be fully k-covered when *exactly* k sensors are deployed in it.

Let C_k be the intersection of k sensing spheres. Using Helly's Theorem (Bollobás, 2006), the maximum volume of the intersection of these k sensing spheres is equal to that of four spheres since $k \geq 4$. However, the maximum intersection (or overlap) volume of four sensing spheres such that the maximum distance between any pair of sensors is equal to r, corresponds to the configuration where the center of each sensing sphere is at distance r from the centers of all other three ones. In this configuration, the edges between the centers of these four spheres form a *regular tetrahedron* and the shape of their intersection volume is known as the *Reuleaux tetrahedron* (Weisstein, 2009b) (Figure 10).

In (Ammari & Das, 2008b), we used Helly's Theorem (Bollobás, 2006) in our analysis of the k-coverage problem and exploited the geometric properties of the Reuleaux triangle to derive a sufficient condition to fully k-cover a two-dimensional field. Note that the Reuleaux triangle of side r, which represents the intersection of three symmetric, congruent disks of radius r, consists of a central regular triangle of side r and three curved regions. More importantly, it has a constant width equal to r (Weisstein, 2009a). We found that a Reuleaux triangle region of width r of a two-dimensional field is guaranteed to be k-covered with exactly k sensors, where r is the radius

of the sensing range of the sensors (Ammari & Das, 2008b). Also, the regular triangle allows a perfect tiling of two-dimensional space. Based on this characterization, we designed an energy-efficient k-coverage configuration protocol for two-dimensional wireless sensor networks (Ammari & Das, 2008b).

Now, we provide some facts why the Reuleaux tetrahedron is not an appropriate solution to our *minimum connected k-coverage* problem. First of all, the Reuleaux tetrahedron does not have a constant breadth whose value is slightly larger than the radius r of the corresponding spheres (Weisstein, 2009b). In contrast to the regular triangle, the regular tetrahedron does not allow a perfect tiling of a three-dimensional space. Indeed, Conway & Torquato (2006) showed that the dihedral angle of a regular tetrahedron is equal to 70.53°, which is not sub-multiple of 360° (Conway & Torquato, 2006). They also gave two arrangements of regular tetrahedra such that five regular tetrahedra packed around a common edge would result in a small gap of 7.36° as shown in Figure 11a, and that twenty regular tetrahedra packed around a common vertex yield gaps that amount to a solid angle of 1.54 steradians as shown in Figure 11b (Conway & Torquato, 2006). This shows that some properties that hold for two-dimensional space are not valid for three-dimensional space. Thus, the extension of the analysis of k-coverage in two-dimensional space to three-dimensional space is not straight-

Figure 11. (a) Five regular tetrahedra about a common edge and (b) twenty regular tetrahedra about a shared vertex (Conway & Torquato, 2006)

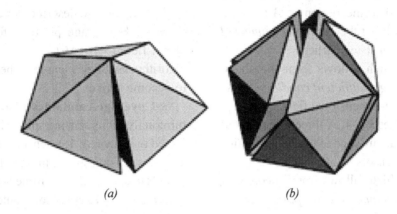

(a)   (b)

Figure 12. (a) two-dimensional projection of a half-sphere and its six slices and (b) a slice

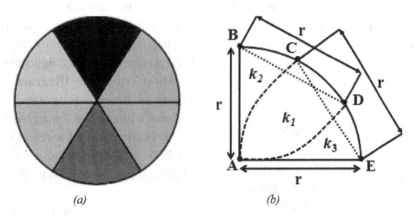

(a)   (b)

forward, and hence another approach should be used (Ammari & Das, 2008a). More precisely, we want to address the following question:

What is the "closest shape" to the Reuleaux tetrahedron that will guarantee energy-efficient k-coverage of a three-dimensional space?

To address this question efficiently, we consider two halves of the sensing sphere of the sensors, i.e., the top half and bottom half. Note that in two-dimensional wireless sensor networks, we divide the sensing disk of the sensors into six overlapping Reuleaux triangles (Ammari & Das, 2008b). By analogy with the two-dimensional analysis given in (Ammari & Das, 2008b), we divide each of

the halves of a sensing sphere into six congruent three-dimensional regions, called *slices*, each of which has three flat faces and one curved face representing an *equilateral spherical triangle* (Figure 12a). Unfortunately, the distance between the point B at the top of a slice and all the points E on the edge of any spherical triangle is larger than r. Thus, a sensor located at B cannot cover any point E and a sensor located at any point E cannot cover B Any sensor located in the region $< A, C, D >$ is able to cover the whole slice as shown in Figure 12b. However, sensors located in the regions $< A, B, C >$ and $< A, D, E >$ cannot cover the entire slice. Thus, if $k_1 = k$, then $k_2 = k_3 = 0$; otherwise, $k_2 = k_3 = k - k_1$. In other words, to guarantee k-coverage of a slice and hence a sensing sphere with a minimum number of sensors, it is necessary that the active sensors be located in the region $< A, C, D >$, thus efficiently solving the *minimum connected k-coverage* problem stated earlier.

Theorem 6, which follows from the above analysis, states a *tight sufficient condition* for *k*-coverage of a three-dimensional field.

Theorem 6: Let $k \geq 4$. A three-dimensional field is *k*-covered if any slice of the field contains at least k active sensors.

Theorem 7, which follows from Theorem 6, computes the *minimum sensor spatial density necessary* to fully *k*-cover a three-dimensional field.

Theorem 7: Let $k > 1$. The minimum sensor spatial density required to guarantee *k*-coverage of a three-dimensional field is computed as

$$\lambda(r, k) = \frac{9\ k}{\pi\ r^3}$$

where r is the radius of the sensing spheres of sensors.

Proof: The volume of a slice is $vol(slice) = \pi\ r^3 / 9$. By Theorem 6, each slice should contain at least k sensors. Thus, *k*-covering a three-dimensional field with a small number of sensors requires that every slice in the field contain exactly k sensors. Thus, under the assumption of uniform sensor distribution, the minimum sensor density to *k*-cover a three-dimensional field is equal to $k/vol(slice)$. This finishes the proof.

Using Theorem 6, Theorem 8 states a sufficient condition to maintain connectivity in three-dimensional *k*-covered wireless sensor networks.

Theorem 8: Let $k > 1$. A three-dimensional *k*-covered wireless sensor network is connected if $R \geq r$, where r and R stand for the radii of the sensing and communication spheres of sensors, respectively.

Protocol Description

In this section, we describe our distributed, randomized *k*-coverage protocol for three-dimensional field (3DIRACC$_k$). First, we present our algorithm enabling a sensor to check its *candidacy* to become active.

***k*-Coverage-candidacy algorithm:** a sensor turns active if its sensing sphere is not *k*-covered. Based on Theorem 1, a sensor randomly decomposes its sensing sphere into twelve slices of side r and checks whether each one of them contains at least k sensors. Each sensor should know the status of its sensing neighbors only to decide whether it is candidate to turn active or not. If any of the twelve slices does not have k active sensors, a sensor is a candidate to become active. Otherwise, it is not. Figure 13 shows the pseudo-code of our *k*-Coverage-Candidacy algorithm.

State Transition Diagram of 3DIRACC$_k$: Figure 14 shows a state transition diagram associated with our distributed *k*-coverage protocol for three-dimensional wireless sensor networks (3DIRACC$_k$). At any time, a sensor can be in one of three states: Ready, Waiting, and Running.

- **Ready state:** a sensor listens to messages and checks its *k*-coverage candidacy to switch to the Running state.
- **Running state:** a sensor is active and can communicate with other sensors and sense the environment.
- **Waiting state:** a sensor is neither communicating with other sensors nor sensing a field, and thus its radio is turned *off*. However, after some fixed time interval, it switches to the Ready state to check its candidacy for *k*-coverage (Figure 13) and receives messages.

At the beginning of their monitoring task, all sensors are in the Running state. Moreover, each sensor chooses randomly and independently of all

Figure 13. k-Coverage-candidacy algorithm

ALGORITHM 2: *k*-Coverage-Candidacy
(* This code is run by each sensor *)
Begin
/* Sensing sphere slicing */
1. Randomly decompose a sensing sphere into twelve slices
/* Localized *k*-coverage candidacy checking */
2. For each slice Do
3.   If it contains *k* awake sensors Then
4.   Skip /* i.e., do nothing */
5.   Else
6.   Return ("candidate")
7.   End
8. End
9. Return ("non-candidate")
End

other sensors a value t_{check} between 0 and t_{check_max} after which it runs the *k-Coverage-Candidacy* algorithm (Figure 13) to check whether it stays active or not (i.e., switch to the Waiting state). Our intuition behind this random selection of t_{check} is to avoid as much as possible higher or lower coverage of any region in a three-dimensional field.

When a sensor runs the *k-Coverage-Candidacy* algorithm and finds out that it is a candidate, it sends out a Notification-RUNNING message to inform all its neighbors. While in the Ready state, a sensor keeps track of its sensing neighbors that are in the Running state. If it finds out that its

sensing area is *k*-covered, it will switch to the Waiting state.

For energy efficiency purposes, a sensor may wish to switch from Running state to Waiting state. For this purpose, a sensor broadcasts a Running-to-Waiting message and waits for some transit time $t_{transit}$. If $t_{transit}$ expires and it has not received any Running-to-Waiting message, it switches to the Waiting state and sends a Notification-Waiting message, where it can stay there for t_{wait} time. When t_{wait} expires, it switches to the Ready state, where it can stay for t_{ready} time. When a sensor in the Ready state receives a Running-to-Waiting

Figure 14. State transition diagram of 3DIRACC$_k$.

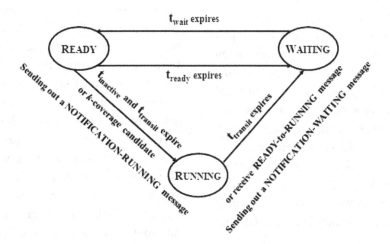

message from its sensing neighbor, it runs the *k-Coverage-Candidacy* algorithm to check whether it needs to turn on.

If a sensor in the Ready state finds out that it has not been active for some $t_{inactive}$ time, it will broadcast a Ready-to-Running message and wait for some $t_{transit}$ time. If $t_{transit}$ expires without receiving any other Ready-to-Running, it will send out a Notification-Running message and switch to the Running state. Otherwise, it stays in the Ready state. a sensor in the Ready state would also apply the same process if it finds out that it has not heard from one of its sensing neighbor within some t_{alive} time. This means that this sensing neighbor has entirely depleted its energy and died. To this end, each sensor in the Running state should broadcast an alIVE message after each t_{active} time. We assume that each sensor remains active for at least t_{active} time.

Performance Evaluation

In this section, we evaluate the performance of $3DIRACC_k$ with a high-level simulator written in the C language. We consider a cubic sensor field of side length 1000 m where all sensors are randomly and uniformly deployed. We consider the same energy model as described in the previous section. All simulations are repeated 200 times and the results are averaged.

Figure 15 plots the sensor spatial density $\lambda(r, k)$ versus the coverage degree k, where the radius r of the sensing range of sensors is equal to 30 m. We observe a close to perfect match between our simulation and analytical results. Notice that $\lambda(r, k)$ increases with k for a fixed r. Indeed, higher coverage degree of a field would require more active sensors. Figure 16 plots $\lambda(r, k)$ versus the radius r of the sensing range of sensors, where the degree of coverage k is equal to 3. We observe that $\lambda(r, k)$ decreases with r for a fixed k. In fact, sensors with larger sensing range would cover more areas and hence less number

of active sensors is required to achieve a certain coverage degree k of a field.

FUTURE TRENDS

We believe that three research areas should be deeply investigated. First, rigorous approaches should focus on heterogeneous sensors with different sensing and communication ranges capabilities. Indeed, real-world sensing applications may require heterogeneous sensors in order to enhance the reliability of the network and extend its lifetime. Moreover, even sensors equipped with identical hardware may not always have the same sensing and communication models. Particularly, the problem of ensuring *k*-coverage of a three-dimensional field needs to be addressed using different tiling strategies of a sphere, assuming that the sensing range of the sensors is represented by a sphere. Indeed, the problems of *k*-coverage and tiling in three-dimensional space seem to be very inter-related. Different models of titling of a sphere could be adapted to solve the problem of *k*-coverage in three-dimensional wireless sensor networks with better bounds on the required sensor spatial density. Then, these results could be extended to account for heterogeneous, three-dimensional wireless sensor networks.

Second, it is worth studying the problem of connected *k*-coverage in mission-oriented wireless sensor networks, where sensors are mobile to accomplish a specific mission that is requested by a central gathering point, such as the sink, at some time. Indeed, most of the existing work focuses on the problem of static, connected *k*-coverage in two-dimensional wireless sensor networks. Therefore, more attention should be given to investigate the problem of guaranteeing mobile *k*-coverage while maintaining network connectivity in mission-oriented wireless sensor networks. In particular, it is necessary to build a unified framework, where *k*-coverage, sensor scheduling, and routing are jointly considered.

Figure 15. vs. k

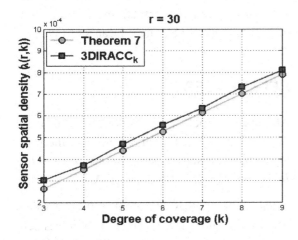

Figure 16. vs. r

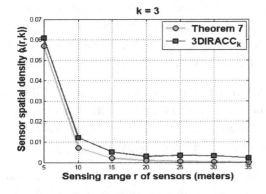

Last, but not least, the problem of determining the minimum number of sensors required to ensure full *k*-coverage of a field should be deeply investigated using more sophisticated approaches, such as percolation theory. These studies should account for the stochastic nature of the sensing and communication ranges of the sensors, which are neither necessary circular nor spherical.

CONCLUSION

Discussion

In this chapter, we have reviewed a sample of approaches for generating connected *k*-coverage

configurations in two-dimensional and three-dimensional wireless sensor networks. Also, we have proposed energy-efficient connected *k*-coverage protocols based on our characterization of *k*-coverage using a geometric analysis of the intersection of *k* sensing disks of the sensors. We have mainly used Helly's Theorem, a fundamental result of convexity theory, which characterizes the intersection of convex sets. For two-dimensional deployment fields, we have exploited the geometric properties of the Reuleaux triangle to compute the number of sensors needed to fully *k*-cover of field. However, for three-dimensional deployment fields, using *Helly's Theorem* (Bollobás, 2006), we find that the extension of the analysis of *k*-coverage in two-dimensional wireless sensor networks to their three-dimensional counterpart is not straightforward. This is due to the fact that some properties that hold for two-dimensional space cannot generalize to two-dimensional space. Precisely, the regular triangle allows a perfect tiling of a two-dimensional space while the regular *tetrahedron* cannot perfectly tile a three-dimensional space. Furthermore, while the *Reuleaux triangle* has a constant width, its three-dimensional counterpart, called *Reuleaux tetrahedron*, does not have a constant breadth. Thus, while *k* sensors deployed in a Reuleaux triangle region of width *r'* in a two-dimensional field guarantee full *k*-coverage of this region, a Reuleaux tetrahedron region of side length *r* in a three-dimensional field is not guaranteed to be fully *k*-covered with exactly sensors deployed in it, where *r'* (respectively, *r*) is the radius of the sensing disk (respectively, sphere) of the sensors. We have proposed a fully distributed *k*-coverage protocol for three-dimensional wireless sensor networks based on the "closest shape" to the Reuleaux tetrahedron.

Concluding Remarks

Although several elegant protocols have been proposed to solve the problem of connected *k*-

coverage in wireless sensor networks as discussed earlier, we believe that more work needs to be done. In particular, the problem of routing on duty-cycled sensors has received little attention in the literature. In particular, joint coverage and geographic forwarding in wireless sensor networks has been overlooked intentionally. This is due to the fact that *all* sensors are assumed to be always *on* during data forwarding. However, this assumption is not valid in real-world applications, where all sensors should not stay *on* all the times to save energy. Furthermore, it has been shown that the sensing ranges of the sensors are irregular and probabilistic (Cao et al., 2005), (Duarte & Hu, 2003, Elfes, 1989). The same results have been found for the communication ranges of the sensors (Zhao & Govindan, 2003, Zhou et al., 2005). In general, the sensing and communications ranges of the sensors do not follow the unit disk and sphere models for two-dimensional and three-dimensional deployment fields, respectively.

Our future work is three-fold. First, we plan to extend our protocols for stochastic sensing and communication models of the sensors while considering both two-dimensional and three-dimensional deployment fields. Second, we focus to complement previous work on joint coverage and routing in wireless sensor networks (Nath & Gibbons, 2007). Precisely, we focus on the design of energy-efficient geographic forwarding on a duty cycled sensors, where every point in a deployment field is covered by at least *k* sensors, in order to promote the applicability of our protocols for real-world sensing applications. Third, we plan to implement our protocols using a sensor testbed to assess their performance.

ACKNOWLEDGMENT

The author would like to thank the editors for their comments, which helped improve the quality and presentation of this chapter. This research is partially supported by the US National Science Foundation (NSF) grant 0917089 and a New Faculty Start-Up Research Grant from Hofstra College of Liberal Arts and Sciences Dean's Office.

REFERENCES

Abrams, Z., Goel, A., & Plotkin, S. (2004). Set *k*-cover algorithms for energy efficient monitoring in wireless sensor networks. In *Proceedings IPSN* (pp. 424-432). Berkeley, CA.

Adlakha, S., & Srivastava, M. (2003). Critical density threshold for coverage in wireless sensor networks. In *Proceedings IEEE WCNC* (pp. 1615-1620). New Orleans, LA:IEEE.

Ai, J., & Abouzeid, A. (2006). Coverage by directional sensors in randomly deployed wireless sensor networks. *Journal of Combinatorial Optimization*, *11*(1), 21–41. doi:10.1007/s10878-006-5975-x

Akyildiz, I. F., Pompili, D., & Melodia, T. (2005). Underwater acoustic sensor networks: research challenges. *Ad Hoc Networks*, *3*, 257–279. doi:10.1016/j.adhoc.2005.01.001

Alam, S., & Haas, Z. (2006). Coverage and connectivity in three-dimensional networks. In *Proceedings ACM MobiCom* (pp. 346-357). Los Angeles, CA:ACM.

Ammari, H. M., & Das, S. K. (2008a). Joint *k*-Coverage and Hybrid Forwarding in Duty-Cycled Three-Dimensional Wireless Sensor Networks. In *Proceedings IEEE SECON* (pp. 170-178). San Francisco, CA:IEEE.

Ammari, H. M., & Das, S. K. (2008b). Clustering-based minimum energy *m*-connected *k*-covered wireless sensor networks. In *Proceedings EWSN*, TPC Best Paper Award [Bologna, Italy.]. *Lecture Notes in Computer Science*, *4913*, 1–16. doi:10.1007/978-3-540-77690-1_1

Ammari, H. M., & Das, S. K. (2009). Fault tolerance measures for large-scale wireless sensor networks. *ACM Transaction on Autonomous and Adaptive Systems, 4*(1), 2:1-2:28.

Bai, X., Kumar, S., Xuan, D., Yun, Z., & Lai, T. H. (2006). Deploying wireless sensors to achieve both coverage and connectivity. In *Proceedings ACM MobiHoc* (pp. 131-142). Florence, Italy: ACM.

Bollobás, B. (2006). *The Art of Mathematics: Coffee Time in Memphis*, Cambridge, UK: Cambridge University Press.

Bulusu, N., Heidemann, J., & Estrin, D. (2000). GPS-less low cost outdoor localization for very small devices. *IEEE Personal Communications Magazine, 7*(5), 28–34. doi:10.1109/98.878533

Cao, Q., Yan, T., Stankovic, J., & Abdelzaher, T. (2005). Analysis of target detection performance for wireless sensor network. In *Proceedings DCOSS*, LNCS 3560 (pp. 276-292). Marina del Rey, CA.

Cardei, M., & Wu, J. (2006). Energy-efficient coverage problems in wireless ad-hoc sensor networks. *Computer Communications, 29*(4), 413–420. doi:10.1016/j.comcom.2004.12.025

Chen, A., Kumar, S., & Lai, T. H. (2007). Designing localized algorithms for barrier coverage. In *Proceedings ACM MobiCom* (pp. 75-86). Montreal, Canada: ACM.

Chen, B., Jameson, K., Balakrishnan, H., & Morris, R. (2002). Span: An energy-efficient coordination algorithm for topology maintenance in ad hoc wireless networks. *ACM Wireless Networks, 8*(5), 481–494. doi:10.1023/A:1016542229220

Conway, J., & Torquato, S. (2006). Tiling, packing, and covering with tetrahedra. In *Proceedings National Academy of Sciences of the United States of America (PNAS), 103*(28), 10612-10617.

Cortes, J., Martinez, S., Karatas, T., & Bullo, F. (2004). Coverage control for mobile sensing networks. *IEEE Transactions on Robotics and Automation, 20*(2), 243–255. doi:10.1109/TRA.2004.824698

Datta, A. K., Gradinariu, M., Linga, P., & Raipin-Parvedy, P. (2005). Self-* distributed query region covering in sensor networks. In *Proceedings IEEE SRDS* (pp. 50-59). Orlando, FL: IEEE.

Datta, A. K., Gradinariu, M., & Patel, R. (2007). Dominating sets-based Self* Minimum Connected Covers of Query Regions in Sensor Networks. New York: Springer.

Duarte, M., & Hu, Y. (2003). Distance based decision fusion in a distributed wireless sensor network. In *Proceedings IPSN* [Palo Alto, CA.]. *Lecture Notes in Computer Science, 2634*, 392–404. doi:10.1007/3-540-36978-3_26

Elfes, A. (1989). Using Occupancy Grids for Mobile Robot Perception and Navigation. *IEEE Computer, 22*(6), 46–57.

Gupta, H., Zhou, Z., Das, S. R., & Gu, Q. (2006). Connected sensor cover: Self-organization of sensor networks for efficient query execution. *IEEE/ACM TON, 14*(1), 55-67.

Huang, C., & Tseng, Y. (2003). The coverage problem in a wireless sensor network. In *Proceedings ACM WSNA* (pp. 115-121). San Diego, CA: ACM.

Huang, C., Tseng, Y., & Lo, L. (2004). The coverage problem in three-dimensional wireless sensor networks. In *Proceedings IEEE Globecom* (pp. 115-121). Dallas, TX: IEEE.

Huang, C., Tseng, Y., & Wu, H. (2007). Distributed protocols for ensuring both coverage and connectivity of a wireless sensor network. *ACM Transactions on Sensor Networks, 3*(1), 1–24. doi:10.1145/1210669.1210674

Kumar, S., Lai, T. H., & Arora, A. (2005). Barrier coverage with wireless sensors. In *Proceedings ACM MobiCom* (pp. 284-298). Cologne, Germany: ACM.

Kumar, S., Lai, T. H., & Balogh, J. (2004). On *k*-coverage in a mostly sleeping sensor network. In *Proceedings ACM MobiCom* (pp. 144-158). Philadelphia: ACM

Kumar, S., Lai, T. H., Posner, M. E., & Sinha, P. (2007). Optimal sleep-wakeup algorithms for barriers of wireless sensors. In *Proceedings IEEE BROADNETS* (pp. 327-336). Raleigh Durham, NC: IEEE.

Lazos, L., & Poovendran, R. (2006a). Coverage in heterogeneous sensor networks. In *Proceedings WiOpt* (pp. 1-10). Boston, MA.

Lazos, L., & Poovendran, R. (2006b). Stochastic coverage in heterogeneous sensor networks. *ACM Transactions on Sensor Networks, 2*(3), 325–358. doi:10.1145/1167935.1167937

Li, X.-Y., Wan, P.-J., & Frieder, O. (2003). Coverage in wireless ad-hoc sensor networks. *IEEE Transactions on Computers, 52*(6), 753–763. doi:10.1109/TC.2003.1204831

Liu, B., Brass, P., & Dousse, O. (2005). Mobility improves coverage of sensor networks. In *Proceedings ACM MobiHoc,* (pp. 300-308). Urbana-Champaign, IL: ACM.

Liu, C., Wu, K., Xiao, Y., & Sun, B. (2006). Random coverage with guaranteed connectivity: Joint scheduling for wireless sensor networks. *IEEE Transactions on Parallel and Distributed Systems, 17*(6), 562–575. doi:10.1109/TPDS.2006.77

Megerian, S., Koushanfar, F., Potkonjak, M., & Srivastava, M. (2001). Exposure in wireless ad-hoc sensor networks. In *Proceedings ACM MobiCom* (pp. 139-150). Rome: ACM.

Megerian, S., Koushanfar, F., Potkonjak, M., & Srivastava, M. (2001). Coverage problems in wireless ad-hoc sensor networks. In *Proceedings IEEE INFOCOM* (pp. 1380-1387). Anchorage, AK: IEEE.

Megerian, S., Koushanfar, F., Potkonjak, M., & Srivastava, M. (2005). Worst and best-case coverage in sensor networks. *IEEE Transactions on Mobile Computing, 4*(1), 84–92. doi:10.1109/TMC.2005.15

Nath, S., & Gibbons, P. B. (2007). Communicating via fireflies: Geographic routing on duty-cycled sensors. In *Proceedings IPSN* (pp. 440-449). Cambridge, MA.

Poduri, S., Pattem, S., Krishnamachari, B., & Sukhatme, G. S. (2006). Sensor network configuration and the curse of dimensionality. In *Proceedings IEEE EmNets*. Cambridge, MA: IEEE.

Pompili, D., Melodia, T., & Akyildiz, I. F. (2006). Deployment analysis in underwater acoustic wireless sensor networks. In *Proceedings ACM WUWNet* (pp. 48-55). Los Angeles: ACM.

Ravelomanana, V. (2004). Extremal properties of three-dimensional sensor networks with applications. *IEEE Transactions on Mobile Computing, 3*(3), 246–257. doi:10.1109/TMC.2004.23

Shakkottai, S., Srikant, R., & Shroff, N. (2005). Unreliable sensor grids: Coverage, connectivity and diameter. *Ad Hoc Networks, 3*(6), 702–716. doi:10.1016/j.adhoc.2004.02.001

Tian, D., & Georganas, N. (2005). Connectivity maintenance and coverage preservation in wireless sensor networks. *Ad Hoc Networks, 3*(6), 744–761. doi:10.1016/j.adhoc.2004.03.001

Weisstein, E. W. (2009a). *Reuleaux Triangle.* Retreived (n.d.). From, http://mathworld.wolfram.com/ReuleauxTriangle.html

Weisstein, E. W. (2009b). Reuleaux Tetrahedron. Retreived (n.d.). From, http://mathworld.wolfram. com/ReuleauxTetrahedron.html

Wu, J., & Li, H. (1999). On calculating connected dominating set for efficient routing in ad hoc wireless networks, In *Proceedings of the Third International Workshop on Discrete Algorithms and Methods for Mobile Computing and Communications* (pp. 7–14). Seattle, WA: ACM.

Xing, G., Wang, X., Zhang, Y., Lu, C., Pless, R., & Gill, C. (2005). Integrated coverage and connectivity configuration for energy conservation in sensor networks. *ACM Transactions on Sensor Networks*, *1*(1), 36–72. doi:10.1145/1077391.1077394

Yang, S., Dai, F., Cardei, M., & Wu, J. (2006). On connected multiple point coverage in wireless sensor networks. *International Journal of Wireless Information Networks*, *13*(4), 289–301. doi:10.1007/s10776-006-0036-z

Yarvis, M., Kushalnagar, N., Singh, H., Rangarajan, A., Liu, Y., & Singh, S. (2005). Exploiting heterogeneity in sensor networks. In *Procedings IEEE Infocom* (pp. 878-890). Miami, FL; IEEE.

Ye, F., Zhong, G., Cheng, J., Lu, S., & Zhang, L. (2003). PEAS: A Robust Energy Conserving Protocol for Long-Lived Sensor Networks. In *Proceedings ICDCS* (pp. 1-10). Providence, RI: IEEE.

Yener, B., Magdon-Ismail, M., & Sivrikaya, F. (2007). Joint problem of power optimal connectivity and coverage in wireless sensor networks. *Wireless Networks*, *13*(4), 537–550. doi:10.1007/s11276-006-5875-0

Zhang, H., & Hou, J. (2005). Maintaining sensing coverage and connectivity in large sensor networks. *Ad Hoc & Sensor Wireless Networks*, *1*(1-2), 89–124.

Zhao, J., & Govindan, R. (2003). Understanding packet delivery performance in dense wireless sensor networks. In *Proceedings ACM SenSys* (pp. 1-13). Los Angeles: ACM.

Zhou, G., He, T., Krishnamurthy, S., & Stankovic, J. (2004). Impact of Radio Irregularity on Wireless Sensor Networks. In *Proceedings MobiSys* (pp. 125-138). Boston: ACM.

Zhou, Z., Das, S., & Gupta, H. (2005). Fault tolerant connected sensor cover with variable sensing and transmission ranges. In *Proceedings IEEE SECON* (pp. 594-604). Santa Clara, CA: IEEE.

KEY TERMS AND DEFINITIONS

Communication Neighbor Set: The *communication neighbor set* $CN(s_i)$ of s_i is a set of all the sensors in its communication range.

Communication Range: The *communication range* of a sensor s_i is a region such that s_i can communicate with any sensor located in this region.

Connectivity: A wireless sensor network is said to be *connected* if all the sensors can communicate with each other directly or indirectly.

Heterogeneous Wireless Sensor Networks: A wireless sensor network is said to be *heterogeneous* if all of its sensors do not have the same storage, processing, battery power, sensing, and communication capabilities.

Interior and Boundary Lenses: An *interior lens* of a cluster is not shared with any of its adjacent clusters while a *boundary lens* is shared by two adjacent clusters.

k-Coverage: Let a be an area of a deployment field. A point $p \in A$ is said to be *covered* if it belongs to the sensing range of at least one sensor. The area a is said to be covered if every point $p \in A$ is covered. a is said to be *k-covered* if each point $p \in A$ belongs to the intersection of sensing ranges of at least k sensors. A wireless sensor network that provides full *k*-coverage of a field is called *k-covered* wireless sensor network, where a *maximum value* of k is called *degree of coverage* of the network.

Sensing Neighbor Set: The *sensing neighbor set* $SN(s_i)$ of s_i is a set of all the sensors located in its sensing range.

Sensing Range: The *sensing range* of a sensor s_i is a region where every event that takes place in this region can be detected by s_i.

Width: The *width* of a closed convex area is the maximum distance between parallel lines that bound it.

Chapter 11
Middleware Support for Wireless Sensor Networks:
A Survey

Tales Heimfarth
Federal University of Rio Grande do Sul, Brazil

Edison Pignaton de Freitas
Federal University of Rio Grande do Sul, Brazil

Flávio Rech Wagner
Federal University of Rio Grande do Sul, Brazil

Tony Larsson
Halmstad University, Sweden

ABSTRACT

Wireless sensor networks (WSNs) are gaining visibility due to several sophisticated applications in which they play a key role, such as in pervasive computing and context-aware systems. However, the evolution of WSN capabilities, especially regarding their ability to provide information, brings complexity to their development, in particular for those application developers that are not familiar with the technology underlying and needed to support WSNs. In order to address this issue and allow the use of the full potential of the sensor network capabilities, the use of a middleware that raises the abstraction level and hides much of the WSN complexity is a promising proposal. However, the development of a middleware for WSNs is not an easy task. Systems based on WSNs have several issues that make them quite different from conventional networked computer systems, thus requiring specific approaches that largely differ from the current solutions. The proposal of this chapter is to address the complexity of middleware made for sensor networks, presenting a taxonomy that characterizes the main issues in the field. An overview of the state-of-the-art is also provided, as well as a critical assessment of existing approaches.

INTRODUCTION

The emerging use of WSNs in different application domains is increasing the importance of this research field. The complexity of issues such as dynamic routing, service discovery, resource management, data fusion, in-network processing, among others, are examples of what is being proposed as research efforts needed in this area. The support for applica-

DOI: 10.4018/978-1-61520-701-5.ch011

tion development through a middleware layer for WSNs has also received a lot of attention recently. A middleware to support the programmability of WSNs is necessary due to the increasing sophistication of the applications running over them. These cooperative sensing related applications require adequate programming abstractions and services to enable a seamless programming.

The main challenge for programming a WSN arises from the large gap between the high-level functional specification of WSN applications and the complexity of the underlying infrastructure. This motivates search and strive for mechanisms that can simplify the programmability of these applications, hiding the mentioned complexity (Wang, Cao, Li & Das, 2008). Complexity arises mainly due to factors such as the dynamic network topology, constrained resources, and low-level sensor programming. These concerns are opposed to the requirements of the applications, such as flexibility, reliability, and reusability.

The complexity of the WSN infrastructure depends largely on which components and configurations that are selected to build it. Moreover, WSNs have a large design space, which adds additional challenges to the development of their applications (Römer, K. & Mattern, F., 2004). When programming applications, the developer must be aware of the configuration of the underlying network. Examples of very important such concerns about the infrastructure when developing WSN applications are: How are the deployment of the WSN to be made? Which mobility pattern the nodes will have? Which resources will be available in each node? To what extent will the network be homogeneous or heterogeneous? What are the characteristics of the communication device used? Is the network always connected or is it setup in a dynamic fashion at need?

Therefore, the middleware used must address the complexity of the WSN infrastructure and be aware of the different configurations of this infrastructure. In this context, useful abstractions to simplify the network programming should be

provided by the middleware, and it should also be able to cover a large part of the WSN design space. However, it is very difficult to provide a homogeneous software platform that supports the development of different applications over different WSN infrastructures.

Key issues that must be addressed by middleware due to the different WSN infrastructures are: energy management, to cope with very small nodes; resource sharing, to use resources efficiently; topology control, due to the different possibilities of deployment; integration of different nodes, in order to cope with the heterogeneity; and adaptability, due to the dynamics of the environment.

From the point of view of the applications, the most important issue of the middleware is the programming model it provides. It should support and make easier the development of applications, while at the same time being as generic as possible. These are often contradictory goals. The abstractions provided to support the programming model are implemented by an underlying set of services.

Since different types of applications may be programmed on the WSN, as area surveillance, object tracking, environmental monitoring, and health care applications, different requirements of the programming model can be recognized. For example, for environmental monitoring a declarative language that allows the expression of queries is a suitable programming model, whereas an event-based programming model is more suitable for event supervision and tracking. It is a big challenge to develop middleware that can support many different types of applications due to the very different requirements that they impose on the programming model.

In this chapter, several state-of-the-art middleware techniques will be presented, taking in account the design space of the WSN infrastructure and the different types of applications. The focus is to provide an insight on how different middleware cope with the infrastructure complexity, giving a suitable programming model to the application developer. Moreover, discussions about how generic the different middleware are will also be presented.

Different surveys exist in the current literature, e.g. (Henricksen & Robinson, 2006); (Kuorilehto, Hannikainen & Hamalainen, 2005), (Molla & Ahamed, 2006), and (Wang, Cao, Li & Das, 2008), as well as publications addressing key issues on developing middleware for WSNs, like (Römer, 2004), (Woo, Madden & Govindan, 2004), and (Yu, Krishnamachari & Prasanna, 2004). Nevertheless, differently from the existing works, our survey presents a novel taxonomy to classify existing middleware solutions. The existing surveys either use a very limited classification of middleware, as in (Henricksen & Robinson, 2006) and (Kuorilehto, Hannikainen & Hamalainen, 2005), or do not use any structure to classify them, for example in (Molla & Ahamed, 2006). The survey (Wang, Cao, Li & Das, 2008) presents the middlewares based on main services supported by them. Our work, differently from others, systematically classifies the existing middleware based on the introduced taxonomy, and the presentation of each middleware is done in a structured way to improve the comparability. Other surveys either analyze just one aspect of the presented middleware or present various aspects in an ad hoc manner. Our survey also improves the existing ones regarding the coverage: eight middlewares are presented in Section 4.

Little work has been done in creating a taxonomy for the WSN middleware technology. The extensive taxonomy presented in (Wang, Cao, Li & Das, 2008) mainly addresses the features of the WSN middleware. Although covering important aspects of WSN middleware, the taxonomy is complex and concentrates on a very generic aspect of the middleware, namely its features. Our taxonomy, on the other hand, focuses on the most important aspect of WSN middleware: the programming abstraction, i.e. the interface between the user and the WSN. Moreover, it is less complex than the taxonomy presented in (Wang,

Cao, Li & Das, 2008), and, finally, different from that work, the presented middlewares' are classified by the taxonomy.

Therefore, the discussions of this chapter are guided by the programming model provided by the different middlewares'. As already stated, in order to better classify the different programming models, a clear and standardized taxonomy is provided, thus guiding a structured discussion about the subject that allows a clear overview of the area.

Since the middleware solutions presented in this chapter are just an initial step on providing a consistent programmability, testing, and deployment environment across the design space, much research is still needed. Therefore, future directions on middleware development are also discussed.

MAIN CHARACTERISTICS AND CHALLENGES OF WSN MIDDLEWARES

WSN Characteristics

The operation of a WSN requires in general methods that are quite different from those used for conventional computer networks. In these networks, the user is interested only in the performed computation, whereas in a WSN the integration of the computation devices with the physical world is the essential concern. This reflects the way of deployment of the sensor nodes. Nodes are spread over an area of interest for example to monitor a certain phenomenon. These sensors can be static or dynamic, moving around the area or even flying over it. Sensing tasks are, ideally, defined at a high-level, and the answers to them can be obtained by a combination of individual contributions of several limited sensors that compose the whole system.

Besides the particular way of operation, WSNs also have certain characteristics that must be addressed in the design of a middleware that aims at supporting them. The first main characteristic is the limited availability of several node resources. More concretely said, WSN nodes are often small devices that have limited energy supply and restricted CPU performance, memory space, and communication bandwidth and range.

The high degree of dynamicity is another important feature found in WSNs. Failures, mobility of nodes or tracked events, and environment obstacles can interfere with and disturb the functionality of the network.

Heterogeneity is also an issue that must be observed. Nodes can be of various types, showing different processing power, amount of memory, available energy, and sensing capabilities. Further most nodes in the network are small and resource constrained, thus bringing the need of harmonizing the interactions and cooperation among different types of nodes.

Regarding the characteristics and operation modes of WSNs, some software design principles are proposed, as the use of distributed algorithms to achieve a global goal of the system and create an adaptive fidelity of algorithms to achieve efficient use of resources.

In traditional networks, selection of nodes is done by using a unique identification, typically a network address. This is because the communication focuses in transferring data between two specific devices. In WSNs, due to the redundant deployment, the user is not interested on getting requested information from a specific node, but much more from a desired region or from nodes that provide a given type of information. The important issue is the information or data and not which node that provides it. This is called data-centric communication, in contrast to conventional address-based communication.

The data-centric or data-driven characteristic of WSNs generates requirements for different routing protocols in which nodes are not targeted by an identification address, but by the data that they provide. In fact, the node itself is not addressed, the importance is focused on the data that it is capable to produce. As an example, in a sensor network that is used to monitor the temperature in a building, the queries are not addressed to nodes "X" and "Y", but to a location with given properties, such as "the conference room in the third floor" or "the place where the temperature is greater than a threshold".

The information retrieval is also singular, as data from several nodes may be combined in order to fulfill the application requirements. The so called "virtual nodes" address this issue, by providing meaningful information as a combination or fusion of simple basic data gathered by correlated (in time and/or space) nodes.

A very important characteristic of WSNs is that they are in many cases deployed in a harsh environment in which the modification of some configuration or the replacement of some software component must be achieved without a physical access. This is the case, for instance, of the ZebraNet project (Juang, Oki, Wang, Martonosi, Peh & Rubenstein, 2002).

Network lifetime is another essential issue in WSNs, due to the fact that batteries cannot be easily replaced in many of the application scenarios. The definition of network lifetime varies: it can go from the time of the first failure up to the moment when the network is disconnected. In the context of this chapter, the lifetime is defined as the time interval during which the network can provide the desired service.

Since sensors are deeply embedded in the physical world, the environment plays an important role in this kind of network. Several changes in the environment may occur, such as changes of weather conditions, movement of obstacles, and node failures. The network must cope with these classes of problems in order to keep its usefulness.

Challenges of WSN Middleware

After highlighting the main characteristics of WSNs, the challenges in the development of a middleware for this domain are presented in this section.

The first challenge in the adoption of a middleware approach aimed for WSNs is to define the basis over which it will be built. In conventional computers, it sits between the operating system and the application. Small sensor nodes do not have this clear separation between application and operating system, and cross-layer optimizations are often used. Many nodes even execute tasks directly in hardware. However, middleware is always supposed to provide functionality of a general nature that simplifies the development network distributed systems and applications.

Looking directly to the goal of middleware for WSNs, it has to support the network development, maintenance, and deployment, as well as the execution of sensing applications. It means that the middleware must: (a) coordinate the nodes, by distributing detailed low-level tasks according to the high-level tasks it receives; (b) perform the merge of individual results issued by distinct nodes; and (c) present the final report to the user. The middleware must also handle the heterogeneity of sensor nodes, providing appropriate abstractions and mechanisms to use different kinds of sensors regarding energy efficiency, robustness, and scalability.

The middleware must also deal with the integration of the WSN to the conventional network with which it is connected. Classical mechanisms and infrastructures do not fit the needs to run in a WSN, but integration must exist between both networks. The WSN middleware must thus address this integration concern as well.

Another important issue is that the middleware must provide mechanisms to inject application knowledge into the infrastructure. This means that the middleware should offer mechanisms for the application to inject support (programs and data) for handling of special needs and be able to adapt itself to such needs. For example, the resource and energy efficiency can be improved by application specific data caching and aggre-

gation in intermediary nodes (Römer, Kasten & Mattern, 2002).

The data-centric communication is a key concern in WSNs. It fits better in an event-based communication paradigm than in a traditional request-reply scheme. A middleware for WSNs should offer this feature.

The high-level dynamicity and adaptability of WSNs require new levels of support for automatic configuration (self-configuration), error handling, and service degradation.

As most embedded systems, WSNs are used in applications in which real-time requirements are present. Therefore, support for time constraints should also be offered.

After this general presentation of the concerns related to the development of middlewares for WSNs, a summary list briefly describes these ideas:

a. **Abstraction Support:** This is one the main goals of the middleware for WSNs, in order to hide the underlying platforms (hardware and operating system), providing a homogeneous view of the entire network;

b. **Data Fusion:** Raw data collected by sensor nodes have to be merged or synthesized in order to provide a high-level and easily understandable format or report;

c. **Resource Constraints:** In order to work under very limited available resources, middlewares for WSNs must be lightweight;

d. **Dynamic Topology:** A key concern is related to the dynamicity of the topology in a sensor network;

e. **Application Knowledge:** In order to provide useful services to the final applications, the middleware has to integrate knowledge about the supported applications;

f. **Programming Paradigm:** Programming paradigms for sensor networks, and also for parts of the middleware for these systems, are quite different from traditional ones;

Figure 1. WSN middleware taxonomy

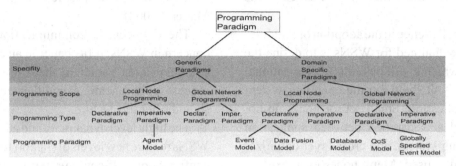

g.   **Adaptability:** Adaptive behavior is a key element in middleware made for WSNs, in order to provide adequate support under operation and environment changes, performance adaptations, among others;

h.   **Scalability:** Middleware for WSNs must support scalability, not only regarding the number of nodes, but also the number of users, must be able to operate over long periods;

i.   **Security:** Security issues in sensor networks, both in data processing and communication, are another key concern because of the applications in which they are used, like health care and military operations;

j.   **QoS Support:** In order to ensure reliable services, middlewares for WSNs have to handle QoS requirements like resource and bandwidth allocation, response time, and deadlines, among others;

k.   **Integration:** Real-world integration is also a challenge, since space and time play crucial roles in sensor networks to identify real-world events and objects and to distinguish them from other ones. Hence, the establishment of common time and space scales (localization) may be an important issue of the middleware;

l.   **Heterogeneity:** The node heterogeneity provides a challenge in the coordination and cooperation among sensors, which have different available resources;

m.   **Data-centric communication support:** Due to the fact that data are the most important entities in the WSN, the middleware must support diverse data-centric mechanisms, e.g. data-based routing.

MIDDLEWARE TAXONOMY

In this section, we present the taxonomy that structures the overview of the state-of-the-art middleware. Although several characteristics may be used to classify WSN middlewares, we decided to base it on the programming paradigm concept. It represents the high-level programming interface that is responsible for hiding the underlying infrastructure to the application developer. The concepts and abstractions offered by the middleware define to which kind of application domain the sensor network is directed. There is an obvious trade-off in the programming paradigm between its expressiveness and the complexity of programming. The less complexity in programming the applications, the less general flexibility and expressiveness is provided by the paradigm.

Our taxonomy is presented in Figure 1. At the first level, we classify middleware programming paradigms based on their specificity: generic and domain specific paradigms. In the generic programming paradigms, different types of WSN applications may be written, as sensor queries, distributed complex in-network processing, and event handling. In the specific programming paradigms, the middleware is specialized for some domain-specific type of application. As an example, the database-based middleware allows the user to specify SQL-like queries that are planned and executed automatically by the middleware. The queries are used to return sensor data and may not be used for generic programming of the sensor network.

The second level in our classification is the Programming Scope, which divides the programming paradigms based on the granularity of the paradigm. In the Local Node Programming model, the developer is responsible for specifying the behavior of individual nodes of the system, including their actions and the cooperation between them. The Global Network Programming approach allows the specification of a single centralized program (global behavior) that will be automatically partitioned and distributed among the sensor nodes.

The Programming Type describes how the desired actions that should be performed by the WSN middleware are implemented: through commands or orders (imperative paradigm) or through a description of the desired results (declarative paradigm). For example, paradigms that describe the types of events (simple or composed) that should be reported by the middleware are classified as declarative, whereas agent-based middleware rely on agents executing commands of the user and therefore belong to the imperative paradigm.

At the bottom level of our taxonomy we have identified different types of middleware, according to the programming paradigm that reflects the main application goals. In the next paragraphs, these different paradigms are presented and situated in our taxonomy.

In Database-Based Middlewares, the sensor network is seen as a network distributed storage where SQL-like queries can be issued and the network then performs a certain task in response. This type of query processing system provides high-level interfaces that allow users to collect and process continuous streams of data. It is an efficient way of implementing monitoring applications without forcing the user to face complex questions like management of multi-hop network topologies or acquiring samples from sensor nodes (Woo, Madden & Govindan, 2004).

Due to the fact that such middleware is specialized in answering sensor queries, it is classified as a domain-specific middleware. Moreover, the user "programs" the network by means of a global query, and, therefore, we classify such a paradigm as Global Network Programming. Additionally, because the SQL-like query is a declarative description of the sample that should be returned by the middleware, the Programming Type of the middleware is classified as based on Declarative Paradigm.

In the class of QoS-Based Middlewares, the QoS requirements coming from the application are used together with the sensor and network information to distribute the tasks among the nodes. The middleware receives the description of the application requirements. These requirements specify in which variables the application is interested in (e.g. blood pressure in a body surveillance application) and which degree of quality the variables should meet. Moreover, the middleware also has as input the sensors' description. This description informs which quality of measurement the sensor can provide for the variables. Based on these data, the middleware can calculate which sets of sensors that satisfy the application QoS requirements for each variable. Therefore, this type of middleware is specialized on QoS sensing applications (domain-specific). Because the user enters the description of the variables and the QoS for the variables of interest and not a program for each node, we classify this type of middleware as Global Network Programming, and its Programming Type is Declarative.

Another Programming Paradigm approach is based on the notion of events. Applications can specify interest in certain state changes of the real world seen as basic events, data conditions or in certain patterns of events (compound events). Upon detection of such an event or event pattern, the sensor nodes send notifications to the interested applications. As already stated, the events are classified in two types: the atomic and the compound ones. An atomic event refers to an event that can be determined merely based on the observation of a sensor, whereas a compound event must be inferred from the detection of other atomic or compound events. The notion of confidence provided by middleware can be used in order to formulate compound events based on degrees of certainty. The middleware may also link an absolute validity interval to the events in order to depict the temporal consistency between the environment and its observed measurement.

In this class, we can have middlewares that allow a global description of the required events, so we say that they follow a Globally Specified Event Model. They are also classified as Global Network Programming model with a Declarative Paradigm, since the user describes the events he/she is interested in. Other middleware implementations just allow a node-based declaration of events, thus they are classified as Local Node Programming.

The Data Fusion Middleware is an example of a domain-specific middleware. A middleware of this category provides support to the programmer by hiding several concerns such as data synchronization, buffer management, and fusion point placement, when developing data fusion applications.

Data aggregation consists of an in-network operation that combines multiple messages coming from different sources into a smaller representation that is either equivalent or represents in a suitable manner the original messages in its contents. It captures the redundancy among data collected by different sensors (Luo, Luo, Liu & Das, 2005).

Since Data Fusion Middlewares are specialized in the data fusion task, we classify them as domain-specific middlewares. Moreover, the user must specify the nodes that are acting as sources

and sinks for each data stream, therefore these middlewares follow a Local Node Programming paradigm. Nevertheless, the fusion points are automatically specified by the middleware, such that this middleware also has some characteristics of the Global Network Programming model. It is a Declarative middleware, since the user has to describe which kind of data fusion is expected in the system.

The Agent-based middlewares use a virtual machine approach to obtain dynamic re-programmability at a reduced cost and they assume that a sensor network system is composed by a common set of services and sub-systems, combined in different ways. The virtual machine offers hardware platform independence and code expressiveness for compactness of the code.

The language that is interpreted by the virtual machine allows a concise description of the composition of services and sub-systems. This approach presents an advantage when compared to the transmission of raw binary code, since the interpreted language produces a smaller code causing less migration overhead. Moreover, it provides a programming model powerful enough to implement any distributed system, while, at the same time, hiding unnecessary low-level details from the application programmer.

Since all kinds of applications and distributed algorithms may be written by this class of middleware, we classify it as a Generic Paradigm. Due to the fact that the user must program all interactions among individual nodes (agents) of the system, this class is identified as a Local Node Programming model. Moreover, the script languages are imperative: commands, like any other imperative language, must be issued by the agents in the system.

After describing the Programming Paradigms that are covered by existing middleware, described in the literature, the next section presents an overview of the state-of-the-art on WSN middlewares.

OVERVIEW OF SOME STATE-OF-THE-ART WSN MIDDLEWARES

In this part of the chapter, a summary of some of the most relevant state-of-the-art middlewares for Wireless Sensor Networks is provided. The selected middlewares are divided according to the taxonomy described in Section 3. Moreover, each middleware is presented highlighting four main aspects: 1) the problems that it proposes to address; 2) the design approach; 3) the system architecture; and 4) the application programming model.

Generic Paradigm: Local Node Programming

In this section, a selection of middlewares that support several types of applications and provide local node programming abstraction are presented. Currently, only middlewares implementing the agent model programming paradigm are representing this group.

Imperative Paradigm: Agent Model

The agent-based approach has being explored by several proposals of middlewares for sensor networks. One of the originators in the use of this approach is Maté (Levis & Culler, 2002), which brought several innovations in using software agents in a middleware for sensor networks. Other proposals added new features and came with additional improvements. The one that showed a great success was Agilla (Fok, Roman & Lu, 05), which is highlighted in this section. This middleware was built on top of a tuple spaces approach based on Linda (Gelernter, 1985), which was an innovation that represented a great improvement in terms of system flexibility and scalability for sensor networks. Furthermore TinyLime, a data-sharing middleware, is also presented in this section.

AGILLA

Agilla (Fok, Roman & Lu, 05) allows users to inject special programs in the WSN by using an abstraction of mobile agents that can migrate across the nodes performing certain application-specific tasks. This migration of agents makes possible the adaptation of the network to run different applications with different demanding requirements.

Problems Addressed: The first main problem addressed by Agilla is that the long deployment time of WSNs may require flexibility in order that they may be modified according to the user requirements that also may change along the time. The other key requirement is the fact that WSNs are generally deployed in highly dynamic environments, such that applications have to be flexible enough in order to continue being useful in changing scenarios. These two aspects are related to each other, but there is a distinction between them, which is the autonomy required by the latter if compared with the former. The former can be successfully addressed with an approach such as Maté (Levis & Culler, 2002), which provides means to perform an update in all nodes of a sensor network. However, the latter requires autonomous behavior of the network in order to tackle the changes that occur in a dynamic operational scenario.

Design Approach: Agilla uses an approach based on Maté (Levis & Culler, 2002), but unlike this one, which divides applications into capsules that are disseminated throughout a network by using flooding, it allows the deployment of applications by injecting new application code using mobile agents. These agents can move around the network nodes in an intelligent way, by simply moving, or cloning themselves to different locations in response to changes that might occur in the environment.

In Agilla an application consists of mobile agents that can perform application-specific tasks. Their coordination is made by the use of

Figure 2. Basic architecture of the Agilla middleware (Adapted from (Fok, Roman & Lu, 05))

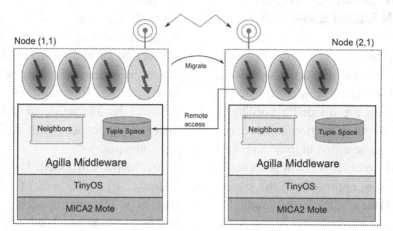

tuple spaces and the nodes are addressed by their location.

System Architecture: The kernel of Agilla consists of a tuple space and a list of node neighbors. The basic architecture of the Agilla middleware with these two components can be seen in Figure 2. Besides these components, the figure also highlights the multiple mobile agents running on top of the middleware.

The tuple space and the list of node neighbors, present in the model, are used to facilitate inter-agent coordination. The tuple space is based on Linda (Gelernter, 1985) and is shared by local agents and it is also remotely accessible, as presented in Figure 2. This approach provides a decoupled style of communication, where one agent can insert a tuple and another one can read or remove it using pattern matching. An acquaintance list is present in every node containing the location of all one-hop neighbor nodes. The middleware is responsible for populating and maintaining this list.

Agilla does not support a global tuple space that spans across multiple nodes due to the constraints related to energy consumption and limitation of bandwidth usage.

Application Programming Model: The applications are represented by scripts that are injected in the network by the use of the mobile

agents. The model used by the agents is based on a stack architecture that allows most instructions to be represented by a single byte. This is an important point to be highlighted, as it is desired that the agents are small enough in order that their movement among the nodes does not induce a great amount of energy consumption.

Agents are written in a low-level assembly-like language similar to that used in Maté. As mentioned above, this simple way to write the agents' code represents an advantage due to its small size, but, on the other hand, it becomes hard to describe more meaningful applications. This aspect of the low-level language is seen by the authors as a weak point of their approach. Moreover, that kind of language is not user friendly, leading to a tedious and error-prone application programming. Recognizing these weaknesses, a proposal to provide a higher-level language to replace the low-level one was done latter on.

Maté

The Maté (Levis, & Culler, 2002) virtual machine is a middleware that enables a wide range of sensor network applications to be composed based on a small set of code capsules, which provides modularity to implement applications. It is a byte-code interpreter that runs on top of TinyOS implemented by a stack-based virtual machine architecture.

Problems Addressed: One of the problems arising when programming applications for WSNs is how to make it simple to perform updates and even install new applications after system deployment. The goal of Maté is to address this problem by providing a flexible way of dynamic (re-)programming a sensor network, and at the same time presenting low system requirements. Moreover, a desired feature is that the messages that carry the new application code and updates should be small.

Design Approach: In order to achieve an easy and dynamic programmability, the approach adopted by Maté is the use of a Virtual Machine that interprets byte-codes, which allows that new applications are installed at runtime, without interference in its operation. Those new applications are intended to be installed in all the network nodes, and, for this purpose, a broadcast message carrying the corresponding byte-code is transmitted across the network. Another feature of the approach is that, once the byte-code is installed, its configuration can be done by parameter adjustments.

Regarding the energy consumption constraint, Maté presents a modular approach to perform updates and dissemination of new codes, by using small capsules containing the code that must be disseminated over the network

System Architecture: Maté is implemented as a single TinyOS component that interprets small pieces of code (the capsules) that can be seen as small agents. Each capsule has 24 instructions of one byte. Larger programs may be composed by several capsules. Each capsule fits in a single network package.

The middleware executes in response to events, which can be related to a timer or to the transmission or reception of a network package. Each event is related to a capsule, and has its own execution context. The virtual machine handles the three execution contexts, for the three types of events above. The execution of the capsules is always triggered by some event.

Application Programming Model: In the interpreted language of Maté, there is a command available that can be used to send the capsules to all neighbors. Upon receiving a capsule, each node tests whether the version is more recent than the current installed one. If positive, the new version is automatically installed. Using this mechanism, it is possible to easily distribute new versions of an existing application as well as new applications. The authors compare the dissemination of code to a virus. It makes possible for a user to enter a query (or other type of application) in a single point of the network, and this application will propagate itself until "infecting" all nodes. A drawback of this approach is that there is no way of selectively choosing the nodes that will receive the new code, as it is propagated to all nodes in the network.

TinyLime

TinyLIME (Curino, Giani, Giorgetta & Giusti, 2005) is a data-sharing middleware based on LIME (Murphy, Picco & Roman, 2001), which in turn adapts and extends for sensor networks the mobile tuple space model presented in Linda (Gelernter, 1985). The main contribution of this work is the operation setting supported by this middleware, which provides self-managed data sharing, supporting response to network changes.

Problems Addressed: The goal of this middleware is to support mobile ad-hoc networks providing an illusion of a shared memory between applications and sensors. Its intention is to avoid a central data collecting point, making possible that mobile nodes move around the area in which the nodes are deployed and by a one-hop communication receive data from the directly reachable nodes.

Design Approach: As TinyLIME is based on Lime, many characteristics are inherited from that middleware. A dynamic tuple space that changes according to the node connectivity is used, which supports node mobility. Connected tuple spaces from different nodes are linked forming a so called federate tuple space. This is in fact a shared space in which agents can move freely. Two important concepts added by Lime are also used: tuple location and reactions. The first one is related to the fact that even if the tuple is accessible by all connected agents, it exists only in one place of the system. The other concept allows an agent to register a special kind of code, called listener, which is executed whenever a tuple matches a particular pattern anywhere in the federation tuple space.

System Architecture: The TinyLIME model was designed and implemented for the Crossbow Mote platform, exploring the TinyOS functionality. The main components and their interactions are: (1) Client Component - A client, like a PDA which is interested in the sensor data, interacts with the sensors through the MoteLimeTupleSpace class, which extends LimeTupleSpace from the LIME API. The format of tuples containing sensor data is predefined and contains four fields (<SensorType, Integer, Integer, Date>), indicating the type of the sensor, the actual sensor reading, how long the mote has been alive, and the time stamp relative to the read value, respectively; (2) Interaction Between Client and Base Station - In the base stations, MoteLimeTupleSpace internally exploits two LIME tuple spaces, one holding data from the sensors and another one for communicating requests from the clients; (3) Base Station to Mote Interaction - The interaction between base station and motes is done by MoteAgent.

Application Programming Model: The operation scenario is different from the one used in traditional centralized models. The proposed alternative does not require multi-hop communication among sensors. Instead, it relies on a one-hop communication between a mobile base station and sparsely distributed ground sensor nodes that do not need to communicate with each other. In this environment, it is also considered the existence of hosts that are clients (e.g. PDAs) that do not have direct access to the sensors, but communicate with the base stations that relay the required data by transmitting the queries to the involved sensors.

In order to support this scenario, TinyLIME uses a transient shared tuple space to store tuples containing the sensed data. The sensor nodes (motes) communicate with a base station when it gets into their communication range. Motes are only seen by TinyLIME when connected to base stations. Each mote has an agent residing in the base station host, with its interface tuple space containing the set of data provided by its sensors (as tuples in the tuple space).

Analysis

In this section, we presented middlewares that use as main programming model the agents, which are autonomous entities residing in the sensor nodes and responsible for some specific task. The Maté middleware is the originator of the approach, in which agents, called capsules, are mobile entities representing queries injected in the network. The middleware uses an assembly-like language for programming and allows communication by means of exchange of messages. This was improved by the Agilla middleware: the tuple space decouples the communication in time and space. Moreover, instead of a simple flooding of the agents, the migration may be controlled by each agent itself and follow some desired policy, improving the efficiency of the middleware.

The TinyLIME middleware also uses tuple spaces to support the communication. But, differently from Agilla, the agents are fixed in the nodes and multi-hop communication is not supported. The operation mode of TinyLIME is not based on injection of agents by a central entity to query the network, but on a mobile base station that collects the data stored in the tuple space. This makes TinyLIME unique in this category of middleware.

Besides the presented middlewares, we can highlight in this category two more middlewares: DAVIM and Sensorware.

DAVIM (Horré, Michiels, Joosen & Verbaeten, 2008) is an adaptive middleware that enables dynamic management of services and isolation between applications simultaneously running on sensor networks. SensorWare (Boulis, Han, Shea & Srivastava, 2007) defines and supports lightweight and mobile control scripts (acting as mobile agents) that allow an efficient use of the computation, communication and sensing resources offered by the sensor nodes. This is done by the use of a service abstraction that can change at runtime by dynamically defining new services.

Sensorware presents a straightforward improvement of Maté. It substitutes the assembly-like language by a high-level scripting language that use and orchestrates existing services to accomplish the objective of the agent.

Domain Specific Paradigm: Local Node Programming

In this section, middleware focusing on domain-specific types of applications and providing local node programming abstraction are presented. All representatives of this category use the declarative paradigm. In addition, the programming abstraction is either based on events or focuses on data fusion.

Declarative Paradigm: Event Model

Although the event model represents a promising paradigm for sensor networks, few representative approaches have adopted it, and it is possible to say that work in this area is still immature. Impala (Liu & Martonosi, 2003) is outstanding in this area, having achieved a great level of maturity and presenting significant results in the deployment of real case studies.

IMPALA

Impala (Liu & Martonosi, 2003) is a middleware for sensor networks that enables application modularity, adaptability, and reparability. This middleware allows software updates to be received via the sensor node's wireless transceivers and to be applied to the running system dynamically. It also provides features to improve system performance, energy efficiency, and reliability. It was developed as a part of a project called ZebraNet (Juang, Oki, Wang, Martonosi, Peh & Rubenstein, 2002), where sensor nodes are placed on free ranging wildlife animals to perform studies of these in their ecosystem.

Problems Addressed: Some basic requirements of the network intended to be used in the ZebraNet project are the long time of deployment and the fact that the application may change during the use of the network, but the change cannot be done with physical access to the nodes. These requirements can be translated into concerns like application adaptability, via updates of the running software. However, in order to not affect the lifetime of the node, the updates cannot have strong impact on the energy consumption.

As the ZebraNet project handles networks with incomplete connectivity, composed by mobile nodes that may not be reached by an update, for instance, the middleware also has to provide interoperability among nodes that may run different versions of protocols, as long as this distinction among protocols does not compromise the system.

Design Approach: Impala is designed in a modular structure, which allows the update of certain parts of the system, without stopping the on going running applications. Another advantage is the reduced consumption of energy in transmitting updates, due to the smaller amount of data contained in the modules' updates, if compared to an entirely monolithic software. This is a feature stressed by the authors, because a monolithic approach tends to be a better solution when constraints such as memory space are relevant and generally provides better performance results. However, as the authors mention, the drawbacks that a monolithic approach bring cancel its benefits, especially when compared to a modular approach as proposed by this middleware, in order to address the adaptability that is required in the ZebraNet project.

Figure 3. Impala framework - layered system architecture (adapted from (Liu & Martonosi, 2003))

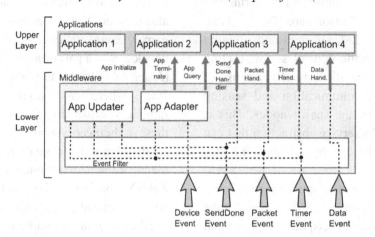

The design rationale of Impala can be summarized as follows: (1) Modularity - the middleware is responsible for switching decisions, and, like that, the applications can be independent and do not need to coordinate with each other; (2) Correctness - programming individual applications is simpler than a single big application; (3) Ease of Updates - Software changes can be done in small pieces of software, the modules; and (4) Energy Efficiency - Instead of transmitting an entirely monolithic program, the updates can be transmitted at the granularity of smaller modules.

System Architecture: The Impala architecture is divided in two layers, which can be seen in Figure 3. The upper layer contains all application protocols and programs. Only one application can run at a time, since there is no support for concurrency in this middleware, which represents a drawback of this approach. The lower layer contains three agents, namely the Application Adapter, the Application Updater, and the Event Filter. The first one is responsible for the adaptations in the application protocols facing different operation conditions. It aims at improving performance, energy efficiency, and robustness. The Application Updater is in charge of reception, propagation, and installation of the updates. The Event Filter captures and dispatches events of the above system units. Five types of events are supported: (1)

Timer Event – it signals when a timer has reached its programmed duration; (2) Packet Event - it signals when a network packet has arrived; (3) Send Done Event - it signals when a packet has been sent or failed to be sent; (4) Data Event - it signals when sensed data from a sensing device are ready to be read; and (5) Device Event - it signals at the detection of a device failure. The events are processed sequentially. Their processing is bound in time in order to avoid delays in handling the other events that might be waiting.

Application Programming Model: Impala uses an event-based programming model. Applications must implement four event handlers: timer handler; packet handler; send done handler, and data handler. The provided user library contains a number of general programming utilities, such as networking utilities that allow applications to send asynchronous messages. Timer utilities are also provided and allow the set up of timers for various purposes, like sending periodic messages, for instance. Impala has global data structures that include a uniform storage image of the sensed data and an execution frame (context) to save the application execution states. The storage of sensed data is composed by data sensed by the local node and data received from the other nodes. It allows applications to have access to changes made by other applications. This feature requires that the applications use the same storage organization. The execution frame saves information such as the use of the network and memory management and is shared by all applications.

Declarative Paradigm: Data Fusion Model

To some extent, almost all middlewares aimed for sensor networks provide some features that can be interpreted as data fusion. However, few approaches really focus on the problem of data fusion as a main concern. DFuse (Ramachandran, Kumar, Wolenetz, Cooper, Agarwalla, Shin, Hutto & Paul, 2003) represents an approach that not only addresses the data fusion problem, but it also proposes a concept to drive application design.

DFUSE

The DFuse middleware (Ramachandran, Kumar, Wolenetz, Cooper, Agarwalla, Shin, Hutto & Paul, 2003) is an architecture for programming of data fusion intensive applications. It supports distributed data fusion with automatic placement and migration of the fusion points. This migration has the goal of maximizing/minimizing some given cost function. This means that the role assignment for each node is decided by the middleware considering the given cost function.

Problems Addressed: The main goal of this middleware is to provide the distribution of the data fusion among the nodes of the network, in order to reduce the power consumption in specific nodes, and in this way enable power savings in the network as a whole, positively affecting the network longevity.

Design Approach: In order to achieve the goal of reducing the energy consumption, the approach adopted is the use of cost functions that will drive the decisions to distribute the tasks among the nodes of the network. The middleware offers the following cost functions: (1) minimize transmission cost without node power considerations; (2) minimize power variance; (3) minimize the ratio of transmission cost to power; and (4) minimize the transmission cost with node power considerations.

The middleware proposal provides an abstraction called *Fusion Channel*, which encapsulates a fusion function, such as an aggregation function or a mathematical filter, providing also data buffering and synchronization features. Instances of this channel can be created and migrated across nodes during system runtime. The channels are also seen as "virtual sensors", as they produce new data based on inputs from sensors or other channels and provide these new data to other channels or actuators as they were data sources like sensors.

System Architecture: The DFuse architecture has two main modules, the Placement Module and the Fusion Module (see Figure 4). It also includes an interface to monitor the resource usage in the node.

Figure 4. DFuse architecture - a high-level view per node. (Adapted from (Ramachandran, Kumar, Wolenetz, Cooper, Agarwalla, Shin, Hutto & Paul, 2003))

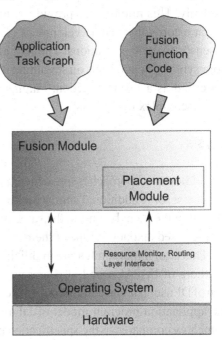

The Placement Module is responsible for applying application-specific cost functions in order to drive the dynamic migration of fusion channels across the nodes. The decision of migrating or not a certain fusion channel is taken locally, taking in account the local resource usage provided by the local monitor and the information of the one-hop neighbors. Based on this information, the cost function is evaluated in order to determine if the cost for hosting the fusion channel would be lower in a neighbor or not.

The Fusion Module is responsible for hosting the fusion functions that operate on the semantic data of the system. It is in fact in this module where the data fusion will take place. The application task graph and the fusion function code, presented in Figure 4, are discussed next.

Application Programming Model: Using the DFuse middleware, the developer is only responsible for implementing the fusion functions and providing the application task graph which provides the data flow information (Sohraby, Minoli & Znati, 2007). The fusion functions and the application task graph are the two inputs of the Fusion Module, as shown in Figure 4. The distribution of these fusion points is made automatically. This middleware provides a Fusion API that affords the easy development of complex sensor fusion applications. The API allows custom synthesis operations on streaming data to be specified as a fusion function, ranging from simple to complex operations.

Analysis

The middleware approaches presented in this section have some additional similarities besides those that make them belong to the same branch in the proposed taxonomy. One of them is modularity. This is an important feature highlighted by the authors of IMPALA and also present in the design of DFuse. The correlation control specified in the application program model of DFuse can be somehow compared to the specification of the

handlers in IMPALA. However, it is possible to say that DFuse presents an improvement in terms of data handling and application specification if compared to IMPALA. Having data fusion as rationale of its design, DFuse raises the abstraction level of the application specification if compared to the event-based approach of IMPALA, as the user thinks in terms of the data he/she desires from the system rather than in how to gather them.

Domain Specific Paradigm: Global Network Programming

In this section, we will focus on the global network programming. Differently from the middlewares presented in the previous sections, the programmer here can specify the behavior of the sensor network as a whole and does not need to program the individual behavior of the nodes.

Declarative Paradigm: Database Model

The database model of sensor networks has become a great success, providing an easy and intuitive way to retrieve information from sensor nodes. There are different proposals that follow this approach, in which TinyDB (Woo, Madden & Govindan, 2004) is outstanding. The great success of TinyDB can be partially attributed to its innovation in using the database model, but also to the "easy to use" application programming features that it provides. Besides TinyDB, this section also briefly introduces the SINA middleware.

TinyDB
TinyDB (Woo, Madden & Govindan, 2004) is a middleware for WSNs that runs on top of TinyOS. TinyDB supports a single "virtual" database table, where each column corresponds to a specific type of sensor. The idea is to provide an abstraction of the sensor network as it were a database. Taking this point of view, the middleware represents a query processing system, which translates a high-level query to low-level commands to the

sensor nodes, in order to retrieve the desired information. This idea frees the user from having to write code in nesC for the sensor nodes, as it has to be done if only pure TinyOS is used as support. The query language is a subset of SQL with some extensions. It supports data aggregation and provides means to optimize this aggregation aiming at reducing the communication needed to answer a given query.

Problems Addressed: A great problem in the use of WSNs is to perform actions that acquires the data from the sensor nodes. Communication is costly, and, besides, not all data gathered by the sensors are needed. Some data aggregation can be done by the nodes in the way the information flows through the network to the data requester user, in order to reduce the amount of total data that arrives at the sink or data requester. These facts motivate the approach proposed by TinyDB, which consists of a middleware that aims at simplifying the request for data by the use of SQL queries and, at the same time, helps to decrease the number of messages exchanged in the network by means of data aggregation.

Design Approach: TinyDB uses an approach that abstracts the sensor network as if it was a distributed data base containing sensor data. The sensor nodes are the data providers, and the interface to request them is a SQL-like query language. As the raw data of all nodes are not necessary to reply to a certain request, for instance in cases where an average value of the observed phenomenon is requested, the strategy adopted is to push the aggregation criteria into the network. The SQL query triggers the aggregation of information in the WSN, and TinyDB supports the distributed aggregation of the data. The *Tiny Aggregation* consists of two phases: (1) a distribution phase, in which the aggregate queries are inserted and propagated in the network; and (2) a collection phase, where the aggregated values are routed from children to parents until arriving at the requester.

The approach adopted by TinyDB supports multiple queries running at the same time in the network. Queries may have different characteristics, such as different sample rates, requiring data that is provided by different types of sensors, etc. The middleware efficiently manages the use of resources needed to respond to the queries, according to the possibilities allowed by the available resources.

System Architecture: The architecture of TinyDB consists of a distributed query processor, which is a software layer that runs in each of the nodes, over the TinyOS sensor network operating system, exploring the services of this particular OS to provide the desired functionality.

In summary, the architecture of a system running TinyDB is the following: a query is generated in a base station, parsed, optimized, and then sent to the network. The query is then disseminated over the nodes in the network and executed. The results are returned back to the base station via the routing tree formed during the query propagation. In order to do that, TinyDB manages the network, controlling the use of the communication link, keeping track of the nodes' neighbors and maintaining routing tables.

Application Programming Model: By using an SQL-like language, the applications consist of a SQL query (i.e. SELECT-FROM-WHERE-GROUPBY clause) that is propagated across the network nodes, in order to perform the request for the data asked by the user. The queries may represent an event-based request for data, by the specification of conditionals (e.g. if a condition occurs, retrieve the specified information), and timing-conditioned requests, in which timing parameters can be specified, such as sampling and reporting intervals. There is also the possibility of performing diagnostic queries, which provide data about the network health status, such as nodes' battery levels. The user can also enforce a physical action in response to a result from a query. This way, a result from a query may trigger the activation of a difference sensor device in the same node, for instance.

The data model supports the data base oriented approach in which the sensor tuples belong to a table, called *sensors*, which has one row per sensor node and per instant in time, with one column per attribute (e.g., temperature, humidity, light, etc.) that represents the measurement the device can produce. Records in this table are materialized only as needed to satisfy a given query. These records are usually stored only for a short period of time or delivered directly out of the network to the requesting base stations. It is possible to apply special functions over the tuples of the sensor table, such as projections and/or transformations, which can be stored at specific points of the network called materialization points.

SINA

SINA (Shen, Srisathapornphat & Jaikaeo, 2001) is a database oriented middleware for sensor networks that facilitates querying, monitoring, and tasking by abstracting the network as a collection of massively distributed objects. It has an execution environment that allows scalable and energy-efficient organization and interaction of sensors.

Problems Addressed: Data access involving more selective acquisition and query in wireless sensor networks are complex problems, especially in applications running in dynamic scenarios with constant changes in the operational environment and network topology. SINA proposes a middleware architecture that abstracts the network such that the user may query the desired information without worrying about underlying details.

Design Approach: SINA supports data accesses by the use of an SQL-like language called SQTL (Sensor Query and Tasking Language) performed as query tasks described by SQTL programming scripts. The information abstraction present in SINA is based on the idea of viewing the network as a collection of datasheets; each of these datasheets contains a collection of attributes of each node. Cells are created in the nodes ac-

cording to requests for information from other nodes. Each created cell is uniquely named and becomes a node attribute.

System Architecture: SINA modules running in each node provide adaptive organization of sensor information, thus facilitating query, event monitoring, and task capability. Sensor nodes are autonomously grouped in clusters, to provide scalability and energy-efficient support. SINA presents three functional components: (1) Hierarchical Clustering - Nodes are grouped in clusters according to their power levels and proximity. These clusters form a hierarchy, recursively providing levels or tiers of clusters; (2) Attribute-based Naming - As an identification of nodes by identifiers is not scalable and causes difficulty to interface with the final user, an attribute-based naming scheme is used; and (3) Location Awareness - It is important that nodes have knowledge about their own physical location as the network can be deployed in large areas. The use of GPS and beacons allows this awareness.

Application Programming Model: SQTL is a procedural scripting language, which is flexible and compact and that act as the programming interface provided by SINA to the applications. SQTL supports primitives for hardware access, location-awareness, and communication. Moreover, it also provides event handling constructions supporting three kinds of events: events generated when a message is received by a sensor node; events triggered periodically by a timer; and events caused by the expiration of a timer.

Declarative Paradigm: QoS Model

The evolution of sensor networks has pushed their use to new kinds of applications, many of them requiring quality of service (QoS) guarantees. However, few approaches already address this issue. MiLAN (Heinzelman, Murphy, Carvalho & Perillo, 2004) presents a design model that aims at harmonizing the different QoS requirements of the applications running in a sensor network.

Figure 5. Overview of a system that employs MiLAN . (Adapted from (Heinzelman, Murphy, Carvalho & Perillo, 2004))

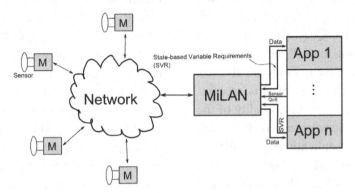

MiLAN

MiLAN (Middleware Linking Applications and Network) (Heinzelman, Murphy, Carvalho & Perillo, 2004) is an adaptive middleware that explores the concept of proactive adaptation in order to respond to the users' needs in sensor networks in which changes occur in the operational environment conditions. It explores the concept of QoS related to the requested data, which allows a definition of a strategy to optimize the data acquisition by directing the aggregation of raw data to provide the required information and the selection of the sensor nodes that may reply to a given request.

Problems Addressed: Due to the fact that WSNs are generally deployed in very dynamic scenarios, a middleware that supports the operation of a network in this kind of scenario has to adapt in response to changes in the environment conditions and in the user requirements. The goal of MiLAN is to address this problem by means of proactive adaptations, which is a before-hand response to the foreseeable needs for adaptation. The final result intended by the use of MiLAN is to meet the QoS requirements established by the user, optimizing application performance in terms of energy consumption.

Design Approach: The main idea in the MiLAN approach is that it receives the requirements from the users (and related QoS), monitors the nodes and network status, and, based

on this information, continuously adapts the network configuration, selecting which sensors that should provide the required data, which sensors should be routers in multi-hop communication, and which of them should play special roles in the network, such as sink or information aggregator.

As the optimization of application performance is an important goal, MiLAN provides support to combine QoS levels established by the users for the different types of information desired, in order to promote an overall optimization of the data retrieving.

System Architecture: In a system that employs MiLAN, each sensor runs a scaled-down version of the MiLAN middleware, as shown in Figure 5. MiLAN does not come with new implementations of network and MAC layers. Instead, it is built on top of the existing protocols. The fulfillment of energy savings and QoS requirements is achieved by varying the parameters of the network, using the available resources provided by the underlying protocols. MiLAN has a different architecture if compared to a traditional middleware that sits between the application and the operating system. Instead, the architecture proposed by MiLAN extends into the network protocol stack. This characteristic allows MiLAN to interoperate with different technologies at different levels of the network protocol stack, depending on the node in which it is installed.

Application Programming Model: In order to retrieve data from the sensor network using MiLAN, application programmers specify variables of interest and the QoS related to each of these variables by using the API provided by the middleware. Moreover, the middleware also has as input descriptions of the sensors. This information makes it possible to infer the quality of the measurement that each sensor can provide for each variable. Basically, the applications transmit their request for information to MiLAN in terms of two entities, the "State-based Variable Requirements" and "Sensor QoS" graphs.

The State-based Variable Requirements are variables of interest to the application and their respective QoS. They are defined as the resulting information that the application wants to gather from the environment. This information can be achieved by MiLAN in different ways, and the definition of how it can provide it is specified in the Sensor QoS graphs.

Declarative Paradigm: Globally Specific Event Model

New applications of sensor networks are requiring the integration of more sophisticated nodes into the network, which have to cooperate with other types of nodes (simpler or even more sophisticated). In order to address the problem of harmonization of this cooperation among heterogeneous nodes, new approaches are emerging, since the traditional ones generally consider a homogeneous network of sensors or very little heterogeneity among them. AWARE (Gil, Maza, Ollero & Marron, 2007) presents a consistent approach to address this issue, integrating sensors embedded in aircraft platforms with current wireless sensor nodes.

AWARE

The AWARE middleware is a part of a European project called AWARE (Autonomous self-deploying and operating Wireless sensor-actuator networks cooperating with AeRial objEcts) (Gil,

Maza, Ollero & Marron, 2007). The goal of this middleware is to provide integration of the information gathered by different types of sensors, including WSNs and mobile robots.

Problems Addressed: The use of different types of sensors in a network mainly brings concerns related to the harmonization of the communication and interaction among these elements. The main problem addressed by AWARE is how to put different kinds of sensors, with different capabilities (in terms of provided data, mobility, and communication), to work together in a way to respond to the user needs of information.

Design Approach: The AWARE platform consists of two networks, a high bandwidth network (HBN) and a low bandwidth network (LBN). The first HBN one is composed of high-end nodes, such as personal computers, cameras, and mobile robots, relying on transmission technology based on IEEE 802.3 and IEEE 802.11. The second network, the LBN, constitutes a WSN composed of low-end nodes, with very limited computing and wireless data transmitting capabilities. These two networks are connected through gateways that are devices capable of communicating with both networks. The middleware has the goal of making the communication between these two networks transparent.

The middleware follows a data-centric approach, which is related to the definition and management of groups and channels. A group defines some environmental condition of interest defined by the user. Whenever the value read by the sensors of a node meets these conditions, the node belongs to that type of group. The definition of a channel is based on the information provided by the user about: (i) which sensor should provide the required data, (ii) sampling rate, (iii) group type, and (iv) group ID to which the channel is associated. Data aggregation is performed through the channel. Command channels are also supported, allowing online changes in the configuration.

System Architecture: The middleware provides a data-centric publish-subscribe approach to access the nodes in the WSN. The application layer is responsible for providing information about the node to the middleware.

A node can join a group when there is a matching of the group conditions. A node joins a group as a member, becoming publisher of any channel to which that group is associated. If there is no group in the vicinity of the node, it creates a group and becomes its leader.

The leader election mechanism avoids conflicts related to duplicated leaders. When a node creates a group, it becomes its leader and begins to send leader beacons that let the other nodes in the vicinity to know about the existence of the group leader. These beacons are rebroadcasted by all nodes in the group. Nodes that do not belong to the group do not retransmit the beacons.

Channels can be associated to more than one group. In order to distinguish information provided by different groups that use the same channel, the data packages include the group ID.

Application Programming Model: AWARE adopts an event-based programming model. The user, which may be a human operator or an application running on a high-end node, describes interest for certain events by defining groups and associating them to channels that denote the information that he/she/it wants to obtain from the specified group.

Analysis

TinyDB presents several features that make it a good choice to manage and interact with a sensor network. It provides an easy way to retrieve information, abstracting the network itself from the user point of view. SINA also has this capability, but using a more sophisticated architectural approach, allowing clustering in the network, which is also hidden from the user. SINA also provides additional features, such as a better energy usage control. However, even if it also uses an SQL-based query language, the interface support provided by TinyDB is more user-friendly. An

additional important database-based middleware, not presented in this text, is Cougar (Bonnet, Gehrke & Seshadri, 2001), which shares several characteristics with TinyDB.

MiLAN and AWARE represent a great evolution if compared with the above middlewares. MiLAN provides an approach that considers application QoS requirements, while AWARE has an additional capability to integrate heterogeneous sensor nodes in the network. These features are neither present in TinyDB nor in SINA.

DISCUSSION

Based on the requirements of WSNs and on the main challenges in the development of adaptive middlewares for WSNs, a critical analysis of the present state-of-the-art efforts in the area can be done, in which the main needs in the area are highlighted.

The features used as a base for the comparison are:

- **Code Mobility:** Support for code mobility, both for update and installation of new services in a node;
- **Flexibility:** Support for network scalability and for managing incoming nodes in the network, as well as the management of the network topology;
- **Node Mobility:** Support for mobile nodes in the network;
- **Node Heterogeneity:** Capability of the middleware to address the needs of both low-end nodes, with few and constrained resources, and more sophisticated sensors, with more powerful computer platforms and advanced resources;
- **Application Knowledge:** The ability of the middleware to respond to the needs of specific applications or groups of applications;
- **Data Fusion:** Support for data aggregation and fusion by the nodes that are in the way

Figure 6. State-of-the-art middleware evaluation

	Code Mobility	Flexibility	Node Mobility	Node Heterogeneity	Application Knowledge	Data Fusion	QoS
AGILLA	3	2	1	1	1	1	0
AWARE	0	2	4	2	1	2	0
IMPALA	2	2	1	0	2	1	0
Maté	3	2	0	0	1	0	0
MiLAN	0	1	0	0	2	2	4
SINA	0	2	2	0	1	1	0
TinyDB	0	1	0	0	1	2	0
TinyLime	4	1	2	2	1	1	0

of data moving from the phenomenon occurrence to the end user that requested the information;

- **QoS:** Support for QoS control that can be provided by the middleware.

By evaluating those features, the analyzed middlewares were graded according to the following levels and presented in Figure 6:

- **4 - Very Good:** Provides a promising approach that addresses the challenges related to the concern;
- **3 - Good:** Provides a support that partially addresses the challenges related to the concern;
- **2 - Average:** Provides a support that addresses a few of the related problems under concern;
- **1 - Weak:** Provides a support that poorly addresses the related concerns;
- **0 - Very Weak:** Minimally addresses or does not address at all the related concerns.

By the information condensed in Figure 6, it is possible to have an idea that there is much to do in order to improve middlewares for WSNs. There are few grades "4" and "3", which means that most of the considered features are poorly addressed by existing middlewares and that they mostly concentrate in just one or two features. As analyzed in Section 2 of this chapter, there is

a need for solutions that combine efforts in all of those features in order to respond to the adaptation needs that are emerging with the new sophisticated applications of wireless sensor networks.

At least two of the analyzed features require special attention, namely "Node Heterogeneity" and "QoS". The first one has mostly low grades in Figure 6, with some proposals partially addressing the issue, but a great need for research in this area can be identified. The second one had extremely low grades in almost all the approaches, except the good grade achieved by MiLAN. It is possible to see that this proposal concentrates its efforts in a QoS solution for WSNs and successfully achieves good results, but it is not integrated with the other needs regarding the other features.

The information provided in this chapter cannot be considered as exhaustive. There are approaches that address the related concerns, as other approaches that deal especially with QoS requirements (see (Chen & Varshney, 2004) for a detailed discussion about approaches that address this specific topic). But, even if they were considered in the presented evaluation, the conclusion would be the same, namely that there is a need to integrate the support for each of the described features in a common middleware platform in order to consider the full range of requirements of new emerging applications.

Some words must also be said about the sensor network programmability features provided by each approach. The low-level programming language presented by Agilla makes the proposal hard to scale in terms of more complex and larger applications, since the development of large applications in such language is tedious and an error-prone task. The capsules used in Maté suffer from the same problem due to the same reasons already mentioned for Agila. IMPALA provides an approach based on the use of handlers that are easy to think about, when developing an application, but still requires that the developer handles low-level concerns. TinyDB, on the other hand, provides a much more user-friendly approach, as the SQL-like language is quite easy to use, especially for those familiar with SQL. The same holds for SINA. The APIs provided by DFuse and MiLAN provide intuitive ways to develop applications, with a more straightforward application development framework. AWARE raises the abstraction level when it comes to define the application, which allows users to think less about the network itself and more in terms of the results they want, which represents an easier way to use the sensor network.

CONCLUSION

This overview of middlewares for WSNs is intended to illustrate the motivations, characteristics, and challenges of such systems. The middlewares are important in order to support the programmability of WSNs, hiding the complexity of the underlying infrastructure and presenting useful abstractions to the application programmer.

The overview has emphasized the difference between the operation modes of WSNs and conventional systems, highlighting that WSNs are essentially data-centric systems, which generate different requirements for the protocols and also for the middleware development.

The capability of offering a suitable abstraction that supports a dynamic environment where nodes have constrained resources is the key challenge in developing WSN middlewares.

In order to better classify the existing middleware approaches for WSNs, a taxonomy focusing on the programming paradigm supported by the different middlewares has been introduced. We decided to use this feature as the main characteristic to classify middleware since it is the interface between the programmer and the system. Moreover, it also defines which kind of applications that can be written for the sensor network. It is important to remark that different programming abstractions define different ways of expressing a given application. Some abstractions are more suitable for a given class of applications and others are more general.

Based on the presented taxonomy, we introduced in this chapter eight different examples of middleware. The presentation of the middlewares was organized in a systematic way (problems addressed, design approach, system architecture, and application programming model), thus allowing comparisons among the different approaches.

We started our presentation with middlewares based on the agent model. In this model, it is usual to use a virtual machine to interpret the code of the agents inserted in the network. Most of the approaches use the mobile agent paradigm, where the queries are inserted in the network as agents and disseminated among the sensor nodes. The main drawback here is the local node programming paradigm: the user must specify the actions of every single agent, running on single nodes, and also think about the possible interactions, at a microscopic level.

In the sequence, the chapter focused on examples of event-based middlewares. In such middlewares, the user can specify a set of events that should be notified. Moreover, some of the middlewares allow dynamic updates and application adaptation besides the basic event-based processing. These are very interesting features but do not overcome the main limitation of this class of middlewares: the state-machine based programming model, which is suitable only for a restricted class of applications.

Data fusion is a very important operation often used by WSN applications. The chapter also presented an example of middleware which specifically supports this operation. This is an extreme example of domain-specific middleware: only data-fusion applications are supported by the middleware.

A very interesting programming paradigm for WSNs is the database model. We have presented in this chapter several examples of this category. Middlewares based on the database model are very efficient in processing sampling queries from the user. Based on the query input, the processing at each node is automatically determined as well as the data aggregation points. The main drawback of the database middlewares is the lack of flexibility: they are mainly designed to process SQL-like queries. Expressing other kinds of applications or even more complex queries may not be possible.

Finally, we focused on the QoS-based middlewares. They can automatically select nodes that meet QoS requirements defined by the user. This sort of middleware is just specialized on collecting data applications.

From the state-of-the-art presentation of this chapter, we come to the conclusion that, although the diverse existing middlewares try to address the challenges presented in the beginning of the chapter using different mechanisms and abstractions, there is a large room for improvements. Improvements are possible because all presented middlewares cover just a subset of the challenges and several of them are specialized only in a specific type of application.

It is desirable to develop a middleware that supports the global network programming paradigm. At the same time, it should be generic enough for different application areas and also support distributed computation. Research efforts are still needed to achieve this ambitious goal.

REFERENCES

Bonnet, P., Gehrke, J. E., & Seshadri, P. (2001). Towards Sensor Database Systems. In *Proceedings of the Second International Conference on Mobile Data Management* (pp.3-14). London: Springer Press.

Boulis, A., Han, C.-C., Shea, R., & Srivastava, M. B. (2007). SensorWare: Programming sensor networks beyond code update and querying. *Pervasive and Mobile Computing, 3*(4), 386–412. doi:10.1016/j.pmcj.2007.04.007

Boulis, A., Han, C.-C., & Srivastava, M. B. (2003). Design and Implementation of a Framework for Efficient and Programmable Sensor Networks. In *Proceedings of The First International Conference on Mobile Systems, Applications, and Services, MobiSys* (pp. 187-200).

Chen, D., & Varshney, P. K. (2004). QoS Support in Wireless Sensor Networks: a Survey. In *Proceedings of the Int'l. Con. On Wireless Networks (ICWN 04)*.

Curino, C., Giani, M., Giorgetta, M., & Giusti, A. (2005). TinyLIME: Bridging Mobile and Sensor Networks through Middleware. In *Proceedings of the 3th IEEE International Conference on Pervasive Computing and Communications* (pp.61-72).

Fok, C. L., Roman, G. C., & Lu, C. (2005). Rapid Development and Flexible Deployment of Adaptive Wireless Sensor Network Applications. In *Proceedings of the 24th International Conference on Distributed Computing Systems* (pp.653-662).

Gelernter, D. (1985). Generative Communication in Linda. *ACM Transactions on Programming Languages and Systems, 7*(1), 80–112. doi:10.1145/2363.2433

Gil, P., Maza, I., Ollero, A., & Marron, P. (2007). Data Centric Middleware for the Integration of Wireless Sensor Networks and Mobile Robots. In *Proceedings of the 7th Conference on Mobile Robots and Competitions*, Paderne, Portugal.

Heinzelman, W. B., Murphy, A. L., Carvalho, H. S., & Perillo, M. A. (2004). Middleware to Support Sensor Network Applications. *IEEE Network*, *18*(1), 6–14. doi:10.1109/MNET.2004.1265828

Henricksen, K., & Robinson, R. (2006). A Survey of Middleware for Sensor Networks: State-of-the-art and Future Directions. In *MidSens '06: Proceedings of the international workshop on Middleware for sensor networks* (pp. 60-65). New York: ACM Press.

Horré, W., Michiels, S., Joosen, W., & Verbaeten, P. (2008). DAVIM: Adaptable Middleware for Sensor Networks. *IEEE Distributed Systems Online*, *9*(1), 1–1. doi:10.1109/MDSO.2008.2

Juang, P., Oki, H., Wang, Y., Martonosi, M., Peh, L. S., & Rubenstein, D. (2002). Energy-Efficient Computing for Wildlife Tracking: Design Tradeoffs and Early Experiences with ZebraNet. In *Proceedings of the 10th International Conference on Architectural Support for Programming Languages and Operating Systems* (pp. 96-107).

Kuorilehto, M., Hännikäinen, M., & Hämäläinen, T. D. (2005). A Survey of Application Distribution in Wireless Sensor Networks. *EURASIP Journal on Wireless Communications and Networking*, *38*(5), 774–788. doi:10.1155/WCN.2005.774

Levis, P., & Culler, D. E. (2002). Maté: a Tiny Virtual Machine For Sensor Networks. In *Proceedings of the 10th International Conference on Architectural Support for Programming Languages and Operating Systems* (pp. 85-95).

Liu, T., & Martonosi, M. (2003). Impala: A Middleware System for Managing Autonomic, Parallel Sensor Systems. In *Proceedings of the 9th ACM SIGPLAN Symposium on Principles and Practice of Parallel Programming* (pp. 107-118).

Luo, H., Luo, J., Liu, Y., & Das, S. K. (2005). Energy Efficient Routing with Adaptive Data Fusion in Sensor Networks. In *Proceedings of the Joint Workshop on Foundations of Mobile Computing* (pp. 80–88).

Madden, S., Franklin, M. J., Hellerstein, J. M., & Hong, W. (2002). TAG: a Tiny Aggregation Service for Ad-hoc Sensor Networks. In *Proceedings of the Symposium on Operating Systems Design and Implementation*.

Molla, M. M., & Ahamed, S. I. (2006). A Survey of Middleware for Sensor Network and Challenges. In *Proceedings of the 2006 International Conference Workshops on Parallel Processing* (pp. 223-228). Washington, USA: IEEE Computer Society.

Murphy, A. L., Picco, G. P., & Roman, G.-C. (2001). LIME: A Middleware for Physical and Logical Mobility. In F. Golshani, P. Dasgupta, & W. Zhao (Eds.), *21st International Conference on Distributed Computing Systems* (pp 524-533).

Ramachandran, U., Kumar, R., Wolenetz, M., Cooper, B., Agarwalla, B., & Shin, J. (2003). Dynamic Data Fusion for Future Sensor Networks. *ACM Transactions on Sensor Networks*, *2*(3), 404–443. doi:10.1145/1167935.1167940

Römer, K. (2004). Programming Paradigms and Middleware for Sensor Networks. In *GI/ITG Workshop on Sensor Networks* (pp. 49-54).

Römer, K., Kasten, O., & Mattern, F. (2002). Middleware Challenges for Wireless Sensor Networks. *ACM SIGMOBILE Mobile Computing and Communication Review (MC2R)*, *6*(4), 59-61.

Römer, K., & Mattern, F. (2004). The Design Space of Wireless Sensor Networks. *IEEE Wireless Communications*, *11*(6), 54–61. doi:10.1109/MWC.2004.1368897

Shen, C., Srisathapornphat, C., & Jaikaeo, C. (2001). Sensor Information Networking Architecture and Applications. *IEEE Personal Communications*, *8*(4), 52–59. doi:10.1109/98.944004

Sohraby, K., Minoli, D., & Znati, T. (2007). *Wireless Sensor Networks - Technology, Protocols and Applications*. New York: Wiley-Interscience.

Tilak, S., Abu-Ghazaleh, N. B., & Heinzelman, W. (2005). *A Taxonomy of Wireless Micro-Sensor Network Models* (Tech. Rep.). New York: University Binghamton, Dept. of Computer Science, System Research Laboratory.

Wang, M. M., Cao, J. N., Li, J., & Das, S. K. (2008). Middleware for Wireless Sensor Networks: A Survey. *Journal of Computer Science And Technology*, *23*(3), 3. doi:10.1007/s11390-008-9135-x

Woo, A., Madden, S., & Govindan, R. (2004). Networking Support for Query Processing in Sensor Networks. *Communications of the ACM*, *47*(6), 47–52. doi:10.1145/990680.990706

Yu, Y., Krishnamachari, B., & Prasanna, V. K. (2004). Issues in Designing Middleware for Wireless Sensor Networks. *IEEE Network*, *18*, 15–21. doi:10.1109/MNET.2004.1265829

KEY TERMS AND DEFINITIONS

Adaptive Middleware: Middleware capable to adapt its behavior according to changes in operation or environment conditions, and/or user requirements.

Data-Centric: Refers to methods that concentrate the focus on the data instead of the individual nodes in a network.

Data Fusion: Technique to combine multiple data sources in a single one.

Middleware: Abstraction layer that lies between application and network, enabling the easy development of complex networking application.

Programming Paradigm: A fundamental style of computer programming, with its abstractions and steps for composing a program.

Virtual Machine: Software implementation of a computer architecture.

Wireless Sensor Networks: A wireless network consisting of spatially distributed autonomous devices using sensors to cooperatively monitor physical or environmental conditions.

Section 3
Security and Privacy

Chapter 12
A Survey on Applied Cryptography in Secure Mobile Ad Hoc Networks and Wireless Sensor Networks

Jianmin Chen
Florida Atlantic University, USA

Jie Wu
Florida Atlantic University, USA

ABSTRACT

Many secure mobile ad hoc networks (MANETs) and wireless sensor networks (WSNs) use techniques of applied cryptography. Numerous security routing protocols and key management schemes have been designed based on public key infrastructure (PKI) and identity-based cryptography. Some of these security protocols are fully adapted to fit the limited power, storage, and CPUs of these networks. For example, one-way hash functions have been used to construct disposable secret keys instead of creating public/private keys for the public key infrastructure. In this survey of MANET and WSN applications we present many network security schemes using cryptographic techniques and give three case studies of popular designs.

INTRODUCTION

This chapter aims to explain how MANET and WSN security design may be improved with a broad knowledge of cryptography. Securing MANETs and WSNs requires consideration of the following factors: dynamic topologies, resource constraints, no infrastructure, and limited physical security. Because WSNs typically have more nodes and less power than MANETs, their security design requires more attention to computational capabilities and memory resources. Much cryptographic, authentication, and authorization research has been conducted into the details of secure routing, key management, and trust management in MANETs and WSNs.

Previous researches have studied attacks and countermeasures in MANETs (Wu & Chen, 2008), key management in MANETs (Wu & Cardei, 2008),

DOI: 10.4018/978-1-61520-701-5.ch012

security locations in WSNs (Srinivasan, 2008), secure routing protocols in MANETs (Pervaiz, 2008), challenges and solutions in wireless security (Lou, 2003), key management schemes in WSNs (Xiao, 2007), and open issues in WSNs (Evans, 2006). To increase network security cryptographic techniques may be applied in different areas of MANETs/WSNs. For example, ID-based cryptography (Shamir, 1984) is used to develop a new certificateless security scheme in MANETs as well as for a security scheme in vehicular ad hoc networks and for other secure routing applications. Case studies of cryptographic techniques in customized MANETs and WSNs will provide the research community with the latest updates in security and performance for MANETs and WSNs. One example of a new foundation for advanced research is a configurable library for elliptic curve cryptography in WSNs called TinyECC (Liu, 2008). Our survey is an effort to promote the use of cryptographic techniques in the ongoing research to better secure MANETs/WSNs.

Our case studies are chosen to discuss symmetric cryptography, public key infrastructure (PKI), identity-based cryptography, threshold cryptography, and batch verification of signatures. After summarizing cryptographic techniques we give an overview of commonly used security designs followed by sections on symmetric cryptographic techniques. Our discussion of the symmetric techniques is based on a case study of LHAP (Zhu & Xu, 2003). Our discussion of the asymmetric techniques, with a special emphasis on composite design, is based on a case study of IKM (Zhang, Liu, Lou & Fang, 2006). Then we discuss how threshold cryptography is used in different cases for secret sharing to make gains in both security and performance. Finally other cryptographic techniques are discussed on the basis of a case study of IBV by Zhang (Zhang, Lu, Ho & Shen, 2008) followed by our presentation of present open issues and future challenges.

CRYPTOGRAPHY TECHNIQUES OF SECURE MANETS/WSNS DESIGN

Security is the combination of processes, procedures, and systems used to ensure confidentiality, authentication, integrity, availability, access control, and non-repudiation.

- **Confidentiality:** The confidentiality is to ensure that information is accessible only to those authorized users or nodes to have access. Since MANETs/WSNs use an open medium, all nodes within the direct transmission range can usually obtain the data. One way to keep information confidential is to encrypt the data. In WSNs confidentiality is employed to protect information from inadvertent disclosure while communicating between one sensor node and another sensor node or between the sensors and the base station. Compromised nodes are a threat to confidentiality if the cryptographic keys are not encrypted and stored in the node.

- **Authentication:** The goal of authentication is to identify a node or a user and to prevent impersonation. In wired networks and infrastructure-based wireless networks it is possible to implement a central authority at a router, base station, or access point. However, there is no central authority in MANETs/WSNs, and it is much more difficult to authenticate an entity. Confidentiality can be achieved via encryption. Authentication can be achieved by using a message authentication code (MAC) (Menezes, Oorschot & Vanstone, 1996).

- **Integrity:** The goal of integrity is to keep a message from being illegally altered or destroyed during transmission. When the data is sent through the wireless medium, the data may be modified or deleted by malicious attackers. When malicious attackers can resend altered data the action is known as a *replay attack*. Integrity can be achieved through hash functions.

- **Non-repudiation:** The goal of non-repudiation is to prevent a message sender from later denying that it has sent the message. The entity which produces a message signature cannot later deny having sent that message. In public key cryptography, a node, *A*, signs the message using its private key. All other nodes verify the signed message by using *A*'s public key, and *A* cannot deny that its signature is attached to the message.

- **Availability:** The goal of availability is to keep the network service or resources available for legitimate users. It ensures the survivability of the network despite malicious incidents. In a WSN, for example, sensor node capturing and denial of service attacks are common problems. Outages may be mitigated by providing alternate routes in the protocols employed by the WSN.

- **Access control:** The goal of access control is to prevent unauthorized use of network services and system resources. Obviously, access control is tied to authentication attributes. Access control is the most commonly needed service in both network communications and individual computer systems.

Cryptography is very strongly tied to mathematics and the number theory. Therefore, creating a new composite cryptographic design is difficult without sound security analysis based on cryptographic reasoning. One way to reach this goal is to learn from others by reviewing current MANET/WSN security schemes and by understanding how cryptographic techniques combine with MANETs/WSNs to provide a security service with reasonable network performance, scalability, storage, and synchronization. Certainly, a security design can be evaluated using different techniques, but our goal is to provide

helpful insight by studying basic cryptographic techniques (as seen in Figure 1) when applied to authentication, trust management, and key management in MANETs/WSNs. Furthermore, we will study several of the commonly-used cryptographic techniques and see how they are employed to deal with different tasks and how to balance the tradeoff between security and performance.

It is a common approach today to use software engineering design patterns to illustrate the design of object-oriented programming. Likewise, cryptographic techniques can be successfully used in different stages of network bootstrap, packet communication, and evaluation factors in the security and performance of MANETs/WSNs. Once these techniques are understood they are easily applied to new designs of these networks.

Overview of Cryptographic Techniques

Choosing which and how often specific cryptographic techniques should be used is difficult. Deciding on network performance evaluation metrics and security analysis techniques is also not easy. The first question may be "when does one use symmetric cryptography and when does one use asymmetric cryptography?" For example, in order to get better performance, a hash key chain may be a better choice than an asymmetric private key for encryption due to dynamic topology changes in some MANETs/WSNs. Specifically, alternative temporary symmetric secret keys (e.g., AES with a 128 bit size key) may be better than asymmetric public keys (e,g., RSA with a 1024 bit size public key).

Many researchers have proposed the use of asymmetric cryptography such as public key infrastructure using RSA (Mehuron, 94) or Elliptic Curve Cryptography (ECC) (Salomaa, 1996) to secure wireless ad hoc network routing protocols (Zhou & Haas, 1999; Yi, Naldurg, & Kravets, 2002; Zapata, 2002). But, considering the ad hoc network computation cost to verify asymmetric

Figure 1. Cryptographic techniques introduction and selected MANET/WSN security schemes applied

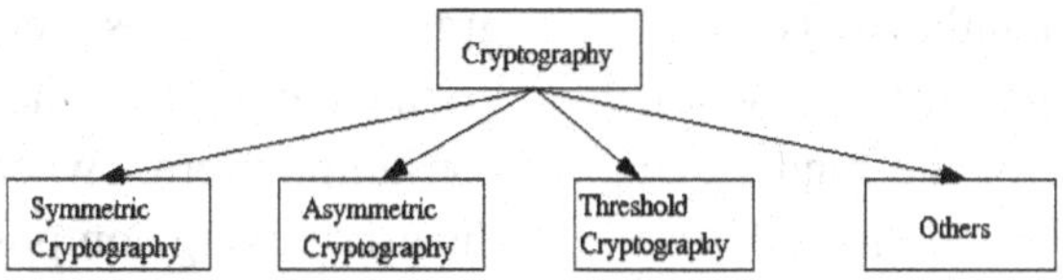

(a) Major components of cryptography applied in MANETs/WSNs.

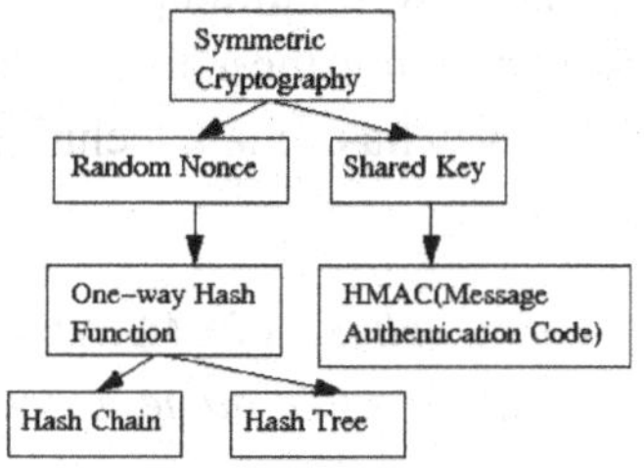

(b) Commonly-used symmetric cryptography techniques and their dependency relationships.

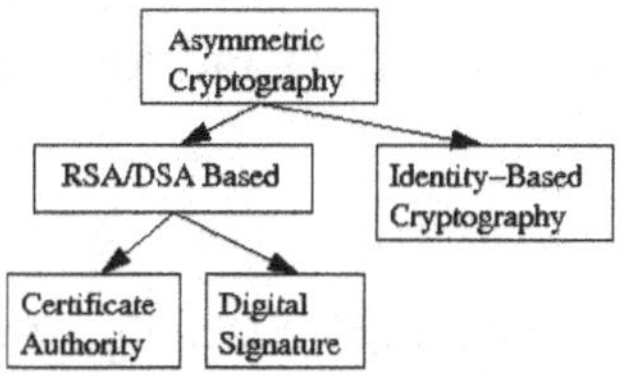

(c) Commonly-used asymmetric cryptography techniques in MANETs/WSNs and their dependency relationships.

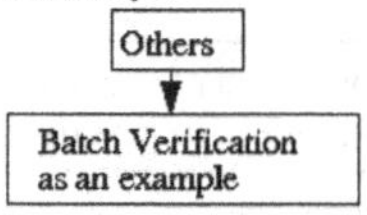

(d) Other techniques in MANETs/WSNs.

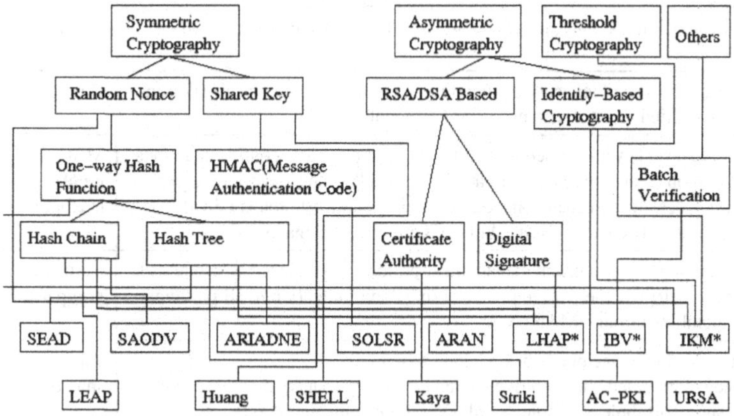

(e) Cryptography techniques used in MANETs/WSNs security schemes. Schemes with * are selected as study cases.

signatures and the frequency of this verification, symmetric keys for encryption and authentication are proposed (Hu, Perrig, & Johnson, 2002; Zhu & Xu, 2003) to secure routing protocols. One of the commonly-used cryptographic techniques is the one-way hash function, from which other techniques (i.e., hash chain, TESLA key, Merkle hash tree and hash tree chain) are derived. The cryptographic techniques used in some MANET/WSN security research work are shown in Figure

1. (Table 1 gives details of each scheme shown in Figure 1.).

Digital signatures, hash functions, and hash functions based on a message authentication code (HMAC) (Menezes, Oorschot, & Vanstone, 1996) are techniques used for data authentication or integrity purposes in securing MANETs/WSNs. A digital signature is usually signed using a private key and can be verified using a public key. In more detail, a public key is protected by the public-key certificate, in which a trusted entity called the certification authority (CA) in public key infrastructure (Menezes, Oorschot, & Vanstone, 1996) vouches for the binding of the public key with the owner's identity. Those cryptographic techniques are used in most security schemes in MANET/WSN design, for example, SOLSR (Clausen, 2003) and ARAN (Sanzgiri, 2002).

It is very challenging to use different cryptographic techniques to deal with different tasks. The good example is in the countermeasure resource consumption error, where the LHAP scheme shows the art of using composite techniques.

Another popular topic of discussion is to determine how to build up MANETs or WSNs and how to maintain the network. For example, the use of one-way hash chain techniques will determine

Table 1. Overview of cryptographic techniques used in security schemes in MANETs/WSNs. (Schemes without a specific name are specified here according to the author's last name, marked with *.)

MANET/WSN Security Scheme	Security Objectives	Cryptographic Techniques
ARAN	Authentication, integrity, and non-repudiation of signaling packets, based on AODV (Perkins, 2001), designed to substitute reactive routing protocols.	Certificate authority, timestamp.
ARIADNE	Authentication and integrity of signal packets, based on the basic operations of DSR (Perkins, 2001).	Symmetric cryptography primitives, hash function and timestamp.
SAODV	Authentication and integrity of signaling packets, a security extension for AODV.	Digital signature and hash chain.
SEAD	Authentication and integrity of signaling packets, based on DSDV (Perkins, 2001), applied to other distance vector protocols.	Hash chain and sequence number.
Huang*	A secure level key infrastructure for multicast to protect data confidentiality via hop-by-hop reencryption and mitigate DoS-based flooding attacks through an intrusion detection and deletion mechanism. The multicast protocol divides a group routing tree into levels and branches in a clustered manner.	MACs and one way sequence number, cluster-based tree as key management.
Kaya*	A dynamic multicast group management protocol is proposed which aims to equally distribute the workload of securing communication to all participating members through MANETs.	Certificate authority and ad hoc group shared key.
LEAP	Source and message one way key chain based authentication and cluster-based shared key in key management to countermeasure wormhole, sinkhole, Sybil, DoS, replay, insider attacks.	Hash chain and cluster-based shared key.
SLSP (Papadimitratos & Haas, 2003)	Authentication, integrity, and non-repudiation of signal packets, extends an intrazone protocol for ZRP (Perkins, 2001).	Certificate authority.
SPAAR (Carter & Yasinsac, 2002)	Authentication, integrity, non-repudiation, and confidentiality, secure position aided ad hoc routing protocol.	Certificate authority and timestamp.
SOLSR	Authentication and integrity of signaling packets.	MACs and timestamp.
SHELL	A cluster-based key management scheme. Each cluster has its own distributed key management entity residing in a-cluster-head node. Therefore, the operational responsibility and key management responsibility are separated, offering better resiliency against node capture.	Group shared key.
LHAP	A hop-by-hop authentication protocol for ad hoc networks.	Digital signature and hash chain.
IKM	Key management to secure mobile ad hoc network, efficient network-wide key update via a single broadcast message.	ID-based cryptography and threshold cryptography.
Striki*	User authentication and Merkle tree-based data authentication in MANETs.	Hash function and hash tree.
IBV	An efficient batch signature verification scheme for vehicular sensor networks.	Batch verification of ID-based signature.

how to bootstrap the network, how to deliver the key chain, how to let nodes join the network, and how the nodes communicate with neighbors and countermeasure attacks. Other cryptographic techniques have to be considered in the design to establish trust relationships and authentication keys in MANETs in order to complement the use of techniques.

Cryptographic techniques are grouped together and associated with each other to support schemes and protocols in MANET/WSN as shown in Figure 1 (a), (b), (c), (d), (e). Cryptography can be categorized into four parts seen in Figure 1 (a); In detail, symmetric key techniques are shown in Figure 1 (b), in which random nonce, shared keys, one-way hash functions, hash chains, hash trees, and message authentication codes are most-commonly-used in MANET/WSN; and as part of Figure 1(e), those symmetric techniques are used for schemes SEAD (Hu, Johnson, & Perrig, 2002), SAODV (Zapata, 2002), ARIADNE (Hu, Perrig, & Johnson, 2002), SOLSR, LEAP (Zhu, Setia, & Jajodia, 2003), Huang (Huang, Buckingham, & Han, 2005), and SHELL (Younis, Ghumman, & Eltoweissy, 2006). Secondly, asymmetric key techniques are presented in Figure 1 (c), in which public/private keys, RSAs, Digital Signature Algorithms (DSA), ID-based cryptography, certificate servers, and digital signatures are commonly-used techniques in MANET/WSN; and as part of Figure 1 (e), those asymmetric techniques are associated to support schemes such as Kaya (Kaya, 2003), ARAN, LHAP, IKM, AC-PKI (Zhang, Liu, Lou, Fang, & Kwon, 2005), and Striki (Striki & Baras, 2004). Third, threshold cryptography is shown in Figure 1 (e) to support part of the IKM scheme, URSA (Luo & Lu, 2004). Last but not least, batch verification based on ID-based signature is shown in Figure 1(d) to represent other cryptographic techniques that are not included in our survey. For example, the IBV scheme. There are many other cryptographic techniques that can be applied in MANETs/WSNs. In the following paragraphs we show a collection of short reviews of cryptographic techniques and a short discussion of selected MANET/WSN security solutions.

- **Symmetric cryptography:** The encryption key is closely related or identical to the decryption key. In practice, keys represent a shared secret between two or more parties that can be used to maintain private communication.

Usually the network can choose a shared secret key to encrypt and decrypt the message once two or more parties have used a public/private key pair to build trust in the hand-shaking stages. This is more feasible and efficient from a computational standpoint than asymmetric key techniques.

- **Random nonce:** In the network, a time-stamp or random number (nonce) is used to make packets fresh and prevent a replay attack (Kaufman, Perlman, & Speciner, 2002). The session key is often generated from a random number. In the public key infrastructure, the shared secret key can be generated from a random number as well.

Cryptographic pseudo random generators typically have a large pool of seed values. The design and implementation of cryptographic pseudo random generators can easily become the weakest point of the system.

- **Shared key:** Less computationally intense symmetric key algorithms are used more often than asymmetric algorithms. In practice, asymmetric algorithms are hundreds of times slower than symmetric key algorithms. The most common are AES, RC4 and IDEA. The disadvantage of shared keys in networks is the requirement of $n(n-1)/2$ shared keys among n nodes in order to have a secure communication between any two nodes.

In wireless sensor networks, some protocols use shared keys. Instead of a shared key for each pair of nodes, called pairwise keys, there may be one shared secret key for the entire network, or a group key for each group or cluster of networks. Lee (2007) (Lee, Leung, Wong, Cao, & Chan, 2007) has a detailed discussion using case studies of five key management protocols: Eschnauer (Eschenauer & Gligor, 2002), Du (Du, 2003), LEAP, SHELL, and Panja (Panja, Madria, & Bhargava, 2006).

- **HMAC message authentication code:** This type of message authentication code is calculated using a hash function in combination with a secret key. Usually in MANETs/WSNs, the hash functions chosen are mostly MD5 or SHA-1. It can also be used to ensure that an unencrypted message retains its original content by calculating the message HMAC using a secret key. For example, see SOLSR, Huang (Huang, Buckingham, & Han, 2005).
- **Hash chain:** It is generated by a successive application of a hash function to a string. Lamport (Menezes, Oorschot & Vanstone, 1996) suggested the use of hash chains as a password protection scheme. Due to the one-way property of secure hash functions, it is impossible to reverse the hash function. A hash chain is a method to produce many one-time keys from a single key, and keys are used in the reversed order of generation. For example, SAODV, ARIADNE, and LEAP are three applications in MANETs/WSNs that use one-way key chains.
- **Hash tree:** It was originally invented to support the handling of many Lamport one-time signatures. At the top of a hash tree there is a top hash or master hash. Nodes higher in the tree are the hashes of their respective children. An example can be found in the MANET/WSN security scheme SEAD.

- **Asymmetric cryptography:** In public key or asymmetric cryptography, there is a pair of public/private keys. The private key is known only to the owner, while the public key is shared with others. One of the earliest public-key cryptographic techniques, known as RSA, was developed in the 1970s. Since then, a large number of encryption, digital signature, key management, and other techniques have been developed in public-key cryptography. Examples include the ElGamal cryptograph system, DSA, and elliptic curve cryptography.
- **Certificate Authority:** A certificate authority is an entity that issues digital certificates for use by other parties. CA is the most important role in many public key infrastructure schemes.

Whether certificate authorities are practical in MANETs/WSNs is a popular topic of debate. But it is wise to take advantage of the CA role if possible even in MANETs/WSNs. Usually network nodes in MANETs trust the CA in the bootstrap stage and can verify the CA's signature. Then, nodes can also verify whether a certain public key does indeed belong to another node, as it is identified in the certificate. For example, ARAN and Kaya (Kaya, 2003) are two applications in MANETs/WSNs that use certificate authority.

- **Digital signature based on RSA/DSA:** The ElGamal signature is based on the difficulty of breaking the discrete log problem. DSA is an updated version of the ElGamal digital signature scheme published in 1994 by FIPS and was chosen as the digital signature standard (DSS) (Mehuron, 94).

Digital signature, using the RSA/DSA algorithm, is popular for authentication or confirming the message's integrity. A digital signature scheme typically consists of three algorithms: a key generation algorithm, a signing algorithm, and a signature verifying algorithm.

In MANETs/WSNs the digital signature is more expensive to compute than a hash function, and digital signatures do not scale well in MANETs/WSNs as the number of nodes grows larger. For example, a digital signature is only performed once in bootstrapping a TESLA key chain in the LHAP scheme.

- **Identity-based cryptography:** This is a type of public-key cryptography. The first identity-based cryptography, developed by Adi Shamir in 1984, uses the identity of the user as a public key. Modern schemes include Boneh/Franklin's pairing-based encryption scheme (Boneh & Franklin, 2001). For example, IKM and AC-PKI schemes are applications that use ID-based cryptography.

- **Batch verification with ID-based signature:** Although there are advantages to ID-based cryptography signature schemes based on pairing, the signature verifications are at least ten times slower than that of DSA or RSA. The batch verification (Yoon, Cheon, & Kim, 2004) of many signatures increases efficiency.

Table 1 lists some security schemes with their security objectives and associated cryptographic techniques.

In general, most surveys have been done on security routing and other specific areas such as key management. Our approach differs in that we concentrate on the cryptography techniques used. We prefer to choose cases that include the latest research in an area, putting different cryptographic techniques under review. The following discussion will focus on cryptographic techniques. Using Figure 1 as the outline, we will go through the discussion from symmetric key techniques to asymmetric key techniques, and from RSA/DSA-based schemes to ID-based cryptography, with some discussion about threshold cryptography. Most of the discussion focuses on three cases

in MANET/WSN security research: the LHAP scheme, the IKM scheme, and the IBV scheme.

Symmetric Key Techniques Applied in MANETs/WSNs

As seen in Figure 1, symmetric key techniques are used in most security MANETs/WSNs schemes. The techniques are random nonce, shared key, one way hash function, message authentication code, hash chain, and hash tree, as seen in Figure 1 (b). These are used so frequently that we must consider the applicable network factors before the techniques are used in a new design.

One way hash chain and TESLA key (Perrig, Canetti, Tygar, & Song, 2000) are faster than the traditional PKI private key calculation. They are used in the design of several security protocols including SAODV, ARIADNE, and LHAP as shown in Figure 1 (e). One-way hash chains are very easy to compute compared to public key distribution, which typically requires central authentication. Thus, in order to achieve the best performance in the network field, we sometimes use hash functions instead of PKI public keys.

Lamport used one-way hash chains for password authentication. In this instance, a one-way hash chain repeatedly applies a one-way hash function starting from a random number. The user picks up the secret key, which is usually a random number. Supposing that the chain length is N, the user runs the hash function N times on the random number. Actually, each hash function value is a key on the chain. In the list of keys the original random number is the most important key because all other secret keys can be calculated via hash function from this number. If a node wants to generate a key chain of size N, it first needs to choose a key, denoted as *seedKey*, which will be the last one used to do the encryption. The one-way hash chains are generated as follows:

$$K(0) = seedKey, K(1) = h(K(0), ..., K(N)) = h(K(N-1))$$

in which h is the one-way hash function. It is infeasible to compute inversely from a one-way hash function. In various standards and applications, the two most-commonly used hash functions are MD5 and SHA-1.

Two commonly-used cryptographic techniques that are used in WSN broadcast authentication are *µTESLA* (Perrig, Szewczyk, Wen, Culler, & Tygar, 2001) and digital signatures. *µTESLA* is considered to be a symmetric cryptography technique, and its variations implement broadcast authentication through delayed disclosure of authentication keys. *µTESLA* keys are based on a symmetric cryptographic hash function, and the operations cost is more efficient even though the network has to be loosely time synchronized and suffers from authentication delays. If digital signatures, such as ECDSA (IEEE, 2006), are used directly for broadcast authentication, they are easily attacked by broadcasting forged packets. The receiving nodes are forced to perform a large amount of unnecessary signature verifications.

To countermeasure DoS attacks when digital signatures are directly used for broadcast authentication, hop-by-hop pre-authentication filters can be used to remove bogus messages before verifying the actual digital signatures. In particular, two filtering techniques, a group-based filter and a key chain-based filter (Dong, Liu, & Ning, 2008), are based on a symmetric cryptographic hash function, hash chain, shared pairwise key, and MAC. When a sender and its neighbor nodes hold a group key in common, an adversary cannot forge messages without compromising the group key. However, a compromised sensor leaks the group key. Alternatively, a sensor node can add a MAC to a broadcast message for each of its neighbor nodes. However, this incurs large communication overhead. Based on the above two simple methods to filter out forged messages, the group-based filter technique has to trade-off communication efficiency with security. Specifically, the group-based filter organizes the neighbor nodes of a sender into multiple groups, which are protected by different keys in a tree structure. In the second filter technique, the key chain-based filter is designed to apply a two-layer filter to deal with the DoS attacks on the verification of signatures and chained keys. On the other hand, one-way key chains feature a simple pre-authentication filter, used by LHAP, which cannot countermeasure the DoS attack because an adversary may claim a key close to the end of the key chain and cause a large amount of unnecessary hash operations. In the two-layer filter the first layer employs a one-way key chain to filter out fake signatures, and the second layer uses existing pairwise keys to prevent a node from conducting unnecessary hash operations.

Key management is a challenging issue in WSNs due to the sensor node's resource constraints. Various key management schemes in WSNs are still based on symmetric key techniques. With varying degrees of key sharing, the key distribution scheme models are generally network keying, pairwise keying, and group keying. For example, in Lee (Lee, Leung, Wong, Cao, & Chan, 2007) the security and operational requirements of WSNs are examined, and five key management protocols - Eschenauer, Du, LEAP, SHELL, and Panja- are reviewed. The key sharing models for WSNs are used to compare the different relationships between the security and operation requirements for WSNs: accessibility, flexibility, and scalability.

Like security, key management in WSNs is comprised of a cross-layered design, which can go from the link layer to the application layer. As an applicable link layer standard in a WSN, IEEE 802.15.4 considers key usage for secure data transmission, but it does not specify how to securely exchange keys. This opens the door to the key management problem that has been the focus of recent research. We sum up the benefits and problems for three models - network keying, pairwise keying and group keying. In network keying the entire network shares a single secret key. The benefits are simple to implement and allow data aggregation and fusion, ease of scale, self-

Table 2. One way hash chain techniques in a variety of MANETs schemes

Secure Routing Protocol	Cryptography Techniques	Network Service Provided
SEAD	One way hash chain	Used on a hop-by-hop basis due to the basic operation of DSV.
ARIANDE	TESLA key	Applied to secure on-demand routing protocols in source-to-destination nature.
LHAP	One way hash chain	Used for traffic packet authentication.
	Merkle hash tree chain	Used to achieve fast hash verification.

organization, flexibility and accessible. However, compromising one node compromises the entire network, losing robustness. In pairwise keying a pairwise model is chosen to allow each specific pair of nodes to share a different key. Hence, the pairwise model has benefits of best robustness and each node is authenticated, but the pairwise model suffers from scalability problems in storage, energy and computation. In addition, the pairwise model is unable to self-organize and is not flexible for addition or removal of nodes. Last but not least, the group model is designed to let each group use a different shared key. It has benefits of allowing multicast and group collaboration, better robustness than network-wide keying, and adjustable scalability with the ability to self-organize within the cluster. On the other hand, the group model lacks efficient storage methods for group keying to the standard of *IEEE 802.15.4*, and is difficult to securely set up. Also, cluster formation information is application-dependent.

So far, we have discussed one-way hash chains, *μTESLA* key, pre-authentication filters on broadcast authentication in MANETs/WSNs, and shared key models in WSNs. Indeed, using symmetric cryptography in networks is a state-of-the-art advancement. Next, we use the case study of the LHAP protocol to enhance the discussion of symmetric cryptography.

Case Study 1: LHAP Protocol

In Figure 1, the three cryptographic techniques that are used in the LHAP protocol are shown as hash chain, hash tree, and digital signature. Taking LHAP as our first case study, we show the advantage of using symmetric cryptographic techniques to handle special network situations in security.

One mechanism employs authentication and ensures that only authorized nodes can inject traffic into MANETs to countermeasure resource

Table 3. LHAP scheme cryptographic techniques customized for different network service

LHAP Cryptographic Techniques	Network Rationale
1024-bit RSA digital signature	The most expensive operation in LHAP, but it is only performed once in bootstrapping a TESLA key chain. Therefore the cost is negligible when amortized over the entire packet.
TESLA key	Used to reduce the number of public key operations for maintaining trust between nodes.
One way hash chain (It is more efficient than HMAC over the message.)	Used to authenticate traffic packets for mainly two reasons: 1: One hash time cost is small compared to the overall end-to-end transmission latency of a packet. 2: Limit network memory used for buffering the received packets, and only authenticate traffic packets to its immediate neighbors to prevent an attacker from launching replay attacks.
Merkle hash tree	Used to support fast hash verification; the maximum number of verifications a receiver has to perform is $O(\log(N))$, where N is the length of a TESLA key chain. The verification process only works for TESLA key chains.

consumption attacks. As a hop-by-hop authentication protocol for MANETs, LHAP resides between the network layer and the data link layer providing a layer of protection that can counter many attacks, including outsider attacks and insider impersonation attacks.

Many security schemes take advantage of the benefits of hash chains. To illustrate, in Table 2, we present the one way hash chain techniques used in different MANET schemes. To trade reduced security for enhanced performance, various cryptographic techniques are customized for different network services in the LHAP scheme, as illustrated in Table 3.

In order to counter a resource consumption attack the LHAP protocol is designed to use authentication of traffic packets to avoid bogus packets. Based on wireless ad hoc network analysis, in cases such as network deployment, nodes joining the network, and a node gaining trust from other nodes in the network, trust management may be based on one way hash traffic key chains and a TESLA key chain. To minimize overhead the node uses an RSA digital signature only for gaining trust while using traffic packet authentication in which keys are generated from the one-way hash chain function. Also, to support fast hash verification, LHAP uses a tree-based authentication scheme, namely the Merkle hash tree.

Through this short case study of LHAP, we have shown that symmetric cryptography can be used creatively in special cases, and that it can be used to compare similar networking schemes. Therefore, cryptography technique studies really do help us to organize security design schemes better.

Asymmetric Cryptographic Techniques Applied in MANET/WSN Security

From Figure 1, we show that asymmetric cryptography is popularly used in the security of MANET/WSN schemes, detail seen in Figure 1

(e). Most public key infrastructure schemes are either based on RSA/DSA or ID-based cryptography, seen in Figure 1 (c). For example, the most popular scheme, ARAN, has been discussed in many surveys (Lou, 2003; Wu, Cardei, & Wu, 2008; Xiao, 2007).

Public key infrastructure in MANETs is a very popular choice securing the networks. Some schemes (Luo & Lu, 2004; Yi, Naldurg, & Kravets, 2002) use a public-key infrastructure to associate public keys with the node's identity. One of PKI's approaches is to pre-load each node with all other nodes's public key certificates prior to network deployment. This approach has two problems: scalability with network size and public key update if needed. Another approach is to use on-demand certificate retrieval, which is not an optimal choice considering communication latency and overhead. Secure routing protocols, such as ARAN, ARIADNE, SEAD, and SPINS (Perrig, Szewczyk, Wen, Culler, & Tygar, 2001), all are based on the assumption that there is pre-existence and pre-sharing of secret and/or public keys for all the nodes in the network. This leaves ad hoc key management and key distribution as an open problem that must be solved.

Several IBC-based certificate-less public-key management schemes for MANETs have been developed by (1) deploying identity-based cryptography (IBC) and threshold secret sharing and (2) by eliminating the assumption of a pre-fixed trust relationship between nodes,. These include Deng, Mukherjee, & Agrawal, 2004; Khalili, Katz, & Arbaugh, 2003; Saxena, Tsudik, & Yi, 2004; Zhang, Liu, Lou & Fang, 2006; and Zhang, Liu, Lou, Fang, & Kwon, 2005. The basic idea is to let some or all network nodes share a network master-key. Some of them (Saxena, Tsudik, & Yi, 2004; Zhang, Liu, Lou, Fang, & Kwon, 2005) use threshold cryptography and collaboratively issue ID-based private keys. The PKI digital signature scheme is widely recognized as the most effective approach for Vehicular Sensor Networks (VSNs) to achieve authentication, integrity, and validity. To avoid scalability problems, the efficient identity-based batch verification scheme (Yoon, Cheon, & Kim, 2004) is proposed, which employs the *batch verification technique* based on IBC (Camenisch, Hohenberger, & Pedersen, 2007). This scheme uses IBC to generate private keys for pseudo identities, so PKI certificates are not needed, and transmission overhead is significantly reduced.

Introduction to Identity-Based Cryptography

In 1984, Shamir proposed the idea behind identity-based encryption. However, there was no workable method to solve the problem until Boneh (2001) invented a practical scheme based on elliptic curves and a mathematical construct called the Weil Pairing.

A bi-linear map is a special mathematical function that makes identity-based encryption work. A bi-linear map is a pairing that has the property: $Pair(a*X, b*Y) = Pair(b*X, a*Y)$.

For identity-based encryption, the operator "*" is used for multiplication of integers with points on elliptic curves. The products, for example $a*X$, are easy to calculate, but the inverse operations, such as finding parameter a from X and value of $a*X$, are practically impossible. The function is one way and practically non-invertible. The concept is actually the same as one-way hash functions; the bi-linear map can be a Weil Pairing.

The following concrete example will more clearly illustrate the pairing technique.

Let p, q be two large primes and E/F_p indicate an elliptic curve $y^2 = x^3 + ax + b$ over the finite field F_p. G_1 is a *q-order* subgroup of the additive group of points of E/F_p, and G_2 is a *q-order* subgroup of the multiplicative group of the finite field $F_{p^2}^*$. The discrete logarithm problem is required to be hard in both G_1 and G_2, which means that it is computationally infeasible to extract the integer x, given $p, q \in G_1$ such that $q = xp$. For

example, a pairing is a map $\psi : G_1 \times G_1 \to G_2$ with the following properties:

- Bilinear property:
  For $\forall P, Q, R, S \in G_1$,
  $\psi(P+Q, R+S) = \psi(P,R)\psi(P,S)\psi(Q,R)\psi(Q,S)$
  And also, for
  $\forall a, b \in Z_q^* = \{a | 1 \le a \le q-1\}$ there is
  $\psi(aP, bQ) = \psi(aP,Q)^b = \psi(P, bQ)^a = \psi(P,Q)^{ab}$

etc.

- Non-degenerate property: If P is a generator of G_1, then $\psi(P,P) \in F_{p^2}^*$ is a generator of G_2.
- Computable property: There is an efficient algorithm to compute $\psi(P,Q)$ for all $P, Q \in G_1$.

A more comprehensive description of how these pairing techniques work can be found in papers (Barreto, Kim, Bynn, & Scott, 2002; Boneh & Franklin, 2001; Boneh, Franklin, 2003).

In our case study, we choose a hybrid cryptography scheme combining threshold cryptography with ID-based cryptography as a certificateless key scheme IKM.

Case Study 2: ID-Based Key Management Scheme – IKM

As seen in Figure 1(e), several cryptographic techniques (including random nonce, one way hash function, threshold cryptography and ID-based cryptography) are used in the IKM scheme. Fig 2 uses cryptographic techniques to break down network initialization in the IKM scheme presenting the design in a comprehensive tree-structure. The complicated design of the network initialization is based on a prototype of the most commonly-used case: one random nonce, one node specific identity, and one hash function to apply node's identity. The IKM scheme may be extended with two random nonce, two sets of identities, two hash functions, and many (up to maximum M)

Figure 2. IKM scheme design network initialization demystified – A threshold cryptography and identity-based cryptography composite design tree structure illustration of parameters. Key_A is network master key for all nodes in network. Key_B is network master key for all network phases. ID_A is node specific identity. ID_B is network phase identity. Func_A is a hash function applied in node's identity in network. Func_B is a hash function applied in phases to generate salts. Phase_1 is network phase salt in first phase in the process of relatively frequent key update. Phase_M is network phase salt in Mth phase which is maximum phase index

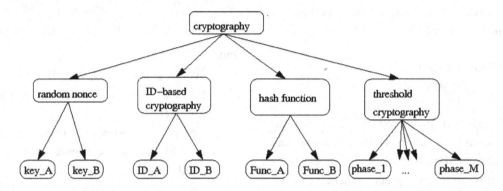

phases. Consider this approach a trial use of the most recent scheme in IBC, through which we encourage readers to employ the "cryptographic techniques used" exercise; ask a series of questions regarding how many cryptographic techniques are used, when to use them, and why to use them. This break-down approach is also a training exercise to encourage us to design a security scheme using the cryptographic techniques used in practical approaches in MANETs/WSNs.

After comparing several IBC-based certificate-less public-key management schemes (Deng, Mukherjee, & Agrawal, 2004; Khalili, Katz, & Arbaugh, 2003; Saxena, Tsudik, & Yi, 2004; Zhang, Liu, Lou, Fang, & Kwon, 2005), IKM solved several issues related to the previous IBC-based key management scheme:

- The security of the whole network is compromised when a threshold number of network nodes who share the network's master key are compromised.
- Significant communication overhead in a large-scale MANET occurs while updating ID-based public/private keys because each node has to contact a threshold number of nodes who share the network master key one by one.

- There is no quantitative argument to prove the advantage of IBC-based public key management schemes over certificate-based cryptography.

One contribution of IKM is to provide a novel construction method of ID-based public/private keys. In IKM each node's public key and private key includes two parts: one is a node specific ID-based element, and the other is a network-wide common element. The node specific ID-based elements are designed to ensure that the compromise of an arbitrary number of nodes does not affect the secrecy of the non-compromised nodes' private keys. With network-wide common key elements a single broadcast message can update the network-wide public/private keys.

Each IKM node has an authentic ID-based public/private key pair and uses the key pair as proof of its group membership. Those key pairs help to implement the mutual authentication, key management, public-key encryption, and digital

Table 4. The cryptography techniques and their functionalities in IKM scheme

Cryptography techniques	Design consideration – IKM scheme functionality.
Random nonce	Used as network master secret keys K_{p_1}, K_{p_2}, in which one constructs a node private key, another constructs a series of network phase private keys.
Hash function	One hash function is used to make a series of network phases. In detail, $salt_i = h(salt_{i-1} + 1)(1 < i \leq M)$, h is a hash function, such as SHA-1.
	Hash function is chosen for ID-based identity application, H_1, which maps arbitrary string to non-zero element in the subgroup G_1.
ID-based cryptography	Node specific element is related to nodes which can join network anytime, and its major concern is to define its public key and private key. For example, node A with identity ID_A has the keys: $< \Gamma_A, \Gamma_A^{-1} >= (H_1(ID_A), K_{p_1}H_1(ID_A))$.
	Network phase specific element is related to phases in different time period, and its public/private key pair is $(H_1(salt_i), K_{p_2}H_1(salt_i))$.
Threshold cryptography	It is used to apply relatively frequent key update to enhance the security. The IKM is composed of a number of continuous, non-overlapping *key update phases*, denoted by p_i for $(1 < i \leq M)$, where M is the maximum possible phase index.

signatures. The IKM key management scheme consists of three phases: key pre-distribution, revocation, and update.

In the key pre-distribution phase a Private Key Generator (PKG), acting as a trusted authority, prepares a set of system parameters and pre-loads every node with certain key contents during network initialization. Next, the PKG distributes its functionality to n distributed authorities which are selected from the overall number of nodes N to enable secure key revocation and update during network operation. The n distributed authorities in IKM are called D-PKGs for convenience.

If a node is compromised, its public key may be explicitly revoked. During network operation, if a node suspects that a peer, say A, has been compromised, the node can send a signed accusation against A to some D-PKGs. When the number of accusations against node A reaches a predefined revocation threshold, denoted by γ, in a certain window, the node A is diagnosed as compromised. The D-PKGs can jointly issue a key revocation against A.

As a common practice, public/private keys of mobile nodes are updated at intervals for various reasons, such as preventing cryptanalysis. IKM also takes this approach and the non-revoked node can update its public key autonomously and its private key via a single broadcast message.

IKM is designed to make the distributed authorities D-PKG's indistinguishable from common nodes via anonymous routing (Zhang, Liu, Lou, Fang, & Kwon, 2005). Because of the shared wireless medium, D-PKGs IDs leak in routing and data packets, making D-PKGs vulnerable to pinpoint attacks.

We will now focus on the basics of IKM related to network initialization, key revocation, key update, and its security analysis. We will also discuss threshold cryptography and how the IKM scheme can benefit from previous threshold cryptography analysis.

Network Initialization

In Table 4, the cryptography techniques used in the IKM scheme are reviewed and matched with the detailed IKM scheme functionality. In Figure 2, IKM scheme network initialization, the main ideas are (1) using threshold cryptography to update the network key for each network phase based on a hash function, and (2) node specific public/private key based on ID-based cryptography.

The PKG does the following three steps to bootstrap the network. First, generate the pairing parameters (p, q, ψ), and select an arbitrary generator W of G_1. Secondly, choose a hash function H_1 that maps arbitrary binary strings to nonzero elements in G_1. Thirdly, choose two distinct random number $K_{p_1}, K_{p_2} \in Z_q^*$ as network master-secrets, and set $W_{p_1} = K_{p_1} W$ and $W_{p_2} = K_{p_2} W$ respectively.

Then, preload parameters $(p, q, \psi, H_1, W, W_{p_1}, W_{p_2})$ to each node; those parameters are public, while network master keys K_{p_1}, K_{p_2} should never be disclosed to any node.

In IKM secret sharing design, only knowledge of K_{p_2} is introduced into the network, and the PKG performs a *(t, n)-threshold* secret sharing of K_{p_2} to avoid the single point of compromise and failure. The random polynomial of threshold cryptograph is $g(x) = K_{p_2} + \sum_{i=1}^{t-1} g_i x^i (\bmod q)$. Distributed authorities D-PKGs are randomly selected from a subset of size n of nodes $(t \leq n < N)$. After that the PKG assigns to each node in D-PKG a secret share computed as $K_{p_2}^V = g(ID_v)$. The design is based on the Lagrange interpolation. The PKG's master secret key K_{p_2} can be obtained from value $g(0)$. However, any subset with size $(t - 1)$ or smaller cannot reconstruct $g(0)$. The PKG also calculates a set of values to enable verifiable secret sharing. The values $\{W_{p_2}^V = K_{p_2}^V W | V \in \Omega\}$ in which Ω is the D-PKG set are preloaded to each D-PKG. Due to the difficulty of solving the DLP in G_1, none of the other D-PKGs can know the secret share $K_{p_2}^V$ of D-PGK V from $W_{p_2}^V$. To make key revocation and update feasible, the IDs of all the D-PKGs are public to each node.

IKM is designed to construct ID-based public/private keys for each node A. The IKM contains a number of continuous, non-overlapping key update phases, in which the ith key update period is denoted by p_i for $(1 < i \leq M)$, where M is the maximum possible phase index. Each phase p_i is associated with a unique binary string called a phase salt, denoted by $salt_i$. The PKG issues a random number $salt_1$ to each node before the deployment of the network, and with a hash function h such as SHA-1, a series of $salt_i$ is generated using $salt_i = h(salt_{i-1} + 1)(1 < i \leq M)$.

There are both node-specific and phase-specific public/private key pairs, and node A's key pair is valid only during phase p_i which is denoted by $< \Gamma_{A,p_i}, \Gamma_{A,p_i}^{-1} >$. Each public key Γ_{A,p_i} and private key Γ_{A,p_i}^{-1} is comprised of a node-specific element and a phase specific element common to all the nodes, both in G_1

$$\Gamma_{A,p_i} := (\Gamma_A, \Gamma_{p_i}) = (H_1(ID_A), H_1(salt_i)),$$

$$\Gamma_{A,p_i}^{-1} := (\Gamma_A^{-1}, \Gamma_{p_i}^{-1}) = (K_{p_1} H_1(ID_A), K_{p_2} H_1(salt_i))$$

At the beginning, the PKG issues $< \Gamma_{A,p_1}, \Gamma_{A,p_1}^{-1} >$ to node A from which $< \Gamma_{A,p_i}, \Gamma_{A,p_i}^{-1} > (1 < i \leq M)$ is originated from the D-PKGs during network operation. $< \Gamma_{p_i}, \Gamma_{p_i}^{-1} >$ is a common public-key and private-key element of phase p_i, and $< \Gamma_A, \Gamma_A^{-1} >$ is a node-specific public-key and private-key elements of node A. The phase p_i public/private key pair changes across key-update phases, while node A public/private key pairs remain the same during the network lifetime and should not be released to node A itself.

Because it is difficult to solve the discrete logarithm problem in the subgroup G_1, it is not possible to calculate the network master secret K_{p_1} and K_{p_2} from an arbitrary number of public/private key pairs. Therefore, IKM has a property which allows it to keep the confidentiality of the node's private key if the node is compromised, regardless of how many key pairs the adversary is able to acquire from compromised nodes. The IKM scheme has more resilience to the compromization of D-PKGs than the conventional key construction method (Saxena, Tsudik, & Yi, 2004; Zhang, Liu, Lou, Fang, & Kwon, 2005).

The IKM scheme allows dynamic node joins at any time. Suppose a new node X joins the network at phase p_i, the PKG only needs to pre-equip X with public system parameters and $< \Gamma_{A,p_i}, \Gamma_{A,p_i}^{-1} >$. Based on the support of a node joining the network at any time, the IKM scheme network size grows without limitation, therefore high network scalability is achieved.

Key Revocation

The IKM scheme includes a key revocation design which has three subprocesses: misbehavior notification, revocation generation, and revocation verification. For our case study, we only show the misbehavior notification; for the other two parts refer to the original paper (Zhang, Liu, Lou & Fang, 2006).

Suppose node B detects node A's misbehavior. Node B generates a signed accusation $\left[ID_A, s_B\right]_{\Gamma_{B,p_i}^{-1}}$ against A, where s_B is a timestamp to countermeasure message replay attacks. If node B sends the revocation message to the D-PKGs, several things must be considered. Since node A may temporarily behave normally, it is not wise for node B to naively flood the accusation. Node A may attempt to lower the number of accusations against it down to the level that is below the predefined revocation threshold γ. Therefore, the IKM scheme takes the approach to let node B unicast the accusation secretly to one of the D-PKGs instead.

During network initialization, the PKG provides each node with a function η that maps each node ID to the IDs of β distinct D-PKGs. Any node A in the network node set, denoted by Λ, is assoiciated with the set $\eta(ID_A) = \{ID_{X_j} | 1 \le j \le \beta, X_j \in \Omega, X_j \ne A\}$. Therefore the node set Λ is divided into n disjoint node sets, each associated with β distributed authorities D-PKGs.

The β determines the tradeoff between resilience to D-PKG compromise and communication overhead. Smaller β leads to a lower related communication overhead, but also to a less resilient network.

Key Update

It is common practice to update keys to countermeasure the cryptanalysis and limit any potential damage from compromised keys. Previous research in MANETs and WSNs provides the related work for updating keys using threshold cryptography, for example, (Zhou & Haas, 1999; Luo & Lu, 2004). In our desire to become expert at application design, it is in our interest to show threshold cryptography applied in different cases and to determine the primary evaluation factors. So in our case study, we show the details of key update in IKM and network analysis.

A new key update phase p_{i+1} starts either because the previous phase p_i times out or because the number of nodes revoked in p_i is not less than the prescribed threshold. In the IKM scheme each node can update its public/private key autonomously. For example, node B uses the following formula: public key case, $\Gamma_{B,p_{i+1}} := (H_1(ID_B), H_1(salt_{i+1}))$ where $salt_i = h(salt_{i-1} + 1)(1 < i \le M)$. From the computation overhead standpoint, there are only two hash operations for node B to compute when updating its public key; private key case, we have $\Gamma_{p_{i+1}}^{-1} = K_{p_2} H_1(salt_{i+1})$. Private key update needs work from t D-PKGs in Ω. In the IKM scheme, the simple way is to assume that $Z \in \Omega$ initiating phase p_{i+1}, but the D-PKGs should take turns balancing their resource usage. Z randomly selects $(t - 1)$ other non-revoked distributed authorities D-PKGs from Ω and sends a request to each one.

The key update method of the IKM design provides the network's self healing capabilities. For example, there is a scenario in which any non-revoked node can recover $\Gamma_{p_j}^{-1}$ for any phase $p_j (j > i)$ if the node did not receive the key-update broadcast message due to MANET mobility, channel errors, and temporary network partitions.

Security Analysis On Threshold Cryptography Used

IKM provides more security than other MANET security schemes using certificate key management CKM (Yi & Kravets, 2003; Zhou & Haas, 1999;) and previous identity based cryptograph IBC-based schemes (Saxena, Tsudik, & Yi, 2004; Zhang, Liu, Lou, Fang, Kwon, 2005 (referred to as o-IKM)).

All these approaches are *(t,n)-threshold* schemes, having the same level of security as long as the *t-limited* assumption holds. The difference is in the worst-case scenario. Table 5 shows the detail. The IKM part of threshold scheme is as secure as conventional certificate based key management CKM's, and it outperforms o-IKM in the worst-case scenario.

After the case study, we will now cover a discussion on threshold cryptography (Shamir, 1979; Desmedt & Frankel, 1989) applied in MANETs and WSNs.

Threshold Cryptography Applied in MANET/WSN Security

In Figure 1 threshold cryptography is shown as a technique used by IKM and URSA schemes in MANET/WSN. Actually the cryptography is widely used in a variety of schemes (Capkun, Buttyan, & Hubaux, 2003; Gouda & Jung, 2004; Kong, Zerfos, Luo, Lu & Zhang, 2001; Luo & Lu, 2004; Saxena, Tsudik, & Yi, 2004; Yi & Kravets, 2003; Zhou & Haas, 1999; Zhang, Liu, Lou & Fang, 2006). We will compare those schemes and discuss the most frequently asked questions when threshold cryptography is applied in MANETs/WSNs.

For detailed background knowledge of threshold cryptography, please refer to the paper (Shamir, 1979; Desmedt & Frankel, 1989). Due to limited space, we don't present the threshold cryptography primitives here. There are several features discussed in research literature, namely verifiable secret sharing (Chor, Goldwasser, Micali, & Awerbuch, 1985; Gennaro, Jarecki, Krawczyk, & Rabin, 1996) and periodical updates on the participants' secret sharing, called proactive secret sharing, (Frankel, Gemmell, MacKenzie, & Yung, 1997; Herzberg, Jarecki, Krawczyk, & Yung, 1995). As applied in MANETs and WSNs, some of them require a trusted centralized authority to bootstrap the secret sharing procedure, while others provide joint secret sharing and do not require any trusted authorities.

Zhou & Haas (1999) used certificate based cryptography (CBC) and *(t,n)-threshold* cryptography in MANET. Let N be the total number of nodes and t, n be the two integers of threshold parameters where $t \leq n > N$. Prior to network deployment the certificate authority CA's public key is furnished to each node, while each node's private key is divided into n shares, each uniquely assigned to one of n chosen nodes. Let us denote them as D-CAs. During network operations any t D-CAs can work together to perform certificate generation and revocation using their secret share, while less than t D-CAs cannot restore the secret key. Yi and Kravets (Yi & Kravets, 2003) proposed that it is better to select more computationally powerful and more physically secure nodes as D-CAs. Both schemes tolerate the compromise of up to $(t-1)$ D-CAs and the failure of up

Table 5. Threshold cryptography worst case comparison: compromised nodes reaches threshold

(t,n)-threshold scheme distributed CAs are compromised	CKM	IKM	o-IKM
Can adversaries construct a secret key? If yes, what key is it?	Yes CA's private key.	Yes One of PKG master secret key.	Yes Same as IKM.
Can adversaries deduce the private key of any non-compromised node?	No	No	Yes
Is overall system security lost?	No	No	Yes

to $(n - t)$ D-CAs according to *(t,n)-threshold* cryptography.

Another application of threshold cryptography in MANETs is URSA (Kong, Zerfos, Luo, Lu & Zhang, 2001; Luo & Lu, 2004), which is a (t, N) threshold scheme where N is the overall number of nodes. URSA provides the network benefit of increased service availability because a certificate can be generated by any t nearby nodes or revoked by any t nearby nodes. The pitfall of this design is that the compromise of any t out of N nodes could break the secret key, that is certificate authorities CA's private key, which leads to loss of overall system security. From the network attacks analysis, several security problems have been studied (Douceur, 2002; Jarecki, Saxena, & Yi, 2004; Narasimha, Tsudik, & Yi, 2003). One major problem is the Sybil (Douceur, 2002) attack, in which an attacker takes as many identities as necessary to collect shares until reaching the threshold after which the CA's private key may be constructed.

Another approach to using threshold cryptography CBC schemes for MANETs is to let each node act as a CA to issue certificates to other nodes (Capkun, Buttyan, & Hubaux, 2003; Gouda & Jung, 2004). This approach is less suitable in MANETs, but it is good for authority-free civilian networks. The IBC-based certificate-less public-key management schemes for MANETs (Deng, Mukherjee, & Agrawal, 2004; Khalili, Katz, & Arbaugh, 2003; Saxena, Tsudik, & Yi, 2004; Zhang, Liu, Lou, Fang, & Kwon, 2005) sometimes use threshold cryptography. Table 6 demonstrates selection criteria applied to threshold cryptography in MANET/WSN. Table 7 is an illustration of the main features and advantages of the various schemes. Table 8 is a collection of threshold cryptography questions and answers seen in papers (Jarecki, Saxena, & Yi, 2004; Narasimha, Tsudik, & Yi, 2003; Zhang, Liu, Lou, & Fang, 2006).

Other Cryptographic Techniques Applied in Security of MANET/WSN

Looking up new research results of applied cryptography and applying them to MANET/WSN security is not as far from reality as it once was. Application only requires time and effort to digest the cryptographic techniques, put together the security analysis, and make up an innovative design.

Of the four categories of cryptography seen in Figure 1 (a), *"others"* is the next topic of discussion. A special cryptography technique called batch verification with ID-based signature (Yoon, Cheon, & Kim, 2004), and its application to the emerging area of vehicular delay tolerant networks, is the third case study in this chapter.

Vehicular sensor networks (VSNs) have been envisioned to be useful in many commercial applications and in road safety systems. It is common practice to apply a digital signature scheme as the countermeasure to attacks and resource abuse for VSNs. Consider the fact that a roadside unit cannot handle receiving a large number of signatures within the short interval, according to the dedicated short range communication broadcast protocol (DSRC). A cryptography technique, called the batch signature verification scheme, based on ID-based cryptography (Fiat, 1989), is applied to communication between vehicles and roadside units. A roadside unit verifies multiple received signatures at the same time to reduce the total verification time dramatically. In VSNs the scheme is designed to employ identity-based cryptography to generate private keys for pseudo identities, achieve conditional privacy preservation, and reduce transmission overhead.

There is an abundance of cryptographic techniques that can be applied in the security of MANETs/WSNs. The latest research in cryptography is advancing so quickly that a new scheme applied to MANET/WSN can dramatically change the performance of that network. For example, the following case study will focus on the vehicular

Table 6. Threshold cryptography criteria applied in MANET/WSN schemes

Certificate authority	Quantities of CA	Asymmetric cryptography	Private key
Selective on network node, not any node in network.	Selective on quantities of CA, not one unique CA.	Selective on RSA/DSA based asymmetric cryptography or IBC based one.	Selective on network-wide element's private key, not node specific element.

Table 7. Threshold cryptography usage in MANET

Scheme information	Main features	Pros/Cons
Zhou & Hass (1999) CBC scheme	Choose n nodes to be D-CAs, each secret key to give n shares, threshold is t, $t \leq n > N$.	Traditional approach
Yi & Kravets (2003) CBC scheme	Certificate authorities selected based on network factors: physical security, computation power, etc.	**Pros**: Consider network factors, thus smart choice from network viewpoint.
URSA (Kong, Zerfos, Luo, Lu & Zhang, 2001; Luo & Lu, 2004) CBC scheme	Each of N nodes is a D-CA, where N is the overall number of nodes.	**Pros**: Increase service availability, any t nearby nodes can provide service. **Cons**: Overall security is decreased, e.g. the Sybil attack.
Multiple CA (Capkun, Buttyan, & Hubaux, 2003; Gouda & Jung, 2004)	Each node acts as a CA.	**Cons**: Less authority available in network.
ID-GAC (Saxena, Tsudik, & Yi, 2004) IBC based	IBC-based access control scheme.	**Pros**: Apply in MANET service availability same as URSA. **Cons**: Overall security is decreased.
IKM, IBC based scheme	There are two parts of public/private keys which are node-specific keys and network-wide common keys; nonetheless, the threshold cryptography only applies to network-wide common element.	**Pros**: Node-specific key elements ensure the secrecy of noncomprised nodes's private key; common key elements enable efficient key updates via a single broadcast message.

sensor network. Improving the batch verification of signatures from linear time to constant time is an algorithm optimization problem in applied cryptography.

Case Study 3: An Identity-Based Batch Verification Scheme IBV

The design of a new security scheme can be very complicated, but for simplicity, we will go through a simple algorithm run time analysis case of the IBV scheme. While the node identity is used in the IKM scheme, pseudo identity is used with network context in the IBV scheme. Multiple batch verification schemes are updated in applied cryptography. Therefore, the IBV scheme is designed to adapt the work from MANETs/WSNs.

The batch verification scheme is designed to handle all the signatures received within a time interval in less time than it would take to verify the same set of signatures independently. There are several batch cryptographic techniques.

Fiat (1989) introduced batch cryptography in 1989, and several other batch schemes (Cha, & Cheon, 2003; Naccache, M'Raihi, Vaudenay, & Raphaeli, 1994; Yoon, Cheon, & Kim, 2004; Zhang, & Kim, 2003; Zhang, Safavi-Naini, & Susilo, 2003) were proposed later. The batch verification scheme (Camenisch, Hohenberger, & Pedersen, 2007) is based on the CL signature scheme and achieves high computational efficiency by not using random oracles. The batch verification scheme operates in constant time rather than linear time. For example, verifying n signatures takes 3 pairing operations instead of $3n$ So, batch verification can be applied to vehicular sensor networks to achieve good scalability.

The Identity-based Batch Verification (IBV) scheme for vehicular traffic related message trans-

Table 8. Design questions related to threshold cryptography technique applied in MANET/WSN

Ideas of threshold cryptography	MANET/WSN network advantage/pitfall analysis
1. Threshold cryptography distributes the ability to decryption or signing etc. service, what is the advantage to use threshold cryptography?	MANET/WSN network has better fault tolerance than non-threshold cryptography, better security.
2. Is secret share verifiable?	If not, it cannot be used in a setting where malicious insiders can exist. It requires a trusted third party to initialize the group during bootstrapping.
3. What is a common problem if MANET has one single trusted authority, one certificate authority CA?	One single trusted authority introduces a single point of failure attack, limited scalability.
4. What are some concerns of configuring MANETs using group authority rather than single one?	Although the group authority can be replicated for better availability, the scalability cannot be addressed by replication alone. Furthermore, unpredicted network faults and partitions complicate placement of group authority "replicas" in the network.
5. Is a fixed threshold policy applicable?	Sometimes it is necessary to reduce the threshold t to motivate the group to operate. A large group of nodes leave the network, resulting in a new smaller group size.
6. How is dynamic group size determined if using dynamic threshold cryptography?	MANET and WSN have distributed, asynchronous and decentralized dynamic group setting. Therefore, every member can send a periodic heart-beat message to the trusted authority to maintain the group size.
7. What cryptography library is commonly used in MANETs?	MIRACL (Chor, Goldwasser, Micali, & Awerbuch, 1985), a standard cryptographic library which is used in IKM.

mission includes four phases: the key generation and pre-distribution phase, the pseudo identity and private key generation phase, the message signing phase, and the batch verification phase.

Key Generation and Pre-Distribution

There are several assumptions about the network. Each vehicle is equipped with a tamper-proof device, and there is trusted authority (TA) which is designed to check the vehicle's identity, and generate and pre-distribute the private master keys of the vehicles. Before the network deploys, the trusted authority sets up the system parameters for all road side and onboard units.

Let G be a cyclic additive group generated by P, and G_T be a cyclic multiplicative group. G and G_T have the same order q which is a big prime number. Let $\psi : G_1 \times G_1 \to G_T$ be a bilinear map.

The trusted authority generates two master keys by randomly choosing $s_1, s_2 \in Z_q^*$ $= \{a \,|\, 1 \le a \le q-1\}$, and computes $P_{pub1} = s_1 P$ $P_{pub2} = s_2 P$ as its public keys.

The tamper-proof device of each vehicle is preloaded with the parameters (s_1, s_2). Each road side unit and vehicle are preloaded with the public parameters $\{G, G_T, q, P, P_{pub1}, P_{pub2}\}$.

Each vehicle is assigned a real identity, denoted as $RID \in G$, and a password, denoted as PWD, where RID uniquely identifies the vehicle, and the PWD is used by the tamper-proof device for authentication.

Pseudo Identity Generation

The tamper-proof device is designed to generate random pseudo identities and corresponding private keys based on identity-based cryptography. The tamper-proof device is designed according to the IBV scheme to be composed of three secure modules: one for authentication, another for pseudo identity generation, and a third for private key generation.

The authentication module protects the tamper-proof device even if it is physically held by the adversary. It authenticates a user's right to use the device's service. In the IBV scheme, the

RID is the vehicle's unique real identity and the password PWD can be generated in various ways. The PWD is generated by a trusted authority TA as the signature of RID.

The pseudo identity generation module is designed to generate a list of random pseudo identities from the authentication RID. Each pseudo identity ID is composed of ID_1 and ID_2. The formula to generate ID_1 and ID_2 is: $ID_1 = rP$, and $ID_2 = RID \oplus H(rP_{pub1})$ where r is a random nonce and r is changed each time so that ID_1 and ID_2 are different for each pseudo ID. $\oplus$ is an Exclusive-OR(XOR) operation. P and P_{pub1} are the public parameters preloaded by the trusted authority TA. ID_1 and ID_2 are used by the private key generation module.

The private key generation module uses identity based cryptography. There are two private keys corresponding to the two pseudo identity IDs, denoted as SK_1 and SK_2. And $SK_1 = s_1 ID_1$ and $SK_2 = s_2 H(ID_1 \| ID_2)$, in which $\|$ is the message concatenation operation.

A vehicle can go through the tamper-proof device using PWD and RID and get a list of pseudo identities $ID = (ID_1, ID_2)$ and the associated private keys $SK = (SK_1, SK_2)$. Note that the pseudo identities and the private keys can be generated offline by the tamper-proof device.

Message Signing

Vehicles can sign a message and send it to the roadside unit. In the IBV scheme, the message signing phase is designed as follows.

Suppose the traffic message, denoted by M_i, is generated by a vehicle, denoted by V_i. V_i uses the tamper-proof device to obtain a pseudo identity $ID^i = (ID_1^i, ID_2^i)$ and the corresponding private key $SK^i = (SK_1^i, SK_2^i)$. The vehicle V_i can compute the signature σ_i of the message M_i, where $\sigma_i = SK_1^i + h(M_i)SK_2^i$. Subsequently, the vehicle V_i sends the final message $\langle ID^i, M_i, \sigma_i \rangle$ to its neighboring roadside unit. These steps are done once every *100-300ms* according to the current

dedicated short range communication broadcast protocol (DSRC).

The signature of the IBV scheme has no need for any signature certificate to be sent along with the message because identity-based cryptography is used. Only a pseudo identity is sent, which has a length of *42* bytes, the sum of the lengths of ID_1^i and ID_2^i. This is much better than the ECDSA signature scheme of *IEEE 609.2* where a 125 byte certificate is contained in the message.

Secondly, the signature of the IBV scheme does not release any real identity information of the vehicle because a pseudo identity is used.

Batch Verification

When a road side unit (RSU) receives a traffic related message from a vehicle in the IBV scheme, the RSU must verify the signature of the message for two reasons: first to ensure the corresponding vehicle doesn't impersonate any other legitimate vehicle, and secondly to prevent the vehicle from disseminating bogus messages. Details of the verification process of the IBV scheme are illustrated in the following single signature verification and batch verification discussion.

Given the system public parameters $\{G, G_T, q, P, P_{pub1}, P_{pub2}\}$ assigned by the trusted authority TA and preloaded on each RSU and vehicle in the network according to the IBV scheme and given the message $\langle ID^i, M_i, \sigma_i \rangle$ sent by the vehicle V_i, then the signature σ_i can be validated by testing if $\Gamma(\sigma_i, P) = \Gamma(ID_1^i, P_{pub1})\Gamma(h(M_i)H(ID_1^i, ID_2^i), P_{pub2})$ as verified below using bi-linear maps bi-linear feature. Therefore, the computation cost for the RSU to verify a single signature is mainly one MapToPoint hash (Boneh, Lynn, & Shacham, 2001), one multiplication, and three pairing operations.

Given n distinct messages denoted as $\langle ID^1, M_1, \sigma_1 \rangle$, $\langle ID^2, M_2, \sigma_2 \rangle, ..., \langle ID^n, M_n, \sigma_n \rangle$ respectively, which are sent by n distinct vehicles denoted as V_1, V_2, ..., V_n, all signatures, denoted as $\sigma_1, \sigma_2, \ldots, \sigma_n$ are valid if $\Gamma(\sum_{i=1}^{n} \sigma_i, P) = \Gamma(\sum_{i=1}^{n} ID_1^i, P_{pub1})\Gamma(\sum_{i=1}^{n} h(M_i)HID^i, P_{pub2})$ in which HID^i denotes $H(ID_1^i \| ID_2^i)$. Detailed verification can be found in the IBV scheme paper (Zhang, Lu, Ho & Shen, 2008).

Batch verification in the IBV scheme reduces the verification delay and the computation cost of verifying n signatures by the RSU to n Map-ToPoint hash, n multiplication, *3n* addition, *n* one-way hash, and *3* pairing operations. Because the computation cost of a pairing operation is much higher than the cost of a MapToPoint hash and a multiplication cost, the verification time for multiple signatures is constant instead of linear with the size of the batch.

Security Analysis

The IBV scheme design is based on ID-based batch verification which can improve efficiency when many signatures must be verified. With the rising interest in pair-based cryptography, much research on identity-based signatures and performance of batch verification of identity-based signatures has been proposed. Here we focus on the IBV scheme security analysis; the basics of the cryptography foundation that supports the IBV scheme. The following three aspects of security analysis will be presented: message authentication, user identity privacy preservation, and traceability by the trusted authority.

- **Message authentication in the IBV scheme:** In review, the IBV signature $\sigma_i = SK_1^i + h(M_i)SK_2^i$ is a one-time identity-based signature. It is impossible to forge an IBV signature without knowing the private key SK_1 and SK_2. The NP-hard computational complexity of the Diffie-Hellman problem in G makes the private key SK_1 and SK_2 derivation from ID_1, P_{pub1}, P and $H(ID_1 \| ID_2)$ infeasible. The Diphantine equation is used to construct the IBV signature σ_i, and it is infeasible

to compute the private key SK_1 and SK_2 from knowledge of σ_i and $h(M_i)$.

- **Identity privacy preservation:** In the design of the IBV scheme preserving identity privacy is implemented using the ElGamal-type ciphertext construction. The real identity RID of a vehicle is used to construct two random pseudo identities ID_1 and ID_2, where $ID_1 = rP$ and $ID_2 = RID \oplus H(rP_{pub1})$, in which r is random number, P and P_{pub1} are public parameters that are preloaded on each roadside and vehicle unit. The master-key (s_1, s_2) is preloaded on each vehicle tamper-proof device. Thus, without the master key (s_1, s_2), it is impossible to get the real identity from the pseudo identity pair. Also, because the pseudo identities (ID_1, ID_2) in each signature are distinct, it is not helpful to compound the series of signatures to get the real identity. In other words, there is no linkability.

- **Traceability by the trusted authority:** In the proposed IBV scheme, the trust authority (TA) can authenticate the signature by using the master key (s_1, s_2) which is preloaded to each tamper-proof device on each vehicle. The value of RID can be computed by evaluating $ID_2 \oplus H(s_1ID_1)$ via the following steps: $ID_2 \oplus H(s_1ID_1) = RID \oplus H(rP_{pub1}) \oplus H(s_1rP)$ $P_{pub1} = s_1P$. From the above two equations, we get $ID_2 \oplus H(s_1ID_1) = RID \oplus H(rs_1P) \oplus H(s_1rP)$ in which $H(rs_1P) \oplus H(s_1rP) = 1$. Thus, we conclude that $ID_2 \oplus H(s_1ID_1) = RID$

OPEN ISSUES AND FUTURE DIRECTIONS

Further MANET/WSN research in industry and academia will make progress as it emphasizes cryptology, with each new cryptographic technique making its own impact in different case studies as well as in overall network security. Among the numerous possibilities are research in vehicular sensor networks, global positioning systems, and new wireless devices. The more cryptographic techniques available to the designer the greater will be the variations possible in the design. For instance, there are several alternatives to ElGamal type ciphertext which can be used to hide the real identity of a vehicle in the IBV scheme.

Some of the research deals with long term effects. Our current research explores more foundational aspects of security, such as the categorization of cryptographic primitives, security routing protocols, broadcast communication, group key and composite key management, and batch verification. Specifically, this chapter's survey of applied cryptography helps overcome the difficulty of understanding complicated security designs.

Other researchers focus on specific, real life problems. For instance, Wu & Chen (2008) investigate attacks and countermeasures within various network layers. Kannhavong (Kannhavong, Nakayama, Nemoto, & Kato, 2007) survey routing attacks and countermeasures against those attacks in MANETs.

Applying improved cryptographic algorithms to the security of MANETs/WSNs has already reduced the computational costs of cryptographic primitive operations and suggested less expensive dedicated cryptographic hardware for the future. For instance, in the IKM evaluation, the IKM's computation cost as an IBC scheme is shown to be less than RSA operations. And, Zhang (2006) pointed out that the Barreto approach can expedite the Tate pairing to be up to 10 times faster than previous methods, although the implementation is still under way.

Much research has been done related to MANET/WSN location privacy. Such techniques as association rules hiding, statistical combinatorics, and data mining can be helpful in the area of ad hoc network privacy. For example, Aggarwal & Yu (2008), address the privacy model and its algorithms against attacks using background knowledge and patterns.

More research is required in the areas of secure routing and key management in MANETs/WSNs. Key management is always a fundamental issue, and cryptographic techniques always play a major role in the handling of keys.

With multiple wireless networks becoming increasingly more important in our daily business lives, it is much easier both to form a MANET/WSN and to expect to run a greater variety of applications on that network. This great variety of possibilities requires network scalability, computer cost, and resource constraints to be considered on a case by case basis. For example, in vehicular sensor networks, power and processing constraints are less important than with MANETs. In addition, the vehicle has temporary infrastructure access via road-side units as seen in the IBV scheme.

Looking ahead, the use of symmetric cryptography and asymmetric cryptography, and their customized usage according to different network stages, will always be a challenge in covering the wide range of network layers in MANETs/WSNs. The current cryptography libraries will expand, and the number of available MANET simulators, and self-developed simulation studies, will increase. Future MANET/WSN security research will explore various ways to reduce complexity and increase abstraction levels as the field moves forward along with all other innovative technologies.

ACKNOWLEDGMENT

This work was supported in part by NSF grants ANI 0073736, EIA 0130806, CCR 0329741, CNS 0422762, CNS 0434533, CNS 0531410, and CNS 0626240.

REFERENCES

Aggarwal, C., & Yu, P. (2008). *Privacy-Preserving Data Mining Models and Algorithms*. New York: Springer Science Business Media, LLC.

Barreto, P., Kim, H., Bynn, B., & Scott, M. (2002). Efficient Algorithms for Pairing-Based Cryptosystems. In *Proceedings of CRYPTO* (pp. 354-368).

Boneh, D., & Franklin, M. (2001). Identify-Based Encryption from the Weil Pairing. In . *Proceedings of CRYPTO, 01*, 213–229.

Boneh, D., & Franklin, M. (2003, March). Identify-Based Encryption from the Weil Pairing. *SIAM Journal on Computing, 32*(3), 586–615. doi:10.1137/S0097539701398521

Boneh, D., Lynn, B., & Shacham, H. (2001). Short signatures from the weil pairing. In . *Proceedings of Asiacrypt, 2248*, 514–532.

Camenisch, J., Hohenberger, S., & Pedersen, M. (2007). Batch verification of short signatures. In . *Proceedings of EUROCRYPT, LNCS, 4514*, 246–263.

Camenisch, J., & Lysyanskaya, A. (2004). Signature schemes and anonymous credentials from bilinear maps. In . *Proceedings of Crypto, LNCS, 3152*, 56–72.

Capkun, S., Buttyan, L., & Hubaux, J. P. (2003, Jan.-Mar.). Self-Organized Public Key Management for Mobile Ad Hoc Networks. *IEEE Transactions on Mobile Computing, 2*(1), 52–64. doi:10.1109/TMC.2003.1195151

Carter, S., & Yasinsac, A. (2002, November). Secure Position Aided Ad hoc Routing Protocol. In *Proceedings of the IASTED International Conference on Communications and Computer Networks* (CCN02).

Cha, J. C., & Cheon, J. H. (2003). An identity-based signature from gap Diffie- Hellman groups. In *Proceedings of Public Key Cryptography* (pp. 18-30).

Chor, B., Goldwasser, S., Micali, S., & Awerbuch, B. (1985). Verifiable secret sharing and achieving simultaneity in the presence of faults. In Proceedings of *FOCS*.

Clausen, T., Adjih, C., Jacquet, P., Laouiti, A., Muhlethaler, A., & Raffo, D. (2003, June). Securing the OLSR Protocol. In *Proceeding of IFIP Med-Hoc-Net*. Dedicated Short Range Communications. (n.d.). *DSRC*. Retrieved (n.d.). from, http://grouper.ieee.org/groups/scc32/dsrc/index.html

Deng, H., Mukherjee, A., & Agrawal, D. (2004, April). Threshold and Identity-Based Key Management and Authentication for Wireless Ad Hoc Networks. In *Proceedings of the Int'l Conf. Information Technology: Coding and Computing (ITCC '04)*.

Desmedt, Y., & Frankel, Y. (1989, August). Threshold Cryptosystems. In . *Proceedings of CRYPTO, 89*, 307–315.

Dong, Q., Liu, D., & Ning, P. (2008). Pre-Authentication Filters: Providing DoS Resistance for Signature-Based Broadcast Authentication in Wireless Sensor Networks. In *Proceedings of ACM Conference on Wireless Network Security* (WiSec).

Douceur, J. R. (2002, March). The Sybil Attack. In *Proceedings of the First Int'l Workshop Peer-to-Peer Systems (IPTPS '02)* (pp. 251-260).

Du, W., et al. (2003). A Pairwise Key Predistribution Scheme for Wireless Sensor Networks. In *Proceedings of the 10th ACM Conf. Comp. Commun. Sec.* (pp. 42-51).

Eschenauer, L., & Gligor, V. D. (2002). A Key-Management Scheme for Distributed Sensor Networks. In *Proceedings of the 9th ACM Conf. Comp. and Commun. Sec.* (pp. 41-47).

Evans, J., Wang, W., & Ewy, B. (2006). Wireless networking security: open issues in trust, management, interoperation, and measurement. [IJSN]. *International Journal of Security and Networks, 1*(1-2), 84–94. doi:10.1504/IJSN.2006.010825

Fiat, A. (1989). Batch RSA. In *Proceedings of Crypto* (pp. 175-185).

Frankel, Y., Gemmell, P., MacKenzie, P. D., & Yung, M. (1997). Optimal resilience proactive public-key cryptosystems. In *Proceedings of FOCS*.

Gennaro, G., Jarecki, S., Krawczyk, H., & Rabin, T. (1996). Robust and efficient sharing of rsa functions. In *Proceedings of CRYPTO*.

Gouda, M. G., & Jung, E. (2004, March). Certificate Dispersal in Ad-Hoc Networks. In *Proceedings of the 24th IEEE Int'l Conf. Distributed Computing Systems* (ICDCS '04).

Herzberg, A., Jarecki, S., Krawczyk, H., & Yung, M. (1995). Proactive secret sharing, or: How to cope with perpetual leakage. In *Proceedings of CRYPTO*.

Hu, Y., Johnson, D., & Perrig, A. (2002). SEAD: Secure Efficient Distance Vector Routing in Mobile Wireless Ad-Hoc Networks. In *Proceedings of the 4th IEEE Workshop on Mobile Computing Systems and Applications (WMCSA'02)* (pp. 3-13).

Hu, Y., Perrig, A., & Johnson, D. (2002). Ariadne: A Secure On-Demand Routing for Ad Hoc Networks. In *Proceedings of MobiCom 2002*, Atlanta.

Huang, J., Buckingham, J., & Han, R. (2005, September). A Level Key Infrastructure for Secure and Efficient Group Communication in Wireless Sensor Networks. In *Proceedings of the 1st Int'l. Conf. on Security and Privacy for Emerging Areas in Commun. Net.* (pp. 249-260).

Ilyas, M. (2003). *The Handbook of Ad Hoc Wireless Networks*. Boca Raton, FL:CRC Press.

Jarecki, S., Saxena, N., & Yi, J. H. (2004, October). An Attack on the Proactive RSA Signature Scheme in the URSA Ad Hoc Network Access Control Protocol. In *Proceedings of the Second ACM Workshop Security of Ad Hoc and Sensor Networks (SASN '04)*.

Kannhavong, B., Nakayama, H., Nemoto, Y., & Kato, N. (October 2007). A Survey of Routing Attacks in Mobile Ad Hoc Networks. In *IEEE Wireless Communications* (pp. 85-91).

Kaufman, C., Perlman, R., & Speciner, M. (2002). *Network Security Private Communication in a Public World*. Upper Saddle River, NJ: Prentice Hall PTR, A division of Pearson Education, Inc.

Kaya, T., et al. (n.d.). Secure Multicast Groups on Ad Hoc Networks. (Oct. 2003). In *Proceedings of ACM SASN '03* (pp. 94-103).

Khalili, A., Katz, J., & Arbaugh, W. (2003, January). Toward Secure Key Distribution in Truly Ad Hoc Networks. In *Proceedings of IEEE Workshop Security and Assurance in Ad Hoc Networks*.

Kong, J., Zerfos, P., Luo, H., Lu, S., & Zhang, L. (2001, November). Providing Robust and Ubiquitous Security Support for Mobile Ad Hoc Networks. In *Proceedings of IEEE Int'l Conf. Network Protocols*.

Lee, J., Leung, V., Wong, K., Cao, J., & Chan, H. (2007, October). Key management issues in wireless sensor networks: current proposals and future developments. In *IEEE Wireless Communications* (pp. 76-84).

Liu, A., & Ning, P. (2008, April). TinyECC: A Configurable Library for Elliptic Curve Cryptography in Wireless Sensor Networks. In *Proceedings of the 7th International Conference on Information Processing in Sensor Networks (IPSN 2008), SPOTS Track* (pp. 245-256).

Lou, W., & Fang, Y. (2003). A Survey of Wireless Security in Mobile Ad Hoc Networks: Challenges and Available Solutions. In Chen, X. Huang, X., & Kluwer, D. (Eds.), *Ad Hoc Wireless Networks* (pp. 319-364). Academic Publishers.

Luo, H., & Lu, S. (2004). URSA: Ubiquitous and Robust Access Control for Mobile Ad-Hoc Networks. *IEEE/ACM Transactions on Networking. Vol. 12 No. 6* (pp. 1049-1063).

Mehuron, W. (1994). Digital Signature Standard (DSS). *U.S. Department of Commerce, National Institute of Standards and Technology (NIST), Information Technology Laboratory (ITL)*. FIPS PEB 186.

Menezes, A., Oorschot, P., & Vanstone, S. (1996). *Handbook of Applied Cryptography*. Boca Raton, FL: CRC Press.

Naccache, D., M'Raihi, D., Vaudenay, S., & Raphaeli, D. (1994). Can D.S.A be improved? complexity trade-offs with the digital signature standard. In . *Proceedings of EUROCRYPT, LNCS, 950*, 77–85.

Narasimha, M., Tsudik, G., & Yi, J. H. (2003, November). On the Utility of Distributed Cryptography in P2P and Manets: The Case of Membership Control. In *Proceedings of. IEEE Int'l Conf. Network Protocols*.

Panja, B., Madria, S. K., & Bhargava, B. (2006). Energy and Communication Efficient Group Key Management Protocol for Hierarchical Sensor Networks, SUTC '06. In *Proceedingsof. IEEE Int'l. Conf. Sensor Networks, Ubiquitous, and Trustworthy Comp.* (pp. 384-393).

Papadimitratos, P., & Haas, Z. J. (2003, January). Secure Link State Routing for Mobile Ad Hoc Networks. In *Proceedings of the 2003 Symposium on Applications and the Internet Workshops (SAINT'03)* (pp.379-383) Washington, DC, USA.

Perkins, C. (2001). *Ad Hoc Networks*. Reading, MA: Addison-Wesley.

Perrig, A., Canetti, R., Tygar, J., & Song, D. (2000). *The TESLA Broadcast Authentication Protocol*. Retrieved (n.d.). from, http://www.ece.cmu.edu/~adrian/projects/tesla-cryptobytes/tesla-cryptobytes.pdf

Perrig, A., Szewczyk, R., Wen, V., Culler, D., & Tygar, D. (2001, July). SPINS: Security protocols for sensor networks. In *Proceedings of Seventh Annual International Conference on Mobile Computing and Networks* (MobiCom).

Pervaiz, M. O., Cardei, M., & Wu, J. (2008). Routing Security in Ad Hoc Wireless Networks. In, HuangS., MacCallumD., & Du, D. (Eds.), *Network Security*. New York: Springer.

Salomaa, A. (1996). *Public-Key Cryptography*. New York: Springer-Verlag.

Sanzgiri, K., Dahill, B., Levine, B., Shields, C., & Belding-Royer, E. (2002). A Secure Routing Protocol for Ad Hoc Networks. In *Proceedings of IEEE International Conference on Network Protocols (ICNP)* (pp. 78-87).

Saxena, N., Tsudik, G., & Yi, J. (Dec. 2004). Identity-Based Access Control for Ad Hoc Groups. In *Proceedings of the Int'l Conf. Information Security and Cryptology*.

Shamir, A. (1979). How to Share a Secret. *Communications of the ACM, 22*(11), 612–613. doi:10.1145/359168.359176

Shamir, A. (1984). Identity Based Cryptosystems and Signature Schemes. In . *Proceedings of CRYPTO, 84*, 47–53.

Srinivasan, A., & Wu, J. (2008). A Survey on Secure Localization in Wireless Sensor Networks. In Furht, B. (Ed.), *Encyclopedia of Wireless and Mobile Communications*. Boca Raton FL: CRC Press, Taylor and Francis Group.

Standard, I. E. E. E. 1609.2. (n.d.). IEEE Trial-Use Standard for Wireless Access in Vehicular Environments - *Security Services for Applications and Management Messages*, 2006.

Striki, M., & Baras, J. (2004, June). Towards Integrating Key Distribution with Entity Authentication for Efficient, Scalable and Secure Group Communication in MANETs. In *Proceedings of IEEE ICC'04, vol.7* pp. 4377-4381.

Wu, B., Cardei, M., & Wu, J. (2008). A Survey of Key Management in Mobile Ad Hoc Networks. In, ZhengJ., ZhangY., & Ma, M. (Eds.), *Hdbk. of Rschon Wireless Security*. Hershey, PA: Idea Group Inc.

Wu, B., Chen, J., Wu, J., & Cardei, J. (2008). A Survey on Attacks and Countermeasures in Mobile Ad Hoc Networks, In Y. Xiao, X. Shen, & D. -Z. Du (Ed.), *Wireless/Mobile Network Security*. New York: Springer.

Xiao, Y., & Rayi, V. (2007). A survey of key management schemes in wireless sensor networks. *Computer Communications*, *30*(11-12), 2314–2341. doi:10.1016/j.comcom.2007.04.009

Yi, S., & Kravets, R. (2003, April). Moca: Mobile Certificate Authority for Wireless Ad Hoc Networks. In *Proceedings of Second Ann. PKI Research Workshop (PKI '03)*.

Yi, S., Naldurg, P., & Kravets, R. (2002). *Security Aware Ad hoc Routing for Wireless Networks*. (Tech. Rep. No. UIUCDCS-R-2002-2290). Illinois, USA: University of Illinois at Urbana-Champaign.

Yoon, H., Cheon, J. H., & Kim, Y. (2004). Batch verification with ID-based signatures. In *Proceedings of Information Security and Cryptology* (pp. 233-248).

Younis, M. F., Ghumman, K., & Eltoweissy, M. (2006). Location-Aware Combinatorial Key Management Scheme for Clustered Sensor Networks. In *IEEE Trans. Parallel and Distrib. Sys., vol.17*, 2006 (pp. 865-882).

Zapata, M. (200). *Secure Ad Hoc On-Demand Distance Vector (SAODV)*. Retreived 2004, from, http://personals.ac.upc.edu/guerrero/papers/draft-guerrero-manet-saodv-01.txt

Zhang, C., Lu, R., Ho, P., & Shen, X. (2008). An Efficient Identity-based Batch Verification Scheme for Vehicular Sensor Networks. In *Proceedings of IEEE INFOCOM*.

Zhang, F., & Kim, K. (2003). Efficient ID-based blind signature and proxy signature from bilinear pairings. In . *Proceedings of ACISP, LNCS, 2727*, 312–323.

Zhang, F., Safavi-Naini, R., & Susilo, W. (2003). Efficient verifiably encrypted signature and partially blind signature from bilinear pairings. In . *Proceedings of Indocrypt, LNCS, 2904*, 191–204.

Zhang, Y., Liu, W., & Lou, W. (2005, March). Anonymous Communications in Mobile Ad Hoc Networks. In . *Proceedings of IEEE INFOCOM*, *05*, 1940–1951.

Zhang, Y., Liu, W., Lou, W., & Fang, Y. (2006, Oct.-Dec.). Securing mobile ad hoc networks with certificateless public keys. In Proceedings of *IEEE Transactions on Dependable and Secure Computing* (Vol. 3, No. 4, pp 386-399).

Zhang, Y., Liu, W., Lou, W., Fang, Y., & Kwon, Y. (2005, May). AC-PKI: Anonymous and Certificateless Public-Key Infrastructure for Mobile Ad Hoc Networks. In *Proceedings of IEEE Int'l Conf. Comm* (pp. 3515-3519).

Zhou, L., & Haas, Z. (1999). Securing Ad Hoc Networks. *IEEE Network Magazine*, *13*(6), 24–30. doi:10.1109/65.806983

Zhu, S., Setia, S., & Jajodia, S. (2003). LEAP: Efficient Security Mechanisms for Large-Scale Distributed Sensor Networks. In *Proceedings of the 10th ACM Conf. Comp. and Commun. Sec.* (pp. 62-72).

Zhu, S., Xu, S., Setia, S., & Jajodia, S. (2003). LHAP: A Lightweight Hop-by-Hop Authentication Protocol for Ad-Hoc Networks. In Proceedings of the *23rd International Conference on Distributed Computing Systems Workshops (ICDCSW '03)*.

KEY TERMS AND DEFINITIONS

Asymmetric Cryptography: In public key or asymmetric cryptography, there is a pair of public/private keys. The private key is known only to the owner, while the public key is shared with others. One of the earliest public-key cryptographic techniques, known as RSA, was developed in the 1970s.

Batch Verification with ID-Based Signature: Although there are advantages to ID-based cryptography signature schemes based on pairing, the signature verifications are at least ten times slower than that of DSA or RSA. The batch verification of many signatures increases efficiency.

Certificate Authority: A certificate authority is an entity that issues digital certificates for use by other parties.

Digital Signature Based on RSA/DSA: The ElGamal signature is based on the difficulty of breaking the discrete log problem. DSA is an updated version of the ElGamal digital signature scheme published in 1994 by FIPS and was chosen as the digital signature standard (DSS). Digital signature, using the RSA/DSA algorithm, is popular for authentication or confirming the message's integrity. A digital signature scheme typically consists of three algorithms: a key generation algorithm, a signing algorithm, and a signature verifying algorithm.

Hash Chain: It is generated by a successive application of a hash function to a string. Due to the one-way property of secure hash functions, it is impossible to reverse the hash function. A hash chain is a method to produce many one-time keys from a single key, and keys are used in the reversed order of generation.

Hash Tree: It was originally invented to support the handling of many Lamport one-time signatures. At the top of a hash tree there is a top hash or master hash. Nodes higher in the tree are the hashes of their respective children.

HMAC Message Authentication Code: This type of message authentication code is calculated using a hash function in combination with a secret key. Usually in MANETs/WSNs, the hash functions chosen are mostly MD5 or SHA-1. It can also be used to ensure that an unencrypted message retains its original content by calculating the message HMAC using a secret key.

Identity-Based Cryptography: This is a type of public-key cryptography. The first identity-based cryptography, developed by Adi Shamir in 1984, uses the identity of the user as a public key. Modern schemes include Boneh/Franklin's pairing-based encryption scheme.

Random Nonce: In the network, a timestamp or random number (nonce) is used to make packets fresh and prevent a replay attack. Cryptographic pseudo random generators typically have a large pool of seed values.

Shared Key: Less computationally intense symmetric key algorithms are used more often than asymmetric algorithms. In practice, asymmetric algorithms are hundreds of times slower than symmetric key algorithms. The most common are AES, RC4 and IDEA.

Symmetric Cryptography: The encryption key is closely related or identical to the decryption key. In practice, keys represent a shared secret between two or more parties that can be used to maintain private communication.

Chapter 13
Privacy and Trust Management Schemes of Wireless Sensor Networks:
A Survey

Riaz Ahmed Shaikh
Kyung Hee University, Korea

Brian J. d'Auriol
Kyung Hee University, Korea

Heejo Lee
Korea University, Korea

Sungyoung Lee
Kyung Hee University, Korea

ABSTRACT

Until recently, researchers have focused on the cryptographic-based security issues more intensively than the privacy and trust issues. However, without the incorporation of trust and privacy features, cryptographic-based security mechanisms are not capable of singlehandedly providing robustness, reliability and completeness in a security solution. In this chapter, we present generic and flexible taxonomies of privacy and trust. We also give detailed critical analyses of the state-of-the-art research, in the field of privacy and trust that is currently not available in the literature. This chapter also highlights the challenging issues and problems.

INTRODUCTION

Security solutions based on cryptography are mainly used to provide protection against security threats, such as fabrication and modification of messages, unauthorized access, etc. For this purpose, assorted security mechanisms such as authentication, confidentiality, and message integrity are used. Additionally, these security mechanisms highly rely on a secure key exchange mechanism [Shaikh et al., 2006a]. However, these cryptography based security mechanisms alone are not capable of pro-

DOI: 10.4018/978-1-61520-701-5.ch013

Figure 1. Relationship between privacy, cryptographic-based security and trust

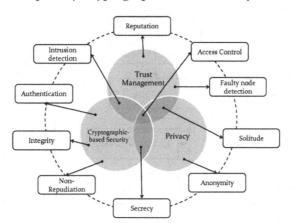

viding robustness, reliability and completeness in a security solution. They can only be achieved by incorporating privacy and trust features as described below.

Privacy features such as route anonymity of the data packets, identity anonymity of nodes and their locations are mainly used to provide protection against security threats such as traffic analysis and eavesdropping. Additionally, these privacy features could also be used to provide protection against security threats such as camouflage [Walters et al., 2006]. Therefore, the incorporation of these privacy features with cryptographic-based security mechanisms add to the degree of completeness of a security solution.

Trust management features, such as reputation is used to provide corresponding access control based on the judgment of the quality of sensor nodes and their services [Walters et al., 2006]. Also, it is used to provide complete reliable routing paths which are free from any malicious, selfish and faulty nodes [Liu et al., 2004]. Therefore, incorporation of trust management features with cryptographic-based security mechanisms help in increasing robustness and reliability of the overall security solution.

The soft relationship between privacy, trust, and cryptographic-based security is shown in Figure 1. This figure illustrates the related aspects of these terms with other commonly found terms used in the security domain. For example, secrecy is a mutual feature of cryptographic-based security and privacy aspects. In order to provide secrecy (also referred to as confidentiality), cipher algorithms (such as AES, DES) are used to prevent disclosure of information from any unauthorized entity. Similarly, an intrusion detection system may need a trust management feature such as reputation as well as a cryptographic-based security feature such as integrity checking to detect any malicious nodes. In like manner, solitude, which is used to isolate a node from the network either willingly or forcefully, is a mutual feature of trust and privacy aspects.

Current research so far seems to intensively focus on the cryptographic-based security aspects of wireless sensor networks. Many security solutions have been proposed such as SPINS [Perrig et al., 2002], TinySec [Karlof et al., 2004], LEAP [Zhu et al., 2003] and LSec [Shaikh et al., 2006b] etc., but surprisingly, less importance is given to privacy and trust issues of wireless sensor networks. Privacy and trust are as important as other security issues and they also contribute in increasing the degree of completeness and reliability of a security solution as discussed above.

In this chapter, we focus on the importance of privacy and trust establishment in wireless sensor networks. In Sections 2 and 3, we present generic and flexible taxonomies of privacy and

Figure 2. Taxonomy of privacy

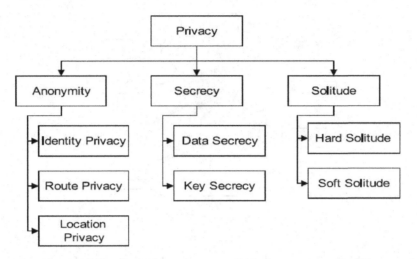

trust respectively. These taxonomies are based on our specific experience with wireless sensor networks. Apart from these taxonomies, these sections also contain a detailed description of the privacy and trust issues of wireless sensor networks. This description is currently not available in the literature [Chan & Perrig, 2003; Djenouri et al., 2005; Perrig et al., 2004; Shaikh et al., 2006c; Walters et al., 2006]. Also, this chapter contains critical analyses of the state-of-the-art research work. Additionally, this chapter also highlights the challenging problems and issues in the field of privacy and trust in wireless sensor networks. Finally, last section concludes the chapter.

PRIVACY

Taxonomy

Privacy generally refers to "ability to control the dissemination of information about oneself" [Anderson, 2001]. In the wireless sensor network domain, so far privacy is mainly provided from anonymity [Misra & Xue, 2006; Ozturk et al., 2004; Wadaa et al., 2004;] and/or secrecy perspective [Karlof et al., 2004; Park & Shin, 2004; Perrig et al., 2002; Zhu et al., 2003]. However, neither anonymity nor secrecy is capable of providing complete privacy. In real life, we observe that complete privacy is gained through three independent but interrelated ways; *anonymity*: when an individual's true identity remains unidentified; *secrecy*: when an individual or a group's information remains protected from disclosure, and *solitude*: when one needs a temporal isolation in which an individual can not serve any request [DeCew & Judith, 2006]. Therefore, in order to achieve full privacy, we need to ensure that all these aspects: anonymity, secrecy, and solitude should be addressed. These three elements are further divided into sub categories as shown in Figure 2.

Anonymity provides three types of privacy protections, identity privacy, route privacy, and location privacy [Zhu et al., 2004].

- **Identity privacy:** No node can get any information about the source and destination nodes. Only the source and destination nodes can identify each other. Also, the source and destination nodes have no information about the real identities of the intermediate forwarding nodes.
- **Route privacy:** No node can predict the information about the complete path (from

source to destination) of the packet. Also, a mobile adversary can not get any clue to trace back the source node either from the contents and/or directional information of the captured packet(s).

- **Location Privacy:** No node can get to know any information about the location (either in terms of physical distance or number of hops) of the sender node except the source, its immediate neighbors and the destination.

Secrecy generally refers to the practice of hiding some information. Information is classified into two categories; one is the secrecy of actual sensed data (also referred as data confidentiality) forwarded by a sensor node to the specific destination and the other is key secrecy that is required to cipher data.

Solitude refers to the condition that a node goes into the state of isolation for a specific period of time. During that interval, the node cannot fulfill jobs nor can it provide services such as packet forwarding to the other nodes. We have categorized solitude into two types. *Soft solitude* refers to the node's decision to be in the solitude state. *Hard solitude* means that other node(s) in a compact or a command node decide to isolate a particular node.

Table 1 gives the classification of proposed privacy schemes (i.e. SAS & CAS [Misra and Xue, 2006], PFR PFR PFR [Ozturk et al., 2004],

PSR PSR PSR [Kamat et al., 2005], SIGF SIGF SIGF [Wood et al., 2006], CEM CEM CEM [Ouyang et al., 2006], GROW GROW GROW [Xi et al., 2006], DIRL DIRL DIRL [Shaikh et al., 2008a]) of wireless sensor networks based on our proposed taxonomy. These schemes are discussed comprehensively in next section.

State-of-the-Art Research

Current research so far sees privacy either from a secrecy perspective or from an anonymity perspective. As mentioned earlier, full privacy consists of three elements: secrecy, anonymity, and solitude. Unfortunately, no solution, in the wireless sensor network domain, can guarantee the triumph of all these three elements in a single solution. In this section, we present the critical analysis of current state-of-the-art research work done so far in the field of privacy in wireless sensor networks.

Anonymity Schemes

In the wireless sensor network domain, some applications demand anonymity, for example, a panda-hunter application [Ozturk et al., 2004]; in which the *Save-The-Panda* organization has deployed sensor nodes to observe the vast habitat for pandas. Whenever any sensor node detects some panda it will make observations, e.g. activity, location, and periodically forward those to the sink node via some multi-path routing strategy.

Table 1. Application of privacy taxonomy

			SAS & CAS	PFR	PSR	SIGF	CEM	GROW	DIRL
Anonymity		Identity	Yes	No	No	No	No	No	Yes
		Route	Depending on routing scheme	Yes	Yes	Yes	Depending on routing scheme	Yes	Yes
		Location	No	Yes	Yes	Yes	Yes	Yes	Yes
Secrecy		Data	Yes	NA	NA	Yes	NA	NA	Yes
		Key	Yes	NA	NA	Yes	NA	NA	Yes
Solitude		Soft	No	No	No	No	No	No	No
		Hard	No	No	No	No	No	No	No

In this scenario, a hunter can try to capture the pandas by back-tracing the routing path until it reaches the source sensor nodes. Therefore, in order to prevent the hunter from back-tracing, the route and location anonymity mechanisms must be enforced. Similarly, in a battlefield application scenario, "the location of a soldier should not be exposed if he initiates a broadcast query" [Xi et al., 2006].

Traditionally, a number of various anonymity schemes have been proposed such as DC-Network [Chaum, 1988], Crowds [Reiter & Rubin, 1998], Onion Routing [Reed et al., 1998], Hordes [Shields & Levine, 2000], ARM [Seys & Preneel, 2006] etc. The most common approach used in these schemes is the employment of cover traffic. Cover traffic represents the dummy packets that are transmitted along with the original packets to the different destinations. In addition to cover traffic, some schemes use pseudonyms for assigning identities to the nodes. The objective of using cover traffic is to make an attacker clueless about the original packet and its destination. This kind of approach is not often suitable for traditional wired networks because it can cause a large amount of traffic overhead. Also these schemes have high computational cost mainly due to encryption and decryption of not only of the original packets but also of the dummy packets as well. These common problems make traditional anonymity schemes especially unsuitable for wireless sensor networks that operate in highly resource constraint environment.

PFR: Ozturk et. al. (2004) proposed a phantom routing scheme for wireless sensor networks that helps in preventing the location of a source node from the attacker. In this scheme, each message reaches the destination in two phases; 1) a walking phase in which the message is unicasted in random fashion within first h_{walk} hops and 2) a message flooding phase in which the message is flooded using the baseline flooding technique. In the first phase, the authors have introduced a bias in the random selection that makes it a directed walk from a pure random walk. The purpose of this approach is to minimize the chances of creating routing loops. However, this approach may incur delays. For example, because of a directed walk, the message may always move away from the base station. Thus, this approach is suitable for the applications that are not time sensitive. The main advantage of this scheme is that source location privacy protection improves as the network size and intensity increase because of the high path diversity. However, if the network size increases, the flooding phase consumes more energy, which in turn reduces the life time of the network.

PSR: Kamat et al. (2005) proposed a phantom single-path routing (PSR) scheme that works in a similar fashion as original phantom routing scheme [Ozturk et al., 2004]. They refer to the earlier one phantom-flood routing (PFR) scheme. The major difference between these two schemes is that, after the walking phase, the packet will be forwarded to the destination via a single path routing strategy such as shortest path routing mechanism. This scheme consumes less energy and requires marginally higher memory (each node need to maintain routing tables) as compared to the phantom-flood routing scheme. The major limitation of this scheme is that it only provides protection against a weaker adversary model.

SAS & CAS: Misra & Xue (2006) proposed two schemes for establishing anonymity in clustered wireless sensor networks. One is called Simple Anonymity Scheme (SAS) and other is called Cryptographic Anonymity Scheme (CAS). Both schemes are based on various assumptions such as sensor nodes are similar, immobile, consist of unique identities, and share pair-wise symmetric keys. The SAS scheme uses dynamic pseudonyms instead of a true identity during communications. Each sensor node needs to store a given range of pseudonyms that are non-contiguous. Therefore, the SAS scheme is memory inefficient. However, the CAS scheme uses keyed hash functions to generate pseudonyms. This makes it more memory efficient as compared to the SAS, but it requires more computation power.

SIGF: Wood et al. (2006) have proposed a configurable secure routing protocol family called Secure Implicit Geographic Forwarding (SIGF) for wireless sensor networks. The SIGF scheme is based on Implicit Geographic Forwarding (IGF) protocol [Blum et al., 2003], in which, a packet is forwarded to the node that lies within the region of a 60° sextant, centered on the direct line from the source to the destination. This approach reduces the path diversity and leads to only limited route anonymity is achieved. The SIGF protocol is mainly proposed by keeping security in mind. That is why some of the privacy aspects have not been covered such as identity privacy. Also, this protocol is unable to provide data secrecy in the presence of identity anonymity. Another drawback of this protocol is that, when there is no trusted node within a forwarding area, it will forward packet to the un-trusted node. So, the reliability of a path is affected.

GROW: Xi et al. (2006) proposed a Greedy Random Walk (GROW) scheme for preserving location of the source node. This scheme works in two phases. In the first phase, the sink node will set up a path through random walk with a node that acts as a receptor. Then the source node will forward the packet(s) towards the receptor in a random walk manner. Once the packet(s) reaches the receptor, it will forward the packet(s) to the sink node through the pre-established path. Here the receptor is acting as a central point between the sink and the source node for every communication session. The selection criteria of trustworthy receptors are not defined.

CEM: Ouyang et al. (2006) proposed a Cyclic Entrapment Method (CEM) to minimize the chance of an adversary to find out the location of the source node. In the CEM, when the message is sent by the source node to the base station, it activates the pre-defined loop(s) along the path. An activation node will generate a fake message and forward it towards the loop and original message is forwarded to the base station via specific routing protocol such as shortest path. Energy consumption in the CEM scheme is mainly dependent upon the number of loops in the path and their size.

DIRL: Shaikh et al. (2008a) have proposed a data privacy mechanism and two identity, route, and location privacy algorithms (IRL and r-IRL) for wireless sensor networks. We refer to this work as DIRL. The unique thing about the DIRL scheme is that it provides data secrecy in the presence of identity privacy. Also, the DIRL scheme assures that all packets will reach their destination by passing through only trusted intermediate nodes. From the memory and energy consumption point of view, the DIRL scheme is not very good in comparison with some other existing schemes such as PSR [Kamat et al., 2005]. However, at the modest cost of energy and memory, the DIRL scheme provides more privacy features (data, identity, route, and location) along with the attributes of trustworthiness and reliability.

Table 2 gives the summary of proposed privacy preserving schemes, i.e. PFR [Ozturk et al., 2004], PSR [Kamat et al., 2005], SAS & CAS [Misra & Xue, 2006], SIGF [Wood et al., 2006], CEM [Ouyang et al., 2006], GROW [Xi et al., 2006] and DIRL [Shaikh et al., 2008a].

Secrecy

Secrecy is generally used to hide the contents of the message from unauthorized access, but it is not used to hide the source and destination identity. Overall, secrecy is achieved through the combination of different security mechanisms such as authentication and confidentiality. Additionally, these security services highly rely on a secure key exchange mechanism [Shaikh et al., 2006a]. Quite recently, many security solutions have been proposed such as SPINS [Perrig et al., 2002], LEAP [Zhu et al., 2003], TinySec [Karlof et al., 2004], LiSP [Park and Shin, 2004], SBKH [Michell & Srinivasan, 2004], LSec [Shaikh et al., 2006b], MUQAMI [Raazi et al., 2007], etc. These provide various security services such

Table 2. Summary of privacy preserving schemes of WSNs

	PFR	PSR	SAS & CAS	SIGF	CEM	GROW	DIRL
Required information for routing	ID of destination	Routing table (e.g. dest. ID, # of hops etc.)	Depending on a routing scheme	Own, dest., & neighborhood locations	Depending on a routing scheme	Routing table (e.g. dest. D, receptor ID etc.)	Own, dest., & neighborhood locations
Transmission Mechanism	1st phase: Point-to-point; 2nd phase: Broadcast	Point-to-point	Depending on a routing scheme	Point-to-point	Point-to-point	Point-to-point	Point-to-point
Decision place for forwarding	1st phase: Transmitter; 2nd phase: Receiver	Transmitter	Depending on a routing scheme	Transmitter	Transmitter	Transmitter	Transmitter
Criteria for forwarding packet to next hop	1st phase: random; 2nd phase: flooding	1st phase: random; 2nd phase: shortest in terms of hops	Depending on a routing scheme	Randomly select any trusted node lies in forwarding region	Depending on a routing scheme	1st phase: random; 2nd phase: Predefined path	Randomly select any trusted node

as authentication, confidentiality, and message integrity. A high level qualitative comparison of these schemes is shown in Figure 3. This figure illustrates that the authentication, confidentiality, and integrity are well accommodated. However others (access control, availability, and non-repudiation) are not. A detailed description of each scheme is given below.

LEAP: Zhu et al. (2003) have proposed the security mechanisms: Localized Encryption and Authentication Protocol (LEAP), and a key management protocol for large scale distributed wireless sensor networks. In order to meet different security requirements, LEAP provides the support of four types of keys for each sensor node: 1) unique secret key that is shared between each sensor node and the base station, 2) pairwise key shared between each pair of neighboring nodes, 3) cluster key shared with multiple neighboring nodes, and 4) a group key that is shared by all the

Figure 3. Comparison of security protocols

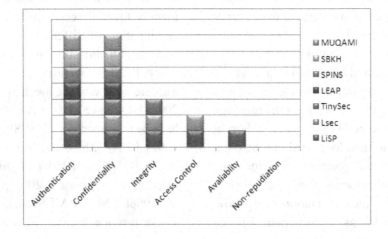

nodes in the network. If a node has d neighbors, it needs to store one individual key, d pairwise keys, d cluster keys and one group key. The authors have employed the uTESLA (Perrig et al., 2002) protocol for broadcast authentication. However, in order to add more security such as inter-node authentication, the authors have used a hop-by-hop authentication strategy in which each node must authenticate the packet before forwarding it to the next hop. For this purpose, each node needs to store a one-way key chain of length L and most recent authenticated key of each neighbor. Therefore, each node needs to store total $3d+2+L$ keys.

SPINS [Description of this protocol has been taken from our published paper [Shaikh et al., 2006]: Perrig et al., (2002) have proposed a security protocols suite called SPINS for wireless sensor networks. SPINS consist of two building blocks, SNEP and uTESLA. SNEP provides data confidentiality, two-party data authentication, and data freshness where as uTESLA provides authenticated broadcast for severely resource constraint environments. For data confidentiality a symmetric encryption mechanism is used in which a secret key called the master key is shared between sensor nodes and the base station. SNEP uses a one-time encryption key that is generated from the unique master key. SNEP uses MAC function for two party authentications and checking data integrity. SPINS is based on a binary security model which means that either it provides maximum security or it does not provide any security. Usage of source routing scheme in SPINS makes the network vulnerable to traffic analysis [Undercoffer et al. 2002].

TinySec: Karlof et al. (2004) have proposed a secure architecture for wireless sensor networks called TinySec. TinySec is the first fully implemented link layer cryptography-based security protocol that provides authentication, integrity, and confidentiality by adding less than 10% of energy, latency, and bandwidth overhead [Shaikh et al., 2006c]. TinySec architecture comprises of

two modes; 1) Authenticated encryption (TinySec-AE) mode, in which TinySec encrypts the payload (data) and authenticate the packet with a MAC. 2) Authentication only (TinySec-AH) mode, in which TinySec authenticates the entire packet with the MAC. TinySec protocol is tightly tied-in with Berkeley TinyOS. Therefore, it can not be used for general sensor network model [Perrig et al., 2004].

LiSP: Park and Shin (2004) have proposed the Lightweight Security Protocol (LiSP) that makes a tradeoff between security and resource consumption through efficient re-keying mechanism. This re-keying mechanism has a number of features such as: efficient key broadcasting, which does not require any retransmissions or acknowledgements; implicit authentication of new keys without incurring any additional overhead; seamless key refreshments; detection and recovery of lost keys. The LiSP protocol does not have any control packets or any type of retransmission that makes it energy efficient and secure against DoS attacks. LiSP provides the support of authentication, confidentiality, data integrity, access control and availability [Shaikh et al., 2006c]. In LiSP, each node need to save atleast eight keys therefore it is memory efficient. Also, the computation cost of LiSP is very low because on average it needs to compute less then three hash computations.

SBKH: Michell & Srinivasan, (2004) have proposed lightweight security protocol called State Based Key Hop (SBKH) for low power devices such as sensor nodes. SBKH achieves authentication, confidentiality, and integrity. In this protocol, two communicating nodes share common knowledge about RC-4 states. These states are used to generate cipher streams. These states remain the same for the pre-defined duration known only to two communicating nodes and will be reinitialized only when the base key changes. This approach gives the benefit of providing less computation overhead as compared to the traditional WEP and WPA 1.0 security solutions where RC-4 states are reinitialized for every packet.

However, the security strength of this scheme is mainly depended on a stronger key management and distribution scheme.

LSec: Shaikh et al., (2006b) have proposed the Lightweight Security (LSec) protocol for wireless sensor networks. LSec provides authentication, access control, confidentiality, and integrity of sensor nodes with simple key exchange mechanism. It works in three phases: 1) Authentication and authorization phase that is performed by using symmetric scheme, 2) Key distribution phase which involves sharing of random secret key by using asymmetric scheme and 3) Data transmission phases which involves transmission of data packets in an encrypted manner. LSec is memory efficient that requires 72 bytes to store keys. Also, it introduces 74.125 bytes of transmission and reception cost per connection.

MUQAMI: Raazi et al., (2007) have proposed a key management scheme for clustered sensor networks called MUQAMI. In MUQAMI, the responsibility of key management is divided among a small fraction of nodes within a cluster. Also, during the normal network operation, this responsibility can be transferred from one node to another with minimal overhead. This eradicates any single point of failure in the network. Also, this scheme is highly scalable and it eradicates all the inter-cluster communication. Lastly, it does not require all nodes to participate in key management, which reduces the security overhead substantially. This scheme is mainly designed for large-scale sensor networks. This scheme is more susceptible to collusion attacks [Moore, 2006] than other schemes such as LEAP+ [Zhu et al., 2006]. Its parameters should be chosen carefully in order to avoid collusion attacks.

Solitude

As we mentioned earlier, so far the concept of solitude is not used for achieving privacy in the wireless sensor networks. The concept of solitude could be applied in different ways. For example,

soft solitude is achieved whenever any node does not want to participate in communication due to any reason such as to preserve energy etc, then that node will broadcast message to all its neighboring nodes. That message contains the node's state change information to solitude state for specific time. Once this message is received by the member nodes, they will no longer consider the solitude node for the purposes of forwarding a packet; virtually considers that the node is an un-trusted node. After the passage of some time, a node's state will reset to original (trusted or un-trusted) state. In order to provide protection against spoofing, receiving node will first perform an Angle of Arrival (AoA) and single strength check [Durresi et al., 2007], which will ensure that the packet was sent by the legitimate source node. Many other AoA-based localization techniques have been specifically proposed for sensor networks such as [Nasipuri & Li, 2002, Rong & Sichitiu, 2006]. Any one of them could be used. The pseudo-code of a Soft Solitude Algorithm (SSA) is given Algorithm 1.

Hard solitude could also be achieved with the help of trust values. If any node is considered to be un-trusted based on its trust value, that node will not be able to participate in a communication for a given period of time. For example, some intrusion detection techniques [Michiardi & Molva, 2002; Buchegger & Boudec, 2002] proposed for ad-hoc networks have the ability to gradually isolate the node(s) in case the node(s) are found to be

Algorithm 1. SSA

```
1: Receive Packet Pkt
2: Get NID = GetNodeID(Pkt)
3: if checkAoA(Pkt) = true then
4: Set timer Δt;
5: Set state = NID_state;
6: while Δt = true do
7: NID_state remain untrusted;
8: end while
9: NID_state = state;
10: else
11: Detect spoof_pkt;
12: end if
```

Figure 4. Taxonomy of trust

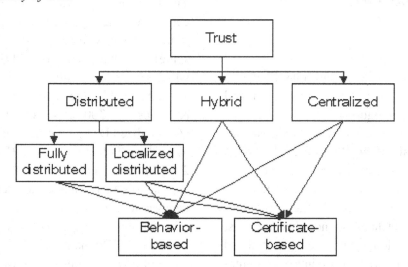

malicious or untrusted. However, those schemes require continuous monitoring and collection of information about intrusions at various places that increases overhead, and make them unsuitable for wireless sensor networks [Bhuse, 2007].

Challenging Issues

The main challenging issue that we are facing here is: "Having limited memory, computational capability and transmission/reception power of wireless sensor networks, is it practically possible to achieve full privacy?" It is not feasible to maintain complete privacy all the time in wireless sensor networks, not only due to the technological constraints but also due to the changing characteristics of application and the network itself. In general, privacy is a dynamic need at every level in wireless sensor networks. Applications, nodes and communication packets require different levels of privacy throughout their operation. Thus, privacy cannot be maintained at the same level all the time and an effective privacy scheme should efficiently cater for the dynamic privacy needs at all levels in wireless sensor networks. Hence, in this type of environment, the best way to achieve privacy is in a flexible and adaptive manner. Here flexible

means that the scheme should support variable levels of privacy and adaptive means that with respect to time and demand, the solution should automatically adjust the required level of privacy. This kind of flexibility and adaptability is currently not available in the presently proposed privacy solutions of wireless sensor networks.

Another challenging problem is to protect various aspects of privacy (as mentioned earlier) against three different ways of privacy disclosure mechanisms [Walters et al., 2006]. One is by traffic analysis [Deng et al., 2004], second is by eavesdropping [Djenouri et al., 2005] and third is by camouflage [Walters et al., 2006]. Traffic analysis means that an attacker can get access to the information like "who is talking to whom?" based on that information, the attacker can infer the role and activities of each sensor node in the network. Eavesdropping means that an attacker can get the information like "what nodes are talking about?" and camouflaging means that an attacker can masquerade (via newly inserted nodes or via compromised nodes) as a normal node to attract the packets to pass through it.

TRUST MANAGEMENT

Taxonomy

Trust management schemes are classified into three categories: centralized, distributed and hybrid as shown in Figure 4.

Centralized trust management (CTM) schemes (e.g. [Blaze et al., 1999; Resnick et al., 2000]) consist of a single globally trusted server that determines the trust values of every node in the network. This gives the benefit of lesser computational overhead at the sensor node because most of the trust calculation is performed at centralized trusted server that has no constraints of computational power and memory. This approach however has the drawbacks of a single point of failure, which makes it least reliable. Also, it suppresses the underlying fact that different nodes may have different trust values about a particular given node [Theodorakopoulos & Baras, 2006]. For large scale sensor networks, centralized trust schemes are not suitable because the total routing cost for the exchange of trust values of a sensor node with the base station is quite energy expensive, especially when the base station is located far from the node. Therefore centralized approach introduces large communication overhead in the sensor network.

Distributed trust management (DTM) schemes (e.g. [Boukerche et al., 2007; Ganeriwal & Srivastava, 2004]) also do not work well for large-scale sensor networks. In the distributed approach, every node locally calculates the trust values of all other nodes in the network that increases the computational cost. Also each node needs to maintain an up-to-date record about the trust values of the entire network in the form of a table. The size of the table is directly proportional to the size of the network which results in a large memory consumption. Each sensor node maintains its own trust record and that gives the benefit of less communication overhead because a node does not need to contact with some centralized server.

The distributed approach is more reliable than the centralized one because it has no single point of failure. In the wireless sensor network domain, some researchers use restricted DTM approach, in which sensor nodes only maintains the trust value about its neighboring nodes only e.g. [Ganeriwal & Srivastava, 2004]. We refer to that approach as a localized DTM approach and the earlier one as a fully DTM approach, e.g. [Boukerche et al., 2007]. The major drawback of the localized DTM approach is that it introduces delay and dependency whenever any node wants to evaluate trust of distant nodes. This is due to the fact that trust is established "dynamically at runtime using the chain of trust relationships between neighboring nodes" [Ganeriwal & Srivastava, 2004].

Hybrid trust management (HTM) schemes (e.g. [Krishna & Maarof, 2003; Shaikh et al., 2006a]) contain the properties of both centralized as well as distributed trust management approaches. The main objective of this approach is to reduce the cost associated with trust evaluation as compared to distributed approaches. This scheme is used with clustering schemes, in which cluster-head acts as a central server for the whole cluster. This approach is more reliable than the centralized one but less reliable than the distributed one. Each node needs to maintain the record of only member nodes, which gives the benefit of less memory consumption than the distributed approach. For intra-cluster communication, nodes need to contact the cluster head. It introduces more communication overhead in the network as compared to the distributed one.

The advantages and disadvantages of all three approaches are summarized in Table 3. All these three trust management approaches are further classified into two categories [Aivaloglou et al., 2007]: certificate-based trust management approach and behavior-based trust management approach. In the certificate-based trust management approach, trust is mainly based on the provision of a valid certificate assigned to a target node by a centralized certification authority or by other

Table 3. Advantages and disadvantages of trust management approaches

	Advantages	**Disadvantages**
Centralized	• Least computational overhead. • Least memory usage.	• Least reliable (single point of failure). • Most communication overhead.
Distributed	• Most Reliable (no single point of failure). • Scalable.	• Most computational overhead. • Most memory usage.
Hybrid	• Less communication overhead than centralized. • Less memory consumption than distributed. • Less computational overhead than distributed. • More reliable and scalable than centralized.	• Large computational overhead then centralized. • Large memory requirement than centralized. • Less scalable and reliable than distributed.

Table 4. Application of trust taxonomy

		Certificate-based	**Behavior-based**
Centralized		-	-
Hybrid		-	GTMS [Shaikh et al.,2009]
		[Aivaloglou et al., 2007]	
Distributed	Fully	ATRM [Boukerch et al., 2007]	-
	Localized	-	PLUS [Yao et al., 2006], RFSN [Ganeriwal & Srivastava, 2004], T-RGR [Liu et al., 2007]

trusted issuer. In the behavior-based trust management approach, an entity calculates the trust values by continuous direct or indirect monitoring of other nodes.

Table 4 gives the classification of proposed trust management schemes of wireless sensor networks based on our proposed trust taxonomy. These schemes are discussed in more comprehensive manner in next section.

State-of-the-Art Research

Research on trust management schemes for wireless sensor networks is in its infancy state. Few schemes have been proposed that are discussed below in chronological order. Our discussion in [Shaikh et al., in press] is extended with additional detail below.

RFSN: Ganeriwal et al. (2004, 2008) have proposed the Reputation based Framework for Sensor Network (RFSN) where each sensor node maintains the reputation for neighboring nodes.

On the basis of that reputation trust values are calculated. The RFSN scheme follows the localized distributed approach and borrows some design features from several existing works in the literature. It uses Bayesian formulation for representing reputation of a node. The RFSN scheme assumes that the node would have enough interactions with the neighbors so that the reputation (beta distribution) can reach to a stationary state. If the mobility is at a higher rate, reputation information will not stabilize and it may degrade its performance. Therefore, this kind of architecture is most suitable for stationary networks as compared to the mobile networks. In the RFSN scheme, nodes are classified into two categories: cooperative and not cooperative. Trust formulation approach of RFSN scheme can not cope with uncertainty situations [Chen et al., 2007]. Also, in their scheme no node is allowed to disseminate bad reputation information. It is resilient against bad-mouthing [Sun et al., 2008] and ballot stuffing attacks [Ganeriwal and Srivastava, 2004] but at the cost of system efficiency, as nodes cannot share bad experiences with each other.

ATRM: Boukerch et al. (2005, 2007) have proposed the Agent based Trust and Reputation Management (ATRM) scheme for wireless sensor networks. The ATRM is based on a clustered wireless sensor networks and calculates trust in a fully distributed manner. Every sensor node holds a local mobile agent that is responsible for administrating trust and reputation of hosting node. ATRM assumes that there is a trusted authority which is responsible for generating and launching mobile agents. It also assumes that mobile agents are resilient against malicious nodes that try to steal or modify information carried by the agent. The major advantage of the ATRM scheme is that they use mobile agents for trust calculation which reduces the bandwidth consumption and time delay. The ATRM scheme work in two phases: 1) Network Initialization phase and 2) Service offering phase. In the first phase, the Agent Launcher (AL) distributes the mobile agents called Trust and Reputation Assessor (TRA) to each node. As long as node has local TRA, it is in service offering phase, in which it is ready to provide trust and reputation management services. This phase is composed of four sub-services: r-certificate acquisition, t-instrument issuance, r-certificate issuance, and trust management routine.

- The *r*-certificate acquisition is pre-transaction service whose objective is to find out the reputation value of the other node. This will be performed by the exchange of certificate request (CertReq) and reply (CertRep) messages. At the end of this service node will decide whether it should start transaction or not.
- The *t*-instrument issuance is a post transaction service whose objective is to evaluate trust value based on the recent context. This will be performed by the exchange of *t*-Instrument issuance (InstrIssument) and acknowledgement (ACK) messages.

- The *r*-Certificate Issuance service is executed periodically by replica TRAs based on the *t*-Instruments of their hosts. Since t- *t*-Instruments are context-specific, therefore in this process single reputation value is calculated based on all context's value.
- The trust management routine is also periodically carried out by every replica TRA to maintain the evaluation table on its hosting node. In each run, this routine will eliminate the any record from the table that is older then specific threshold time.

PLUS: Yao et al. (2006) have proposed Parameterized and Localized trUst management Scheme (PLUS) for sensor networks security. The authors adopt a localized distributed approach and trust is calculated based on either direct observations or indirect observations. Trust calculation mechanism involves the combination of six parameters: 1) ordering, 2) integrity checking, 3) confidentiality checking, 4) responsibility checking, 5) positivity checking and 5) cooperative checking. The involvement of so many parameters makes this scheme less generic and more complex. For example in 'ordering', node checks whether the packet forwarded by node *i* is really coming from the base station or not. For this purpose, they assume that all the important control packets generated by the base station must contain hashed sequence number (HSN). Based on that HSN it performs checking. If the check is passed then the trust value of the forwarding node will increase. The involvement of the HSN in control packets introduces two problems: 1) it increases the size of the packet that results in higher consumption of the transmission and reception power, 2) it increases the computational cost at the sensor node because sensor node needs to verify the control packet that contains the HSN. Also, in 'positivity' checking case, judge node monitors the suspected node *i* whether the node has participated in the exchange of opinions as well as whether it has sent report measurement to the base station with an appropriate frequency. This parameter forces the sensor nodes to remain in promiscuous mode all the time. In the PLUS scheme, node is classified into four categories: 1) Distrust (untrustworthy), 2) Minimal (low trust), 3) Average (common trustworthy), and 4) Good (trustworthy). However the mechanism of computing boundaries of four trust levels is missing.

T-RGR: Liu et al. (2007) have proposed a very simple trust management scheme for Resilient Geographic Routing (T-RGR) scheme. Their trust algorithm works in a localized distributed manner, in which each node monitors the behavior of the one-hop neighbors. If neighboring node successfully forwards the packet it will increase the trust value by a constant parameter, δt, and if it drops the packet then the source node will decrease its trust value by another constant parameter, Δt. If the trust value of a particular node is greater than the predefined threshold value, then the node will be considered as a trusted node, otherwise it will be un-trusted. In their paper, the authors do not mention the mechanism to calculate those three constant parameters that make their scheme non-adaptive. The main advantage of their scheme is that it is not only simple and easy to implement but it also consume less memory and energy. The main problem in their scheme is that each node only relies on its direct monitoring for the calculation of a trust value. This makes their scheme vulnerable to collaborative attacks.

FTSN: Aivaloglou et al. (2007) have proposed Flexible Trust establishment Framework for Sensor Networks (FTSN) but it is still in initial phases. The unique thing about the FTSN is that it combines the features of certificate-based and behavior-based trust establishment approaches. Some subset of nodes in the network performs certificate–based trust evaluation and some subset of nodes, called supervision nodes in the network; perform behavior-based trust evaluation. A certificate validation is performed locally and is distributed before the deployment of the sensor nodes in the field. These certificates are signed by offline trust management authorities. Since this scheme is based on pre-deployment knowledge, so it is suitable for static sensor network environment. Nodes are either classified into trusted or un-trusted. Support of un-certain evidence is not available in this framework.

GTMS: Shaikh et al. (2009) have proposed lightweight Group-based trust management scheme (GTMS) for clustered wireless sensor networks. The unique thing about the GTMS scheme is that in contrast to traditional trust management approaches, which always focus on trust values of individual users, the GTMS scheme evaluates the trust of a group of users. That group based approach gives the benefit of less memory consumption. GTMS calculates the trust value based on direct or indirect observations. Direct observations represent the number of successful and unsuccessful interactions and indirect observations represent the recommendations of trusted peers about a specific node. For example, a sender will consider an interaction as successful if the sender receives an assurance that the packet is successfully received by the neighbor node and that node has forwarded the packet towards the destination in an unaltered fashion.

The GTMS works with two topologies. One is the intra-group topology where distributed trust management is used. The other is inter-group topology where centralized trust management approach is employed. For the intra-group network, each sensor that is a member of the group, calculates individual trust values for all group members. Based on the trust values, a node assigns one of the three possible states: 1) trusted, 2) un-trusted or 3) un-certain to other member nodes. After that, each node forwards the trust state of all the group member nodes to the CH. Then, centralized trust management takes over. Based on the trust states of all group members, a CH detects the malicious node(s) and forwards a report to the base station. On request, each CH also sends trust values of other CHs to the base station. Once this information reaches the base station, it assigns one of the three possible states to the whole group. On request, the base station will forward the current state of a specific group to the CHs. This methodology helps to drastically reduce the cost associated with trust evaluation of distant nodes. [Shaikh et al.,2009]

GTMS is intrusion-tolerant and provides protection against malicious, selfish and faulty nodes. Authors have provided detailed theoretical and simulation-based analysis and evaluation from the perspective of security resiliency, communication overhead, memory overhead and energy consumption analysis. Results show that GTMS scheme is lightweight and more suitable for large-scale wireless sensor networks.

Table 5 gives a qualitative comparison of the proposed schemes based on number of different parameters as discussed below:

- **Trust-based on direct observations:** Represents the trust value that is calculated based on the personal interaction experience with other nodes and/or via monitoring of nodes which reside inside its radio range.
- **Trust-based on indirect observations:** Represents the value that is obtained from the recommendations of the peer nodes.
- **Trust levels:** Depending on the scope and functionality, various trust management schemes provide support for different trust levels. Minimally, we can classify the nodes into the two categories of trusted and un-trusted.

- Dependency on routing scheme: There are various routing schemes that have been proposed for wireless sensor networks. If a proposed trust management scheme is independent of any specific routing strategy then that scheme is considered to be a generic scheme.

Challenging Issues

The main research problem is "How to establish dynamic trust relationships in large scale wireless sensor networks?" This problem needs to be investigated within two network environments:

- **Static wireless sensor network environment:** In which both sensor nodes and the base station are stationary and each sensor node has a unique identity.
- **Mobile wireless sensor network environment:** In which sensor nodes and base station both are mobile. Sensor nodes may or may not have unique identities.

General research challenges that arise in the design of trust management schemes for both the network environments are:

- How to make our trust management scheme lightweight? Here lightweight means scheme should consume less energy, low memory and less computation power.
- How to make our trust management scheme resilient against security threats?

Table 5. Comparative features of trust management schemes

	RFSN	ATRM	PLUS	GTMS	T-RGR	FTSN
Trust-based on direct observations	Yes	Yes	Yes	Yes	Yes	Yes
Trust-based on indirect observations	Yes	No	Yes	Yes	No	Yes
Trust levels	2	-	4	3	2	2
Dependency on routing scheme	Any	Any clustered based RS	PLUS_R	Any clustered based RS	Any geographic RS	Any

If any node in the network has been compromised by an intruder and starts sending false information then the other nodes in the network should be able to detect that.

- How to make our trust management scheme flexible and robust?

By flexible we mean that the trust management scheme should be independent of any topology of sensor networks. By robust we mean that it should be reliable enough to provide accurate trust values in a timely manner. The specific issue that is related to the mobile wireless sensor network environment is, "How to maintain the trust level when a node is moving among different clusters of the network?"

In general, trust calculation at each node, in fact, measures the confidence in node reliability. Ideally network traffic conditions such as congestion, bandwidth, etc should not affect the trust attached to a node. Assume that node A sends data to node B, but because of packet loss due to congestion, packets do not reach node B successfully. In this case, node A will think that node B is not cooperating and not providing the required service. So node A will reduce the trust level of node B. This is a very challenging problem that how such intermittent failures which occur due to bad network parameters can be filtered automatically and the trust actually reflects the correct cooperative metric of node B. So far, not much focus has been given on this issue.

There are many application scenarios in which sensor nodes do not have unique identities or the identities should remain hidden for achieving anonymity in wireless sensor networks [Misra and Xue, 2006; Olariu et al., 2005]. So the challenging problem is: without knowing identities, how to establish and maintain trust between communicating nodes? In order to calculate trust, various schemes keep the track of past behavior of other nodes, here the issues are:

- Node should keep the record of how many past interactions?

- What weight should be given to old interactions and very recent interactions? and
- What weight should be assigned to the direct observations and to the indirect observations?

CONCLUSION

Current research so far focuses on the security issues of wireless sensor networks. Although many survey papers are available in the security domain of wireless sensor networks, but we did not find any work in the literature which discusses the privacy and trust issues of wireless sensor networks in detail. In this chapter, we have given critical analysis of the current state-of-the-art research work done so far in the field of privacy and trust of wireless sensor network domain. We also presented generic and flexible taxonomies of privacy and trust that are based on our own research experience with wireless sensor networks. At the end, we also highlighted the challenging issues and problems of privacy and trust that need to be resolved.

REFERENCES

Aivaloglou, E., Gritzalis, S., & Skianis, C. (2007). Towards a flexible trust establishment framework for sensor networks. *Telecommunication Systems*, 35(3), 207–213. doi:10.1007/s11235-007-9049-x

Anderson, R. J. (2001). *Security Engineering: A Guide to Building Dependable Distributed Systems*. Indianapolis, IN: Publisher Wiley.

Bhuse, V. S. (2007). Lightweight intrusion detection: A second line of defense for unguarded wireless sensor networks. *PhD thesis, Dept. of Comp. Sci., Western Michigan University*.

Blaze, M., Feigenbaum, J., Ioannidis, J., & Keromytics. (1999). A. The keynote trust management system. In *RFC2704*.

Blum, B., He, T., Son, S., & Stankovic, J. (2003). *IGF: A state-free robust communication protocol for wireless sensor networks* (Tech. Rep. CS-2003-11). Dept. of Comp. Sci. University of Virginia, USA.

Boukerche, A., & Li, X. (2005). An agent-based trust and reputation management scheme for wireless sensor networks. *48th annual IEEE Global Telecommunications Conference* (pp. 1857–1861). St. Louis, MO: IEEE Press.

Boukerche, A., & Li, X., & EL-Khatib, K. (2007). Trust-based security for wireless ad hoc and sensor networks. *Computer Communications*, *30*(11-12), 2413–2427. doi:10.1016/j.comcom.2007.04.022

Buchegger, S., & Boudec, J. L. (2002). Performance analysis of the CONFIDANT protocol. In *Proceedings of the 13th ACM Symp. on Mobile Ad Hoc Networking and Computing* (pp. 226–236). Lausanne, Switzerland: ACM Press.

Chan, H., & Perrig, A. (2003). Security and privacy in sensor networks. *Computer*, *36*(10), 103–105. doi:10.1109/MC.2003.1236475

Chaum, D. L. (1988). The dinning cryptographers problem: unconditional sender and recipient untraceability. *Journal of Cryptographyg*, *1*(1), 65–75.

Chen, H., Wu, H., Zhou, X., & Gao, C. (2007). Reputation-based trust in wireless sensor networks. *Int. Conference on Multimedia and Ubiquitous Engineering*, (pp. 603–607), Korea: IEEE Computer Society.

DeCew & Judith. (2006). Privacy. In Zalta, E. N., (ed.), *The Stanford Encyclopedia of Philosophy (Fall 2006 Edition)*, Stanford, CA: Metaphysics Research Lab, CSLI, Stanford University.

Deng, J., Han, R., & Mishra, S. (2004). Countermeasures against traffic analysis attacks in wireless sensor networks (Tech. Report CU-CS-987-04), Comp. Sci. Dept, University of Colorado, Boulder.

Djenouri, D., Khelladi, L., & Badache, A. (2005). A survey of security issues in mobile ad hoc and sensor networks. *IEEE Communications Surveys and Tutorials*, *7*(4), 2–28. doi:10.1109/COMST.2005.1593277

Durresi, A., Paruchuri, V., Durresi, M., & Barolli, L. (2007). Anonymous routing for mobile wireless ad hoc networks. *International Journal of Distributed Sensor Networks*, *3*(1), 105–117. doi:10.1080/15501320601069846

Ganeriwal, S., Balzano, L. K., & Srivastava, M. B. (2008). Reputation-based framework for high integrity sensor networks. *ACM Transaction on Sensor Networks*, *4*(3), 1–37. doi:10.1145/1362542.1362546

Ganeriwal, S., & Srivastava, M. B. (2004). Reputation-based framework for high integrity sensor networks. *ACM Security for Ad-hoc and Sensor Networks*, (pp. 66–67). New York: ACM Press.

Kamat, P., Zhang, Y., Trappe, W., & Ozturk, C. (2005). Enhancing source-location privacy in sensor network routing. In *Proceedings of the 25th IEEE Int. conf. on Distributed Computing Systems*, (pp. 599–608), Columbus, OH: IEEE Computer Society.

Karlof, C., Sastry, N., & Wagner, D. (2004). TinySec: a link layer security architecture for wireless sensor networks. In Proceedings of the *2nd Int. Conf. on Embedded networked sensor systems*, (pp. 162–175), Baltimore, MD:ACM Press.

Krishna, K., & bin Maarof, A. (2003). A hybrid trust management model for MAS based trading society. *The Int. Arab Journal of Information Technology*, *1*(1), 60–68.

Liu, K., Abu-Ghazaleh, N., & Kang, K.-D. (2007). Location verification and trust management for resilient geographic routing. *Journal of Parallel and Distributed Computing*, *67*(2), 215–228. doi:10.1016/j.jpdc.2006.08.001

Liu, Z., Joy, A. W., & Thompson, R. A. (2004). A dynamic trust model for mobile ad hoc networks. In *Proceedings of the 10th IEEE Int. Workshop on Future Trends of Distributed Computing Systems*, (pp. 80–85), Suzhou, China: IEEE Computer Society.

Michell, S., & Srinivasan, K. (2004). State based key hop protocol: a lightweight security protocol for wireless networks. In *Proceedings of the 1st ACM international Workshop on Performance Evaluation of Wireless Ad Hoc, Sensor, and Ubiquitous Networks*, (pp. 112-118). Venezia, Italy: ACM Press.

Michiardi, P., & Molva, R. (2002). CORE: A collaborative reputation mechanism to enforce node cooperation in mobile ad hoc networks. In *Proceedings of the 6th IFIP conf. on communications and multimedia security*, (pp. 107–121), Portoroz, Slovenia: Kluwer Academic Publishers.

Misra, S., & Xue, G. (2006). Efficient anonymity schemes for clustered wireless sensor networks. *International Journal of Sensor Networks, 1*(1/2), 50–63. doi:10.1504/IJSNET.2006.010834

Moore, T. (2006). A Collusion Attack on Pairwise Key Predistribution Schemes for Distributed Sensor Networks. In *Proceedings of the 4th Annual IEEE International Conference on Pervasive Computing and Communications Workshops (PERCOMW'06)*, pp. 251-255. IEEE Computer Society.

Nasipuri, A., & Li, K. (2002). A directionality based location discovery scheme for wireless sensor networks. In Proceedings of the *1st ACM international Workshop on Wireless Sensor Networks and Applications*, (pp. 105-111). Atlanta, Georgia, USA: ACM Press.

Olariu, S., Xu, Q., Eltoweissy, M., Wadaa, A., & Zomaya, A. Y. (2005). Protecting the communication structure in sensor networks. *International Journal of Distributed Sensor Networks, 1*(2), 187–203. doi:10.1080/15501320590966440

Ouyang, Y., Le, Z., Chen, G., Ford, J., & Makedon, F. (2006). Entrapping adversaries for source protection in sensor networks. *2006 Int. Sym. on a World of Wireless, Mobile and Multimedia Network*, (pp. 23–34), Buffalo, NY: IEEE Computer Society.

Ozturk, C., Zhang, Y., & Trappe, W. (2004). Source-location privacy in energy-constrained sensor network routing. In *Proceedings of the 2nd ACM workshop on Security of Ad hoc and Sensor Networks*, (pp. 88–93), Washington, DC: ACM Press.

Park, T., & Shin, K. G. (2004). LiSP: A lightweight security protocol for wireless sensor networks. *ACM Transaction on Embedded Computing Sys., 3*(3), 634–660. doi:10.1145/1015047.1015056

Perrig, A., Stankovic, J., & Wagner, D. (2004). Security in wireless sensor networks. *Communications of the ACM, 47*(6), 53–57. doi:10.1145/990680.990707

Perrig, A., Szewczyk, R., Tygar, J. D., Wen, V., & Culler, D. E. (2002). SPINS: security protocols for sensor networks. *Wireless Networks, 8*(5), 521–534. doi:10.1023/A:1016598314198

Raazi, S., Khan, A., Khan, F., Lee, S., & Song, Y.-J. (2007). MUQAMI: A locally distributed key management scheme for clustered sensor networks. In Etalle, S. and Marsh, S., (eds), *Int. Federation for Infor. Proc.*, (pp. 333–348). Boston: Springer.

Reed, M. G., Syverson, P. F., & Goldschlag, D. M. (1998). Anonymous connections and onion routing. *IEEE Journal on Selected Areas in Communications, 6*(4), 482–494. doi:10.1109/49.668972

Reiter, M. K., & Rubin, A. D. (1998). Crowds: Anonymity for web transactions. *ACM Transactions on Information and System Security, 1*(1), 66–92. doi:10.1145/290163.290168

Resnick, P., Zeckhauser, R., Friedman, E., & Kuwabara, K. (2000). Reputation systems: Facilitating trust in internet interactions. *Communications of the ACM, 43*(12), 45–48. doi:10.1145/355112.355122

Rong, P., & Sichitiu, M. L. (2006). Angle of Arrival Localization for Wireless Sensor Networks. In *Proceedings of the 3rd Annual IEEE Communications Society Conference on Sensor and Ad Hoc Communications and Networks (SECON '06)*, (pp. 374-382), Reston, VA: IEEE Communications Society.

Seys, S., & Preneel, B. (2006). ARM: Anonymous routing protocol for mobile ad hoc networks. In *Proceedings of the 20th Int. conf. on Advanced Information Networking and Applications*, (pp. 33-37), Vienna Austria: IEEE Press.

Shaikh, R. A., Jameel, H., d'Auriol, B. J., Lee, S., Song, Y.-J., & Lee, H. (2008a). Network level privacy for wireless sensor networks. In *Proceedings of the 4th International Conference on Information Assurance and Security (IAS 2008)*, (pp. 261-266), Naples, Italy: IEEE Computer Society.

Shaikh, R. A., Jameel, H., d'Auriol, B. J., Lee, S., Song, Y.-J., & Lee, H. (2009). Group-based Trust Management Scheme for Clustered Wireless Sensor Networks. *IEEE Transactions on Parallel and Distributed Systems, 20*(11), 1698-1712. doi:10.1109/TPDS.2008.258

Shaikh, R. A., Jameel, H., Lee, S., Rajput, S., & Song, Y. J. (2006a). Trust management problem in distributed wireless sensor networks. In *Proceedings of the 12th IEEE Int. Conf. on Embedded Real Time Computing Systems and its Applications*, (pp. 411-414). Sydney, Australia: IEEE Computer Society.

Shaikh, R. A., Lee, S., Khan, M. A. U., & Song, Y. J. (2006b). LSec: Lightweight security protocol for distributed wireless sensor network. In *11th IFIP Int. Conf. on Personal Wireless Comm., LNCS 4217*, (pp. 367-377), Albacete, Spain: Springer-Verlag.

Shaikh, R. A., Lee, S., Song, Y. J., & Zhung, Y. (2006c). Securing distributed wireless sensor networks: Issues and guidelines. In *Proceedings of the IEEE Int. Conf. on Sensor Networks, Ubiquitous, and Trustworthy Computing- vol. 2 - Workshops*, pp. 226-231, Taiwan: IEEE Computer Society.

Shields, C., & Levine, B. N. (2000). A protocol for anonymous communication over the internet. In *Proceedings of the 27th ACM conf. on Computer and communications security*, (pp. 33-42), Athens, Greece: ACM Press.

Sun, Y. L., Zhu, H., & Liu, K. J. R. (2008). Defense of trust management vulnerabilities in distributed networks. *IEEE Communications Magazine, 46*(2), 112-119. doi:10.1109/MCOM.2008.4473092

Theodorakopoulos, G., & Baras, J. S. (2006). On trust models and trust evaluation metrics for ad hoc networks. *IEEE Journal on Selected Areas in Communications, 24*(2), 318-328. doi:10.1109/JSAC.2005.861390

Undercoffer, J., Avancha, S., Joshi, A., & Pinkston, J. (2002). Security for Sensor Networks. Paper presented at *2002 CADIP Research Symposium*, Baltimore, MD.

Wadaa, A., Olariu, S., Wilson, L., Eltoweissy, M., & Jones, K. (2004). On providing anonymity in wireless sensor networks. In *Proceedings of the 10th Int. conf. on Parallel and Distributed Systems*, (pp. 411-418), California, USA: IEEE Computer Society.

Walters, J. P., Liang, Z., Shi, W., & Chaudhary, V. (2006). Wireless sensor network security: A survey. In Xiao, Y., (ed.), *Security in Distributed, Grid, and Pervasive Computing*, (pp. 367-410). CRC Press.

Wood, A. D., Fang, L., Stankovic, J. A., & He, T. (2006). SIGF: a family of configurable, secure routing protocols for wireless sensor networks. In *Proceedings of the 4th ACM workshop on Security of ad hoc and sensor networks*, (pp. 35-48). Alexandria, VA: ACM Press.

Xi, Y., Schwiebert, L., & Shi, W. (2006). Preserving source location privacy in monitoring-based wireless sensor networks. In *Proceedings of the Parallel and Distributed Processing Symposium (IPDPS 2006)*, Rhodes Island, Greece: IEEE Computer Society.

Yao, Z., Kim, D., & Doh, Y. (2006). PLUS: Parameterized and localized trust management scheme for sensor networks security. In *Proceedings of the 3rd IEEE Int. Conf. on Mobile Adhoc and Sensor Systems*, (pp. 437–446), Vancouver, Canada: IEEE Computer Society.

Zhu, B., Wan, Z., Kankanhalli, M. S., Bao, F., & Deng, R. H. (2004). Anonymous secure routing in mobile ad-hoc networks. In *Proceedings of the 29th IEEE Int. conf. on Local Computer Networks*, (pp. 102–108), Tampa, USA: IEEE Computer Society.

Zhu, S., Setia, S., & Jajodia, S. (2003). LEAP: efficient security mechanisms for large-scale distributed sensor networks. In *Proceedings of the 10th ACM Conf. on Computer and Comm. security*, (pp. 62–72). NY: ACM Press.

Zhu, S., Setia, S., & Jajodia, S. (2006). LEAP+: Efficient security mechanisms for large-scale distributed sensor networks. *ACM Trans. Sensor Networks*, 2(4), 500–528. doi:10.1145/1218556.1218559

KEY TERMS AND DEFINITIONS

Centralized Trust Management: A single globally trusted server determines the trust values of every node in the network.

Distributed Trust Management: Every node locally calculates the trust values of all other nodes in the neighborhood or network.

Hard Solitude: Means that other nodes in a compact or a command node decide to isolate a particular node.

Identity Privacy: No node can get any information about the source and destination nodes. Only the source and destination nodes can identify each other. Also, the source and destination nodes have no information about the real identities of the intermediate forwarding nodes.

Location Privacy: No node can get to know any information about the location (either in terms of physical distance or number of hops) of the sender node except the source, its immediate neighbors and the destination.

Route Privacy: No node can predict the information about the complete path (from source to destination) of the packet. Also, a mobile adversary can not get any clue to trace back the source node either from the contents and/or directional information of the captured packet(s).

Soft Solitude: Refers to the node's decision to be in the solitude state.

Solitude: Refers to the condition that a node goes into the state of isolation for a specific period of time. During that interval, the node cannot fulfill jobs nor can it provide services such as packet forwarding to the other nodes.

Trust: Represents the level of confidence on other entity.

Chapter 14
Distributed Group Security for Wireless Sensor Networks

Juan Hernández-Serrano
Universitat Politècnica de Catalunya, Spain

Juan Vera-del-Campo
Universitat Politècnica de Catalunya, Spain

Josep Pegueroles
Universitat Politècnica de Catalunya, Spain

Miguel Soriano
Universitat Politècnica de Catalunya, Spain

ABSTRACT

Wireless sensor networks (WSNs) are made up of large groups of sensor nodes that usually perform distributed monitoring services. These services are often cooperative and interchange sensitive data, so communications within the group of sensor nodes must be secured. Group key management (GKM) protocols appeared, and were broadly studied, in order to ensure the privacy and authentication throughout the group life. However, GKM for WSNs is already challenging due to the exposed nature of wireless media, the constrained resources of sensor nodes, and the need of ad-hoc self-organization in many scenarios. In this chapter we present the basis of GKM and its state-of-the art for WSNs. We analyze the current non-resolved topics and we present a GKM proposal that solves some of these topics: it minimizes both the rekeying costs when the group membership changes and the routing cost within the group.

INTRODUCTION

Wireless sensor networks (WSN) comprise a wide range of devices that autonomously intercommunicate offering a wide variety of services: fire detection, fleet management, health monitoring, etc. Very often, WSNs are deployed in hostile environ-ments, like battlefields, where there is a presence of enemies or potential attackers and hence there is a necessity for providing them with security. And not only military networks have strong security requirements. Unfortunately, there is an increasing risk of terrorist attacks against civil and crowded places that may have a WSN deployed, such as hotels, hospitals or train stations.

DOI: 10.4018/978-1-61520-701-5.ch014

Under the circumstances, ensuring the security in WSN becomes critical. However, the "special" nature of the WSNs makes this task specially challenging. First of all, sensor nodes are often "low-end" devices with constrained resources and hence the use of well-known but expensive security algorithms (e.g. asymmetric cryptography) is not feasible or it must be minimized. Second, WSNs are in many cases unattended self-organized ad-hoc networks and so the security management must be assumed within the group of sensors without the presence of a fixed powerful infrastructure which could lead the security. And third, the wireless media makes the physical layer very accessible for an attacker, which can jam, inject or modify link layer packets without difficulty and which can easily compromise and spoof a sensor node.

In short, the widespread use of such networks produces new security challenges: there is a necessity for securing the communications within the group of sensors and it must be tackled in a self-configuration manner with active cooperation of the sensor nodes. If in addition we take into account the constrained resources of sensor nodes, we can also assert that the cost of security management must be minimized and thus shared out by the group members. That is to say, securing group communications becomes mandatory and its implementation must be distributed tackled and with an overall low cost per sensor node.

The rest of this chapter is organized as follows. First, we present the necessary background on group security and then we, second, continue with its particularization to WSNs. Within this section we analyze the current issues on developing group security for WSNs and we present state-of-the art solutions. Next, we detail an own proposal of distributed group security for WSNs and we propose future directions in order to enhance the current group security systems. Finally, we present our concluding remarks.

BACKGROUND ON GROUP SECURITY

In cooperation frameworks, such as WSNs, data is shared by incumbents and, in many cases, carries sensible information. Therefore, data must be protected, but just against non group members and not within the group membership, and this is what is called group security. Group security is, thus, targeted to provide group privacy and group authentication: data is protected from outsiders and the only sources of communication are the members of the group. As a result, it is merely based on the use of a common shared secret called the session key or group key. This key allows every group member to: 1) send encrypted data; 2) decrypt received data, and 3) authenticate itself as a group member since the knowledge of the session key guarantees that it belongs to the group. As only the current group members ought to know the session key, such key must be updated every time the membership of the group changes. *Group key management* (GKM) is the branch of knowledge that studies the generation and updating of the keying material used for securing the group during its whole life (Wallner, Harder & Agee, 1998).

Traditionally the update of keying material or *rekeying* is performed by a centralized trusted third party called the *key server* (KS). Within the simplest approach, the KS delivers updated keys through individual secure channels with every member (Wallner et al., 1998). This method is both non-scalable and infeasible for large groups since a single peer-to-peer connection is needed for each of the group members. Therefore, the rekeying complexity problem is O(N), with N the number of group members, and the KS has to wait for the delivery of N messages to actually change the session key. To mitigate this scalability problem two main branches of GKM have been studied: delegation and tree-based approaches; although combination of both is possible.

Figure 1. GKM based on delegation of tasks

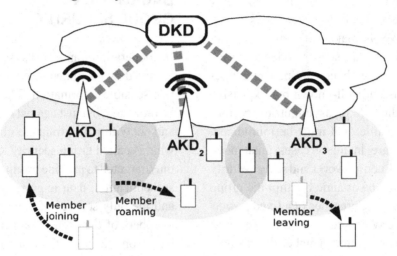

GKM Based on Delegation of Tasks

A common approach for designing a scalable GKM service is to adopt a hierarchical structure, and a number of proposed key management algorithms have adopted such an approach (Hardjono, Cain & Doraswamy, 2000; Hardjono, Cain & Monga, 1998; Mittra, 1997). Broadly, these re-keying algorithms operate by hierarchically dividing the key management domain into smaller administratively scoped areas on either a physical or logical basis. The domain is managed by a *domain key distributor* (DKD), and each area by an *area key distributor* (AKD) which assumes most of the KS functions (delegated from the DKD) within the area. Intra-area rekeying algorithms are defined to locally (within an area) manage member joinings and leavings. Inter-area rekeying algorithms are defined to handle the security relations when members roam between areas. Joining, leavings and roamings are graphically represented in figure 1.

Group security is independently managed by the AKD in every area and there are different area (subgroup) keys within the domain. As a result, when a group member in area *i* wants to communicate with another member in area *j*, either it must know a shared secret (a domain or group

key) or a "translation" between areas becomes mandatory. Whether there is a domain key or on the contrary there is a need of a translation is the main difference between the delegation approaches to GKM.

Many GKM contributions use translation between areas assuming the basis of Iolus framework (Mittra, 1997), where the AKDs are in charge of such translation. In Iolus the AKDs are called *group security intermediaries* (GSI) and the DKD is the *group security controller* (GSC). The domain is split into smaller areas in a geographical basis and the GSI are strategically positioned between areas so they know the area keys of at least two areas. When a message must cross between areas, it is routed through the GSI that first decrypts the message with the incoming area key and then delivers the message to the next area encrypting it with the outgoing area key.

Other GKM proposals based on delegation derive from Intra-Domain Group Key Management Protocol (Hardjono et al., 1998) and specially from the Group Key Management Architecture (Baugher, Canetti, Dondeti & Lindholm, 2002) and avoid "translation" by using a common shared group key or domain key called the *data encryption key* (DEK). In such proposals the DKD creates a DEK and securely sends it to all AKDs. Each

Figure 2. A logical KEK tree

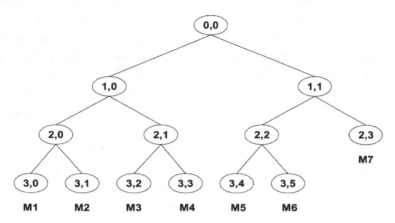

AKD is in charge of securely distributing (and redistributing) this DEK to its members. How the rekeying within an area is carried out and how it affects other areas (as the domain key is probably updated) determines the different inter-area rekeying algorithm approaches in the literature. A good summary of them can be found in (Zhang, DeCleene, Kurose & Towsley, 2002).

GKM Based on Tree-Based Approaches

The most successful proposal for reducing the rekeying problem order is the use of logical key encryption keys (KEK) tree hierarchies. Many GKM rekeying algorithms, such as (Balenson, McGrew & Sherman, 1999; Harney & Harder, 1999; Harney & Muckenhirn, 1997; Pegueroles, Rico-Novella, Hernández-Serrano & Soriano, 2003b; Wallner et al., 1998), use these trees since they substantially improve the efficiency in terms of bandwidth and latency. In this kind of methods, the KS builds a KEK tree only known to him and assigns a set of these KEKs to each of the members of the multicast group. This set of keys corresponds to the path from the tree's leafs - where the members are - to its root. When a member leaves or joins a group only the KEKs belonging to that member need to be changed. Then new keys are delivered to the remaining

members and the tree is reconstructed using the unchanged keys.

For example, in figure 2 a logical KEK tree is shown. A KEK is associated to every node in the tree; e.g. the KEK associated to node (0,0) is denoted as $K_{0,0}$. The KS knows all the KEKs in the tree and every member knows the KEKs in the tree path from its leaf to the tree root. Therefore, e.g. member M1 knows the keys $K_{3,0}, K_{2,0}, K_{1,0}$ and $K_{0,0}$, and notice that $K_{2,0}$ is shared by M1 and M2; $K_{1,0}$ is shared by M1, M2, M3 and M4; and $K_{0,0}$ is shared by all members. Now, suppose that e.g. M4 leaves the group. The compromised KEKs are $K_{3,3}, K_{2,1}, K_{1,0}$ and $K_{0,0}$ and thus the KS update them with the new ones: $K'_{3,3}, K'_{2,1}, K'_{1,0}$ and $K'_{0,0}$. Now, instead of sending an encrypted message per member with the necessary updated KEKs, the KS can only send: $K'_{0,0}$ and $K'_{1,0}$ encrypted with $K_{2,0}$, that only M1 and M2 can decrypt; $K'_{2,1}$ and $K'_{1,0}$ encrypted with $K_{3,2}$, that M3 will read; and $K'_{0,0}$ with $K_{1,1}$, that will be read by M5, M6 and M7. In short, the cost of rekeying is reduced from $O(N)$ to $O(L)$, with N the number of members and L the depth of the tree, and this a extremely valuable property for systems in where message sending is a very costly process, such as WSNs.

As the number of necessary messages for rekeying depends on L, tree-based algorithms are more efficient when L gets its minimum value. And this is the case of a balanced tree (all leafs at the same level or between at most two adjacent levels) where the depth of the tree is exactly $\log_2 N$. The need for balanced trees have been well studied and many protocols (Li, Yang, Gouda & Lam, 2001; Pegueroles, Rico-Novella, Hernández-Serrano & Soriano, 2003a) have tackled with it.

DISTRIBUTED GROUP SECURITY IN WSNS

In the following section we analyze the current issues and problems found when developing group security protocols for WSNs. Next, we present an overview of the state-of-the art GKM solutions targeted to such networks including our proposal of distributed group security for WSNs. Finally, we discuss the potential future research directions.

Issues, Controversies, Problems

As said before, ensuring security in WSN is specially challenging due to the following features of such networks:

- Sensor nodes are, in many cases, "low-end" devices with constrained resources. This characteristic forces the researchers to find GKM solutions with as little cost as possible. As a result GKM for WSNs must minimize the necessary transmitted messages for rekeying and reduce or just avoid the use of costly cryptographic operations such as asymmetric primitives.
- The wireless medium is specially exposed to attackers and thus the probability of node compromising is higher than in classical wired scenarios. A sensor node can be easily kidnapped by and attacker which can easily obtain its secrets (Hartung, Balasalle & Han, 2005). Therefore, any security algorithm for WSNs has to be resilient against the compromised sensor nodes.

- WSNs often are unattended self-organized ad-hoc networks. In such cases there isn't the presence of a fixed powerful infrastructure which could lead the group security. As a consequence the group security management must be distributed among the sensor nodes that cooperate to deal with it.

The need of being resilience and the lack of a centralized key server makes group security for WSNs even more challenging. In traditional GKM, the KS shares keys (KEKs) others than the group key with the group members. As a result, when the group key is compromised due e.g. to a node capture, the key server uses these KEKs to isolate the node and update the group key. However, when no central KS is assumed, the only way of protecting links from the attacker and negotiating a new group key is the use of shared keys between pair of nodes. Therefore in WSNs it is necessary to have a GKM system able to tackle with two different problems:

- **Global keying:** The global keying problem refers to the need of a common shared key or group key. This is the traditional only task of GKM since, as stated above, it provides confidentiality and group authentication, but its merely usage is not feasible for WSNs since it doesn't provide resilience at all when there is a lack of a centralized KS.
- **Pairwise keying:** The pairwise keying problem is related to establish different shared keys between every pair of nodes within the group. The use of these keys is necessary as it provides the only method to protect links when the group key is compromised and there is a lack of a centralized KS (distributed proposals).

From all the above we can conclude that a new case of study arises when referring to GKM over WSNs: the pairwise keying. As a consequence most of researchers in group security for WSNs has stressed on this topic.

Note that the pairwise keying extreme solution is that every node must store a shared key with every other node in the network. However, this is not a scalable solution due to the memory constraints. Because of this, security researchers have worked into find a more optimal solution, such as pairwise keying only with every direct neighbor. Several approaches have been presented and we present a short of them in next section.

When talking about the global keying problem, two main approaches can be detected: 1) "classical" centralized GKM solutions relying on a fixed server or base station, and 2) the "new" distributed GKM proposals where every member contributes to manage the group security. The first approach just consists on the application of known efficient GKM techniques to WSNs. However, the second approach is a new open topic and little research has been carried out so far on the application of distributed GKM to WSNs. An overview of state-of-the-art global keying solutions for WSNs is also presented in next section.

Solutions and Recommendations

Following we present and discuss state-of-the art solutions for both the pairwise keying and the global keying problem. Next we present an own contribution to GKM for WSNs that solves many of the open challenges and especially the ones from the global keying problem that it is less studied by the researchers.

Pairwise Keying

Pairwise keying refers to establish keys between pairs of nodes in order to protect the communication links, which are specially exposed in wireless media. Such pairwise keys are used to secure data hop-by-hop at the link layer. Ideally, there had to be at least a pairwise key for any pair of neighboring nodes. There are two ways of dealing

with this problem: the pre-deployment and the post-deployment approach. The former is based on the pre-distribution of the pairwise keys, and the second one on the negotiation of such keys after the network deployment.

Pre-Deployment Approach

Pre-deployment approaches are based on the pre-distribution of pairwise keys among the network nodes. Since the pairwise keys are needed to guarantee secure hop-by-hop routing, each node must share pairwise keys with at least its neighbors. Assuming that there isn't a post-deployment knowledge there is no way to decide which nodes will be neighbors, and thus, as a naïve solution, every node is pre-loaded with a secret shared key with any other node in the network. However this solution presents scalability risks since, for a number N of group members, a total of $N \cdot (N-1)/2$ keys must be initially created and every node must store $N-1$ keys. The need of obtaining a balance between the probability of sharing a key and the necessary stored keys was introduced in (Eschenauer & Gligor, 2002) as the "key pool" paradigm. In this paradigm, a certain subset of not necessarily different keys from a common "key pool" is pre-distributed to every sensor node. Therefore, only the nodes sharing a key, or a certain set of keys (Chan, Perrig & Song, 2003), can interchange messages securely. As the compromise of a member only reveals some of the system keys, network resilience is higher than in the naïve approach. Moreover, the fact of knowing just a subset of the total key avoids the scalability risk.

The quest of better and better pairwise key pre-distribution schemes which provide a high probability of sharing a key between neighbors for the smallest possible subset of pre-loaded keys in every member is one of the research hot topics today and many contributions have been presented with such a purpose. The source of these proposals is the following variant of the birthday paradox (Menezes, Oorschot & Vanstone, 1997):

Given a set S with k elements; two subsets S_1 and S_2 with m_1 and m_2 elements, respectively, from S are randomly chosen; the probability of $S_1 \cap S_2 \neq 0$ can be expressed as in 1.

$$K_{i,j} = f(g(K_{i+1,2j}), g(K_{i+1,2j+1})) \qquad \begin{array}{l} 0 \leq i < L \\ 0 \leq j < 2^i \end{array}$$

$$K_{i,j} = random(i,j) \qquad \begin{array}{l} i = L \\ 0 \leq j < 2^i \end{array}$$

$$(1)$$

It was demonstrated in (Menezes et al., 1997) that the expression value in 1 is greater than ½ for k big enough and $m_1 = m_2 \approx k^{\frac{1}{2}}$. From that assessment it is easily deduced that for a total set of k keys it is enough to pre-distribute a subset of $k^{\frac{1}{2}}$ keys to every member in order to guarantee a probability of at least a 50% of two random nodes sharing a key. For the shake of clarity, in figure 3 we represent the expression in 1 with $k=900$ and $m_1=m_2$. The reader can appreciate how the probability of sharing a key is nearly one with a relatively small pre-loaded subset of the keys.

The application of such properties to a basic scheme of random pairwise key pre-distribution in WSNs was first presented in (Eschenauer & Gligor, 2002), but many other proposals share the same mathematical background with more or less modifications. These proposals randomly choose key chains from a known set and distribute them to the members. The fact of randomly choose the keys is why these approaches are referred in the literature as probabilistic proposals. For example, a q-composite key scheme was presented in (Chan & Perrig, 2003). Basis of this scheme is that a secure link can be only established when both involved nodes share q keys instead of just one key. Obviously the probability of sharing q keys is lower than sharing just one key, that as previously stated is quite high, and hence the number of operative neighbors is reduced, but on the other hand more network resilience is provided since a compromised node spoofing a link communication must know the q shared keys instead of just one. Other example is (Q. Huang, Cukier, Kobayashi, Liu & Zhang, 2003) which

Figure 3. Probability of two members sharing at least one key from a pool of k=900 keys and a variable number m of pre-loaded keys in both members

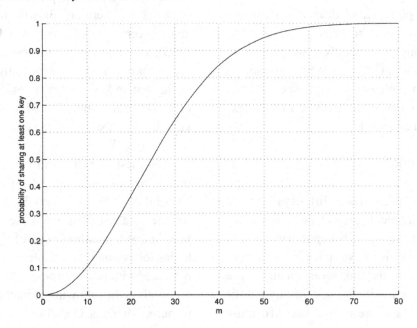

reduces the number of necessary pre-loaded keys by using two-dimensional key pools. And many other probabilistic contributions need at least to be mentioned here such as the ones in (D. D. Hwang, Lai & Verbauwhede, 2004; J. Hwang & Kim, 2004; Pietro, Mancini & Mei, 2003; Zhu, Setia & Jajodia, 2006).

To further enhance the security of key pre-distribution, it is also usual in the literature to use deterministic processes alone or together with probabilistic ones. For example, in (Du, Deng, Han & Varshney, 2004; D. Huang, Mehta, Medhi & Harn, 2004; Liu & Ning, 2003a; Liu, Ning & Du, 2008; Yu & Guan, 2005) probabilistic processes are complemented with positioning systems which help to decide how to establish the secure links; (Camtepe & Yener, 2004; Lee & Stinson, 2004a) use block design and combinatorials to guarantee a shared key between any two neighbor nodes; and others, such as (Blundo et al., 1998; Du et al., 2005; Lee & Stinson, 2004b; Liu & Ning, 2003b, 2003c), use techniques which are similar to Blom key matrices (Blom, 1984) or threshold-based schemes.

Post-Deployment Approach

Post-deployment approaches rely on the negotiation of the necessary pairwise keys just after the network deployment. The simplest model (Anderson, Chan & Perrig, 2004) allows nodes to agree the pairwise keys within their neighborhood in "clear" because the treat level at this point is, in most scenarios, very low. To enhance the security, in (Du et al., 2005) all the nodes are pre-loaded with a temporal common shared key that initially allows them to securely negotiate pairwise keys with their neighbors. In SPINS (Perrig, Szewczyk, Tygar, Wen & Culler, 2002), a trusted third party is used to securely negotiate the neighbor pairwise keys just after the network deployment. Other traditional approach would be the use of any known key agreement protocol, such as Diffie-Hellman (Diffie & Hellman, 1976). However, key agreement protocols are based on

asymmetric cryptographic and hence, as previously stated, its use in WSNs must be avoided or minimized due to resource constraints.

In order to achieve a better understanding of post deployment approaches we detail next the basic behavior of the SPINS proposal, which have been widely accepted in the literature. Within this proposal, Perrig et al. presented a node-to-node key agreement constructed from symmetric-key algorithms and hence with an overall low cost per node. The proposed symmetric protocol is based on a hierarchical centralized infrastructure and, therefore, uses the base station as a trusted agent for key setup. The authors avoid the use of any pre-deployment scheme beyond a shared secret between any sensor node and the base station. The protocol is detailed below.

Suppose that the node A needs to establish a shared secret session key SK_{AB} with node B. Initially, A and B do not share any secret and thus they use a trusted third party S, which is the base station. Both member A and B share a secret key with S: K_{AS} and K_{BS} respectively. The sequence of messages of the protocol is the following:

$$A \rightarrow B: N_A, A$$
$$B \rightarrow S: N_A, N_B, A, B, MAC(K_{BS}, N_A | N_B | A | B)$$
$$S \rightarrow A: \{SK_{AB}\}_{KAS}, MAC(K'_{AS}, N_A | B | \{SK_{AB}\}_{KAS})$$
$$S \rightarrow B: \{SK_{AB}\}_{KBS}, MAC(K'_{BS}, N_A | A | \{SK_{AB}\}_{KBS})$$

The nonces N_A and N_B ensure strong key freshness to both A and B. Confidentiality is ensured through encryption with the keys K_{AS} and K_{BS} of the established session key SK_{AB}, and message authentication through the MAC using keys K'_{AS} and K'_{BS}.

As the reader can easily extract, the post-deployment approaches rely on the presence of a trusted third party in order to secure the establishment of the pairwise keys and thus cannot be applied in isolated WSN scenarios without a fixed presence of a powerful base station (or any kind of power node).

Global Keying

The global keying problem is the "classical" challenge of GKM and it refers to establish a common shared secret or group key that provides:

- **Group confidentiality:** Data is encrypted/decrypted with the group key, and thus it is protected from external access
- **Group authentication:** The mere fact of knowing the group key identifies a user as a current group member

Obviously, the group keys must be only known by the current members of the group in order to maintain the group security. Therefore, it becomes mandatory to update the group key every time the group membership changes in order to handle such a requirement. How the group key is created and updated determines the different proposals until now.

In WSN the global keying problem has been poorly tackled. In fact, solutions either rely on the application of the "traditional" centralized GKM techniques or are just limited extensions of the pairwise keying solutions. Centralized GKM techniques rely on a fixed security server/manager that it is in charge of creating/updating the keying material. Their application to WSNs just relies on the mere presence of a powerful fixed node (the base station) and the main challenge is to take care about sensor constraints in terms of available security primitives and power consumption. However, there are many scenarios where a WSN lacks of a fixed infrastructure and a need of a distributed self-organizing solution arises.

Solutions coming from the pairwise keying, such as (Chan & Perrig, 2003; Du et al., 2004; Eschenauer & Gligor, 2002; J. Hwang & Kim, 2004; D. D. Hwang et al., 2004; Pietro et al., 2003; Zhu et al., 2006), focus on generating the initial group key from the pre-distributed keys. Group members use the pairwise keys to agree or contribute to a common shared secret or global key. These approaches do not need a fixed central entity and thus they are more suitable for many WSN scenarios, since they provide a distributed self-organized method to generate a group key. Nevertheless, they only focus on the creation of the group key and leave its update aside. As a consequence, the lack of a rekeying scheme will result in a repetition of the initial creation every time the group membership changes; and, as the initial creation of the secure group with its consequent common shared secret is a heavy process, this is a very costly process for common constrained resources sensor nodes, not to mention how worse the situation becomes when the group is very large. As the state-of-the-art proposals do not provide efficient solutions for the necessary rekeying when the group membership changes due to addition or loss/compromise of nodes, therefore distributed GKM techniques focusing on rekeying efficiency must be provided.

Next we present techniques which show how the deployed sensor group can on the one hand autonomously manage its self-security in a distributed manner, as well as on the other hand, still achieve the rekeying performance of the centralized solutions.

Considering that the more energy consumption activity in WSNs is the transmission of a message over a wireless channel, any GKM proposal for WSNs must be designed to minimize the number of messages that has to be transmitted for rekeying. It is accepted by the scientific community that the more efficient GKM schemes in terms of rekeying in the literature, are the ones based on logical key trees that can reduce the number of necessary transmitted messages to only $O(\log_2 N)$ with N the number of members of the group. However, such proposals do not fulfill the requirement of being decentralized since they rely on fixed centralized security servers. As a consequence of that, distributing the tree-based GKM proposals is challenging but still only a few contributions have been proposed:

- LKHW (Di Pietro, Mancini, Law, Etalle & Havinga, 2003) is a tree-based GKM proposal that relies on hierarchical networks where there is a presence of powerful nodes that deal with the security. Therefore, it cannot then be considered for peer-to-peer sensor groups in which all nodes have similar capabilities.
- TGDH (Kim, Perrig & Tsudik, 2004) is another similar approach that creates/manages a logical tree of KEKs by means of Diffie-Hellman key agreements. However, since it relies on non-symmetric cryptography, it is not well suited for common low-end sensors.

We have proposed in (Hernández-Serrano, 2008) a GKM protocol that uses logical KEK trees distributively constructed from contributions of all group members with an overall cost per member very low and with the mere use of symmetric cryptography. Next section details such proposal.

A DISTRIBUTED GKM PROPOSAL FOR WSNS

In (Hernández-Serrano, Pegueroles & Soriano, 2008) we presented a distributed GKM proposal based on logical KEK trees. This proposal fulfills two of the main design objectives of GKM used in WSNs: 1) it minimizes the cost of updating keying material in terms of transmitted messages, and 2) it achieves the former objective by implementing a distributed GKM system based on peer relations in which every group member contributes. The protocol was formalized and evaluated from both analytical and simulation results. From the evaluation, we demonstrated that the protocol do not just achieve the rekeying efficiency of the centralized tree-based GKM protocols but also distributes the necessary transmitted messages between the group members.

Such results, originally targeted to MANETs, are also especially relevant for large dynamic sensor groups. However, they could still be improved in terms of reducing computational cost and sharing information between layers (cross-layering). Therefore, we redefined our proposal with two main changes (Hernández-Serrano, 2008):

- We replaced the asymmetric cryptography used in (Hernández-Serrano et al., 2008) with symmetric cryptography to reduce the computational cost of the new protocol. As a result, source authentication (pair of private and public keys) is always replaced by group authentication (the group key $K_{0,0}$).
- We added a cross-layer system to the routing layers, which enables the system to maintain the necessary routes using only the messages for creating/maintaining the secure group and avoids unnecessary network flooding.

These changes are designed to reduce the consumption of resources: the first compensates for the restrictions on low-end devices in WSNs, and the second derives from the fact that the original protocol in (Hernández-Serrano et al., 2008) only requires that nodes know their neighbors so that the knowledge could be expanded step by step to eventually cover every group member. If this knowledge is used and it is provided to upper routing layers, each node is able to determine all its required routes and the costly route discovery process (unnecessary flooding) does not need to be executed.

The logical KEK tree used in this proposal is generated from cooperation of the sensor nodes and thus it is created in a bottom-top manner: every KEK is generated as a function of the KEKs associated to the underlying nodes. Consequently, we define the key associated to a node (i,j) as in expression 2. With i, the depth (row) in the tree; j, the column position within row i (see the logical tree of KEKs in figure 2); L, the depth of the tree; $g()$, an unidirectional function that it is used to blind the keys; $f()$, a combinatorial function; and *random*(), a random function used to create the keys associated to the leaf nodes (the group members).

$$\delta = \frac{N}{M \times M} \cdot \pi \cdot R^2 \qquad (2)$$

As in every tree based GKM scheme, each member must be able to calculate all of the tree keys from the leaf node where it is placed to the root of the tree. Consequently, every member must know both its key and the blinded keys of every sibling node in its path to the root. For example, consider the tree figure 2, M1 must know its own KEK $K_{3,0}$ and the blinded KEKs $g(K_{3,1})$, $g(K_{2,1})$, $g(K_{1,1})$, and now M1 can calculate its necessary tree KEKs as: $K_{2,0}=f(g(K_{3,0}), g(K_{3,1}))$, $K_{1,0}=f(g(K_{2,0}), g(K_{2,1}))$ and $K_{0,0}=f(g(K_{1,0}), g(K_{1,1}))$.

Next we present how the logical tree of KEKs is created by cooperation of the group senor nodes and how the global key is updated when the group membership changes.

Initial Creation of the Group

Since the new protocol uses symmetric cryptography, during its initial execution all group nodes must be preloaded with an initial symmetrical key, which is the current $K_{0,0}$. The system therefore uses group authentication in which the members only need to know the current session key. Group authentication is sufficient if we assume that all members are trusted and hence attacks from insiders are not possible.

After the WSN deployment the sensor nodes agree pairwise keys with all its neighboring nodes by means of the common shared secret $K_{0,0}$. This approximation, which works in a similar way as the one in (Du et al., 2005) does, solves the problem of establishing pairwise keys. Consequently we focus now on how to solve the global keying problem by means of a logical KEK tree which also makes rekeying as efficient as possible.

The initial creation of the tree works as follows. First of all, each node publishes its weight to its neighbors by sending a local broadcast message. Therefore, once it has been completed each node should know its one-hop neighbors. Once every node knows its neighborhood, the nodes start the association by pairs. Association is just the generation of a shared key by means of interchanging the blinded keys of both incumbents as in expression 2.

After association, a key is created and shared by both incumbents, and the node with highest weight is chosen as the leader of the pair. In the next step, pairs start the association forming groups of four nodes. Consequently, a new key is created and shared by all of the four members of the group. Next, groups of four nodes start the association into sets of eight nodes with a new shared key. The association process is repeated time and again until an only logical key tree is created. In order to know how groups of nodes associate each other we defined a priority function that assigns a unique weight to every node. Every group is leaded by what we call the *tree leader* (TL), which is the node with highest weight within the group. We defined the group with the highest association priority as the one with the lowest tree depth, the highest weight (sum of member weights) and the leader with highest weight; in this order.

More in detail, the association process between two trees is just the interchange of the blinded root keys of both trees. A new logical tree is created by establishing a new root node. The two associating trees hang from this new root and, thus every member can generate the new root key as in expression 2.

For the sake of clarity, we next summarize the overall behavior of the protocol with an example. Let us suppose a sensor deployment such as the one in figure 4. The numbers represent each member weight; and the lines between members represent direct visibility at the link layer. Tables 1, 2, 3 and 4 show the network hops between a source node (row) and a destination node (column) during the different iterations of the protocol.

Figure 4. Establishment of the logical tree of KEKs (secure group)

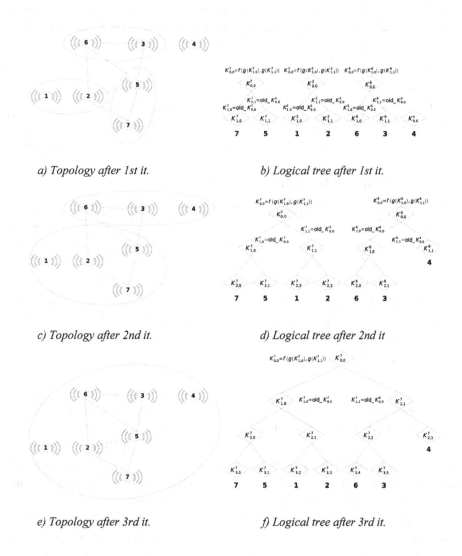

a) Topology after 1st it. b) Logical tree after 1st it.

c) Topology after 2nd it. d) Logical tree after 2nd it

e) Topology after 3rd it. f) Logical tree after 3rd it.

Table 1. Initial table of routes

Node IDs	1	2	3	4	5	6	7
1	0	1	-	-	-	-	-
2	1	0	-	-	1	1	1
3	-	-	0	1	1	1	-
4	-	-	1	0	-	-	-
5	-	1	1	-	0	1	1
6	-	1	1	-	1	0	-
7	-	1	-	-	1	-	0

Table 2. Table of routes after first iteration

Node IDs	1	2	3	4	5	6	7
1	0	1	-	-	-	-	-
2	1	0	2(6)	-	1	1	1
3	-	-	0	1	1	1	2(5)
4	-	-	1	0	-	2(3)	-
5	-	1	1	-	0	1	1
6	-	1	1	2(3)	1	0	2(5)|3(3)
7	-	1	-	-	1	2(5)	0

Table 3. Table of routes after second iteration

Node IDs	1	2	3	4	5	6	7
1	0	1	-	-	-	-	2(2)
2	1	0	2(6)	-	1	1	1
3	-	-	0	1	1	1	2(5)
4	-	-	1	0	-	2(3)	-
5	-	1	1	-	0	1	1
6	-	1	1	2(3)	1	0	2(5)|2(2)|3(3)
7	-	1	-	-	1	2(5)|2(2)	0

Table 4. Table of routes after the third iteration

Node IDs	1	2	3	4	5	6	7
1	0	1	-	-	-	4(2)	2(2)
2	1	0	2(6)	-	1	1	1
3	-	-	0	1	1	1	2(5)|3(6)
4	-	-	1	0	-	2(3)	4(3)
5	-	1	1	-	0	1	1
6	-	1	1	2(3)	1	0	2(5)|2(2)|3(3)
7	-	1	-	-	1	2(5)|2(2)	0

The identifier of the first node in the route path is shown in parenthesis. A dash is used to denote no known route.

First of all, each node publishes its weight to its neighbors by sending a local broadcast message. Therefore, after the broadcast each node should know its one-hop neighbors. At this point, the route tables are as shown in table 1.

The first iteration of the protocol proceeds as follows (see figures 4a and 4b). The members that are known to be the heaviest in their neighborhoods are 4, 6 and 7. The remaining members wait for the following decisions: 1 waits for 2; 2 for 7; 3 for 6; and 5 for 7. This first iteration is carried out as follows:

- The association requests are sent: 4 requests association with 3; 6 with 5; and 7 with 5.
- 3 does not answer 4 because it is waiting for the decision of 6.

- 5 confirms the request received from 7. Therefore, 5 and 7 are associated in the tree [7,5] with 7 as the TL and weight of the new tree of 7+5=12. Both report the action to their neighboring leaders, which in this case are their vicinity:
  ○ 5 reports to 2, 6 and 3. In the report 5 includes its distance in hops to 7. Now both 6 and 3 know a two-hop route to 7 via 5. Node 2 does not store this path because it already knows a shorter one (direct visibility).
  ○ 7 informs 2 that it has already been associated in this phase and sends updated tree information to 2
- 2 now waits for decision of the next heaviest unassociated neighbor, which in this case is 6.
- 6, which was waiting for a response from 5, is now the heaviest unassociated node in its neighborhood. Therefore, 6 requests

association with the next heaviest unassociated neighbor, 3.

- 3 answers 6 affirmatively and hence 6 and 3 are associated in the tree [6, 3] and inform their neighboring leaders about the new created tree:
  - 6 informs 2 and 5 that it has already been associated in this phase.
  - 3 informs 4 and 5 that it has already been associated in this phase. Now both 4 and 5 know a two-hop route to 6 via 3. Node 5 does not update its routing table because it already has a shorter path to 6 (direct visibility).
- 2, which was waiting for a response from 6, is now the heaviest unassociated node and requests association with 1.
- 1 answers 2 affirmatively and thus 2 and 1 are associated in the tree [2, 1] and they inform their neighboring leaders:
  - 2 informs 6, 5 and 7.
  - 1 does not need to inform any other node because its only neighbor is 2.
- Finally, 4, which had requested association with 3, assumes that it cannot associate with any other member because it has already received a report from 3, and informs its neighbor that it is 3.

At this point, the subphase of associations of the first iteration has finished. Now, the neighboring trees are updated and associated nodes inform their leaders of the nodes that they know:

- 3 informs 6 that it can reach 7 in two hops and 4 in one hop. 6 now knows how to reach tree leader 4 in two hops via 3. It also knows an alternative three-hop route to 7 via 3.
- 5 informs 7 that it can reach 6 in one hop and 2 in one hop. 7 now knows how to reach tree leader 6 in two hops via 5. It does not store the route to 2 because it already has direct visibility of this node.

- 1 informs 2 that it does not know any other nodes.

Now, first iteration is over and next one (see figures 4b and 4c) can be started. The known routes are shown in table 2. At this moment, the trees with highest priority in its neighborhood are [4] and [7,5]. The rest waits: [6,3] waits for [4]; and [2,1] for [7,5]. Next steps are:

- [4] requests association with [6,3]; and [7,5] with [6,3]. [6,3] answers yes to [4], both form a new tree [6,3,4] (depth=2 and weight=13) and report to their neighbors.
- Now [7,5] is the tree with the minimum depth and greatest weight in its neighborhood. Therefore it requests association with [1,2]. [1,2] answers affirmative, both form a new tree [7,5,1,2] and report to their neighbor tree [6,3,4].

Once the association subphase of the second iteration is finished, every new ex-TL reports the route cost in network hops to the TLs that it currently knows to its new TL. As consequence, the route table after the second iteration is the one in table 3.

Finally, in iteration 3 (see figures 4e and 4f) [6,3,4] is associated with [7,5,1,2] and a unique logical KEK tree is generated. Member 7 (the tree leader) finds out that there are not any neighbor tree and the protocol is over. Once again new ex-TL reports the route cost in network hops to the TLs that it currently knows to its new TL.

Now the group of sensors has created a logical KEK tree (see figure 4f) and they all share a new global key $K_{0,0}$. The protocol is now finished and the routing table is as shown in table 4. As expected, the final routing table shows that each node can find a route to its sibling TLs. These sibling TLs are the TLs of the subtrees hanging from the direct underlying nodes from every node in its path to the root that are not actually in this path.

Figure 5. Depth of the created logical KEK tree

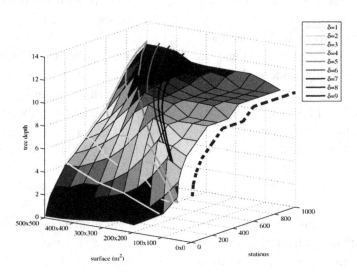

- 7 knows how to reach 5 (leader of [5]), 2 (leader of [2.1]) and 6 (leader of [6, 3, 4])
- 5 knows how to reach 7 (leader of [7]), 2 (leader of [2.1]) and 6 (leader of [6, 3, 4])
- 1 knows how to reach 2 (leader of [2]), 7 (leader of [7.5]) and 6 (leader of [6, 3, 4])
- 2 knows how to reach 1 (leader of [1]), 7 (leader of [7.5]) and 6 (leader of [6, 3, 4])
- 6 knows how to reach 3 (leader of [3]), 4 (leader of [4]) and 7 (leader of [7, 5, 1, 2])
- 3 knows how to reach 6 (leader of [6]), 4 (leader of [4]) and 7 (leader of [7, 5, 1, 2])
- 4 knows how to reach 6 (leader of [6, 3]) and 7 (leader of [7, 5, 1, 2])

As we will detail later, the routes to the sibling TLs are the only necessary unicast routes when rekeying and hence, as the creation of the tree has provided such information (cross-layer info), we can avoid a costly route discovery protocol.

So far, we have only explained how the routing table and the logical tree key are generated simultaneously from the contributions of all group members. The creation of the tree is probably more costly than just agreeing a group key but this is justified by its use for handling the necessary rekeyings because of new joins or members leaving the group. The use of the logical KEK tree

for rekeying is detailed in next section. Anyway, the overall distributed creation of the logical KEK tree has been evaluated in (Hernández-Serrano, 2008). Among the parameters evaluated by simulation we present next the most important ones: first, the depth of the generated tree (see figure 5), and second, the mean and maximum of transmitted messages per node in order to create the tree (see figure 6). The former is important to check whether the created tree is balanced or not and therefore whether it will be efficient in terms of rekeying. The second one is very important since transmitting a message is by far the more costly operation for a sensor node.

In order to understand the presented results we define next the normalized density δ as the density relative to the station range. Assumed a circular range area of $\pi \cdot R^2$, a square deployment area of $M \times M$, and a total number of sensor nodes deployed of N, δ is denoted as in expression 3.

$$\overline{M} = \frac{1}{N}\left(N + N\mu + \frac{N}{2}\mu + \frac{N}{4}\mu + \ldots + \mu\right) = 1 + \mu \sum_{i=0}^{\log_2 N - 1}\left(\frac{1}{2}\right)^i = 1 + \mu\frac{\frac{1}{N} - 1}{\frac{1}{2} - 1}$$
$$\lim_{N \to \infty} \overline{M} = 1 + 2\mu$$

(3)

The normalized density δ gives us a reliable approximation of the number of stations within the range area of a specific station (including it-

Figure 6. Sent messages per node during key tree creation

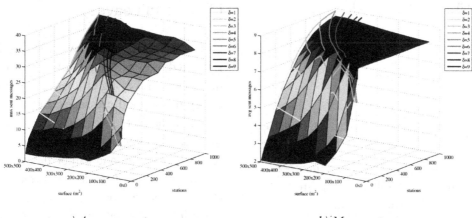

a) *Average sent messages*   b) *Max. sent messages*

self). It was demonstrated in (Hernández-Serrano, 2008) that in order to avoid (with high probability) unlinked sensor nodes (without stations in their ranges) δ must be greater than 5 stations per range area and hence only results for $\delta \geq 5$ have to be taken into account.

Results in figures 5 and 6 were obtained from 30 evaluations used for specific areas from 100x100 to 500x500 m². The number of stations varies from 10 to 1000 and the station range radius is 20 meters. On the figures there are always represented different δ values as color lines. As stated before the results have to be considered only for $\delta \geq 5$.

On figure 5 it is observed that, as expected, the tree depth increases logarithmically with the number of stations. It is also interesting to note that the protocol generates logical trees that are very close to being balanced (depth of

$$\Pr\left\{S_1 \cap S_2 \neq 0\right\} = 1 - \frac{(k - m_1)!(k - m_2)!}{k!(k - m_1 - m_2)!} \quad ,$$

represented as a black dashed line). In fact, the created trees only differ from the ideal balanced tree by one or two levels. From this result we verified that the created tree is balanced and thus efficient for its purpose that it is the rekeying.

Balanced trees are achieved by using protocol design guidelines that prioritize trees with the least depth for association. This ensures that the smallest trees do not remain without association for more rounds than necessary and that differences in tree depth cannot increase beyond one or two levels. Unbalancing occurs when trees with different depths associate.

Figure 6a shows the average number of messages sent by a node during the creation of the logical KEK tree. It can be clearly seen that the protocol converges to an upper limit of eight messages. We can justify such result with a simple analytical inspection. At the beginning of the protocol each of the N nodes sends a message (publishes its weight) and then the nodes start associations until construct a single logical KEK tree. Since associations are configured in pairs, the number of tree leaders is reduced by half in each iteration. If we denote μ as the average number of messages sent by a TL during each iteration, we can denote the average number of messages sent by a node as in expression 4. If we apply the limit for large values of N to this expression, we obtain the expression in 5, which corresponds to the simulated value when $\mu \approx 3.5$, that is to say that with this proposal each node must send a mean of only 8 messages during the creation of the logical tree of KEKs independently of the size of the group.

$$K_{i,j} = f(g(K_{i+1,2j}), g(K_{i+1,2j+1})) \qquad \begin{array}{l} 0 \le i < L \\ 0 \le j < 2^i \end{array}$$

$$K_{i,j} = random(i,j) \qquad \begin{array}{l} i = L \\ 0 \le j < 2^i \end{array}$$

$$\qquad\qquad\qquad\qquad\qquad\qquad (4)$$

$$\delta = \frac{N}{M \times M} \cdot \pi \cdot R^2 \qquad\qquad (5)$$

Figure 6b shows the maximum number of messages sent by a member of the group for different areas and numbers of nodes. We can deduce from the figure that the maximum number of sent messages increases with $\log_2 N$. Note that the node that sends the most messages is the final tree leader; it will therefore send messages during each of the $\log_2 N$ iterations. Since μ messages are sent per iteration, we can determine that the node will send $\log_2 1000 \cdot 3.5 \approx 35$ messages for 1000 members and assuming $\mu \approx 3.5$, which matches the simulation results.

At this point it is useful to recall that, with a trivial solution, the key server sends N messages to establish the group key and N messages for re-keying; with the centralized tree-based GKM protocols, the KS uses N messages to establish the group key and $O(\log_2 N)$ messages for re-keying; with our proposal, the node that sends the most messages only sends $\mu \cdot \log_2 N$ when the group key is established and (as with any other tree-based algorithm and as detailed in next section) $O(\log_2 N)$ messages for the re-keying process. However, with our protocol every node in the group sends an average of at most 8 messages due to the distribution of the GKM tasks and regardless of the group size. This is probably the most valuable property of this contribution since it allow us to assert that this proposal is perfectly scalable since the creation of the logical KEK tree is a very low cost process regardless of the number of deployed sensors.

Management of Group Dynamism (Joins and Leaves)

So far, we have shown how the logical KEK tree is created; now we detail how to use this tree to improve the rekeying efficiency in terms of necessary transmitted messages. Let us begin with case of node that leaves the group.

A leaving event is produced by node compromise or failure and can be detected in one of two ways: by using an *intrusion detection system* (IDS); or when a node detects a broken connection when it tries to send a package to the following hop. However, since we are unable to determine whether the node has failed down or simply does not respond because it has been compromised, the rekeying process must be executed to ensure that the group security after the leaving.

The leaving process is carried out as follows. First of all, the nodes that detect the leaving member remove all of the paths that pass through it from their routing tables and send a broadcast message to notify the *rekeying master* (RM) of the leaving event. The RM is the node hanging from the same node of the tree as the compromised node or, if it does not exist, the leader of the sub-tree hanging from the first sibling node of the leaving member in a leaf-root way. Just after being aware of the leaving event, the RM also removes the affected routes and initiates the leaving process. It first deletes every spare node and consequently moves itself or its subtree. Then, the RM updates its key and regenerates all the keys in its path to the root. Next, the RM sends to its sibling TLs the needed new blinded keys to reconstruct the tree. After receive the blinded keys, each involved leader sends them to the rest of members in its subtree encrypted with the subtree root key. And finally every member in the tree regenerates the tree keys in its path to the root.

Note that the communication between the RM and its sibling TLs must be unicast and secure. Nevertheless, since the group key is compromised, hop-by-hop security is used by means of the pre-

Figure 7. Prob. of losing a route

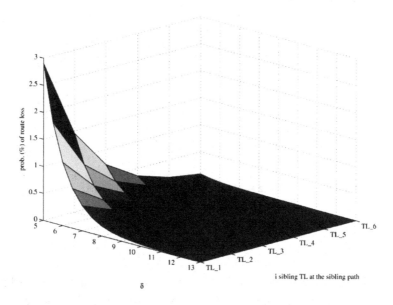

negotiated pairwise keys that are not known by the leaving member.

We now return to the previous operation example as it was after the creation of the logical tree of KEKs and routing table (see figure 4). Let us assume that member 3 detects that member 6 no longer responds. Member 3 is the RM and must therefore initiate a leaving process for member 6 by updating the compromised tree keys, irrespective of whether member 6 has failed or been compromised. In this case, the compromised keys are the ones associated with nodes (3,4), (2,2), (1,1) and (0,0). First, the RM removes the remaining tree nodes (3,4) and (3,5), replaces itself at node (2,2), and generates a new key $K'_{2,2}$. Next, the RM sends to its sibling TLs (7 and 4) the blinded keys that they need to reconstruct their required tree keys, that is to say:

- The RM, which is member 3, sends to 4 the blinded key $g(K'_{2,2})$. After receiving the key, 4 reconstructs $K'_{1,1}$ as $f(g(K'_{2,2}),g(K_{2,3}))$ and the new group key $K'_{0,0}$ as $f(g(K_{1,0}),g(K'_{1,1}))$.
- 3 sends to 7 the blinded keys $g(K'_{1,1})$.

7 reconstruct the group key as $K'_{0,0}$ $=f(g(K_{1,0}),g(K'_{1,1}))$ and also send $g(K'_{1,1})$ encrypted with the non-compromised $K_{1,0}$ to its tree members 5, 1 and 2 in order to allow them to also reconstruct the new group key.

The message between 3 and 4 is secured with the pairwise key pre-negotiated between both members. The message between 3 and 7 follows the path 3-5-7 and hence is first secured with the pairwise key between 3 and 5 and then with the pairwise key between 5 and 7.

The process to handle new member joins works in a similar way and hence it is not detailed here. The only difference is that the RM is the group member receiving the join request.

As the reader can have already deduced, with this proposal we achieve the same rekeying efficiency as the GKM protocols based on logical trees of KEKs do: we have reduced the necessary rekeying messages to $O(\log_2 N)$ with N the number of members of the group. Moreover there is not an only entity in charge of sending these messages since there are many candidates for the RM, and

Figure 8. Example of a leaving process

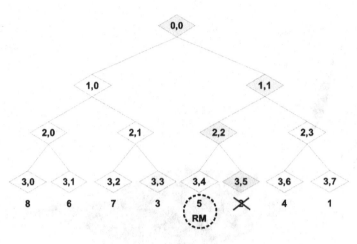

thus the rekeying is shared. Compared with the naïve solution the improvement is even better since, as stated before, the number of rekeying messages is $O(N)$ since the key server establishes one secure channel with every group member in order to send him the new group key. Moreover, the naïve solution is completely centralized and relies on the use of a powerful node or base station. Compared with the distributed pairwise approaches in the literature, our proposal provides methods to efficiently manage new member joins and leaves. With the pairwise approaches a whole new creation of the group should be re-run with the consequent cost and time. Nevertheless with our proposal just a few $O(\log_2 N)$ of sent messages is needed.

As the cost of rekeying is similar to centralized GKM protocols ($O(\log_2 N)$ transmitted messages), we now assess the effectiveness of the cross-layer mechanism. Specifically, we determined by simulation the probability that a member loss will remove a possible route from the RM to any of its sibling TLs (see figure 7). If this occurs, the RM has to execute a route discovery protocol and assume the corresponding cost.

In figure 7, the sibling TLs are enumerated from the leaf to the root. That is to say, TL_1 is the TL of the subtree hanging from the node which is underlying node from the first node in the RM path to the root that are not actually in

this path. For example, according to the member loss in figure 8, TL_1 is member 4, which is the TL of the subtree hanging from node (2,3), and TL_2 is member 8, which is the TL of the subtree hanging from node (1,0).

Figure 7 shows that for $\delta > 7$ the probability of no finding a route due to a member loss is less than 0.5%, even when the communication is with the nearest TL, which is with high probability the closest one in network hop and thus the one with less potential known routes. We can use this information to determine how to deploy the network according to a policy-defined required probability. For example, if we want the probability route loss to be no higher than 0.1%, we will have to force a deployment with $\delta \geq 9$; that is to say, if we have an area of 5km^2 and a station range radius R of 20 meters, we can obtain from expression 3 the minimum recommended number of stations to deploy as $N=(\text{area} \cdot \delta)/(\pi \cdot R^2) \approx 38.810$ sensor nodes. In any case, when a route is not found it does not mean that communication cannot be established but that a unicast route discovery protocol such as AODV must be executed. Since the route discovery is done by flooding the network, it is a costly process that must be as far as possible avoided. The cross-layer mechanism in our proposal significantly reduces the probability of executing this process.

FUTURE TRENDS

Until now we have presented a state-of-the-art of GKM for WSNs. Many proposals have been detailed and it could be assert that, despite of needing further enhancements, they on the whole fulfill the main design objectives. However many closely related branches need further work.

Any good GKM protocol provides with the methods to update the group key whenever a node is compromised or becomes malicious. Nevertheless, execution of the leaving process is a consequence of a previous detection of this node and thus a parallel system for finding the malicious user is necessary. *Intrusion detection systems* (IDS) are valuable tools for detecting such compromised nodes, as well as many other attacks. In WSNs, IDSs can detect which nodes are suspicious or malicious, and provide with this information to other protocols of the node (e.g. routing, aggregation). Due to the gaps in the existing research, a distributed IDS architecture for WSNs is in great demand. The IDS architecture will have to fulfill all the following properties (Roman, Lopez & Gritzalis, 2008): full network coverage (cover all the information flow of the network), simplicity (use mainly simple components, statistics and mechanisms), and adaptability (possibility to include new detection mechanisms and existing research, or to exclude those detection mechanisms that are not needed).

Considering the cryptography advances in WSNs, the advent of new public key primitives for sensor nodes (Malan, Welsh & Smith, 2004) opens a new, uncharted area in the field of key infrastructures. Public key cryptography could help on the secure creation of pairwise keys after network deployment or in on-the-fly manner. However, it is necessary to develop specific primitives which are specially designed for highly constrained devices such as sensor nodes.

CONCLUSION

In this chapter we have studied and discussed the development of group security schemes for WSNs. The group security is, within the simplest approach, merely based on the use of a common shared secret or group key that provides both group authentication and confidentiality. Since the group key must be only known to the current members, a group key update or rekeying becomes necessary whenever the group membership changes. The GKM is the branch of knowledge that studies how to generate and efficiently update the group key. As the GKM is far from being a new research line, we first introduced the necessary background and the most important contributions to GKM in the literature. From these contributions we stress the GKM protocols based on logical trees of KEK that achieve the best performance in terms of reducing the necessary messages for rekeying.

Once presented the "classical" background on GKM we have discussed the current state-of-the-art proposal specifically targeted to WSNs. These proposals are focused into two main objectives: generation of the global key and generation of pairwise keys between nodes. The former is related to establish and update the global key or group key (common shared secret). The second one refers to the generation of shared keys between pair of nodes and it is needed to secure the communication whenever the group key is compromised and there is a lack of a central key server (distributed proposals). Unfortunately, state-of-the-art proposals focus on the initial establishing of pairwise and global keys but do not take into account the rekeying cost generated by members joining and leaving events. Therefore we present an own GKM proposal for WSNs that is designed to reduce the rekeying cost when the group membership changes. To extend the life of the nodes and reduce management costs, the proposal incorporates a cross-layer mechanism into the routing protocol that reduces the route discovery cost. We have presented analytical and simulation results of the initial establishment of the secure group. These results probe that the protocol has no deadlocks, that the logical key tree formed tends to be balanced; that the spent bandwidth in terms of sent messages only grows with the logarithm of the number of members of the group; and that the secure group can be established with enough routing data to avoid most of the necessary route discovery process. With all these results we can conclude that protocol is perfectly suited for large sensor groups. This is, to the best of our knowledge, the only solution suitable for distributed WSNs that achieves the same efficiency level as the centralized solutions for the rekeying with only a limited overhead during the initial creation of the group with respect to the state-of-the-art pairwise approaches.

Finally, we have argue the future trends of GKM for WSNs that includes the developing of distributed IDS and further work on public key cryptography for resource constrained devices such as sensor nodes.

REFERENCES

Anderson, R., Chan, H., & Perrig, A. (2004, October). Key infection: smart trust for smart dust. In *ICNP 2004: Proceedings of the 12th IEEE International Conference on Network Protocols*, (pp. 206-215).

Balenson, D. M., McGrew, D., & Sherman, A. (1999, 15 March). Key management for large dynamic groups: One-way function trees and amortized initialization. In *IRTF SMUG meeting*. (Internet Draft <draft-balenson-groupkeymgmt-oft-00.txt>).

Baugher, M., Canetti, R., Dondeti, L., & Lindholm, F. (2002, 28 February). *Group key management architecture*. (Internet Draft -draft-ietf-msec-gkmarch-02.txt-).

Blom, R. (1984). An optimal class of symmetric key generation systems. In *Proceedings of Eurocrypt 84*.

Blundo, C., Santis, A. D., Vaccaro, U., Herzberg, A., Kutten, S., & Yong, M. (1998). Perfectly secure key distribution for dynamic conferences. *Information and Computation, 146*(1), 1–23. doi:10.1006/inco.1998.2717

Camtepe, S. A., & Yener, B. (2004). Combinatorial design of key distribution mechanisms for wireless sensor networks. *Computer Security (ESORICS 2004). Lecture Notes in Computer Science, 3193*, 293–308.

Chan, H., & Perrig, A. (2003). Security and privacy in sensor networks. *IEEE Computer Magazine, 36*(10), 103–105.

Chan, H., Perrig, A., & Song, D. (2003, May). Random key predistribution schemes for sensor networks. In *SP '03: Proceedings of the 2003 IEEE symposium on Security and Privacy (pp. 197–213)*. Washington, DC: IEEE Computer Society.

Di Pietro, R., Mancini, L., Law, Y. W., Etalle, S., & Havinga, P. (2003, October). Lkhw: a directed diffusion-based secure multicast scheme for wireless sensor networks. In *International conference on parallel processing workshops* (pp. 397-406).

Diffie, W., & Hellman, M. E. (1976). New directions in cryptography. *IEEE Transactions on Information Theory, 22*(6), 644–654. doi:10.1109/TIT.1976.1055638

Du, W., Deng, J., Han, Y. S., & Varshney, P. K. (2004). A key management scheme for wireless sensor networks using deployment knowledge. *IEEE INFOCOM, 1*, 586–597. doi:10.1109/INFCOM.2004.1354530

Du, W., Deng, J., Han, Y. S., Varshney, P. K., Katz, J., & Khalili, A. (2005). A pairwise key predistribution scheme for wireless sensor networks. *ACM Transactions on Information and System Security, 8*(2), 228–258. doi:10.1145/1065545.1065548

Eschenauer, L., & Gligor, V. D. (2002). A key-management scheme for distributed sensor networks. In *CCS '02: Proceedings of the 9th ACM conference on Computer and Communications Security* (pp. 41-47). New York: ACM Press.

Hardjono, T., Cain, B., & Doraswamy, N. (2000, August). *A framework for group key management for multicast security.* (Internet Draft <draft-ietf-ipsec-gkmframework-03.txt>).

Hardjono, T., Cain, B., & Monga, I. (1998, November*). Intra-domain group key management protocol.* (Internet Draft <draft-ietf-ipsec-intragkm-00.txt>).

Harney, & Harder. (1999, March). *Logical Key Hierarchy protocol (LKH).* (Internet Draft. <draft-harney-sparta-lkhp-sec-00>).

Harney, & Muckenhirn. (1997). *Group key management protocol architecture.* IETF RFC2094.

Hartung, C., Balasalle, J., & Han, R. (2005, January). *Node compromise in sensor networks: The need for secure systems* (Tech. Rep.). University of Colorado at Boulder.

Hernández-Serrano, J. (2008, July). *Contribución a la seguridad de grupo en redes inalámbricas avanzadas.* Unpublished doctoral dissertation, Universitat Politècnica de Catalunya, Spain.

Hernández-Serrano, J., Pegueroles, J., & Soriano, M. (2008, November). Shared self-organized GKM protocol for MANETs. [JISE]. *Journal of Information Science and Engineering, 24*(6), 1629–1646.

Huang, D., Mehta, M., Medhi, D., & Harn, L. (2004). Location-aware key management scheme for wireless sensor networks. In *Proceedings of the 2nd ACM workshop on security of ad hoc and sensor networks.* New York: ACM Press.

Huang, Q., Cukier, J., Kobayashi, H., Liu, B., & Zhang, J. (2003). Fast authenticated key establishment protocols for self-organizing sensor networks. In *WSNA '03: Proceedings of the 2nd ACM international conference on Wireless Sensor Networks and Applications* (pp. 141-150). New York: ACM Press.

Hwang, D. D., Lai, B.-C. C., & Verbauwhede, I. (2004, July). Energy-memory-security tradeoffs in distributed sensor networks. In Proceedings of *LNCS third international conference on ad-hoc, mobile, and wireless networks (adhoc-now)* (pp. 70-81). Vancouver, Canada: Springer.

Hwang, J., & Kim, Y. (2004). Revisiting random key pre-distribution schemes for wireless sensor networks. In *SASN '04: Proceedings of the 2nd ACM workshop on Security of Ad hoc and Sensor Networks* (pp. 43-52). New York: ACM Press.

Kim, Y., Perrig, A., & Tsudik, G. (2004). Tree-based group key agreement. *ACM Transactions on Information and System Security, 7*(1), 60–96. doi:10.1145/984334.984337

Lee, J., & Stinson, D. (2004a). *A combinatorial approach to key pre-distributed sensor networks.* Retrieved (n.d.). from, http://www.cacr.math.uwaterloo.ca/dstinson/

Lee, J., & Stinson, D. (2004b). *Deterministic key pre-distribution schemes for distributed sensor networks.* Retrieved (n.d.). from, http://www.cacr.math.uwaterloo.ca/dstinson/

Li, X. S., Yang, Y. R., Gouda, M. G., & Lam, S. S. (2001). Batch rekeying for secure group communications. In *WWW '01: Proceedings of the 10th international conference on World Wide Web* (pp. 525–534). New York: ACM Press.

Liu, D., & Ning, P. (2003a). Efficient distribution of key chain commitments for broadcast authentication in distributed sensor networks. In *Preceedings of the 10th annual network and distributed system security symposium* (pp. 263–276).

Liu, D., & Ning, P. (2003b). Establishing pairwise keys in distributed sensor networks. In *CCS '03: Proceedings of the 10th ACM conference on Computer and Communications Security* (pp. 52–61). New York: ACM Press.

Liu, D., & Ning, P. (2003c). Location-based pairwise key establishments for static sensor networks. In *SASN '03: Proceedings of the 1st ACM workshop on Security of Ad hoc and Sensor Networks* (pp. 72–82). New York: ACM Press.

Liu, D., Ning, P., & Du, W. (2008). Group-based key predistribution for wireless sensor networks. *ACM Trans. Sen. Netw., 4*(2), 1–30. doi:10.1145/1340771.1340777

Malan, D. J., Welsh, M., & Smith, M. D. (2004). A public-key infrastructure for key distribution in TinyOS based on elliptic curve cryptography. In *Proceedings of 1st IEEE Communications Society Conference on Sensor and Ad Hoc Communications and Networks* (Secon'04).

Menezes, A. J., Oorschot, P. C. V., & Vanstone, S. A. (1997). *Handbook of applied cryptography.* Boca Raton, FL: CRC Press.

Mittra, S. (1997). Iolus: a framework for scalable secure multicasting. *SIGCOMM Comput. Commun. Rev., 27*(4), 277–288. doi:10.1145/263109.263179

Pegueroles, J., Rico-Novella, F., Hernández-Serrano, J., & Soriano, M. (2003a). *Adapting GDOI for balanced batch-lkh (<draft-ietf-msec-gdoi-batch-lkh-00.txt>. Internet Draft).

Pegueroles, J., Rico-Novella, F., Hernández-Serrano, J., & Soriano, M. (2003b). Improved LKH for batch rekeying in multicast groups. In *Proceedings of IEEE international conference on Information Technology Research and Education* (ITRE). New Jersey, USA.

Perrig, A., Szewczyk, R., Tygar, J. D., Wen, V., & Culler, D. E. (2002). SPINS: security protocols for sensor networks. *Wireless Networks, 8*(5), 521–534. doi:10.1023/A:1016598314198

Pietro, R. D., Mancini, L. V., & Mei, A. (2003). Random key-assignment for secure wireless sensor networks. In *SASN '03: Proceedings of the 1st ACM workshop on Security of Ad hoc and Sensor Networks* (pp. 62-71). New York: ACM Press.

Roman, R., Lopez, J., & Gritzalis, S. (2008). Situation awareness mechanisms for wireless sensor networks [security in mobile ad hoc]. *IEEE Communications Magazine, 46*(4), 102–107. doi:10.1109/MCOM.2008.4481348

Wallner, D., Harder, E., & Agee, R. (1998). *Key management for multicast: issues and architectures.* RFC 2627.

Yu, Z., & Guan, Y. (2005, April). A key pre-distribution scheme using deployment knowledge for wireless sensor networks. In *IPSN '05: Proceedings of the 4th international symposium on Information Processing in Sensor Networks* (p. 261-268). Piscataway, NJ: IEEE Press.

Zhang, C., DeCleene, B., Kurose, J., & Towsley, D. (2002). Comparison of inter-area rekeying algorithms for secure wireless group communications. *Performance Evaluation, 49*(1-4), 1–20. doi:10.1016/S0166-5316(02)00120-7

Zhu, S., Setia, S., & Jajodia, S. (2006). LEAP+: Efficient security mechanisms for large-scale distributed sensor networks. *ACM Trans. Sen. Netw., 2*(4), 500–528. doi:10.1145/1218556.1218559

KEY TERMS AND DEFINITIONS

Centralized: Within this chapter a protocol is centralized when there is an only top level entity in charge of its management. This is the case of traditional approaches to GKM that rely on a fixed infrastructure which is the KS.

Cross-Layer: Within the context of this chapter, cross-layer refers to useful data that break the boundaries of the OSI layer model with the purpose of reducing costs.

Distributed: In a distributed network the management is shared by several entities. When these entities are not a subset but all the network entities then we talk about fully-distributed networks. Fully-distributed networks are often based on peer-to-peer (P2P) relations where users communicate in an ad-hoc manner.

GKM: Group Key Management. Branch of knowledge closely related to securing group communications that studies how to generate the group keying material and how to update it when necessary (rekeying) with focus on efficiency and cost reduction.

KS: Key Server. In centralized GKM approaches it refers to the entity that is in charge of managing the group security.

Rekeying: The rekeying is the process of updating the keying material. In centralized GKM approaches is carried out by the Key Server (KS).

Siblings: The sibling nodes of a member in a logical tree of KEKs are the direct underlying nodes from every node in its path to the root that are not actually in this path. The set of sibling nodes is denoted as the sibling path. Thus, the sibling trees of a member are the trees hanging from the nodes of its sibling path.

Chapter 15
Jamming Attacks and Countermeasures in Wireless Sensor Networks

Yan-Qiang Sun
National University of Defense Technology, China

Xiao-Dong Wang
National University of Defense Technology, China

ABSTRACT

Guaranteeing security of the sensor network is a challenging job due to the open wireless medium and energy constrained hardware. Jamming style Denial-of-Service attacks is the transmission of radio signals that disrupt communications by decreasing the signal to noise ratio. These attacks can easily be launched by jammer through, either bypassing MAC-layer protocols or emitting a radio signal targeted at blocking a particular channel. In this chapter, we survey different jamming attack models and metrics, and figure out the difficulty of detecting and defending such attacks. We also illustrate the existed detecting strategies involving signal strength and packet delivery ratio and defending mechanisms such as channel surfing, mapping jammed region, and timing channel. After that, we explore methods to localize a jammer, and propose an algorithm Geometric-Covering based Localization. Later, we discuss the future research issues in jamming sensor networks and corresponding countermeasures.

INTRODUCTION

The broadcast nature of wireless networks makes them particularly vulnerable to radio interference, which prevents the normal communications (Akyildiz IF, 2002; N. Ahmed, 2005). This interference or jamming can destroy wireless transmission and may occur either by means of unintentional interference or collision at the receiver side or intentional attacks. The jamming attack can be easily launched since it can be implemented by simply listening to the open medium and broadcasting in the same frequency band as the sensor networks (Y.W. Law, 2005, Feb).

In order to cope with this kind of Denial-Of-Service style attack, many strategies and techniques have been developed. The traditional method is to use the sophisticated physical layer technologies such as DSSS (Direct Sequence Spread Spectrum) and FHSS (Frequency Hopping Spread Spectrum),

DOI: 10.4018/978-1-61520-701-5.ch015

Figure 1. An example of Jamming attack in WSN

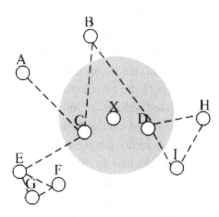

which have been widely used in military communication (Y.W. Law, 2005, November). However, it can be too costly for the energy and frequency constrained sensor networks. So many kinds of evasion strategies have been researched, such as wormhole-based anti-jamming techniques (M. Cagalj, 2007), channel surfing (W. Xu, 2007) and timing channel (W. Xu, 2008).

In this chapter, we survey issues related to jamming sensor networks by examining both the attack and defend sides of the problem. We first give two kinds of jamming attack models and two widely used metrics. Then, we would like to introduce approaches to detecting jamming attack and defending mechanisms. After that, we explore methods to localize a jammer, and propose an algorithm Geometric-Covering based Localization. At last, we analyze the future trends of jamming attacks and corresponding countermeasures in wireless sensor networks.

JAMMING ATTACK MODELS AND METRICS

Jamming attack is a kind of Denial of Service attack, which prevents other nodes from using the channel to communicate by occupying the channel that they are communicating on (W. Xu, 2005). We define the jammer in wireless sensor network as an entity who is purposefully trying to interfere with the physical transmission and reception of wireless communications. A typical scenario of jamming attack is shown in figure 1. The normal nodes C and D has been jammed by the malicious node X, so the communications between the jammed nodes(C, D) and the normal nodes (A, B, E, H, I) are disrupted.

Jamming Attack Models

There are many different attack strategies that a jammer can perform in order to interfere with other wireless communications. It is hard and impractical to cover all the possible attack models that might exit (A. D. Wood, 2007), so we list two kinds of jamming attacks that have been widely researched and proven to be effective in disrupting wireless communication in wireless sensor networks.

1. Active jammer

    The active jammer tries to block the channel irrespective of the traffic pattern on the channel.

    ○   Constant jammer

    The constant jammer continually emits radio signals and can be implemented using either a waveform generator that continuously sends radio signals [11] *or a normal wireless device that continually sends out random bits to the channel without following any MAC-layer etiquette* (W. Xu, 2005).

    ○   Deceptive jammer

    The deceptive jammer constantly injects regular packets to the channel without any gap between subsequent packet transmissions, instead of sending out random bits (W. Xu, 2005). As a result, a normal communicator will be deceived into believing there is a legitimate packet and be duped to remain in the receive state (W. Xu, 2005).

    ○   Random jammer

    Instead of continuously sending out a radio signal, *random jammer alternates between sleeping and jamming. Specifically, after jamming for*

t_j units of time, it turns off its radio, and enters a "sleeping" mode. It will resume jamming after sleeping for t_s time, t_s and t_s can be either random or fixed values (W. Xu, 2005).

2. Reactive jammer

The three models mentioned above are active jammers, which are usually effective because they make the channel busy all the time. However, it is relatively easy to detect because of the property of "online". For the reactive jammer, it is not necessary to jam the channel when nobody is communicating with. Instead, the jammer keeps quiet when the channel is not busy, but starts transmitting radio signals as soon as it senses activity on the channel (W. Xu, 2005). It is specifically targets the reception of a message.

Wenyuan Xu (2005) implemented the above four jammer models using Berkeley Motes that employ a ChipCon CC1000 RF transceiver and use TinyOS as the operating system. They disabled channel sensing and back off operations to bypass the MAC protocol, so that the jammer can blast on the channel irrespective of other activities that are taking place.

Jamming Metrics

The purpose of the jammer is to block the legitimate wireless communication. *A jammer can achieve this goal either by preventing a real data source from sending out a packet, or by preventing the reception of legitimate packets* (W. Xu, 2005). Here, we assume A and B denote two legitimate wireless nodes, and X to be a jammer. We use the following two metrics that have been defined By Wenyuan Xu (2005) to measure the effectiveness of a jammer.

- **Packet Send Ratio (PSR)**: *The ratio of packets that are successfully sent out by a legitimate traffic source compared to the number of packets it intends to send out at the MAC layer* (W. Xu, 2005). For example, suppose node A has a packet to send,

in the MAC protocol employed by Mica2, the channel must be sensed as being in an idle state for at least some random amount of time before A can send out a packet. Further, different MAC protocols have different definitions on an idle channel . They may compare the signal strength measured with a fixed threshold, or just adapt the threshold based on the noise level on the channel. A radio interference attack may cause the channel to be sensed as busy, causing A's transmission to be delayed. *If too many packets are buffered in the MAC layer, the newly arrived packets will be dropped* (W. Xu, 2005). It is also possible that a packet stays in the MAC layer for too long, resulting in a timeout and packets being discarded. If A intends to send out n messages, but only m of them go through, the PSR is $\dfrac{m}{n}$ (W. Xu, 2005). The PSR can be easily measured by a wireless device by keeping track of the number of packets it intends to send and the number of packets that are successfully sent out (W. Xu, 2005).

- **Packet Delivery Ratio (PDR)**: *The ratio of packets that are successfully delivered to a destination compared to the number of packets that have been sent out by the sender* (W. Xu, 2005). Even after the packet is sent out by A, B may not be able to decode it correctly, due to the interference introduced by X. Such a scenario is an unsuccessful delivery. The PDR may be measured at the receiver B by calculating the ratio of the number of packets that pass the CRC check with respect to the number of packets (or preambles) received. PDR may also be calculated at the sender A by having B send back an acknowledge packet. In either case, if no packets are received, the PDR is defined to be 0 (W. Xu, 2005).

Problem Definition

There problems have been abstracted:

1) Identify whether an attack is being performed or not
   - Determine what type of jamming attack is being performed;
   - Could the network be experiencing node failure or congestion?
2) Overcome the Attack
   Network of nodes must perform some kind of method to overcome the effects of the attack.
3) Locate a jammer
   Whether can we get the physical location of jammers instead of using the evasion approaches after detecting the existence of jamming attacks, in particular, how to locate a jammer?

JAMMING DETECTION

Many wireless sensor networks are susceptible to jamming attacks. To ensure the availability of sensor communications, the jamming detection mechanisms must be developed that are easy to scale, distributed and have low false positives (M. Li. 2007). It is the first step to build a secure and robust wireless sensor networks, and it is also challenging because jammers can employ different jamming attack models.

Utility Metrics

Wood and Stankovic (2003) applied heuristics to determine whether the current node was experiencing non-transient interference that we called jamming. They expanded the definition of jamming to include any kind of denial-of-service condition in which the utility of the communication channel dropped below a certain threshold. Then they broadened the jamming model to include mobility, pulse jamming, and even link-layer jamming, all based on the jamming impact on the local ability

to communicate(C. Karlof, 2003). The idea they proposed is that below this utility threshold, the nodes in sensor networks are unable to communicate effectively enough for long enough to accomplish given tasks. So the value of utility threshold is necessarily dependent on the purpose of the sensor networks. And factors which may impact this utility metric include (Wood, 2003):

- *Repeated inability to access wireless channel*
- *Bad framing*
- *Checksum failures*
- *Illegal values for addresses or other fields*
- *Protocol violations (e.g., missing ACKs)*
- *Excessive received signal ratio*
- *Low signal-to-noise ratio*
- *Repeated collisions*
- *Duration of condition*

These data may be obtained from the local radio hardware, MAC layer, or other available sensors as mentioned by Wood and Stankovic (2003). Due to the complexity of jamming detection and its sensitivity to the deployment environment, the authors constructed it as a separate module, where the mapping module is provided for mapping service in order to defend the jamming attacks, being discussed later in the jamming defending secton.

Detecting Jamming Attacks

Wenyuan Xu (2006) presented several measurements that could be employed by wireless devices for the purpose of detecting jamming attacks. They did lots of experiments to verify the effectiveness of the measurements and also figured out some scenarios where these measurements may not be effective in detecting a jamming attack, even in fact could lead to false detections. At last, they concluded that statistics built upon individual measurements may cause false decisions, and it is impossible to detecting all kinds of jamming attacks just relying on single measurement. So two improved detection strategies were developed, which were both built upon the fundamental assumption that communicating parties should have some basis for knowing what their characteristics should be if they were not jammed, and consequently could use this as a basis for differentiating jammed scenarios from mere poor link conditions.

• Signal Strength

Since the signal strength distribution may be effected because of the presence of a jammer, it is seemingly a natural measurement to be used to detect the jamming attacks in wireless sensor networks. In order to illustrate the effect a jammer would have on the received signal strength, Wenyuan Xu (2006) presented results of several experiments conducted with the MICA2 Mote platform (described in more detail by W. Xu (2005)). By looking at raw time series data, it is clear that any statistic solely based on a moving average of the RSSI values would be hard pressed to discriminate between a normal traffic scenario and a reactive jammer scenario. Furthermore, the shape of the RSSI time series for normal traffic scenarios and the reactive jammer are too similar to rely on spectral discrimination techniques for discrimination. Further analysis of these methods and the difficulties associated with using signal strength readings can be found in the paper written by W. Xu (2005). Overall, these results suggest the following important observation: *we may not be able to use simple statistics, such as average signal strength or energy, to discriminate jamming scenarios from normal traffic scenarios because it is not straightforward to devise a threshold that can separate these two scenarios* (W. Xu, 2005).

• Carrier Sensing Time

It seems possible to make use of carrier sensing time as a measurement to determine whether one of the nodes in sensor networks is jammed. The reasonability behind this is that a jammer can prevent a legitimate source from sending out packets because the channel might appear constantly busy to the source. In this case, it is very natural for one to keep track of the amount of time it spends waiting for the channel to become idle, i.e. the carrier sensing time, and compare it with the sensing time during normal traffic operations to determine whether it is jammed (W. Xu, 2005). W. Xu (2005) explored this possibility. They observed that using carrier sensing time is suitable when the following two conditions are true: the jammer is non-reactive or non-random, and the underlying MAC protocol determines whether a channel is idle by comparing the noise level with a fixed threshold. If these two conditions are true, carrier sensing time is an efficient way to discriminate a jammed scenario from a normal ill-functioning scenario, such as congestion, because the sensing time will be bounded, although large, in a congested situation, but unbounded in a jammed situation. Overall, carrier sensing time alone cannot be used to detect all the jamming scenarios (Wenyuan Xu, 2006).

• Packet Delivery Ratio (PDR)

PDR may be used to detect the jamming attacks in wireless sensor networks, because the jammer can effectively disrupt transmissions, leading to a much lower PDR. The PDR can be measured in the following two ways: either by the sender, or by the receiver (W. Xu, 2006). At the sender side, the PDR can be calculated by keeping track of how many acknowledgements it receives from the receiver. At the receiver side, the PDR can be calculated using the ratio of the number of packets that pass the CRC check with respect to the number of packets (or preambles) received (W. Xu, 2005). Wenyuan Xu said: "*Unlike signal strength and carrier sensing time, PDR must be measured during a specified window of time where a baseline amount of traffic is expected. If no packet is received over that time window, then the PDR within that window is zero*". For example, a powerful jammer leads to a very poor PDR (close to 0) which indicates that PDR may be a good measurement in detecting jamming attacks.

Table 1. The capability of coping with different attack models (W. Xu, 2005)

	Signal strength		Carrier sensing time	Packet delivery ratio
	Average	Spectral discrimination		
Constant jammer	x	√	√	√
Deceptive jammer	x	√	√	√
Random jammer	x	x	x	√
Reactive jammer	x	x	x	√

Wenyuan Xu (2006) investigated how much PDR degradation can be caused by non-jamming normal network dynamics, such as congestion or failures at the sender side. Their studies showed that even in a highly congested situation where a raw traffic rate of 19.38 kb/s is offered to MICA2 radio whose maximum bandwidth capacity is 12.364 kb/s at a 100 percent duty cycle, the PDR measured by the receiver is still around 78 percent (Wenyuan Xu, 2006). As a result, a simple threshold mechanism based on the PDR value can be used to differentiate a jamming attack, regardless of the jamming model, from a congested network condition. Wenyuan Xu depicted that: "*Although PDR is quite effective in discriminating jamming from congestion, it is not as effective for other network dynamics, such as a sender battery failure, or a sender moving out of a receiver's communication range, because these dynamics can result in sudden PDR drop in much the same way as a jammer does*". Specifically, if the sender's battery drains out, it stops sending packets, so the corresponding PDR will be zero (W. Xu, 2005).

- Jamming detection with consistency checks

The signal strength, carrier sensing time and packet delivery ratio are basic statistics for detecting jamming attacks in wireless sensor networks. Can these basic statistics differentiate between jamming scenario from a normal scenario including congestion? The answer is no! No single measurement is capable of detecting all kinds of jamming attacks. Table 1 shows the capability of

coping with different attack models among the three main basic measurements. From the table we may find that PDR is a relative good statistic, but cannot deal with hardware failure.

So multimodal strategies were proposed by (W. Xu, 2005), in which PDR and signal strength readings are combined together. The reasonability behind this is that: In a normal scenario with no interference, a high signal strength value corresponds to a high PDR. However, if the signal strength is low (i.e., the strength of the signal is comparable to noise levels), the PDR will also be low (W. Xu, 2006). On the other hand, a low PDR does not necessarily imply a low signal strength: it may be that all of a node's neighbors have died (perhaps from consuming battery resources or device faults), or the node is jammed. The key observation here is that in the first case, the signal strength is low, which is consistent with a low PDR measurement, while in the jammed case, the signal strength should be high, which contradicts the fact that the PDR is low (Wenyuan Xu, 2006). This is summarized in table 2.

Based on these observations, Wenyuan Xu (2005) defined a multimodal consistency check, which mainly depends on a (PDR, SS) (SS: Signal Strength) look-up table that is built empirically:

- *Measure (PDR, SS) during a guaranteed time of non-interfered network*
- *Divide the data into PDR bins, calculate the mean and variance for the data within each bin*

Table 2. Packet delivery ratio and signal strength

Observed PDR	Observed Signal Strength	Typical Scenarios
PDR=0 (no preamble is received)	Low signal strength	Non jammed: neighbor failure, neighbor absence, neighbors being blocked, etc.
PDR=0 (no preamble is received)	High signal strength	Node jammed
PDR low (packets are corrupted)	Low signal strength	Non jammed: neighbor being faraway
PDR low (packets are corrupted)	High signal strength	Node jammed

- *Get the upper bound for the maximum SS that would have produced a particular PDR value during a normal case*
- *Partition the (PDR, SS) plane into a jammed-region and a non-jammed region*

Figure 2 shows a typical process of detecting whether a jamming attack happens based on the multimodal consistency check method (Wenyuan Xu, 2006). To illustrate how such a detection scheme might operate, they presented the results of their investigation, which was conducted using MICA2 Motes. They varied source-receiver separation for four different jammers. The PDR and SS readings were averaged over a sufficient time window to remove anomalous fluctuations

(e.g. hardware-related or fading-related variations) (Wenyuan Xu, 2006). They found the 99 percent SS confidence levels for different regions, and defined the jammed-region to be the region of (PDR, SS) that is above the 99 percent signal strength confidence intervals and whose PDR values are less than 65 percent (Wenyuan Xu, 2006).

JAMMING DEFENSE STRATEGIES

In this section, we will introduce mainly five defense strategies of jamming sensor networks. It contains: Mapping Service (Wood, 2003), Channel Surfing (W. Xu, 2004, 2007), Spatial Retreat (W. Xu, 2004), Wormhole-based Anti-Jamming (C. Karlof, 2003) and Timing Channel (W. Xu, 2008).

Mapping Jammed Region

In order to defend the jamming attacks, Wood and Stankovic (2003) proposed a mapping service for wireless sensor networks that can provide the following benefits:

- *Feedback to routing and directory services*
- *An effective abstraction at a higher-level than local congestion, failed neighbors, and broken routes*
- *Support for avoiding the region by network-controlled vehicles, military assets, emergency personnel*
- *Reports to a base-station for further jamming localization*

Figure 2. A typical process of detecting whether a jamming attack happens

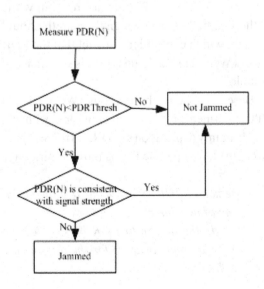

- *Aid to power management strategies for nodes inside and around jammed regions*

The basic idea: Nodes near the border of a jammed region notify neighbors outside of the region of jamming. These nodes form groups and use a lightweight, low-state management mechanism to coalescing groups and mapping the extent of the jammed region. Bridge members aid neighborhoods of low connectivity. An eager eavesdropping strategy provides forward and backward information diffusion among mapping members.

Channel Surfing

The basic idea of channel surfing proposed by Wenyuan Xu (2004, 2007) is inspired by frequency hopping techniques, but it operates at the link layer in an on-demand fashion. That is, if we are blocked at a particular channel, we can resume our communication by switching to a "safe" channel. Generally, in channel surfing, the nodes that detect themselves as jammed should immediately switch to another orthogonal channel and wait for opportunities to reconnect to the rest of the network. After they lose connectivity to their neighbors that refer to as boundary nodes, the neighbors will discover the disappearance of the jammed nodes and temporally switch to the new channel to search for them. If the lost neighbors are found on the new channel, the boundary nodes will participate in rebuilding the connectivity of the entire work (W.Xu, 2007).

However, it is not easy to implement these basic techniques in the distributed computing environments as reliably coordinating multiple devices switching to new channels faces the usual challenges: asynchrony, latency, and scalability, which depicted by Wenyuan Xu (2004, 2007). To cope with these challenges, Wenyuan Xu (2007) proposed the following channel surfing variations: coordinated channel switching and spectral multiplexing. In coordinated channel switching the

entire network changes its channel to a new channel. The scheme begins with the jammed nodes detecting they are jammed. Boundary nodes, which are not jammed but are neighbors of jammed nodes, will detect the absence of their neighbors on the original channel and probe the next channel to see if their neighbors are still nearby. If a node detects beacons on the new channel, it will switch back to the original channel and transmit a broadcast message informing the entire network to switch to the new channel. As Wenyuan Xu (2007) has said:" *One advantage of this scheme is its simplicity, and the success of performing such strategy doesn't depend on the likelihood that each individual node can detect the loss of its neighbors but the quick formation of network connectivity in spite of the radio interference*." However, performing a coordinated channel switching across an entire network incurs significant latency as the scale of the network increases; as a result, the network may be in an unstable phase where some devices are on an old channel while others are waiting on the new channel. To alleviate the latency problem, spectral multiplexing was proposed by Wenyuan Xu (2007). In spectral multiplexing only jammed regions switch channels, and the boundary nodes of a jammed region serve as relay nodes between different spectral zones.

Problems of the channel surfing is, in sensor networks, nodes commonly only have 1 single communication frequency, further more, messages are flooded across the entire network to inform nodes of the channel change, it may be too costly.

Spatial Retreat

The basic idea is: The nodes physically move out of the effectively jammed area. So the strategy of spatial retreat is suitable for the mobile sensor networks, in which nodes have mobility and are equipped with GPS or other simple localization. When the jammed nodes detect that it has been jammed, it choose a random direction to evacuate from jammed area. If no nodes are within its radio range, it moves along the boundary of the jammed area until it reconnects to the rest of the network. Therefore, the process of spatial retreat contains two phases (Wenyuan Xu, 2004):

• Escape phase

The nodes located within the jammed area move to "safe" regions, and stay connected with the rest of the network,

• Reconstruction phase

The mobile nodes move about to achieve uniform network coverage, thereby preventing the jammer from partitioning the network.

However, there are some challenges that the spatial retreat has to face. First, the ability to move nodes can be costly and difficult to control. Second, moving nodes can possibly break communication with neighboring nodes. And even more, global repositioning scheme may be difficult, which needs to define position of all nodes to avoid the jamming attack.

Wormhole-Based Anti-Jamming

Wormholes (Y.Hu, 2003) were so far considered to be a threat in wireless sensor networks. In a wormhole attack, an adversary tunnels messages received in one part of the network over a low latency link and replays them in a different part. The simplest instance of this attack is a single node situated between two other nodes forwarding messages between the two of them, and wormhole attacks more commonly involve two distant malicious nodes colluding to understate their distance from each other by relaying packets along an out-of-bound channel available only to the attacker (C. Karlof, 2003). However, if the attacker performs this tunneling honestly and reliably, no harm is done; the attacker actually provides a useful service in connecting the network more efficiently. Using this observation, M. Cagalj (2007) proposed wormhole-based techniques to defend jamming attacks. They defined a path

which called probabilistic wormhole to guarantee that, for the given randomly located jammer, there is a positive probability that a sensor node residing in the jammed region of an attacker forms a path from such a region to the area not affected by jamming (Cagalj, 2007). Three mechanisms were presented and analyzed: 1) wired pairs of sensor nodes, 2) coordinated frequency-hopping pairs, and 3) uncoordinated channel-hopping pairs of nodes.

• Wormholes via wired pairs of sensor nodes

In this solution, M. Cagalj (2007) proposed augmenting a wireless sensor network with a certain number of pairs of sensor nodes that are each connected through a wire. These connected nodes are also equipped with wireless transceivers, just like regular sensor nodes. As shown in figure 3, the connected pair (1, 2) creates a link resistant to jamming from the jammed region. When node 1 senses the presence of the jammer, it makes use of the wire channel to communicate with its peer node 2. As the wired channel between node 1 and 2 are not affected by the jammer, the message sent by node 1 is successfully received by node 2.

• Wormholes via frequency hopping pairs

The solution discussed above is based on pairs of nodes connected through wires, which has the obvious drawback that it requires wires to be deployed in the field. In this solution, two types of sensor nodes are distinguished based on coordinated frequency hopping. The first type is regular node equipped with an ordinary single-channel radio. The second type is sensor node equipped with two radios: the regular radio and a radio with frequency-hopping (FH) capability (M. Cagalj, 2007). The thing that should be mentioned is that only a certain number of FH-enabled nodes are deployed because of the FH radios' substantial overhead on sensor nodes in wireless networks. The goal of this solution is to ensure that, with a high probability, FH pairs form at least one wormhole in the presence of jamming attack.

- Wormholes via uncoordinated channel-hopping

The solution based on the coordinated FH pairs still requires a certain level of synchronization between the FH nodes that make a pair. In this solution, the authors explored the feasibility of a completely uncoordinated channel-hopping approach, which tries to create probabilistic wormholes by using sensor nodes that are capable of hopping between radio channels that ideally span a large frequency band. The major difference between channel-hopping and frequency-hopping is that, with the former one, an entire packet is transmitted on a single channel. In other words, with channel-hopping, sensor nodes hop between different channels in a much slower way (per packet basis), as compared to the classical frequency –hopping (e.g., Bluetooth) (M. Cagalj. 2007).

Timing Channels

The jamming defense strategies discussed above try to repair network connectivity in the presence of jammer. However, these evasion strategies are costly and limited to overcome a broadband jammer with unlimited power. Wenyuan Xu (2008) proposed an alternative to such evasion methods. It was motivated by the Tai Chi Chuan philosophy: A strength of four ounces can defeat a force of a thousand pounds. The main idea is to create a low bit-rate overlay that exits on top of the conventional physical/link layers and that would survive in the presence of a persistent broadband jammer. This low-rate channel is named as timing channel which can exit in the presence of jamming attacks. The 4-Ounce Overlay exits between data link and network layers, and involves the creation of a timing channel, framing, error correction, and authentication services (Wenyuan Xu, 2008)

This method is based on the possibility of jamming detection and mapping jammed regions, which have been proved possible through use of network utility metrics, such as packet delivery ratio with received signal strength (Wenyuan Xu,

2005). Once a jamming attack happens in sensor network, the legitimate Sender and Receiver will start with creating a timing channel. To form the timing channel, Receiver must detect failed packets transmitted by Sender. Regardless of the type of jammer employed, detecting packet failures is generally possible as long as the Sender is able to employ transmissions power levels comparable to the jammer's interference power (Wenyuan Xu, 2008).

LOCATE THE JAMMER

In section 3 and 4, we surveyed the detecting and defending mechanisms against jamming attacks. However, sometimes we need to know the position of the jammer as accurate as possible. For instance, one can deal with a jammer by localizing it and destroy it through human intervention. Additionally, the location of jammer provides important information for network operations in various layers (Hongbo Liu, 2009). For instance, a routing protocol can choose a route that does not traverse the jammed region to avoid wasting resources due to failed packet delivery. Nevertheless, localizing a jammer is not an easy job. First, jammers are not complied with localization

Figure 3. Wormholes via wired pairs of sensor nodes

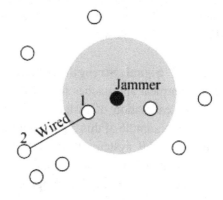

protocols. Most existing localization protocols need special hardware, e.g., GPS or ultrasound transmitter to measure the time difference of arrival. Second, we lack the feasible technique to differentiate the jamming signal from the legal signal. Finally, the proposed localization methods should not require extensive communication due to the energy-constrained sensors. To address these challenges, we first survey the existing localization algorithms, and find that Centroid Localization (CL) and Weighted Centroid Localization (WCL) can be used to locate a jammer, but the location accuracy of these two methods is very sensitive to nodes densities. Hongbo Liu (2009) proposed a Virtual Force Iterative Localization algorithm, but the cost of computation is high. Thus, we propose a novel algorithm, Geometry-Covering Localization (GCL), which uses the knowledge of computing geometry, especially the convex hull problem. Simulation results showed that GCL can achieve higher accuracy than CL, and has time complexity of $O(nlogn)$, which is proper to sensor networks.

Range-Based Localization

There has been a hot topic in the field of sensor localization, most of which can be grouped into two categories: range-based approaches and range-free approaches. Range-based algorithms involve estimating distance to beacons based on the measurement of various physical properties, which are able to provide fine-grained localization. A summary of two most popular range-based algorithms are given below:

- **Received Signal Strength (RSS):** One of the simplest approaches that have been used for estimation of distances between nodes. An example of this type of approach is the RADAR system (P. Bahl & V. N. Padmanabhan, 2000) where signal strength from static nodes is used to roughly track mobile indoor nodes.

- **Time Difference of Arrival (TDOA):** Range between nodes can be estimated far more accurately by estimating the time difference between transmission and reception of slow traveling signals such as ultrasound or acoustic waves. The obvious shortcoming of this approach is the need for nodes to have additional hardware for dealing with such slow traveling signals (U. Bischoff, 2006; Y. Chen, 2006).

Apparently, these two methods can not be applied to localize jammers due to the disrupted communication signals under the jamming attack scenarios.

Range-Free Localization

In contrast to the range-based algorithms, the range-free approaches focus on using a minimal amount of information to derive a location estimate (T. He, 2003).

- **Proximity:** Proximity measurement may be the simplest of all location techniques. It only needs to make a decision on whether two nodes are within radio range of each other. The decision can be made based on whether any packets are received at all, or can be based on a threshold approach requiring some consistent connectivity over time (P. Wilson, 2007).

- **Centroid Localization (CL):** Centroid Localization utilizes position information of all neighboring nodes, which are located within the transmission range of the targeted node. For example, assume that there are N neighboring nodes $\{(X_1, Y_1), (X_2, Y_2), ..., (X_n, Y_n)\}$. Then we can estimate the position of the target node by:

$$(\bar{X}_{target}, \bar{Y}_{target}) = (\frac{\sum_{k=1}^{N} X_k}{N}, \frac{\sum_{k=1}^{N} Y_k}{N}).$$

An enhance version of CL is Weighed Centroid Localization (WCL) (J. Blumenthal, 2007), which adds weight value into the process of estimating target node position. For example, the distance between the target node and its neighbors can be used as the metric. Then the target node can be estimated by:

$$(\bar{X}_{target}, \bar{Y}_{target}) = (\frac{\sum_{k=1}^{N} w_k X_k}{\sum_{k=1}^{N} w_k}, \frac{\sum_{k=1}^{N} w_k Y_k}{\sum_{k=1}^{N} w_k}).$$

Both the CL and WCL can be used to locate the jammer. However, these two methods are susceptible to the density and distribution of the wireless sensor networks.

Virtual Force Iterative Localization

To increase accuracy, Hongbo Liu (2009) proposed the Virtual Force Iterative Localization (VFIL) algorithm which iteratively estimates the jammer's location by utilizing the network topology. VFIL begins with a coarse estimation of the jammer's position. For instance, CL can be leveraged to perform initial position estimation, and then the jammer's position is re-estimated iteratively until the result is close to the true position. The goal of VFIL is to search for an estimation of the jammed region that can cover all the jammed nodes whereas does not contain any boundary nodes. To achieve this object, two virtual forces were defined, one is Pull Force F_{pull}^{i} generated by a jammed node i that is outside of the jammed region, another is Push Force F_{push}^{j} generated by a boundary node j that is located inside of the jammed region. F_{pull}^{i} and F_{push}^{j} were defined as normalized vectors that point to/from the estimated jammer's position:

$$F_{pull}^{i} = [\frac{X_i - \bar{X}_0}{\sqrt{(X_i - \bar{X}_0)^2 + (Y_i - \bar{Y}_0)^2}}, \frac{Y_i - \bar{Y}_0}{\sqrt{(X_i - \bar{X}_0)^2 + (Y_i - \bar{Y}_0)^2}}]$$

$$F_{push}^{j} = [\frac{X_0 - X_j}{\sqrt{(X_0 - X_j)^2 + (Y_0 - Y_j)^2}}, \frac{Y_0 - Y_j}{\sqrt{(X_0 - X_j)^2 + (Y_0 - Y_j)^2}}]$$

where $(\bar{X}_0, \bar{Y}_0)$ is the estimated position of the jammer, (X_i, Y_i) is the position of the jammed node and (X_j, Y_j) is the position of a boundary node. Then a joint force F_{joint} as the combination of all F_{pull}^{i} and F_{push}^{j} based on the formula of force synthesization was defined:

$$F_{joint} = \frac{\sum_{i \in J} F_{pull}^{i} + \sum_{j \in B} F_{push}^{j}}{|\sum_{i \in J} F_{pull}^{i} + \sum_{j \in B} F_{push}^{j}|},$$

where J is the set of jammed nodes that are located outside of the estimated jammed region, and B is the set of boundary nodes that are located within the estimated jammed region. At each iteration, the F_{joint} can be used as guider to make sure that VFIL achieve the goal.

Geometric-Covering Based Localization

In this part, we propose an algorithm: Geometric-Covering based Localization (GCL) which uses the knowledge of computing geometry, especially the convex hull problem.

Network Assumptions

- **Network model**
  Devising a generic approach that works across all kinds of sensor networks is impractical. We consider a wireless sensor network over a large area. As a initial work, we aim to tailor our proposal to a category of sensor networks with the following characteristics.
    - **Stationary**. We assume that once deployed, the location of each sensor node remains unchanged.
    - **Location-Aware.** Each node knows its location coordinates and its neighbors' locations. This is a reasonable

assumption as many applications require location services, and the neighbor table can be maintained by most routing protocols.

○ **Jamming Detection.** We focus on localizing a jammer after it has been detected. So we do not consider the process of detecting jamming attack. And we utilize the detecting method (Wenyuan Xu, 2005) that provided a consistency checking.

- **Jamming model**

In section 2.1, we have introduced many different jamming attack strategies that a jammer can perform in order to interfere with or jam wireless communications. However, the consequences of different jammers are the same: the normal communications are disrupted and the nodes cannot communicate with their neighbors. In general, as shown in figure 4, we divide network nodes into two categories under jamming condition, jammed and unaffected nodes. And also, we assume a static jammer which has an isotropic effect, e.g., the jammed region can be modeled as a circular region centered at the jammer's location as shown in figure 4.

GCL Algorithm

In this section, we present our algorithm Geometric-Covering based Localization. We note that the localization can be performed at a special node or a central management unit.

- **Preliminary Knowledge**

Convex hull. The convex hull of a set of Q of points is the smallest convex polygon P for which each point in Q is either on the boundary of P or in its interior. We denote the convex hull of Q by $CH(Q)$.

○ **Smallest circle covering.** The smallest circle covering, simply speaking, is the problem of finding the smallest circle that completely contains a

set of points. Formally, given a set Q of n planar points, find the circle C of smallest radius such that all points in Q are contained in either C or its boundary.

- **Algorithm**

In case of locating a jammer, the neighbor nodes of the jammer are jammed nodes. Therefore, in order to computing the position of the jammer, GCL collects all coordinates of jammed nodes. Assume that there are N jammed nodes, denoted by $Q=\{(X_1,Y_1),(X_2,Y_2),...,(X_n,Y_n)\}$. Then the localization steps in GCL are summarized as follows.

Step 1: Compute the convex hull of the jammed nodes set Q, denoted by

$$CH(Q) = \{p_1, p_2, \cdots, p_m\}$$

Step 2: Calculate the diameter of $CH(Q)$, denoted by l, in which p_i, p_j is the endpoint. The midpoint of l is denoted by $o(k)$ as the centre of the circle and $|l|/2$ by $r(k)$ as the radius of the centre. If $d(p_v, o(k)) \leq r(k), (v \in 1,2,\cdots m, v \neq i,j)$ then $o(k)$ is the estimated jammer's position. If not, go to step 3

Step 3: Calculate the distance between every vertex of $CH(Q) = \{p_1, p_2, \cdots, p_m\}$ and l, denoted

Figure 4. Jamming attack scenario in GCL

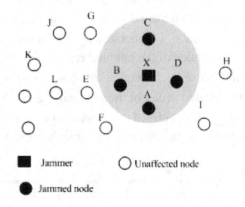

by $d(p_u, l)$, where $p_u \in \{p_1, p_2, \cdots p_m\}$.

Step 4: Calculate $maxd(p_u, l)$, and the corresponding u is denoted by k.

Step 5: Computing the perpendicular bisector of l and $\overline{p_i p_k}$ respectively. And these two perpendicular bisectors intersect at $o(k)$, which is used as the centre of the circle. And the distance $d(p_i, o(k))$ between p_i and $o(k)$, denoted by $r(k)$, is treated as the radius of the centre.

Step 6: If $d(p_v, o(k)) \leq r(k), (v \in 1, 2, \cdots m, v \neq i, j, k)$, then $o(k)$ is the estimated jammer's position.

Else if

$$d(p_{k'}, o(k)) > r(k),$$
$$(k' \in 1, 2, \cdots m, k' \neq i, j, k)$$
$\& p_{k'}$ and p_k are at the same side of l

Then replace p_k with $p_{k'}$.

Repeat step 5 and 6 until k' does not exit.
Else if

$d(p_{k'}, o(k)) > r(k) \ \& \ \& p_k$ and
$p_{k'}$ (or p_k) are not at the same side of l
$\& \ \& \ min(\angle p_i p_{k''} p_{k'})(k'' \in 1, 2, \cdots m, k'' \neq i, j, k)$

Then draw a circle by the point $p_i, p_{k'}$, and $p_{k''}$.
End

GCL Algorithm Analysis

Correctness

In the GCL algorithm, we first calculate the convex hull of the jammed nodes, then all the jammed nodes will fall inside or on the boundary of $CH(Q)$. Consequently, the problem of localizing a jammer becomes a typical computing geometry question: Find a smallest covering circle to make every jammed node covered, and the centre of the circle will be the estimated position of the jammer. The correctness of smallest covering problem has been verified by Peide Zhou (1995). Thus, we have shown the correctness of GCL algorithm.

Complexity

The time complexity of calculating the convex hull is $O(nlogn)$ (Peide Zhou, 1995). And rest computations of GCL are all less than $O(nlogn)$, so the time complexity of GCL is $O(nlogn)$.

Figure 5. Sensor network scenario for simulation

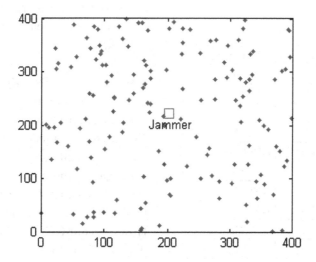

Figure 6. Impact of nodes density to jammer localization

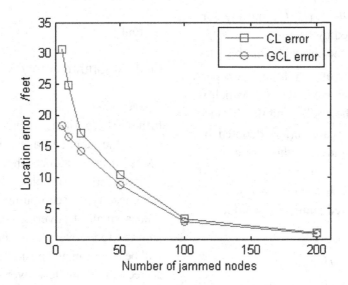

Simulation Evaluation

We simply simulate a wireless sensor network scenario in a square field with a size of 400 feet x 400 feet using MATLAB 7.1 as shown in figure 5. Sensor nodes are randomly distributed in this area with a transmission range of 40 feet. We evaluate the performance of locating the jammer by using GCL and CL under various network densities. In order to study the impact of sensor network densities, we place the jammer at the centre of

the simulation area so that the jammer can be surrounded by multiple network nodes.

Accuracy

First, we implement the CL and GCL algorithm in the experiment to evaluate the accuracy of localizing a jammer with a transmission range of 60 feet and 10 jammed nodes around. We run 10 times of localization using each algorithm to obtain the estimated position of the jammer as shown in table 3.

From the table above, we found that, in most cases, our GCL algorithm has achieved higher accuracy when localizing a jammer compared with the CL.

Impact of Node Density

Next we study the impact of sensor nodes densities to the location performance. As shown in figure 6, we varied the number of jammed nodes (5, 10, 20, 50, 100, and 200). We observed that both the CL and GCL are more or less sensitive to the sensor network densities. In general, when the density of nodes is higher, the localization error is smaller. However, our GCL achieves better performance than the CL algorithm.

Table 3. Accuracy of GCL and CL

CL	GCL	True Position
(233.7, 182.8)	(225.7,196.0)	(200,200)
(174.3,198.3)	(203.8,197.1)	(200,200)
(243.3,210.3)	(226.6,183.7)	(200,200)
(167.1,235.6)	(176.4,226.3)	(200,200)
(147.4,216.6)	(174.7,203.7)	(200,200)
(248.2, 213.7)	(206.7,190.9)	(200,200)
(157.4,222.9)	(175.5,199.7)	(200,200)
(236.8, 245.9)	(181.5,207.3)	(200,200)
(179.9,151.8)	(208.4,194.2)	(200,200)
(179.1,181.8)	(196.7,195.3)	(200,200)

FUTURE RESEARCH ISSUES

There are still many research issues in jamming wireless sensor networks, such as:

- Identification of MAC and network layer protocol employed by just sniffing the radio traffic
- Effects of jamming on deployment quality measure
- Differentiation of jamming from network congestion or sensor failures
- Designing jammer-resistant MAC and network layers
- Cross-layer protocol research to resist jamming

CONCLUSION

In this chapter, we survey issues related to jamming style Denial-of-Service attack in sensor networks by examining both the attack and defend sides of the problem. We illustrate different jamming attack models that might be used to disrupt sensor networks and metrics that can be used as measurements of jamming attacks. Then, we list out methods that can be employed by the sensor network in order to detect the presence of jamming, such as signal strength, carrier sensing time and packet delivery ratio, and conclude that basic measurement alone is not sufficient for classifying the presence of a jammer, and a consistency check detection method is needed. In defending section, we list five strategies for coping with jamming: mapping jammed region, channel surfing, spatial retreat, wormhole-based anti-jamming and timing channel. In case of localizing a jammer, we explore the typical range-based and range-free localization algorithm, and then proposed a Geometric-Covering Localization method to locate a jammer in static wireless sensor networks. At last, we discuss the future trends of jamming attacks and corresponding countermeasure in wireless sensor networks.

REFERENCES

Ahmed, N., Kanhere, S., & Jha, S. (2005). The Holes Problem in Wireless Sensor Networks: A Survey. *Mobile Computing and Communications Review ACM, 9*(2), 102–110.

Akyildiz, I. F., Su, W., Sankarasubramaniam, Y., & Cayirci, E. (2002). A survey on sensor networks. *IEEE Communications Magazine, 40*(8), 102–114. doi:10.1109/MCOM.2002.1024422

Bahl, P., & Padmanabhan, V. N. (2000). *RADAR: An in-building RF-based user location and tracking system.* Paper presented at the IEEE International Conference on Computer Communications.

Bischoff, U., Strohbach, M., Hazas, M., & Kortuem, G. (2006). *Constraint-based distance estimation in ad-hoc wireless sensor networks.* Paper presented at the European Workshop on Sensor Networks.

Blumenthal, J., Grossmann, R., Golatowski, F., & Timmermann, D. D. (2007). *Weighted centroid localization in zigbee-based sensor networks.* Paper presented at the IEEE International Symposium on Intelligent Signal Processing.

Karlof, C., & Wagner, D.(2003). Secure routing in wireless sensor networks: Attacks and countermeasures. In *Proceedings of the First IEEE International Workshop on Sensor Network Protocols and Applications* (pp. 293-315). Anchorage, AK: ACM Press.

Cagalj, M., Capkun, S., & Hubaux, J. P. (2007, January). Wormhole-Based Anti-jamming Techniques in Sensor Network. *IEEE Transactions on Mobile Computing, 6*(1), 1–15. doi:10.1109/TMC.2007.250674

Chen, Y., Kleisouris, K., Li, X., Trappe, W., & Martin, R. P. (2006). The robustness of localization algorithms to signal strength attacks: a comparative study. In *Proceedings of International Conference on Distributed Computing in Sensor Systems* (DCOSS, pp 546-563).

He, T., Huang, C., Bium, B. M., Stankovic, J. A., & Abdelzaher, T. (2003). *Range-free localization schemes for large scale sensor networks*. Paper presented at the ACM Mobile Computing Conference.

Hu, Y., Perrig, A., & Johnson, D. (2003). *Packet Leashes: A Defense against Wormhole Attacks in Wireless Ad Hoc Networks*. Paper presented at the IEEE International Conference on Computer Communications.

Law, Y. W., Hartel, P., den Hartog, J., & Havinga, P. (2005, Feb). Link-layer Jamming Attacks on S-MAC. In *Proceedings of. 2nd European Workshop on WSN* (pp. 217-225).

Law, Y. W., van Hoesel, L., Doumen, J., Hartel, P., & Havinga, P. (2005, November). Energy-Efficient Link-Layer Jamming Attacks against Wireless Sensor Networks MAC Protocols. *Proceedings of third ACM workshop on Security of ad hoc and sensor networks* (pp. 76-88).

Li, M., Koutsopoulos, I., & Poovendran, R. (2007). *Optimal Jamming Attacks and Network Defense Policies in Wireless Sensor Networks*. Paper presented at the IEEE International Conference on Computer Communications.

Liu, H., Xu, W., Chen, Y., & Liu, Z. (2009). Localizing Jammers in Wireless Networks. *To appear in Proceedings of IEEE PerCom International Workshop on Pervasive Wireless Networking (IEEE PWN) (Held in conjunction with IEEE PerCom)*. Texas, USA.

Polastre, J., Hill, J., & Culler, D. (2004). Versatile low power media access for wireless sensor networks. In *Proceedings of the 2nd international conference on Embedded networked sensor systems* (pp. 95-107). Anchorage, AK: ACM Press.

Wilson, P., Prashanth, D., & Aghajan, H. (2007). *Utilizing RFID signaling scheme for localization of stationary objects and speed estimation of mobile objects*. Paper presented at the IEEE International Conference on RFID.

Wood, A. D., Stankovic, J. A., & Son, S. H. (2003). *JAM: A Jammed-Area Mapping Service for Sensor Networks*. Paper presented at the meetings of 24th IEEE International Real-Time System Symposium. California, USA

Wood, A. D., Stankovic, J. A., & Zhou, G. (2007). *DEEJAM: Defeating Energy-Efficient Jamming in IEEE 802.15.4-based Wireless Networks*. Paper presented at the 4th Annual IEEE Communications Society Conference on Sensor, Mesh and Ad Hoc Communications and Networks. San Diego, CA.

Xu, W., Ma, K., Trappe, W., & Zhang, Y. (2006). Jamming Sensor Networks: Attacks and Defense Strategies. *IEEE Networks Special Issue on Sensor Networks, 20*, 41–47.

Xu, W., Trappe, W., & Zhang, Y. (2004). Channel Suffering and Spatial Retreats: Defenses against Wireless Denial of Service. *Proceedings of. 2004 ACM Workshop. Wireless Security* (pp. 80-89). ACM Press.

Xu, W., Trappe, W., & Zhang, Y. (2007, April). *Channel Surfing: Defending Wireless Sensor Networks from Interference*. Paper presented at the International Conference on Information Processing in Sensor Networks (IPSN 2007).

Xu, W., Trappe, W., & Zhang, Y. (2008, April). Anti-jamming Timing Channels for Wireless Networks. *Proceedings of 2008 ACM Workshop. Wireless Security* (pp. 203-213). ACM Press.

Xu, W., Trappe, W., Zhang, Y., & Wood, T. (2005, May). *The feasibility of Launching and Detecting Jamming Attacks in Wireless Networks*, Paper presented at the 6th ACM International Symposium on Mobile Ad Hoc Networking and Computing.

Zhou, P. (1995). Algorithms for Some problems in Geometric Covering. *Journal of Beijing Institute of Technology*, *15*(5), 21–25.

KEY TERMS AND DEFINITIONS

Channel Surfing: The nodes that detect themselves as jammed should immediately switch to another orthogonal channel and wait for opportunities to reconnect to the rest of the network

Convex Hull: The convex hull of a set of Q of points is the smallest convex polygon P for which each point in Q is either on the boundary of P or in its interior. We denote the convex hull of Q by $CH(Q)$.

DSSS: Direct Sequence Spread Spectrum

FHSS: Frequency Hopping Spread Spectrum

Jamming Attack: A kind of Denial of Service attack, which prevents other nodes from using the channel to communicate by occupying the channel that they are communicating on

Mapping Service: Nodes near the border of a jammed region notify neighbors outside of the region of jamming

Packet Delivery Ratio: The ratio of packets that are successfully delivered to a destination compared to the number of packets that have been sent out by the sender

Signal Strength: The value of received signal strength indicator

Smallest Circle Covering: The smallest circle covering, simply speaking, is the problem of finding the smallest circle that completely contains a set of points. Formally, given a set Q of n planar points, find the circle C of smallest radius such that all points in Q are contained in either C or its boundary.

Timing Channel: A low bit-rate overlay that exits on top of the conventional physical/link layers and that would survive in the presence of a persistent broadband jammer

Wormhole-Based Anti-Jamming: A path which called probabilistic wormhole to guarantee that, for the given randomly located jammer, there is a positive probability that a sensor node residing in the jammed region of an attacker forms a path from such a region to the area not affected by jamming

Section 4
Practices and Applications

Chapter 16
Visualizations of Wireless Sensor Network Data

Brian J. d'Auriol
Kyung Hee University, Korea

Sungyoung Lee
Kyung Hee University, Korea

Young-Koo Lee
Kyung Hee University, Korea

ABSTRACT

Wireless sensor networks can provide large amounts of data that, when combined with pre-processing and data analysis processes, can generate large amounts of data that may be difficult to present in visual forms. Often, understanding of the data and how it spatially and temporally changes as well as the patterns suggested by the data are of interest to human viewers. This chapter considers the issues involved in the visual presentations of such data and includes an analysis of data set sizes generated by wireless sensor networks and a survey of existing wireless sensor network visualization systems. A novel model is presented that can include not only the raw data but also derived data indicating certain patterns that the raw data may indicate. The model is informally presented and a simulation-based example illustrates its use and potential.

INTRODUCTION

Wireless Sensor Networks are quickly realizing a potential to support large and ultra-large scale data sensor, gathering and processing applications. Applications suitable for such wireless sensor networks include ubiquitous and quickly deployable systems that can meet the anytime and anywhere demands for quickly obtaining information about

the environment, processing that information and then presenting that information to human communities to facilitate better understanding about the environment. The latter includes the visualization of the sensor information and is the main focus of this chapter.

There are many types of user communities that may be interested in the information obtained via sensor networks. Very broadly, these would include scientists, policy and decision makers, educators and general public interests. The first two types of

DOI: 10.4018/978-1-61520-701-5.ch016

communities are often involved in modeling and seek to understand the sensor obtained information as observations in the context of these underlying models; or, as in the case of the policy and decision makers, base professional decisions upon this understanding. Educators are primarily interested in facilitating the learning process and may use visualizations in two ways, either by considering the sensor acquired information singularly, or as combined with the underlying models. General public interests however would often be satisfied by merely the sensor acquired information. The visualization model described here incorporates both of these visualization levels and therefore suggests its wide-scope application potential.

There are many issues involved in the visual presentation of wireless sensor network acquired information to broad audiences. Some of data related issues include: large and ultra-large scale deployments, high frequency data acquisition rates, and, multiple imagery and multimedia streams. The presentation of information will also depend upon the needs of the user communities and in particular the selection of the information level appropriate for those needs. In particular, the decision makers may require presentations to afford sufficient depth of understanding in time-critical applications. Since the latter imposes additional requirements, the focus of this chapter emphasizes the visualization of wireless sensor network information for presentation to decision makers to facilitate understanding leading to effective decisions in time-critical situations.

The objectives of this chapter are three-fold. Firstly, to discuss issues about the potential large data set sizes generated by wireless sensor network. Secondly, to survey existing wireless sensor network visualization systems. And, thirdly, to present a new visualization model that can accommodate large data set sizes and address the limitations of existing visualization systems.

BACKGROUND

The primary purpose of sensor networks is to acquire information about some environment. Sensor data is obtained both spatially and temporally, and for purposes of this chapter, is assumed to be transmitted to a computational base station for pre-processing and visual displaying. The first part of this section discusses the significant large data set sizes that wireless sensor networks impose upon visualization systems based upon a simple analysis. The second part discusses several wireless sensor applications in context of current day realistic data set sizes. And the third part discusses several existing sensor visualization systems.

Characteristics and Properties

The ideal maximum amount of information available for a visualization is limited by the sensor communication bandwidth. Two communication technologies can be used. Radio Frequency based systems have bandwidths in the 40 kbps and 250 kbps ranges (Polastre, 2004), although, newer systems may be capable of somewhat higher rates. Free space optical based technology is newly emerging and can support data rates in the order of 10 gbps or higher (see d'Auriol et al., (2009) for further discussion). The kilo bits per second range is sufficient to support typical environmental data sensing such as acceleration, temperature or humidity; but not high definition imagery nor video; whereas, the giga bits per second range can support both. Assuming an eight-bit short word representation for environmental type data; then, a 40 kpbs data rate can deliver 5120 values per second, a 250 kpbs data rate can deliver 32,000 values per second and a 10 gbps data rate can deliver over 1.3 billion values per second. And, assuming a 1280 by 720 pixel, 24-bit color image (without compression); then, a 10 gbps data rate can deliver 485 images per second.

It is unlikely that the ideal maximums truly represent the realistic maximum information available for visualizations. In general, actual data transmission rates depend upon many other factors including sensor sampling rates, power utilization requirements, application requirements, and communication traffic properties; all of which could reduce the amount of information available to visualizations. There are additionally other operations such as data aggregation and data re-sampling (e.g. for downsizing) which could further reduce the amount of information. However, at the same time, derived information obtained from processing the sensor acquired or 'raw' data can be combined with the sensor data thus increasing the amount of information available to visualizations. In general, the amount of data used in a visualization depends upon these and other factors so as to support the extent of the human viewers' requirements and needs. Let us consider an available information modification factor (for brevity, this will be shortened to the term 'modifier' in the rest of this chapter) as a percentage of the ideal maximums; for realistic systems, the modifier will likely be quite low.

Visualization metrics define various measurable aspects of a visualization. Loosely, *visual density* can be considered to be a measure of how much data is displayed in a single visualization. At the extremes, the density is minimal for a 'blank' visual and is maximal if the information is encoded and presented as a single pixel. Usually, a single information item in a visualization requires many pixels for representation. Additionally, since sensor networks have distributed nodes, it is likely that the visualization would consist of multiple sensor nodes placed on the screen thereby further reducing the screen area available per sensor node.

The following analysis assumes a 1400 by 1050 color pixel output device. A standard character size of 12 by 8 pixels suggests a small but sufficiently recognizable visual primitive. Assuming one data value is mapped to one visual primitive and without regard for specific screen coordinate placements, then the visual density can be calculated given the amount of information obtained from the sensor network. Figure 1 illustrates this analysis: consider the three visual primitive sizes of 100, 400 and 700 pixels with increasing amounts of information from 1000 to 10,000 items at the modifier set to 0.5; then, for the 100 primitive size, there is enough space on the screen to represent this data, but when the primitive size quadruples, more screen space is devoted to each primitive and the density reaches one just before 8000 data values. This analysis, for

Figure 1. A visual density model: visual primitive sizes of 100, 400 and 700 for the number of information items from 1000 to 10,000 at a modifier set to 0.5

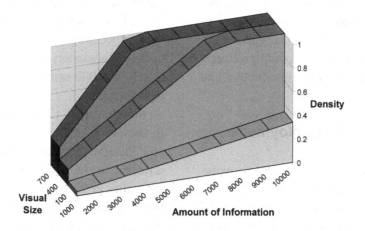

the 40 kbps data rate, with a primitive size of 100 and modifier set to 0.5 suggests that a maximum of five nodes can be viewed simultaneously; and with the modifier set to 0.1, suggests that 28 nodes can be so viewed.

Clearly, wireless sensor networks impose very demanding requirements upon visualizations. The simple analysis here indicates that low data rate and small scale sensor network deployments may be accommodated in visualizations; however, neither moderate nor large scale deployments can be. Visualization operations such as zooming, scrolling or panning could be used for moderate or large scale visualization applications. However, doing so places some additional requirements on providing navigational context information, and may depend upon the inherent relationships of the data itself (an obvious hierarchy here is the spatial placement of the sensor nodes where scrolling and panning would allow applications with many nodes to be represented and zooming would allow drilling down into the information content of a single node). Often, it may be very useful to isolate one or two parameters in a visualization thereby reducing the data requirements. There is in fact a large body of literature that is concerned with the visual presentation of large amounts of information and those techniques may also be of use as well. However, the implication of the afore analysis for large and ultra large wireless sensor network deployments (e.g. in the order of hundreds to tens of thousands of nodes interconnected by high bandwidth radio frequency or optical networks) is that a 'traditional' approach to visualization is problematic to providing a clearly understandable 'picture' of the information and its meaning to human viewers.

Analysis of Existing Deployments

This section briefly surveys several recent wireless sensor network applications in terms of the visualization presentation requirements. Several examples of real-world wireless sensor deploy-

ments suggest that past deployments had supported relatively low amounts of acquired data and that current deployments support more modest amounts of acquired data. Furthermore, several applications either directly indicate the need or benefit of incorporating underlying models for prediction purposes. For other applications, we believe that the incorporation of a model would provide enhanced benefit.

Mainwaring et al. (2004) discuss a wildlife habitat monitoring application (The Great Duck Island study). Their primary visualization needs include both the visual presentation of data as well as patterns indicated by this data. Thirty two sensors are deployed. The sensor data includes five essential scalars as well as desired additional scalar and vector data with data encoding sizes estimated between eight and 16 bits. The required sampling rates are significantly more modest than the maximums considered earlier and are based on a time scale of minutes and hours. Szudziejka et al. (2003) mentions that over one million sensor readings over a period of about five months were collected: "making it difficult to analyze the data".

Lédeczi et al. (2005) discuss an application for countersniper detection in urban combat zones. Their primary interest is the detection of sniper activity with associated geographical visualization. Fifty-six to 60 sensors are deployed. They indicate that sensor data can be comprised of seven scalars or vectors, although, in their work, they use only a subset of these parameters. Powerful local processing at the node is available. Two of these parameters are sampled up to approximately 100,000 samples per second at a 12 bit representation.

Stoianov et al. (2007) discuss an application for monitoring leaks and other anomalies in water pipelines. Much of their visualization needs are reflected in the detection and identification of anomalies in the water flow system. The data set includes several scalars and vectors. The required sampling rates vary depending on the specific data in the order of 1000 to 1500 samples every five minutes with a transmission capability of up to 600 samples per second.

More recent work indicate more demanding amounts of data: Chen and Chou (2008) describe a wireless system capable of supporting 50 to 100 streams at 500 samples per second; and, Barrenetxea et al. (2008) indicate on the order of megabytes of sensor acquired data available for visualizations.

Basha et al. (2008) describe an application methodology that includes connecting an underlying model useful for prediction with the raw data sensing. They also survey many other comparable systems and applications noting the absence of model predictive capabilities. Predictive models as well as augmented visualizations appear in Hull et al. (2006).

This brief survey illustrates the increasing size, availability, heterogeneity and demand of wireless sensor network acquired data as testbed applications give way to more broad 'real-world' deployed systems. For the most part, visual presentations of the information in these surveyed works rely on standard plots (for example, accelerometer data is mostly presented in two and three graph multiplots (see d'Auriol et al., (2008) for a detailed discussion), although, several applications incorporate map-based visualizations.

Past and present day wireless sensor network systems provide specific manageable data that is suitable for standard types of visualizations; however, the augmented demands for larger deployable systems in more complex application environments as indicated by the ideal communications maximums and the incorporation of underlying models studied here pose significant visualization manageability issues for even near-future deployments. Furthermore, an emerging theme noticed in some of the surveyed works includes generalizable approaches that reduce specific application, system or environment fine tuning of sensor and network parameters. Lastly, predictions such as in (p. 122, Wessner,

2006) suggest the continuing fast expansion in sensor-based systems.

Existing Visualizations

There are more than a few visualization environments, frameworks or systems that have been developed over the past years. (This observation is in stark contrast with the impression given by some of the recent publications in this area.) This is not surprising in that, as wireless sensor networks continue to transition to more complex real-world deployments, the complexity of the network as well as the sensing environment also continue to grow; thereby, driving a need for better visualization tools in order to deal with increased information content.

Visualizations of wireless sensor networks fall into three broad categories: visualizations of the network operational conditions (Network), visualizations of the sensed data (Sensed Data), and visualizations that combine network and sensed data (Hybrid). A survey of several existing environments, frameworks and systems using these categories for classification is given below. In some cases, the distinction between the Network and Hybrid categories is made based on the primary purpose and clearly dominate visualization capability of the particular system. In addition, visualization environments, frameworks and systems may be fixed, that is, the systems designer pre-selects the types of allowable visualizations, partially extendable, that is, the user may select from a wide-range of parameterizable options, or flexibly extendable, that is, the user may develop scripts as plug-ins.

Network

Visualizations aimed at the network operational conditions are often useful for two main reasons, first, to develop, test and debug sensor deployments, either in-situ or by simulation; and second, to monitor deployed network status and health..

Many of these systems also incorporate limited per node visualization of sensed data, often, associated with textual labels on a graph-based topology display or trend graphs of sensor data. Some systems are flexibly extendable, apparently providing support for additional visualizations, perhaps including visualizations of sensed data (however, at the time of this review, none of these systems provide much evidence of such application extension to sensed data).

The Emview tool, a part of the EmStar system (Girod et al., 2004), the ISEE sensor network monitoring environment (Ivester and Lim, 2006) is view tool, and the Sensor Network Analyzer (SNA) by DaintreeNetworks (Daintree, 2008) are examples of visualization systems that are primarily aimed at visualizing network operational information and provide very limited or no capability for sensed data visualization.

NetTopo is a recent simulator and visualizer for wireless sensor networks (Shu et al., 2008) that contains a testbed visualization module primarily providing network topology visualizations useful for analyzing network algorithms. The visualization display is subdivided into three regions: a display canvas, a node property display and a message display for use in logging and debugging. The authors indicate that visualizations of sensed data are also available via defining wrapper functions to obtain the sensed data, although it appears that some of this data is exported to other standard graphing applications.

TinyViz is part of the TinyOS mote simulator (TOSSIM) (Levis et al., 2003). This is a framework that manages the event and command interface to TOSSIM. Visualization is accomplished via plug-ins. A set of basic plug-ins are available and users may implement their own for specific purposes. The primary purpose of the available visualizations is aimed at network operational data which is displayed as a graph, although some basic plug-in are provided to display sensor values and contouring. Other plug-ins may be user defined allowing TinyViz

to perhaps provide some additional visualizations of simulated sensed data.

Sensed Data

Whereas the general properties of wireless sensor networks beg a graph-based topology display, the domains of the environments sensed by sensor networks are specific. Broadly, general methods can be applied to the sensed data which share degrees of commonality amongst the data properties or specific methods can be applied which construct specific visualization models or systems for the specific data requirements.

Scattered data methods combined with Voronoi diagram abstractions are used by Szudziejka et al. (2003) to visualize temperature information obtained from the Great Duck Island study. Due to the properties of the sensed data, their visualizations are animation-based.

The augmented reality visual interface system proposed by Claros et al. (2007) combines visualizations of the sensed data with visualizations of the sensor physical environment. A visualization of the sensed data is firstly rendered and subsequently transformed into an image with graphical tags. This transformed data is used by the augmented reality application to position and display the visualization images onto a real-world scene; thereby, providing three dimensional environment scene contexts to the visualization.

WiseObserver (Castillo et al., 2008) provides a number of sensed data visualizations including evolution charts that plot graphs of sensed data over time; interpolation maps and evolution videos that provides spatial color mapping, contouring, etc. of selected data; and report generation that provides document along with text information. The windows graphical user interface also allows for multiple views to be displayed, thereby providing some comparative capability.

A more domain restricted sensed data visualizations include CarTel (Hull et al., 2006) which makes use of map-based visualizations to provide location information.

The sensed data visualization approach adopted by Fan et al. (2004) makes use of the GIS Geographic Resources Analysis Support System (GRASS) to provide map-based visualizations.

Hybrid

Hybrid systems provide visualizations of both network operational conditions and sensed data. In some cases, dual visualization approaches effectively provide for each independent of each other, in other cases, a combined visualization can be defined. The latter, whilst useful in understanding the conditions of the network in the context of the sensed data, may lead to increased confusion about understanding the implications of the sensed data in the context of the environment being monitored.

SpyGlass is a wireless sensor network visualizer (Buschmann et al., 2005) that provides information for use in sensor network debugging, evaluation and understanding of the software. Within this focus, sensed data can be visualized. The visualization component of SpyGlass consists of a graphical user interface that is subdivided into three regions: a display canvas, a sidebar for tree-structured textual information about the network and a message display for use in debugging. The canvas itself is three-layered and provides for background imaging, graph-based relational information between nodes displayed and node-based detail information about a node. Plug-ins can be defined for each of these layers thereby allowing specific visualizations to be defined as needed.

Octopus is a visualization and control tool for wireless sensor networks specifically designed for TinyOS 2.x together with a limited number of mote devices (Jurdak, 2008). Its graphical user interface incorporates two types of pre-defined visualizations: a network map for graph-based topology display, and a network chart for sensed data versus time curve plotting.

SNAMP provides a multi-view visualization framework for wireless sensor networks (Yang et al., 2006) that provides multiple views: topology, packets, measurement and sensing chart. The first three pertain to network operational data; the latter, to sensor data. The front-end visualizer allows the incorporation of user defined visualizations to the software.

In-Situ real-time visualization for difficult-to-work-in-environments is described in Selavo et al., 2006. The architecture for SeeMote device is presented, in particular, its LCD and LED buttons, and, visualizations of network operational data as well as sensed data are shown. Visualizations are limited due to the low-resource usage intention of the SeeMote device. New visualizations can be developed via scripting that are based on a limited number of visual outputs (e.g. lines and filled boxes, text, menus, and color).

The Mote-View (Crossbow, 2007) from crossbow Company is a commercial tool that incorporates visualization of the wireless sensor network (e.g. node status and network topology as well as the sensed data. For the latter, a set of pre-defined data level visualizations are provided via menu selection (data, charts, histogram, scatterplot and topology) together with a per node user selection dialog (which also displays some network status information). Mainly, these visualizations provide details about the sensed values per selected nodes. Additionally, there also are some limited comparison and statistical visualizations. Specifically: the 'data' visualization provides tabular detail of the sensed values, the 'chart' visualization provides for plotting per node (maximum 24 nodes) the data over time (maximum three sensor types, i.e. three graphs), the 'histogram' displays simple statistical distribution of single sensor (maximum 24 sensors) data, the 'scatterplot' displays two sensor readings against each other for a selected set of nodes, and the 'topology' provides for a node topology graph superimposed on a background, either a bit-map of the user's choice, or a colorized gradient of a selected sensor data. Related software, the Surge Network viewer also from the same company, provides similar although reported less visualization capability (see the discussion in (Buschmann et al., 2005)).

Summary

Almost all of the existing visualization systems and approaches surveyed above provide visualizations at the data visualization level; and leave the understanding and interpretation of that data to the viewers (although, the singular approach of Szudziejka et al. (2003), based on general scattered data methods, may have a broader scope).

Many of the visualization capabilities provide for visualizations of the wireless sensor network itself – for development, testing and debugging or for in-situ operational monitoring. This observation suggests two things: first, that, despite the intense research, development and deployment of wireless sensor networks, there continues to be real or perceived challenges to successful deployments that motive the continued development of these types of visualization systems, and second, that application deployments may not have yet reached sufficient deployment maturity necessary to motivate corresponding intense research efforts to provide effective visualizations of the sensed data. In many cases, visualizations of sensed data is well provided by systems that also well provide for network environment data visualizations.

The visualizations surveyed here are often informative for small sized networks; however, its usefulness for large-scale applications is less certain. In some cases, the graphical user interface provides standard panning or scrolling capabilities, however, with apparently little or no navigational context information available nor other more widely available virtual camera features (e.g. projection, zooming). As such, for the most part, these visualization systems represent typical, low-fidelity, and abstract visual representations of the information that suggest their unsuitability for large-scale applications. The singular approach of Claros et al. (2007), however, specifically addresses the fidelity and context issues.

In addition, some of the older or commercial software are either systems or vendor dependent making it difficult to adopt widely. Newer systems provide greater flexibility. In addition, intended future work on a number of these newer systems include further visualization developments (e.g. three dimensional visualization support).

MULTIPLE LEVEL VISUALIZATION

The Multiple Level Visualization (MLV) model is classified as a Hybrid model in terms of the categorical classification of the previous section since it defines a singular model that is equally applicable to either network operational data or sensed data. However, since the MLV model includes additional elements, its semantics are not found in any of the surveyed models, hence, a part of this model also lies outside of this classification. Although substantively different in approach, the work of Szudziejka et al. (2003) is closely related to the MLV model in that both aim at general methods widely applicable in different networks or for different applications; also, the work of Claros et al. (2007) is closely related in that both three dimensional environment scene context is provided. The MLV model is formally presented in (d'Auriol, 2009). However, various earlier aspects are presented in (d'Auriol, 2006; d'Auriol et al., 2006; d'Auriol et al., 2007). The presentation of the model here is semi-formal to allow for easier reading and understanding. The MLV model is based on the alternative approach of connecting an underlying model with the sensor acquired data. We suggest that various features of our model may be applicable to much of the afore mentioned surveyed work; and, by virtue of its alternative basis, may be able to partially address the large data size issues. Lastly, the idea that an underlying model supports the observations provided by wireless sensor networks has been previously mentioned (see for example, (Sect. 7.5, Zhao et al., 2004)), although, the afore mentioned survey does not indicate such incorporation into sensor network visualizations.

In general, the information obtained from a wireless sensor network has two fundamental properties: structure and value. Structure refers to the x, y, and z coordinates of the physical sensor placement, its GPS coordinates or some other placement location coordinates. Value refers to the measured or observed information obtained by the sensors. Values may either be defined in discrete or continuous space and have associated minimum, maximum and normal operating ranges, for example, temperatures inside of a living room or the voltage and frequency of power lines. These properties have often been noted elsewhere in the literature; see for example (Brodlie, 1992; Ware, 2004). In much of the visualization literature, structured information is referred to as data.

Sensor data is obtained from a single sensor at different times and hence it is an ordered set of values. Let $D^*_k=(D_1, D_2, ...)$ denote this ordered set for the kth sensor and where D_i denotes all of the sensor data at some ith time. Each sensor obtains a data vector consisting of structured and value components. Let $D=(d^s_0, d^s_1, ..., d^s_{m1-1}, d^v_0, d^v_1, ..., d^v_{m2-1})$ for m1 coordinates and m2 values and each d^v_i is defined on some interval representing the range of the sensed information. D^s denotes the structure subset while D^v denotes the value subset. A data level visualization is any visualization of D^*_k.

Data level visualizations are very commonly found in both the research and popular literature (see the previous section). By itself it can be very useful in facilitating understanding about spatial and temporal environment changes reflected by the sensor acquired data. However, the semantics of the environment comes from human understanding about the environment; in this sense the sensor acquired data are stand-alone entities without predefined semantics. Data level visualizations tend to be straightforward using well-known techniques such as coloring or contouring on a map (see the previous section).

However, in many situations and environments, there exists some underlying model that either may describe these spatial-temporal changes or predict such changes. Often, in science, an objective is the discovery of such models; in engineering, the design of systems based on such models. For policy and decision makers, the predictive capability of such models can be used as the basis for decisions. In some cases, an underlying model is either difficult to develop or is not known. Nevertheless, in all of these cases, an underlying model provides semantics for D^*_k. For many sensor network systems, the sensors are placed so as to provide observations about some underlying model.

A typical dynamic systems model determines a state space, often continuous, that represents the states of the variables in that system (see (Dorf, 1974)). This continuous state space can be discretized and thus represented by a specific type of finite state machine called an Orthogonal Organized Finite State Machine (OOFSM) (d'Auriol, 2006). Consider a one dimensional system: a collection of temperature sensors where one may discretize this system in sub-ranges say of ten degrees; or, frequency sensors of a power line where one may discretize in sub-ranges say of [58,60), [60,60] and (60,62] Hz. These discretized states can further represent nominal operative conditions, exceptional conditions or abnormal conditions of the system. In general, each state space variable represents an orthogonal parameter and hence very high dimensional OOFSMs can be defined; for example, even small power grid models may have dimensions of several hundred variables. In general, finite state machines have been used to model dynamic systems, see for example (Blouin, 2003; Cassandras & LaFortune, 1999; Jodogne, 2002; Marchand et al., 2000).

More formally, an OOFSM represents a lattice partitioned, and therefore a discretized, state space of a dynamic system and is defined by the tuple M=(Y, L, VY) (the notation is greatly simplified here, see (d'Auriol, 2009; d'Auriol, 2006) for a

Figure 2. Illustration of a two dimensional OOFSM with a uniform partitioning, 16 states and two uniform regions (source: modified from d'Auriol, 2006)

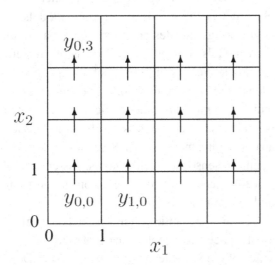

detailed mathematical presentation). The lattice partitioning L applied to an n dimension state space leads to a set of discretized states Y and in general defines a set of partition boundaries of the state space. A trajectory of the state space is the evolution of a set of state space parameters from one state to another. In terms of the OOFSM, the evolution is noted as the state to state transition across a partition boundary. In general, there exists a set of possible trajectories, many of which will intersect the same boundary and effectively reduce to a single state to state transition; however, many others may intersect other partition boundaries thereby defining multiple possible state to state transitions. The region field set, VY, denotes the union of all of these state to state transitions across the OOFSM. And a uniform region field is a collection of states all which have the same region field (subsets of VY can be uniform). A formal proof of the OOFSM's representation of such a system's state space is given in (d'Auriol, 2009; d'Auriol, 2006). Figure 2 illustrates this for a two dimensional system of four discretized states in each dimension; here, there are two

uniform region fields with the first being null (terminal states) associated with the `top row' and the second being `up-wards only' associated with the remaining states; the lower-case 'y's indicate specific states in Y. Diagonal transitions are disallowed; however, changing the resolution of the partitioning will often reduce or eliminate such transitions.

If a dynamic system is known, than the dimension and variables of the state space are known and L can be determined based upon computational or other requirement. The system also may provide predictions about how the states may evolve thus deriving VY. Sensor data represent observations about this system (if observable) at a particular point in space and time and either confirms the prediction of the model or does not. Specifically, sensor data determine specific states and changes between sensor data determine VY_s. When $VY_s = VY_s$, then the sensor data confirms the model; however, when $VY_s \neq VY_s$, then the sensor data suggests some abnormal condition that may be outside of the model. In both cases, the semantics of the model extend, although by different degrees, to cover the situation or events records by the sensors.

However, for the case where there is no dynamic system or it is unknown, the sensor acquired data can still be used to determine the OOFSM. Specifically, let L either be applied to D^s or D^v. The former implies that the OOFSM's structure is based on the physical space of the sensors and that transitions through this space reflect relationships between the values provided by the sensors. The latter implies that the OOFSM's structure is based on the observable state space variables. In this case, the semantics is similar to that when an underlying model is known, albeit without the ability to compare with model predictions.

An extended example is now discussed. A grid of 5 by 5 by 5 temperature sensors is simulated for a particular room location. A known underlying model for the temperature distribution in this room is assumed and therefore the visualizations

Figure 3. Top view of temperature state space

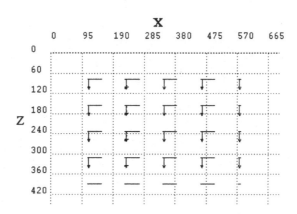

Figure 4. Front view of temperature state space

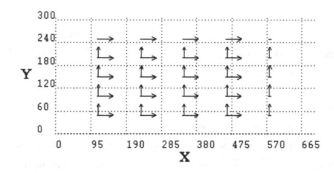

discussed seek to identify normal, unusual or abnormal environment conditions between the predictions of the model and the simulated sensor recorded values. The selected room location is part of the uLCRC (ubiquitous Lifecare Research Center) located at Kyung Hee University Global Campus in Korea. The uLCRC is a long-term academic, corporate and government consortium which aims at monitoring daily life of human behaviors and activities as well as providing proactively context-aware health related services via various types of sensors in an integrated environment. The uLRC consists of three rooms however, only the main office room is modeled in this simulation. The room contains a single air conditioning unit located in a corner and is modeled as a point source of cool air. There is also a single door, located at a different corner that provides

an entrance from the hallway to the facility. In the simulation, the opening of the door assumes that warm air is introduced to the room. Figure 7 shows a cut-away of this room: camera images are texture mapped to rectangles representing the room's walls, the air conditioning unit is shown at the back-right of the room, the door, not shown, is located at the front-right of the room. A simple linear air current model is used in the simulation; for more realistic simulations, a standard thermal convention model could be used. The simulation determines the expected air temperatures at each coordinate of the temperature sensors.

Since the simulation includes a known model, the state space is three dimensional representing the x, y and z coordinates of the sensor locations (this would also be appropriate for the partitioning of the structured data components in the unknown

Figure 5. Side view of temperature state space

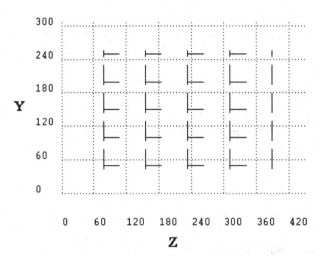

model case). An OOFSM is therefore determined based upon the partitioning of this physical space placing each sensor within a single state. The objective function $T_i > T_j$ for temperatures T_i and T_j obtained from two neighboring state-based sensors determines the state to state transitions (this function is derived from the underlying model's semantics, namely, about the temperature distributions, although, it would also be appropriate for the partitioning of the value data components in the unknown model case). Figures 3 – 6 show the top, front, side and three dimensional views of these

OOFSM transitions as arrows (all visualizations are done in AVS/Express). In general, the color of each arrow represents the sensor acquired values (however, neither color nor grayness is included in these figures). Other than the edge states, the uniformness of the region is apparent. These visualizations show expected behaviors of the temperature system in the room.

Figure 7 shows the combined data and underlying model level visualizations for this simulation embedded in the three dimensional room scene. There are three elements of data level visualiza-

Figure 6. General view showing the three dimensional temperature state spaces

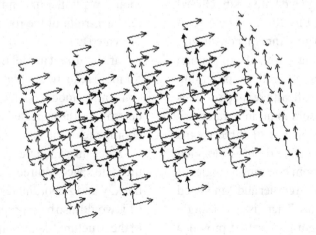

Figure 7. Combined data and model level visualizations

tion incorporated here: the colors of the arrows represent the temperature values, three isosurfaces together with the orthoslice show the temperature distribution throughout the room. Note that, by itself, the data level visualizations do not incorporate the semantics of the underlying model, that is, the precise possible trajectories of the temperature distribution are not evident. However, when combined with the underlying model visualizations of the state space, the semantics of warmer to cooler air transitions are indeed evident. Nevertheless, since the sensor observations confirm the model predictions, the additional semantics provided by the state to state transitions may not provide much in the way of additional advantage in understanding the temperature distribution.

The next part of the simulation introduces a heat source as the external door is opened. Although a similar thermal convention model could be used to model this event; and a combined model could be developed to model the interaction of both events, in general, we may assume that some unpredictable event could cause a change in the state space observations which do not correspond with the predictions of the underlying model. Let us consider this assumption in the following discussion.

Figure 8 shows the data level visualization corresponding to the assumed abnormal condition of a heat source in the front-right of the room. Comparing with the previous figure, the isosurfaces are significantly changed in this part of the room. However, the visualization itself does not provide any clear indication of an abnormal condition. Indeed, it would be left to the human viewer to decide based upon experience and/or knowledge that the isosurface shape in this figure shows some abnormality.

Figure 9 shows the state space visualization corresponding with the underlying model. Note that the front-right state to state transitions form a clearly identifiable region that has different behavior from the rest of the figure. Both the regional localization and the regional behavior are evident from this type of visualization. In general, a rich visualization is potentially available when both visualization levels are included.

The previous section introduced a simple model based on visual density calculations to illustrate the size and scope of sensor acquired information. Here, this model is applied to determine how well the dual level model presented here may address this issue. First, since the data level visualizations rely upon the same set of visual

Figure 8. Data level visualization of abnormal conditions

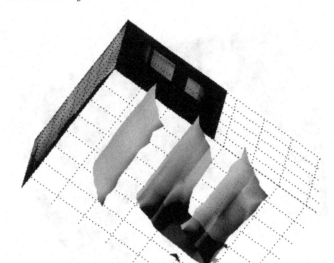

primitives as assumed previously, there may not be any savings without some further manipulation. Second, as visual primitives go, arrows take up very few pixels and may be closely aligned, that is, many more arrows could be utilized than the previous analysis would suggest. However, the incorporation of a large set of like arrows could increase the visual clutter in the visualization and thereby detract from the overall benefit provided by this model. In many cases, individual state behaviors are not of interest; rather, it is the region's size and behavior that is much more interesting. Although the presentation of the model in this chapter does not illustrate this, it is possible to compress uniform regions into a single 'superstate' like representation; of course, any data level visualizations would also require corresponding transformations. Doing so addresses the first issue in that less specific data points are used in the data level visualization and address the second as fewer arrows of the same orientation are incorporated into the figure. Zooming can be used to drill into specific regions.

There are two final comments about the potential application of this multilevel visualization model. First, the state space of the underlying model may be very large (almost certainly will

be much greater than the three dimensions of this extended example). A subsequent model is needed in which to provide either state space reductions or state space navigation so that the high dimensional OOFSM can be explored. Such a model has been considered in (d'Auriol et al., 2006). Second, the state space of the underlying model may not overlap with the physical space of the sensor network environment, as was the case in the extended example. In general, the state space describes a model of a system that is embedded in the physical space; then, the overall parameter space could be combined and again, the model in (d'Auriol et al., 2006) may be used as well.

CONCLUSION

Visualizations of wireless sensor networks and data obtained from these networks are very important to both understanding the operational characteristics of the networks and the behavior and 'meaning' of the sensed data. A simple categorical-based classification is introduced in this chapter in order to distinguish visualization systems that are mainly intended for visualizations of network operational conditions from those that

Figure 9. Abnormal region of behavior

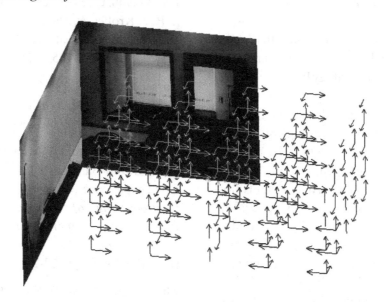

are mainly intended for visualizations of sensed data; and from those that provide for both. A number of visualization environments, frameworks and systems that have been proposed in the past years are classified accordingly. The survey reveals that many of the visualizations of network operational data are typical (i.e., graph-based, node-labeled, or chart-based) and low-fidelity; although some may be under development to provide higher quality visualizations. The survey also reveals that some of the visualizations aimed at sensed data are of higher quality.

Wireless sensor networks can also provide large amounts of data that when combined with pre-processing and data analysis processes can generate large amounts of data that may be difficult to present in visual forms. A simple analysis based upon the maximum amount of information that can be delivered from the sensor network together with a survey of several wireless sensor network applications suggest that near future sensor deployments could generate more information than can be accommodated by typical visualizations. The surveyed existing systems aimed at visualizations of network operational conditions are informative for small scale networks; many of which are scalable in terms of network size. However, as

networks grow larger, many of the visualizations in these systems may not scale adequately due to issues such as context and navigation. Even those systems aimed at visualizations of sensed data may also have scalability issues. The generation of large amounts of data from wireless sensor networks continues to pose challenges.

This chapter introduces a novel model called the Multiple Level Visualization (MLV) model that is developed to address some of the aforementioned limitations and provide more advanced and higher-fidelity visual display. The MLV model combines visualizations of either the sensed data or network operational data with that of an underlying model that describes the semantics of the data. It is the inclusion of the underlying model that constitutes the unique direction of this model. An extended example illustrates the MLV model (although this example only covers visualizations of sensed data). An application of this model for large and ultra large scale sensor deployments that includes zooming, navigation and other visualization features and capabilities could provide a solution to address some of the issues inherent in the visualization of information from these types of sensor deployments.

ACKNOWLEDGMENT

This research was supported by the MKE (Ministry of Knowledge Economy), Korea, under the ITRC (Information Technology Research Center) support program supervised by the IITA (Institute of Information Technology Advancement)"(IITA-2008-C1090-0801-0002) and by the MIC (Ministry of Information and Communication), Korea, Under the ITFSIP (IT Foreign Specialist Inviting Program) supervised by the IITA (Institute of Information Technology Advancement, C1012-0801-0003. Also, this work is financially supported by the Ministry of Education and Human Resources Development (MOE), the Ministry of Commerce, Industry and Energy (MOCIE) and the Ministry of Labor (MOLAB) through the fostering project of the Lab of Excellency.

REFERENCES

Barrenetxea, G., Ingelrest, F., Schaefer, G., & Vetterli, M. (2005-2006, 2007, 2008, November). The hitchhiker's guide to successful wireless sensor network deployments. In *Proceedings of the 6th ACM Conference on Embedded Network Sensor Systems*. Raleigh, NC, USA, November 05 - 07, 2008 (SenSys '08 pp. 43-56). New York: ACM

Basha, E. A., Ravela, S., & Rus, D. (2005-2007, 2008, November). Model-based monitoring for early warning flood detection. In *Proceedings of the 6th ACM Conference on Embedded Network Sensor Systems*, Raleigh, NC, (SenSys '08, pp. 295-308). New York:ACM

Blouin, S. (2003). Finite-state Machine Abstractions of Continuous Systems. Unpublished PhD thesis, Chemical Engineering Department,Queens University, Kingston, Canada.

Brodlie, K., Carpenter, L. A., Earnshaw, R. A., Gallop, R., Hubbolt, R., Mumford, A. M., et al. (1992). *Scientific Visualization: Techniques and Applications*. New York: Springer-Verlag.

Buschmann, C., Pfisterer, D., Fischer, S., Fekete, S. P., & Kröller, A. (2005). SpyGlass: A Wireless Sensor Network visualizer. *ACM SIGBED Review*, *2*(1), 1–6. doi:10.1145/1121782.1121784

Cassandras, C. G., & Lafortune, S. (1999). *Introduction to Discrete Event Systems*. Bostone, MA: Kluwer Academic Publishers.

Castillo, J. A., Ortiz, A. M., López, V., Olivates, T., & Orozco-Barbosa, L. (2008). WiseObserver: a real experience with wireless sensor networks. In *Proceedings of the 3nd ACM Workshop on Performance Monitoring and Measurement of Heterogeneous Wireless and Wired Networks* (PM2HW2N '08,pp. 23-26). New York: ACM.

Chen, C., & Chou, P. H. (2008, April 22-24). EcoDAQ: A Densely Distributed, High Bandwidth Wireless Data Acquisition System. In *Proc. of the 7th Intl. Conf. on Information Processing in Sensor Networks*. Information Processing In Sensor Networks (pp. 545-546). IEEE Computer Soc., Wash., DC.

Claros, D., De Haro, M., Dominguez, M., de Trazegnies, C., Urdiales, C., & Sandoval, F. (2007). Augmented Reality Visualization Interface for Biometric Wireless Sensor Networks, In *LNCS 4507 Computational and Ambient Intelligence* (pp. 1074-1081).

Crossbow (2007). *MoteView Users Manual, Revision A*. Retrieved May 2007 from, (PN: 7430-0008-05) http://www.xbow.com

d'Auriol, B. J. (2006). A Finite State Machine Model to Support the Visualization of Complex Dynamic Systems, In *H. Arabnia, The 2006 International Conference on Modeling, Simulation and Visualization Methods* (MSV'06, pp. 304-310)

d'Auriol, B. J. (2009). (paper in preparation). *Multilevel Visualization*.

d'Auriol, B. J., Carswell, P., & Gecsi, K. (2006). A TransDimension Visualization Model for Complex Dynamic System Visualizations, In *H. Arabnia, The 2006 International Conference on Modeling, Simulation and Visualization Methods* (MSV'06. pp. 318-324).

d'Auriol, B. J., Kim, J., Lee, S. Y., & Lee, Y. K. (2007). Orthogonal Organized Finite State Machine Application to Sensor Acquired Information, In V. Malyshkin (Ed.), *9th International Conference on Parallel Computing Technologies* (PaCT-2007): LNCS 4671. Parallel Computing Technologies (pp. 111-118). Pereslavl-Zalessky, Russia: Springer.

d'Auriol, B. J., Nguyen, T., Pham, T., Lee, S. Y., & Lee, Y.-K. (2008, July 14-17). Viewer Perception of Superellipsoid-Based Accelerometer Visualization Techniques, In *Proceedings of the 2008 International Conference on Modeling, Simulation and Visualization Methods* (MSV'08, pp. 129-135), Las Vegas, NV.

d'Auriol, B. J., Niu, Y., Lee, S. Y., & Lee, Y. K. (2009). The Plasma Free Space Optical Model for Ubiquitous System, In *Proceedings of the 3rd International Conference on Ubiquitous Information Management and Communication* (pp. 446-455).

Daintree Networks. (2008), Sensor Network Analyzer (SNA): *Standard Edition (Product Data Sheet)*. Retrieved (n.d.). from, http://www.daintree.net

Dorf, R. C. (1974). *Modern Control Systems, Second Edition*. Reading, MA: Addison Wesley.

Fan, F., & Biagioni, E. S. (2004). An Approach to Data Visualization and Interpretation for Sensor Networks, In *Proceedings of the 37th Hawaii International Conference on System Sciences* (pp. 1-9).

Girod, L., Elson, J., Cerpa, A., Stathopoulos, T., Ramanathan, N., & Estrin, D. (2004) EmStar: a Software Environment for Developing and Deploying Wireless Sensor Networks, In *Proceedings of the USENIX General Track* (pp. 283-296).

Hull, B., Bychkovsky, V., Zhang, Y., Chen, K., Goraczko, M., Miu, A., et al. (2006, October 31-Novermber 3). CarTel: a distributed mobile sensor computing system. In *Proceedings of the 4th international Conference on Embedded Networked Sensor Systems* (Boulder, CO., SenSys '06,pp. 125-138). New York:ACM

Ivester, M., & Lim, A. (2006). Interactive and Extensible Framework for Execution and Monitoring of Wireless Sensor Networks, In *Proceedings of the First International Conference on Communication System Software and Middleware* (Comsware 2006. pp. 1-10).

Jodogne, S. (2002). Orthogonal finite-state representations. In *Sixth Meeting of the ADVANCE Project*, Edinburgh: Springer.

Jurdak, R. Ruzzelli., A. & Boivineau, S. (2008) *Octopus: User Documentation*. Retreived (n.d.). from, http://www.csi.ucd.ie/content/octopus-dashboard-sensor-networks-visual-control

Kosterev, D., Taylor, C., & Mittelstadt, W. (1999). Model validation for the august 10, 1996 WSCC system outage. *IEEE Transactions on Power Systems*, *14*(3), 967–979. doi:10.1109/59.780909

Lédeczi, Á., Nádas, A., Völgyesi, P., Balogh, G., Kusy, B., & Sallai, J. (2005, November). Countersniper system for urban warfare. *ACM Trans. Sen. Netw.*, *1*(2), 153–177. doi:10.1145/1105688.1105689

Levis, P., Lee, N., Welsh, M., & Culler, D. (2003). TOOSIM: Accurate and Scalable Simulation of Entire TinyOS Applications. In *Proceedings of the First ACM Conference on Embedded Networked Sensor Systems* (SenSys 2003, pp. 126-137).

Mainwaring, A., Polastre, J., Szewczyk, R., Culler, D., & Anderson, J. (2002). *Wireless Sensor Networks for Habitat Monitoring, in WSNA'02*, Atlanta, GA:ACM Press.

Marchand, H., Boivineau, O., & Lafortune, S. (2000). Optimal control of discrete event systems under partial observation (Tech. Rep. CGR-00-10).

Polastre, J., Szewczyk, R., Sharp, C., & Culler, D. (2004, August 22-24). *The Mote Revolution: Low Power Wireless Sensor Network Devices*, Presented at Hot Chips 16, A Symposium on High Performance Chips, Stanford Memorial Auditorium, CA, USA.

Selavo, L., Zhao, G., & Stankovic, J. (2006, October). SeeMote: In-Situ Visualization and Logging Device for Wireless Sensor Networks (BaseNets 2006, pp. 1-9).

Shu, L. Wu. C., Zhang, Y., Chen, J., Wang, L., & Hauswirth, M. (2008). NetTopo: Beyond Simulator and Visualizor for Wireless Sensor Networks, In *Proceedings of the Second International Conference on Future Generation Communication and Networking* (FGCN 2008, pp. 17-20).

Stoianov, I., Nachman, L., & Madden, S. (2007). PIPENET: A Wireless Sensor Network for Pipeline Monitoring, In. *IPSN'07*, Cambridge, MA: ACM Press.

Szudziejka, V., Kreylos, O., & Hamann, B. (2003). Visualization of environmental data generated by wireless sensor networks, In Copsey, D., (ed.), *Proceedings of the 2003 UC Davis Student Workshop on Computing* (TR CSE-2003-24, pp. 40-41).

Ware, C. (2004). *Information Visualization Perception for Design*. San Francisco: Morgan Kaufmann Publishers.

Wessner, C. W. (2006). *The Telecommunications Challenge: Changing Technologies and Evolving Policies*, Report of a Symposium. Washington DC: The National Academies Press.

Yang, Y., Xia, P., Huang, L., & Zhou, Q. Xu, Y., & Li, X. (2006). SNAMP: A Multi-sniffer and Multi-view Visualization Platform for Wireless Sensor Networks, In *Proc. of the 2006 IST IEEE Conf. on Industrial Electronics and Applications* (pp. 1-4).

Zhao, F., & Guibas, L. J. (2004). *Wireless sensor networks: an information processing approach*. San Francisco: Morgan Kaufmann.

KEY TERMS AND DEFINITIONS

Data Level Visualization: Visualization aimed at displaying the values and patterns of the sensed data, may be combined with derived data visualizations, that is, visualizations of pre-processed sensed data.

Multiple Level Visualization (MLV) Model: New visualization model that combines data level and underlying model level visualizations so as to provide underlying model semantics coupled with standard data visualizations of the sensed environment.

Orthogonal Organized Finite State Machine (OOFSM): A special finite state machine abstraction used to represent the state-space transitions and state-space regions of behavior of the underlying model.

Underlying Model: Dynamic system model composed of state-space parameters either observable or not which provides semantics about the sensed environment; observable parameters are sensed by the wireless sensor network.

Underlying Model Level Visualization: Visualization aimed at displaying the state-space transitions and behavior described by the underlying model.

Visualization: Displaying information appropriately to facilitate human understanding leading to decision making about the sensed environment; usually, pictorial or graphical displays.

Wireless Sensor Networks: Networks of sensor nodes capable of acquiring sensed information about the environment and communicating that information via wireless data links to base stations.

Chapter 17

Sink Mobility in Wireless Sensor Networks:
From Theory to Practice

Natalija Vlajic
York University, Canada

Dusan Stevanovic
York University, Canada

George Spanogiannopoulos
York University, Canada

ABSTRACT

The use of sink mobility in wireless sensor networks (WSN) is commonly recognized as one of the most effective means of load balancing, ultimately leading to fewer failed nodes and longer network lifetime. The aim of this chapter is to provide a comprehensive overview and evaluation of various WSN deployment strategies involving sink mobility as discussed in the literature to date. The evaluation of the surveyed techniques is based not only on the traditional performance metrics (energy consumption, network lifetime, packet delay); but, more importantly, on their practical feasibility in real-world WSN applications. The chapter also includes sample results of a detailed OPNET-based simulation study. The results outline a few key challenges associated with the use of mobile sinks in ZigBee sensor networks. By combining analytical and real-world perspective on a wide range of issues concerning sink mobility, the content of this book chapter is intended for both theoreticians and practitioners working in the field of wireless sensor networks.

INTRODUCTION

The conventional wireless sensor network (WSN) architecture, as described in the majority of the literature, assumes the existence of a large number of miniature battery-powered sensor devices scattered

over an area of interest and organized in an ad-hoc communication manner. The primary goal of the wireless sensors is to gather relevant data from the environment and, subsequently, to route the gathered data to a central processing node, commonly referred to as sink. The sink is generally assumed to have far superior capabilities than the 'ordinary' sensors, i.e.

DOI: 10.4018/978-1-61520-701-5.ch017

Figure 1. Conventional WSN architecture: single static sink with many-to-one multi-hop communication

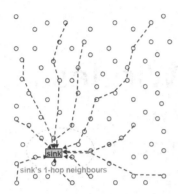

nodes, and it serves as a gateway point to the end user of the system. In the networking literature, this type of systems comprised of multiple sending and one superior receiving station are known as *many-to-one* systems.

From the communication perspective, the conventional many-to-one WSNs rely on multi-hop forwarding, i.e. routing, to deliver data to the sink. (In a WSN employing multi-hop routing, data is passed from the source node to the sink by being relayed by intermediate sensors, as illustrated in Figure 1.) Two most commonly recognized benefits of multi-hop routing are simplicity and energy efficiency (Wang, Hossam, & Xu, 2006). Namely, the physical-layer setup of a multi-hop WSN is pretty straightforward, as all nodes employ the same fixed radio range, regardless of their location and/or distance from the sink. In addition to being fixed, the nodes' radio range is also assumed to be relatively short, i.e. needs to ensure communication between immediate topological neighbours only, and as such has a conserving effect on the nodes' energy supplies. (Energy conservation is often considered to be the most important parameter in the design and operation of WSNs.)

In addition to the above advantages, there are also some serious challenges associated with the use of multi-hop routing in the conventional

wireless sensor systems of many-to-one type, including:

1. **Funnelling Effect**: In many-to-one multi-hop WSNs, the nodes in the sink's one-hop vicinity are required to 'funnel' (forward) data on behalf of all other, more distant, nodes (see Figure 1). Clearly, if such a network is to support a traffic-intense application with intervals of time in which the data collection rate dominates the data forwarding rate, e.g. video-based target tracking, congestion may start to build up at the bottleneck nodes of the sink's one-hop neighbourhood. Consequently, the incidents of packet dropping and/or retransmission become more frequent, leading to increasingly degraded network performance.

2. **Hot-Spot Problem**: By being required to forward a disproportionately higher volume of traffic compared to other nodes in the network, the sink's one-hop neighbours - also known as the 'hot-spot' nodes - tend to exhaust their energy and 'die' earlier relative to other nodes. Unless adequately dealt with, through measures that reduce the load and prevent the failure of the sink's one hop neighbours, the hot-spot problem can lead to a complete isolation of the sink node, resulting in the failure of the entire network.

Based on the above, it should be clear that the funnelling effect and hot-spot problem are fundamentally related to each other – both appear as side-effects of multi-hop routing in WSNs of many-to-one type, and the former phenomenon can aggravate the later. Nevertheless, there are several key differences between the two phenomena:

a. The hot-spot phenomenon refers to the possibility of a topological collapse (complete sink isolation) due to a highly skewed distribution of routing load among the network nodes. Though generally more pronounced under conditions of high traffic intensity, this phenomenon occurs regardless of the type of WSN application – even in networks of low traffic intensity, the nodes in the sink's vicinity will exhaust their energy earlier relative to other nodes. The funnelling effect, on the other hand, refers to a quality of service (QoS) aspect of network performance, i.e. level of congestion or packet dropping, and it occurs only under conditions of high traffic intensity.

b. The defence against these two phenomena asks for different type of countermeasures. For example, the most effective way of dealing with the funnelling effect is through the use of MAC-level scheduling (Ahn, Hong, Miluzzo, Campbell, & Cuomo, November, 2006) or data-compression techniques (see Petrovic, May, 2003). On the other hand, effective dealing with the hot-spot problem often requires some degree of topological change in the network.

Given its higher prevalence and potentially more detrimental effect on the network operation, the hot-spot problem and its respective countermeasures are the main topic of this book chapter. More specifically, the book chapter focuses on the use of mobile sink(s) as currently the most superior approach to solving the hot-spot problem in many-to-one multi-hop WSNs. The main sections of the book chapter are organized as follows:

- In the section "Defence against the Hot-spot Problem", we provide a general overview of known techniques for combating the hot-spot problem, and we demonstrate the superior effectiveness of the approaches involving sink mobility.

- In the section "Deployment Strategies Involving Sink Mobility", we introduce the reader to a wide range of theoretical ideas concerning the use of mobile sinks

and, subsequently, we identify those ideas and solutions with the highest chances of being utilized in practice.

- In the section "Multi-hop Data Delivery with Predictable/Fixed Trajectory Model: Analytical Evaluation", we conduct a detailed theoretical analysis of the mobile sink deployment strategy that has been identified as the most realistic in the presented literature overview - multi-hop data delivery with predictable sink trajectory. Specifically, we evaluate different types of predictable trajectories (outer periphery, mid periphery, mid-cross, diagonal) using three different performance measures: network lifetime, total energy consumption, and packet delay. The results of this analysis reveal that from the perspective of network lifetime, the outer peripheral and diagonal trajectories outperform the other two types of trajectories and are also far superior over the use of a static sink. On the other hand, from the perspective of overall energy consumption, the static sink strategy appears to be optimal.

- In the section "Sink Mobility in ZigBee WSNs: Simulation-based Evaluation", we present sample results of our OPNET-based study aimed at examining the real-world applicability of the analytical findings from the previous two sections. The presented results suggest that, while our theoretical findings are generally applicable, one has to be aware of a few inherent challenges arising from the use of a mobile sink over a static sink in ZigBee-based sensor networks.

DEFENCE AGAINST THE HOT-SPOT PROBLEM

To date, the following techniques for combating the negative effects of the hot-spot phenomenon in the conventional many-to-one multi-hop WSNs have been introduced and discussed in the literature:

a. Deployment of Additional Sensors

An obvious way to minimize the effects or mitigate the worst-case consequences of the hot-spot phenomenon is by bringing a new set of sensors into the sink's immediate vicinity, as suggested by (Yang, Xiao, & Chen, April, 2007). Namely, the deployment of new sensors enables the following to be achieved: 1) reduce the load on, and thus extend the lifetime of, the nodes already operating in the sink's one-hop neighbourhood; and 2) prevent the isolation of the sink node in the case that the previously operating nodes fail.

Despite its intuitive appeal, the practical feasibility of this approach is very questionable, especially in WSNs that need to operate in remote, rescue, or hostile environments. In these WSNs, the deployment of new nodes would likely require that these nodes be dropped by air or brought in by robots, thus rendering the whole approach costly and/or imprecise, i.e. of uncertain effectiveness. (For example, the aerial deployment cannot guarantee that the sensors would fall exactly where intended - within the sink's one-hop neighbourhood.)

b. Adjustment of Transmission Power

Another possible way of dealing with the hot-spot phenomenon is by allowing the second- or third- hop nodes to extend their radio range and communicate directly with the sink (see Akkaya, Younis, & Youssef, Positioning of base stations in wireless sensor networks, 2007). Similar to (a), this approach can be used as both: 1) a preventive measure - to reduce the load on, and extend the lifetime of, the sink's first-hop neighbours, and 2) a reactive measure - to keep the sink connected to the rest of the network following the failure of the sink's first-hop neighbours.

The major drawback of this approach is the fact that, for example, a 2-time increase in the radio range of a sensor node implies an increased demand of 2^a times on the nodes energy supply, where $a \geq 2$ represents the path-loss exponent of the given environment (Rappaport, 2001). Thus, by conducting direct communication with the sink, the second- and third- hop nodes would place extraordinary pressure on their energy supplies, ultimately exposing themselves to the risk of very rapid failure – even more rapid than the failure of the sink's first-hop neighbours.

c. Use of Multiple Static Sinks

The use of multiple static sinks as an approach to combating the host-spot phenomenon has been discussed by (Kalantari & Shayman, April, 2006; Meng, Wu, Xu, Jeong, Lee, & Lee, August, 2007; Wan, Eisenman, Campbell, & Crowcroft, November, 2005). The most obvious advantage of substituting one sink with multiple static sinks – assuming the sinks are strategically distributed across the network and the sensors forward their collected data to the nearest sink – is in splitting and more even spreading of routing load among the sensor nodes. Specifically, in the presence of multiple static sinks, each sink as well as its respective one-hop neighbours would receive only a fraction of the overall traffic load. This, consequently, would reduce the probability of network congestion and/or individual node failure.

One should be aware, however, that the multiple sink strategy provides only a partial defence against the hot-spot phenomenon, since the nodes located in the vicinity of each sink would still consume their energy earlier than other nodes (Khan, Gansterer, & Haring, 2007). Hence, in a traffic-intense network with relatively few sinks, the isolation of sink nodes and partition of the sensor field would still be likely, just as in the single sink scenario. Here, it is also worth noting that the effectiveness of the multiple sink strategy can generally be improved be increasing the number of deployed sinks. Nevertheless, the budgeting constraints of most real-world applications would allow the purchase of a (very) limited number of sink stations, thus limiting the overall effectiveness of the multiple-sink strategy.

d. Use of a Mobile Sink

The idea of using a mobile sink in place of a static sink as a defence against the hot-spot problem is quite intuitive: continuous change of the sink's residing location implies continuous change of the sink's one-hop neighbourhood, and as such can prevent that any particular group of nodes be exposed to a load excessively higher than the load of others. Of course, the actual effectiveness of deploying a mobile sink is ultimately determined by the shape of the sink's trajectory. Clearly, more 'diverse' trajectories can be expected to have better load-balancing properties and, consequently, be more effective against the hot-spot phenomenon. (By diverse, we mean a trajectory that spans over a sufficiently large number of different points within the network field.)

From the practical point of view, there are two possible ways of obtaining a *mobile* sink:

1. By mounting the hardware of a static sink (including its wireless module and antenna) on top of a mobile robot. In this case, the movement of the robot can be pre-programmed or controlled remotely by the network operator.

2. By placing a static sink in the backpack of a rescue crew or on an emergency vehicle (Akkaya, Younis, & Youssef, Sink repositioning for enhanced performance in wireless sensor networks, 2005). The movement of such a sink is clearly constrained by the movement of the respective carrier and, consequently, is appropriate for use in only a limited number of applications.

The key advantages of the mobile sink approach to combating the hot spot problem over the other earlier mentioned approaches are:

*   **Superior distribution of traffic and routing load:** Generally speaking, approaches (a) and (b) can provide remedy against the hot spot problem, but they posses very limited load balancing capability – with these techniques, the nodes in the sinks immediate vicinity continue to experience disproportionately higher load than the other nodes in the network. Consequently, the risk of node failure and sink isolation remains high. Compared to (a) and (b), approach (c) exhibits much better load balancing capability, as it manages to split and distribute the traffic across the network. However, the actual effectiveness of approach (c) is closely tied to the number of deployed (static) sinks. Now, relative to approach (c), strategy (d) is equivalent to a consecutive use of a very large if not an indefinite number of static sinks. (The operation of a single moving sink could be seen as the operation of a large number of static sinks, each operating at a different time (Ma, Chen, & Salomaa, 2007.) Hence, in terms of load balancing, the effectiveness of the mobile sink strategy (d) potentially far surpasses the effectiveness of the multiple static sink strategy (c).

*   **Relatively low cost:** The cost of purchasing a mobile robot, which would be sufficiently reliable and robust for deployment in rough outdoor terrain (e.g. Pioneer or Kephera type robot), is in the range of a few thousand US dollars. This is considerably less than the cost of deploying an airplane (as, in some cases, would be required by approach (a)) or the cost of purchasing a large number of static sinks (as required by approach (c)).

In the following section, we will look at various strategies of deploying a mobile sink in WSNs of many-to-one type, as proposed in the literature to date. We will also discuss the inherent pros and cons of each of these strategies, paying special attention to their real-world applicability.

DEPLOYMENT STRATEGIES INVOLVING SINK MOBILITY

The existing WSN literature can generally be grouped into two broad categories: theoretical and application-centric works (Raman & Chebrolu, ACM SIGCOMM, July 2008 issue). It is somewhat disappointing to observe that one particular subgroup of WSN literature – the works on the subject of sink mobility – largely falls in the first of the two categories (theoretical works), and as such " ... *make no mention, or only a passing mention, of the specific application scenarios in which the presented ideas are applicable*" (Raman & Chebrolu, 2008, p. 75).

In the view of the above critique, the purpose of this section is to provide the reader with a comprehensive as well as a realistic perspective on the topic of sink mobility. Specifically, in the first subsection, the reader is familiarized with various theoretical ideas concerning the use of mobile sinks, as found in the literature. Each of these ideas is discussed not only from the perspective of their expected performance; but, more importantly, from the perspective of their practical feasibility in real-world WSNs. In the subsequent subsection, the ideas and solutions with the highest chance of being utilized in practice are explicitly identified.

Taxonomy of Deployment Scenarios Involving Mobile Sink(s)

The existing WSN works on the subject of sink mobility can be categorized based on the following two criteria (Ekici, Gu, & Bozdag, 2006): (1) assumed type of node-to-sink data delivery, and (2) assumed type of mobile sink trajectory. A detailed look at each of the two criteria and their respective subcategories is provided below.

Types of Node-to-Sink Data Delivery

The term *node-to-sink data delivery* generally refers to the way the sensory readings travel, i.e. are transmitted, from the respective node to the sink. Types of node-to-sink data delivery most commonly studied in the literature, in the context of sink mobility, include: single-hop, multi-hop, and limited multi-hop.

Single-Hop Data Delivery
Single-hop data delivery, as the name implies, assumes direct one-hop communication between a sensor and the sink, and as such eliminates the need for inter-sensor communication (i.e. data exchange). In WSNs deploying a single static sink, this form of data delivery should generally be avoided – to be able to communicate directly with the sink, each sensor node would have to extend its radio range over multiple if not tens of hops, thus putting itself at the risk of rapid battery exhaustion (recall the discussion from previous section). In WSNs deploying a mobile sink, on the other hand, the implementation of single-hop data delivery is much less demanding or detrimental, as it does not require that the sensors' radio range be changed. Instead, the sensors can now wait for the mobile sink to enter their existing unchanged transmission range, at which point they can upload their respective readings. Note, in the remainder of this chapter, the term *single-hop* will be used to refer to this later and more realistic variant of direct sensor-to-sink data delivery, involving fixed short-range sensor nodes and a moving sink, as discussed in (Chakrabarti, Sabharwal, & Aazhang, August, 2006; Chatzigiannakis, Kinalis, & Nikoletseas, October, 2006; Ekici, Gu, & Bozdag, 2006). (An illustration of this type of delivery is provided in Figure 2).)

Advantages of single-hop delivery: Given that the single-hop delivery relieves the ordinary sensors from relaying/routing of other sensors' packets, individual sensor lifetime as well as the lifetime of the entire network is expected to increase.

Figure 2. Examples of single-hop deployment scenarios involving sink mobility

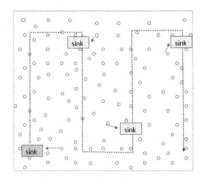

(a) Single-hop delivery with random mobility	*(b) Single-hop delivery with predictable mobility*

Disadvantages of single-hop delivery: The extraordinary demands placed upon the mobile sink are the most obvious drawback of this delivery scheme. Namely, the effectiveness of the single-hop delivery depends on the mobile sink's ability to visit most, i.e. every single, static sensor in the network. Doing so will not only consume a large amount of the sink's energy, but in some cases may not even be feasible due to the nature or inaccessibility of the underlying terrain. (For example, the terrain may abound in obstacles or be under the enemy observation, thus preventing the robot from visiting all of the network nodes). Another serious drawback of the single-hop delivery is the fact that in order to upload their sensor readings, the sensors must wait to be visited by the sink. As a consequence, in a network that covers a wide geographic area and/or comprises a large number of sensors, a significant delay may occur between the time a sensory reading was taken and the time the reading was uploaded to the sink.

Multi-Hop Data Delivery

Multi-hop is the de-facto form of data delivery in the conventional many-to-one WSNs, as already discussed in Introduction. In WSNs that deploy this form of delivery, the sensors are able relay their readings to the sink (via other nodes) in real-time, i.e. as soon as the readings have been taken (see Figures 3.a) and 3.b)).

Advantages of multi-hop delivery: In general, the delays introduced by the multi-hop delivery are much shorter than the ones introduced by the single-hop delivery, since wireless transmission of data takes much less time than the physical locomotion of a robot. This may not be true only under conditions of extreme traffic congestion, resulting in excessive wireless transmission delays. Another advantage of the multi-hop delivery is the fact that the operation of the respective network is not fundamentally dependant on the type of mobile sink trajectory – a multi-hop WSN will continue to function, and the sensors' readings are able to reach the sink, even if the sink decides to become fully stationary. The same, clearly, cannot be said for the one-hop data delivery; in which case the mobile sink's ability to move around the network is the key precondition for the normal operation of the network.

Disadvantages of multi-hop delivery: As discussed previously, many-to-one multi-hop WSNs are generally vulnerable to the hot-spot and funnelling-effect problems. (Though, we have also shown that the first of the two problems can be very well controlled through the use of mobile sinks.) Another issue associated with the use of multi-hop data delivery is a potentially much higher communication load carried by individual sensor nodes, not only in the sink's neighbourhood but across the entire network, compared

Figure 3. Examples of multi-hop deployment scenarios involving sink mobility

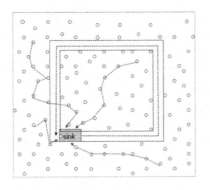

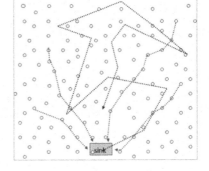

(a) Multi-hop delivery with random mobility   *(b) Multi-hop delivery with predictable mobility*

to the load carried in the case of single-hop data delivery. Namely, the nodes of a multi-hop WSN are responsible for forwarding their own as well as data from other sensors. As a result, the average life expectancy of individual nodes and of the entire network is shorter than in networks that deploy the single-hop delivery scheme.

Limited Multi-Hop Data Delivery

Limited multi-hop data delivery is a hybrid approach that combines the features of single-hop and multi-hop node-to-sink data delivery techniques. Specifically, the limited multi-hop assumes the existence of several designated (static) anchor-nodes responsible for collecting the sensory readings from other (ordinary) nearby nodes via multi-hop transmissions. Subsequently, the anchor nodes get visited by the mobile sink, one by one, at which point the single-hop delivery is used to upload the previously collected readings from the anchor nodes to the sink. (For more see (Kansal, Somasundara, Jea, Srivastava, & Estrin, June, 2004) and (Khan, Gansterer, & Haring, 2007).)

Advantages of limited multi-hop delivery: By reducing the average hop distance traveled by packets carrying the nodes' sensory readings (instead of being routed to the sink, the packets

are now routed to the nearest anchor point, thus passing through fewer intermediate sensors), the limited multi-hop attempts to reduce the 'per-node' load and increase the network lifetime, compared to that observed under pure multi-hop delivery. Similarly, by reducing the number of nodes that need to be visited by the mobile sink (the sink now visits only a few designated anchor nodes), the limited multi-hop technique attempts to reduce the locomotion demands on the mobile sink as well as the delay between taking and uploading of sensory readings, compared to that observed under pure single-hop delivery.

Disadvantages of limited multi-hop delivery. In its essence, the limited multi-hop delivery is just a 'compromise solution' – while providing improvement in one network-performance metric it inevitably deteriorates the network performance in another metric. For example, when compared to the pure single-hop data delivery, the limited multi-hop provides better packet delay but worse node/network lifetime. When compared to the pure multi-hop data delivery, the limited multi-hop provides better node/network lifetime but worse packet delay. Or, put another way, in each performance category (packet delay, network lifetime, network's overall energy consumption), the limited multi-hop emerges as the 'mid-range' performer.

Types of Mobile Sink Trajectories

In the context of this book chapter, the term *mobile sink trajectory* will refer to the physical path that a mobile sink takes/follows while moving through the network field. Types of mobile sink trajectories most commonly studied in the literature include: random, predictable or fixed, controlled or optimized. Before we take a detailed look at each of the three types of trajectories, it is worth noting that the type of sink trajectory and the model of data delivery deployed in a WSN are not directly related to, or constrained by, each other. Theoretically it is possible to combine almost any type of sink trajectory with any type of data delivery (see Table 1).

Random Trajectory
As the name implies, the trajectory of a randomly moving sink comprises a sequence of segments of arbitrary length and direction (see Figures 2.a)

and 3.a)). The mobile sink's speed along each segment, and the pause time between movements along different segments, can also be arbitrary, although these two conditions do not have to hold to satisfy the randomness requirement. Examples of research works on the use of random trajectory in WSNs involving a mobile sink include (Shah, Roy, Jain, & Brunette, May, 2003) and (Shakya, Zhang, Zhang, & Lampe, May, 2007).

Advantages of random trajectory: In multi-hop WSNs, the random trajectory is generally very effective in defending against the hot-spot problem: randomized change of the sink's residing location implies randomized distribution of the traffic and routing load across the network, and as such minimizes the probability that any of the network nodes exhausts its energy and dies much earlier than other nodes.

Disadvantages of random trajectory: In single-hop WSNs, random trajectory may imply long delays between the time a sensory reading is taken and the time the respective sensor is visited (the reading is uploaded) by the sink. More importantly, in both single- and multi- hop WSNs,

Table 1. Types of data delivery in many-to-one WSNs involving a mobile sink and respective/expected performance (the performance quantifiers, e.g. generally short vs. generally long, are relative to each other)

	TYPE OF NODE-TO-SINK DATA DELIVERY	**Multi-hop**	**Single-hop**	**Limited Multi-hop (Hybrid)**
CATEGO-RIES	**TYPE OF SINK TRAJECTORY**	**Random:** (Cheng et al. 2008) **Predictable:** (Luo et al. 2006), (Tacconi et al. 2007) **Controlled:** (Bi et al. 2007), (Vincze et al. 2005)	**Random:** (Chatzigiannakis et al. 2006), (Shah et al. 2003) **Predictable:** (Chakrabarti et al. 2006) **Controlled:** (Somasundara et al. 2004), (Ekici et al. 2006)	**Predictable:** (Khan et al. 2007) **Controlled:** (Kansal et al. 2004)
PERFOR-MANCE	**MOBILITY/ENERGY DEMANDS ON SINK**	generally limited	generally high	limited
	DELAY	generally short	generally long	medium
	NETWORK LIFETIME / MAX PER-NODE LOAD	generally short	generally long	medium
	OVERALL ENERGY CONSUMPTION	generally large	generally small	medium

the feasibility of the random sink trajectory is highly questionable. Namely, most real-world WSN environments abound in obstacles (e.g. walls, furniture, indoor machinery, and vegetation, rocks, ponds, in outdoor environments). Hence, in such environments, it seems very unlikely if not impossible that a sink be able to arbitrarily move through the network space – in just any direction, to any distance, following a straight line.

Predictable or Fixed Trajectory

In contrast to the random trajectory, the predictable/fixed trajectory is fully deterministic – with this type of trajectory, the sink is expected to continuously follow the same path through the network (see Figures 2.b) and 3.b)). The predictable/fixed trajectory is typically forced upon the mobile sink by the nature of the physical terrain and/or presence of obstacles in the environment.

An example of a research study that deals with this type of trajectory is (Tacconi, Carreras, Miorandi, Chiti, Casile, & Fantacci, March, 2007). The study looks at the use of a WSN to track free parking spots and provide this information to the cars circling the parking lot. Here, cars are envisioned as mobile sinks (i.e. end recipients and users of the gathered data) which, by the nature of parking-lot design, are expected to follow a well-defined, deterministic trajectory.

Advantages of predictable/fixed trajectory: Real-world WSN environments, especially ones found in urban and industrial settings, are known to abound in small- and large- scale obstacles. In these environments, a trajectory confined to a limited number of path-segments appears to be most likely, thus making the predictable/fixed trajectory a far more realistic model of sink movement than the other two types of trajectories.

Disadvantages of predictable/fixed trajectory: In multi-hop WSNs, the predictable/fixed trajectory is generally less effective against the hot-spot phenomenon than the other two types of trajectories. Namely, by moving along a limited (and possibly predefined) set of path-segments,

the sink's ability to distribute the traffic and routing load across a multi-hop WSN is also limited. Moreover, in single-hop WSNs, the sink's inability to reach all areas of the network field may fundamentally jeopardize the network's ability to function.

Controlled or Optimized Trajectory

In the case of controlled/optimized trajectory, the sink's path is a function of a particular network variable (e.g. current state of nodes' batteries, intensity and/or direction of traffic flows, etc.), and the path gets continuously adjusted to ensure optimal network performance with regard to one or more performance metrics. Examples illustrating the concept of controlled/optimized trajectory include:

a. A multi-hop WSN in which the mobile sink receives regular updates on the state of nodes' batteries. Based on the received updates, the mobile keeps adjusting its position so as to minimize the routing load on the nodes with critically low battery levels, as discussed by (Bi, Sun, Ma, Li, Khan, & Chen, 2007).

b. A one-hop WSN for pollution monitoring, in which nodes deployed in different areas of the network operate at different sampling rates. The sink's visitation of nodes, i.e. path through the network, is controlled by these sampling rates, with the aim to minimize the probability that any of the sensor nodes' buffers overflow (Somasundara, Ramamoorthy, & Srivastava, December, 2004).

Other examples of works dealing with mobile sink trajectories of controlled/optimized type are (Ekici, Gu, & Bozdag, 2006), (Luo, Panchard, Piorkowski, Grossglauser, & Hubaux, June, 2006) and (Vincze & Vida, October, 2005).

Advantages of controlled/optimized trajectory: The earlier discussion has shown that the nature and physical shape of the random and predictable trajectories are not, in any direct way, related the state of and/or events occurring inside the respective WSN. On the other hand, the physical shape of the optimized trajectory is meant to be fully controlled from inside the network, with the goal to optimize one or more aspects of the network's performance. Hence, there should be no question, from the perspective of network related benefit(s), that the effectiveness of the controlled/optimized trajectory will always far surpass the effectiveness of the other two types of sink trajectories.

Disadvantages of controlled/optimized trajectory: In most of the literature on the subject of optimized mobile sink trajectory, the shape of the trajectory is assumed to be exclusively controlled by, i.e. from inside, the network. Nevertheless, as already mentioned, the existence of obstacles in real-world WSN environments is likely to constrain the movement of the sink, possibly preventing it from following the (theoretically) optimal trajectory. To our knowledge, there has been no previous study on the effects of environmental constraints and obstacles on the network performance under the controlled/optimized sink trajectory. Hence, the actual real-world effectiveness of this type of trajectory remains very much an open question.

Table 1 summarizes the preceding discussion concerning different models of data delivery and mobile sink trajectory. Specifically, column-wise, the table outlines the key performance characteristics of multi-hop, single-hop and limited multi-hop data delivery, relative to each other, in four different performance categories: mobility and/or energy demands on sink, packet delay, network lifetime, overall energy consumption. The table also specifies the types of sink trajectories that are feasible under each data delivery model (second row), as well as provides examples of relevant research works in each respective category.

Mobile-Sink Deployment Scenario Close to Reality

In the previous section, the reader has been introduced to a wide range of theoretical models and ideas concerning the use of mobile sink(s) in the conventional many-to-one WSNs. The goal of this section is to identify those models and ideas most applicable in the real world. To be able to do so, we begin our discussion with a brief survey of seven real-world wireless sensor network systems, whose design and/or operation has been reported in the literature:

1. **Microsoft's Data Centers HVAC System:** (Gohring, 2008) The system employs a ZigBee-based sensor network for temperature and humidity monitoring inside a Microsoft's data center. The collected information is then used to control fan speeds and air-conditioning inside the center with the goal to reduce the energy cost and increase the data center's revenue.

2. **Smart Car Parking System:** (Mick, 2008) The system utilizes a large-scale WSN to alert drivers of available parking spaces in the greater San Francisco area. The advantages of the proposed systems include: decreased car fuel cost for consumers, reduced pollution on city streets, etc.

3. **Semiconductor Plant and North Sea System:** (Krishnamurthy et al., 2005) The system involves the use of WSNs in a semiconductor fabrication plant and onboard an oil tanker in the North Sea to ensure product quality as well as efficient and safe operation.

4. **Structural Monitoring System Wisden:** (Xu et al., 2004) The system utilizes a *multi-hop* WSN to collect and store vibration measurements (caused by earthquakes, wind and passing vehicles) inside large structures/ buildings for the purpose of damage detection and localization.

5. **Volcano Monitoring System:** (Werner-Allen et al., 2006) The system employs a *multi-hop* sensor network for detection and monitoring of interesting volcanic activity at Reventador.

6. **FireWxNet System:** (Hartung et al., 2006) The system involves the use of a *multi-hop* WSN for monitoring weather conditions and fire prevention in rugged wild land environments.

7. **Vehicle Tracking System:** (He et al., 2004) The system utilizes a *multi-hop* WSN for the detection and tracking of moving vehicles in missions that involve a high element of risk for human personnel.

Although not exhaustive, the above examples are fairly representative of a large number of WSN systems operating (or envisioned operating) in the real-world, and as such can help in setting the ground for a more realistic evaluation of the theoretical ideas previously discussed.

The original implementation of all of the above mentioned systems has been based on the traditional many-to-one model of WSNs, assuming the use of a single static sink. At the same time, in all seven of the mentioned systems, the extension of network lifetime has been identified as one of the most important goals of their design and operation. Consequently, this allow us to argue that each of these systems would likely benefit from the use of a mobile/moveable sink in place of a static sink, since increased network lifetime is known to be associated with the use of mobile sinks (see first two sections of this chapter). We should also stress, however, that the actual level of benefit obtained from the use of a mobile sink would ultimately depend on the type, i.e. appropriateness, of data delivery and sink trajectory deployed.

Following the above, we further postulate: if in the enlisted or similar systems one was to contemplate the use of a mobile sink, the implementation would likely ask for *multi-hop* node-to-sink type of data delivery accompanied with *predictable/fixed*

sink trajectory. Namely, in all seven systems (with the possible exception of 4)), the sensory readings need to be delivered to the main processing centre, i.e. sink, with minimal delays, so as to enable a prompt and adequate action in case of computer-overhead, fire, or unsafe conditions. As discussed earlier, the multi-hop data delivery has the best chances of meeting this requirement (see Table 1). Also, in all seven of the mentioned systems, the underlying physical environments abound in obstacles (e.g. walls, furniture, machinery in the indoor systems 1), 3) and 4); rocks, craters, ponds in the natural outdoor systems 5) and 6); buildings and parks in the urban outdoor system 2)), thus suggesting the *predictable/fixed trajectory* as the only admissible course of movement for the mobile sink.

In the closing of this section, we would like to note the following: even though *multi-hop* data delivery accompanied with *predictable/fixed* sink trajectory appears to be the most likely mode of mobile sink deployment in the real world, one should not exclude the possibility that there are (i.e. will be) situations in which the use of other types of data delivery or sink trajectory would be more appropriate. Due to the page limit of this book chapter, we omit such 'possible but less likely' application scenarios from further discussion, and concentrate exclusively on the analysis of the *multi-hop* data delivery with a *predictable/fixed* trajectory deployment model.

MULTI-HOP DATA DELIVERY WITH PREDICTABLE/FIXED TRAJECTORY MODEL: ANALYTICAL EVALUATION

The discussion in the previous section has shown that, when it comes to the use of mobile sinks in real-world WSNs, *multi-hop* data delivery accompanied with *predictable* sink trajectory emerges as the most likely deployment model. To date, only a handful of published works have dealt with this particular model of mobile sink deployment (e.g. Luo, Panchard, Piorkowski, Grossglauser, & Hubaux, June, 2006 and Tacconi, Carreras, Miorandi, Chiti, Casile, & Fantacci, March, 2007). Moreover, each of those works has been very limited in its scope, focusing on just one particular aspect of network performance. For example, (Luo, Panchard, Piorkowski, Grossglauser, & Hubaux, June, 2006) look at network lifetime, while (Tacconi, Carreras, Miorandi, Chiti, Casile, & Fantacci, March, 2007) at packet latency as only relevant performance measures.

While keeping the limitations of the earlier works as well as the constraints of real-world environments in mind, this section offers a comprehensive analysis of the *multi-hop data delivery with predictable sink trajectory* model of operation. This analysis considers not only one but all relevant performance metrics, while offering results applicable to a wide range of scenarios. The content of the section is organized as follows. In the following subsection, the main network assumptions are stated and the choice of performance metrics is justified. In the subsequent subsection, the key analytical findings are introduced to the reader.

Analytical Preliminaries

Network Assumptions

The wireless sensor network model used in the proceeding analysis is based on the following assumptions:

- The network field covers a square-shaped area (see Figure 4), and as such resembles the layout of either of these: a floor of a building or an industrial plant, a parking lot, a man-made park, etc.
- The sensor nodes are distributed in a semi-random fashion through the network field – one randomly positioned node per one small square-shaped cell (see Figure 5). There are a total of *nxn* nodes in the

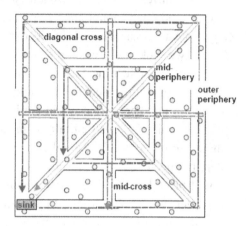

Figure 4. A WSN over a square-shaped field with constrained mobile sink trajectories

network. From the communication perspective, the nodes form a so-called virtual grid (Vlajic & Xia, June 2006), in which each node is able to directly communicate with eight of its topological neighbours. As for the nodes' sensing function, every node collects and transmits one packet of data once every T time units.

- One mobile sink is deployed in the network. The sink's radio range is equal to the radio range of ordinary static sensors. The sink's speed is relatively slow compared to the frequency of the sensors' reporting – it takes T (or a multiple of T) time units for the mobile sink to move between two sensor nodes, i.e. to move through one square-shaped cell shown in Figure 5. This implies that during its residence in each cell, the mobile sink receives at least one packet of data from each of the sensors.
- The sink's movement is constrained to several possible paths, each of rather regular geometric form. Our analysis focuses on four specific paths: outer periphery, mid-periphery, diagonal cross, and mid-cross, as illustrated in Figure 4. (Note, in reality, the predictable trajectory of a mobile sink could be constrained by various objects

and/or conditions and, consequently, can take very different shapes. Obviously, in a single study, we cannot analyze or even enumerate all possible cases of trajectories. Instead, we look at a few simple and very generic predictable trajectories which, in this or slightly modified form, could be found in a number of real-world environments. We also believe that, although not exhaustive in terms of the number of considered trajectories, the presented analysis will provide a good foundation which the reader will be able to build upon and/or customize to any specific real-world scenario.

- The routing protocol employed in the network is a version of 'shortest-path', in which the next-hop neighbour minimizes both: the hop- and geographical- distance to the sink.

Performance Metrics

The three performance metrics that will be considered in the proceeding analysis include:

a. **Total Energy Consumption:** This performance metric is intended to reflect the overall (i.e. cumulative) energy consumption of all nodes in the network. Signal transmission is known to be the most costly aspect of a sensor node's operation. Each node engages in transmission not only when sending its own, but also when relaying/routing other nodes' data. Accordingly, a straightforward way to approximate the total energy consumption in a multi-hop WSN (see Figure 3) is by calculating the overall number of hops that all the packets, i.e. bits, generated in the network traverse. (In the remainder of this chapter, we will refer to this value as *cumulative hop count*.) From the practical point of view, total energy consumption is

the key performance metric in WSNs where a frequent replacement of sensors' batteries is possible, yet we are interested in minimizing the operational cost of the network, most of which comes from the regular replacement of batteries.

b. **Network Lifetime:** This metric, commonly used in the WSN literature, is intended to reflect the time span from the network's initial deployment to the first loss of coverage (Luo, Panchard, Piorkowski, Grossglauser, & Hubaux, June, 2006). As such, network lifetime can alternatively be defined as the 'time until the first node dies'. The easiest to capture indicator of this metric is the maximum per-node load, where a node's load corresponds to the number of packets sent from or routed through the given node. Clearly, the network setup that minimizes the maximum node load is the one that will ensure the maximum network lifetime. From the practical point of view, network lifetime is a critical performance metric in applications where frequent replacement of sensors' batteries is not possible, and the operator of the network is interested in minimizing the probability of a node (i.e. network) failure.

c. **Average Packet Delay:** This metric has a clear significance in WSN applications where the timeliness of received information is a concern. Under the assumption of negligible propagation, processing and queuing delay, the average packet delay turns out to be simply the average transmission (i.e. routing) delay, which can be approximated through the following: 1) calculate the overall number of hops traversed by all the packets observed in the network, and 2) divide this number by the actual packet count. Please note that the first of the two values (cumulative hop count) is also used in estimating the network's total energy consumption. Thus, if the actual packet count is known and as such represents a simple constant, the cumulative hop count can serve as a good indicator of both - total energy consumption and average packet delay.

Analytical Results

In this section, we compare the performance of the four selected predictable mobile-sink trajectories (periphery, mid-periphery, diagonal, and mid-cross) with respect to the three selected performance metrics (network lifetime, total energy consumption, and average packet delay). In addition to analyzing the four different strategies of mobile sink deployment relative to each other, we also compare these strategies against the deployment of a single static sink.

Total Energy Consumption

As previously indicated, a relatively simple way to approximate the total energy consumption in a multi-hop WSN is through the use of *cumulative hop count* metric (Vlajic & Xia, June 2006). Namely, if we annotate *the energy consumed by a sensor for transmission of one bit of data (over one hop)* with E_{T-bit}, *the energy consumed by a sensor for reception of one bit of data (over one hop)* with E_{R-bit}, *the number of bits sent by all nodes over all hops in the network (i.e. cumulative bit hop count)* with CHC, and if we assume that every bit-transmission is followed by a corresponding bit-reception, then the total energy consumption in the network (E) turns out to be:

$$E = CHC \cdot (E_{T-bit} + E_{R-bit}) \text{ [Joules]} \qquad (4.1)$$

Here, E_{T-bit} [Joules/bit-hop] and E_{R-bit} [Joules/bit-hop] are considered to be identical for all nodes in the network, as all nodes employ the same radio range and share the same/similar hardware characteristics. The two parameters can easily

Figure 5. Static sink deployment strategy

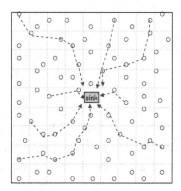

be determined from the nodes' specifications. In contrast to E_{T-bit} and E_{R-bit}, CHC [bit-hops] is a network-wide and much less predictable parameter – its value depends not only on the amount of traffic originally generated by the sensors, but also on the configuration of routing paths in the network, which are primarily influenced by the geographical position of the sink. Given the preceding discussion and form of (3.1), it is evident that by knowing CHC one can easily approximate E, and vice versa.

Now, to estimate the network's total energy consumption that would arise from the deployment of a mobile sink moving along a particular trajectory, over a given interval of time, the following needs to be done:

1. Derive a closed-form expression for the network's cumulative bit-hop count (CHC), assuming the sink's fixed residence at one particular point, i.e. cell, along the observed trajectory.
2. Derive a closed-form expression for network's total cumulative hop count (TCHC), obtained by summing all CHCs over all cells along the observed trajectory that get visited by the sink within the given interval of time. The relationship between TCHC and CHC can be represented as:

$$TCHC = \sum_{over\ all\ visited\ cells} CHC[bit - hops]$$
$$(4.2)$$

In order to enable a fair comparison of the static and different sink mobility scenarios, we will assume in all cases that the observed interval of time is equivalent to the sink's visitation of 4*(n-1) cells. Note, in a network of size *nxn*, the sink visits exactly 4*(n-1) cells while making one full revolution along the peripheral, diagonal, or mid-cross trajectories (see Figures 6-8).

a. **Total Energy Consumption in a Static Sink Scenario:** In the static sink scenario, we consider the sink to be positioned in the centre of the network field, as shown in Figure 5. Now, to find the respective expression for CHC, let us observed that under this strategy and in one interval of time T, there will be 1+(1*4+4) nodes (i.e. packets generated) at a 1-hop distance from the sink. (Recall, we assume that each sensor generates one packet of data in one interval of time T. Also, in the mobile sink scenarios, we assume T to be the sink's residence time at one cell.) Furthermore, there will be (3*4+4) packets generated at a 2-hop distance from the sink, and passing a 2-hop distance to reach the sink; 3(5*4+4) packets generated at and passing a 3-hop distance to reach the sink, etc. Consequently, the expression for CHC and the respective TCHC take the following form:

$$CHC_{static}(n,) =$$

$$1 + \sum (2 \cdot i - 1) \cdot 4 + 4) \cdot i = 1 + \frac{n^3}{3} - \frac{n}{3} \quad (4.3)$$

$$TCHC_{static}(n,) = 4 \cdot (n-1) \cdot CHC_{static}(n,) \quad (4.4)$$

b. **Total Energy Consumption in Scenario involving a Mobile Sink with Peripheral Trajectory:** To find a general analytical expression for the cumulative hop count (CHC) observed in a WSN of size *nxn* assuming the sink's location at an arbitrary cell *k* along the outer peripheral trajectory (see Figure 6), let us look at the following special cases:

Figure 6. Peripheral mobile sink trajectory

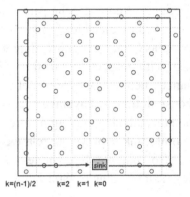

Figure 7. Diagonal trajectory

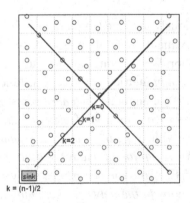

Figure 8. Mid-cross trajectory

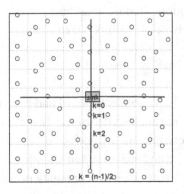

1. When $k=0$ (the mid bottom cell, as shown in Figure 5), there will be 5+1 packets generated at a 1-hop distance (and passing this distance to reach the sink), 9 packets generated at a 2-hop distance, 13 packets generated at a 3-hop distance, ..., $4*(n-1)/2 +1$ packets generated at a distance of $(n-1)/2$ hops, and then sets of n packets at distances of $(n-1)/2+1$, $(n-1)/2+2$, .., $(n-1)$ hops, respectively. Based on this, we obtain the following expression for $CHC_{periphery}(n,k=0)$.

$$CHC_{periphery}(n, k = 0)$$
$$= 1 + \sum_{i=1}^{\frac{n-1}{2}-k} (4 \cdot i + 1) \cdot i + \sum_{i=\frac{n+1}{2}+k}^{n-1} n \cdot i \qquad (4.5)$$

2. When $k=(n-1)/2$ (one of the corner locations), there will be 3+1 packets generated at a 1-hop distance, 5 packets generated at a 2-hop distance, .., $(2n-1)$ packets generated at a $(n-1)$-hop distance. Based on this, the following expression for $CHC(n,k=(n-1)/2)$ is obtained.

$$CHC_{periphery}\left(n, k = \frac{n-1}{2}\right) = 1 + \sum_{i=1}^{n-1}(2 \cdot i + 1) \cdot i$$
$$(4.6)$$

3. Through a procedure similar to (1) and (2), the following expression for CHC under an arbitrary k is derived:

$$CHC_{periphery}(n,k) =$$
$$1 + \sum_{i=1}^{\frac{n-1}{2}-k}(4 \cdot i + 1) \cdot i + \sum_{i=\frac{n-1}{2}-k+1}^{\frac{n+1}{2}+k-1}\left(2 \cdot i + \frac{n+1}{2} - k\right) \cdot i + \sum_{i=\frac{n+1}{2}+k}^{n-1} n \cdot i$$
$$(4.7)$$

Now, to obtain the expression for the total cumulative hop count (TCHC), we sum CHCs over all cells visited by the sink while moving along the outer peripheral trajectory, as shown below. (Note, in the case of outer peripheral trajectory, there are four corner cells where $k=(n-1)/2$, four central cells where $k=0$, and eight instances of every other type of cell $(k=0, .., (n-1)/2-1)$.) The resulting equation is provided below.

$$TCHC_{periphery}(n) =$$
$$8 \cdot \sum_{k=1}^{\frac{n-1}{2}-1} CHC_{periphery}(n,k) + 4 \cdot CHC_{periphery}(n,0) + 4 \cdot CHC_{periphery}\left(n, \frac{n-1}{2}\right)$$
$$(4.8)$$

c. **Total Energy Consumption in Scenario involving a Mobile Sink with Diagonal Trajectory:** By following a procedure analogous to the one presented above, the following expressions for CHC and TCHC under the diagonal trajectory (see Figure 7) are obtained:

$$CHC_{diagonal}(n,k) = 1 + \sum_{i=1}^{\frac{n-1}{2}-k} 8 \cdot i^2 + \sum_{i=\frac{n-1}{2}-k+1}^{\frac{n+1}{2}+k-1}(2 \cdot i + n - 2 \cdot k) \cdot i$$
$$(4.9)$$

$$TCHC_{diagonal}(n) =$$
$$8 \cdot \sum_{k=1}^{\frac{n-1}{2}-1} CHC_{diagonal}(n,k) + 4 \cdot CHC_{diagonal}(n,0) + 4 \cdot CHC_{diagonal}\left(n, \frac{n-1}{2}\right)$$
$$(4.10)$$

(d) **Total Energy Consumption in Scenario involving a Mobile Sink with Mid-Cross Trajectory:** Again, by following a procedure analogous to the one presented in section (b), the following expressions for CHC and TCHC under the mid-cross trajectory (see Figure 8) are derived:

$$CHC_{mid-cross}(n,k)=$$

$$1+\sum_{i=1}^{\frac{n-1}{2}-k}8\cdot i^2+\sum_{i=\frac{n-1}{2}-k+1}^{\frac{n+1}{2}-1}(4i+n-2k)\cdot i+\sum_{i=\frac{n+1}{2}}^{\frac{n+1}{2}+k-1}n\cdot i$$

$$(4.11)$$

$$TCHC_{mid-cross}(n)=$$

$$8\cdot\sum_{k=1}^{\frac{n-1}{2}-1}CHC_{mkd-cross}(n,k)+4\cdot CHC_{mid\,cross}(n,0)+4\cdot CHC_{mid\,cross}\left(n,\frac{n-1}{2}\right)$$

$$(4.12)$$

(e) **Total Energy Consumption in Scenario involving a Mobile Sink with Mid-Peripheral Trajectory:** Finally, the expressions for CHC and TCHC under the mid-peripheral trajectory are provided below,

$$CHC_{mid-periphery}(n,p,k)=$$

$$1+\sum_{i=1}^{p}8\cdot i^2+\sum_{i=p+1}^{\frac{n+1}{2}-k-1}(4i+1+2\cdot p)\cdot i+\sum_{i=\frac{n+1}{2}-k}^{\frac{n+1}{2}+k-1}\left(2i+\frac{n+1}{2}-k+p\right)\cdot i+\sum_{i=\frac{n+1}{2}+k}^{n-p-1}n\cdot i$$

$$(4.13)$$

$$TCHC_{mid-periphery}(n,p)=\frac{\frac{n-1}{2}}{\frac{n-1}{2}-p}\cdot$$

$$\left[8\cdot\sum_{k=1}^{\frac{n-1}{2}-p-1}CHC_{mkd-periphery}(n,p,k)+4\cdot CHC_{mid-periphery}(n,p,0)+4\cdot CHC_{mid-periphery}\left(n,p,\frac{n-1}{2}-p\right)\right]$$

$$(4.14)$$

where p∈{0,..,(n-2)/2} represents the offset of the mid-peripheral from the outer peripheral trajectory in terms of hops (see Figure 9).

(f) **Comparison of the Total Cumulative Hop Counts:** The numerical values of TCHCs, for the static sink and the four different mobile sink deployment strategies, under varying network sizes, are plotted in Figure 10. The figure shows that in terms of total cumulative hop count and, consequently, in terms of total energy consumption, the static sink strategy outperforms all four of the mobile

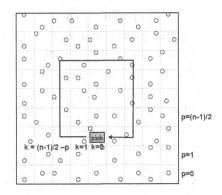

Figure 9. Mid-peripheral trajectory

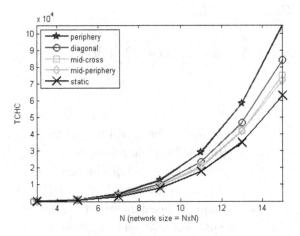

Figure 10. Total cumulative hop counts (TCHCs) under different sink-deployment strategies

sink deployment strategies. Although this may appear surprising and contrary to the discussion presented earlier in the chapter, such a result, in fact, could have been anticipated. Namely, on average, the location of the sink in the center of the network field induces the shortest sensor-to-sink routes and, as such, naturally results in a smaller hop count (i.e. smaller overall energy consumption) relative to the other examined strategies. Furthermore, when suggesting the possible superiority of mobile- over static- sink deployment, our earlier argument referred to the hot-spot and network lifetime problems, not the network's overall energy consumption.

Before closing this section, we would like to note that among the four mobile-sink trajectories (according to Figure 10), the mid-cross and mid-periphery offer noticeably better performance in terms of the overall energy consumption than the other two types of trajectories.

Average Packet Delay

As explained earlier in this section, (total) cumulative hop count could also serve as a good indicator of the average packet delay occurring in the network. Hence, our general conclusions from the total cumulative hop count analysis apply here as well.

Network Lifetime / Maximum Node Load

As stated previously in this section, in order to determine the lifetime of a WSN, one needs to: 1) know the precise load of every sensor in the network (i.e. number of packets routed through this node), and 2) identify the node that experiences the maximum load. Now, it turns out that finding a closed-form expression for individual node load in a network of the type as considered in our analysis is highly challenging if not impossible. Therefore, we have resorted to coding a MATLAB routine (see Appendix 1), and evaluating the per-node load numerically. The results of this evaluation are presented in Figure 11.

It is interesting to observe that the results shown in Figure 11 are very much opposite to the results shown in Figure 10 – in networks of size 6x6 and larger, the static sink is now the worst and the mobile-sink with peripheral trajectory the best performing strategy. The superiority of the mobile sink deployment strategies is especially pronounced in networks of larger sizes. For example, in a network comprising 11x11 nodes, the maximum node load in case of the mobile sink with the peripheral or diagonal trajectory is 3 times smaller, i.e. the network lifetime is 3 times

Figure 11. Network lifetime (i.e. <u>maximum</u> node load) under different sink-deployment strategies

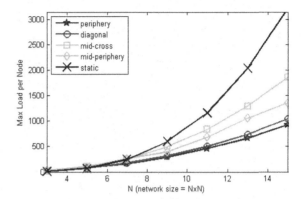

longer, than that of the static sink deployment strategy. As such, these results fully confirm our earlier theoretical postulates.

In order for the reader to better understand and validate the above results, the actual per-node loads in a network of size 9x9 under various sink deployment strategies are presented in Figure 12. The graphs depict the fundamental difference in the performance of the static-sink and various mobile-sink strategies with regard to the distribution of routing load. Specifically, in the cases of the static sink and the mobile sink with mid-cross and mid-peripheral trajectory (Figure 12.a), 12.b) and 12.c)) most of the load gets concentrated at/near the centre of the network. On the other hand, the strategies involving the mobile sink with diagonal and peripheral trajectory are evidently much better at distributing the routing load across the network grid (see Figures 12.d) and 12.e)). By shifting some of the routing load towards the edges of the network, the two later strategies manage to reduce the load at/near the network's centre and thus increase the lifetime of the respective nodes, ultimately producing a positive effect on the lifetime of the entire network.

As a final note, we would like to mention that in an earlier study on the use of a mobile sink following a predictable trajectory (Luo, Panchard, Piorkowski, Grossglauser, & Hubaux, June, 2006),

Figure 12. Total loads per node in a WSN comprising of 9×9 nodes

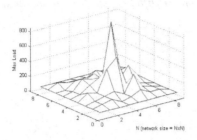

(a)   Load in static scenario in 9 ×9 grid

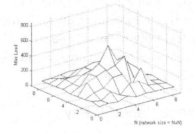

(b)   Load in mid-cross trajectory scenario in 9×9 grid

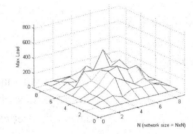

(c)   Load in mid-peripheral trajectory scenario in 9×9 grid

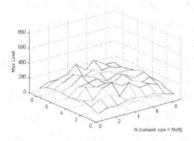

(d) Load in peripheral trajectory scenario in 9×9 grid

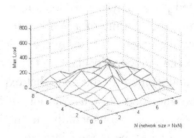

(e) Load in diagonal trajectory scenario in 9×9 grid

the superiority of outer-periphery over other types of trajectories, from the perspective of network lifetime, was also observed. We would also like to point out that there are a few fundamental differences between our study and that by (Luo, Panchard, Piorkowski, Grossglauser, & Hubaux, June, 2006):

1.   In (Luo, Panchard, Piorkowski, Grossglauser, & Hubaux, June, 2006), the authors look at WSNs of extreme sizes (800 nodes), whereas our study concentrates on more realistic smaller-size WSNs. (According to Raman & Chebrolu (2008), most real-world WSNs comprise only a few tens of sensors. In those rare instances of larger-size WSNs, comprising hundreds or thousands of nodes, there is typically " ... *no need to have one single network, but rather form a few smaller (independently operating) ones.*")

2.   The network model employed in (Luo, Panchard, Piorkowski, Grossglauser, & Hubaux, June, 2006) assumes a circular shaped WSN with nodes distributed according to a Poisson process, whereas our study looks at WSNs of randomized virtual-grid type.

3. In (Luo, Panchard, Piorkowski, Grossglauser, & Hubaux, June, 2006), no explicit mention of the diagonal trajectory is made. On the other hand, in our study, the diagonal trajectory emerges as an almost equally effective alternative to the outer peripheral trajectory (see Figure 11). In fact, upon considering its performance in all categories (total energy consumption, packet delay, network lifetime), the diagonal trajectory appears to be overall superior than the outer periphery – better in the first two, and as effective in the last category.

SINK MOBILITY IN ZIGBEE WSNS: SIMULATION-BASED EVALUATION

In the final section of this book chapter, we present the results of our OPNET-based simulation study. The study is aimed at evaluating the applicability of the theoretical findings presented in previous two sections to real-world WSNs that deploy ZigBee (Alliance Z., 2008) as the underlying network technology. The results suggest that, while the theoretical findings are generally applicable, one also needs to be aware of a few inherent challenges arising from the use of mobile sinks in ZigBee-based WSNs.

Overview of ZigBee

The IEEE 802.15.4 / ZigBee suite of standards is considered the 'technology of choice' for applications involving sensor networks, due to their ability to ensure reliable, low-power and cost-effective communication. In recent years, the technology has been gaining wide industrial acceptance, one proof of this being the use of ZigBee to support WSNs in Microsoft's data centres (Gohring, 2008).

The ZigBee standard (Alliance Z., 2008) defines 3 possible types of network device: ZigBee coordinator, ZigBee router, and ZigBee end-device.

- Sigsbee coordinator is responsible for starting the network and choosing the key network parameters. There is one single coordinator in any given ZigBee network. In WSNs deploying ZigBee, the role of network coordinator is typically given to the sink node.

- ZigBee routers are devices capable of routing data. In the case of ZigBee-based WSNs, (ordinary) sensor nodes assume the role of network routers, in addition to performing the sensing function.

- ZigBee end-devices are devices with no routing capability - end-devices rely on their parents (coordinator or routers) to transmit/route their packets. In WSNs of conventional architecture, the role of end-device is not directly suited for either type of nodes – the sink or sensors.

The ZigBee standard also defines 2 possible types of network topology: tree and mesh.

- The ZigBee tree topology employs a simple routing algorithm that relies on the established parent-child associations. Namely, through ZigBee's hierarchical addressing scheme, each parent device knows the addresses of its children. Hence, when a router receives a packet, it examines the destination address of the packet. If the given address happens to be in its own address block, the router then forwards the packet to the appropriate child. Otherwise, the router forwards the packet to its parent. The procedure gets recursively repeated until the final destination is reached.

- The ZigBee mesh topology implements a proactive routing algorithm which allows each routing capable device to forward packets not only to its parent, but also to other neighbouring routers. According to this algorithm, *route requests* are broadcasted on-demand to all neighbouring nodes.

The routes are, subsequently, constructed based on *route replies* received from the neighbouring nodes and/or the destination. The mesh routing algorithm also includes a route repair mechanism in order to allow the reconstruction of broken paths.

Due to its much greater flexibility, which is essential for the implementation of the shortest-path routing model, the mesh topology is far more suited for applications involving conventional WSNs.

Simulation Preliminaries

Simulation Tool

The simulation results presented in this section are generated using OPNET Modeler (OPNET Modeler Inc., 2008). OPNET Modeler is a commercial software tool with a wide-spread use in the networking industry. For the purposes of our analysis, OPNET's ZigBee model has been customized to include the Battery Module (The ART-WiSe framework, 2008), which is required for monitoring of nodes' energy consumption. The given battery module is based on the specification of MICAz and TelosB motes (MicaZ Datasheet Specs, 2008).

Network Parameters

The network dimensions (virtual grid sizes) deployed in our simulation study are representative of the virtual grid sizes utilized in the theoretical analysis of the previous section. Specifically, the study looks at the performance of ZigBee-based WSNs of the following four sizes:

5×5 + 1 (25 ZigBee routers/sensors and 1 mobile sink)

7×7 + 1 (49 ZigBee routers/sensors and 1 mobile sink)

9×9 + 1 (81 ZigBee routers/sensors and 1 mobile sink)

11×11 + 1 (121 ZigBee routers/sensors and 1 mobile sink)

Other important simulation parameters are enlisted below:

- Dimensions of virtual-grid cells (see Figure 5) are set to 30m × 30m, with a single ZigBee router randomly positioned inside each cell.
- Nodes' transmission range is set to 80m - long enough to allow each ZigBee router to communicate with routers in 8 neighbouring cells.
- In order to simulate the *shortest hop distance* routing, the network(s) deploy ZigBee's mesh topology.
- The mobile sink's velocity is set to be constant, with the mobile sink spending a total of 10 minutes in each grid cell.
- The implementation of sink's (re)association procedure, required due to its mobility, is based on the following principles: the sink starts dropping packets when it moves beyond the transmission range of its current 'point of attachment'. Once the percentage of dropped packets exceeds **50%**, the sink un-joins its current and attempts to find a new 'point of attachment.
- Every sensor node generates one packet of data every 5 minutes (packet inter-arrival time = 300 seconds), starting at a randomly picked initial packet generation time.
- The size of individual data packets is 400 bits.
- The overall simulation time is set to 21600 seconds (i.e. 6 hours).

Performance Metrics

Generally speaking, the analytical results presented in the previous section could be called hypothetical – the respective analysis was concerned with network performance at the level of user-data routing while assuming idealized conditions at the lower levels

Figure 13. Maximum node load simulation results

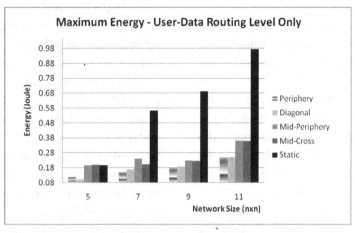

a)   *User-data routing level*

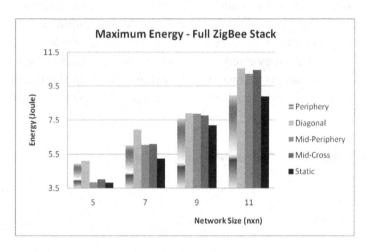

b)   *Across all layers*

of the OSI stack: network management, MAC, and physical layer. More specifically, all energy-related costs associated with the routing of control-data and any packet/transmission delays possibly induced at the MAC and physical layer were ignored. There were two main reasons for this: 1) to simplify the mathematical aspect of the analysis, and 2) to make the results more general - the lower-level parameters/conditions are uniquely determined by the actual protocols employed in the network and can significantly vary from one technology to another. (For example, ZigBee / IEEE 802.15.4 parameters/conditions are very much different from those found in WiFi / IEEE 802.11 or Bluetooth networks.) In order to compensate for this possible deficiency of the derived theoretical results, our simulation-based study of ZigBee-based WSNs looks at both: network performance at the user-data routing level only, as well as network performance across all layers of the OSI stack.

Simulation Results

In this section, the most significant results of our OPNET simulation study are presented. Also, the key differences between the simulation results and the derived analytical results from the previous section are outlined.

Maximum Node Load

The simulation results concerning the maximum node load, at the level of user-data routing and across the full ZigBee stack, are shown in Figure 13.a) and 13.b) respectively.

Now, the results of Figure 13.a) are pretty much in line with the analytical results shown in Figure 11, with the maximum node load under the static sink deployment being the highest, and the maximum node load under the mobile sink with the diagonal and outer-peripheral trajectories generally being the lowest. As for the maximum node load across the full ZigBee stack shown in Figure 13.b), these results turn out to be considerably different from those of Figure 11. Namely, according to Figure 13.b), in the networks of sizes 5x5, 7x7, and 9x9, the static sink strategy emerges as the best performer, only to be surpassed by the mobile sink with the outer-peripheral trajectory

strategy in the 11x11 network scenario. These results imply that, in ZigBee-based WSNs involving a mobile sink, the energy cost arising from the management of sink's mobility (repeated re-association procedure and respective route maintenance) is so significant that it completely negates the earlier identified theoretical advantage of using a mobile sink. Or, put another way, for anybody interested in extending the lifetime of a ZigBee-based WSN through the use of a mobile sink, these results are an indication of how critical the minimization of control overhead induced by the sink's mobility can be.

As a part of a recent study, we have integrated several known techniques for reduction of Zig-Bee's mobility-related control overhead into our simulation model; subsequently, we have examined the effects of these techniques on the performance of the four mobile sink deployment strategies. The results of this study are very encour-

Figure 14. Simulation Results in a WSN comprising of 7×7 nodes

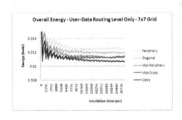
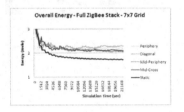
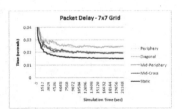

| *a) Overall energy consumption at user-data routing level* | *b) Overall energy consumption across all ZigBee layers* | *c) Average packet delay* |

Figure 15. Simulation Results in a WSN comprising of 9×9 nodes

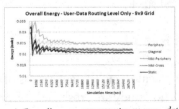
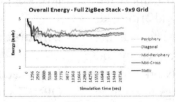
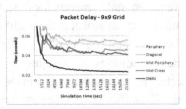

| *a) Overall energy consumption at user - data routing level* | *b) Overall energy consumption across all ZigBee layers* | *c) Average packet delay* |

aging – with the given techniques, the maximum node load under the mobile sink with the outer-peripheral trajectory was reduced well beyond that observed under the static sink deployment strategy, in networks of all sizes. These results will soon be reported in a full paper.

Overall Energy Consumption and Packet Delivery Delay

The simulation results concerning: 1) the overall energy consumption at the user-data routing level, 2) the overall energy consumption across all ZigBee layers, and 3) the average packet delay, in networks of sizes 7x7 and 9x9, are shown in Figures 14 and 15 respectively. (The simulation results for networks of sizes 5×5 and 11×11 follow the same performance pattern and are omitted due to space limitations.)

The obtained results are generally in line with the respective theoretical results presented in Figure 10, with the static sink strategy being the best and the mobile-sink with the outer-peripheral trajectory being the worst performer. It is interesting to observe, however, that when the energy-related performances of different deployment strategies are compared against each other, the superiority of the static sink scenario over the mobile deployment strategies is much more pronounced in the category of 'energy consumption across all ZigBee layers' (Figure 14.b)) than in the category of 'energy consumption at the user-data routing level' (Figure 14.a)). Again, as in the case of the maximum node load results, this can be explained by the fact that compared to the four deployment strategies involving a mobile sink, the static sink strategy achieves a much smaller control overhead, thus ultimately consuming far less energy.

CONCLUSION

In this book chapter, we have examined various theoretical and practical issues concerning the use of a mobile sink in a WSN of many-to-one multi-hop type. The discussion and presented results have shown that from the analytical point of view, a longer network lifetime is the first and most obvious benefit associated with the use a mobile sink. However, the extent of this benefit will very much depend on the actual type and shape of the sink's trajectory. We have also seen that when it comes to real-world WSNs, specifically ZigBee-based WSNs, the theoretical results may not fully apply – the use of a mobile sink may provide no benefit over the use of a static sink, unless there is a way to control the amount of mobility-related overhead traffic. The development and implementation of techniques for reduction of mobility-related overhead in ZigBee-based WSNs deploying a mobile sink remains an open field of study.

REFERENCES

Ahn, G. S., Hong, S. G., Miluzzo, E., Campbell, A. T., & Cuomo, F. (November, 2006). Funneling-MAC: a localized, sink-oriented MAC for boosting fidelity in sensor networks. In *Proceedings of The 4th ACM Conference on Embedded Networked Sensor Systems (SenSys 2006)*, (pp. 293-306). Boulder, CO.

Akkaya, K., Younis, M., & Youssef, W. (2005). Sink repositioning for enhanced performance in wireless sensor networks. *Elsevier Computer Networks Journal, 49*, 512–542.

Akkaya, K., Younis, M., & Youssef, W. (2007). Positioning of base stations in wireless sensor networks. *IEEE Communications Magazine, 45*(4), 96–102. doi:10.1109/MCOM.2007.343618

Alliance, Z. (2006). *ZigBee Alliance*. Retrieved October, 2006 from, http://www.zigbee.org/imwp/download.asp?ContentID=14128

Bi, Y., Sun, L., Ma, J., Li, N., Khan, I. A., & Chen, C. (2007, November). HUMS: An Autonomous Moving Strategy for Mobile Sinks in Data-gathering Sensor Networks. *EURASIP Journal on Wireless Communications and Networking*.

Chakrabarti, A., Sabharwal, A., & Aazhang, B. (2006, August). Communication Power Optimization in Sensor Network with a Path-Constrained Mobile Observer. [TOSN]. *ACM Transactions on Sensor Networks, 2*(3), 297–324. doi:10.1145/1167935.1167936

Chatzigiannakis, I., Kinalis, A., & Nikoletseas, S. (October, 2006). Sink mobility protocols for data collection in wireless sensor networks. In *Proceedings of the International Workshop on Mobility Managementand Wireless Access*, (pp. 52-59). Terromolinos, Spain.

Cheng, Z., Perillo, M., & Heinzelman, W. B. (2008). General network lifetime and cost models for evaluating sensor network deployment strate-gies. *IEEE Transactions on Mobile Computing, 7*(4), 484–497. doi:10.1109/TMC.2007.70784

Ekici, E., Gu, Y., & Bozdag, D. (2006). Mobility-Based Communication in Wireless Sensor Net-works. *IEEE Communications Magazine, 44*(7), 56–62. doi:10.1109/MCOM.2006.1668382

Gohring, N. (2008, July). *Networkworld*. Retrieved July, 2008 from http://www.networkworld.com/news/2008/072908-microsoft-shows-off-data-center-monitoring.html

Hartung, C., Han, R., Seielstad, C., & Holbrook, S. (June, 2006). FireWxNet: A MultiTiered Monitoring Weather Conditions in Wildland Fire Environments. *International Conference on Mobile Systems, Applications and Services (MobiSys 2006)*, (pp. 28-41). Uppsala, Sweden.

He, T., Krishnamurthy, S., Stankovic, J. A., Abdelzaher, T., Luo, L., Stoleru, R., et al. (June, 2004). Energy-Efficient Surveillance System Using Wireless Sensor Networks. *International Conference on Mobile Systems, Applications and Services (MobiSys 2006)*, (pp. 270-283). Boston, MA.

Kalantari, M., & Shayman, M. (April, 2006). De-sign Optimization of Multi-Sink Sensor Networks by Analogy to Electrostatic Theory. *IEEE Wireless Communications and Networking Conference*, (pp. 431-438). Las Vegas, USA.

Kansal, A., Somasundara, A., Jea, D., Srivastava, M., & Estrin, D. (June, 2004). Intelligent Fluid Infrastructure for Embedded Networks. *2nd ACM USENIX International Conference on Mobile Systems, Applications, and Services (MobiSys '04)*. Boston, USA.

Khan, M. I., Gansterer, W. N., & Haring, G. (2007). Congestion Avoidance and Energy Efficient Rout-ing Protocol for Wireless Sensor Networks with a Mobile Sink. *Journal of Networks, 2*(2).

Krishnamurthy, L., Adler, R., Buonadonna, P., Chhabra, J., Flanigan, M., Kushalnagar, N., et al. (Nov 2005). Design and Deployment of Industrial Sensor Networks: Experiences from a Semicon-ductor Plant and the North Sea. *ACM Conference on Embedded Networked Sensor Systems (SenSys 2005)*, (pp. 64-75). San Diego, CA.

Luo, J., Panchard, J., Piorkowski, M., Grossglaus-er, M., & Hubaux, J.-P. (June, 2006). MobiRoute: Routing towards a Mobile Sink for Improving Lifetime in Sensor Networks. *2nd IEEE/ACM International Conference on Distributed Comput-ing in Sensor.* San Francisco, CA.

Ma, J., Chen, C., & Salomaa, J. P. (2007). WSN for Large Scale Mobile Sensing. *Journal of Signal Processing Systems, 51*(3).

Meng, M., Wu, X., Xu, H., Jeong, B., Lee, S., & Lee, Y. (August, 2007). Energy Efficient Routing in Multiple Sink Sensor Networks. *Fifth International Conference on Computational Science and Applications*, (pp. 561-564). Kuala Lumpur, Malaysia.

MicaZ Datasheet Specs. (2008). *Crossbow Technology*. Retrieved November, 11, 2008 from, http://www.xbow.com/Products/Product_pdf_files/Wireless_pdf/MICAz_Datasheet.pdf

Mick, J. (2008, July). *DailyTech*. Retrieved July, 2008 from, http://www.dailytech.com/San+Francisco+Pioneers+Smartphonecoordinated+Parking/article12364.htm

OPNET Modeler Inc. (2008, November). *OPNET Inc*. Retrieved November 11, 2008, from, http://www.opnet.com/solutions/network_rd/modeler.html

Petrovic, D. S. (May, 2003). Data funneling: routing with aggregation and compression for wireless sensor networks. *The First IEEE International Workshop on Sensor Network Protocols and Applications, 2003.*, (pp. 156-162). Anchorage, AK.

Raman, B., & Chebrolu, K. (2008). Censor Networks: A Critique of "Sensor Networks". *ACM SIGCOMM Computer Communication Review*, 38(3), 75–78. doi:10.1145/1384609.1384618

Rappaport, T. S. (2001). *Wireless Communications: Principles and Practice* (2nd Edition ed.). Upper Saddle River, NJ: Prentice Hall PTR.

Shah, R. C., Roy, S., Jain, S., & Brunette, W. (May, 2003). Data MULEs: Modeling a Three-tier Architecture for Sparse Sensor Networks. *IEEE Workshop on Sensor Network Protocols and Applications (SNPA)*, (pp. 30 - 41). Anchorage, AK.

Shakya, M., Zhang, J., Zhang, P., & Lampe, M. (May, 2007). Design and optimization of wireless sensor network with mobile gateway. *21st International Conference on Advanced Information Networking and Applications Workshops*, (pp. 415-420). Niagara Falls, Canada.

Somasundara, A. A., Ramamoorthy, A., & Srivastava, M. B. (December, 2004). Mobile Element Scheduling for Efficient Data Collection in Wireless Sensor Networks with Dynamic Deadlines. *25th IEEE International Real-Time Systems Symposium (RTSS)*, (pp. 296 - 305). Lisbon, Portugal.

Tacconi, D., Carreras, I., Miorandi, D., Chiti, F., Casile, A., & Fantacci, R. (March, 2007). *Mobile applications in disconnected WSN: System architecture, routing framework, applications* (Tech. Rep.). University of Florence, Florence.

The, A. RT-WiSe framework. (2008, May). *IPP Hurray*. Retrieved May, 2008 from, http://www.hurray.isep.ipp.pt/art-wise/

Vincze, Z., & Vida, R. (October, 2005). Multi-hop Wireless Sensor Networks with Mobile Sink. *ACM Conference on Emerging Network Experiment and Technology (CoNEXT'05)*. Toulouse, France.

Vlajic, N., & Xia, D. (June 2006). Wireless Sensor Networks: To Cluster or Not To Cluster? *IEEE International Symposium on a World of Wireless and Multimedia Networks*. Niagara Falls, Buffalo-NY.

Wan, C., Eisenman, S. B., Campbell, A. T., & Crowcroft, J. (November, 2005). Siphon: overload traffic management using multi-radio virtual sinks in sensor networks. In *Proceedings of the 3rd international conference on Embedded networked sensor systems (SenSys '05)*, (pp. 116-129). San Diego, CA.

Wang, Q., Hossam, H., & Xu, K. (2006). A Practical Perspective on Wireless Sensor Networks. In Mahgoub, I. & Ilyas, M. (Eds.) *Smart Dust: Sensor Network Applications, Architecture, and Design.* CRC Press, Taylor & Francis Group.

Werner-Allen, G., Lorincz, K., Johnson, J., Lees, J., & Welsh, M. (2006). Fidelity and yield in a volcano monitoring sensor network. In *OSDI*, (pp. 381 - 396). Seattle, WA.

Xu, N., Rangwala, S., Chintalapudi, K., Ganesan, D., Broad, A., Govindan, R., et al. (Nov. 2004). A Wireless Sensor Network For Structural Monitoring. *ACM Conference on Embedded Networked Sensor Systems (SenSys 2004)*, (pp. 13-24). Baltimore, MD.

Yang, G., Xiao, M., & Chen, C. (April, 2007). A Simple Energy-Balancing Method in RFID Sensor Networks. *2007 IEEE International Workshop on Anti-counterfeiting, Security, Identification*, (pp. 306-310).

KEY TERMS AND DEFINITIONS

Controlled Trajectory: In a controlled/optimized trajectory, the sink's path is a function of a particular network variable (e.g. current state of nodes' batteries, intensity and/or direction of traffic flows, etc.), and the path gets continuously adjusted to ensure optimal network performance with regard to one or more performance metrics.

Hot-Spot Problem: The hot-spot problem refers to sink node's one-hop neighbours being required to forward a disproportionately higher volume of traffic compared to other nodes in the network. The sink's one-hop neighbours - also known as the 'hot-spot' nodes - tend to exhaust their energy and 'die' earlier relative to other nodes. Unless adequately dealt with, through measures that reduce the load and prevent the failure of the sink's one hop neighbours, the hot-spot problem

can lead to a complete isolation of the sink node, resulting in the failure of the entire network.

Limited Multi-Hop Data Delivery: Limited multi-hop data delivery is a hybrid approach that combines the features of single-hop and multi-hop node-to-sink data delivery techniques. Specifically, the limited multi-hop assumes the existence of several designated (static) anchornodes responsible for collecting the sensory readings from other (ordinary) nearby nodes via multi-hop transmissions. Subsequently, the anchor nodes get visited by the mobile sink, one by one, at which point the single-hop delivery is used to upload the previously collected readings from the anchor nodes to the sink.

Multi-Hop Data Delivery: Multi-hop communication allows for utilization of several/multiple hops for communication between a sensor and the sink. Packets are forwarded by intermediary nodes in the network, located between a sensor and the sink.

Network Lifetime: This metric, commonly used in the WSN literature, is intended to reflect the time span from the network's initial deployment to the first loss of coverage. As such, network lifetime can alternatively be defined as the 'time until the first node dies'. The easiest to capture indicator of this metric is the maximum per-node load, where a node's load corresponds to the number of packets sent from or routed through the given node. Clearly, the network setup that minimizes the maximum node load is the one that will ensure the maximum network lifetime.

Predictable Trajectory: The predictable/fixed trajectory is fully deterministic type of mobility– with this type of trajectory, the sink is expected to continuously follow the same path through the network. The predictable/fixed trajectory is typically forced upon the mobile sink by the nature of the physical terrain and/or presence of obstacles in the environment.

Random Trajectory: As the name implies, the trajectory of a randomly moving sink comprises a sequence of segments of arbitrary length and direction. The mobile sink's speed along each segment, and the pause time between movements along different segments, can also be arbitrary, although these two conditions do not have to hold to satisfy the randomness requirement.

Single-Hop Data Delivery: Single-hop data delivery, as the name implies, assumes direct one-hop communication between a sensor and the sink, and as such eliminates the need for inter-sensor communication (i.e. data exchange).

APPENDIX

The ensuing two tables display two MATLAB functions utilized in obtaining the maximum load under different sink-deployment strategies. The code was developed assuming the same network parameters and assumptions as outlined in the section "Multi-hop Data Delivery with Predictable/Fixed Trajectory Model: Analytical Evaluation".

1. **RandGrid(n)** – takes the dimensions of a desired virtual square grid (of size $n \times n$) and generates random locations for sensor nodes inside each cell of the grid (Table 2).
2. **LoadEval(trajectory, p, n)** – utilizing **RandGrid(n)** creates the virtual grid and calculates the maximum load among the nodes in the grid. The results shown in Figure 11 were generated by running this function with 300 different seeds for each set of input parameters (Table 3).

Table 2. RandGrid() MATLAB function

```
% RandGrid function takes the dimensions of a desired virtual square grid (NxN)
% and generates random locations for sensor nodes inside each cell of the given grid.
%
function [X, Y] = RandGrid(N)
%
% Usage - Possible input parameters:
% N is an odd natural number representing the grid dimension
%
%--------------------------------------------------------------------------------------
X = zeros(N,N);
Y = zeros(N,N);
% Loop thru the entire grid and randomly place a sensor within the area of each cell of the grid.
for i = 1:1:N
for j = 1:1:N
X(i, j) = (j - 1) + rand;
Y(i, j) = (i - 1) + rand;
end
end
```

Table 3. LoadEval() MATLAB function

```
function [maxTotalLoad] = LoadEval(trajectory, p, N)
% This function calculates the total load per node and returns the maximum load among all nodes
% in the virtual grid of size N x N
%
% Usage - Possible input parameters:
% static sink: trajectory = 0
% outer periphery: trajectory = 1
% mid-periphery: trajectory = 2
% diagonal cross: trajectory = 3
% mid-cross: trajectory = 4
%
% p is mid-periphery hop distance offset from the outer periphery
%
% N is an odd natural number representing the grid dimension
%
%----------------------------------------------------------------------
close all;

% Maximum total load
maxTotalLoad = 0;
% initialize load matrix
TotalLoad = zeros(N,N);
% initialize trajectory for visualization purposes
Trajectory = zeros(N,N);
% mesh grid points for graphs
[x,y] = meshgrid(1:1:N, 1:1:N);
%----------------------------------------------------------------------

% randomize node locations
X = zeros(N,N);
Y = zeros(N,N);
[X, Y] = RandGrid(N);


%----------------------------------------------------------------------

if (trajectory < 0 || trajectory > 4)
error('Invalid trajectory selected')
elseif (N < 1)
error('Invalid grid dimension selected')
end

% ReportingCounter is used to annotate the number of reporting cycles, i.e. the number of
% different locations visited by the sink.
% Number of cells/locations on the outer periphery = 8*radius = 8*(N-1)/2.
reportingCounter = 0;

% collect load until at least (8*(N-1)/2) cells/locations have been visited
while (reportingCounter < (8*((N-1)/2)))
for h = 1:1:N
for g = 1:1:N
% check if sink is located on a valid path for specific trajectory selected in the input
if ((trajectory == 0) && (h == ((N+1)/2)) && (g == ((N+1)/2)))
%valid static sink location, at network centre
validSinkLocation = true;
elseif ((trajectory == 1) && (h == 1 || g == 1 || h == N || g == N))
% valid sink location for outer periphery trajectory
validSinkLocation = true;
elseif ((trajectory == 2) ...
&& ((h == (p+1) && g >= (p+1) && g <= (N-p)) ...
|| (h == (N-p) && g >= (p+1) && g <= (N-p)) ...
|| (g == (p+1) && h > (p+1) && h < (N-p)) ...
```

```
|| (g == (N–p) && h > (p+1) && h < (N–p))))
% valid sink location for mid-periphery trajectory
validSinkLocation = true;
elseif ((trajectory == 3) && ((h == g) || (g == (N-h+1))))
% valid sink location for diagonal-cross trajectory
validSinkLocation = true;
elseif ((trajectory == 4) && ((h == ((N+1)/2)) || (g == ((N+1)/2))))
% valid sink location for mid-cross trajectory
validSinkLocation = true;
else
% current position in the grid is invalid sink location for trajectory selected in
% the input
validSinkLocation = false;
end
% Sink location is valid, so calculate the load for all nodes at this sink's location
% (I,J) are used to annotate coordinates of current sink location.
if (validSinkLocation == true)
I = h;
J = g;
% current sink location
sinkLocation = [X(I, J), Y(I, J)];
% mark that sink was at this location
Trajectory(I, J) = Trajectory(I, J) + 1;
% increment reporting rounds counter
reportingCounter = reportingCounter + 1;

% Determine nodes' loads with regard to current sink location by:
% 1) Looping through entire grid,
% 2) Finding all nodes at hop distance k, starting from the largest possible k,
% 3) Determining the best forwarding neighbour for the given node,
% 4) Transferring all of the given node's load to the best (closest) neighbour.
LOAD = ones(N,N);

for k = (N-1):(-1):2
for p = 1:1:N
for q = 1:1:N
% current node is at hop distance k from current sink location
if (max(abs(I - p), abs(J - q)) == k)
nodeLocation = [X(p, q), Y(p, q)];
bestDist = norm(sinkLocation - nodeLocation);
bestNeighbor = [p, q];
% look at immediate neighbours (with smaller hop distance)
% select the node geographically closest to the sink
for i = (-1):1:1
for j = (-1):1:1
if ((abs(i) == 1 || abs(j) == 1) ...
&& (p+i) >= 1 && (q+j) >= 1 ...
&& (p+i) <= N && (q+j) <= N)
neighborLocation = [X(p+i, q+j), Y(p+i, q+j)];
dist = norm(sinkLocation - neighborLocation);
% found a closer neighbour
if (dist < bestDist)
bestDist = dist;
bestNeighbor = [p+i, q+j];
end
end
end
end
% forward data to neighbour with the closest distance to the sink,
% i.e. increase its load
LOAD(bestNeighbor(1), bestNeighbor(2))...
```

```
= LOAD(bestNeighbor(1), bestNeighbor(2))+ LOAD(p, q);
end
end
end
end
% updated the total load by adding the load accumulated during the current reporting
% cycle
TotalLoad = TotalLoad + LOAD;
end
end
end
end

%------------------------------------------------------------------------
% Output and Present Total and Maximum Total Load

disp('Total Load:')
disp(TotalLoad)
figure(2);
mesh(x, y, Trajectory);
xlabel('The dimension of the network')
figure(3);
mesh(x, y, TotalLoad);
xlabel('The dimension of the network')
maxTotalLoad = max(max(TotalLoad));
disp(sprintf('Maximum Load per Node: %d\n', maxTotalLoad));
```

Chapter 18

Network–Wide Broadcast Service in Wireless Sensor Networks

Feng Wang
Simon Fraser University, Canada

Jiangchuan Liu
Simon Fraser University, Canada

ABSTRACT

Network-wide Broadcast is one of the most fundamental services in wireless sensor networks (WSNs). It facilitates sensor nodes to propagate messages across the whole network, serving a wide range of higher-level operations and thus being critical to the overall network design. A distinct feature of WSNs is that many sensor nodes alternate between the active state and the dormant state, so as to conserve energy and extend the lifetime of the network. Unfortunately, the impact of such cycles has been largely ignored in existing network-wide broadcast implementations that adopt the common assumption of all sensor nodes being active all over the whole broadcast process. In this chapter, we first provide a brief survey on previous research works on network-wide broadcast services. We then revisit the network-wide broadcast problem by remodeling it with active/dormant cycles and showing the practical lower bounds for the time and message costs, respectively. We also propose an adaptive algorithm named RBS (Reliable Broadcast Service) for dynamic message forwarding scheduling in this context, which enables a reliable and efficient broadcast service with low delay. The performance of the proposed solution is evaluated under diverse network configurations. The results suggest that the proposed solution is close to the lower bounds of both time and forwarding costs, and it well resists to the network size and wireless loss increases.

INTRODUCTION

Network-wide broadcast is one of the most fundamental services in wireless sensor networks

DOI: 10.4018/978-1-61520-701-5.ch018

(WSNs) (Akyildiz, Su, Sankarasubramaniam & Cayirci, 2002). It facilitates sensor nodes to propagate messages across the whole network, serving a wide range of higher-level operations: During the networking configuration, control messages may be broadcast from the sink to all the sensor nodes; For data collection, the interest or query messages may be broadcast within the network; Upon observing an event, a sensor node may broadcast a message to coordinate with other sensor nodes for tracking the event and storing the sensed data; to name but a few. Hence, implementing an effective network-wide broadcast service is critical to the overall performance optimization of a WSN.

Flooding and gossiping (Akyildiz, Su, Sankarasubramaniam & Cayirci, 2002) are two commonly used broadcast approaches, though their basic forms are known inefficient. If we assume that all sensor nodes in the network are active during the broadcast process (referred to as *all-node-active assumption*), ideally every sensor node needs to receive and forward the broadcast message at most once. Significant efforts thus have been made toward enhancing the efficiency of the basic flooding or gossiping, while retaining their robustness in the presence of the error-prone transmissions (Guo, 2004; Kyasanur, Choudhury & Gupta, 2006; Stann, Heidemann, Shroff & Murtaza, 2006).

The all-node-active assumption is valid for wired networks and for many conventional multi-hop wireless networks. It however fails to capture the uniqueness of energy-constrained wireless sensor networks. In a WSN, the sensor nodes are often alternating between the *dormant* state and the *active* state (Gu, Hwang, He & Du, 2007; Liu, Zhao, Cheung & Guibas, 2004; Wang, Xing, Zhang, Lu, Pless & Gill, 2003; Yan, He & Stankovic, 2003); in the former, they go to sleep and thus consume little energy, while in the latter, they actively perform sensing tasks and communications, consuming significantly more energy (e.g., 56 *mW* for IEEE802.15.4 radio plus 6 to 15 *mW* for Atmel ATmega 128L micro-controller and possible sensing devices on a MicaZ sensor mote). Define *duty-cycle* as the ratio between the active period and the full active/dormant period. A low duty-cycle WSN clearly has a much longer lifetime for operation, but breaks the all-node-active assumption. In such a network, if the

number of sensor nodes is very small, it may be possible to wake up all sensor nodes for message broadcast through global synchronization with the customized active/dormant schedules. For larger scale WSNs, however, synchronization itself remains an open problem. More importantly, the duty-cycle schedules are often optimized for the given application or deployment, and a broadcast service accommodating the schedules is thus expected for the cross-layer optimization of the overall system.

In this chapter, we revisit the network-wide broadcast problem in low duty-cycle WSNs. Their scale, together with their application/deployment-specific duty-cycles, renders the all-node-active assumption impractical. This in turn introduces a series of new challenges toward implementing the network-wide broadcast service. From a local viewpoint, since the neighbors of a node are not active simultaneously, a node would have to forward a message multiple times at different time instances; From a global viewpoint, since the topology is time-varying with no persistent connectivity, if not well-planned, the latency for a broadcast message to reach all sensor nodes can be significantly prolonged. The error-prone wireless links further aggravate these problems. The experiments later in this chapter have shown that, for ultra low duty-cycles (less than 0.4), a conventional broadcast strategy would simply fail to cover all the sensor nodes within an acceptable timeframe.

To this end, we remodel the network-wide broadcast problem in this new context, seeking a balance between efficiency and delay with reliability guarantees. Based on the model, we propose an adaptive algorithm named RBS to dynamically schedule message forwardings, which does not depend on any global synchronization and maintains only the local topology information. The algorithm exploits overhearing to further reduce the message costs while still keeping the broadcast robust, timely and reliable. We conduct extensive simulations to evaluate the proposed algorithm.

The results demonstrate that the proposed algorithm is robust, reliable and close to the practical lower bounds in terms of the time and messages used for the network-wide broadcast.

The remainder of this chapter is organized as follows. We first give a brief survey on previous related research works. Following the survey, we revisit the network-wide broadcast problem with the consideration of low duty-cycle. We then propose the algorithm design for RBS in the fourth section to solve this problem. In the fifth section, we conduct extensive simulations to evaluate the proposed algorithm. Then some further discussions on practical issues are given in the sixth section. And we conclude this chapter and give some directions of our future work in the last section.

BACKGROUND AND RELATED WORK

Generally, the background of our work is highly related to two areas of the literature researching on wireless sensor networks, namely, network-wide broadcast and low duty-cycle networks. In the following two subsections, we briefly investigate prior works in these two areas and discuss how they related to our work proposed in this chapter.

Network-Wide Broadcast

There have been numerous studies on network-wide broadcast in wired networks and in wireless ad hoc networks (Deering, 1989; Floyd, Jacobson, Liu, McCanne & Zhang, 1997; Ni, Tseng, Chen & Sheu, 1999). Flooding and gossiping (Akyildiz, Su, Sankarasubramaniam & Cayirci, 2002) are two commonly used network-wide broadcast approaches. In flooding, each sensor node forwards the received broadcast message until the message reaches its maximum hop count. This approach provides high robustness against wireless communication losses and high reliability

for message deliveries. It however causes many duplicate messages being forwarded and thus leads to a significant amount of unnecessary energy consumptions. On the other hand, in gossiping, received broadcast messages are only forwarded with some pre-defined probability. By theoretical analysis, a threshold probability exists for a given topology and wireless communication loss rate, so that the whole network can be covered with a high probability. Thus by setting the pre-defined probability just above the threshold, a great amount of duplicate messages can be avoided. Nevertheless, the pre-defined probability is very sensitive to the changes of the network topology and wireless communication losses, which often leads to unsatisfactory reliability for message deliveries in practice.

Ideally, if without wireless communication losses, every sensor node needs to receive and forward the broadcast message at most once. Thus though their basic forms are known inefficient, significant efforts have been made toward enhancing the efficiency of the flooding or gossiping, while retaining their robustness in the presence of error-prone transmissions.

In Guo's work (2004), an algorithm based on timing heuristic was proposed to reduce redundant message forwardings in the basic flooding as well as to extend the lifetime of the network. To suppress duplicate forwardings, a sensor node only schedules a message forwarding when it receives a broadcast message for the first time. Also a short latency named FDL (Forwarding-node Declaration Latency) is introduced before a node forwards a message, and if a forwarding for the same message is overheard, the node cancels its forwarding to further reduce duplicate forwardings. To extend the lifetime of the network, for a sensor node u, its FDL is computed based on its residual energy $E_t(u)$, specifically, by the following equation

$$FDL(u) = T \cdot (1 - \frac{E_t(u)}{E_{ref}(u)}) + t_D(u), \qquad (1)$$

where T is a timing constant, $t_D(u)$ is the maximum delay related to signal processing, transceiver switching and so forth at the potential forwarding nodes other than u, and E_{ref} is the maximum energy capacity of a battery. As a result, each time that several neighboring nodes receive a broadcast message, only the node with the highest residual energy and thus the shortest FDL will forward the message. Other nodes by overhearing will suppress their own forwardings to save the energy so that the lifetime of the network is extended.

Smart Gossip (Kyasanur, Choudhury & Gupta, 2006), on the other hand, extends the basic gossiping to minimize the forwarding overhead while still keeping a reasonable reliability. Different from the basic gossiping that uses the same static forwarding probability for all sensor nodes, the authors proposed to dynamically adapt the forwarding probability on each sensor node to its local topology and the originator of the broadcast message. Specifically, based on where the forwarded broadcast message comes from and who is its last forwarder, a node's neighbors are divided into three sets, namely, *parent*, *child* and *sibling*. The neighbors in the parent set are those that the node depends on to receive the first forwarded message; the neighbors in the child set are those that depend on the node to receive the first forwarded message; and the remaining neighbors are in the sibling set. Given an expected network delivery ratio τ, the required per-hop delivery ratio τ_{hop} can be estimated by the following equation

$$(\tau_{hop})^{\delta} = \tau \tag{2}$$

where δ is the estimated diameter of the network. Thus for a node with K neighbors in its parent set, the required forwarding probability ($p_{required}$) for each parent neighbor can be estimated using the inequation

$$(1 - p_{required})^{K} < (1 - \tau_{hop}). \tag{3}$$

Each node then collects $p_{required}$ from all its child neighbors and uses the maximum as its own forwarding probability. Also, the three sets and $p_{required}$ on each node are computed periodically based on the recent message forwarding history, so as to make the forwarding probability adaptive to the network dynamics (e.g., node failure).

A more recent work is RBP (Robust Broadcast Propagation) (Stann, Heidemann, Shroff & Murtaza, 2006), which extends the flooding-based approach and targets for high reliability broadcast. It lets each sensor node do forwarding when receiving the broadcast message for the first time. Then by overhearing, a node can quickly identify the percentage of its neighbors that have successfully received the message. Based on this percentage and the local density (i.e., the number of neighbors), a node determines whether to retransmit the message, where the principle is that for a low density, the message will be retransmitted until a high receiving percentage is achieved, while for a high density, a moderate receiving percentage is enough to stop further retransmissions. To counter wireless losses, explicit ACKs will be sent to the nodes that are heard rebroadcasting a message several times. In addition, if a node finds itself highly depending on another node to receive broadcast messages, the link between them is deemed as an important link. And the downstream node will then notify the upstream node to increase the number of retransmissions so as to improve the probabilities of message deliveries.

To enhance the reliability one step further, Levis, Patel, Culler and Shenker (2004) proposed an approach with perfect broadcast reliability for the code redistribution and update propagation. To keep the codes updated, each sensor node transmits a summary of its codes if it has not heard a few of other sensor nodes do so. When receiving a code summary from its neighbor, a node compares the received summary with its own summary. If the neighbor's summary is old, the node then sends its new code to the neighbor. And if the neighbor's summary is newer, the node

Table 1. A summary of different network-wide broadcast solutions

Network-wide Broadcast Solution	Basic Approach	Balance Energy Consumption	Topology-Aware	Reliability	Delay	Message Cost
Guo, 2004	Flooding	Yes	No	Moderate	Moderate	Very Low
Kyasanur, Choudhury & Gupta, 2006	Gossiping	No	Yes	High	Low	Very Low
Stann, Heidemann, Shroff & Murtaza, 2006	Flooding	No	Yes	Very High	Low	Low
Levis, Patel, Culler & Shenker, 2004	Flooding	Yes	No	Perfect	Moderate	Low

retransmits its own summary so as to trigger the neighbor to send the new code. Otherwise, a node counts the number of summaries received within one time interval, if the number exceeds a threshold, the node suppresses its own summary transmission so as to save energy. And to balance the energy costs, within each time interval, a node randomly picks its summary transmission time by following a uniform distribution. Moreover, the length of a time interval is set to a lower bound when a summary of new codes is received, so as to accelerate the code update propagations. After that, the length of each next interval will be the double of the current one until it reaches an upper bound, which can further help to reduce the energy costs.

To give a clear overview, we list the network-wide broadcast solutions mentioned previously in Tab. 1. Different from these solutions, the work proposed in this chapter considers a more realistic scenario with sensor nodes alternating between the active state and the dormant state to save energy. In this scenario, above previous works may fail or suffer from poor performance due to the invalidation of the *all-node-active* assumption.

Low Duty-Cycle Networks

There have been recent works investigating low duty-cycle in wireless sensor networks (Gu & He, 2007; Miller, Sengul & Gupta, 2005). Among them, PBBF (Probability-Based Broadcast Forwarding) (Miller, Sengul & Gupta, 2005) con-

sidered the scenario that the MAC layer works in a low duty-cycle pattern, where the radios of all sensor nodes synchronously turn to the active mode for possible communications and turn to the dormant mode to save energy. In this scenario, the authors introduced two parameters p and q by which the tradeoff among the reliability, latency and energy consumption can be tuned. Specifically, the parameter p, is the probability that a node forwards a broadcast message *immediately* without ensuring that any of its neighbors is active. And the parameter q, is the probability that for a given node and a give time instant when it is supposed to be dormant, the node instead stays in the active state with the expectation that it might be a receiver of an *immediate forwarding*. In addition, each sensor node only forwards a broadcast message when it receives the message for the first time. Thus with a high value of p and q, more nodes will forward the message immediately and more nodes will receive the message more quickly due to staying in the active states instead of the dormant states, thus low latency is achieved with the costs of extra energy consumption. On the other hand, with a high value of p and a low value of q, low latency can still be achieved but now with the costs of low reliability. And with a low value of p and a high value of q, energy consumption is sacrificed for high reliability. Based on these variations, the authors conducted analysis and suggested that in practice, a possible value region of p and q can first be determined based on the reliability required by the application, then p and q can be finely tuned within the region for a better tradeoff between the latency and energy consumption. The duty-cycle pattern considered in this chapter, however, is in the application layer and greatly different from the one aforementioned for the MAC layer. In addition, such a MAC layer duty-cycle pattern highly depends on the global synchronization to work properly, which also causes problems in real-world deployments.

DSF (Dynamic Switch-based Forwarding) (Gu & He, 2007) is proposed for data forwarding from any sensor node to the sink in low duty-cycle wireless sensor networks, where the main idea is to let a sensor node dynamically select the next-hop forwarder among its neighbors based on their duty-cycle schedules and other metrics such as the link loss and end-to-end delay. In particular, given a time range that a message must be forwarded within (otherwise the message will be dropped), all its neighbors that will turn active within this time range form a forwarding sequence ordered by the time to turn active. During a message forwarding, a node will try to transmit the message to the first neighbor in the sequence when the neighbor turns active. If the transmission succeeds, the message forwarding is done; otherwise the node will try the remaining neighbors in the sequence one by one at the time they become active until a transmission succeeds or all neighbors are tried. And the data forwarding problem can be modeled as selecting a subsequence of neighbors from the sequence such that by only trying to forward the message to the neighbors in this subsequence, some expected metric goal can be optimized. For example, to maximize the expected delivery ratio (*EDR*), given any subsequence S_n^e, where a message to be forwarded from node e and a total number n of its neighbors are in the subsequence. The overall probability $P^e(i)$ that a transmission by node e is successful at the i-th forwarder (after i-1 failures) can be calculated by the following equation

$$P^e(i) = [\prod_{j=1}^{i-1}(1 - p_{ej})]p_{ei}, \tag{4}$$

where p_{ei} is the success ratio of a round-trip transmission (DATA and ACK) between node e and the i-th forwarder in S_n^e. And the expected delivery ratio of node e can then be obtained by the following recursive equation

$$EDR_e(S_n^e) = \sum_{i=1}^{n} P^e(i) EDR_i. \tag{5}$$

Thus the optimal subsequence that maximizes EDR_e can be computed by a dynamic programming algorithm. And for each sensor node, *EDR* can then be computed hop-by-hop starting from the sink, where EDR_{sink}=1. In the paper, the authors also discussed how to optimize other metrics such as the expected end-to-end delay and expected energy consumption. However, the general approaches used in the paper are only suitable for data deliveries from a few source nodes to the sink, where the computational complexity and the number of states maintained on each sensor node increase linearly with the number of receivers. As a result, such approaches can not be directly used for the network-wide broadcast service since all the sensor nodes except for the broadcast message originator now become receivers.

Summary

Although many efforts have been done in the literature of network-wide broadcast, most of the prior works depend on the all-node-active assumption to work properly and efficiently. On the other hand, very few of the research works on low duty-cycle wireless sensor networks focused on the network-wide broadcast service in the application layer. To this end, our work complements them by considering the scenario of providing a network-wide broadcast service in low duty cycle wireless sensor networks, which calls for novel solutions to ensure broadcast messages are successfully delivered to all destinations.

PROBLEM STATEMENT

In this section, we revisit the broadcast problem with the consideration of low duty-cycle, where sensor nodes may alternate between the active state and the dormant state either before or during a network-wide broadcast process, and the active-dormant patterns may not be pre-determined. We assume that every sensor node at its turning active (or after a timeout) will try to send a wake-up message to notify its neighbors how long it will stay actively. By this means, a node during an active period, may find several neighbors turning active later than itself and thus decide whether to forward a received broadcast message to them. It is worth noting that this assumption is not mandatory in practice, since there may be other traffic between sensor nodes, which may already indicate a node is active. We will further discuss how to relax this assumption in the sixth section.

Also, we divide time equally into small discrete time units. Within a time unit, each node sends at most one message so as to reduce the possibility of collision. Thus the network-wide broadcast problem can be modeled as follows: for each sensor node having received a broadcast message, how to find out a forwarding schedule which guarantees that all its neighbors receive the message while keeping the total costs of time units and messages as few as possible.

For ease of clarification, we first consider a motivative toy case shown in Fig. 1, where sensor node 0 needs to broadcast a message to all other sensor nodes from 1 to 6 which are all dormant at the beginning. Suppose node 0 keeps active during the broadcast process and there is no wireless loss, then it is easy to find out the schedule that node 0 waits until all its neighbors wake up and then forwards the broadcast message uses the smallest number of forwarded messages to accomplish the broadcast, i.e., only one message is forwarded. However, since the active-dormant patterns of the nodes may be significantly different from each other, it may take a very long time

for node 0 to wait for all its neighbors becoming active together. In the worst, if there is no overlap among the active periods of node 0's neighbors, the time to accomplish the broadcast becomes infinity, i.e., the strategy does not work. On the other hand, one may figure out another schedule where node 0 forwards the broadcast message as soon as it finds out a neighbor is turning active. By this means, the time used for the broadcast is bounded by the time when the last neighbor turns active. However, comparing with the previous schedule, this schedule in the worst case may use six times number of forwarded messages to finish the broadcast.

Therefore, in general, there is a time/message cost trade-off in the scheduling problem. The lower bound of the time cost can be achieved by letting all sensor nodes forward a broadcast message as soon as they receive it for the first time or find any neighbor turns active. And the lower bound of the message cost can be achieved by letting every sensor node defer the broadcast

Figure 1. An example to illustrate broadcast under low duty-cycle. The double-arrowed line indicates two sensor nodes can communicate with each other if both are active

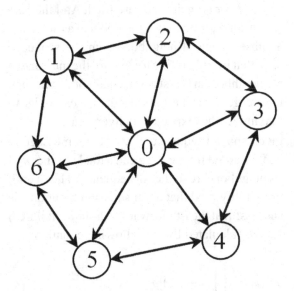

message forwarding until all its neighbors are active. It is worth noting that under such a situation, since all the neighbors are active from the view point of an individual node, previous broadcast algorithms based on the all-node-active assumption might still work. However, as we will show in the fifth section, such approaches suffer long propagation time and may fail under extremely low duty-cycle. In next section, we propose an algorithm that is both time and message efficient for reliable broadcast service in large-scale low duty-cycle WSNs.

ALGORITHM DESIGN FOR RBS

In this section, we give the details of the proposed algorithm for RBS. Generally, the proposed algorithm schedules message forwardings in a way that balances the total number of messages forwarded and the time used for broadcast. Specifically, a sensor node exploits overhearing to dynamically adjust its own forwarding schedule so as to adapt to the active/dormant patterns and forwarding schedules of its neighbors, and only issues a message forwarding if it finds out otherwise some neighbor may miss the message due to turning dormant and can not be contacted until next time the neighbor becomes active.

However, to achieve such a scheduling algorithm and make it efficient, there are some issues to be considered. First, since a node may turn active and dormant before and during the broadcast process, even to know whether a node has received the broadcast message may become difficult. If all sensor nodes are active, it is easy to use overhearing to know that a node has received the message. However, a node now may be dormant and not able to overhear when its neighbor forwards the broadcast message. Besides, simply forwarding a broadcast message when a neighbor is going to turn dormant may cause the forwarding redundant if the neighbor has already been forwarded the message. Thus how

to identify and avoid such situations as many as possible also needs to be considered. In addition, the wireless losses may further complicate above mentioned issues. In the remainder of this section, we propose our algorithm for reliable broadcast service under low duty-cycle, and show how to solve these challenges.

Definitions

Before moving into the detail, we first give some definitions that will be used later in this chapter. Given a sensor node with its neighbors, *ReceivingSet*, shortened as *RcvSet*, is the set of neighbors whose having received the broadcast message has been known by the node. RcvSet is a subset of the neighbors that have received the broadcast message. And if RcvSet is equal to the set of all the neighbors, this indicates that all the neighbors have received the broadcast message. For example, in Fig. 1, if node 0 knows node 1, 2 and 3 have received the broadcast message, then these three nodes are in node 0's RcvSet. However, if node 4 has received the broadcast message from node 3, which is currently not known by node 0, node 4 is not in node 0's RcvSet.

Another set that will be used is CoverSet. We call a node's active neighbor is *covered*, if during the node's current active period, it knows a broadcast message has been sent within its neighbor's communication range. Also we consider a neighbor is covered if it is known to have received the broadcast message. And a node's *CoverSet* is consisted of all its covered neighbors. For example, in Fig. 1, if node 0 is active and knows node 1, 2 and 3 are active, then after node 0 sends the broadcast message, node 1, 2 and 3 are in node 0's CoverSet. It is worth noting that due to the existence of the wireless communication losses, a node's RcvSet may only be a subset of its Cover-Set, since a node covered by a broadcast message may not actually receive the message.

Besides, we use *WakeUpSet* to track a node's neighbors turning active and dormant. A neighbor is added into this set when a wake-up message from the neighbor is received. And the neighbor is removed from this set as it goes to the dormant state.

Also, for ease of exposition, we consider the case that every time one message is broadcast over the network. It is easy to be extended to support multiple messages by maintaining RcvSet and CoverSet for each message and deleting these sets after all the neighbors have received the corresponding message. A more comprehensive description of the extension to handle multiple messages simultaneously can be found in the work of Wang & Liu (2007).

Algorithm Description

In our algorithm, each sensor node maintains the identifications of its neighbors. And for each of its neighbors, the node also maintains the identifications of the neighbor's neighbors. We assume these identifications can be sorted into a determined order, such as IP or MAC addresses sorted in the increasing order. Thus by using a bitmap, an RcvSet can be compressed into several bytes and easily piggy-backed with the broadcast or wake-up messages.

In addition, we assume that each message to be broadcast over the network has a sequence number, which can uniquely identify the message. Also the sequence number is assigned to each broadcast message continuously in the increasing order[1]. For a sensor node, there are two important sequence numbers. One is called the last received sequence number (shortened as *LastRcvNO*), which is the largest number among the numbers of all the broadcast messages successfully received by this node. The other is called current sequence number (shortened as *CurrentNO*), which is the number of the broadcast message known by the node to be on the broadcast process. Also, a node can piggy-back its LastRcvNO in its wake-up message to facilitate the updating of the RcvSet and CoverSet of its neighbors, which will be further

explained later with the algorithm. In case of one broadcast message a time, the difference between the two sequence numbers is at most one, where the LastRcvNO is one less than the CurrentNO, indicating the node knows a new message is under broadcast from its neighbors' wake-up messages, but has not received the message yet. Otherwise, the two numbers are always equal.

Fig. 2 shows the flow chart of the main algorithm that a sensor node would do in each time unit during its active periods. Based on the received/overheard wake-up and broadcast messages, the algorithm dynamically schedules message forwarding to adapt to the active/dormant patterns and forwarding schedules of its neighbors, which will be elaborated in the following paragraphs.

Wake-Up Message Processing

On receiving a wake-up message, a node adds the neighbor that sent the message into its WakeUpSet and sets a countdown timer according to the neighbor's active time indicated in the message. When the timer counts back to zero, the node removes the neighbor from its WakeUpSet. Also, the neighbor's LastRcvNO and its RcvSet are piggy-backed in the wake-up message. If both the node's LastRcvNO and CurrentNO are less than the neighbor's LastRcvNO, which means the neighbor must have received a broadcast message that has not arrived in the node yet, the node sets its CurrentNO equal to the neighbor's lastRcvNO, creates a RcvSet and a CoverSet according to the new CurrentNO and adds the neighbor and other neighbors appearing in the neighbor's RcvSet into the two sets. Otherwise, if the node's CurrentNO is the same as the neighbor's LastRcvNO, the node then updates its RcvSet and CoverSet by adding the neighbor and other neighbors in the neighbor's RcvSet; and if the neighbor's LastRcvNO is less than its own, which means the neighbor has not received the new broadcast message, the node updates its CoverSet by removing the neighbor from the set.

Figure 2. The flow chart for the operations of a sensor node in each time unit during its active periods

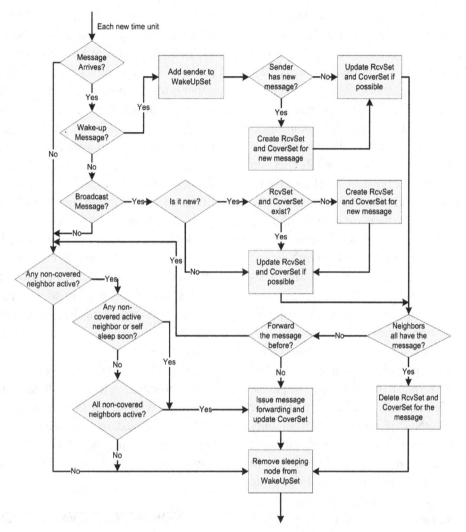

For example, in Fig. 1, on receiving a wake-up message from node 1, which contains node 2 and 6 in its RcvSet and will be active for 60 time units, node 0 adds node 1 into its WakeUpSet and sets a countdown timer so as to remove node 1 from the set after 60 time units. If both its LastRcvNO and CurrentNO are less than node 1's LastRcvNO, node 0 knows node 1 has received a broadcast message that itself has not received yet, then sets its CurrentNO equal to node 1's LastRcvNO and creates a RcvSet and a CoverSet for the new CurrentNO and adds node 1, 2 and 6 into these new sets so that it will not waste messages to cover

these nodes later when it receives the new broadcast message. Otherwise, if node 0's CurrentNO is the same as node 1's LastRcvNO, then node 0 adds node 1, 2 and 6 into its RcvSet and CoverSet. In both cases, by adding node 1, 2 and 6 into the RcvSet, even node 0 has not overheard messages from node 2 and 6, it can still quickly build up its RcvSet and reduce the chances to waste messages to cover node 2 and 6. And in case that node 1's LastRcvNO is less than its own, node 0 removes node 1 from its CoverSet so that it can schedule a message forwarding to node 1 later.

Broadcast Message Processing

On receiving a broadcast message from a neighbor, a node first compares the message's sequence number with its CurrentNO. If its CurrentNO is less, which means the node receives a new broadcast message, it sets both its CurrentNO and LastRcvNO equal to the message's sequence number, creates a RcvSet and a CoverSet according to the new CurrentNO, adds the neighbor and other neighbors appearing in the neighbor's RcvSet into both sets, and adds all its active neighbors (in its WakeUpSet) also covered by this message into the CoverSet. Otherwise, the node directly updates its current RcvSet and CoverSet if the message's sequence number is equal to its CurrentNO, and sets its LastRcvNO equal to the message's sequence number if its LastRcvNO is less.

Again we use Fig. 1 as an example. Suppose node 0 receives a broadcast message from node 1. If node 0's LastRcvNO is less than the message's sequence number, node 0 sets both its CurrentNO and LastRcvNO equal to the message's sequence number and creates an RcvSet and CoverSet for the new CurrentNO. If node 1's RcvSet only contains node 2, then node 2 is added into node 0's RcvSet and CoverSet. And if node 6 is in node 0's WakeUpSet, then it is added into node 0's CoverSet. By this means, node 0 may avoid send duplicate message to node 6 unless node 6 has not received node 1's message and later notifies node 0 by piggy-backing its LastRcvNO in other messages to node 0. If the message is not new to node 0, node 0 directly updates its RcvSet and CoverSet. Also node 0 sets its LastRcvNO equal to the message's sequence number if its LastRcvNO is smaller.

Dynamic Forwarding Scheduling

If there is any updating in a node's RcvSet, it will check whether all its neighbors have received the broadcast message. If so, the node considers the broadcast for this message is done and deletes the corresponding RcvSet and CoverSet. Otherwise, the node checks its WakeUpSet and CoverSet to see whether there is any non-covered active neighbor. If so, the node further checks whether itself or any such neighbor will turn dormant soon or whether all the non-covered neighbors are currently active. If either is true, it then issues a message forwarding. After the message forwarding, the node updates its CoverSet by adding these neighbors. It is worth noting that to accelerate the propagation of the broadcast message, a node also forwards the message after it receives the message for the first time, unless it finds out that all its neighbors have already received the message.

Besides the operations performed each time unit during active periods, on turning active or after a timeout, a sensor node also resets its CoverSet equal to its RcvSet so that the neighbors previously covered but having not received the broadcast message will get a chance to be forwarded later.

From above description, it is easy to find out that a broadcast process will terminate by our algorithm with all nodes receiving the broadcast message. In next section, we can see that RBS actually performs very close to practical lower bounds in terms of the time and message costs.

PERFORMANCE EVALUATION

Settings

We conduct extensive simulations to evaluate our algorithm. To control the active/dormant pattern of each sensor node easily, we build a customized simulator, where the sensor nodes' turning active and dormant periods are randomly generated with pre-determined average values based on the assigned duty cycle. The average period for a sensor node's turning active and dormant once is set to 50 time units. Thus for a 0.4 duty cycle, the average active period is 20 time units and the average dormant period is 30 time units. Also we adopt the wireless communication loss model used by the work of Kyasanur, Choudhury, & Gupta (2006), where the messages are randomly dropped based on a specified packet error probability.

Also for comparison, we implemented two algorithms based on the time and message lower bounds discussed in the third section. The first one is called *Time-Optimal*, which forwards a broadcast message as soon as the message is received for the first time or a received wake-up message indicates that its sender has not received the broadcast message yet. As discussed in the third section, this algorithm achieves the lower bound of the time costs and thus is time optimal. The other algorithm is called *Message-Optimal*, which only forwards the broadcast message when all its non-covered neighbors become active. And to guarantee it achieving the lower bound of the message costs and accelerate its execution time, the algorithm has the real-time information about which neighbor is active and which neighbor has not received the broadcast message yet.

In addition, we use the following two metrics to measure RBS's performance. One is the total number of messages forwarded during the whole broadcast process. The other is the time units taken for 99% of sensor nodes receiving the broadcast message. The reason for not choosing 100% is that in spite of their small number, there may be nodes of extremely low degree in the randomly generated topologies. In extremely low duty cycle, it may take enormous time for these nodes and their neighbors to wake up together so that the broadcast message can be transmitted. Thus although RBS does terminate with 100% reliability, using the time to achieve 100% reliability may not reflect the algorithm's real performance on message propagation.

With these two metrics, we conduct simulations to see how the performance of the three algorithms changes with the number of sensor nodes, the wireless communication loss or different duty-cycles. Specifically, we consider a sensing field which is a square of 200 by 200 m^2. And the wireless communication range of a sensor node is set to 10 m. We randomly generated 10 random topologies for each number of sensor nodes starting from 800 to 2000 with an increment of 200. And each data point shown in the results is the average of 10 topologies with 10 runs on each topology.

Simulation Results

Scalability with Node Number

We first evaluate how the performance of the three algorithms changes with different number of sensor nodes, where the wireless loss rate is set to 0.3 and the whole WSN works under 0.4 duty-cycle. Fig. 3 and Fig. 4 show the results. It is clear to see that the time costs used for RBS to achieve 99% reliability are very close to the time costs of Time-Optimal and much less than that of Message-Optimal. At the same time, the total number of messages forwarded by RBS is much less than that of Time-Optimal and becomes close to the message costs of Message-Optimal when the scale of the network grows large (note that *Y*-axis starts from 500). Also with the number of nodes increased, both the time costs used for RBS and Time-Optimal are in a decreasing trend. This is because increasing the number of nodes lets a node have more neighbors, which gives it more chances to be covered by the forwarded messages during the same period. As a contrast, the time costs used for Message-Optimal fluctuate and are generally in an increasing trend. The reason behind is that due to the randomly generated active/dormant patterns, the increasing number of neighbors takes a node much more time to wait for all the non-covered neighbors becoming active together. For the total messages forwarded for broadcast, both RBS and Message-Optimal increase linearly with the number of nodes, while Time-Optimal increases non-linearly and much faster than the other two algorithms due to the increasing possibility that a node forwards the broadcast message prematurely on hearing a wake-up message from a non-covered neighbor.

Figure 3. Time units used to achieve 99% reliability with different number of sensor nodes (0.4 duty-cycle, 0.3 wireless communication loss rate)

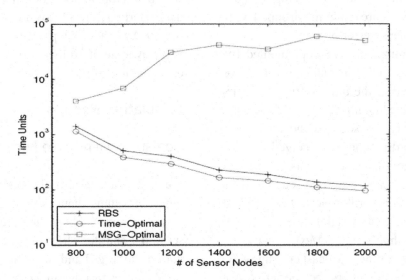

Robustness Against Wireless Loss

Next, we fix the number of nodes to 2000 and compare the three algorithms under different wireless communication loss rates with sensor nodes working in 0.4 duty-cycle. The results are shown in Fig. 5 and Fig. 6. Still we find that RBS outperforms Message-Optimal and is very close to Time-Optimal in terms of the time used to achieve 99% reliability. And for the number of forwarded messages, RBS works close to Message-Optimal and much better than Time-Optimal. In addition, in Fig. 5, the time costs taken by RBS and Time-Optimal increase slightly when the wireless loss

Figure 4. Total number of messages used with different number of sensor nodes (0.4 duty-cycle, 0.3 wireless communication loss rate)

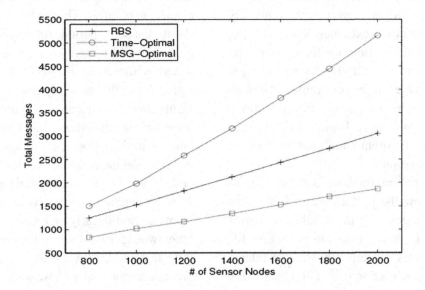

Figure 5. Time units used to achieve 99% reliability under different wireless communication loss rate (0.4 duty-cycle, 2000 sensor nodes)

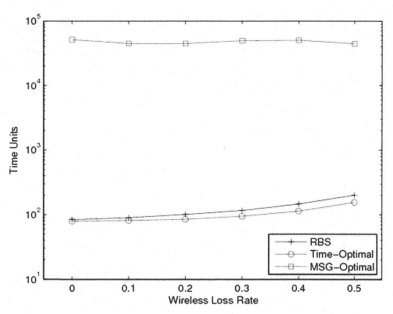

rate grows, but the time taken by Message-Optimal has no obvious trend with the change of the wireless loss rate. This is because for Message-Optimal, most of the time costs are spent in waiting for all the non-covered neighbors becoming active together, which is mainly determined by the duty-cycle. Since the duty-cycle has not changed, such waiting time will not change much either. In Fig. 6, an anti-intuitive phenomenon is that the messages forwarded by Time-Optimal decrease as the wireless loss rate increases. A close look into the simulations reveals that since Time-Optimal forwards the broadcast message as soon as it receives a wake-up message from a non-covered neighbor, the increasing of the wireless loss rate reduces the number of successfully delivered wake-up messages and as a result the number of pre-maturely forwarded broadcast messages has also been reduced. Also in Fig. 6, both the messages forwarded by RBS and Message-Optimal increase with the increasing of the wireless loss rate, however, RBS increases much more slowly than Message-Optimal, which shows that RBS is robust against the wireless losses.

Adaptability to Duty-Cycle

Also, we conduct simulations to evaluate how RBS performs under different duty-cycles. We set the wireless communication loss rate to 0.3 and again use 2000 sensor nodes. During the simulations, we find that Message-Optimal may not terminate when the duty-cycle becomes much lower. Thus for these extremely low duty-cycles, only the results of RBS and Time-Optimal are presented. Fig. 7 and Fig. 8 show the results of the three algorithms. We can see that the time costs used by Message-Optimal grow exponentially when the duty-cycle decreases, while the time costs used by RBS and Time-Optimal increase much more slowly and RBS performs very close to Time-Optimal. In addition, the total number of messages forwarded by RBS is close to the total number of messages forwarded by Message-Optimal and better than that of Time-Optimal in all cases.

In summary, RBS achieves a near optimal performance: Its time cost is very close to that of the Time-Optimal strategy, and the total number of forwarded messages is close to that of the

Figure 6. Total number of messages used under different wireless communication loss rate (0.4 duty-cycle, 2000 sensor nodes)

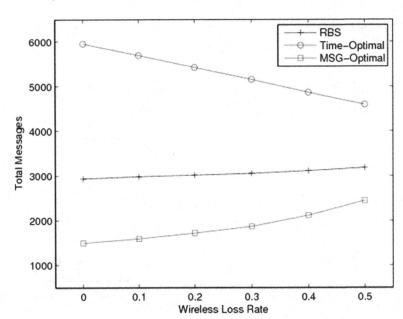

Message-Optimal strategy. In addition, the time and message costs of RBS both scale well with the size of the network. RBS is also robust against wireless communication losses and copes well with different duty-cycles.

FURTHER DISCUSSION ON PRACTICAL ISSUES

Till now, we have proposed RBS and evaluated it by thorough simulations where the main network traffics are broadcast messages. In practice, the network-wide broadcast service may work together with other services, such as sensing, routing and in-network processing. Thus by overhearing and piggy-backing several bytes containing the remaining active time, the broadcast message sequence number and RcvSet bitmap to the communication traffics from other services, the performance of RBS can be further improved. Specifically, the overhearing makes RBS more sensitive to neighbors' waking up in

spite of the loss of the wake-up messages. And piggy-backing with other communication traffic helps the RcvSet bitmaps to be exchanged more timely, thus a node can know its neighbors have already received the broadcast message in time and avoid redundant forwardings.

Another practical issue is to provide flexible reliability. So far we have struck to achieve the perfect reliability for the network-wide broadcast service. This may be not mandatory in some applications, where it may be preferred to trade a small portion of reliability off with the great reduction in the number of forwarded broadcast messages. In such cases, RBS can be easily modified to terminate when $x\%$ neighbors of a sensor node are in its RcvSet. And x can be determined locally based on the topology information by approaches such as the one proposed by Stann, Heidemann, Shroff, & Murtaza (2006). Yet another way is to let RBS terminate after all the neighbors not in the RcvSet have been covered at least several times, which can also be determined by numerical analysis and will be done as one of our future works.

Figure 7. Time units used to achieve 99% reliability under different duty-cycle (0.3 wireless communication loss rate, 2000 sensor nodes)

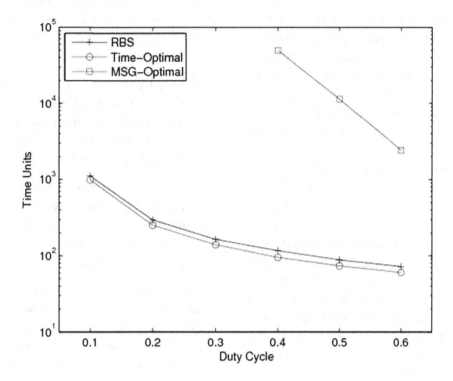

Figure 8. Total number of messages used under different duty-cycle (0.3 wireless communication loss rate, 2000 sensor nodes)

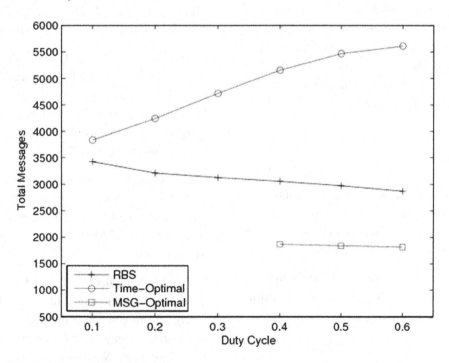

CONCLUSION AND FUTURE WORK

In this chapter, we revisited the network-wide broadcast problem in wireless sensor network by considering a more practical scenario, where sensor nodes work in low duty-cycle to save energy and extend the lifetime of the whole network. Under low duty-cycle, sensor nodes may turn active and dormant before and during a broadcast process. The active/dormant pattern is often application-specific and depends on the deployed environment. All these make previous broadcast algorithms designed for the all-node-active scenarios either failed or inefficient. To solve this problem, we first remodeled the network-wide broadcast problem with the consideration of low duty-cycle. Based on the model, we proposed an adaptive algorithm, which dynamically schedules message forwardings and achieves high reliability and near optimal performance in terms of the time and message costs. This was further verified by our extensive simulations.

Next we would like to implement RBS on real sensor networks as a general service and conduct experiments to see how RBS performs with background traffics and how it interacts with other services. We are also interested in using probabilistic methods to model the tradeoff between costs and reliability, thus extending RBS to support customized reliability requirements from different applications.

REFERENCES

Akyildiz, I. F., Su, W., Sankarasubramaniam, Y., & Cayirci, E. (2002). A Survey on Sensor Networks. *IEEE Communications Magazine, 40*(8), 102–114. doi:10.1109/MCOM.2002.1024422

Deering, S. (1989). *Scalable Multicast Routing Protocol*. Unpublished doctoral dissertation, Stanford University.

Floyd, S., Jacobson, V., Liu, C., McCanne, S., & Zhang, L. (1997). A Reliable Multicast Framework for Light-weight Sessions and Application Level Framing. *IEEE/ACM Transactions on Networking, 5*(6), 784-803.

Gu, Y., & He, T. (2007). Data Forwarding in Extremely Low Duty-Cycle Sensor Networks with Unreliable Communication Links. In *Proceedings of the ACM International Conference on Embedded Networked Sensor Systems* (pp. 321-334).

Gu, Y., Hwang, J., He, T., & Du, D. H.-C. (2007). μSense: A Unified Asymmetric Sensing Coverage Architecture for Wireless Sensor Networks. In *IEEE International Conference on Distributed Computing Systems* (pp. 8-8).

Guo, X. (2004). Broadcasting for Network Lifetime Maximization in Wireless Sensor Networks. In *Proceedings of the Annual IEEE Communications Society Conference on Sensor and Ad Hoc Communications and Networks* (pp. 352-358).

Kyasanur, P., Choudhury, R. R., & Gupta, I. (2006). Smart Gossip: An Adaptive Gossip-based Broadcasting Service for Sensor Networks. In *Proceedings of the IEEE International Conference on Mobile Adhoc and Sensor Systems* (pp. 91-100).

Levis, P., Patel, N., Culler, D., & Shenker, S. (2004). Trickle: A Self-Regulating Algorithm for Code Propagation and Maintenance in Wireless Sensor Networks. In *Proceedings of the USENIX Symposium on Networked Systems Design and Implementation* (pp. 2-2).

Liu, J., Zhao, F., Cheung, P., & Guibas, L. (2004). Apply Geometric Duality to Energy-efficient Non-local Phenomenon Awareness using Sensor Networks. *IEEE Wireless Communications, 11*(8), 62–68.

Miller, M., Sengul, C., & Gupta, I. (2005). Exploring the Energy-Latency Tradeoff for Broadcasts in Energy-Saving Sensor Networks. In *Proceedings of the IEEE International Conference on Distributed Computing Systems* (pp.17-26).

Ni, S.-Y., Tseng, Y.-C., Chen, Y.-S., & Sheu, J.-P. (1999). The Broadcast Storm Problem in a Mobile Ad Hoc Network, In *Proceedings of the Annual ACM International Conference on Mobile Computing and Networking* (pp. 151-162).

Stann, F., Heidemann, J., Shroff, R., & Murtaza, M. Z. (2006). RBP: Robust Broadcast Propagation in Wireless Networks. In *Proceedings of the ACM International Conference on Embedded Networked Sensor Systems* (pp. 85-98).

Wang, F., & Liu, J. (2007). *RBS: A Reliable Broadcast Service for Large-Scale Low Duty-Cycled Wireless Sensor Networks* (Tech. Rep.). Burnaby, British Columbia: Simon Fraser University, School of Computing Science.

Wang, X., Xing, G., Zhang, Y., Lu, C., Pless, R., & Gill, C. (2003). Integrated Coverage and Connectivity Configuration in Wireless Sensor Networks. In *Proceedings of the ACM International Conference on Embedded Networked Sensor Systems* (pp. 28-39).

Yan, T., He, T., & Stankovic, J. (2003). Differentiated Surveillance for Sensor Networks. In *Proceedings of the ACM International Conference on Embedded Networked Sensor Systems* (pp. 51-62).

KEY TERMS AND DEFINITIONS

Active State: The state where a sensor node is fully functioning. In this state, a sensor node may perform tasks such sensing the ambient environment, processing data or communicating with neighboring sensor nodes.

Dormant State: The state where a sensor node turns off most of its functioning components so as to greatly reduce energy consumption. In this state, often a count down timer is used so as to wake the sensor node up when its dormant state is finished.

Duty-Cycle: In a wireless sensor network, sensor nodes are often designed to be able to switch between the active state and the dormant state. Duty-cycle is the ratio between the active period and the full active/dormant period.

Forwarding: The communication mode that a host transmits a message to its neighboring (one-hop) hosts. Due to the broadcast nature of the wireless communication, a message forwarded by a sensor node may be received by all its neighboring sensor nodes.

Network-Wide Broadcast: The communication mode that a host in a network sends a message to all the other hosts in the same network.

Reliability: In the network-wide broadcast scenario, it means the ratio of hosts that have successfully received the broadcast message to all the hosts that are supposed to receive the message when the broadcast process finishes.

Wireless Sensor Network: The network that contains sensor nodes for sensing the ambient environment. Each sensor node may also perform tasks such as processing data and communicating with neighboring sensor nodes. The communication is often done by wireless communications. And often there is a base station in the network, where sensing data are sent for further utilization.

ENDNOTE

[1] In this chapter we assume that each time only one message is broadcast network-wide. Thus when another message needs to be broadcast, its originator can simply set the sequence number as the previous (sent/received) broadcast message's sequence number plus one. In case that multiple messages are broadcast simultaneously from different originators, the sequence numbers can be first indexed by the ID of the originator and maintained separately. A more comprehensive description to handle multiple messages simultaneously can be found in the work of Wang & Liu (2007).

Chapter 19
Description and Analysis of an Indoor Positioning System that Uses Wireless ZigBee Technology

Roberto Vazquez
University of Valladolid, Spain

Javier Herrero
HC Technologies, Pza Santa María, Spain

Daniel Herrero
HC Technologies, Pza Santa María, Spain

Jaime Gómez
University of Valladolid, Spain

ABSTRACT

Along this chapter, an indoor positioning system that uses wireless ZigBee technology is presented and evaluated. In this system, mobile wireless devices measure the level of the received signal from reference nodes, which are placed in well-known positions. With this information, the position is estimated and presented to the user in a graphical way. A precision study is presented, being this study done in function of positions and numbers of reference nodes. It is also analyzed the presence of obstacles in the system, and a study of maximum distance inter nodes that allow positioning with a minimum of quality in the results is done.

INTRODUCTION

The positioning systems arose years ago with the aim of being able at any moment to know the situation of mobile objects and being able to realize their pursuit. Those systems were very complex and expensive at the time because they required the newest technology of the time and therefore they were not accessible for the great majority of the people and organizations. Nowadays, and due to the development of the technologies and the

DOI: 10.4018/978-1-61520-701-5.ch019

electronics, these systems have evolved a lot and are used daily in multiple sectors of the society (such as auto motion, industrial automatism or people pursuit). This height in telecommunications and this type of applications has been translated in multitude of location, positioning and pursuit systems, with diverse technologies associated, generating a wide range of possibilities.

At the moment, there is a group of positioning systems very interesting. They can locate in interiors using radio frequency signals. In order to reach the objective of the location Bluetooth (Labiod et al., 2007), WLAN (Labiod et al., 2007), ZigBee (ZigBee Standards Organization, 2008; Labiod et al., 2007) or RFID technology (Ni et al., 2003) can be used. For such aim, Chipcon Company (now absorbed by Texas Instruments) commercialized the chip CC2431 that, besides being a ZigBee transceiver, included a microprocessor of 8051 family and a hardware location engine (Aamodt, 2006). With those chips, a wireless location system can be mounted in interiors (although also it is valid for mixed exteriors or surroundings) based on radio signals, and more concretely, in the study of the received signals strength. With these measures, an estimation of the position of mobile devices can be estimated with the help of several perfectly located nodes which work as a reference.

They are the methods to carry this estimation out based on the received signal strength. One of them, the simplest one, consists of making a map with the signal powers that would arrive at a group of located receivers. They cover the surface where it is the location is needed, with a certain overlapping. Thus, comparing the received signal levels from the mobile devices with the measures previously stored, the position can be estimated.

Bahl et al., presented in March of 2000 a work called `RADAR: An In-Building RF-based User Location and Tracking System', in which RADAR system is described (Bahl et al., 2000). As indicated in the title, RADAR is a location system based

on radio frequency whose purpose is the location and pursuit of objects or movable people in interiors. RADAR is based on the engraving and processing of the information of signal power of the system in different stations located in strategic positions. These radio stations covered the area where the system was working with some overlapping to be able to provide more trustworthy and precise results. The measured data from previous experimentation with a propagation model for the signal are combined to determine the position of the object or subject to locate.

The second method consists of placing the fixed receivers and eliminating the previous measures to be stored in the data base. These fixed receivers, or a central node, take the level from the signal received from the mobile device and based in a model of attenuation of the signal (or following another algorithm) they will compute the distance between every fixed node and a moving body, being possible to make a location.

In 2006 Masashi et al., published 'Indoor Location System using RSSI measurement of Wireless sensor network based on ZigBee standard' (Sugano et al.). As its name indicates, during this study a location system was implemented based on the measurement of the power of the received signals using wireless nodes with ZigBee technology. The implemented system sends packages that at least, include their identity to three sensorial nodes that measure the RSSI and are in a well-known position. Obtained these values, fixed nodes send the data to a central node which computes them and estimates the position of the movable nodes from these measured data.

The third method is a similar idea, but decentralizing the system. Fixed nodes send packages to movable nodes, and they receive different signal power levels from every reference node. These packages contain the real position of the reference nodes, and taking the data, movable devices are able to estimate their own positions.

Apart from these three works, a high number of studies about positioning systems had been done in scientific literature. It could be read Tannuri & Morishita (2006), Wallbaum (2007), Ohyama & Kobayashi (1996), Mosshammer et al. (2008), Heidari & Pahfavan (2004) and Matteo et al. (2005) among others.

Based on these principles, a location system was built. It had a reference node network, and fixed nodes gave their well-known position cyclical to a movable device. They were equipped with an adjustable hardware location engine. Then, with received positions and signal strengths associated it could estimate its own position.

LOCATION METHODS

Global Positioning System, or GPS is the most popular positioning method. They are GPS and GLONASS. These systems are sustained in a satellites network orbiting around the Earth. Those satellites communicate with terrestrial receivers and using their signals we can consider their position. GPS has prevailed round the world over GLONASS, but both are operative. Although both are based on a same idea, they are not compatible among them. However precision is very good, with an allowable error of about 2,5 meters, being possible to reduce it to approximately 1 meters using techniques like the DGPS in which a differential correction of the measurement is realized. The European Union in association with the European Space Agency and other countries is investing a lot of money and work in an alternative system for civil uses called Galileo. This system will be compatible and interoperable with GPS and will improve the precision.

Talking about local location, in smaller scenarios, the alternatives are much diverse. There are techniques that uses from Odometry to the Artificial Vision, Interferometry, Radars, Ultrasounds, arrays of antennas or microphones, Radiofrequency signals... There are a lot of commercial systems that can be used, and multitude of new ones are being developed because the implantation cost of a system is not as elevated as in the case of location by satellite.

The most interesting systems for the case of this study is the wireless outdoor location system (although also it is valid indoor or in mixed scenarios) based on radiofrequency signals, and more concretely based on the power of the received signals. With these measures, an estimation of the position of a movable device between several perfectly located nodes that serve as reference can be performed.

There are three methods that exist to perform this estimation based on the power of the received signal. One of them, the simplest one, consists of making a map with the signal powers that would arrive from the movable device to a group of located receivers that cover the surface where the location wants to be performed, with a certain overlapping. Thus, comparing the received signal levels from the movable device with the measures previously stored, the position can be considered. The second method consists of placing the fixed receivers and eliminating the mapping of measures to be stored in the data base. These fixed receivers, or a central node, will take the level of the signal received from the movable device and according to a signal attenuation model or to the way it has been implemented, they will compute the data and consider the position of the moving body. The third method is a similar idea, but decentralizing it and emitting from the fixed nodes and receiving different power levels of signals in the movable device. This movable device will consider its own position.

ZIGBEE TECHNOLOGY

ZigBee term describes a standard wireless protocol for personal area networks or WPAN (ZigBee Standards Organization, 2008). ZigBee is different from other wireless standards since it has been designed to support a diverse market of applications that require low cost and low consumption, with more sophisticated connectivity than previous wireless systems which existed until ZigBee appeared. It arises like an alternative to the existing systems in the same frequencies, with a low data transmission and low cycle of connectivity.

ZigBee is a standard of hardware and software based on the standard IEEE 802.15.4. This important standard defines the hardware and the software, which has been described in the terms of networking, like physical layers (PHY), and control of access layer (MAC). ZigBee adds to them several possibilities in network layer, which makes the networks possible to expand them easily. It also defines the possible alternatives in application layer, which is the last one specified by ZigBee.

In order to guard this set of solutions that form the ZigBee technology in 2002 ZigBee Alliance was created. This organization, without profit spirit, is an enterprise association with many companies from the sector, mainly semiconductors, devices and microelectronics systems manufacturers, interested in this technology to leave ahead and constitutes a real standard that assures the interoperability of their products. Also, they organize sessions destined to make advises and presenting the different solutions related to ZigBee. A specification updated to year 2006 exists where the entire referring to ZigBee is.

The most outstanding applications of ZigBee are:

- **Home and building automation:** Security, alarms, reading of water meters, gas, electricity, control of illumination, control of accesses, irrigation control, domotic applications in which is not desired or it is not possible to create a thread framework of communications, etc.
- **Sanitary attention:** Monitoring of patients and equipment for the health (fitness).
- **Industrial control:** Temperature sensor, process control, remote instrumentation, etc. In addition, due to its low speed

of data transmission and its nature of low consumption, also enters the markets of the remote control for the consumption electronics, and devices like keyboards, mice and joysticks.

CHIPCON'S SYSTEM ON CHIP CC2431

Chipcon was a Norwegian company dedicated to Microelectronics that developed a chips family, the CC24XX, which worked with ZigBee technology. Later, Texas Instruments absorbed the Nordic company, but it maintained these products. For this chapter, chip CC2431 is from special interest (Aamodt, 2006).

CC2431 is a chip that follows the philosophy 'System-on-a-chip' or 'system on chip' (SoC or SOC), which means that all the components of the electronic system are integrated in a unique integrated circuit or chip. SoCs, will support digital signals, analogical and even mixed, and some times they have hardware to communicate by radiofrequency.

Chipcon created the emitting-receiving ZigBee CC2430. CC2430 represents the second generation of ZigBee chips of the company Chipcon and is a real SoC solution, combining the leader transceiver CC2420 of 2.4GHz that follows IEEE to 802.15.4 and a microcontroller 8051. 8051 is a microcontroller developed by Intel in 1980. It is a very popular microcontroller. The Core of 8051 is used in more than 100 microcontrollers from more than 20 independent manufacturers such as Atmel, Scythes Semiconducting, Philips, Winbond ..., among others. The official denomination of Intel for this family of microcontrollers is MCS 51.

Once it was verified that Chips would have a place in the market and due to the great commercial success that they achieved, Chipcon realized that the same Chips could be used for positioning systems by using the parameter RSSI of ZigBee, related to the power of the received signal. They raised the creation of a new chip which allowed the handling of this variable to make an estimation of the location with an easy process. Then the CC2431 arose. It is an identical SoC chip to the CC2430 adding a new hardware block that works as a location engine. This hardware location engine needs at least three measures of signal level with their three respective coordinates of the signal origins, and can make an estimation of its own position.

Therefore, CC2431 is a compact and small chip. It's only necessary to add a power feeding system, an adaptation filters system to an antenna that works in the rank of frequencies of ZigBee, and all the sensors, switches, indicators, etc. which are desired to use. You can make some robust and simple devices with a low cost that make possible the creation of wireless systems, useful for indoor and outdoor location and with a very low consumption.

COMPLETE DESCRIPTION OF THE SYSTEM

The system used in the present study is the evaluation kit CC2431DK, from Chipcon company, with some modifications at software level, and a personal computer Acer Aspire 1351LC computer where were running software applications.

Hardware

There are microelectronic boards that work as reference nodes and movable device to be located. These boards as well are made with three fundamental pieces: SOC_BB, CC2431 EM and a 2,4 GHz antenna. SOC_BB simply is a board where the necessary batteries (two 1,5V batteries) and switches and some buttons are integrated. In addition it includes connectors in which we are able to set the device as reference node or on the opposite way, set it as movable device, and an adapter to CC2431 EM board. CC2431 EM

Figure 1. CC2431 EM

(Evaluation Module) board contains the adapter towards SOC_BB board from where it receives the necessary feeding, the chip CC2431 and all the necessary circuitry that make possible the system works, as well as the adapter for the antenna, that is the third piece that forms each final devices that are going to form the network ZigBee. The appearance is shown in Figure 1.

The other basic hardware piece of the system is the SmartRF04EB board, to which one CC2430 EM board is connected, which is identical to CC2431 EM but with one CC2430 chip instead CC2431 since it is not necessary to contain the location hardware engine that CC2431 chip incorporates. This board is the centre of the system and will receive all the information interchanged by nodes of the ZigBee network, as well as the information sent by movable devices in broadcast way. It contains multitude of peripheral associates as a screen, switches, LEDs or different I/O adapter. For this study the unique peripheral used is the USB connector to communicate the board with a personal computer. Figure 2 shows a SmartRF04EB board.

Figure 2. SmartRF04EB

Software

In the laptop there is an application running called General Chipcon Packet Sniffer CC2431. This application picks up from port USB all the ZigBee messages that the SmartRF04EB board receives in broadcasting from the ZigBee network mounted to perform the location. These data allows the application to make a graphical representation of the estimated position of a movable device. Also, with this application, it is possible to modify the parameters of the fixed reference nodes (to fit the position where they are at the moment) and modify the parameters of the movable devices that let the system make a better estimation of the position with hardware engine (to fit variables A and N will the system adapt the real conditions of attenuation of the real scenario).

As it has already been let intuit, the SmartRF04EB board is programmed to redirect the information. Every data arriving to the board via ZigBee will be given to the computer application to be treated properly. Every data arriving from the PC to the board will be sent towards the ZigBee network in order to make the devices set their parameters.

Reference nodes contain a program in low consumption state. It listen to the network and it only perform some actions in case they receive a message to renew the position that have associate in his memory or a request asking to send in broadcast their stored position.

Finally, movable devices include a program that maintains the device in low consumption listening to the network. In case they receive a message asking them to modify their parameters A and N (necessary for a correct adjustment of the hardware location engine) stored in his memory they will perform the actions. Periodically, movable nodes send in broadcast a message towards reference nodes soliciting their position, to which they respond. With this answer the moving body can obtain not only the position of near reference nodes, but also obtains the measurement of the received signals power from the RSSI variable. They use them with the hardware location engine to calculate their own position. Then, they send a message to the network with this estimation and return to low consumption state.

General Vision of the System

The system is a wireless location system which is able to operate in indoors and outdoors, with a low cost and low consumption. The algorithm of location of the system is based on the value of RSSI. This value is related to the power of the received signal and therefore it decreases according the distance grows. The greater advantage of this system (aside from the low consumption and price) is that the task of estimation of the position of the movable devices is realized by devices to locate, decentralizing the system and making possible to use a simpler central node (it only have to send messages to set parameters and receive messages with estimated locations). Also, and thanks to this fact, the system will be easily expandable. For this study, more messages in the central node are received because they contribute with interest information and allow us to appreciate the operation of the algorithm.

The location engine can use up to 16 reference nodes in each calculation. Each reference node supplies its Cartesian coordinates X and Y (location is realized in 2-D) and the measured RSSI for each received package. With the aid of the transmission parameters A and N, the movable device is able to determine an estimation of its own position.

In order to be able to perform the location of these devices, a 2-D mesh is needed. A location could be perform in 3-D, but always with supplementary software develop. The point (X=0, Y=0) will be always taken as reference, not being possible to use negative measures. The recommendations of the manufacturers indicate that for a good operation of the system it is enough to place a reference node by each 100 m², in other words, to place the reference nodes spaced 10 m in both Cartesian coordinate axes.

Reference nodes then, are placed in static and well-known positions stored in their internal memories. Its main function is to give reference packages to the movable devices when they are requested, providing its position in the packages. Movable devices extract the value of RSSI from these packages. Movable nodes however, communicate with the nearest reference nodes, collecting their coordinates and the power of the signal received from each one. Then, and thanks to the hardware location engine, they calculate their own positions. Central node communicate with the network, and can reshape the positions stored by the fixed nodes as well as the parameters A and N stored in movable nodes.

Program flows of the reference nodes and movable devices are shown respectively in Figure 3 and Figure 4.

The application is configured to make SmartRF04EB board to request periodically the reference nodes position and it requests it too to the movable devices to initiate the location process. In this way, the configuration of the mesh can be appreciated in the PC. It is able to change it in case of being erroneous or moving some reference nodes, and periodically shows the situation of each moving body in the screen.

OBJECTIVES

With the case of study that is presented we try to observe, in an objective and critical way, the behaviour of the system described before under different circumstances, being able of obtaining important conclusions on the precision of the system. The system is proved changing several variable parameters. They are:

Figure 3. Reference node's program flow

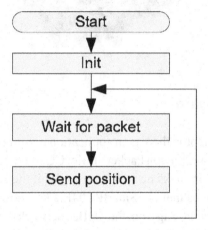

Figure 4. Movable device's program flow

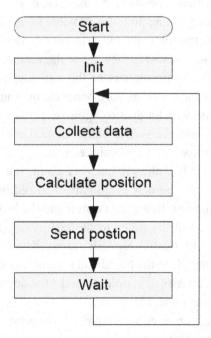

- The number of implied nodes of reference in the location. In this experience it is tried to verify how a major or less number of incoming signals to a movable device, which uses for the estimation of its own position, affects.
- The disposition of the reference nodes in order to observe the variation of benefits when the shape of the mesh of fixed nodes of the system changes.
- The distance between reference nodes. The variation of the nodal separation in the mesh of the system makes possible to analyze the influence of this distance in the precision of the movable devices.
- Existence or not of obstacles, walls or objects that can interfere with and attenuate the signals.

Besides it, in this study an experiment that verifies the reach of the system with a minimum benefits and quality is made.

EXPOSITION OF THE STUDY

The study is based on the observation of the estimation that a movable device makes of its own position under different circumstances, modifying the state of the factors that the part take in the calculation of that estimation.

In order to be able to draw trustworthy conclusions was established a reference frame to be possible to compare the results that are obtained in the different experiments. That's the reason why firstly a general analysis of precision is made in a scenario that fulfils the recommendations of the manufacturer of CC2431, realizing an observation during 3 minutes intervals in 4 selected physical points that had been selected for their special interest. Later, the test scenario modifies as far as some of the parameters described in section 7.1, obtaining new measures of the precision and stability of the system that now can be compared

with the initial situation. Then, interesting conclusions can be extracted.

Also, a small experiment was made about the distance that the system can reach with minimum benefits of quality and stability in the calculated position, measuring the estimation of the position in an axis and moving away further and further the movable device from the used reference nodes.

In order to finalize the study, the influence of the existence of objects and walls in the system was verified, something that in the previous experiments did not vary the results because experiments were realized in a clear of obstacles area.

General Analysis of Precision in the Positioning

In this study, tests of precision with the positioning system was tried to be performed. For this purpose we will start with a base scenario that is described next and it will be modified searching for conclusive data. Therefore, in this 7.1 point, we are going to establish a reference frame and we will reach some precision and stability measures that we will compare later in the following sections.

Test Scenario

The basic scenario corresponds with the sports center (Figure 5 and Figure 6) of Apostolado del Sagrado Corazón school, located in the Old Way of Simancas in Valladolid, Spain.

The futsal ground inside the sport center was used of reference for the measures, whose measures correspond with Figure 7. For this first test the recommendations of Texas Instruments were followed, placing the fixed reference nodes spaced 10 meters one from next one.

As indicated in Figure 8, 1 meter outside the field the receiver of the control centre (the application) was placed acting like Cartesian coordinate origin. A square mesh with 9 reference nodes was created. This mesh was 20 x 20

Figure 5. School's sport center where test were made

meters long (same as futsal ground). A node was placed in each one of the four corners of half the field, one in the centre, another one in centre of the goal, and three intermediate reference nodes. These distances made possible that all the ZigBee messages can arrive at the SmartRF04EB board and let have a good visualization of everything that was happening in the network in real time at the PC screen.

Immediately, measurements of the precision were taken.

Figure 6. PC connected to SmartRF04EB

Figure 7. Futsal field

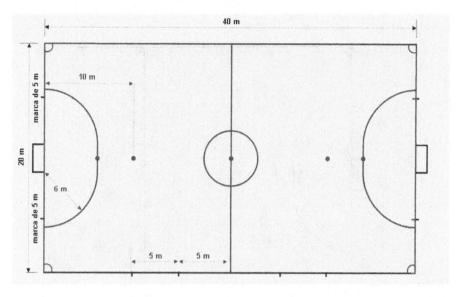

Results

The system was started up. First, the Smart-RF04EB board was connected via USB to the personal computer in order to be recognized and all suitable controllers were installed to make possible the communications between the PC and the microelectronic board. This installation is automatic. That is the reason why a user does not have to realize difficult steps to set up the hardware.

Once the board was recognized by the computer, the application CC2431 Location Engine, developed by the Chipcon Company (at the moment absorbed by Texas Instruments), was launched. This application will be the one which regulate the operation of the system giving the precise orders to the SmartRF04EB board, and receiving from it all the traffic of interest in the network of nodes. Also, the application show by the screen of the computer a graphical visualization of the reference node positions. Same happens with the position of possible movable devices in the area where location system works. The application let the administrator to set the position of the reference nodes, to modify the parameters that

control the performance of the hardware location engine in movable devices, to ask for the position of a concrete movable device... etc.

With the board connected to the computer and the application launched correctly, then only last to create the reference node network that have been described previously. They were placed in its correct physical position, and they were put one by one into operation, readjusting the position stored in its memory using the application before switching on the next one. In this way, the squared network of 9 reference nodes was created. According to they were connected, was appraised in the application that the reference nodes were appearing in the screen exactly in the position which they had saved last time they had been used, becoming the readjustment compulsory to adapt it to the new network.

When the network was created, a locatable movable device was introduced inside the network. It was connected, and initial tests were made to fit the parameters A and N that have influence on the behaviour of the location engine. The parameter A is obtained measuring the value of RSSI 1 m ahead of a reference node, and in theory this measurement should be the same in all directions.

Figure 8. Reference test scenario

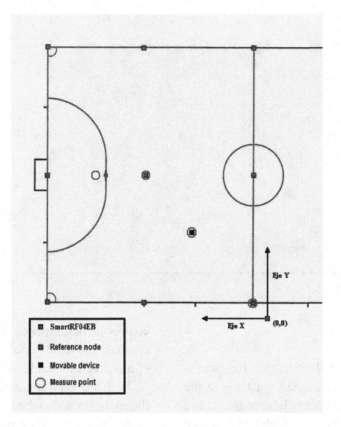

As we know, used antennas are not isotropic and. That is the reason why an average measurement with 4 points was done with one of the reference nodes (as it is recommended from Texas Instruments), obtaining an approximated value of 40 for A. Of course, this parameter would not change in next experiments. For the case of N it is more complicated. N is related to a variable called in the application n_index and describes how the signal is attenuated with the distance, as is appraised in Figure 9. This will be influenced by the thickness of possible walls, the material which was used to make them, etc. It has to be calculated experimentally, but it is very complicated to get an optimal value (almost impossible).

Therefore, knowing by empirical studies realized by experts in scenarios without any obstacles (same as our case of study) that a value of n_index located between 15 and 25 offer good results, parameter N of the movable device was adjusted to 3,875, which corresponds with a value of n_index of 20.

Immediately, the measurements of the precision of the system were started. It is necessary to be conscious that by construction, the system has an intrinsic error of 0,25 meters. The reason is that coordinates can be fit with this precision (the minimum separation between two positions in an axis is of 0,5 meters), and hardware location engine offers the result of the estimation of the position using the same system,, with values of coordinates each half meter. Therefore, the limitations of the system do that the minimum difference between two measures cannot be inferior to 50 centimeters, being created that error that cannot be eliminated.

Four physical points were taken to realize the measurement of the precision. These are (1,1),

(6,6), (11,11) and (15,11) by different reasons. The point (1,1), is the corner of the node mesh next to the SmartRF04EB board, and in that location it is going to be placed a reference node in this experiment as in the next 4 sections, which causes a good datum point. The point (6,6), on the other hand, always is going to be inside reference nodes mesh that will be used, but with different conditions, just like it happens to the point (11,11). However the point (16,11) was considered interesting because in a section it will be outside the fixed nodes mesh and an this case was considered attractive to know how the behavior of the system would be.

The answer in the application was not immediate while the movable device moved in the network. It is due to the frequency which this device uses to ask for reference nodes positions and calculate its own position. After a small time the device was graphically printed in the screen in a stable position and their coordinates were showed. Mean values obtained are resumed in Table 1.

As shown in the results, the estimation under the described circumstances is very good. It is important to emphasize that the measures had a small temporary fluctuation, because some packages for different reasons could be lost, or affected by some noise affecting to the RSSI used for the

estimation. These values described in the Table 1, correspond with the value that after a period of observation of 3 minutes by each position was most stable, indicating than the fluctuations did not alter the measures in this case over 0,5 meters (± 25 centimetres).

Analysis of the Influence of the Number of Reference Nodes

Once obtained initial data that will be used to establish a comparison, in this test, the precision of the system will be measure varying the number of reference nodes. It is tried to see the way it affects that a movable device communicates with a more or less reference nodes that send packages with their position and make possible to get a RSSI value to estimate its position.

Test Scenario

For the accomplishment of the precision measurements in this case previous scene was modified slightly. It was continued with the recommendation of the manufacturer of the chips to place reference nodes spaced 10 meters in a network, but several of them were disconnected. As we know, the minimum of reference nodes needed to be able to

Figure 9. Ideal signal attenuation with distance

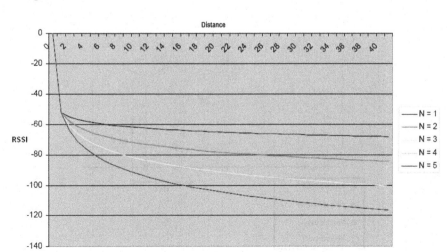

carry out the estimation of the position in a movable device is 3, and it was decided to maintain a squared reference node network, but only with 4 nodes. Therefore, the other 5 reference nodes were switched off, staying in operation nodes located in the points (1,1), (11, 1), (1,11) and (11,11), as appraised in Figure 10.

With the new test scene operative, to the visualization of the system operation under these circumstances was made.

Results

When the 5 reference nodes that have been indicated were switched off, as it is logical, it is thought that the measures will be worse because the movable device is going to have less signal sources which send their packages obtaining reference coordinates and power of signals values. Then, in view of the results that are in Table 1 this is certain, but not in all cases.

Table 1. Reference measures

Real Position	Estimated Position	Error-X	Error-Y
(1,1)	(1,1)	0	0
(6,6)	(6.5,5.5)	0.5	0.5
(11,11)	(11.5,11)	0.5	0
(16,11)	(15.5, 11)	0.5	0

Table 2. Results using less reference nodes

Real Position	Estimated Position	Error-X	Error-Y
(1,1)	(1,1)	0	0
(6,6)	(5.5,5.5)	0.5	0.5
(11,11)	(11.5,11)	0.5	0
(16,11)	(14, 10)	2	1

As it is appraised on the results, in the point (1,1) there is not a big variation because movable device is located right with a reference node

Figure 10. Reference node network with only 4 nodes

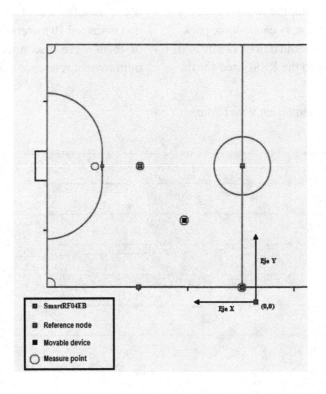

Figure 11. Reference node network changing its geometry

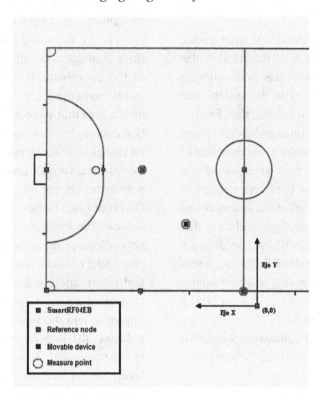

alongside. In the point (6,6), however, there is a bigger error against previous case. This is due to the signals that before came from the now dull nodes that contributed with a very little information to the movable device when it was in (1,1), but were more important in this point. Even so, the fluctuations of the coordinates were a little more frequent than in reference case. In the point (11,11), it is similar to the situation in point (1,1). Movable device is next to a reference node, so the other don't affect too much to measurement. However, in the point (16,11), the variation is more important. First of all, this point is now outside the nodes mesh that is being used, and therefore is remoter of the nodes that provide it the information, and so the signals arrive more attenuated, and values of RSSI, are more similar between different nodes and fluctuate more. Therefore, not only the measurement is worse, as it was possible to be verified, the measurement was less stable.

The quality of the system with less nodes was affected mainly according to we approached the edges the fixed nodes network, and if the movable device continued moving away beyond the limits of the mesh, then it is when the precision experimented a remarkable decrease. As a conclusion that can be obtained from this test, it is important to fit the design of reference nodes mesh to the area where it is desired to locate to a moving body.

Analysis of the Influence of the Reference Nodes Disposition

The scenario described on 7.1 has been taken as a reference for the study, and we tried in this experiment to analyze the influence of the disposition of the reference nodes or, in other words, the influence to have a reference nodes network that is not squared in this case.

Test Scenario

The scene of tests that is needed to analyze the influence of the disposition of the fixed nodes varies of the previous experiences in the number of involved reference nodes in the system, but also in the geometry of the mesh that they form. It was continued with the recommendation to place a fixed nodes spaced 10 meters in the direction of each coordinate axis, and they were placed in a rectangular mesh formed by 6 reference nodes. In this way, fixed nodes placed in the line opposed to the computer in the football ground were disconnected. Then those nodes whose coordinate Y was 21 meters were switched off. Reference nodes involved in these measurements had coordinates (1,1), (11,1), (21,1), (1,11), (11,11) and (21,11), as it is appraised in Figure 11.

With this disposition, the measures were realized to verify the operation of the system under these conditions.

Results

The previous scenario from 7.2 was used and then the 2 new nodes that were needed in addition for this section were put into operation. An improvement of the behaviour of the system was anticipated, mainly in the measurement point that before was outside the reference node mesh and now it was inside the network. Results are shown in Table 3.

In the point (1,1) there is no remarkable variation because as it has happened always, is

Table 3. Results using a rectangular mesh

Real Position	Estimated Position	Error-X	Error-Y
(1,1)	(1,1)	0	0
(6,6)	(5.5,5.5)	0.5	0.5
(11,11)	(11.5,11)	0.5	0
(16,11)	(15.5, 10.5)	0.5	0.5

a reference node just alongside, influencing the movable device to make a great estimation. In the point (6,6), the small error of the previous sections continues, but the fluctuation of the value of the coordinate X was reduced, being more stable, something logical if we thought that both new nodes that were added to the previous section essentially contribute with more information on this axis. Once again, in the point (11,11) the measurement is good and it does not almost change with time. On the other hand, as anticipated, in the point (16,11) there is a remarkable improvement of the location. This is due to that pair of new reference nodes with value of coordinate X equal to 21 which contribute with some important information to the movable device while it calculated its position. The considered position improves and the fluctuation of the measures is reduced, although as anticipated, the estimation is less stable than in the initial case that has been taken from reference of the measures.

Once again it is clear that to fit the design of the meshes to the geometry of the area where is desired to locate a moving body is important. It is very interesting to realize that the devices which are going to be located should not approach too much to the edges of the networks if the estimation of the position is desired to be precise and stable.

Analysis of the Influence of the Distance Between Nodes I

In the present section the influence of approximate all the reference nodes to each other is verified. The geometry squared stays for the mesh of 9 fixed nodes but, however, they are spaced only 5 meters now in both Cartesian axes. This experience tries to demonstrate the differences between the recommendations of separate 10 meters the reference nodes and this particular case.

Test Scenario

For the accomplishment of the precision measurements in this circumstance the initial scene of the study in section 7.1 was taken modifying it again. The 9 reference nodes were used. They were placed in a square network but the distance among them was reduced to only 5 meters. Therefore, the reference nodes were placed in the points (1,1), (6,1), (11, 1), (6,1), (6,6), (6,11), (11,1), (11,6) and (11,11), as appraised in Figure 12. Their new well known position was stored in their memory using the application running on the PC.

When the scenario was operative we could begin with the analysis of the results that were obtained with the system in this case.

Results

During the accomplishment of this experiment, we had 9 reference nodes in the test scene. Since they are spaced 5 meter one from another one, the mesh conserved its squared geometry. The distance that separate nodes is reduced exactly to half distance that was used in 7.1 to perform the tests. This way, movable devices placed in an inner point the reference nodes mesh receive signals with greater power than in reference case. In Figure 12 the results obtained in this situation are shown.

At first, the most frequent results are good, not very different from those from the reference case. In point (1,1), as it has happened in all the

Table 4. Results reducing distance between nodes

Real Position	Estimated Position	Error-X	Error-Y
(1,1)	(1,1)	0	0
(6,6)	(6,5.5)	0	0.5
(11,11)	(11.5,11)	0.5	0
(16,11)	(15, 10.5)	1	0.5

experiments, the measurement is good due to the extreme proximity to one of the reference nodes. In the point (11,11), the same happens, it exists a fixed node in that position, causing a very good and very stable estimation. On the opposite, at point (6,6), although the mean measurement is good, the estimation is not stable enough as it would be desired. The explanation can be found in the randomness of the power of the received signal. The power of the signals which contains the packages from an origin to a destiny is very variable with time, and only the mean value turns out a trustworthy variable to make estimations. Added to it, the problem that arises here is that the difference between the powers of arrival signals is not very high because all reference nodes (except the one placed in (6,6)) are all near (6,6) And if several packages of an origin arrives with a little different power from which they should have, an important variation in the estimation of the position will be displayed. In order to verify this, to watch out what happens placing the movable device in inner points of mesh was decided, and it was verified that the estimation of the position was not very trustworthy, although the mean error was small. On the other hand, in the measurement point (16,11) the situation is different. Once again this point remains outside the network, like in section 7.2, although now there are more fixed nodes and they contribute with information to the movable device. The calculated average position is better than the one obtained in 7.2 and the variation between measures is something smaller.

The average quality of the system in this occasion was good, but it was more influenced by the different noises that altered the power of the packages. Again, if the movable device were placed outside the reference nodes mesh, the estimation of the position was worse, and its quality decreased with the distance to the network. From this study it is important to emphasize that placing the nodes very closed to each is not good because of differences between the powers of the received signals are reduced, bringing about a

Figure 12. Reference nodes mesh reducing space between nodes

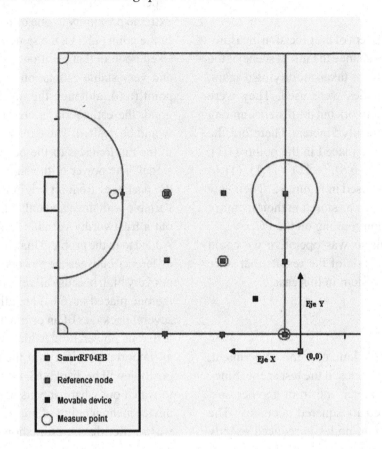

greater uncertainty by the noises and increasing the cost of the system as well.

Analysis of the Influence of the Distance Between Nodes II

In contrast to section 7.4, in this section the influence in the precision of the positioning increasing the distance between reference nodes is studied. In fact, and due to the size of the enclosure where the experiments were realized, this study is a hybrid experience between moving away the reference nodes and the experiment realized in section 7.2., maintaining the squared geometry of the fixed nodes mesh, reducing the number of nodes to 4. At this time, how affects not to respect the Texas Instruments recommendations is studied, separating 10 meters reference nodes, being conscious

that there is a smaller number of reference nodes, but also knowing that they would be very remote of the zones where measures are going to be taken and consequently its influence would be small.

Test Scenario

For the accomplishment of the precision measurements in this chance the 4 reference nodes were placed in the corners of the fixed nodes mesh of section 7.1, disconnecting the other five. A squared with 4 node mesh was built, separated 20 meters in Cartesians axis. Fixed reference nodes then were located in the points of coordinates (1,1), (21, 1), (1,21) and (21,21), as schematically showed in Figure 13.

It was the time to see the answer of the system in these circumstances.

Figure 13. Reference nodes mesh increasing the space between nodes

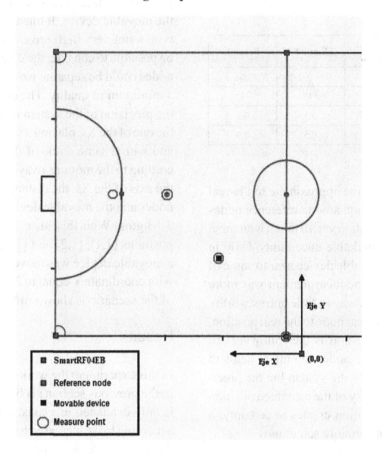

Results

As commented previously, in this experiment there are two important variations. One of them is the distance between reference nodes that have been increased until the double in relation to the case taken as reference in section 7.1. The second important change is due to the space limitations of the sport centre where the tests were realized, and it derivates in the use of only 4 reference nodes to the location of the movable device. This is going to influence, but we are going to assume that the distance among the further nodes and the positions where the movable device will be placed is sufficiently high to consider this influence as very low. Table 5 shows the realized measurements.

Same as in previous cases, at point (1,1), the measurement was very good. It was better; the estimation in that point was the most stable of

all the cases, since the signal coming from the other nodes of reference were much weaker than in other occasions. Something similar happens in the case of the measures in (6,6) and (16,11). In these two cases, the average measures are turned aside towards the nodes that they have nearer, but it is important to pay attention that the error is not as great as it would be expected. However, measures varied much with time, oscillating on the average position in a pair of meters in some occasion. The case of the point (11,11) is special because it is the midpoint between the 4 reference nodes that were assets. It was appraised that the estimation of movable device of its own position was approximated to the real position, but again, this measurement varied much with time respect to that average position.

The system in this case was not very precise. Although it is certain that the calculated aver-

Table 5. Results increasing the distance between nodes

Real Position	Estimated Position	Error-X	Error-Y
(1,1)	(1,1)	0	0
(6,6)	(5,5)	1	1
(11,11)	(10.5,10.5)	0.5	0.5
(16,11)	(16.5, 10.5)	0.5	0.5

age positions were quite approximate to the real positions, the separation among reference nodes caused that the estimation varied much with time, giving rise to a remarkable uncertainty. Only in the cases when a movable device near to some of the reference nodes the measurement was more stable, although it is always a little untrustworthy and nearer from the near node to the real position. One important conclusion is the following. To separate the reference nodes too much helps to make lower the cost of the system but the precision decay and stability of the estimated position is worse. This separation creates uncertainty, a non permissible uncertainly sometimes.

Reach Analysis

This section is different from previous sections. Whereas before how good is the precision of the system was evaluated checking the estimation made in the movable devices in different circumstances, now, in this section, we want to check how increasing the distances between a movable device and some reference nodes that let the movable device make an estimation of its position with a minimum of quality in the approach.

Test Scenario

In this event, only the 3 strictly necessary reference nodes that a movable device needs to collect the minimum data for the calculation of its approximate position are going to be used The idea is to perform a verification of the maximum distance that let signals arrives from the reference nodes to the movable device. It must consider its position with a not very high error. In this way it would be possible to consider the distance that reference nodes could be separate working the system with a minimum of quality. The experience measures the precision of the system in an axis, in this case the one of the Xs, placing 3 very close fixed nodes and with a same value of the coordinate Y. According to the moving away of movable device in the axis of the Xs, the distance between the fixed nodes and the movable device is equalled for the 3 lengths. With this idea, 3 reference nodes in positions (1,1), (1,2) and (1,3) were placed, and a movable device was moved in the straight line with coordinate Y equal to 2 meters. The schema of the scenario is shown in Figure 14.

Results

In this experiment the work dynamics carried out in the previous sections is broken. The precision is only evaluated in x-axis. This is thus because according to we increased to the distance between the movable device and the group of fixed nodes throughout the imaginary straight line Y = 2, the angle forming with the nodes is going to reduce, observing the three nodes with a very small distance among them in y-axis.

With lower distances than 5 meters, the average measurement was exact in x-axis, with a very high stability. This is logical because nodes are closed to the movable device. That is the reason why the power received by the moving body is high from the 3 positions. It was continued moving away the movable device at distance intervals of 2.5 meters. When the distance increased to 7.5 meters to the nodes, the average estimation of the position continued being correct, although in the stability of the measurement the effects of the distance noticed slightly introducing a small uncertainty of maximum (observed) 0.5 meters. When increasing the distance to 10 meters, most of the time it was verified that the average

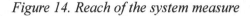

Figure 14. Reach of the system measure

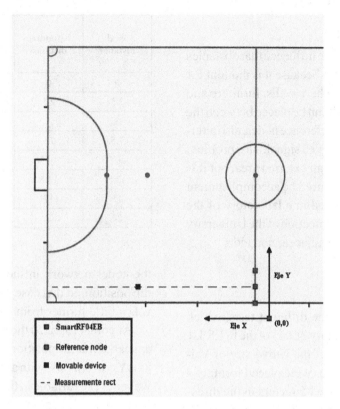

measurement already had an error of 0,5 meters, considering the movable device a coordinate X of 11,5 meters instead of 11 meters. The losses of packages and the arrival of signals with smaller power of the hoped one already were begun to notice slightly. This fact can be appreciated in Figure 9 where it is shown that a small variation of the signal level introduces a variation in the considered distance. The stability also noticed the effects of the loss of a greater number of packages and the variation of the power, with errors of up to 1 meter sometimes. When placing the moving body in the point (13,5, 2), was to hope that the benefits make worse a little more. It also happened with an average estimation of 14,5 meters for the coordinate X, and one variation of this more remarkable position, although with a observed maximum error of 1 meter again. When the distance was 15 meters from the reference nodes, the average calculated coordinate X already was 18,

with a very small stability and errors of 2 meters sometimes. The last point of measurement was taken in (21,2), located at 20 meters from distance of the reference nodes (double the recommended distance by the manufacturer). The estimation of the position was very bad, varying between observed measures up to 12 meters. The results are printed in Table 6 6.

Therefore, from this experiment the conclusion that can be extracted is that for a distance of about 15 meters, a trustworthy system can be obtained (as long as the benefits do not require a greater precision). This comes to reinforce the previous experiments on the distance between nodes, being demonstrated that if the reference nodes of a network are separated too much, the influence on the system is high and the estimations of the position of the different movable devices are bad and with much uncertainty.

Analysis of the Influence of Obstacles

It is important to know the influence that obstacles could have in the system because it is thought for indoors scenarios too, where walls, furniture and other objects exist and it can be placed between the device to locate and the reference nodes, also altering the power of the received signal. This problem also can be displayed in an external area, but it is less problematic. Therefore, the accomplishment of this test was developed in a laboratory of the E.T.S.I.T of Telecommunication of the University of Valladolid and some adjacent corridors.

Test Scenario

For the elaboration of the different tests, which were made in the laboratory 2L013 of the E.T.S.I.T of Telecommunication of the University of Valladolid, 4 reference nodes were placed forming a square nodes mesh spaced 5 meters in the direction of the Cartesian coordinates, as centered as possible in the laboratory in the laboratory. The coordinates were set to (20, 20), (25,20), (20,25) and (25,25) to be able to move the movable device around them and could be shown the estimated position in the application of the personal computer (no negative values allowed). While the device to locate was inside the laboratory, the experience would be equal to the realized in section 7.2 but with smaller number of nodes. When movable device moves away outside the laboratory, how they affect the intermediate obstacle, like the door and walls, was shown.

Results

Inside the laboratory, locating a movable device was verified that the situation was the same as case in section 7.2, as it is logical. However, it is remarkable what happened when the movable device was moved outside the laboratory. It was verified that the intermediate walls and the door between the new position of the moving body and

Table 6. System reach results

Real distance	Estimated distance	Average error	Maximum error from estimation distance
0	0	0	0
2.5	2.5	0	0
5	5	0	0
7.5	7.5	0	0.5
10	10.5	0.5	1
12.5	14.5	2	1
15	17	2	2
20	26	6	7

the nodes network influenced the estimation of the position. In this case, the considered position was a little further from the network.

At point (15,22.5) the average estimation that appeared was the position (17,22). The error in the axis Y corresponds with a small deviation, but that was not very variable in time, since the y-axis was in parallel to the walls that influenced over the signals, not influencing in that coordinate. However, in coordinate X a reduction in the precision of the measurement is appraised in about 2 meters by excess, and with a greater variability, arriving even at 3 meters respect to real position. In the case of the point (15,15), the situation is worse because now the walls influence in both axes. Movable device considers an average position (17,17), with similar variations to the experimented ones previously in the x-axis.

These errors are due to the attenuation that undergoes the signals because of the walls of the laboratory, but they are possible to be avoided. As was commented before, the parameters A and N used in the movable device to fit the hardware location engine response can be modified to correct these errors. Increasing a little N, it is possible to correct in part the error caused for existence of intermediate walls in the system.

The problem arises when in the same surrounding there are zones without obstacles and without walls and zones where there are objects, being able to locate a movable device in both zones. Then it will be necessary to look for a solution testing with different values from A and N and selecting the pair which make the system works better.

RESULTS SUMMARY

Here we have the most outstanding results of the experiments in a summary, being commented the most important aspects. It started from a reference frame that respected the indications of Texas Instruments as far as the separation between reference nodes and using a squared geometry for the mesh of reference nodes. The system started with a mesh of 9 fixed nodes in a square that were extended throughout x-axis and y-axis. Then, precision measurements were settled and comparing with other tests some conclusions could be extracted.

Reducing the number of reference nodes, but respecting the symmetry of the network and the distances between nodes, it was verified that the quality of the system was reduced according to the approaching of the edges of the mesh. Precision and stability of the measures especially decreased when the movable device continued moving away

beyond the limits established for the mesh.

Changing the geometry used for the network of fixed nodes (rectangular in this case) but respecting the distance between the nodes used previously, it was clearly observed that it is very important to fit the design of the mesh to the shape of the area where is desired to locate a moving object. This demonstrates measures improve enough respect to the previous case, mainly in the zones where the network was not covered with fixed nodes before.

Later a squared geometry was took again for the network of reference nodes, and then the distance among them was vary, not following then the recommendations of the chip CC2431 manufacturer. When approaching reference nodes to each other in the middle of the previous distance, the system gave a good answer in general terms, although the noise affected with a small fluctuation in the measures when the movable device was inside mesh, nevertheless, and as it is logical, outside the mesh the precision was not very good (although case improved with respect to the second one). However, when moving away the nodes to 20 meters of distance, the system lost precision in general terms, with great variation of estimation between measures for a physical position with time.

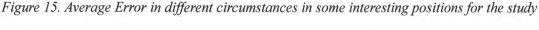

Figure 15. Average Error in different circumstances in some interesting positions for the study

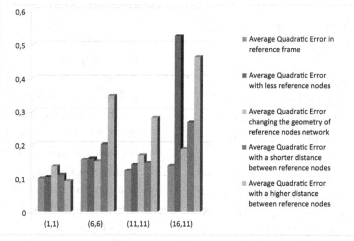

The influence of the objects and obstacles that can be found in the zone where the location can be performed was checked. These obstacles attenuate the power of the signals, and therefore alter the estimations that the movable devices calculate. In order to avoid this as far as possible, the parameters that control the hardware location engine can be regulated, but this solution must be calculated for each case in particular.

From results took along the experience, Figure 15 has been elaborated. It shows in a graphic the conclusions developed previously.

Finally, the reach of the system is important to emphasize. Evidently, the greater distance the smaller power. When the power received is small, the influence of the loss of packages and noise is greater, obtaining a result worse than in shorter distances to the reference nodes. The experiment showed that for distances up to 15 meters the precision is very bad and with a too high margin of uncertainty. Then system can is not considered working correctly. For distances smaller than 10 meters, the system behaves with good benefits. Then, for those intermediate distances between the 10 and 15 meters it is when the precision requirements in different applications is necessary to evaluate, to know if the system is valid or not. It is shown in Table 7:

CONCLUSION

The most important conclusions obtained from the study, are the following:

- The system is a wireless system with low cost, an easy installation and maintenance, and it can be handled by any person with minimum knowledge in applied computer science.
- The internal precision of the system incurs errors of 25 centimetres as a minimum in case of badly considering a position. It is caused because the measures in the coordinate axes

Table 7. Measures for distance with the system

Real distance	Average estimated distance	Average estimated error	Maximum error from estimated distance
0	0	0	0
2.5	2.5	0	0
5	5	0	0
7.5	7.5	0	0.5
10	10.5	0.5	1
12.5	14.5	2	1
15	17	2	2
20	26	6	7

are expressed in meters, with the possibility of spacing them 0.5 meters.

- The network of reference nodes which have to be installed in the zone where is desired to realize the location of movable devices must adjust as much as possible in its geometry to the shape of this area, avoiding found the mobile devices in the limits of the covered area, in which case the uncertainty and the precision would be worse.
- The separation among reference nodes must not being farer beyond 10 meters to support a good operation of the system. With a smaller separation, the improvement is not substantial with respect to the increase of price that it would suppose. Placing them very separated however, money saving can be achieve but the benefits can decay too much when a precise and trustworthy system is required.
- The existence of walls and other obstacles influence the system, but this interference can be compensated modifying the parameters which regulate the operation of the hardware location engine. Once again, this regulation depends on a particular scene, and has to be adjusted from experimental tries and based on the results that are desired to obtain.

REFERENCES

Aamodt, K. (2006). CC2431 Location Engine. *Application Note AN042*, Texas Instruments

Bahl, P., & Padmanabhan, V. N. (2000, March). RADAR: An In-Building RF-based User Location and Tracking System. In *Proceedings. of the IEEE Infocom 2000, The Conference on Computer Communications*. Tel-Aviv, Israel.

Heidari, M. & Pahfavan, K. (2004, June). *Performance Evaluation of Indoor Geolocation Systems Using PROPSim Hardware and Ray Tracing Software*. Paper presented at the 2004 International Workshop on Wireless Ad-Hoc Networks. Oulu, Finland

Labiod, H., Afifi, H., & de Santis, C. (2007). *Wi-Fi, Bluetooth, ZigBee and WiMax*. New York: Springer.

Matteo, Z., Giuseppe, A., Claudia, C., & Mario, F. (2005, May). *Performance evaluation of a differential-GPS ground station for high accurate satellite navigation services*. Paper presented at the 2004 IEEE 59th Vehicular Technology Conference. Milan.

Mosshammer, R. R., Eickhoff, R., Huemer, M., & Weigel, R. (2008). System topologies and performance evaluation of the RESOLUTION embedded local positioning system. *Elektrotechnik & Informationstechnik, 125*(10), 347–352. doi:10.1007/s00502-008-0572-6

Ni, L. M., Liu, Y., Lau, Y. C., & Patil, A. P. (2003). LANDMARC: Indoor Location Sensing Using Active RFID. *Wireless Networks, 10*(6), 701–710. doi:10.1023/B:WINE.0000044029.06344.dd

Ohyama, S., & Kobayashi, A. (1996). Local positioning system by means of enclosing signal field. *Sensors and Actuators, A54*, 457–463.

Sugano, M., Kawazoe, T., Ohta, Y., & Murata, M. (2006). *Indoor Localization System Using RSSI Measurement of Wireless Sensor Network Based on ZigBee Standard*. The IASTED International Conference on Wireless Sensor Networks (WSN 2006), Banff, Canada.

Tannuri, E. A., & Morishita, E. A. (2006). Experimental and numerical evaluation of a typical dynamic positioning system. *Applied Ocean Research, 28*, 133–146. doi:10.1016/j.apor.2006.05.005

Wallbaum, M. (2007). A priori error estimates for wireless local area network positioning systems. *Pervasive and Mobile Computing, 3*, 560–580. doi:10.1016/j.pmcj.2007.02.002

ZigBee Standards Organization. (2008) *ZigBee Specifications*. Retrieved march 20, 2008, from http://www.zigbee.org

KEY TERMS AND DEFINITIONS

Hardware Location Engine: A hardware device that calculates the position of an object using only a few parameters that receives as inputs.

Positioning System: Mechanism for determining the location of an object in space. Technologies for this task exist ranging from worldwide coverage with meter accuracy to workspace coverage with sub-millimeter accuracy.

Radio Communication: The transmission of signals by modulation of electromagnetic waves with frequencies below those of visible light. Electromagnetic radiation travels by means of oscillating electromagnetic fields that pass through the air and the vacuum of space. Information is carried by systematically changing (modulating) some property of the radiated waves such as amplitude, frequency, or phase.

Received Signal Strength: In telecommunications particularly in radio, it is the magnitude of the electric field received from a wireless emitter at a reference point.

Wireless Communication: Transfer of information over a distance without the use of electrical conductors or "wires". The distances involved may be short (a few meters as in television remote control) or long (thousands or millions of kilometers for radio communications). When the context is clear the term is often shortened to "wireless".

Wireless communication is generally considered to be a branch of telecommunications.

ZigBee Alliance: A group of companies which maintain and publish the ZigBee standard in order to maintain it.

ZigBee: Low-cost low-power, wireless mesh networking standard based on the IEEE 802.15.4-2003 standard for wireless personal area networks (WPANs).

Chapter 20
Sensor Web:
Integration of Sensor Networks with Web and Cyber Infrastructure

Tomasz Kobialka
University of Melbourne, Australia

Rajkumar Buyya
University of Melbourne, Australia

Peng Deng
University of Melbourne, Australia

Lars Kulik
University of Melbourne, Australia

Marimuthu Palaniswami
University of Melbourne, Australia

ABSTRACT

As sensor network deployments grow and mature there emerge a common set of operations and transformations. These can be grouped into a conceptual framework called Sensor Web. Sensor Web combines cyber infrastructure with a Service Oriented Architecture (SOA) and sensor networks to provide access to heterogeneous sensor resources in a deployment independent manner. In this chapter we present the Open Sensor Web Architecture (OSWA), a platform independent middleware for developing sensor applications. OSWA is built upon a uniform set of operations and standard data representations as defined in the Sensor Web Enablement Method (SWE) by the Open Geospatial Consortium (OGC). OSWA uses open source and grid technologies to meet the challenging needs of collecting and analyzing observational data and making it accessible for aggregation, archiving and decision making.

DOI: 10.4018/978-1-61520-701-5.ch020

INTRODUCTION

Sensor networks are persistent computing systems composed of large numbers of sensor nodes. These nodes communicate with one another over wireless low-bandwidth links and have limited processing capacity. They work together to collect information about their surrounding environment, which may include temperature, light or GPS information. As sensor networks grow and their ability to measure real-time information in an accurate and reliable fashion improves, a new research challenge, how to collect and analyze recorded information, presents itself.

Deployment scenarios for sensor networks are countless and diverse, sensors may be used for military applications, weather forecasting, tsunami detection, pollution detection and for power management in schools and office buildings. In many of these cases the software management tools for data aggregation and decision making are tightly coupled with each application scenario. However, as these systems grow and mature, a set of common data operations and transformations begin to emerge. All application scenarios will need to query a sensor network and retrieve some observational data. Some scenarios may require information from historic queries be stored in a repository for further analysis. They may require regular queries to be scheduled and automatically dispatched without external operator intervention. Scenarios may need to share information among themselves to aid in decision making tasks. For example, a tsunami warning system may rely on water level information from two geographically distributed sets of sensors developed by competing hardware vendors. These requirements present significant challenges in resource interoperability, fault tolerance and software reliability. A solution to these emerging challenges is to implement a set of uniform operations and a standard representation for sensor data which will fulfill the software needs of a sensor network regardless of the application or deployment scenario.

A Service Oriented Architecture (SOA) allows us to describe, discover and invoke services from heterogeneous platforms using XML and SOAP standards. Services can be defined for common operations including data aggregation, scheduling, resource allocation and resource discovery. We can exploit these properties by combing sensors and sensor networks with a SOA to present sensors as important resources which can be discovered, accessed and where applicable, controlled via the World Wide Web. We refer to this combination of technologies as the Sensor Web. Taking this concept a step further, when interlinked, geographically distributed services form what is called a Sensor Grid which is a key step in the integration of sensor networks and the distributed computing platforms of SOA and Grid Computing. The integration of Sensors Networks with the cyber infrastructure of Grid Computing brings several benefits to the community. The heavy load of information processing can be moved from sensor networks to the backend distributed systems. This separation is beneficial because it reduces the energy and power needed by the sensors, allowing them to concentrate on sensing and sending information. Cross-organizational collaboration is streamlined, because geographically distributed resources can be accessed over common Grid protocols. Data produced by heterogeneous resources can be combined with the aid of common XML formats, eliminating data incompatibility issues.

Figure 1 demonstrates an abstract vision of the Sensor Web; various sensors and sensor nodes form a web view and are treated as available services to all the users who access the Web. A researcher wishing to predict whether a tsunami is going to occur may query the entire Sensor Web and retrieve the response either from real-time sensors that have been registered on the web or from historical data in database. Data from all sources can be aggregated and used by modeling or visualization tools to aid in tsunami prediction. This can be shared among collaborative parties

Figure 1. Abstract vision of the Sensor Web

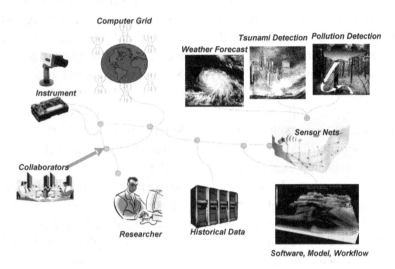

which may run algorithms or transformations over the raw data with the aid of grid resources. In this way, individual resources can be coupled together to perform complex tasks which where not previously achievable.

Driven by the growing demand for data sharing among geographically distributed heterogeneous sensor networks the Open Geospatial Consortium (OGC) (Open Geospatial Consortium, 2008), a geospatial standards authority, has defined the Sensor Web Enablement (SWE) method. SWE includes specifications of interfaces and encodings that enable discovering, accessing, and obtaining sensor data as well as sensor-processing services. These specifications form the blueprint upon which the Sensor Web architecture can be developed. In this chapter we present an implementation of the SWE method which we refer to as Open Sensor Web Architecture (OSWA). We explore the technologies and challenges that have arisen from our experiences with implementing the OSWA. A key aim of which is to provide a software infrastructure that simplifies the task of application development for heterogeneous wireless sensor networks. We present a critical analysis of the proposed standards developed for Sensor Web by the OGC including the challenges

and benefits of working with standards bodies. We introduce the descriptions of core services and encodings which form the SWE, including Sensor Model Language, Observations and Measurements, Transducer Model Language, Sensor Observation Service, Sensor Planning Service, Web Notification Service and Sensor Alert Service. We describe the design and architecture for each of the core services including the challenges and solutions in developing services for heterogeneous sensor hardware and operating system resources. We provide an analysis of a SunSPOT sensor network deployment using a gesture recognition application deployed onto OSWA which includes design and implementation details and results. Finally we conclude by proposing our vision for the future growth of Sensor Web and our OSWA.

RELATED WORK

The integration of sensor networks and grid computing into a sensor grid was initially outlined by Tham and Buyya (2005). Tham and Buyya introduced some early work in the field by proposing the possible implementation of distributed information fusion and distributed autonomous decision-making algorithms. Gaynor et al. (2004) presents a data-collection-network approach to over come the technical problems of integrating resource constrained wireless sensors into grid applications. This takes the form of network infrastructure with a grid API to access heterogeneous sensor resources, referred to as Hourglass. Reichardt (2005) introduced the Sensor Web Enablement (SWE) method which is an important step in connecting sensor networks with web and cyber infrastructure. The method consists of a set of standard services and encoding which can be used to build a framework for discovering and interacting with web-connected sensors and for accessing sensor networks over the web.

52North (Simonis, 2004) is an initiative supported by the Institute for Geoinformatics at the University of Munster, Germany. 52North has developed an open source software set based on the SWE method. They have developed a set of Java web services based on the specifications and data encodings as well as several SWE clients capable of communicating with services and visualizing observational data. Services developed by 52North are deployed as standard Web Services, and the focus of this project is on geospatial data. Sensor observations are retrieved from a geographic information systems (GIS) database called PostGIS and encoded in SWE descriptions. PostGIS acts as an interface between the service and the sensor systems. Interfaces are defined so that new sensor databases or sources can be easily integrated into the architecture.

The GeoICT group at York University (Tao et al. 2004) has built an OGC SWE compliant Sensor Web infrastructure. They have developed a Sensor Web client capable of visualizing geospatial data, and a set of stateless Web Services called GeoSWIFT. The GeoSWIFT Sensing Server implements all the interfaces of a typical observation service and is capable of communicating with Webcams. They have also created an initial Registry Service.

Microsoft has released the MSR SenseWeb Project (Suman, Jie & Feng 2006) which allows users to publish their sensor data on a portal web site. Microsoft has implemented its own XML ontology along with a set of querying and tasking mechanisms. The ontology is influenced by encodings introduced in the SWE method. Support is provided for sensors running TinyOS and devices such as webcams. Microsoft is not affiliated with the OGC Consortium and there is no support for Linux based operating systems. The current application of SenseWeb is limited to publishing data, with little support for post processing, although it is likely that this will change as the project matures.

OSWA is an implementation of the SWE method that extends the typical Web Service interface definitions by implementing them as Stateful Web Services called WSRF. To our knowledge there are no other published SWE implementations which use WSRF. WSRF opens the door for OSWA services to communicate with data and computational grid resources. OSWA supports heterogeneous sensor resources on TinyDB, SunSPOT, TinyOS and Linux. Implementations such as GeoSWIFT and 52North typically support one or two sensor operating systems, although they include constructs to extend these. Microsoft's SenseWeb Project includes support of TinyOS, but not Linux. No other SWE method implementations support the same diversity of operating systems as OSWA. In OSWA we have introduced a caching method into the service responsible for communicating with the sensor networks, this is a novel feature which improves performance and has not been implemented by any other research groups. There are many research groups working on sensor node middleware solutions. This is middleware which resides on top of the sensor operating system. Some notable projects include MiLAN (Heinzelman et al., 2004), Agilla (Fok, Roman & Lu, 2005), DSWare (Li, Son & Stankovic, 2003) and MagnetOS (Barr et al., 2002). It is our intention to expand OSWA into the sensor operating system level and provide a lower level middleware solution. Future research opportunities include developing a specific service for this purpose.

SENSOR WEB ENABLEMENT

As sensor network deployments grow obstacles begin to arise as an outcome of connecting heterogeneous sensor resources and sharing observational data. A research challenge presents itself in how to collect and analyze observational data from heterogeneous sensor networks and make it accessible for aggregation, archiving and decision making.

The Sensor Web Enablement (SWE) method is defined by the Open Geospatial Consortium (OGC); it includes specifications for interfaces, protocols and encodings which enable implementation of interoperable and scalable service-oriented networks of heterogeneous sensor systems and client applications (Botts, Percivall, Reed, & Davidson, 2007). OSWA is an implementation of the SWE method which consists of the following XML encodings and interfaces:

1. **Sensor Model Language (SensorML):** A set of standard models and XML schema for describing sensor systems and processes
2. **Observations and Measurements Schema (O&M):** A set of standard models and XML schema for describing physical phenomena observed by sensor systems
3. **Transducer markup Language (TML):** A XML schema and encoding for describing real-time streaming data recorded by transducers
4. **Sensor Observations Service (SOS):** A web service interface definition for requesting observations from sensor networks and observation repositories
5. **Sensor Planning Service (SPS):** A web service interface definition for scheduling

and planning observational requests to sensor networks
6. **Web Notification Service (WNS):** A web service interface definition for the transmission of messages between SWE services
7. **Sensor Alert Service (SAS):** A web service interface definition for publishing and subscribing to alerts from sensors

Services and encodings presented in the SWE method are decoupled from any particular deployment scenario. Interfaces are defined in such a manner that services responsible for performing independent tasks can co-ordinate with each other to complete a common goal. When coupled together services form a middleware layer which is capable of meeting the complex demands of a heterogeneous multi-user system.

The implementation of service interfaces, based on a common set of standards, has many advantages. Research groups or commercial companies are free to design their own service implementations with the confidence that services will be capable of interacting with one another. A group at the University of Melbourne can build a SPS while another group in Europe builds a SOS. Service descriptions can be published on the World Wide Web so that both groups can use each other's resources, mutually benefiting both teams. Standard models for data encoding such as SensorML, O&M and TML allow data to be shared among implementations and encourage collaboration among SWE implementations. The use of XML as a basis for the schemas allows for platform independence, software can be developed to run on Linux or Microsoft platforms with the confidence that there will be no data incompatibilities. The rich semantic capability of XML is well suited for data exchange, and capable of meeting growing needs in data encapsulation.

The feedback of user experience and contribution of ideas to standards bodies is an important step in developing a community and promoting

Figure 2. Sensor Web enablement service interaction

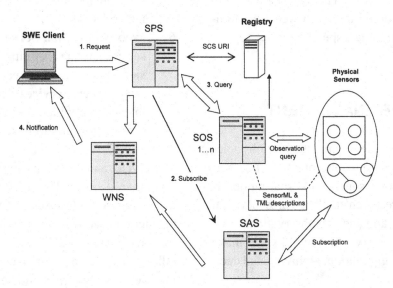

broader adoption of standards among researchers and businesses. Standards can only mature if they are underpinned by practical experience. However, in an emerging technology such as Sensor Web where feedback from deployment experience may be quite high, this presents an interesting challenge. Development efforts of pioneering adopters who have invested in early standards may seem in vain, as their systems can quickly cease to conform to the most recent standards release. The growth of XML tools is one technology which can ease this burden. A variety of tools exist which facilitate the generation of code from XML schemas. This it aids the developer by reducing some of the tedious works. Ultimately, however, it is important that researchers, developers and standards bodies work together to foster a strong community which can meet these challenges.

Moodley & Simonis (2006) raise some drawbacks on the SWE approach. SWE does not have a formal conceptual model that links all the services and encodings together. This complicates the task of combing data with different granularities of time, space and measured phenomena. The encodings lack explicit semantics, so it is difficult to discover if two or more sets of observed

phenomena are related. SensorML is an attempt at partially meeting this requirement. However, although it can be used to describe the sensors themselves, it does not provide a semantic description of the sensor and the phenomenon that it measures.

A vision for Sensor Web is to have service components working together to execute a user request and achieve a common goal as illustrated in Figure 2. A SWE enabled client is interested in retrieving observational data from a set of physical sensor nodes. The client knows the physical location of the sensors it is interested in, the duration for which it is interested in reading the observational data for (1 hour, 1 day, 1 week, etc.) and the observational data (light, acceleration, temperature, etc.) it requires. The client constructs a request containing this information and sends it to the SPS. The SPS then determines the Universal Resource Indicator (URI) of the appropriate SOS instance by querying a registry of available services. When a SOS instance comes online it automatically registers its capabilities with the Registry. If the client requires alerts, the SPS subscribes to a SAS, if events described in the alert request occur the SAS will automatically inform the WNS, which will perform an action or communicate this information back to the client. Once the subscriptions have been dealt with the SPS will query the appropriate SOS instance which will send the request to a sensor network and retrieve the observational result. The SPS will notify the WNS that the request has been completed; the WNS will forward the location of the observation data and the outcome of the plan to the SWE client. The client can then collect the observational data.

The SWE presents a framework of service descriptions and XML schemas for communication protocols. A research challenge lies in the design and architecture of the services and specifications in a robust, efficient, platform independent and secure manner. In OSWA we attempt to tackle this challenge, using the SWE method as a basis upon which to build robust platform independent middleware.

OPEN SENSORWEB ARCHITECTURE

The OSWA is an implementation of the SWE method. The various components defined for OSWA are outlined in Figure 3. We can identify four core layers namely Fabric, Services, Development and Application. Fundamental services are provided by low-level components whereas higher-level components provide tools for creating applications and management of the lifecycle of data captured through sensor networks.

We use the SWE specifications as a blueprint upon which to base our Service layer. It is important to feed back real-world deployment experience into the design and architecture of services. Ultimately deployment experience should drive standards improvement, although, given the nature of standards this often a lengthy process. It is through the early embrace of emerging technologies that we can demonstrate their advantages that can then lead to improvements in standards. With this in mind, we have implemented the Service layer as a set of WSRF services. The move to

Figure 3. High level view of OSWA

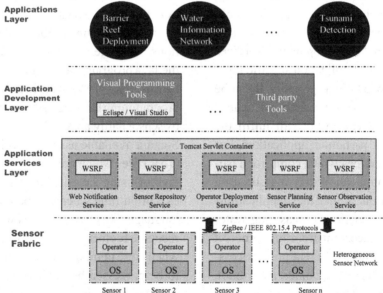

WSRF grew out of a need to support on going queries which persist over time to services. These queries require state information which is difficult to implement in traditional Web Services. WSRF changes the dynamics of the SWE framework because services are no longer passive sources for data to be pulled from. They are active data sources which push data out to clients following a publish-subscribe paradigm. In future work we plan to explore the introduction of additional services, such as an Operator Deployment Service which can facilitate the deployment of application specific operators onto the physical sensor networks. Operators could communicate with one another to form an overlay network which would be hardware transparent and capable of enforcing efficient network communication, fault tolerance, resource management and discovery, code management and energy saving schemes.

A key aim of the OSWA is to provide a software infrastructure that simplifies the task of application development for heterogeneous wireless sensor networks. Once services have been deployed we can further abstract the details of the services into interfaces which we couple together to form an API, this forms the basis of the Application Development layer. Developers can then use the API to build and deploy sensor applications, define relationships between services and build job scheduling schemes through an interactive GUI. A challenge in the development of services is to decouple as much application specific logic from the service code base as possible. It can be difficult to develop services which are neutral to the deployment scenario but still fulfill the idiosyncrasies of a particular application. For example, a set of services which comprise a tsunami monitoring application may also be used for pollution detection. Both of these applications may have the commonality of measuring water temperature or displacement from the same set of sensors but have quite different post processing, scheduling and result outcomes. A common approach is to express these idiosyncrasies using

a XML model. Although this often introduces additional computational processing time which may not be acceptable in real time applications, it is important that XML models can provide the semantic descriptions necessary to encapsulate this information.

The Sensor Fabric layer includes the Operating System and application code deployed onto physical sensors which allows them to record observations and network among themselves. Currently it is up to developers to program and deploy applications at the Fabric layer. This is not an ideal situation as it requires the programmer to directly interface with the sensors and manage the storage, processing, recording and transport of observation data. Furthermore physical access to individual sensors is required, making it difficult to update code on large numbers of remote sensing nodes. A solution to this problem is to deploy a sensor node middleware onto the sensors themselves. The middleware acts as an interface to the underlying operating system and provides code management, allocation and migration facilities. It is our intention to expand OSWA into the fabric layer and provide a multi tier middleware solution. The Operator Deployment Service is a step in this direction.

Technologies such as Java, Tomcat, XML, SOAP and Web Services provide great opportunities for developing platform independent applications but come with a cost. OSWA is written in the Java programming language. Java was chosen because it is a platform independent object oriented programming language, software libraries released by sensor hardware vendors such as Crossbow and Sun are available as Java Archived Repositories (JAR) files which are simple to use. A disadvantage of using Java is that it is not as fast in its execution time when compared to a lower-level language like C. Java comes with a memory and resource footprint which may affect performance when large numbers of simultaneous requests are to be processed by services. However, constant improvements in JVM technologies mean this situation can only improve with time. The platform heterogeneity and ease of programming outweigh any disadvantages associated with Java.

A challenge of OSWA is how to support ongoing sensor queries which persist over time to heterogeneous sensor networks. Traditional Web Services are stateless, making it difficult to create and maintain persistent relationships between services. Stateful Web Services provide access to data values that persist over time and evolve as a result of Web Service interactions. The Web Service Resource Framework (WSRF) defines conventions for managing state so that applications discover, inspect and interact with stateful resources in standard and interoperable ways, as defined by the OASIS standards body. Java WS Core is a component of the Globus Toolkit, a set of software components for building distributed systems and it is a popular Grid middleware platform. WSRF is underlined by a notification-subscription interaction pattern. A client subscribes to a service resource and if

any changes of state occur on that resource, the service will automatically notify the client of the changes. This eliminates the need for a client to poll the service for changes, as is typical from Web Services, thus reducing the network traffic among services and improving performance. The introduction of WSRF is a key step forward in evolving sensor web technologies into a sensor grid. Services are deployed on an Apache Tomcat container. Tomcat is a servlet container which provides an environment for java code to run on. It is written in Java and is a stable and free technology maintained by volunteers.

SOAP is used as the communication protocol between clients and services; SOAP relies on XML as its message protocol and HTTP for negotiation and transportation. The use of XML comes with a processing burden. Transformations need to be performed between the data views of XML and Java object. Managing these relationships manually can be cumbersome and error prone. One solution is to automatically generate Java objects

Figure 4. Conceptual model for SensorML processes (Botts, 2007)

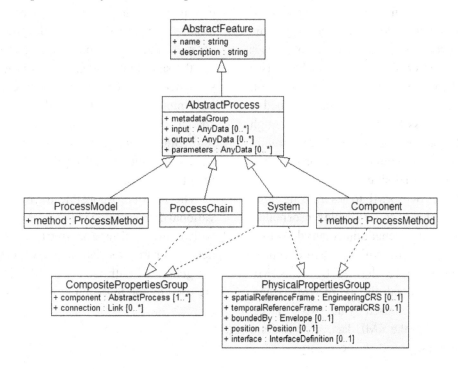

from XML schema using a Java-to-XML binding framework like XMLBeans.

For the remainder of this section we discuss each of the components defined in SWE method, introduce the architecture of these components as implemented by us in the OSWA, and explore the relationships between services and encodings.

Sensor Model Language

SensorML is used to describe the processes and processing components associated with the measurement and post-measurement transformation of observations (Botts, 2007). A process is any entity that takes an input, applies a set of well-defined methods, and results in an output. Processes can be linked together into executable process chains which describe the mapping from input to output between components. This conceptual model for processes in SensorML is illustrated in Figure 4. Process chains are useful in deriving high-level information, which is not otherwise attainable from a single process. For example, a process chain could include the retrieval of raw observational results and the on-demand processing of those results into more meaningful output. SensorML is particularly useful in describing sensor systems and in the processing and analysis of sensor observations. Observations recorded by sensor systems and encoded in the O&M specification can be encoded within SensorML and described as a SensorML process. SensorML is robust enough to handle the processing of data from virtually any sensor whether mobile, in-situ or remotely sensed, or active or passive.

The SOS uses SensorML to describe the capabilities and metadata of any available sensor nodes. The SPS accepts user scheduling plans described as SensorML processes which it then executes. In OSWA Java objects defined by SensorML models are derived with the aid of XMLBeans. These objects are then used by the two services, to encode or decode the XML data. A SWE client may also use these objects as necessary.

Observations and Measurements Schema

The O&M schema is an encoding for observations and measurements retrieved from a sensor network by the SOS. The purpose of the O&M is to alleviate the need for sensor-specific or research independent data formats for describing data retrieved from sensor networks. An observation is any event which has a value that describes some phenomenon. The term measurement is used to identify a numeric quantity associated with the observation. The phenomenon is a property of an identifiable object, which is the feature of interest of the observation (Cox, 2007). For example, if a sensor network is deployed in a room to measure the light intensity, the observed property would be lux, the photometric unit for describing illuminance and the feature of interest would be light. An Observation model identifies the real-world observation target for which observations are made; an extract of this model is illustrated in Figure 5. This includes the value of the observed property and may include a description of the process used to generate the result. Using our light example the value recorded in an office might be 320 lux, the process could be the procedure used for recording light. Constructs exist for describing the sampling time and result time for a time sequence of observations. An observation may have metadata associated with it, such as a geospatial location. Observations can be composed into collections, which share some commonalities such as the same sampling time or the same feature of interest.

The SOS is responsible for returning observational data encoded in the O&M specification, which can be real-time data retrieved from a sensor network or archived data. In OSWA, O&M objects are generated with the aid of XMLBeans, these are used by the SOS and by SWE clients.

Figure 5. O&M model extract (Na, 2007)

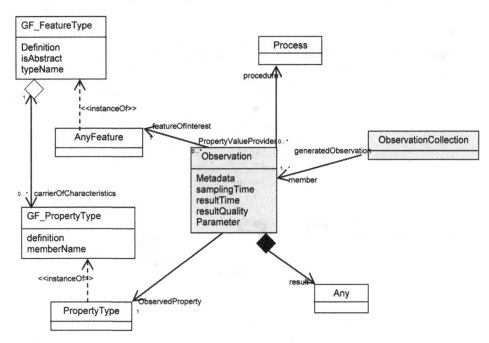

Transducer Markup Language

TML defines a set of models which are used in describing the data captured from transducers, along with methods for communicating real-time sensor data. TML includes information necessary for the post-processing of data by the eventual recipient. A transducer is typically a group of devices which can capture real-time data from multiple phenomena. Transducers work in two ways, they can sense data or data can be sent to them to produce some sort of predetermined result. An advantage of TML is that it makes it possible to share data across application domains. It can be used for retrieving data from live sources or archived sources.

TML is a recent addition to the SWE set of encodings, in our OSWA we are yet to implement it and have no access to a transducer device. The SOS is primary responsible for returning the sets of transducer results encoded in TML. It is the responsibility of the SOS to communicate with the transducer device.

Sensor Observation Service

The SOS is a service responsible for forwarding requests to the sensor network and retrieving the recorded observational results. It acts as an intermediary between the client and real time or archived sensor observation data. It provides a common interface to communicating with sets of heterogeneous networks and archived data sources. The SOS communicates with the sensor network via a base station node which acts as a bridge between the service and sensor nodes. Observational results retrieved from the sensors are returned to the client encoded in the O&M specifications. Metadata describing the sensor platform (hardware capacity, sensor types) is returned in the SensorML encoding. SWE enabled clients can connect directly to the SOS to retrieve near real-time data, or for complex queries the SPS can facilitate lifecycle management and coordinate data retrieval from multiple SOS instances simultaneously.

The SOS is composed of three core operations *DescribeSensor, GetObservation* and *Get-*

Figure 6. Architecture of the SOS

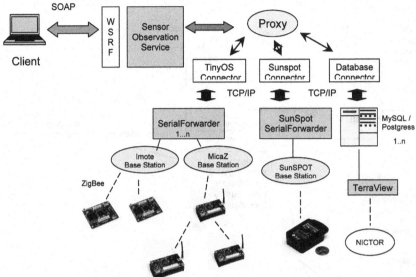

Capabilities. GetObservation is responsible for returning observations from a sensor network, *GetCapabilities* returns metadata information about the SOS service and *DescribeSensor* is responsible for returning metadata information about the sensor nodes. Other operations include *RegisterSensor*, and *InsertObservation*, which are used to support transactions along with six enhanced operations including *GetResult, GetFeaturesOfInterest, GetFeaturesOfInterestTime, DescribeFeatureOfInterest, DescribeObservationType* and *DescribeResultModel*.

In OSWA we implement all the core operations described in the specification, they include *GetObservation, DescribeSensor* and *GetCapabilities*. These interfaces provide sensor descriptions and observational data from a heterogeneous set of sensor networks which include TinyOS running on Mica2, MicaZ and Imote2, NICTOR sensors developed by NICTA running Linux and SunSPOT sensors running Java. The architecture of the SOS is illustrated in Figure 6.

A client connects to the SOS interface through the Globus WSRF library. The client can be a user connecting with a SWE client or a service, such as the SPS, initiating the connection on behalf of a user executable plan. Once the connection has been negotiated calls to the *DescribeSensor* or *GetCapabilities* operations are directed through a proxy class to a database connector. The database connector communicates with a PostgreSQL database to retrieve metadata information describing the sensor hardware (*DescribeSensor*) and the offerings that are available from the SOS (*GetCapabilities*). This information is encoded in SensorML and returned to the client, which will use it to determine if the SOS service is capable of fulfilling an observational request. The client will send a *GetObservation* request to retrieve observational data from the sensor network. The request will contain a SQL-like syntax, information encapsulated in the syntax may include; the sensor network type (vendor), the location of the network, the observed phenomenon (light, temperature, acceleration, etc.), a threshold value (only values temperature values greater than 0 degrees Celsius), the duration for which to sense the data, the update frequency for observations, and the network ID's of sensors to be queried. This information may vary with each application context. The proxy will distribute the query request to the appropriate network connector. It is the responsibility of the connector to communicate with the base station and retrieve the observational results. In most cases this is facilitated by a daemon which forwards queries to the serial port that the base station is connected to. The observational data recorded by the sensors is then published on a TCP/IP port and is available for the connector class to retrieve. In some cases, such as for the NICTOR sensors, this interface occurs via a database. NICTORS are unique in that they publish their observational results directly to a MySQL database. Once the observation data has been collected it is encoded into the O&M specification and returned to the client.

As part of a continued effort to enhance the performance of the SOS we have introduced a cache mechanism into the SOS architecture. A bottleneck of the SOS has been the inability of sensor networks to handle more than one query at a time, without some special operators or middleware deployed onto the sensors. When an observational query is sent to a sensor network that query must return a result before a consecutive query can be fulfilled. In a system with multiple concurrent users, all users are interested in an immediate response, which can lead to a major performance bottleneck. To overcome this bottleneck, a cache mechanism has been developed that consists of a two-level cache chain incorporated with query aggregation rules and a partial matching scheme

Figure 7. Architecture of cache mechanism

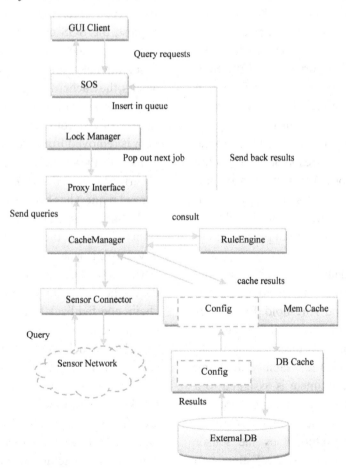

to improve accuracy and performance. The cache mechanism handles the parsing of cached queries and the predicting of results for current queries. Query results are stored in a local cache, incoming queries are checked against historical ones and if the query strings are similar and lie within a timeout the cached results are returned instead of sending a query to the sensor network.

The key components of the Cache mechanism are illustrated in Figure 7: where the cache interfaces with the proxy and connector components of the SOS architecture.

A CacheManager maintains a cache chain, which gives orders of precedence to the available caches. Upon receiving an observation request the CacheManager checks with the RuleEngine to determine if it should query the Cache. The RuleEngine maintains a series of parameters, which are designed to improve the accuracy of the Cache in order to maximize the cache hit rate. The query SQL string is used as a key for caches. The CacheManager checks each Cache in the cache chain and returns a hit if the SQL string exists. If the RuleEngine determines that the CacheManager is unlikely to retrieve a cache hit, i.e. if the Cache is full, or entries are expired or don't match the key, control is returned to the CacheManager which redirects the query to the physical sensor network. When the sensor network returns observational results the proxy will forward these to the CacheManager which will update the cache chain along with parameters in the RuleEngine.

The RuleEngine analyses the observational results and makes changes to the tolerance parameters of Estimate and Threshold. Estimate is a numeric value given to the rate of change observed in the environment by the sensor network. A small Estimate is given to a rapidly changing environment; a larger estimate is assigned to a stable environment. The Estimate is determined by analyzing the difference between consecutively recorded observations returned by the sensor network. The RuleEngine uses the Estimate to determine if a query should be checked against the Cache or not. If the current time exceeds the last update time of a cached result plus the Estimate, then the request is redirect to the sensor network, because the environment is changing fast, thus any observation cached will already be out of date. This ensures that the Cache is queried only in circumstances where we are confident that a cache hit is as close as possible to a correct reflection of the physical environment. The Estimate is initialized in the configuration file and dynamically changed by the RuleEngine at runtime to reflect the changing environment. The Threshold is a dynamically changing parameter that adapts to the cache size, frequency of entries being cached and the values of entries.

The Caches are strung together to form a cache chain. Typically, the memory cache takes precedence over the database and is limited in size. Upon receiving an incoming observation request, the SOS calls the CacheManager. If a cache is hit, the cached result is returned as the observational result to the client. If a cache miss occurs, the CacheManager will insert the observation request into a queue of observational requests to be retrieved from the sensor network. When an item is removed from the queue, a second check is made by the CacheManager to determine if any observational results have been cached while the current request has been waiting in the queue. If a cache hit occurs, the result is returned to the client. If there is a miss, the SensorProxy will query the physical sensor network. The observational results retrieved are then written by the CacheManager to the Cache, following the precedence of the cache chain, and fed back to the RuleEngine. A cache hit can occur if two cache keys (query strings) are "similar", which means that they are exactly the same or their values lie within the tolerance Threshold. The responsibility of determining whether two keys are similar is given to the Comparer. If a cache miss occurs, the Comparer can still use existing cached entries to achieve a cache hit by using partial matching schemes. When writing a new entry to the cache, if a cache entry already exists with the same key, the new observational results will replace the stale data. Otherwise, a new key entry is allocated and a new result entry is placed into the cache. If the cache is full, we employ two eviction strategies, Least Recently Used (LRU) and least rank. Although other more complex cache eviction strategies exist, LRU is our primary choice because it works well in an environment where there is a high temporal locality of reference in the workload (that is, when most recently reference objects are most likely to be referenced again in the near future). After being stored in the cache, the result is feedback to the RuleEngine.

Sensor Planning Service

The SPS is responsible for providing a high level planning, scheduling, tasking, collection, processing, archiving of requests for all services. A SWE client can submit a SensorML encoded plan to the

SPS, the plan must contain the observation request, location of the sensors, duration of the request and any other relevant metadata or post-measurement processing requirements. The SPS is responsible for discovering available SOS instances from a registry of services and capabilities, processing and scheduling the plan, managing subscription requests to the SAS and forwarding notifications to the WNS. Use of the SPS should be limited to circumstances where observational data from more than one SOS instances is required, where the connection duration may persist over some period of time, and where post-processing, such as data aggregation from several sensor networks or archived observation sources is required. We define these types of requests as complex observational requests.

In a similar fashion to the SOS there are both mandatory and optional operations which are required to be implemented. *GetCapabilities*, *DescribeTasking*, *Submit*, and *DescribeResultAccess* are all mandatory operations. *GetCapabilities* is responsible for returning metadata information

Figure 8. Architecture of the SPS

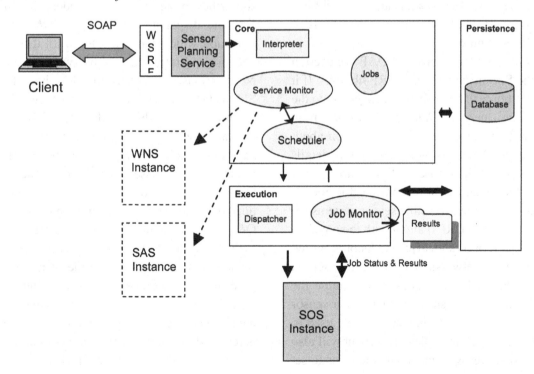

regarding the capabilities of the SPS. *DescribeTasking* returns information about all parameters, which need to be set by a client, to perform a *Submit* request. The *Submit* operation submits the user plan for scheduling and execution by the SPS. *DescribeResultAccess* returns the SOS location that the SPS communicates with in order to access observational data from a particular sensor. Optional operations which may also be implemented include *GetFeasability*, *GetStatus*, *Update* and *Cancel*. *GetFeasability* provides feedback to the client on the feasibility of executing the plan, this includes checking the validity of the parameters, and locating a SOS instance and checking the instance can fulfill the request. *GetStatus* returns the current status of the request. *Update* allows the client to update a previously submitted plan and *Cancel* terminates a plan.

In the OSWA we implement all the mandatory and optional operations. The architecture of our SPS is based heavily on earlier design work our team did in developing a Grid resource broker called Gridbus. The Gridbus Broker is a scheduler for distributed data-intensive applications on global grids (Venugopal, Nadiminti, Gibbins, Buyya, 2008). The architecture of the SPS is illustrated in Figure 8.

A client will initialize a SOAP connection with the SPS using Globus WSRF libraries. If it is the first time that the client is connecting to the SPS it may query the *GetCapabilities* operation in order to retrieve metadata information about the service. The operation may return details describing the hardware and software of the server, the organization responsible for operating the server, the accessible sensor systems (in the form of SOS URI's) along with the physical location and observational phenomenon recorded by the sensor networks. The client may send a *DescribeResultAccess* request to determine the SOS location responsible for a particular sensor it is interested in. This data can then be used in the construction of the plan. The client will also need to discover what parameters it needs to set

in order to perform a *Submit* operation. It is the responsibility of the *DescribeTasking* operation to provide this information. The client uses *DescribeTasking* response to construct a user plan that will contain all the information necessary to execute a *GetObservation* operation on a SOS instance, along with any pre-processing, post measurement processing, archiving, notification, duration and any other tasking it wishes to perform on the observational data.

When the client performs a *Submit* operation an interpreter decodes the user plan and constructs a job. A Job is an object that encapsulates all the content described in the user plan. If the user plan specifies the client to be notified of the Job completion via the WNS (email, SMS, Instant Message, phone call), the SPS registers the Job with the WNS. The state of the Job at anytime throughout its lifecycle is maintained by a Hibernate database. The Job is placed in a queue and scheduled for execution. A Service Monitor thread sits in the background and discovers any new SWE service instances that may be accessible on the network. When the Job is ready for execution a dispatcher subscribes to the SOS instance identified in the Job, the dispatcher calls the *GetObservation* operation which communicates with the sensor network and retrieves the observational results. Notifications are sent by the SOS back to the SPS as the Job executes, these are forwarded to the Job Monitor which updates the Job state in the database. Whenever the SPS receives a *GetStatus* request it retrieves the current state of the job from the database and returns it to the client. Once the observation is completed the SOS returns the results to the Job Monitor encoded in the O&M specification. The results are then written to the local file system and if any post processing requirements are included in the plan these are performed. Post-processing may include error checking, performing additional calculations or data transformations. Upon completion the data is returned to the client, for more complex plans which persist over time, or which may require

Figure 9. SAS overview (Simonis, 2007)

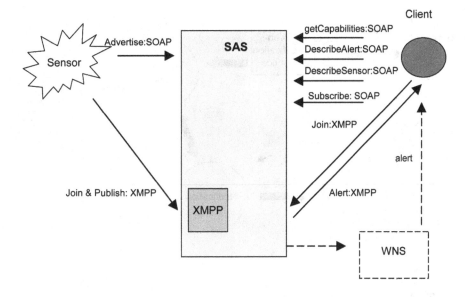

alternative communication means (SMS, email etc..) the WNS is notified which in turn notifies the client over the clients preferred protocol.

Sensor Alert Service

The SAS specification provides an interface for sensor nodes to advertise and publish alerts. Clients can subscribe to data that matches specific criteria, for example when the battery is low or if an observation value is returned above or below a threshold value. When this data becomes available the SAS notifies the client. Intelligent sensors can connect to the SAS and make their resources available to clients for subscription. The SAS uses the Extensible Messaging and Presence Protocol (XMPP), a decentralized open XML-based protocol targeted at near real-time communication, to publish sensor data.

The SAS specification outlines ten operations that can be requested by a client and performed by a SAS server. The required operations include *GetCapabilities*, *Subscribe*, *CancelSubscription*, *RenewSubscription*, *DescribeAlert*, and *DescribeSensor*. The remainder of the operations are optional, they include *GetWSDL*, *Advertise*,

CancelAdvertisement and *RenewAdvertisement*.

The SAS is a new addition to the SWE method and therefore it is yet to be implemented in the OSWA. We will briefly describe the proposed functionality of the interfaces, which are illustrated in Figure 9.

GetCapabilities returns metadata describing the abilities of the SAS implementation. The *Subscribe* operation allows clients to subscribe to the advertised capabilities. *CancelSubscription* terminates the subscription and *RenewSubscription* restarts the subscription. *DescribeAlert* returns the structure of the data observed by a particular sensor. This includes the physical phenomenon being observed and format of the recorded data. Upon receiving a return value the client has enough information to *Subscribe* to the alert. *GetWSDL* returns the WSDL description of the SAS interface. *Advertise* allows sensors to advertise their capabilities to a SAS instance. Sensors or data producers calling *Advertise* will be added to the sensor pool and are available to clients for subscription. *CancelAdvertisement* allows the advertising sensor source to terminate the relationship and be removed from the pool. *RenewAdvertisement* restarts the advertisement.

Figure 10. The architecture of the WNS as implemented in OSWA

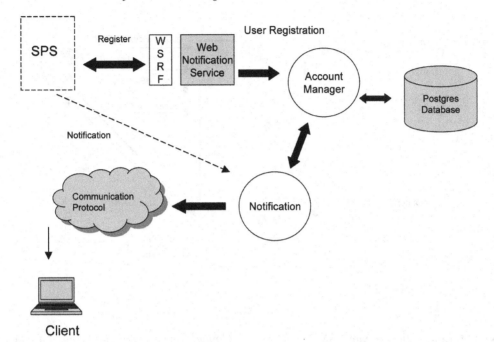

Web Notification Service

The WNS is an asynchronous messaging service whereby users can subscribe and receive notifications, over one of several protocols, on any interesting phenomena that may occur in any SWE service. Any service can call the WNS to send a notification. The WNS handles two notification methods, one-way, where notifications from services are forwarded to the client and two-way where a response is expected from the client. A variety of communication clients can be programmed into the WNS model, including email or SMS. This allows for users to program their mobile devices to accept notifications describing processing errors or completed SPS plan requests.

Mandatory operations defined for the WNS include *GetCapabilities*, *RegisterUser*, and *DoNotification*. Optional operations are *DoCommunication* and *DoReply*. *GetCapabilities* works in a similar fashion to previously mentioned services, returning metadata about the WNS. *RegisterUser*

allows a client to register to receive notifcations and *DoNotification* initiates the notification of the registered user. *DoCommunication* is called to initiate the communication with the user, and *DoReply* accepts a user response to a two-way notification.

In the OSWA we implement the mandatory operations described in the WNS specification. The mandatory operations are used to perform one-way communication whilst optional operations are only required for two-way notification. The architecture of the WNS is illustrated in Figure 10.

Clients can discover the capabilities of the WNS by calling the *GetCapabilities* operation, this returns the available communication protocols implemented by the WNS. When a client calls the *RegisterUser* operation on the WNS the user is assigned a registration ID by the Account Manager which is then stored in a Postgres database. The client case can be any SWE service, however it will typically by the SPS, as this is responsible for managing the scheduling of user plans. When some interesting event occurs the SPS will send a *DoNotification* request to the WNS. This is handled by a Notification class which discovers the user details from the Account Manager and notifies an end-user client with an appropriate communication protocol. In the OSWA we implement email as the preferred protocol, although an interface exists so virtually any protocol can be easily added.

In the following section we present the problem of gesture recognition and build and deploy a gesture recognition application using the OSWA to access real-time observational data produced by SunSPOT sensors, transform the observational results and visualize the data.

CASE STUDY: A GESTURE RECOGNITION APPLICATION

From GUIs to multi-touch surface pads, speech to gesturing, the ways we interact with computers are diversifying more than ever before. To demonstrate the usability of the OSWA we implemented a prototype arm gesture recognition system, trying to free the user from the keyboard and mouse and incorporate a more natural gesture user interface utilizing sensors, machine learning and sensor web. This requires the user to hold a sensor node in their hand and perform a gesture with their arm. Each unique gesture has a different semantic meaning, this may include a letter of the alphabet, moving to the next slide in a presentation or opening and closing a browser window. In this section we introduce the problem of gesture recognition; we outline the idiosyncrasies and challenges in building a gesture recognition system. We introduce the software components which we need to meet these challenges and use the SOS to collect and process gesture data, which we forward to a SWE client for visualization.

Human motion is an inherent continuous event and difficult to predict. Theoretically, the human motion recognition problem is similar to voice recognition which is well studied. The main difference is that human motion occurs in three dimensional space, which requires measurements to be recorded for at least three axes. To recognize human gestures first we need to capture gesture data and transmit it for further processing. This raw data can then be analyzed by recognition algorithms in order to extract some useful meaning or content.

SunSPOT (Sun Small Programmable Object Technology) is an open source software package and hardware sensor node developed by Sun Microsystems. Developers can customize both virtual machine source code and circuit board design to meet their own special requirements, using Java to write applications and deploy them on the physical devices. SunSPOT devices come with a light sensor, temperature sensor, and accelerometer integrated onto the sensor board.

There are two main challenges in for gesture recognition. The first challenge is segmentation, i.e., how to identify the beginning and end of a motion in a multi-attribute data stream. The second challenge is to recognize the segmented

Figure 11. Variations in acceleration data from the X, Y and Z axis produced by similar motions

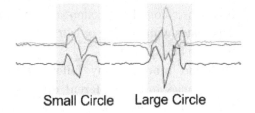

Small Circle   Large Circle

Figure 12. Variations in acceleration data from the X, Y and Z axis produced by similar gestures with different durations

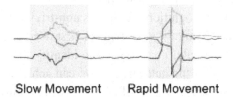

Slow Movement   Rapid Movement

stream with a high level of accuracy. To fulfill these requirements there are several challenges which need to be addressed (Li, Zheng, & Prabhakaran, 2007):

- **Similar motions may look different:** Due to variance in speed and direction, similar motions can have variations in a multi-attribute data stream. Figure 11, illustrates the raw acceleration data produced by gestures of a small (10 centimeter) diameter circle and a larger diameter (60 centimeters) circle.

- **Similar motions vary in duration:** Different people may perform the same gesture in different ways. Even the same person can not perform exactly the same gesture at the same speed twice. The sensor sampling rate may differ as well. This data series is illustrated in Figure 12.

- **Similar motions may have different meanings:** Illustrated in Figure 13 are the accelerometer readings from three gestures with similar motions, but with different semantic meanings. Complete motions are concatenated by brief transitions, and the motion candidates in a stream can contain these transitions. Hence, the difference between a complete motion and motion candidates with missing or extra segments needs to be captured.

- **Different motions may follow similar trajectories but in different directions:** For example Figure 14 illustrates a clockwise circle and a counter clockwise circle which follow a similar trajectory but may produce two different results.

These challenges show that a solution to the gesture recognition problem is non-trivial. For a gesture recognition system we first need to segment the data, i.e., we need to identify the start and end of the data stream. We can achieve this manually with the SunSPOT nodes by holding and

Figure 13. Variations in acceleration data from the X, Y and Z axis produced by the motion of three similar gestures with unique character outcomes

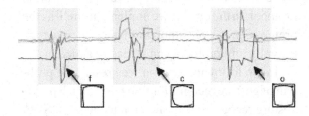

Figure 14. Variations in acceleration data from X, Y & Z axis produced by similar trajectories in different directions

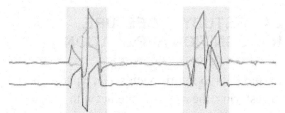

Clockwise Circle   Counter-clockwise Circle

releasing a button to explicitly mark the beginning and end of a gesture. To recognize the human motion in the stream we can use a Hidden Markov Model (HMM) (Baum & Petrie, 1966). Hidden Markov is defined as a set of states of which one state is the initial state, a set of output symbols, and a set of state transitions. Each state transition is represented by the state from which the transition starts, the state to which transition moves, the output symbol generated, and the probability that the transition is taken. HMM are especially known for their applications in temporal pattern recognition such as speech (Rabiner, 1989). In the context of gesture recognition, each state could represent a set of possible hand positions. The HMM which holds the highest probability of state transitions could be determined as the user's most likely gesture. HMM need to be trained before they can be used for recognition. It is important to determine the appropriate number

Figure 15. Architecture of the gesture recognition system with relation to existing SOS components

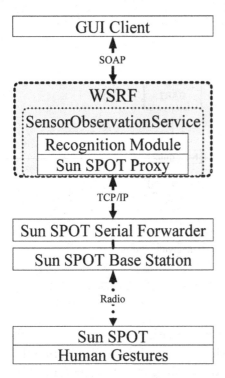

Figure 16. X, Y and Z axes on Sun SPOT accelerometer (SunSPOT, 2008)

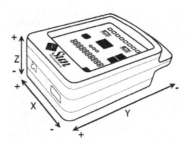

of states for each gesture to maximize accuracy and performance.

We implemented the gesture recognition system in the SOS component of the OSWA. We only use the SOS because we are interested in near-real time observational data from the sensor network. We have no need to schedule the data so we don't use the SPS. Likewise there are no notifications to be sent so we don't use the WNS. The architecture of gesture recognition system with relation to the SOS and its components is illustrated in Figure 15.

We develop a small application in Java which we deploy on the SunSPOT module, the application uses the onboard accelerometer to capture arm movement and forward it to a base station node. Acceleration data from all X, Y and Z axes (Figure 16) is significant as is tilt on all 3 of these axes. These 6 parameters are later analyzed by the Recognition Module. The base station simply acts

as a relay, forwarding packets to the SunSPOT sensor and forwarding the observational results back to a serial port.

One problem with directly using the serial port is that only one application can interact with at any time. A solution to this is the Sun SPOT Serial Forwarder which opens a packet source and lets applications connect over a TCP/IP socket.

To interface with the SOS we implement a connector (SunSPOTConnector) and a recognition (Recognizer) module. The architecture of these components with relation to existing SOS components is illustrated in Figure 17. The SunSPOTConnector interfaces with the SunSPOT Serial Forwarder. The recognition module is invoked by the SunSPOTProxy to analyze the data series observed by the SunSPOT sensors. The SunSOPTObservationFormatter encodes the observational result returned from the sensors into the O&M format, which is later returned to the SWE client. The Recognizer performs the HMM transformation on the raw observation data. Two open source components are used by the Recognizer, the Gesture and Activity Recognition Toolkit (GART) (GART, 2008) and the Hidden Markov Model Toolkit (CU-HTK) ("HTK Speech Recognition", 2008). GART is a prototyping toolkit for the rapid creation of gesture-based applications, developed by the Contextual Computing Group at the Georgia Institute of Technology. It attempts to minimize the complexity of underlying machine learning algorithms and encapsulates functions

Figure 17. Architecture of the SunSPOT proxy and connector

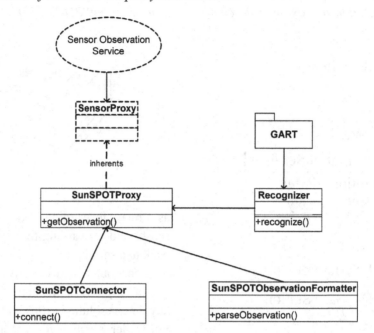

provided by CU-HTK. The CU-HTK is a portable toolkit for building and manipulating HMM, it is primarily used in speech recognition research. CU-HTK was developed in partnership with the Machine Intelligence Laboratory at Cambridge University and Microsoft.

Prior running the experiment, we use the SunSPOT sensors to produce a segmented sample of acceleration data for the HMM. We do this by running a small host side application and repeating a set of predefined training gestures. The resulting training data is maintained as a set of XML files that hold all raw gesture samples along with their names and configuration arguments. All consecutive observational data produced during the use of the system is used to improve on the initial gesture library and build up an experience set. This training need only be done once, during execution the CU-HTK loads the experience set from disk and compares it to the recorded gesture data. The identified gesture data is encapsulated in the O&M encoding by SunSOPTObservation-Formatter and returned to the SWE client

The GUI client is a simple SWE client deployed as a Java desktop application that interfaces with

user. The client uses WSRF to connect to the SOS and when a gesture is identified it prints the result in a text box. The gesture result consist of a letter of the alphabet which mapped to a particular motion recorded by the sensors. Figure 18 is a screenshot of system in action. For illustration purposes the GUI client, SOS instance and Sun SPOT Serial Forwarder are running on a single machine. One gesture is performed and the raw acceleration data series is printed out in the console. The identified gesture is returned and printed in GUI client.

To evaluate the performance of the gesture training and recognition system, test cases were chosen from EdgeWrite ("EdgeWrite Text", 2008) a unistroke text entry method developed by University of Washington. Its benefits include increased physical stability, tactility, accuracy, and the ability to function with minimal sensing. Instead of testing a whole character set in this experiment, we selected two sets of characters. As illustrated in Figure 19, each set has its own characteristics. In set S1, the shape and track of each gesture is unique. The Recognizer should be able to identify these characters easily. In set S2 the six gestures share similarities among each

Figure 18. Gesture recognition system in action

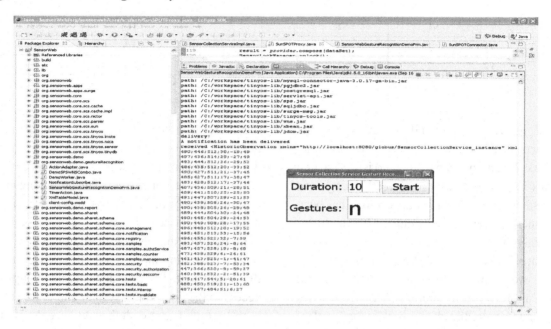

Figure 19: Two character sets

other. When compared to S1 we expect the level of recognition accuracy for S2 to be lower. We want to know the difference of accuracy between these two gesture sets.

Because the HMM is a supervised learning algorithm more training samples will lead to a higher recognition accuracy. So for each gesture set we train two versions of the sample sets:

- **Version 1(V1):** 5 samples per gesture. User specific, trained by one developer
- **Version 2(V2):** 10 samples per gesture. Non user specific (relative), trained by two individuals each taking turns in producing gestures

Compared with V1, V2 contains more training samples, so it should get higher accuracy result.

When we run each of these sample sets on the two gesture sets we get a total of four training sets, these are illustrated in Table 1.

10 gestures were performed on each of the training sets to evaluate the performance of the system. Table 2 depicts a summary of the accuracy after the experiment. Generally, the recognition accuracy of S1 is higher than S2, which is to be expected because characters in S1 are unique. V2 achieves higher accuracy for both character sets because it contains more training samples than V1.

Tables 3, 4, 5 & 6 are confusion matrices which expand on the recognition accuracy detail for each of the four training tests in Table 2. A confusion matrix is used in machine learning to illustrate correct and incorrect classification results. For each confusion matrix the x-axis represents the

Table 1. Improved Sample matrix

	S1	S2
V1	N1	N3
V2	N2	N4

Table 2. Summary of accuracy

	S1	S2
V1	85%	75%
V2	96%	78%

actual gesture performed and the y-axis represents the recognized gesture by system. For example, in Table 3, cell **nn** (shadowed area) is 70% which means 70% of gesture **n** was correctly recognized by system. On the other hand, gesture **n** has been incorrectly recognized as **a** and **g** at 20% and 10% respectively.

Statistically, the HMM recognition engine works well with a minimum of 75% accuracy. Using more training samples, it easily reaches 96%. As we predicted, the higher the level of training samples, the greater the level of accuracy we can expect from the system. The more motion is recorded about one particular user, the more the system can recognize their actions. In this example we use letters of the alphabet because

they offer a suitable amount of complexity in gesture variance. The accuracy of results produced by alphabet set gives us confidence in the performance of our system. The core concept of gesture recognition can be expanded to a variety of deployment scenarios. A user can use perform hand gestures during a presentation which may result in changing slides, or initializing a multimedia component. The user does not need physical access to the computing system and is free to be mobile and interact with the audience during the course of the presentation.

In this experiment, we successfully implemented a gesture recognition system using Sun-SPOT sensors, a machine learning algorithm and OSWA. The SOS was used to retrieve real time

Table 3. Confusion matrix of N1

85%	a	g	n	t	x	s
a	70%	30%	0	0	0	0
g	0	100%	0	0	0	0
n	20%	10%	70%	0	0	0
t	0	20%	0	80%	0	0
x	0	10%	0	0	90%	0
s	0	0	0	0	0	100%

Table 4. Confusion matrix of N2

96%	a	g	n	t	x	s
a	100%	0	0	0	0	0
g	0	100%	0	0	0	0
n	0	0	100%	0	0	0
t	0	0	0	100%	0	0
x	10%	0	0	0	90%	0
s	0	10%	0	0	0	90%

Table 5. Confusion matrix of N3

75%	o	c	f	g	q	Ø
o	70%	0	0	0	20%	10%
c	0	30%	0	50%	0	10%
f	0	0	90%	10%	0	0
g	0	0	0	100%	0	0
q	10%	0	0	40%	60%	0
Ø	0	10%	0	0	0	100%

Table 6. Confusion matrix of N4

78%	o	c	f	g	q	Ø
o	80%	0	0	0	20%	0
c	0	90%	10%	0	0	0
f	0	0	90%	0	10%	0
g	0	0	0	40%	60%	0
q	10%	0	0	20%	80%	0
Ø	0	10%	0	0	10%	90%

observation data from the SunSPOT sensors. The raw observation results were transformed using a Gesture Recognition and a HMM toolkit and forwarded to a SWE client for visualization. Additional development work which was required to implement the application consisted of: introducing an application specific Connector class to the SunSPOT proxy interface, a data formatter class to encapsulate the application specific acceleration data, a recognition module to perform post-processing on the raw observational data, and the visualization capabilities in the form of a text box on the SWE client. Besides these modifications the SOS provided all the necessary architecture components to fulfill our objective. With the additional components added to the architecture we could run our experiment and measure the accuracy of the gesture recognition algorithm. The algorithm saw improved performance in accuracy with the help of more training samples, reaching 96% accuracy at its peak. The combination of OSWA and gesture recognition has the potential to free the user physical access to a computing system and provide them with an accurate alternative.

CONCLUSION AND FUTURE WORKS

In this chapter we have introduced Sensor Web and the OGC SWE method. Sensor Web provides a conceptual framework where geographically distributed services can provide access to heterogeneous sensor resources regardless of the deployment scenario. The SWE method outlines a set of common data description formats and service interfaces which when implemented can realize the vision of a Sensor Web. Application independent data description formats are important for sharing data from heterogeneous sensor resources among independent deployment scenarios. A common set of Service descriptions encourages the development of services by research organizations and businesses to communicate with one another in

order to achieve cross-organizational collaboration, mutually benefiting stakeholders. OSWA is one implementation of the SWE Method which implements services as stateful web services using WSRF. OSWA is developed in Java and implements all the mandatory operations defined for the SOS, WNS and SPS, along with encodings for SensorML and O&M schema. The SOS provides access to a set of heterogeneous sensors and sensor operating systems including hardware developed by Crossbow running TinyOS, SunSPOT's running Java and NICTOR sensors running Linux. The SPS is built on the architecture of the Gridbus Broker, a mature broker application. A gesture recognition application has been presented in order to demonstrate the functionality of the SOS, a major OSWA component. This case study illustrates the ability of OSWA to meet the needs of almost any deployment scenario. Future developments which we intend to commit resources to include:

- An operator service, capable of hiding hardware implementation details from users. Users could use the operator service to automatically deploy and update applications on the sensor nodes without the need to physically access the sensor network. An overlay network would be deployed on the sensors which would be hardware transparent and capable of fulfilling the demands of the operator service. The overlay network would manage energy efficiency, security and automatic network configuration.
- A GUI IDE providing access to service operations and allowing users to visually construct applications, service plans and sensor deployments. Users could drag-and-drop GUI elements which would result in the generation of code that could automatically be deployed on the sensor network by the operator service. Users could visually construct SPS plans and describe service interactions.
- Data driven workflows, which could be deployed on the overlay network and across

services. Sensor observations could automatically trigger service interactions and perform complex tasks.

• An implementation of the SAS along with the TML encoding.

REFERENCES

Barr, R., Bicket, J. C., Dantas, D. S., Du, B., Kim, T. W. D., Zhou, B., & Sirer, E. G. (2002). On the Need for System-Level Support for Ad hoc and Sensor Networks. *Operating Systems Review*, *36*(2), 15. doi:10.1145/509526.509528

Baum, L. E., & Petrie, T. (1966). Statistical Inference for Probabilistic Functions of Finite State Markov Chains. *Annals of Mathematical Statistics*, *37*(6). doi:10.1214/aoms/1177699147

Botts, M. (2007). *OpenGIS Sensor Model Language (SensorML) Implementation Specification,* Open Geospatial Consortium Inc. Retrieved November 17, 2008, from http://portal.opengeospatial.org/files/?artifact_id=21273

Botts, M., Percivall, G., Reed, C., & Davidson, J. (2007). *OGC Sensor Web Enablement: Overview And High Level Architecture*, Open Geospatial Consortium Inc. Retrieved November 3, 2008, from http://portal.opengeospatial.org/files/?artifact_id=25562

Cox, S. (2007). *Observations and Measurements – Part 1 – Observations schema*, Open Geospatial Consortium Inc. Retrieved November 5, 2008, from http://portal.opengeospatial.org/files/?artifact_id=22466

EdgeWrite. (2008) *EdgeWrite Text Entry*. Retrieved June 6, 2008, from http://depts.washington.edu/ewrite

Fok, C., Roman, G., & Lu, C. (2005). Mobile agent middleware for sensor networks: An application case study. In *Proceedings of. The 4ᵗʰ int Conf. Information Processing in Sensor Networks* (pp. 382-387), UCLA, Los Angles.

GART. (2008) *GART*. Retrieved June 6, 2008 from http://wiki.cc.gatech.edu/ccg/projects/gt2k/gt2k.

Gaynor, M., Moulton, S., Welsh, M., LaCombe, E., Rowan, A., & Wynne, J. (2004). Integrating WSN with the Grid . *IEEE Internet Computing*, *8*, 32–39. doi:10.1109/MIC.2004.18

Heinzelman, W. B., Murphy, A. L., Carvalho, H. S., & Perillo, M. A. (2004). Middleware to support sensor network applications. *IEEE Network*, *18*(1), 6–14. doi:10.1109/MNET.2004.1265828

HTK. (2008) *HTK Speech Recognition Toolkit*. Retrieved June 6, 2008, from http://htk.eng.cam.ac.uk

Li, C., Zheng, S. Q., & Prabhakaran, B. (2007) Segmentation and recognition of motion streams by similarity search..*ACM Trans. Multimedia Comput. Commun. Appl., 3*(3).

Li, S., Son, S., & Stankovic, J. (2003). Event Detection services using data service middleware in distributed sensor sensor networks. In *Proceedings of The 2ⁿᵈ Int. Workshop Information Processing in Sensor Networks* (pp.502-517), Palo Alto, CA.

Moodley, D., & Simonis, I. (2006). *New Architecture for the Sensor Web: the SWAP-Framework*, Semantic Sensor Networks Workshop 2006, 5th International Semantic Web Conference ISWC 2006, Athens, GA.

Na, A. (2007). *Sensor Observation Service*, Open Geospatial Consortium Inc. Retrieved November 25, 2008, from http://portal.opengeospatial.org/files/index.php?artifact_id=26667&passcode=xk3nxmxma23st1y6g6hh

Open Geospatial. (2008) *Open Geospatial Consortium Inc.* Retrieved 15 December 2008, from http://www.opengeospatial.org

Rabiner, L.,R., (1989) *A tutorial on hidden Markov models and selected applications in speech recognition.* InProceedings of the IEEE, *77*(2).

Sawada, H., Hashimoto, S. (1997*)* Gesture recognition using an acceleration sensor and its application to musical performance control. *Electronics and Communications in Japan, 80*(5).

Simonis, I. (2004). Sensor Webs: A RoadMap. In *Proceedings. of the 1st Goettinger GI and Remote Sensing Days*, Institute for Geoinformatics, University of Muenster.

Simonis, I. (2007). *OGC Sensor Alert Service Implementation Specification*, Open Geospatial Consortium Inc. Retrieved November 25, 2008, from http://portal.opengeospatial.org/files/index.php?artifact_id=24780&version=1&format=pdf

Suman, N., Jie, L., & Feng, Z. (2006). *Challenges in Building a Portal for Sensors World-Wide.* Paper presented at the First Workshop on World-Sensor-Web: Mobile Device Centric Sensory Networks and Applications, Boulder, CO.

SunSpotWorld. (2008) *SunSpotWorld - Home of Project Sun SPOT.* Retrieved June 6, 2008, from http://www.sunspotworld.com

Tao, V., Liang, S., Croitoru, A., Haider, Z., & Wang, C. (2004). GeoSWIFT: Open Geospatial Sensing Services for Sensor Web. In: Stefanidis, A., Nittel, S. (eds), *GeoSensor Networks* (pp.267-274), Boca Raton, FL: CRC Press.

Tham, C. K., Buyya, R. (2005). SensorGrid: Integrating sensor networks and grid computing. *CSI Communications 29*24-29.

Venugopal, S., Nadiminti, K., Gibbins, H., & Buyya, R. (2008). Designing a Resource Broker for Heterogeneous Grids. *Software, Practice & Experience*, *38*(8), 793–825. doi:10.1002/spe.849

KEY TERMS AND DEFINITIONS

Observations and Measurements: A set of standard models and XML schema as defined in the Sensor Web Enablement method by the Open Geospatial Consortium for describing physical phenomena observed by sensor systems.

Sensor Alert Service: A web service interface definition as defined in the Sensor Web Enablement method by the Open Geospatial Consortium for publishing and subscribing to alerts from sensors.

SensorML: A set of standard models and XML schema as defined in the Sensor Web Enablement method by the Open Geospatial Consortium for describing sensor systems and processes.

Sensor Observation Service: A web service interface definition as defined in the Sensor Web Enablement method by the Open Geospatial Consortium for requesting observations from sensor networks and observation repositories.

Sensor Planning Service: A web service interface definition as defined in the Sensor Web Enablement method by the Open Geospatial Consortium for scheduling and planning observational requests to sensor networks.

Sensor Web: The combination of sensor networks and a service oriented architecture, so that sensors are viewed as resources which can be controlled and accessed over the World Wide Web.

TML: An XML schema and encoding as defined in the Sensor Web Enablement method by the Open Geospatial Consortium for describing real-time streaming data recorded by transducers.

Web Notification Service: A web service interface definition as defined in the Sensor Web Enablement method by the Open Geospatial Consortium for the transmission of messages between SWE services.

Compilation of References

Aamodt, K. (2006). CC2431 Location Engine. *Application Note AN042*, Texas Instruments

Abbas, C. B., Gonzalez, R., Cardenas, N., & Villalba, L. J. (2007). A proposal of a Wireless Sensor Network Routing Protocol. *IFIP International Federation for Information Processing, Volume 245, Personal Wireless Communications*. 410–422. Boston: Springer.

Aboelaze, M., & Aloul, F. (2005, March). *Current and Future Trends in Sensor Networks: A Survey.* Paper presented at the Second IFIP International Conference on Wireless and Optical Communications Networks (WOCN), Dubai, UAE.

Abrams, Z., Goel, A., & Plotkin, S. (2004). Set *k*-cover algorithms for energy efficient monitoring in wireless sensor networks. In *Proceedings IPSN* (pp. 424-432). Berkeley, CA.

Adamou, M., Lee, I., & Shin, I. (2001, December). *An energy efficient real-time medium access control protocol for wireless ad-hoc networks.* Paper presented at WIP session of IEEE Real-time systems symposium, London.

Adlakha, S., & Srivastava, M. (2003). Critical density threshold for coverage in wireless sensor networks. In *Proceedings IEEE WCNC* (pp. 1615-1620). New Orleans, LA: IEEE.

Aggarwal, C., & Yu, P. (2008). *Privacy-Preserving Data Mining Models and Algorithms.* New York: Springer Science Business Media, LLC.

Ahmed, N., Kanhere, S., & Jha, S. (2005). The Holes Problem in Wireless Sensor Networks: A Survey. *Mobile Computing and Communications Review ACM, 9*(2), 102–110.

Ahn, G. S., Hong, S. G., Miluzzo, E., Campbell, A. T., & Cuomo, F. (November, 2006). Funneling-MAC: a localized, sink-oriented MAC for boosting fidelity in sensor networks. In *Proceedings of The 4th ACM Conference on Embedded Networked Sensor Systems (SenSys 2006)*, (pp. 293-306). Boulder, CO.

Aho, A., & Johnson, S. (1976). Optimal Code Generation for Expression Trees. *Journal of the ACM, 23*(3), 488–501. doi:10.1145/321958.321970

Ai, J., & Abouzeid, A. (2006). Coverage by directional sensors in randomly deployed wireless sensor networks. *Journal of Combinatorial Optimization, 11*(1), 21–41. doi:10.1007/s10878-006-5975-x

Aivaloglou, E., Gritzalis, S., & Skianis, C. (2007). Towards a flexible trust establishment framework for sensor networks. *Telecommunication Systems, 35*(3), 207–213. doi:10.1007/s11235-007-9049-x

Akkaya, K., & Younis, M. (2003, May 19-22). An energy-aware QoS routing protocol for wireless sensor networks. In *Proceedings of the 23rd International Conference on Distributed Computing Systems Workshops, 2003.* (pp 710-715).

Akkaya, K., & Younis, M. (2004). Energy-aware delay-constrained routing in wireless sensor networks. *International Journal of Communication Systems, 17*, 663–687. doi:10.1002/dac.673

Akkaya, K., & Younis, M. (2005). A survey on routing protocols for wireless sensor network. *Journal on*

Ad Hoc Networks (Elsevier), 3, 325–349. doi:10.1016/j. adhoc.2003.09.010

Akkaya, K., Younis, M., & Youssef, W. (2005). Sink repositioning for enhanced performance in wireless sensor networks. *Elsevier Computer Networks Journal, 49*, 512–542.

Akkaya, K., Younis, M., & Youssef, W. (2007). Positioning of base stations in wireless sensor networks. *IEEE Communications Magazine, 45*(4), 96–102. doi:10.1109/ MCOM.2007.343618

Akyildiz, I. F., & Erich, P. (2006). Stuntebeck.Wireless underground sensor networks: Research challenges. *Journal on Ad Hoc Networks (Elsevier), 4*, 669–686. doi:10.1016/j.adhoc.2006.04.003

Akyildiz, I. F., Pompili, D., & Melodia, T. (2005). Underwater acoustic sensor networks: research challenges. *Ad Hoc Networks, 3*, 257–279. doi:10.1016/j. adhoc.2005.01.001

Akyildiz, I. F., Su, W., Sankarasubramaniam, Y., & Cayirci, E. (2002b). A survey on sensor networks. *IEEE Communications Magazine, 40*, 102–116. doi:10.1109/ MCOM.2002.1024422

Akyildiz, I. F., Su, W., Sankasubramaniam, Y., & Cayirci, E. (2002a). Wireless Sensor Networks: A Survey. *Journal of Computer Networks, 38*, 393–422. doi:10.1016/ S1389-1286(01)00302-4

Alam, S., & Haas, Z. (2006). Coverage and connectivity in three-dimensional networks. In *Proceedings ACM MobiCom* (pp. 346-357). Los Angeles, CA:ACM.

Alazzawi, L. K., Elkateeb, A. M., & Ramesh, A. (2008, March). *Scalability Analysis for wireless sensor networks routing protocols.* Paper presented at 22nd International Conference on Advanced Information Networking and Applications, GinoWan, Okinawa, Japan

Alex, H., Kumar, M., & Shirazi, B. (2004). MidFusion: middleware for information fusion in sensor network applications. In *Proceedings of International Conference on Intelligent Sensors, Sensor Networks and Information (ISSNIP)*. Australia.

Ali, M., Saif, U., Dunkels, A., Voigt, T., Römer, K., & Langendoen, K. (2006). Medium access control issues in sensor networks. *SIGCOMM Computer Communication Review, 36*(2), 33–36. doi:10.1145/1129582.1129592

Alippi, C., Anastasi, G., Galperti, C., Mancini, F., & Roveri, M. (2007). Adaptive Sampling for Energy Conservation in Wireless Sensor Networks for Snow Monitoring Applications. *IEEE International Conference on Mobile Adhoc and Sensor Systems.* (pp.1-6).

Al-Karaki, J. N., & Kamal, A. E. (2004, December). Routing Techniques in Wireless Sensor Network: a Survey. *IEEE Wireless Communications., 11*(6), 6–28. doi:10.1109/MWC.2004.1368893

Alliance, Z. (2006). *ZigBee Alliance.* Retrieved October, 2006 from, http://www.zigbee.org/imwp/download. asp?ContentID=14128

Alsharabi, N., Fa, L. R., Zing, F., & Ghurab, M. (2008). Wireless sensor networks of battlefields hotspot: Challenges and solutions In *Proceedings from WiOPT'08: The 6th International Symposium on Modeling and Optimization in Mobile, Ad Hoc, and Wireless Networks and Workshops*, Berlin, Germany.

Amirijoo, M., & Son, S. H., & Hansson, J. (2007. June). *QoD adaptation for achieving lifetime predictability of WSN nodes communicating over satellite links.* Paper presented in Fourth International Conference on Networked Sensing Systems (INSS).

Ammari, H. M., & Das, S. K. (2008a). Joint *k*-Coverage and Hybrid Forwarding in Duty-Cycled Three-Dimensional Wireless Sensor Networks. In *Proceedings IEEE SECON* (pp. 170-178). San Francisco, CA:IEEE.

Ammari, H. M., & Das, S. K. (2008b). Clustering-based minimum energy *m*-connected *k*-covered wireless sensor networks. In *Proceedings EWSN*, TPC Best Paper Award [Bologna, Italy.]. *Lecture Notes in Computer Science, 4913*, 1–16. doi:10.1007/978-3-540-77690-1_1

Ammari, H. M., & Das, S. K. (2009). Fault tolerance measures for large-scale wireless sensor networks. *ACM Transaction on Autonomous and Adaptive Systems, 4*(1), 2:1-2:28.

Andersin, M., Rosberg, Z., & Zander, J. (1996). Gradual removals in cellular PCS with constrained power control and noise. *Wireless Networks*, 2(1), 27–43. doi:10.1007/BF01201460

Anderson, R. J. (2001). *Security Engineering: A Guide to Building Dependable Distributed Systems*. Indianapolis, IN: Publisher Wiley.

Anderson, R., Chan, H., & Perrig, A. (2004, October). Key infection: smart trust for smart dust. In *ICNP 2004: Proceedings of the 12th IEEE International Conference on Network Protocols*, (pp. 206-215).

Andrew, H., Maja, J. M., & Gaurav, S. S. (2002a). An Incremental Self-Deployment Algorithm for Mobile Sensor Networks. *Autonomous Robots Special Issue on Intelligent Embedded Systems*, 13, 113–126.

Andrew, H., Maja, J. M., & Gaurav, S. S. (2002b). Mobile Sensor Network Deployment using Potential Fields: A Distributed, Scalable Solution to the Area Coverage Problem. *Proceedings of the 6th International Symposium on Distributed Autonomous Robotics Systems* (pp. 299-308). Fukuoka, Japan.

Ansari, J., Ang, T., & Mähönen, P. (2009). Spectrum Agile Medium Access Control Protocol for Wireless Sensor Networks. *Demonstrated at ACM SIGCOMM*. Barcelona, Spain.

Ansari, J., Pankin, D., & Mähönen, P. (2008). Radio-triggered wake-ups with addressing capabilities for extremely low power sensor network applications. In *Proceedings from PIMRC'08: The 19th Personal, Indoor and Mobile Radio Communications Symposium*, Cannes, France.

Ansari, J., Pankin, D., & Mähönen, P. (2008c). Demo abstract: Radio-triggered wake-ups with addressing capabilities for extremely low power sensor network applications. Presented at EWSN'08: *The 5th European Conference on Wireless Sensor Networks*, Bologna, Italy.

Ansari, J., Riihijärvi, J., Mähönen, P., & Haapola, J. (2007). Implementation and Performance Evaluation of nanoMAC: A Low-Power MAC Solution for High Density Wireless Sensor Networks. *Int. Journal of Sensor Networks*, 2(5), 341–349. doi:10.1504/IJSNET.2007.014361

Ansari, J., Zhang, X., & Mähönen, P. (2007a). Demo abstract: Multi-radio medium access control protocol for wireless sensor networks. In *Proceedings from SenSys'07: The 5th ACM Conference on Embedded Networked Sensor Systems*, Sydney, Austrailia.

Ansari, J., Zhang, X., & Mähönen, P. (2008a). Multi-radio Medium Access Control Protocol for Wireless Sensor Networks. In *Proceedings from WEWSN'08: The 1st International Workshop on Energy in Wireless Sensor Networks*, Santorini, Greece.

Ansari, J., Zhang, X., & Mähönen, P. (2008b). Poster Abstract: Traffic Aware Medium Access Control Protocol for Wireless Sensor Networks. *Presented at EuroSys'08*. Glasgow, Scotland.

Aquino, A. L., Figueiredo, C. M., Nakamura, E. F., Frery, A. C., Loureiro, A. A., & Fernandes, A. O. (2008, March). *Sensor Stream Reduction For Clustered Wireless Sensor Networks*. Paper presented at ACM Symposium on Applied Computing 2008 (SAC 2008), Fortaleza, Brazil.

Arampatzis, T., Lygeros, J., & Manesis, S. (2005, June). *A Survey of Applications of Wireless Sensors and Wireless Sensor Networks*. Paper presented at the 13th IEEE Mediterranean Conference on Control and Automation, Limassol, Cyprus.

Aurenhammer, F. (1991). Voronoi Diagrams – A Survey Of A Fundamental Geometric Data Structure. *ACM Computing Surveys*, 23, 345–405. doi:10.1145/116873.116880

Avago Technologies. (n.d.). *HSMS-285x-series surface mount zero bias schottky detector diodes data sheet*. Retrieved December 02, 2008, from, http://www.avagotech.com

Avvenuti, M., Corsini, P., Masci, P., & Vecchio, A. (2006). Increasing the Efficiency of Preamble Sampling Protocols for Wireless Sensor Networks. In *Proceedings from IWCMC'05: The First Mobile Computing and Wireless Communication International Conference*, Amman, Jordan.

Babu, S., Motwani, R., Munagala, K., Nishizawa, I., & Widom, J. (2004). *Adaptive Ordering of Pipelined Stream Filters.* Paper presented at the ACM SIGMOD International Conference on Management of Data, Paris.

Baccelli, F., Bambos, N., & Chan, C. (2006). Optimal Power, Throughput and Routing for Wireless Link Arrays. In *25th IEEE International Conference on Computer Communications, Joint Conference of the IEEE Computer and Communications Societies (INFOCOM)*. Barcelona, Spain: IEEE.

Bachir, A., Barthel, D., Heusse, M., & Duda, A. (2006). Micro-frame preamble MAC for multihop wireless sensor networks. In *Proceedings from ICC '06: The IEEE International Conference on Communications 2006*, Istanbul, Turkey.

Bahl, P., & Padmanabhan, V. N. (2000, March). RADAR: An In-Building RF-based User Location and Tracking System. In *Proceedings. of the IEEE Infocom 2000, The Conference on Computer Communications*. Tel-Aviv, Israel.

Bai, X., Kumar, S., Xuan, D., Yun, Z., & Lai, T. H. (2006). Deploying wireless sensors to achieve both coverage and connectivity. In *Proceedings ACM MobiHoc* (pp. 131-142). Florence, Italy:ACM.

Baker, C. R., Armijo, K., Belka, S., Benhabib, M., Bhargava, V., Burkhart, N., et al. (2007, May). *Wireless sensor networks for home health care.* Paper presented at the International Conference on Advanced Information Networking and Applications Workshops (AINAW), Ontario, Canada.

Balenson, D. M., McGrew, D., & Sherman, A. (1999, 15 March). Key management for large dynamic groups: One-way function trees and amortized initialization. In *IRTF SMUG meeting.* (Internet Draft <draft-balenson-groupkeymgmt-oft-00.txt>).

Bannach, D., Lukowicz, P., & Amft, O. (2008). Rapid Prototyping of Activity Recognition Applications. *IEEE Pervasive Computing / IEEE Computer Society [and] IEEE Communications Society, 7*(2), 22–31. doi:10.1109/MPRV.2008.36

Baroudi, U. (2007). EQoSA: Energy And QoS Aware MAC For Wireless Sensor Network. In *Proceeding of the 9th International Symposium on Signal Processing and its Applications, IEEE, 1*(3), 1306-1309

Barr, R., Bicket, J. C., Dantas, D. S., Du, B., Kim, T. W. D., Zhou, B., & Sirer, E. G. (2002). On the Need for System-Level Support for Ad hoc and Sensor Networks. *Operating Systems Review, 36*(2), 15. doi:10.1145/509526.509528

Barrenetxea, G., Ingelrest, F., Schaefer, G., & Vetterli, M. (2005-2006, 2007, 2008, November). The hitchhiker's guide to successful wireless sensor network deployments. In *Proceedings of the 6th ACM Conference on Embedded Network Sensor Systems*. Raleigh, NC, USA, November 05 - 07, 2008 (SenSys '08 pp. 43-56). New York: ACM

Barreto, P., Kim, H., Bynn, B., & Scott, M. (2002). Efficient Algorithms for Pairing-Based Cryptosystems. In *Proceedings of CRYPTO* (pp. 354-368).

Basha, E. A., Ravela, S., & Rus, D. (2005-2007, 2008, November). Model-based monitoring for early warning flood detection. In *Proceedings of the 6th ACM Conference on Embedded Network Sensor Systems*, Raleigh, NC, (SenSys '08, pp. 295-308). New York:ACM

Batcher, K., & Papachristou, C. (1999). Instruction randomization self test for processor cores. In *Proceedings of the VLSI Test Symposium* (pp. 34-40).

Baugher, M., Canetti, R., Dondeti, L., & Lindholm, F. (2002, 28 February). *Group key management architecture.* (Internet Draft -draft-ietf-msec-gkmarch-02.txt-).

Baum, L. E., & Petrie, T. (1966). Statistical Inference for Probabilistic Functions of Finite State Markov Chains. *Annals of Mathematical Statistics, 37*(6). doi:10.1214/aoms/1177699147

Behzad, A., & Rubin, I. (2007). Optimum integrated link scheduling and power control for multihop wireless networks. *IEEE Transactions on Vehicular Technology, 56*(1), 194–205. doi:10.1109/TVT.2006.883734

Bentley, J. L., Sleator, D. D., Tarjan, R. E., & Wei, V. K. (1986). A locally adaptive data compression scheme. *Communications of the ACM, 24*(4), 320–330. doi:10.1145/5684.5688

Bhardwaj, M., & Chandrakasan, A. (2002). Bounding the lifetime of sensor networks via optimal role assignments. In *Proceedings of INFOCOM 2002.*

Bhatnagar, S., Deb, B., & Nath, B. (2001, September). *Service Differentiation in Sensor Networks.* Paper presented at Fourth International Symposium on Wireless Personal Multimedia Communications.

Bhatti, S., Carlson, J., Dai, H., et al. (2005). MANTIS OS: an embedded multithreaded operating system for wireless micro sensor platforms. *Mobile Networks and Applications archive, 10*(4), 563-579

Bhuse, V. S. (2007). Lightweight intrusion detection: A second line of defense for unguarded wireless sensor networks. *PhD thesis, Dept. of Comp. Sci., Western Michigan University.*

Bi, Y., Sun, L., Ma, J., Li, N., Khan, I. A., & Chen, C. (2007, November). HUMS: An Autonomous Moving Strategy for Mobile Sinks in Data-gathering Sensor Networks. *EURASIP Journal on Wireless Communications and Networking.*

Bischoff, U., Strohbach, M., Hazas, M., & Kortuem, G. (2006). *Constraint-based distance estimation in ad-hoc wireless sensor networks.* Paper presented at the European Workshop on Sensor Networks.

Blake, S., Black, D., Carlson, M., Davies, E., Wang, Z., & Weiss, W. (1998, December). *An Architecture for Differentiated Services.* RFC 2475, IETF.

Blaze, M., Feigenbaum, J., Ioannidis, J., & Keromytics. (1999). A. The keynote trust management system. In *RFC2704.*

Blom, R. (1984). An optimal class of symmetric key generation systems. In *Proceedings of Eurocrypt 84.*

Blouin, S. (2003). Finite-state Machine Abstractions of Continuous Systems. Unpublished PhD thesis, Chemical Engineering Department, Queens University, Kingston, Canada.

Blum, B., He, T., Son, S., & Stankovic, J. (2003). *IGF: A state-free robust communication protocol for wireless sensor networks* (Tech. Rep. CS-2003-11). Dept. of Comp. Sci. University of Virginia, USA.

Blumenthal, J., Grossmann, R., Golatowski, F., & Timmermann, D. D. (2007). *Weighted centroid localization in zigbee-based sensor networks.* Paper presented at the IEEE International Symposium on Intelligent Signal Processing. Karlof, C., & Wagner, D. (2003). Secure routing in wireless sensor networks: Attacks and countermeasures. In *Proceedings of the First IEEE International Workshop on Sensor Network Protocols and Applications* (pp. 293-315). Anchorage, AK: ACM Press.

Blundo, C., Santis, A. D., Vaccaro, U., Herzberg, A., Kutten, S., & Yong, M. (1998). Perfectly secure key distribution for dynamic conferences. *Information and Computation, 146*(1), 1–23. doi:10.1006/inco.1998.2717

Bollobás, B. (2006). *The Art of Mathematics: Coffee Time in Memphis*, Cambridge, UK: Cambridge University Press.

Boneh, D., & Franklin, M. (2001). Identify-Based Encryption from the Weil Pairing. In *Proceedings of CRYPTO, 01,* 213–229.

Boneh, D., & Franklin, M. (2003, March). Identify-Based Encryption from the Weil Pairing. *SIAM Journal on Computing, 32*(3), 586–615. doi:10.1137/S0097539701398521

Boneh, D., Lynn, B., & Shacham, H. (2001). Short signatures from the weil pairing. In *Proceedings of Asiacrypt, 2248,* 514–532.

Bonnet, P., Gehrke, J. E., & Seshadri, P. (2001). Towards Sensor Database Systems. In *Proceedings of the Second International Conference on Mobile Data Management* (pp.3-14). London: Springer Press.

Borbash, S. A., & Ephremides, A. (2006). Wireless link scheduling with power control and SINR constraints. *IEEE Transactions on Information Theory, 52*(11), 5106–5111. doi:10.1109/TIT.2006.883617

Botts, M. (2007). *OpenGIS Sensor Model Language (SensorML) Implementation Specification,* Open Geospatial Consortium Inc. Retrieved November 17, 2008, from http://portal.opengeospatial.org/files/?artifact_id=21273

Botts, M., Percivall, G., Reed, C., & Davidson, J. (2007). *OGC Sensor Web Enablement: Overview And High Level Architecture,* Open Geospatial Consortium Inc. Retrieved November 3, 2008, from http://portal.opengeospatial.org/files/?artifact_id=25562

Boukerche, A., & Li, X. (2005). An agent-based trust and reputation management scheme for wireless sensor networks. *48th annual IEEE Global Telecommunications Conference* (pp. 1857–1861). St. Louis, MO: IEEE Press.

Boukerche, A., & Li, X., & EL-Khatib, K. (2007). Trust-based security for wireless ad hoc and sensor networks. *Computer Communications, 30*(11-12), 2413–2427. doi:10.1016/j.comcom.2007.04.022

Boukerche, A., Oliveira, H. A. B., Nakamura, E. F., & Loureiro, A. A. F. (2007). Localization Systems for Wireless Sensor Networks. *IEEE Wireless Communications, 14*(6), 6–12. doi:10.1109/MWC.2007.4407221

Boulis, A., Han, C.-C., & Srivastava, M. B. (2003). Design and Implementation of a Framework for Efficient and Programmable Sensor Networks. In *Proceedings of The First International Conference on Mobile Systems, Applications, and Services, MobiSys* (pp. 187-200).

Boulis, A., Han, C.-C., Shea, R., & Srivastava, M. B. (2007). SensorWare: Programming sensor networks beyond code update and querying. *Pervasive and Mobile Computing, 3*(4), 386–412. doi:10.1016/j.pmcj.2007.04.007

Braden, R., Zhang, L., Berson, S., Herzog, S., & Jamin, S. (1997, September). *Resource reSerVation Protocol (RSVP) version 1, Functional Specification,* RFC 2205, IETF.

Brahme, D., & Abraham, J. A. (1984). Functional testing of microprocessors. *IEEE Transactions on Computers,* (C-33), 475–485. doi:10.1109/TC.1984.1676471

Brar, G., Blough, D., & Santi, P. (2006). Computationally efficient scheduling with the physical interference model for throughput improvement in wireless mesh networks. In M. Gerla, C. Petrioli, & R. Ramjee (Ed.), *Proceedings of the 12th annual international conference on Mobile computing and networking (MOBICOM)* (pp.2-13). Los Angeles: ACM.

Brodlie, K., Carpenter, L. A., Earnshaw, R. A., Gallop, R., Hubbolt, R., Mumford, A. M., et al. (1992). *Scientific Visualization: Techniques and Applications.* New York: Springer-Verlag.

Buchegger, S., & Boudec, J. L. (2002). Performance analysis of the CONFIDANT protocol. In *Proceedings of the 13th ACM Symp. on Mobile Ad Hoc Networking and Computing* (pp. 226–236). Lausanne, Switzerland: ACM Press.

Buettner, M., Yee, G., Anderson, E., & Han, R. (2006). X-MAC: A short preamble MAC protocol for duty-cycled wireless networks. In *Proceedings from Sensys'06: The 4th ACM Conference on Embedded Networked Sensor Systems,* Boulder, CO.

Bulusu, N., Heidemann, J., & Estrin, D. (2000). GPS-less low cost outdoor localization for very small devices. *IEEE Personal Communications Magazine, 7*(5), 28–34. doi:10.1109/98.878533

Burkhart, M., Rickenbach, P., von Wattenhofer, R., & Zollinger, A. (2004). Does Topology Control Reduce Interference? In J. Murai, C. E. Perkins, & L. Tassiulas (Eds.), *Proceedings of the 5th ACM Interational Symposium on Mobile Ad Hoc Networking and Computing (MOBIHOC)* (pp.9-19).Tokyo: ACM.

Burrell, J., Brooke, T., & Bechwith, R. (2004). Vineyard computing: sensor networks in agricultural production. *IEEE Pervasive Computing / IEEE Computer Society [and] IEEE Communications Society, 3*(1), 38–45. doi:10.1109/MPRV.2004.1269130

Buschmann, C., Pfisterer, D., Fischer, S., Fekete, S. P., & Kröller, A. (2005). SpyGlass: A Wireless Sensor Network visualizer. *ACM SIGBED Review, 2*(1), 1–6. doi:10.1145/1121782.1121784

Caccamo, M., Zhang, L. Y., Sha, L., & Buttazzo, G. (2002, December). An Implicit Prioritized Access Protocol for Wireless Sensor Network. In *Proceedings of the IEEE Real-Time Systems Symposium.*

Cagalj, M., Capkun, S., & Hubaux, J. P. (2007, January). Wormhole-Based Anti-jamming Techniques in Sensor Network. *IEEE Transactions on Mobile Computing, 6*(1), 1–15. doi:10.1109/TMC.2007.250674

Callaway, E. H. (2004). *Wireless sensor networks: architectures and protocols.* Boca Raton, FL: Auerbach Publications.

Camenisch, J., & Lysyanskaya, A. (2004). Signature schemes and anonymous credentials from bilinear maps. In . *Proceedings of Crypto, LNCS, 3152,* 56–72.

Camenisch, J., Hohenberger, S., & Pedersen, M. (2007). Batch verification of short signatures. In . *Proceedings of EUROCRYPT, LNCS, 4514,* 246–263.

Camtepe, S. A., & Yener, B. (2004). Combinatorial design of key distribution mechanisms for wireless sensor networks. *Computer Security (ESORICS 2004). Lecture Notes in Computer Science, 3193,* 293–308.

Cao, Q., Yan, T., Stankovic, J., & Abdelzaher, T. (2005). Analysis of target detection performance for wireless sensor network. In *Proceedings DCOSS,* LNCS 3560 (pp. 276-292). Marina del Rey, CA.

Capkun, S., Buttyan, L., & Hubaux, J. P. (2003, Jan.-Mar.). Self-Organized Public Key Management for Mobile Ad Hoc Networks. *IEEE Transactions on Mobile Computing, 2*(1), 52–64. doi:10.1109/TMC.2003.1195151

Cardei, M., & Wu, J. (2006). Energy-efficient coverage problems in wireless ad-hoc sensor networks. *Computer Communications, 29*(4), 413–420. doi:10.1016/j.comcom.2004.12.025

Carle, J., & Simplot, D. (2004). Energy Efficient Area Monitoring by Sensor Networks. *IEEE Computer, 37*(2), 40–46.

Carlos, F., Hernandez, G., & Pablo, H. (2007). Ibarguen-goytia-Gonzalez and etc. Wireless Sensor Networks and Applications: a Survey. *IJCSNS International Journal of Computer Science and Network Security, 7*(3), 264–273.

Carter, S., & Yasinsac, A. (2002, November). Secure Position Aided Ad hoc Routing Protocol. In *Proceedings of the IASTED International Conference on Communications and Computer Networks* (CCN02).

Cassandras, C. G., & Lafortune, S. (1999). *Introduction to Discrete Event Systems.* Bostone, MA: Kluwer Academic Publishers.

Castillo, J. A., Ortiz, A. M., López, V., Olivates, T., & Orozco-Barbosa, L. (2008). WiseObserver: a real experience with wireless sensor networks. In *Proceedings of the 3nd ACM Workshop on Performance Monitoring and Measurement of Heterogeneous Wireless and Wired Networks* (PM2HW2N '08, pp. 23-26). New York: ACM.

Cha, J. C., & Cheon, J. H. (2003). An identity-based signature from gap Diffie- Hellman groups. In *Proceedings of Public Key Cryptography* (pp. 18-30).

Chafekar, D., Kumar, V. S., Marathe, M., Parthasarathy, S., & Srinivasan, A. (2007). Cross-layer latency minimization for wireless networks using SINR constraints. In E. Kranakis, E. M. Belding, & E. Modiano (Eds.), *Proceedings of the 8th ACM Interational Symposium on Mobile Ad Hoc Networking and Computing (MOBIHOC)* (pp.110-119). Montreal, Quebec: ACM.

Chakrabarti, A., Sabharwal, A., & Aazhang, B. (2006, August). Communication Power Optimization in Sensor Network with a Path-Constrained Mobile Observer. [TOSN]. *ACM Transactions on Sensor Networks, 2*(3), 297–324. doi:10.1145/1167935.1167936

Chalermek, I., Ramesh, G., Deborah, E., John, H., & Fabio, S. (2003). *Directed diffusion for wireless sensor networking* (Vol. 11, pp. 2-16). Piscataway, NJ: IEEE Press.

Chan, H., & Perrig, A. (2003). Security and privacy in sensor networks. *IEEE Computer Magazine, 36*(10), 103–105.

Chan, H., Perrig, A., & Song, D. (2003, May). Random key predistribution schemes for sensor networks. In

SP '03: Proceedings of the 2003 IEEE symposium on Security and Privacy (pp. 197–213). Washington, DC: IEEE Computer Society.

Chang, N., Kim, K., & Lee, H. G. (2000). Cycle-accurate energy consumption measurement and analysis: Case study of ARM7TDMI. In *Proceedings of the International Symposium on Low Power Electronics and Design* (pp. 185-190).

Chang, R. S., & Wang, S. H. (2008). Self-Deployment by Density Control in Sensor Networks. *IEEE Transactions on Vehicular Technology, 57*(3), 1745–1755. doi:10.1109/TVT.2007.907279

Chatzigiannakis, I., Kinalis, A., & Nikoletseas, S. (October, 2006). Sink mobility protocols for data collection in wireless sensor networks. In *Proceedings of the International Workshop on Mobility Managementand Wireless Access*, (pp. 52-59). Terromolinos, Spain.

Chaum, D. L. (1988). The dinning cryptographers problem: unconditional sender and recipient untraceability. *Journal of Cryptophyg, 1*(1), 65–75.

Chellappan, S., Wenjun, G., Xiaole, B., Dong, X., Bin, M., & Kaizhong, Z. (2007). Deploying Wireless Sensor Networks under Limited Mobility Constraints. *IEEE Transactions on Mobile Computing, 6*(10), 1142–1157. doi:10.1109/TMC.2007.1032

Chen, A., Kumar, S., & Lai, T. H. (2007). Designing localized algorithms for barrier coverage. In *Proceedings ACM MobiCom* (pp. 75-86). Montreal, Canada: ACM.

Chen, B., Jameson, K., Balakrishnan, H., & Morris, R. (2002). Span: An energy-efficient coordination algorithm for topology maintenance in ad hoc wireless networks. *ACM Wireless Networks, 8*(5), 481–494. doi:10.1023/A:1016542229220

Chen, C., & Chou, P. H. (2008, April 22-24). EcoDAQ: A Densely Distributed, High Bandwidth Wireless Data Acquisition System. In *Proc. of the 7th Intl. Conf. on Information Processing in Sensor Networks*. Information Processing In Sensor Networks (pp. 545-546). IEEE Computer Soc., Wash., DC.

Chen, C., & Ma, J. (2006). MEMOSEN: Multi-radio Enabled MObile Wireless SEnsor Network. In *Proceedings from AINA'06: The IEEE 20th International Conference on Advanced Information Networking and Applications*, Vienna, Austria.

Chen, D., & Varshney, P. K. (2004). QoS Support in Wireless Sensor Networks: A Survey. In *Proceedings of the International Conference on Wireless Networks (ICWN 2004)*, Las Vegas.

Chen, H., Wu, H., Zhou, X., & Gao, C. (2007). Reputation-based trust in wireless sensor networks. *Int. Conference on Multimedia and Ubiquitous Engineering*, (pp. 603–607), Korea: IEEE Computer Society.

Chen, L., & Dey, S. (2001). Software-based self-testing methodology for processor cores. *IEEE Transactions on Computer-Aided Design of Integrated Circuits and Systems, 20*, 369–280. doi:10.1109/43.913755

Chen, L., & Heinzelman, W. B. (2005, March). QoS-aware Routing Based on Bandwidth Estimation for Mobile Ad Hoc Networks, *IEEE Journal on Selected Areas of Communication, Special Issue on Wireless Ad Hoc Networks, 23*(3).

Chen, S., & Nahrstedt, K. (1999, August). Distributed Quality-of-Service Routing in ad-hoc Networks. *IEEE Journal on Selected Areas in Communications, 17*(8).

Chen, S., & Yang, N. (2006). Congestion Avoidance Based on Lightweight Buffer Management in Sensor Networks. *Parallel and Distributed Systems . IEEE Transactions, 17*(9), 934–946.

Chen, W., Chen, L., Chen, Z., & Tu, S. (2005). A Realtime Dynamic Traffic Control System Based on Wireless Sensor Network. In *Proceedings from ICCP'05: The 34th International Conference on Parallel Processing Workshops*, Washington D.C.

Chen, Y., Kleisouris, K., Li, X., Trappe, W., & Martin, R. P. (2006). The robustness of localization algorithms to signal strength attacks: a comparative study. In *Proceedings of International Conference on Distributed Computing in Sensor Systems* (DCOSS, pp 546-563).

Chenard, J.-S., Zilic, Z., Chu, C. Y., & Popovic, M. (2005). Design methodology for wireless nodes with printed antennas. In *Proceedings of the Design Automation Conference* (pp. 291-296).

Cheng Tien, E., & Ruzena, B. (2004). *Congestion control and fairness for many-to-one routing in sensor networks.* Paper presented at the Proceedings of the 2nd international conference on Embedded networked sensor systems, Baltimore, MD.

Cheng, Z., Perillo, M., & Heinzelman, W. B. (2008). General network lifetime and cost models for evaluating sensor network deployment strategies. *IEEE Transactions on Mobile Computing, 7*(4), 484–497. doi:10.1109/TMC.2007.70784

Cheung, H., & Gupta, S. K. (1996). A BIST methodology for comprehensive testing of RAM with reduced heat dissipation. In *Proceedings of the International Test Conference* (pp. 386-395).

Chiang, M. W., Zilic, Z., Radecka, K., & Chenard, J.-S. (2004). Architectures of increased availability wireless sensor network nodes. In *Proceedings of the International Test Conference* (pp. 1232-1241).

Chieh-Yih, W., Andrew, T. C., & Lakshman, K. (2002). *PSFQ: a reliable transport protocol for wireless sensor networks.* Paper presented at the Proceedings of the 1st ACM international workshop on Wireless sensor networks and applications. Atlanta, GA

Chieh-Yih, W., Shane, B. E., & Andrew, T. C. (2003). CODA: congestion detection and avoidance in sensor networks. Paper presented at the *Proceedings of the 1st international conference on Embedded networked sensor systems.* Los Angelas

Chieh-Yih, W., Shane, B. E., Andrew, T. C., & Jon, C. (2005). Siphon: overload traffic management using multiradio virtual sinks in sensor networks. Paper presented at the *Proceedings of the 3rd international conference on Embedded networked sensor systems.*

Chipcon, A. S. (2004). *CC2420 data sheet* (revision 1.4). Oslo, Norway: Chipcon AS.

Chong, C. Y., & Kumar, S. P. (2003). Sensor Networks: Evolution, Opportunities, and Challenges. *Proceedings of the IEEE, 91*(8), 1247–1256. doi:10.1109/JPROC.2003.814918

Chonggang, W., Sohraby, K., Bo, L., Daneshmand, M. A. D. M., & Yueming Hu, A. Y. H. (2006). A survey of transport protocols for wireless sensor networks. *Network, IEEE, 20*(3), 34–40. doi:10.1109/MNET.2006.1637930

Chonggang, W., Sohraby, K., Lawrence, V., Bo Li, A. B. L., & Yueming Hu, A. Y. H. (2006). *Priority-based Congestion Control in Wireless Sensor Networks.* Paper presented at the Sensor Networks, Ubiquitous, and Trustworthy Computing, 2006. IEEE International Conference.

Chor, B., Goldwasser, S., Micali, S., & Awerbuch, B. (1985). Verifiable secret sharing and achieving simultaneity in the presence of faults. In Proceedings of *FOCS.*

Chu, X., Kobialka, T., Durnota, B., & Buyya, R. (2006). *Open Sensor Web Architecture: Core Services.* Paper presented at the Intelligent Sensing and Information Processing.

Claros, D., De Haro, M., Dominguez, M., de Trazegnies, C., Urdiales, C., & Sandoval, F. (2007). Augmented Reality Visualization Interface for Biometric Wireless Sensor Networks, In *LNCS 4507 Computational and Ambient Intelligence* (pp. 1074-1081).

Clausen, T., Adjih, C., Jacquet, P., Laouiti, A., Muhlethaler, A., & Raffo, D. (2003, June). Securing the OLSR Protocol. In *Proceeding of IFIP Med-Hoc-Net.* Dedicated Short Range Communications. (n.d.). *DSRC.* Retreived (n.d.). from, http://grouper.ieee.org/groups/scc32/dsrc/index.html

Conway, J., & Torquato, S. (2006). Tiling, packing, and covering with tetrahedra. In *Proceedings National Academy of Sciences of the United States of America (PNAS), 103*(28), 10612-10617.

Correia, L. H., Macedo, D. F., dos Santos, A. L., & Nogueira, J. M., (2005, April). *Issues on QoS Schemes in Wireless Sensor Networks.* (Tech. Rep. RT.DCC.004/2005). Universidade Federal de Minas Gerais DCC: Brasil, Belo Horizonte.

Cortes, J., Martinez, S., Karatas, T., & Bullo, F. (2004). Coverage control for mobile sensing networks. *IEEE Transactions on Robotics and Automation, 20*(2), 243–255. doi:10.1109/TRA.2004.824698

Cox, S. (2007). *Observations and Measurements – Part 1 – Observations schema*, Open Geospatial Consortium Inc. Retrieved November 5, 2008, from http://portal.opengeospatial.org/files/?artifact_id=22466

Crossbow (2007). *MoteView Users Manual, Revision A.* Retrieved May 2007 from, (PN: 7430-0008-05) http://www.xbow.com

Curino, C., Giani, M., Giorgetta, M., & Giusti, A. (2005). TinyLIME: Bridging Mobile and Sensor Networks through Middleware. In *Proceedings of the 3th IEEE International Conference on Pervasive Computing and Communications* (pp.61-72).

d'Auriol, B. J. (2006). A Finite State Machine Model to Support the Visualization of Complex Dynamic Systems, In *H. Arabnia, The 2006 International Conference on Modeling, Simulation and Visualization Methods* (MSV'06, pp. 304-310)

d'Auriol, B. J. (2009). (paper in preparation). *Multilevel Visualization.*

d'Auriol, B. J., Carswell, P., & Gecsi, K. (2006). A TransDimension Visualization Model for Complex Dynamic System Visualizations, In *H. Arabnia, The 2006 International Conference on Modeling, Simulation and Visualization Methods* (MSV'06. pp. 318-324).

d'Auriol, B. J., Kim, J., Lee, S. Y., & Lee, Y. K. (2007). Orthogonal Organized Finite State Machine Application to Sensor Acquired Information, In V. Malyshkin (Ed.), *9th International Conference on Parallel Computing Technologies* (PaCT-2007): LNCS 4671. Parallel Computing Technologies (pp. 111-118). Pereslavl-Zalessky, Russia: Springer.

d'Auriol, B. J., Nguyen, T., Pham, T., Lee, S. Y., & Lee, Y.-K. (2008, July 14-17). Viewer Perception of Superellipsoid-Based Accelerometer Visualization Techniques, In *Proceedings of the 2008 International Conference on Modeling, Simulation and Visualization Methods* (MSV'08, pp. 129-135), Las Vegas, NV.

d'Auriol, B. J., Niu, Y., Lee, S. Y., & Lee, Y. K. (2009). The Plasma Free Space Optical Model for Ubiquitous System, In *Proceedings of the 3rd International Conference on Ubiquitous Information Management and Communication* (pp. 446-455).

Dabrowski, J. (2003). BiST model for IC RF-transceiver front-end. In *Proceedings of the International Symposium on Defect and Fault Tolerance in VLSI Systems* (pp. 295-302).

Daintree Networks. (2008), Sensor Network Analyzer (SNA): *Standard Edition (Product Data Sheet).* Retrieved (n.d.). from, http://www.daintree.net

Dam, T. V., & Langendoen, K. (2003). An adaptive energy-efficient MAC protocol for wireless sensor networks. In *Proceedings from Sensys'03: The 1st International Conference on Embedded networked sensor systems,* Los Angeles.

Das, A. K., Marks, R. J., II, Arabshahi, P., & Gray, A. (2005). Power controlled minimum frame length scheduling in TDMA wireless networks with sectored antennas. In *24th IEEE International Conference on Computer Communications, Joint Conference of the IEEE Computer and Communications Societies (INFOCOM).* Miami, FL: IEEE.

Datta, A. K., Gradinariu, M., & Patel, R. (2007). Dominating sets-based Self* Minimum Connected Covers of Query Regions in Sensor Networks. New York: Springer.

Datta, A. K., Gradinariu, M., Linga, P., & Raipin-Parvedy, P. (2005). Self-* distributed query region covering in sensor networks. In *Proceedings IEEE SRDS* (pp. 50-59). Orlando, FL:IEEE.

Deb, B., Bhatnagar, S., & Nath, B. (2003). *ReInForM: reliable information forwarding using multiple paths in sensor networks.* Paper presented at the Local Computer Networks, 2003. LCN '03. In Proceedings of 28th Annual IEEE International Conference.

Deb, B., Bhatnagar, S., & Nath, B. (2003a, September). *Information Assurance in Sensor Networks.* Paper presented at 2nd International ACM Workshop on Wireless Sensor Networks (WSNA), San Diego CA.

DeCew & Judith. (2006). Privacy. In Zalta, E. N., (ed.), *The Stanford Encyclopedia of Philosophy (Fall 2006 Edition)*, Stanford, CA: Metaphysics Research Lab, CSLI, Stanford University.

Deering, S. (1989). *Scalable Multicast Routing Protocol*. Unpublished doctoral dissertation, Stanford University.

Delicato, F. C., Pires, P. F., Rust, L., Pirmez, L., & Ferreira, J. (2005). *Reflective middleware for wireless sensor networks*. Paper presented at 20th Annual ACM Symposium on Applied Computing (ACM SAC), USA.

Delin, K. A. (2002). The Sensor Web: A Macro-Instrument for Coordinated Sensing. *Sensors, 2,* 270–285. doi:10.3390/s20700270

Demirkol, I., Ersoy, C., & Alagoz, F. (2006). MAC Protocols for Wireless Sensor Networks: a Survey. *IEEE Communications Magazine, 44*(4), 115–121. doi:10.1109/MCOM.2006.1632658

Deng, H., Mukherjee, A., & Agrawal, D. (2004, April). Threshold and Identity-Based Key Management and Authentication for Wireless Ad Hoc Networks. In *Proceedings of the Int'l Conf. Information Technology: Coding and Computing (ITCC '04)*.

Deng, J., Han, R., & Mishra, S. (2004). Countermeasures against traffic analysis attacks in wireless sensor networks (Tech. Report CU-CS-987-04), Comp. Sci. Dept, University of Colorado, Boulder.

Dermibas, M. (2005). *Wireless sensor networks for monitoring of large public buildings (Technical Report)*. Buffalo, NY: University at Buffalo, Department of Computer Science and Engineering.

Deshpande, A., Guestrin, C., Madden, S., Hellerstein, J. M., & Hong, W. (2004). *Model-Driven Data Aquisition in Sensor Networks*. Paper presented at the International Conference on Very Large Data Bases.

Deshpande, A., Guestrin, C., Madden, S., Hellerstein, J. M., & Hong, W. (2005). *Exploiting Correlated Attributes in Acuqisitional Query Processing*. Paper presented at the IEEE International Conference on Data Engineering.

Desmedt, Y., & Frankel, Y. (1989, August). Threshold Cryptosystems. In . *Proceedings of CRYPTO, 89,* 307–315.

Di Pietro, R., Mancini, L., Law, Y. W., Etalle, S., & Havinga, P. (2003, October). Lkhw: a directed diffusion-based secure multicast scheme for wireless sensor networks. In *International conference on parallel processing workshops* (pp. 397-406).

Diffie, W., & Hellman, M. E. (1976). New directions in cryptography. *IEEE Transactions on Information Theory, 22*(6), 644–654. doi:10.1109/TIT.1976.1055638

Djenouri, D., Khelladi, L., & Badache, A. (2005). A survey of security issues in mobile ad hoc and sensor networks. *IEEE Communications Surveys and Tutorials, 7*(4), 2–28. doi:10.1109/COMST.2005.1593277

Djukic, P., & Valaee, S. (2006). *WLC12-4: Reliable and Energy Efficient Transport Layer for Sensor Networks.* Paper presented at the Global Telecommunications Conference, 2006. GLOBECOM '06. IEEE.

Dong, Q., Liu, D., & Ning, P. (2008). Pre-Authentication Filters: Providing DoS Resistance for Signature-Based Broadcast Authentication in Wireless Sensor Networks. In *Proceedings of ACM Conference on Wireless Network Security* (WiSec).

Dorf, R. C. (1974). *Modern Control Systems, Second Edition*. Reading, MA: Addison Wesley.

Douceur, J. R. (2002, March). The Sybil Attack. In *Proceedings of the First Int'l Workshop Peer-to-Peer Systems (IPTPS '02)* (pp. 251-260).

Dousse, O., Mannersalo, P., & Thiran, P. (2004). Latency of wireless sensor networks with uncoordinated power saving mechanisms. In *Proceedings of the ACM MobiHoc 2004*.

Du, W., Deng, J., Han, Y. S., & Varshney, P. K. (2004). A key management scheme for wireless sensor networks using deployment knowledge. *IEEE INFOCOM, 1,* 586–597. doi:10.1109/INFCOM.2004.1354530

Du, W., Deng, J., Han, Y. S., Varshney, P. K., Katz, J., & Khalili, A. (2005). A pairwise key predistribution scheme for wireless sensor networks. *ACM Transactions on Information and System Security*, *8*(2), 228–258. doi:10.1145/1065545.1065548

Du, W., et al. (2003). A Pairwise Key Predistribution Scheme for Wireless Sensor Networks. In *Proceedings of the 10th ACM Conf. Comp. Commun. Sec.* (pp. 42-51).

Du, X. J., & Chen, H. H. (2008). Security in Wireless sensor networks. *IEEE Wireless Communications*, *15*(4), 60–66. doi:10.1109/MWC.2008.4599222

Duarte, M., & Hu, Y. (2003). Distance based decision fusion in a distributed wireless sensor network. In *Proceedings IPSN* [Palo Alto, CA.]. *Lecture Notes in Computer Science*, *2634*, 392–404. doi:10.1007/3-540-36978-3_26

Dunkels, A., Grönvall, B., & Voigt, T. (2004, November). *Contiki - a Lightweight and Flexible Operating System for Tiny Networked Sensors.* Paper presented at Proceedings of the 29th Annual IEEE International Conference on Local Computer Networks, Tampa, FL.

Durresi, A., Paruchuri, V., Durresi, M., & Barolli, L. (2007). Anonymous routing for mobile wireless ad hoc networks. *International Journal of Distributed Sensor Networks*, *3*(1), 105–117. doi:10.1080/15501320601069846

EdgeWrite. (2008) *EdgeWrite Text Entry.* Retrieved June 6, 2008, from http://depts.washington.edu/ewrite

Edith, C. H. N., Yangfan, Z., Michael, R. L., & Jiangchuan, L. (2006). *Reliable Reporting of Delay-Sensitive Events in Wireless Sensor-Actuator Networks.* Paper presented at the Mobile Adhoc and Sensor Systems (MASS), 2006 IEEE International Conference.

Ekici, E., Gu, Y., & Bozdag, D. (2006). Mobility-Based Communication in Wireless Sensor Networks. *IEEE Communications Magazine*, *44*(7), 56–62. doi:10.1109/MCOM.2006.1668382

ElBatt, T., & Ephremides, A. (2002). Joint scheduling and power control for wireless ad-hoc networks. In *21st IEEE International Conference on Computer Communications, Joint Conference of the IEEE Computer and Communications Societies (INFOCOM)*. New York: IEEE.

Elfes, A. (1989). Using Occupancy Grids for Mobile Robot Perception and Navigation. *IEEE Computer*, *22*(6), 46–57.

El-Hoiydi, A., Decotignie, J. D., Enz, C., & Roux, E. L. (2003). Poster abstract: WiseMAC, an ultra low power MAC protocol for the WiseNET wireless sensor network. In *Proceedings from Sensys'03: The 1st International Conference on Embedded networked sensor systems*, Los Angeles, CA, USA.

Elson, J., & Estrin, D. (2001, April). *Time Synchronization for Wireless Sensor Networks.* Paper presented at the IEEE International Parallel and Distributed Processing Symposium (IPDPS), San Francisco.

Errol, L. L., & Guoliang, X. (2007). Relay Node Placement in Wireless Sensor Networks. *IEEE Transactions on Computers*, *56*(1), 134–138. doi:10.1109/TC.2007.250629

Eschenauer, L., & Gligor, V. D. (2002). A Key-Management Scheme for Distributed Sensor Networks. In *Proceedings of the 9th ACM Conf. Comp. and Commun. Sec.* (pp. 41-47).

Evans, J., Wang, W., & Ewy, B. (2006). Wireless networking security: open issues in trust, management, interoperation, and measurement. [IJSN]. *International Journal of Security and Networks*, *1*(1-2), 84–94. doi:10.1504/IJSN.2006.010825

Faller, N. (1973). An adaptive system for data compression. *Record of the 7th Asilomar Conference on Circuits, Systems, and Computers* (pp. 593-597).

Fan, F., & Biagioni, E. S. (2004). An Approach to Data Visualization and Interpretation for Sensor Networks, In *Proceedings of the 37th Hawaii International Conference on System Sciences* (pp. 1-9).

Fan, Y., Chen, A., Songwu, L., & Zhang, L. (2001). A scalable solution to minimum cost forwarding in large sensor networks. *Tenth International Conference on Computer Communications and Networks.* (pp.304-309).

Fan, Y., Gary, Z., Songwu, L., & Lixia, Z. (2005). Gradient broadcast: a robust data delivery protocol for large scale sensor networks (Vol. 11, pp. 285-298). The Netherlands: Kluwer Academic Publishers.

Fang, W.-W., Qian, D.-P., & Liu, Y. (2008). *Transmission control protocols for wireless sensor networks*. Ruan Jian Xue Bao/Journal of Software, *19*(6), 1439-1451.

Felemban, E., Lee, C. G., Ekici, E., Boder, R., & Vural, S. (2005). *Probabilistic QoS guarantee in reliability and timeliness domains in wireless sensor networks.* Paper presented at the INFOCOM 2005. 24th Annual Joint Conference of the IEEE Computer and Communications Societies, Hongwei, Z., Anish, A., Young-ri, C., & Mohamed, G. G. (2007). *Reliable bursty convergecast in wireless sensor networks* (Vol. 30, pp. 2560-2576). Oxford, UK: Butterworth-Heinemann.

Felemban, E., Lee, C., Ekici, E., Boder, R., & Vural, S. (2005, March). Probabilistic QoS guarantee in Reliability and Timeliness Domains in Wireless Sensor Network. In [Miami, FL.]. *Proceedings of the IEEE INFOCOM, 4*, 2646–2657.

Fiat, A. (1989). Batch RSA. In *Proceedings of Crypto* (pp. 175-185).

Fiedler, M. (1973). Algebraic connectivity of graphs. *Czechoslovak Lathematics Journal*, 298-305.

Fischer, S., Petrova, M., Mähönen, P., & Vöcking, B. (2007) Distributed Load Balancing Algorithm for Adaptive Channel Allocation for Cognitive Radios. In *Proceedings from CrownCom'07: The Second International Conference on Cognitive Radio Oriented Wireless Networks and Communications*, Orlando, USA.

Floyd, S., Jacobson, V., Liu, C., McCanne, S., & Zhang, L. (1997). A Reliable Multicast Framework for Lightweight Sessions and Application Level Framing. *IEEE/ACM Transactions on Networking, 5*(6), 784-803.

Fok, C. L., Roman, G. C., & Lu, C. (2005). Rapid Development and Flexible Deployment of Adaptive Wireless Sensor Network Applications. In *Proceedings of the 24th International Conference on Distributed Computing Systems* (pp.653-662).

Fok, C., Roman, G., & Lu, C. (2005). Mobile agent middleware for sensor networks: An application case study. In *Proceedings of. The 4th int Conf. Information Processing in Sensor Networks* (pp. 382-387), UCLA, Los Angeles.

Foster, I. (2005, May 6). Service-Oriented Science. *Science, 308*, 814–817. doi:10.1126/science.1110411

Frankel, Y., Gemmell, P., MacKenzie, P. D., & Yung, M. (1997). Optimal resilience proactive public-key cryptosystems. In *Proceedings of FOCS*.

Fu, L., Liew, S., & Huang, J. (2008). Joint power control and link scheduling in wireless networks for throughput optimization. In *Proceedings IEEE ICC*. Beijing, China: IEEE.

Fu, L., Liew, S., & Huang, J. (2009). Power controlled scheduling with consecutive transmission constraints: complexity analysis and algorithm design. In *28th IEEE International Conference on Computer Communications, Joint Conference of the IEEE Computer and Communications Societies (INFOCOM)*. Rio de Janeiro, Brazil: IEEE.

Fussen, M. (2004). *Sensor networks: interference reduction and possible applications*. Diploma Thesis, Distributed Computing Group, ETH Zurich, Switzerland.

Gage, D. W. (1992). Command control for many-robot systems. *The Nineteenth Annual AUVS Technical Symposium* (pp. 22-24). Huntsville, Alabama, USA, Garey, M.R., & Johnson, D.S. (1979). *Computers and Intractability: A Guide to the Theory of NP-Completeness*. New York: W.H Freeman and Co.

Gallager, R. G. (1978). Variations on a theme by huffman. *IEEE Transactions on Information Theory, 24*(6), 668–674. doi:10.1109/TIT.1978.1055959

Ganeriwal, S., & Srivastava, M. B. (2004). Reputation-based framework for high integrity sensor networks. *ACM Security for Ad-hoc and Sensor Networks*, (pp. 66–67). New York: ACM Press.

Ganeriwal, S., Balzano, L. K., & Srivastava, M. B. (2008). Reputation-based framework for high integrity sensor networks. *ACM Transaction on Sensor Networks, 4*(3), 1–37. doi:10.1145/1362542.1362546

Gao, T., Massey, T., Selavo, L., Welsh, M., & Sarrafzadeh, M. (2007). *Participatory User Centered Design Techniques for a Large Scale Ad-Hoc Health Information System.* Paper presented at the HealthNet.

Gao, Y., Hou, J. C., & Nguyen, H. (2008). Topology control for maintaining network connectivity and maximizing network capacity under the physical model. In *27th IEEE International Conference on Computer Communications, Joint Conference of the IEEE Computer and Communications Societies (INFOCOM)*. Phoenix, AZ: IEEE.

GART. (2008) *GART*. Retrieved June 6, 2008 from http://wiki.cc.gatech.edu/ccg/projects/gt2k/gt2k.

Gaynor, M., Moulton, S., Welsh, M., LaCombe, E., Rowan, A., & Wynne, J. (2004). Integrating WSN with the Grid. *IEEE Internet Computing*, *8*, 32–39. doi:10.1109/MIC.2004.18

Gelernter, D. (1985). Generative Communication in Linda. *ACM Transactions on Programming Languages and Systems*, *7*(1), 80–112. doi:10.1145/2363.2433

Gennaro, G., Jarecki, S., Krawczyk, H., & Rabin, T. (1996). Robust and efficient sharing of rsa functions. In *Proceedings of CRYPTO*.

Gibbons, P. B., Karp, B., Ke, Y., Nath, S., & Seshan, S. (2003). IrisNet: An Architecture for a World-Wide Sensor Web. *IEEE Pervasive Computing / IEEE Computer Society [and] IEEE Communications Society*, *2*(4), 22–33. doi:10.1109/MPRV.2003.1251166

Gil, P., Maza, I., Ollero, A., & Marron, P. (2007). Data Centric Middleware for the Integration of Wireless Sensor Networks and Mobile Robots. In *Proceedings of the 7th Conference on Mobile Robots and Competitions*, Paderne, Portugal.

Girod, L., Elson, J., Cerpa, A., Stathopoulos, T., Ramanathan, N., & Estrin, D. (2004) EmStar: a Software Environment for Developing and Deploying Wireless Sensor Networks, In *Proceedings of the USENIX General Track* (pp. 283-296).

Gohring, N. (2008, July). *Networkworld*. Retrieved July, 2008 from, http://www.networkworld.com/news/2008/072908-microsoft-shows-off-data-center-monitoring.html

Goldsmith, A. J., & Wicker, S. B. (2002). Design challenges for energy-constrained ad hoc wireless networks.

IEEE Wireless Communications, *9*(4), 8–27. doi:10.1109/MWC.2002.1028874

Golomb, S. W. (1966). Run-length encodings. *IEEE Transactions on Information Theory*, *12*(3), 399–401. doi:10.1109/TIT.1966.1053907

Gong, H., Liu, M., Mao, Y., Chen, L., & Xie, L. (2005) Traffic adaptive MAC protocol for wireless sensor network. In *Proceedings from ICCNMC'05: The 2005 International Conference on Computer Networks and Mobile Computing*, Zhangjiajie, China.

Gouda, M. G., & Jung, E. (2004, March). Certificate Dispersal in Ad-Hoc Networks. In *Proceedings of the 24th IEEE Int'l Conf. Distributed Computing Systems (ICDCS '04)*.

Goussevskaia, O., Halldorsson, M., Wattenhofer, R., & Welzl, E. (2009). Capacity of arbitrary wireless networks. In *28th IEEE International Conference on Computer Communications, Joint Conference of the IEEE Computer and Communications Societies (INFOCOM)*. Rio de Janeiro, Brazil: IEEE.

Goussevskaia, O., Oswald, Y. A., & Wattenhofer, R. (2007). Complexity in geometric SINR. In E. Kranakis, E. M. Belding, & E. Modiano (Eds.), *Proceedings of the 8th ACM Interational Symposium on Mobile Ad Hoc Networking and Computing (MOBIHOC)* (pp.110-119). Montreal, Quebec: ACM.

GreenOrbs. Retrieved, from http://www.greenorbs.org/

Gu, L., & Stankovic, J. (2005). Radio-Triggered Wake-Up for Wireless Sensor Networks. *Real-Time Systems*, *29*, 157–182. doi:10.1007/s11241-005-6883-z

Gu, L., & Stankovic, J. A. (2006). *t-kernel: Providing Reliable OS Support to Wireless Sensor Networks*. Paper presented at the ACM Conference on Embedded Networked Sensor Systems, Boulder, CO.

Gu, Y., & He, T. (2007). Data Forwarding in Extremely Low Duty-Cycle Sensor Networks with Unreliable Communication Links. In *Proceedings of the ACM International Conference on Embedded Networked Sensor Systems* (pp. 321-334).

Gu, Y., Hwang, J., He, T., & Du, D. H.-C. (2007). μSense: A Unified Asymmetric Sensing Coverage Architecture for Wireless Sensor Networks. In *IEEE International Conference on Distributed Computing Systems* (pp. 8-8).

Guo, X. (2004). Broadcasting for Network Lifetime Maximization in Wireless Sensor Networks. In *Proceedings of the Annual IEEE Communications Society Conference on Sensor and Ad Hoc Communications and Networks* (pp. 352-358).

Gupta, G., & Younis, M. (2003). Fault-tolerant clustering of wireless sensor networks. *2003 IEEE Wireless Communications and Networking, 3,* 1579-1584.

Gupta, H., Zhou, Z., Das, S. R., & Gu, Q. (2006). Connected sensor cover: Self-organization of sensor networks for efficient query execution. *IEEE/ACM TON, 14*(1), 55-67.

Gupta, P., & Kumar, P. R. (2000). The capacity of wireless networks. *IEEE Transactions on Information Theory, 46*(2), 388–404. doi:10.1109/18.825799

Hadim, S., & Mohamed, N. (2006). Middleware Challenges and Approaches for Wireless Sensor Networks. *IEEE Distributed Systems Online, 7(3),* March 2006.

Hajek, B. E., & Sasaki, G. H. (1988). Link scheduling in polynomial time. *IEEE Transactions on Information Theory, 34*(5), 910–917. doi:10.1109/18.21215

Halder, A., Bhattacharya, S., Srinivasan, G., & Chatterjee, A. (2005). A system-level alternate test approach for specification test of RF transceivers in loopback mode. In *Proceedings of the International Conference on VLSI Design* (pp. 289-294).

Halldorsson, M., & Wattenhofer, R. (2009). Wireless communication is in APX. In *36th International Colloquium on Automata, Languages and Programming (ICALP)*, Rhodes, Greece: Springer-Verlag.

Han, C., Rengaswamy, R. K., Shea, R., Kohler, E., & Srivastava, M. (2005, June). *SOS: A dynamic operating system for sensor networks.* Paper presented at the Third International Conference on Mobile Systems, Applications, And Services (Mobisys), Seattle, WA.

Hardjono, T., Cain, B., & Doraswamy, N. (2000, August). *A framework for group key management for multicast security.* (Internet Draft <draft-ietf-ipsec-gkmframe-work-03.txt>).

Hardjono, T., Cain, B., & Monga, I. (1998, November*). Intra-domain group key management protocol.* (Internet Draft <draft-ietf-ipsec-intragkm-00.txt>).

Hariri, S., & Huitlu, H. (1995). Hierarchical modeling of availability in distributed systems. *IEEE Transactions on Software Engineering,* (21): 50–56. doi:10.1109/32.341847

Harney, & Harder. (1999, March). *Logical Key Hierarchy protocol (LKH).* (Internet Draft. <draft-harney-sparta-lkhp-sec-00>).

Harney, & Muckenhirn. (1997). *Group key management protocol architecture.* IETF RFC2094.

Hartung, C., Balasalle, J., & Han, R. (2005, January). *Node compromise in sensor networks: The need for secure systems* (Tech. Rep.). University of Colorado at Boulder.

Hartung, C., Han, R., Seielstad, C., & Holbrook, S. (June, 2006). FireWxNet: A MultiTiered Monitoring Weather Conditions in Wildland Fire Environments. *International Conference on Mobile Systems, Applications and Services (MobiSys 2006),* (pp. 28-41). Uppsala, Sweden.

He, T., Huang, C., Bium, B. M., Stankovic, J. A., & Abdelzaher, T. (2003). *Range-free localization schemes for large scale sensor networks.* Paper presented at the ACM Mobile Computing Conference.

He, T., Krishnamurthy, S., Stankovic, J. A., Abdelzaher, T., Luo, L., Stoleru, R., et al. (June, 2004). Energy-Efficient Surveillance System Using Wireless Sensor Networks. *International Conference on Mobile Systems, Applications and Services (MobiSys 2006),* (pp. 270-283). Boston, MA.

He, T., Stankovic, J. A., Lu, C., & Abdelzaher, T. (2003, May). SPEED: A stateless protocol for realtime communication in sensor networks. In *Proceedings of the International Conference on Distributed Computing Systems,* Providence, RI.

He, T., Stankovic, J. A., Lu, C., & Abdelzaher, T. F. (2005). A Spatiotemporal Protocol for Wireless Sensor Network. *IEEE Transactions on Parallel and Distributed Systems, 16*(10), 995–1006. doi:10.1109/TPDS.2005.116

He, Y., Li, M., Liu, Y., Zhao, J., Huang, W., & Ma, J. (2008, May 27). *Collaborative Query Processing among Heterogeneous Sensor Networks.* Paper presented at the HeterSANET co-located with the ACM International Symposium on Mobile Ad Hoc Networking and Computing, Hong Kong, China.

Heidari, M. &Pahfavan, K. (2004, June). *Performance Evaluation of Indoor Geolocation Systems Using PROP-Sim Hardware and Ray Tracing Software.* Paper presented at the 2004 International Workshop on Wireless Ad-Hoc Networks. Oulu, Finland

Heinzelman, W. B., Murphy, A. L., Carvalho, H. S., & Perillo, M. A. (2004). Middleware to support sensor network applications. *IEEE Network, 18*(1), 6–14. doi:10.1109/MNET.2004.1265828

Heinzelman, W. R., Chandrakasan, A., & Balakrishnan, H. (2000). Energy-Efficient Communication Protocol for Wireless Microsensor Networks. *Hawaii international Conference on System Sciences: Vol. 8.*

Heinzelman, W., Chandrakasan, A., & Balakrishnan, H. (2002). An application-specific protocol architecture for wireless microsensor networks. *IEEE Transactions on Wireless Communications, 1*(4), 660–670. doi:10.1109/TWC.2002.804190

Heinzelman, W., Kulik, J., & Balakrishnan, H. (1999, August). Adaptive Protocols for Information Dissemination in Wireless Sensor Networks. In *Proceedings of the 5th ACM/IEEE Mobicom Conference*, Seattle, WA.

Hely, D., Bancel, F., Flottes, M. L., & Rouzeyre, B. (2006). Secure scan techniques: A comparison. In *Proceedings of the International On-Line Testing Symposium* (pp. 119-124).

Henricksen, K., & Robinson, R. (2006). A Survey of Middleware for Sensor Networks: State-of-the-art and Future Directions. In *MidSens '06: Proceedings of the international workshop on Middleware for sensor networks* (pp. 60-65). New York: ACM Press.

Henricksen, K., & Robinson, R. (2006, December). *A Survey of Middleware for Sensor Networks: State-of-the-Art and Future Directions.* Paper presented at the ACM MidSens, Melbourne, Australia.

Hernández-Serrano, J. (2008, July). *Contribución a la seguridad de grupo en redes inalámbricas avanzadas.* Unpublished doctoral dissertation, Universitat Politècnica de Catalunya, Spain.

Hernández-Serrano, J., Pegueroles, J., & Soriano, M. (2008, November). Shared self-organized GKM protocol for MANETs. [JISE]. *Journal of Information Science and Engineering, 24*(6), 1629–1646.

Herzberg, A., Jarecki, S., Krawczyk, H., & Yung, M. (1995). Proactive secret sharing, or: How to cope with perpetual leakage. In *Proceedings of CRYPTO.*

Hill, J., & Culler, D. (2002). Mica: A Wireless Platform For Deeply Embedded Networks. *IEEE Micro, 22*, 12–24. doi:10.1109/MM.2002.1134340

Horré, W., Michiels, S., Joosen, W., & Verbaeten, P. (2008). DAVIM: Adaptable Middleware for Sensor Networks. *IEEE Distributed Systems Online, 9*(1), 1–1. doi:10.1109/MDSO.2008.2

Hou, Y. T., Shi, Y., & Sherali, H. D. (2006). Optimal base station selection for anycast routing in wireless sensor networks. *IEEE Transactions on Vehicular Technology, 55*(3), 813–821. doi:10.1109/TVT.2006.873822

HTK. (2008) *HTK Speech Recognition Toolkit.* Retrieved June 6, 2008, from http://htk.eng.cam.ac.uk

Hu, Y., Johnson, D., & Perrig, A. (2002). SEAD: Secure Efficient Distance Vector Routing in Mobile Wireless Ad-Hoc Networks. In *Proceedings of the 4th IEEE Workshop on Mobile Computing Systems and Applications (WMCSA'02)* (pp. 3-13).

Hu, Y., Perrig, A., & Johnson, D. (2002). Ariadne: A Secure On-Demand Routing for Ad Hoc Networks. In *Proceedings of MobiCom 2002*, Atlanta.

Hu, Y., Perrig, A., & Johnson, D. (2003). *Packet Leashes: A Defense against Wormhole Attacks in Wireless Ad Hoc Networks.* Paper presented at the IEEE International Conference on Computer Communications.

Hua, Q.-S. (2009). *Scheduling wireless links with SINR constraints*. PhD Thesis, The University of Hong Kong.

Hua, Q.-S., & Lau, F. C. M. (2006). The scheduling and energy complexity of strong connectivity in ultra-wideband networks. In E. Alba, C.-F. Chiasserini, N. B. Abu-Ghazaleh, & R. Lo Cigno (Eds.), *Proceedings of the 9th International Symposium on Modeling Analysis and Simulation of Wireless and Mobile Systems (MSWiM)* (pp.282-290). Terromolinos, Spain: ACM.

Hua, Q.-S., & Lau, F. C. M. (2008). Exact and approximate link scheduling algorithms under the physical interference model. In M. Segal, & A. Kesselman (Eds.), *Proceedings of the DIALM-POMC Joint Workshop on Foundations of Mobile Computing* (pp.45-54). Toronto, Canada: ACM.

Hua, Q.-S., Wang, Y., Yu, D., & Lau, F. C. M. (2009a). Set multi-covering via Inclusion-Exclusion. *Theoretical Computer Science, 410*(38-40), 3882–3892. doi:10.1016/j.tcs.2009.05.020

Hua, Q.-S., Yu, D., Lau, F. C. M., & Wang, Y. (2009b). Exact algorithms for set multicover and multiset multicover problems. In Y. Dong, D.-Z. Du, O. H. Ibarra (Eds.), *Proceedings of the 20th International Symposium on Algorithms and Computation (ISAAC)* (pp. 34-44). Hawaii: Springer.

Huang, C. F., & Lin, M. (2004). Network coverage using low duty-cycled sensors: random & coordinated sleep algorithms. *Proceedings of the 3rd international symposium on Information processing in sensor networks* (pp. 433-442). Berkeley, California, USA.

Huang, C., & Tseng, Y. (2003). The coverage problem in a wireless sensor network. In *Proceedings ACM WSNA* (pp. 115-121). San Diego, CA: ACM.

Huang, C., Tseng, Y., & Lo, L. (2004). The coverage problem in three-dimensional wireless sensor networks. In *Proceedings IEEE Globecom* (pp. 115-121). Dallas, TX: IEEE.

Huang, C., Tseng, Y., & Wu, H. (2007). Distributed protocols for ensuring both coverage and connectivity of a wireless sensor network. *ACM Transactions on Sensor Networks, 3*(1), 1–24. doi:10.1145/1210669.1210674

Huang, D., Mehta, M., Medhi, D., & Harn, L. (2004). Location-aware key management scheme for wireless sensor networks. In *Proceedings of the 2nd ACM workshop on security of ad hoc and sensor networks*. New York: ACM Press.

Huang, H., Andréa, W. R., & Michael, S. (2005). Dynamic Coverage in Ad-Hoc Sensor Networks. *Mobile Networks and Applications, 10*(1-2), 9–17. doi:10.1023/B:MONE.0000048542.38105.99

Huang, J., Buckingham, J., & Han, R. (2005, September). A Level Key Infrastructure for Secure and Efficient Group Communication in Wireless Sensor Networks. In *Proceedings of the 1st Int'l. Conf. on Security and Privacy for Emerging Areas in Commun. Net.* (pp. 249-260).

Huang, Q., Cukier, J., Kobayashi, H., Liu, B., & Zhang, J. (2003). Fast authenticated key establishment protocols for self-organizing sensor networks. In *WSNA '03: Proceedings of the 2nd ACM international conference on Wireless Sensor Networks and Applications* (pp. 141-150). New York: ACM Press.

Huffman, D. A. (1952). A method for the construction of minimum-redundancy codes. In *. Proceedings of the IRE, 40*(9), 1098–1101. doi:10.1109/JRPROC.1952.273898

Hull, B., Bychkovsky, V., Zhang, Y., Chen, K., Goraczko, M., Miu, A., et al. (2006, October 31-Novermber 3). CarTel: a distributed mobile sensor computing system. In *Proceedings of the 4th international Conference on Embedded Networked Sensor Systems* (Boulder, CO., SenSys '06,pp. 125-138). New York:ACM

Hull, B., Jamieson, K., & Balakrishnan, H. (2004). *Mitigating congestion in wireless sensor networks*. Cambridge MA: MIT Computer Science and Arti_cial Intelligence Laboratory

Hwang, D. D., Lai, B.-C. C., & Verbauwhede, I. (2004, July). Energy-memory-security tradeoffs in distributed sensor networks. In Proceedings of *LNCS third international conference on ad-hoc, mobile, and wireless networks (adhoc-now)* (pp. 70-81). Vancouver, Canada: Springer.

Hwang, J., & Kim, Y. (2004). Revisiting random key pre-distribution schemes for wireless sensor networks. In *SASN '04: Proceedings of the 2nd ACM workshop on Security of Ad hoc and Sensor Networks* (pp. 43-52). New York: ACM Press.

Ibrahim, A. S., Seddik, K. G., & Liu, K. J. R. (2007). Improving Connectivity via Relays Deployment in Wireless Sensor Networks. *IEEE Global Telecommunications Conference*, (pp. 1159–1163).

Ibriq, J., & Mahgoub, I. (2004, July). *Cluster-based routing in wireless sensor networks: Issues and challenges.* Paper presented at International Symposium on Performance Evaluation of Computer and Telecommunication Systems, San Jose, CA.

Ikikardes, T., Hofbauer, M., Kaelin, A., & May, M. (2007). A robust, responsive, distributed tree-based routing algorithm guaranteeing n valid links per node in wireless ad-hoc networks. In *Proceedings from ISCC '07: The 2007 IEEE International Symposium on Computers and Communications.* Aveiro, Portugal.

Ilyas, M. (2003). *The Handbook of Ad Hoc Wireless Networks.* Boca Raton, FL:CRC Press.

Institute of Electrical and Electronics Engineers. (2006). *Wireless medium access control (MAC) and physical layer (PHY) specifications for low rate wireless personal area networks (LRWPANs)* (IEEE standard 802.15.4).

Intanagonwiwat, C., Govindan, R., & Estrin, D. (2000). Directed diffusion: a scalable and robust communication paradigm for sensor networks. *6th Annual international Conference on Mobile Computing and Networking.* (pp. 56-67).

Intanagonwiwat, C., Govindan, R., Estrin, D., Heidemann, J., & Silva, F. (2003). Directed diffusion for wireless sensor networking. *IEEE/ACM Transactions on Networking, 11*(1), 2-16

ITU-T. Telecommunication Standardization Sector of International Telecommunication Union. ITU-T Recommendation E.800. (1994, August).*Telephone Network and ISDN Quality of Service, Network Management and Traffic Engineering: Terms and definitions related to quality of service and network performance including dependability. RetrievedAugust20, 2008,* from http://www.itu.int/rec/T-REC-E.800-199408-I/en

Ivester, M., & Lim, A. (2006). Interactive and Extensible Framework for Execution and Monitoring of Wireless Sensor Networks, In *Proceedings of the First International Conference on Communication System Software and Middleware* (Comsware 2006. pp. 1-10).

Iyengar, R., Kar, K., & Banerjee, S. (2005). Low-coordination Topologies for Redundancy in Sensor Networks. *Proceedings of the 6th ACM International Symposium on Mobile Ad Hoc Networking and Computing* (pp. 332-342), Urbana-Champaign, IL, USA.

Iyer, R., & Kleinrock, L. (2003). *QoS Control for Sensor Networks.* Paper presented at the IEEE International Communications Conference (ICC 2003), May 11-15. Anchorage, AK.

Iyer, Y. G., Gandham, S., & Venkatesan, S. (2005). STCP: a generic transport layer protocol for wireless sensor networks. Paper presented at the Computer Communications and Networks, 2005. ICCCN 2005. In *Proceedings of the 14th International Conference.*

Jangeun, J., & Sichitiu, M. L. (2003). *Fairness and QoS in multihop wireless networks.* Paper presented at the Vehicular Technology Conference, 2003. VTC 2003-Fall. 2003 IEEE 58th.

Jarecki, S., Saxena, N., & Yi, J. H. (2004, October). An Attack on the Proactive RSA Signature Scheme in the URSA Ad Hoc Network Access Control Protocol. In *Proceedings of the Second ACM Workshop Security of Ad Hoc and Sensor Networks (SASN '04).*

Jodogne, S. (2002). Orthogonal finite-state representations. In *Sixth Meeting of the ADVANCE Project*, Edinburgh: Springer.

Joo, C., Lin, X., & Shroff, N. B. (2008). Understanding the capacity region of the greedy maximal scheduling algorithm in multi-hop wireless networks. In *27th IEEE International Conference on Computer Communications, Joint Conference of the IEEE Computer and Communications Societies (INFOCOM).* Phoenix, AZ: IEEE.

Joshi, B. S., & Hosseini, S. H. (1998). Efficient algorithms for microprocessor testing. In *Proceedings of the Annual Reliability and Maintainability Symposium* (pp. 100-104).

Juang, P., Oki, H., Wang, Y., Martonosi, M., Peh, L. S., & Rubenstein, D. (2002). Energy-Efficient Computing for Wildlife Tracking: Design Tradeoffs and Early Experiences with ZebraNet. In *Proceedings of the 10th International Conference on Architectural Support for Programming Languages and Operating Systems* (pp. 96-107).

Jurdak, R. (2007). *Wireless ad hoc and sensor networks: A cross-layer design perspective*. New York: *Springer Publishers*.

Jurdak, R. Ruzzelli., A. & Boivineau, S. (2008) *Octopus: User Documentation*. Retreived (n.d.). from, http://www.csi.ucd.ie/content/octopus-dashboard-sensor-networks-visual-control

Kalantari, M., & Shayman, M. (April, 2006). Design Optimization of Multi-Sink Sensor Networks by Analogy to Electrostatic Theory. *IEEE Wireless Communications and Networking Conference*, (pp. 431-438). Las Vegas, USA.

Kamat, P., Zhang, Y., Trappe, W., & Ozturk, C. (2005). Enhancing source-location privacy in sensor network routing. In *Proceedings of the 25th IEEE Int. conf. on Distributed Computing Systems*, (pp. 599–608), Columbus, OH: IEEE Computer Society.

Kannhavong, B., Nakayama, H., Nemoto, Y., & Kato, N. (October 2007). A Survey of Routing Attacks in Mobile Ad Hoc Networks. In *IEEE Wireless Communications* (pp. 85-91).

Kansal, A., Somasundara, A., Jea, D., Srivastava, M., & Estrin, D. (June, 2004). Intelligent Fluid Infrastructure for Embedded Networks. *2nd ACM USENIX International Conference on Mobile Systems, Applications, and Services (MobiSys'04)*. Boston, USA.

Karl, H. (2003). *A short survey of wireless sensor networks* (Technical Report TKN-03-018). Technical University Berlin, Telecommunication Networks Group.

Karlof, C., Sastry, N., & Wagner, D. (2004). TinySec: a link layer security architecture for wireless sensor networks. In Proceedings of the *2nd Int. Conf. on Embedded networked sensor systems*, (pp. 162–175), Baltimore, MD: ACM Press.

Kaufman, C., Perlman, R., & Speciner, M. (2002). *Network Security Private Communication in a Public World*. Upper Saddle River, NJ: Prentice Hall PTR, A division of Pearson Education, Inc.

Kay, J., & Frolik, J. (2004). *Quality of Service Analysis and Control for Wireless Sensor Networks*. Paper presented at the 1st IEEE International Conference on Mobile Ad-hoc and Sensor Systems. October 24-27. Fort Lauderdale, FL.

Kaya, T., et al. (n.d.). Secure Multicast Groups on Ad Hoc Networks. (Oct. 2003). In *Proceedings of ACM SASN '03* (pp. 94-103).

Khalili, A., Katz, J., & Arbaugh, W. (2003, January). Toward Secure Key Distribution in Truly Ad Hoc Networks. In *Proceedings of IEEE Workshop Security and Assurance in Ad Hoc Networks*.

Khan, M. I., Gansterer, W. N., & Haring, G. (2007). Congestion Avoidance and Energy Efficient Routing Protocol for Wireless Sensor Networks with a Mobile Sink. *Journal of Networks*, *2*(2).

Khoche, A., Volkerink, E., Rivoir, J., & Mitra, S. (2002). Test vector Compression Using EDA-ATE synergies. In *Proceedings of the VLSI Test Symposium* (pp. 97-102).

Kim, S., Pakzad, S., Culler, D., Demmel, J., Fenves, G., Glaser, S., & Turon, M. (2007). Health Monitoring of Civil Infrastructures Using Wireless Sensor Networks. In *Proceedings from IPSN'07: The 6th international conference on Information processing in sensor networks*, New York.

Kim, Y., Perrig, A., & Tsudik, G. (2004). Tree-based group key agreement. *ACM Transactions on Information and System Security*, *7*(1), 60–96. doi:10.1145/984334.984337

Klues, K., Hackmann, G., Chipara, O., & Lu, C. (2007). A component based architecture for power-efficient media access control in wireless sensor networks. In *Proceedings from Sensys'07: The 5th international conference on Embedded networked sensor systems,* Sydney, Austrailia.

Knuth, D. E. (1985). Dynamic huffman coding. *Journal of Algorithms, 6*(2), 163–180. doi:10.1016/0196-6774(85)90036-7

Kohvakka, M., Arpinen, T., Hännikäinen, M., & Hämäläinen, T. D. (2006). High-performance multi-radio wireless sensor networks platform. In *Proceedings from REALMAN'06: The 2nd international workshop on Multi-hop ad hoc networks: from theory to reality,* New York.

Kong, J., Zerfos, P., Luo, H., Lu, S., & Zhang, L. (2001, November). Providing Robust and Ubiquitous Security Support for Mobile Ad Hoc Networks. In *Proceedings of IEEE Int'l Conf. Network Protocols.*

Kosterev, D., Taylor, C., & Mittelstadt, W. (1999). Model validation for the august 10, 1996 WSCC system outage. *IEEE Transactions on Power Systems, 14*(3), 967–979. doi:10.1109/59.780909

Koushanfar, F., Potkonjak, M., & Sangiovanni-Vincentelli, A. (2004). Fault tolerance in wireless sensor networks. In I. Mahgoub and M. Ilyas (Eds.), *Handbook of Sensor Networks* (section VIII), Boca Raton, FL: CRC Press.

Kozat, U. C., Koutsopoulos, I., & Tassiulas, L. (2006). Cross-layer design for power efficiency and QoS provisioning in multi-hop wireless networks. *IEEE Transactions on Wireless Communications, 5*(11), 3306–3315. doi:10.1109/TWC.2006.05058

Kranitis, N., Paschalis, A., Gizopoulos, D., & Xenoulis, G. (2005). Software-based self-testing of embedded processors. *IEEE Transactions on Computers, 54*, 461–475. doi:10.1109/TC.2005.68

Kranitis, N., Paschalis, A., Gizopoulos, D., & Zorian, Y. (2002). Effective software self-test methodology for processor cores. In *Proceedings of Design, Automation and Test in Europe* (pp. 592-597).

Krishna, K., & bin Maarof, A. (2003). A hybrid trust management model for MAS based trading society.

The Int. Arab Journal of Information Technology, 1(1), 60–68.

Krishnamachari, B., & Ordonez, F. (2003). *Analysis of energy-efficient, fair routing in wireless sensor networks through non-linear optimization.* Paper presented at the Vehicular Technology Conference, 2003. VTC 2003-Fall. 2003 IEEE 58th.

Krishnamachari, B., Estrin, D., & Wicker, S. (2002, June). *Modeling Data-Centric Routing in Wireless Sensor Networks.* Paper presented at the IEEE INFOCOM, New York.

Krishnamurthy, L., Adler, R., Buonadonna, P., Chhabra, J., Flanigan, M., Kushalnagar, N., et al. (2005). Design and deployment of industrial sensor networks: experiences from a semiconductor plant and the north sea. In *Proceedings of the 3rd international conference on Embedded networked sensor systems SenSys '05,* (pp 64-75). ACM Press.

Krishnamurthy, L., Adler, R., Buonadonna, P., Chhabra, J., Flanigan, M., Kushalnagar, N., et al. (Nov 2005). Design and Deployment of Industrial Sensor Networks: Experiences from a Semiconductor Plant and the North Sea. *ACM Conference on Embedded Networked Sensor Systems (SenSys 2005),* (pp. 64-75). San Diego, CA.

Krohn, A., Beigl, M., Decker, C., & Riedel, T. S. (2007, June). *Collaborative Time Synchronization in Wireless Sensor Networks.* Paper presented at the International Conference on Networked Sensing Systems, Braunschweig, Germany.

Krstic, A., Lai, W.-C., Cheng, K.-T., Chen, L., & Dey, S. (2002). Embedded software-based self-test for programmable core-based designs. *IEEE Design & Test of Computers, 19*(4), 18–27. doi:10.1109/MDT.2002.1018130

Kulik, J., Heinzelman, W., & Balakrishnan, H. (2002). Negotiation-based protocols for disseminating information in wireless sensor networks. *Wireless Networks Journal, 8*(2/3), 169–185. doi:10.1023/A:1013715909417

Kumar, R. R. H., Cao GH, etc. (2006). *Congestion aware routing in sensor networks.* Retreived (n.d.). from, http://nsrc.cse.psu. edu/tech_report/NAS-TR-0036-2006.pdf

Kumar, S., Lai, T. H., & Arora, A. (2005). Barrier coverage with wireless sensors. In *Proceedings ACM MobiCom* (pp. 284-298). Cologne, Germany: ACM.

Kumar, S., Lai, T. H., & Balogh, J. (2004). On *k*-coverage in a mostly sleeping sensor network. In *Proceedings ACM MobiCom* (pp. 144-158). Philadelphia: ACM

Kumar, S., Lai, T. H., Posner, M. E., & Sinha, P. (2007). Optimal sleep-wakeup algorithms for barriers of wireless sensors. In *Proceedings IEEE BROADNETS* (pp. 327-336). Raleigh Durham, NC: IEEE.

Kuorilehto, M., Hännikäinen, M., & Hämäläinen, T. D. (2005). A Survey of Application Distribution in Wireless Sensor Networks. *EURASIP Journal on Wireless Communications and Networking, 38*(5), 774–788. doi:10.1155/WCN.2005.774

Kyasanur, P., Choudhury, R. R., & Gupta, I. (2006). Smart Gossip: An Adaptive Gossip-based Broadcasting Service for Sensor Networks. In *Proceedings of the IEEE International Conference on Mobile Adhoc and Sensor Systems* (pp. 91-100).

Labiod, H., Afifi, H., & de Santis, C. (2007). *Wi-Fi, Bluetooth, ZigBee and WiMax*. New York: Springer.

Lala, P. K. (2001). *Self-checking and fault-tolerant digital design* (1ˢᵗ ed.). San Francisco: Morgan Kaufmann.

Lalis, S., & Karipidis, A. (2002). *An Open Market-Based Architecture for Distributed Computing.* Paper presented at the IEEE International Parallel & Distributed Processing Symposium.

Langendoen, K. (2007). Medium Access Control in Wireless Networks. *Volume II: Practice and Standards,* 1st ed. New York: Nova Science Publishers.

Laura, G., Andrew, T. C., & Sergio, P. (2005). CONCERT: aggregation-based CONgestion Control for sEnsoR neTworks. Paper presented at the *Proceedings of the 3rd international conference on Embedded networked sensor systems.*

Law, Y. W., Hartel, P., den Hartog, J., & Havinga, P. (2005, Feb). Link-layer Jamming Attacks on S-MAC. In *Proceedings of. 2nd European Workshop on WSN* (pp. 217-225).

Law, Y. W., van Hoesel, L., Doumen, J., Hartel, P., & Havinga, P. (2005, November). Energy-Efficient Link-Layer Jamming Attacks against Wireless Sensor Networks MAC Protocols. *Proceedings of third ACM workshop on Security of ad hoc and sensor networks* (pp. 76-88).

Lazos, L., & Poovendran, R. (2006a). Coverage in heterogeneous sensor networks. In *Proceedings WiOpt* (pp. 1-10). Boston, MA.

Lazos, L., & Poovendran, R. (2006b). Stochastic coverage in heterogeneous sensor networks. *ACM Transactions on Sensor Networks, 2*(3), 325–358. doi:10.1145/1167935.1167937

Lédeczi, Á., Nádas, A., Völgyesi, P., Balogh, G., Kusy, B., & Sallai, J. (2005, November). Countersniper system for urban warfare. *ACM Trans. Sen. Netw., 1*(2), 153–177. doi:10.1145/1105688.1105689

Lee, J., & Stinson, D. (2004a). *A combinatorial approach to key pre-distributed sensor networks.* Retrieved (n.d.). from, http://www.cacr.math.uwaterloo.ca/dstinson/

Lee, J., & Stinson, D. (2004b). *Deterministic key pre-distribution schemes for distributed sensor networks.* Retrieved (n.d.). from,http://www.cacr.math.uwaterloo.ca/dstinson/

Lee, J., Leung, V., Wong, K., Cao, J., & Chan, H. (2007, October). Key management issues in wireless sensor networks: current proposals and future developments. In *IEEE Wireless Communications* (pp. 76-84).

Lee, T. H., Lin, J. C., & Su, Y. T. (1995). Downlink power control algorithms for cellular radio systems. *IEEE Transactions on Vehicular Technology, 44*(1), 89–94. doi:10.1109/25.350273

Lelewer, D. A., & Hirschberg, D. S. (1987). Data compression. *ACM Computing Surveys, 19*(3), 261–296. doi:10.1145/45072.45074

Levis, P., & Culler, D. E. (2002). Maté: a Tiny Virtual Machine For Sensor Networks. In *Proceedings of the 10th International Conference on Architectural Support for Programming Languages and Operating Systems* (pp. 85-95).

Levis, P., Lee, N., Welsh, M., & Culler, D. (2003). TOOSIM: Accurate and Scalable Simulation of Entire TinyOS Applications. In *Proceedings of the First ACM Conference on Embedded Networked Sensor Systems* (SenSys 2003, pp. 126-137).

Levis, P., Madden, S., Polastre, J., Szewczyk, R., Whitehouse, K., Woo, A., et al. (2005). In W. Weber, J. M. Rabaey, and E. Aarts (Eds.) *TinyOS: An operating system for wireless sensor networks* (pp.115-148), Ambient Intelligence. New York: Springer-Verlag.

Levis, P., Patel, N., Culler, D., & Shenker, S. (2004). Trickle: A Self-Regulating Algorithm for Code Propagation and Maintenance in Wireless Sensor Networks. In *Proceedings of the USENIX Symposium on Networked Systems Design and Implementation* (pp. 2-2).

Lewis, F. L. (2004). Wireless Sensor Networks. In Cook, D.J. and Das, S.K. (Eds.), *Smart Environments: Technologies, Protocols, and Applications* (pp. 11-46). New York.John Wiley.

Li, C., Zheng, S. Q., & Prabhakaran, B. (2007) Segmentation and recognition of motion streams by similarity search..*ACM Trans. Multimedia Comput. Commun. Appl., 3*(3).

Li, M., & Liu, Y. (2007a). *Rendered Path: Range-Free Localization in Anisotropic Sensor Networks with Holes.* Paper presented at the ACM International Conference on Mobile Computing and Networking.

Li, M., & Liu, Y. (2007b). *Underground Structure Monitoring with Wireless Sensor Networks.* Paper presented at the IEEE International Conference on Information Processing in Sensor Networks.

Li, M., Koutsopoulos, I., & Poovendran, R. (2007). *Optimal Jamming Attacks and Network Defense Policies in Wireless Sensor Networks.* Paper presented at the IEEE International Conference on Computer Communications.

Li, M., Yan, T., Ganesan, D., Lyons, E., & Shenoy, P. (2007). *Multi-user Data Sharing in Radar Sensor Networks.* Paper presented at the ACM Conference on Embedded Networked Sensor Systems, Sydney, Australia.

Li, S., Li, S., Liao, X., Peng, S., Zhu, P. A. Z. P., & Jiang, J. A. J. J. (2007). Credit based Fairness Control in Wireless Sensor Network. *Paper presented at the Software Engineering, Artificial Intelligence, Networking, and Parallel/Distributed Computing, 2007. SNPD 2007. Eighth ACIS International Conference.*

Li, S., Son, S., & Stankovic, J. (2003). Event Detection services using data service middleware in distributed sensor sensor networks. In *Proceedings of The 2nd Int. Workshop Information Processing in Sensor Networks* (pp.502-517), Palo Alto, CA.

Li, X. S., Yang, Y. R., Gouda, M. G., & Lam, S. S. (2001). Batch rekeying for secure group communications. In *WWW '01: Proceedings of the 10th international conference on World Wide Web* (pp. 525–534). New York: ACM Press.

Li, X. Y., Wan, P. J., & Frieder, O. (2003). Coverage in wireless ad hoc sensor networks. *IEEE Transactions on Computers, 52*(6), 753–763. doi:10.1109/TC.2003.1204831

Li, Y., & Ephremides, A. (2007). A joint scheduling, power control, and routing algorithm for ad hoc wireless networks. *Ad Hoc Networks, 5*(7), 959–973. doi:10.1016/j.adhoc.2006.04.005

Li, Z., Shen, H., & Alsaify, B. (2008). Integrating RFID with Wireless Sensor Networks for Inhabitant, Environment and Health Monitoring. *14th IEEE International Conference on Parallel and Distributed Systems.* (pp.639-646).

Lian, J., Naik, K., Chen, L., Liu, Y., & Agnew, G. (2007). Gradient Boundary Detection for Time Series Snapshot Construction in Sensor Networks. *IEEE Transactions on Parallel and Distributed Systems.*

Liang, H.-G., Hellebrand, S., & Wunderlicht, H.-J. (2001). Two-dimensional test data compression for scan-based deterministic BIST. In *Proceedings of the International Test Conference* (pp. 894-902).

Lim, A. (2002). Support for reliability in self-organizing sensor networks. In *Proceedings of the 5th International Conference on Information Fusion* (pp 973-980) IEEE Press, vol. 2, 2002.

Lin, G., & Xue, G. (1999). Steiner tree problem with minimum number of Steiner points and bounded edge-length. *Information Processing Letters, 69*, 53–57. doi:10.1016/S0020-0190(98)00201-4

Lionel, M. N., & Liu, Y. H. (2008, June). *China's National Research Project on Wireless Sensor Networks.* Paper presented at the International Conference on Ubiquitous and Trustworthy Computing (SUTC), Taichung, Taiwan.

Liu, A., & Ning, P. (2008, April). TinyECC: A Configurable Library for Elliptic Curve Cryptography in Wireless Sensor Networks. In *Proceedings of the 7th International Conference on Information Processing in Sensor Networks (IPSN 2008), SPOTS Track* (pp. 245-256).

Liu, B., Brass, P., & Dousse, O. (2005). Mobility improves coverage of sensor networks. In *Proceedings ACM MobiHoc,* (pp. 300-308). Urbana-Champaign, IL: ACM.

Liu, C. L., & Layland, J. W. (1973). Scheduling algorithms for multiprogramming in hard real time environment. *Journal of the ACM, 20*(1), 46–61. doi:10.1145/321738.321743

Liu, C., Wu, K., Xiao, Y., & Sun, B. (2006). Random coverage with guaranteed connectivity: Joint scheduling for wireless sensor networks. *IEEE Transactions on Parallel and Distributed Systems, 17*(6), 562–575. doi:10.1109/TPDS.2006.77

Liu, D., & Ning, P. (2003a). Efficient distribution of key chain commitments for broadcast authentication in distributed sensor networks. In *Preceedings of the 10th annual network and distributed system security symposium* (pp. 263–276).

Liu, D., & Ning, P. (2003b). Establishing pairwise keys in distributed sensor networks. In *CCS '03: Proceedings of the 10th ACM conference on Computer and Communications Security* (pp. 52–61). New York: ACM Press.

Liu, D., & Ning, P. (2003c). Location-based pairwise key establishments for static sensor networks. In *SASN '03: Proceedings of the 1st ACM workshop on Security of Ad hoc and Sensor Networks* (pp. 72–82). New York: ACM Press.

Liu, D., Ning, P., & Du, W. (2008). Group-based key predistribution for wireless sensor networks. *ACM Trans. Sen. Netw., 4*(2), 1–30. doi:10.1145/1340771.1340777

Liu, H., Xu, W., Chen, Y., & Liu, Z. (2009). Localizing Jammers in Wireless Networks. *To appear in Proceedings of IEEE PerCom International Workshop on Pervasive Wireless Networking (IEEE PWN) (Held in conjunction with IEEE PerCom).* Texas, USA.

Liu, J., Zhao, F., Cheung, P., & Guibas, L. (2004). Apply Geometric Duality to Energy-efficient Non-local Phenomenon Awareness using Sensor Networks. *IEEE Wireless Communications, 11*(8), 62–68.

Liu, K., Abu-Ghazaleh, N., & Kang, K.-D. (2007). Location verification and trust management for resilient geographic routing. *Journal of Parallel and Distributed Computing, 67*(2), 215–228. doi:10.1016/j.jpdc.2006.08.001

Liu, T., & Martonosi, M. (2003). Impala: A Middleware System for Managing Autonomic, Parallel Sensor Systems. In *Proceedings of the 9th ACM SIGPLAN Symposium on Principles and Practice of Parallel Programming* (pp. 107-118).

Liu, Z., Joy, A. W., & Thompson, R. A. (2004). A dynamic trust model for mobile ad hoc networks. In *Proceedings of the 10th IEEE Int. Workshop on Future Trends of Distributed Computing Systems,* (pp. 80–85), Suzhou, China: IEEE Computer Society.

Locher, T., von Rickenbach, P., & Wattenhofer, R. (2008). Sensor networks continue to puzzle: selected open problems. In S. Rao, M. Chatterjee, P. Jayanti, C. S. R. Murthy, & S. K. Saha (Eds.): *9th International Conference on Distributed Computing and Networking (ICDCN)* (pp.25-38). Kolkata, India: Springer.

Lockheed Martin Corporation. (2006). *Buoyant Wire Antenna System, Submarine Communications.* Marion, MA: Lockheed Martin Corporation.

Lou, W., & Fang, Y. (2003). A Survey of Wireless Security in Mobile Ad Hoc Networks: Challenges and Available Solutions. In Chen, X. Huang, X., & Kluwer, D. (Eds.), *Ad Hoc Wireless Networks* (pp. 319-364). Academic Publishers.

Lu, C., Blum, B. M., Abdelzaher, T. F., Stankovic, J. A., & He, T. (2002). RAP: A Real-Time Communication Architecture for Large-Scale Wireless Sensor Networks. In *Proceedings of the IEEE Real-Time and Embedded Technology and Applications Symposium (RTAS 2002)*, San Jose, CA, September 2002.

Lu, G., Krishnamachari, B., & Raghavendra, C. S. (2004, April). An Adaptive Energy-Efficient and Low-Latency MAC for Data Gathering in Wireless Sensor Networks. In *Proceedings of the 18th International Parallel and Distributed Processing Symposium (IPDPS'04)*, 26-30 April 2004. (pp 224)

Lu, K. J., Qian, Y., Rodriguez, D., Rivera, W., & Rodriguez, M. (2007, November). *Wireless Sensor Networks for Environmental Monitoring Applications: A Design Framework*. Paper presented at the IEEE Global Telecommunications Conference, Washington, DC.

Luo, H., & Lu, S. (2004). URSA: Ubiquitous and Robust Access Control for Mobile Ad-Hoc Networks. *IEEE/ACM Transactions on Networking. Vol.12 No.6* (pp. 1049-1063).

Luo, H., Luo, J., Liu, Y., & Das, S. K. (2005). Energy Efficient Routing with Adaptive Data Fusion in Sensor Networks. In *Proceedings of the Joint Workshop on Foundations of Mobile Computing* (pp. 80–88).

Luo, J., Liu, H. Z., Li, R. F., & Bao, L. (2008) QoS-Oriented Asynchronous Clustering Protocol in Wireless Sensor Networks. In *Proceeding of the IEEE Intl Conference on Wireless Communication, Networking and Mobile Networking (WICOM)*, Dalian, China, Oct. 12-14, 2008.

Luo, J., Panchard, J., Piorkowski, M., Grossglauser, M., & Hubaux, J.-P. (June, 2006). MobiRoute: Routing towards a Mobile Sink for Improving Lifetime in Sensor Networks. *2nd IEEE/ACM International Conference on Distributed Computing in Sensor*. San Francisco, CA.

Lutfiyya, H., Molenkamp, G., Katchabaw, M., & Bauer, M. (2001). *Issues in managing soft QoS requirements in distributed systems using a policy-based framework*. Paper presented at 2nd International Workshop on Policies for Distributed Systems and Networks.

Ma, J., Chen, C., & Salomaa, J. P. (2007). WSN for Large Scale Mobile Sensing. *Journal of Signal Processing Systems, 51*(3).

MacKie-Mason, J. K., Osepayshvili, A., Reeves, D. M., & Wellman, M. P. (2004, June 3-7). *Price Prediction Strategies for Market-Based Scheduling*. Paper presented at the 14th International Conference on Automated Planning & Scheduling, Whistler, British Columbia, Canada.

Madden, S. R., Franklin, M. J., Hellerstein, J. M., & Hong, W. (2005). TinyDB: An Acquisitional Query Processing System for Sensor Networks. *ACM Transactions on Database Systems, 30*(1), 122–173. doi:10.1145/1061318.1061322

Madden, S., Franklin, M. J., Hellerstein, J. M., & Hong, W. (2002). TAG: a Tiny Aggregation Service for Ad-hoc Sensor Networks. In *Proceedings of the Symposium on Operating Systems Design and Implementation*.

Mainwaring, A., Polastre, J., Szewczyk, R., Culler, D., & Anderson, J. (2002). *Wireless Sensor Networks for Habitat Monitoring, in WSNA'02*, Atlanta, GA:ACM Press.

Malan, D. J., Welsh, M., & Smith, M. D. (2004). A public-key infrastructure for key distribution in TinyOS based on elliptic curve cryptography. In *Proceedings of 1st IEEE Communications Society Conference on Sensor and Ad Hoc Communications and Networks* (Secon'04).

Manjeshwar, A., & Agrawal, D. P. (2001). TEEN: a routing protocol for enhanced efficiency in wireless sensor networks. *Parallel and Distributed Processing Symposium*. (pp. 2009-2015).

Manjeshwar, A., & Agrawal, D. P. (2002). APTEEN: A Hybrid Protocol for Efficient Routing and Comprehensive Information Retrieval in Wireless Sensor Networks. *Parallel and Distributed Processing Symposium*. (pp. 195-202).

Mao, G., Fidan, B., & Anderson, B. D. O. (2007). Wireless sensor network localization techniques. *Journal of Computer and Telecommunications Networking, 51*(10), 2529–2553.

Mao, Y., Chan, E., Chen, G., & Wu, J. (2006). Energy Efficient Fractional Coverage Schemes for Low Cost

Wireless Sensor Networks. *26th IEEE International Conference on Distributed Computing Systems Workshops.* (pp. 79-79).

Marchand, H., Boivineau, O., & Lafortune, S. (2000). Optimal control of discrete event systems under partial observation (Tech. Rep. CGR-00-10).

Maroti, M., Simon, G., Ledeczi, A., & Sztipanovits, J. (2004). Shooter localization in urban terrain. *IEEE Computer, 37,* 60–61.

Masri, W., & Mammeri, Z. (2007, September). *Middleware for Wireless Sensor Networks: A Comparative Analysis.* Paper presented at the IFIP International Conference on Network and Parallel Computing Workshops, Dalian, China.

Matteo, Z., Giuseppe, A., Claudia, C., & Mario, F. (2005, May). *Performance evaluation of a differential-GPS ground station for high accurate satellite navigation services.* Paper presented at the 2004 IEEE 59th Vehicular Technology Conference. Milan.

Maxim, A. B., & Gaurav, S. S. (2002). *Multi-robot Dynamic Coverage of a Planar Bounded Environment* (Technical report). U.S.A., University of Southern California, Robotic Embedded Systems Laboratory.

Megerian, S., Koushanfar, F., Potkonjak, M., & Srivastava, M. (2001). Exposure in wireless ad-hoc sensor networks. In *Proceedings ACM MobiCom* (pp. 139-150). Rome: ACM.

Megerian, S., Koushanfar, F., Potkonjak, M., & Srivastava, M. (2001). Coverage problems in wireless ad-hoc sensor networks. In *Proceedings IEEE INFOCOM* (pp. 1380-1387). Anchorage, AK: IEEE.

Megerian, S., Koushanfar, F., Potkonjak, M., & Srivastava, M. (2005). Worst and best-case coverage in sensor networks. *IEEE Transactions on Mobile Computing, 4*(1), 84–92. doi:10.1109/TMC.2005.15

Meguerdichian, S., Koushanfar, F., Potkonjak, M., & Srivastava, M. (2001). Coverage Problems in Wireless Ad-Hoc Sensor Networks. *Proceedings of IEEE Infocom,* (pp. 1380-1387).

Meguerdichian, S., Koushanfar, F., Qu, G., & Potkonjak, M. (2001b). *Exposure in Wireless Ad-hoc Sensor Networks. Paper* presented at Mobile Computing and Networking, (pp. 139-150).

Mehuron, W. (1994). Digital Signature Standard (DSS). *U.S. Department of Commerce, National Institute of Standards and Technology (NIST), Information Technology Laboratory (ITL).* FIPS PEB 186.

Menezes, A. J., Oorschot, P. C. V., & Vanstone, S. A. (1997). *Handbook of applied cryptography.* Boca Raton, FL: CRC Press.

Menezes, A., Oorschot, P., & Vanstone, S. (1996). *Handbook of Applied Cryptography.* Boca Raton, FL: CRC Press.

Meng, M., Wu, X., Xu, H., Jeong, B., Lee, S., & Lee, Y. (August, 2007). Energy Efficient Routing in Multiple Sink Sensor Networks. *Fifth International Conference on Computational Science and Applications,* (pp. 561-564). Kuala Lumpur, Malaysia.

Mhatre, V. P., Rosenberg, C., Kofman, D., Mazumdar, R., & Shroff, N. (2005). A minimum cost heterogeneous sensor network with a lifetime constraint. *IEEE Transactions on Mobile Computing, 4*(1), 4–15. doi:10.1109/TMC.2005.2

MicaZ Datasheet Specs. (2008). *Crossbow Technology.* Retrieved November, 11, 2008 from, http://www.xbow.com/Products/Product_pdf_files/Wireless_pdf/MICAz_Datasheet.pdf

Michell, S., & Srinivasan, K. (2004). State based key hop protocol: a lightweight security protocol for wireless networks. In *Proceedings of the 1st ACM international Workshop on Performance Evaluation of Wireless Ad Hoc, Sensor, and Ubiquitous Networks,* (pp. 112-118). Venezia, Italy: ACM Press.

Michiardi, P., & Molva, R. (2002). CORE: A collaborative reputation mechanism to enforce node cooperation in mobile ad hoc networks. In *Proceedings of the 6th IFIP conf. on communications and multimedia security,* (pp. 107–121), Portoroz, Slovenia: Kluwer Academic Publishers.

Mick, J. (2008, July). *DailyTech*. Retrieved July, 2008 from, http://www.dailytech.com/San+Francisco+Pionee rs+Smartphonecoordinated+Parking/article12364.htm

Miller, M., Sengul, C., & Gupta, I. (2005). Exploring the Energy-Latency Tradeoff for Broadcasts in Energy-Saving Sensor Networks. In *Proceedings of the IEEE International Conference on Distributed Computing Systems* (pp.17-26).

Mini, R., Loureiro, A., & Nath, B. (2006). Energy Map Construction for Wireless Sensor Network under a Finite Energy Budget. In *Proceedings of the Seventh ACM/IEEE MSWiM'04*, October 4-6, 2004, Venezia, Italy.

Misra, S., & Xue, G. (2006). Efficient anonymity schemes for clustered wireless sensor networks. *International Journal of Sensor Networks*, *1*(1/2), 50–63. doi:10.1504/ IJSNET.2006.010834

Mittra, S. (1997). Iolus: a framework for scalable secure multicasting. *SIGCOMM Comput. Commun. Rev.*, *27*(4), 277–288. doi:10.1145/263109.263179

Molla, M. M., & Ahamed, S. I. (2006). A Survey of Middleware for Sensor Network and Challenges. In *Proceedings of the 2006 International Conference Workshops on Parallel Processing* (pp. 223-228). Washington, USA: IEEE Computer Society.

Moodley, D., & Simonis, I. (2006). *New Architecture for the Sensor Web: the SWAP-Framework*, Semantic Sensor Networks Workshop 2006, 5th International Semantic Web Conference ISWC 2006, Athens, GA.

Moore, T. (2006). A Collusion Attack on Pairwise Key Predistribution Schemes for Distributed Sensor Networks. In *Proceedings of the 4th Annual IEEE International Conference on Pervasive Computing and Communications Workshops (PERCOMW'06)*, pp. 251-255. IEEE Computer Society.

Moscibroda, T. (2007). The worst-case capacity of wireless sensor networks. In T. F. Abdelzaher, L. J. Guibas, & M. Welsh (Eds.), *Proceedings of the 6th International Conference on Information Processing in Sensor Networks (IPSN)* (pp.1-10). Cambridge, MA: ACM.

Moscibroda, T., & Wattenhofer, R. (2006). The complexity of connectivity in wireless networks. In *25th IEEE International Conference on Computer Communications, Joint Conference of the IEEE Computer and Communications Societies (INFOCOM)*. Barcelona, Spain: IEEE.

Moscibroda, T., Oswald, Y. A., & Wattenhofer, R. (2007). How optimal are wireless scheduling protocols? In *26th IEEE International Conference on Computer Communications, Joint Conference of the IEEE Computer and Communications Societies (INFOCOM)* (pp. 1433-1441). Anchorage, AK: IEEE.

Moscibroda, T., Wattenhofer, R., & Weber, Y. (2006). Protocol design beyond graph-based models. In Proceedings 5th Workshop on Hot Topics in Networks (HotNets-V) (pp.25-30). Irvine, California: ACM.

Moscibroda, T., Wattenhofer, R., & Zollinger, A. (2006). Topology control meets SINR: the scheduling complexity of arbitrary topologies. In S. Palazzo, M. Conti, & R. Sivakumar (Eds.), *Proceedings of the 7th ACM International Symposium on Mobile Ad Hoc Networking and Computing (MOBIHOC)* (pp.310-321). Florence, Italy: ACM.

Mosshammer, R. R., Eickhoff, R., Huemer, M., & Weigel, R. (2008). System topologies and performance evaluation of the RESOLUTION embedded local positioning system. *Elektrotechnik & Informationstechnik*, *125*(10), 347–352. doi:10.1007/s00502-008-0572-6

Mozer, M. C. (2004). *Lessons from an Adaptive House In Smart environments: Technologies, protocols, and applications*. Hoboken, NJ: J. Wiley & Sons.

Mulmuley, K. (1994). *Computational Geometry: An Introduction Through Randomized Algorithms*. New Jersey, USA: Prentice-Hall.

Murphy, A. L., Picco, G. P., & Roman, G.-C. (2001). LIME: A Middleware for Physical and Logical Mobility. In F. Golshani, P. Dasgupta, & W. Zhao (Eds.), *21st International Conference on Distributed Computing Systems* (pp 524-533).

Na, A. (2007). *Sensor Observation Service*, Open Geospatial Consortium Inc. Retrieved November 25,

2008, from http://portal.opengeospatial.org/files/index.php?artifact_id=26667&passcode=xk3nxmxma23st1y6g6hh

Naccache, D., M'Raihi, D., Vaudenay, S., & Raphaeli, D. (1994). Can D.S.A be improved? complexity trade-offs with the digital signature standard. In . *Proceedings of EUROCRYPT, LNCS, 950*, 77–85.

Narasimha, M., Tsudik, G., & Yi, J. H. (2003, November). On the Utility of Distributed Cryptography in P2P and Manets: The Case of Membership Control. In *Proceedings of. IEEE Int'l Conf. Network Protocols*.

Nasipuri, A., & Li, K. (2002). A directionality based location discovery scheme for wireless sensor networks. In Proceedings of the *1st ACM international Workshop on Wireless Sensor Networks and Applications*, (pp. 105-111). Atlanta, Georgia, USA: ACM Press.

Nath, S., & Gibbons, P. B. (2007). Communicating via fireflies: Geographic routing on duty-cycled sensors. In *Proceedings IPSN* (pp. 440-449). Cambridge, MA.

Nath, S., Liu, J., & Zhao, F. (2007). SensorMap for Wide-Area Sensor Webs. *IEEE Computer Magazine, 40*(7), 90–93.

nesC: A Programming Language for Deeply Networked Systems. Retrieved, from the World Wide Web: http://nescc.sourceforge.net/

Ni, L. M., Liu, Y., Lau, Y. C., & Patil, A. P. (2003). LANDMARC: Indoor Location Sensing Using Active RFID. *Wireless Networks, 10*(6), 701–710. doi:10.1023/B:WINE.0000044029.06344.dd

Ni, S.-Y., Tseng, Y.-C., Chen, Y.-S., & Sheu, J.-P. (1999). The Broadcast Storm Problem in a Mobile Ad Hoc Network, In *Proceedings of the Annual ACM International Conference on Mobile Computing and Networking* (pp. 151-162).

Nikolaidis, S., & Laopoulos, T. (2001). Instruction-level power consumption estimation embedded processors low-power applications. *International Workshop on Intelligent Data Acquisition and Advanced Computing Systems: Technology and Applications* (pp. 139-142).

Nojeong, H., & Pramod, K. V. (2003). An Intelligent Deployment and Clustering Algorithm for a Distributed Mobile Sensor Network. *Proceedings of the 2003 IEEE International Conference on Systems, Man & Cybernetics*, vol.5 (pp. 4576-4581).

OceanSense. Retrieved, from the World Wide Web: http://www.cse.ust.hk/~liu/Ocean/

Ohyama, S., & Kobayashi, A. (1996). Local positioning system by means of enclosing signal field. *Sensors and Actuators, A54*, 457–463.

Olariu, S., Xu, Q., Eltoweissy, M., Wadaa, A., & Zomaya, A. Y. (2005). Protecting the communication structure in sensor networks. *International Journal of Distributed Sensor Networks, 1*(2), 187–203. doi:10.1080/15501320590966440

Oldewurtel, F., Ansari, J., & Mähönen, P. (2008). Cross-layer design for distributed source coding in wireless sensor networks. In *Proceedings from SENSORCOMM'08 The Second International Conference on Sensor Technologies and Applications*, Cap Esterel, France.

Open Geospatial. (2008) *Open Geospatial Consortium Inc.* Retrieved 15 December 2008, from http://www.opengeospatial.org

OPNET Modeler Inc. (2008, November). *OPNET Inc.* Retrieved November 11, 2008, from, http://www.opnet.com/solutions/network_rd/modeler.html

Oppermann, I., Stoica, L., Rabbachin, A., Shelby, Z., & Haapola, J. (2004). UWB wireless sensor networks: UWEN-a practical example. *IEEE Communications Magazine, 42*, 527–532. doi:10.1109/MCOM.2004.1367555

Otis, B. (2005). *Ultra-low power wireless technologies for sensor networks*. Ph.D. dissertation, University of California, Berkeley, USA.

Ouyang, Y., Le, Z., Chen, G., Ford, J., & Makedon, F. (2006). Entrapping adversaries for source protection in sensor networks. *2006 Int. Sym. on a World of Wireless, Mobile and Multimedia Network*, (pp. 23–34), Buffalo, NY: IEEE Computer Society.

Ozev, S., Orailoglu, A., & Olgaard, C. V. (2002). Multi-level testability analysis and solutions for integrated bluetooth transceivers. *IEEE Design & Test of Computers, 19*(5), 82–91. doi:10.1109/MDT.2002.1033796

Ozsu, M. T., & Valduriez, P. (1999). *Principles of Distributed Database Systems* (second ed.)New York: Prentice Hall.

Ozturk, C., Zhang, Y., & Trappe, W. (2004). Source-location privacy in energy-constrained sensor network routing. In *Proceedings of the 2nd ACM workshop on Security of Ad hoc and Sensor Networks*, (pp. 88–93), Washington, DC: ACM Press.

Paavola, M., (2007, December). *Wireless Technologies in Process Automation – Review and an Application Example.* (Report A No 33), Hawai, USA: University of Oulu. Control Engineering Laboratory.

Pan, V. Y., & Chen, Z. Q. (1999). The complexity of the matrix eigenproblem. In *Proceedings of the 31st Annual ACM Symposium on Theory of Computing (STOC)* (pp. 507-516). Atlanta, GA: ACM.

Panja, B., Madria, S. K., & Bhargava, B. (2006). Energy and Communication Efficient Group Key Management Protocol for Hierarchical Sensor Networks, SUTC '06. In *Proceedings of. IEEE Int'l. Conf. Sensor Networks, Ubiquitous, and Trustworthy Comp.* (pp. 384-393).

Papadimitratos, P., & Haas, Z. J. (2003, January). Secure Link State Routing for Mobile Ad Hoc Networks. In *Proceedings of the 2003 Symposium on Applications and the Internet Workshops (SAINT'03)* (pp.379-383) Washington, DC, USA.

Park, T., & Shin, K. G. (2004). LiSP: A lightweight security protocol for wireless sensor networks. *ACM Transaction on Embedded Computing Sys., 3*(3), 634–660. doi:10.1145/1015047.1015056

Paschalis, A., & Gizopoulos, D. (2005). Effective software-based self-test strategies for on-line periodic testing of embedded processors. *IEEE Transactions on Computer-Aided Design of Integrated Circuits and Systems, 24*, 88–99. doi:10.1109/TCAD.2004.839486

Paul, J. L. (2000, July). *Smart Sensor Web: Web-based exploitation of sensor fusion for visualization of the tactical battlefield.* Paper presented at the Third International Conference on Information Fusion, Paris, France.

Pegueroles, J., Rico-Novella, F., Hernández-Serrano, J., & Soriano, M. (2003a). *Adapting GDOI for balanced batch-lkh (<draft-ietf-msec-gdoi-batch-lkh-00.txt>.* Internet Draft).

Pegueroles, J., Rico-Novella, F., Hernández-Serrano, J., & Soriano, M. (2003b). Improved LKH for batch rekeying in multicast groups. In *Proceedings of IEEE international conference on Information Technology Research and Education* (ITRE). New Jersey, USA.

Pepe, A., Borgman, C. L., Mayernik, M., & Wallis, J. C. (2007). *Knitting a Fabric of Sensor Data Resources.* Paper presented at the the ACM Workshop on Data Sharing and Interoperability on the World-wide Sensor Web, Cambridge, MA.

Percivall, G., & Reed, C. (2006). OGC Sensor Web Enablement Standards. *Sensors & Transducers Journal, 71*(9), 698–706.

Perillo, M., & Heinzelman, W. (2003, May). Providing Application QoS Through Intelligent Sensor Management. In *Proceedings of the 1st IEEE International Workshop on Sensor Network Protocols and Applications (SNPA '03).*

Perkins, C. (2001). *Ad Hoc Networks.* Reading, MA: Addison-Wesley.

Perrig, A., Canetti, R., Tygar, J., & Song, D. (2000). *The TESLA Broadcast Authentication Protocol.* Retreived (n.d.). from, http://www.ece.cmu.edu/~adrian/projects/tesla-cryptobytes/tesla-cryptobytes.pdfPerrig, A., Szewczyk, R., Wen, V., Culler, D., & Tygar, D. (2001, July). SPINS: Security protocols for sensor networks. In *Proceedings of Seventh Annual International Conference on Mobile Computing and Networks* (MobiCom).

Perrig, A., Stankovic, J., & Wagner, D. (2004). Security in wireless sensor networks. *Communications of the ACM, 47*(6), 53–57. doi:10.1145/990680.990707

Perrig, A., Szewczyk, R., Tygar, J. D., Wen, V., & Culler, D. E. (2002). SPINS: security protocols for sensor networks. *Wireless Networks, 8*(5), 521–534. doi:10.1023/A:1016598314198

Perrig, A., Szewczyk, R., Tygar, J. D., Wen, V., & Culler, D. E. (2002). SPINS: security protocols for sensor networks. *Wireless Networks, 8*(5), 521–534. doi:10.1023/A:1016598314198

Pervaiz, M. O., Cardei, M., & Wu, J. (2008). Routing Security in Ad Hoc Wireless Networks. In, HuangS., MacCallumD., & Du, D. (Eds.), *Network Security.* New York: Springer.

Peterson, L. L., & Davie, B. S. (2003). *Computer Networks: A Systems Approach,* 3rd Edition. San Francisco: Morgan Kaufmann Publishers Inc.

Petrova, M., Wu, L., Mähönen, P., & Riihijärvi, J. (2007). Interference measurements on performance degradation between colocated IEEE 802.11g/n and IEEE 802.15.4 networks. In *Proceedings from ICN'07: The 6th International Conference on Networks,* Sainte-Luce, Martinique.

Petrovic, D. S. (May, 2003). Data funneling: routing with aggregation and compression for wireless sensor networks. *The First IEEE International Workshop on Sensor Network Protocols and Applications, 2003.,* (pp. 156-162). Anchorage, AK.

Pietro, R. D., Mancini, L. V., & Mei, A. (2003). Random key-assignment for secure wireless sensor networks. In *SASN '03: Proceedings of the 1st ACM workshop on Security of Ad hoc and Sensor Networks* (pp. 62-71). New York: ACM Press.

Pillai, S. U., Suel, T., & Cha, S. (2005). The perron-frobenius theorem and some of its applications. *IEEE Signal Processing Magazine, 22*(2), 62–75. doi:10.1109/MSP.2005.1406483

Poduri, S., Pattem, S., Krishnamachari, B., & Sukhatme, G. S. (2006). Sensor network configuration and the curse of dimensionality. In *Proceedings IEEE EmNets.* Cambridge, MA: IEEE.

Polastre, J. (2003). *Sensor Network Media Access Design,* (Technical Report), University of California, Berkeley, USA.

Polastre, J., Hill, J., & Culler, D. (2004). Versatile low power media access for wireless sensor networks. In *Proceedings from Sensys'04: The second international conference on Embedded networked sensor systems,* Baltimore, USA.

Polastre, J., Szewczyk, R., Sharp, C., & Culler, D. (2004, August 22-24). *The Mote Revolution: Low Power Wireless Sensor Network Devices,* Presented at Hot Chips 16, A Symposium on High Performance Chips, Stanford Memorial Auditorium, CA, USA.

Pompili, D., Melodia, T., & Akyildiz, I. F. (2006). Deployment analysis in underwater acoustic wireless sensor networks. In *Proceedings ACM WUWNet* (pp. 48-55). Los Angeles: ACM.

Popa, L., Raiciu, C., Stoica, I., & Rosenblum, D. S. (2006). Reducing Congestion Effects in Wireless Networks by Multipath Routing. Paper presented at the Network Protocols, 2006. ICNP '06. *Proceedings of the 2006 14th IEEE International Conference.*

Prabhakar, B., Uysal-Biyikoglu, E., & Gamal, A. E. (2001). Energy-efficient transmission over a wireless link via lazy packet scheduling. In *Proceedings of the IEEE InfoCom, 2001.*

Puccinelli, D., & Haenggi, M. (2005). Wireless sensor networks: applications and challenges of ubiquitous sensing. *IEEE Circuits and Systems Magazine, 5*(3), 19–31. doi:10.1109/MCAS.2005.1507522

Qu, W., Li, K., Masaru, K., & Takashi, N. (2007). *An Efficient Method for Improving Data Collection Precision in Lifetime-adaptive Wireless Sensor Networks.* Paper presented at IEEE International Conference on Communications ICC '07. 24-28 June. (pp 3161-3166) Glasgow, UK.

Raazi, S., Khan, A., Khan, F., Lee, S., & Song, Y.-J. (2007). MUQAMI: A locally distributed key management scheme for clustered sensor networks. In Etalle, S. and Marsh, S., (eds), *Int. Federation for Infor. Proc.,* (pp. 333–348). Boston: Springer.

Rabiner, L.,R., (1989) *A tutorial on hidden Markov models and selected applications in speech recognition.* In Proceedings of the IEEE, *77*(2).

Raghavendra, C. S., Kumar, V. K. P., & Hariri, S. (1988). Reliability analysis in distributed systems. *IEEE Transactions on Computers, 37,* 352–358. doi:10.1109/12.2173

Raginsky, D., & Estrin, D. (2002). Rumor routing algorithm for sensor networks. *1st ACM international Workshop on Wireless Sensor Networks and Applications.* (pp. 22-31).

Rajendran, V., Obraczka, K., & Garcia-Luna-Aceves, J. (2003). Energy efficient, collision-free medium access control for wireless sensor networks. In *Proceedings from Sensys'03: The first international conference on Embedded networked sensor systems*, Los Angeles, CA.

Ramachandran, U., Kumar, R., Wolenetz, M., Cooper, B., Agarwalla, B., & Shin, J. (2003). Dynamic Data Fusion for Future Sensor Networks. *ACM Transactions on Sensor Networks, 2*(3), 404–443. doi:10.1145/1167935.1167940

Ramamurthi, V., Reaz, A. S., Dixit, S., & Mukherjee, B. (2008). Link scheduling and power control in wireless mesh networks with directional antennas. In *Proceedings IEEE ICC*. Beijing, China: IEEE.

Raman, B., & Chebrolu, K. (2008). Censor Networks: A Critique of "Sensor Networks". *ACM SIGCOMM Computer Communication Review, 38*(3), 75–78. doi:10.1145/1384609.1384618

Rappaport, T. S. (2001). *Wireless Communications: Principles and Practice* (2nd Edition ed.). Upper Saddle River, NJ: Prentice Hall PTR.

Ravelomanana, V. (2004). Extremal properties of three-dimensional sensor networks with applications. *IEEE Transactions on Mobile Computing, 3*(3), 246–257. doi:10.1109/TMC.2004.23

Reddy, S., Chen, G., Fulkerson, B., Kim, S. J., Park, U., Yau, N., et al. (2007). *Sensor-Internet Share and Search - Enabling Collaboration of Citizen Scientists.* Paper presented at the ACM Workshop on Data Sharing and Interoperability on the World-wide Sensor Web, Cambridge, MA.

Reed, M. G., Syverson, P. F., & Goldschlag, D. M. (1998). Anonymous connections and onion routing. *IEEE Journal on Selected Areas in Communications, 6*(4), 482–494. doi:10.1109/49.668972

Reiter, M. K., & Rubin, A. D. (1998). Crowds: Anonymity for web transactions. *ACM Transactions on Information and System Security, 1*(1), 66–92. doi:10.1145/290163.290168

Rentala, P., Musunui, R., & Gandlham, S. (2002). Survey on Sensor Networks (Technical Report UTDCS-33-02). Dallas, TX:University of Texas at Dallas.

Resnick, P., Zeckhauser, R., Friedman, E., & Kuwabara, K. (2000). Reputation systems: Facilitating trust in internet interactions. *Communications of the ACM, 43*(12), 45–48. doi:10.1145/355112.355122

Rhee, I., Warrier, A., Aia, M., & Min, J. (2005). Z-MAC: a hybrid MAC for wireless sensor networks, In *Proceedings from Sensys'05: The third international conference on Embedded networked sensor systems*, San Diego, USA.

Rice, R. F. (1979). *Some practical universal noiseless coding techniques* (JPL Technical Report No. 79-22). Pasedena, CA: Jet Propulsion Laboratory.

Ringwald, M., & Römer, K. (2005). BitMAC: A Deterministic, Collision-Free, and Robust MAC Protocol for Sensor Networks. In *Proceedings from EWSN'05: The Second IEEE European Workshop on Wireless Sensor Networks,* Istanbul, Turkey.

Ringwald, M., & Römer, K. (2008). Poster abstract: BurstMAC low idle overhead and high throughput in one MAC protocol. Presented at EWSN'08: *The 5th European Confererence on Wireless Sensor Networks,* Bologna, Italy.

Rodrigo, F., Sylvia, R., Jerry, Z., Cheng Tien, E., David, C., Scott, S., et al. (2005). Beacon vector routing: scalable point-to-point routing in wireless sensornets. Paper presented at the *Proceedings of the 2nd conference on Symposium on Networked Systems Design \& Implementation - Volume 2.*

Roman, R., Lopez, J., & Gritzalis, S. (2008). Situation awareness mechanisms for wireless sensor networks [security in mobile ad hoc]. *IEEE Communications Magazine, 46*(4), 102–107. doi:10.1109/MCOM.2008.4481348

Römer, K. (2004). Programming Paradigms and Middleware for Sensor Networks. In *GI/ITG Workshop on Sensor Networks* (pp. 49-54).

Römer, K., & Mattern, F. (2004). The Design Space of Wireless Sensor Networks. *IEEE Wireless Communications, 11*(6), 54–61. doi:10.1109/MWC.2004.1368897

Römer, K., Kasten, O., & Mattern, F. (2002). Middleware Challenges for Wireless Sensor Networks. *ACM SIGMOBILE Mobile Computing and Communication Review (MC2R), 6*(4), 59-61.

Rong, P., & Sichitiu, M. L. (2006). Angle of Arrival Localization for Wireless Sensor Networks. In *Proceedings of the 3rd Annual IEEE Communications Society Conference on Sensor and Ad Hoc Communications and Networks (SECON '06)*, (pp. 374-382), Reston, VA: IEEE Communications Society.

Ruiz, L. B., Siqueira, I. G., & Oliverira, L. B. (2004). *Fault management in event-driven wireless sensor networks.* Paper presented at 7th ACM/IEEE Int. Symposium on Modeling, Analysis and Simulation of Wireless and Mobile Systems, Italy.

Ruzzelli, A. G., O'Hare, G., Jurdak, R., & Tynan, R. (2006). Advantages of Dual Channel {MAC} for Wireless Sensor Networks. In *Proceedings from COMSWARE '06: The First International Conference on COMmunication System softWAre and MiddlewaRE*, New Delhi, India.

Salomaa, A. (1996). *Public-Key Cryptography.* New York: Springer-Verlag.

Salomon, D. (2004). *Data compression: The complete reference (3rd ed.).* New York: Springer.

Sankarasubramaniam, Y., Akan, O. B., & Akyildiz, I. F. (2003). *ESRT: Event-to-Sink Reliable Transport in Wireless Sensor networks.* Paper presented at MobiHoc 2003, Annapolis, Maryland, June 2003.

Santi, P. (2005). *Topology control in wireless ad hoc and sensor networks.* Chichester, UK: John Wiley and Sons.

Santosh, K., & Ten, H., Lai, & József, B. (2008). On k−coverage in a mostly sleeping sensor network. *Wireless Networks, 14*(3), 277–294. doi:10.1007/s11276-006-9958-8

Sanzgiri, K., Dahill, B., Levine, B., Shields, C., & Belding-Royer, E. (2002). A Secure Routing Protocol for Ad Hoc Networks. In *Proceedings of IEEE International Conference on Network Protocols (ICNP)* (pp. 78-87).

Sawada, H., Hashimoto, S. (1997) Gesture recognition using an acceleration sensor and its application to musical performance control. *Electronics and Communications in Japan, 80*(5).

Saxena, N., Tsudik, G., & Yi, J. (Dec. 2004). Identity-Based Access Control for Ad Hoc Groups. In *Proceedings of the Int'l Conf. Information Security and Cryptology.*

Schmid, S., & Wattenhofer, R. (2006). Algorithmic models for sensor networks. In *Proceedings 20th International Parallel and Distributed Processing Symposium (IPDPS).* Rhodes Island, Greece: IEEE.

Schurgers, C., & Srivastava, M. B. (2001, October). *Energy Efficient Routing In Wireless Sensor Networks.* Paper presented at the IEEE Military Communications Conference, Washington, D.C.

Schurgers, C., Aberhorne, O., & Srivastava, M. B. (2001) *Modulation scaling for energy-aware communication systems.* Paper presented at ISLPED 2001, (pp. 96–99).

Schurgers, C., Tsiatsis, V., Ganeriwal, S., & Srivastava, M. (2002). Optimizing sensor networks in the energy-latency-density design space. *IEEE Transactions on Mobile Computing*, 70–80. doi:10.1109/TMC.2002.1011060

Selavo, L., Zhao, G., & Stankovic, J. (2006, October). SeeMote: In-Situ Visualization and Logging Device for Wireless Sensor Networks (BaseNets 2006, pp. 1-9).

Semiconductor Industry Association. (2005). *International technology roadmap for semiconductors: Test and test equipment. Computer, 37*(1), 47–56. doi:10.1109/MC.2004.1260725

SensorMap. Retrieved, from http://atom.research.microsoft.com/sensormap

Seung-Jong, P., Ramanuja, V., Raghupathy, S., & Ian, F. A. (2004). A scalable approach for reliable downstream data delivery in wireless sensor networks. Paper presented at the *Proceedings of the 5th ACM international symposium on Mobile ad hoc networking and computing.*

Seys, S., & Preneel, B. (2006). ARM: Anonymous routing protocol for mobile ad hoc networks. In *Proceedings of the 20th Int. conf. on Advanced Information Networking and Applications*, (pp. 33–37), Vienna Austria: IEEE Press.

Shah, R. C., Roy, S., Jain, S., & Brunette, W. (May, 2003). Data MULEs: Modeling a Three-tier Architecture for Sparse Sensor Networks. *IEEE Workshop on Sensor Network Protocols and Applications (SNPA)*, (pp. 30 - 41). Anchorage, AK.

Shaikh, R. A., Jameel, H., d'Auriol, B. J., Lee, S., Song, Y.-J., & Lee, H. (2008a). Network level privacy for wireless sensor networks. In *Proceedings of the 4th International Conference on Information Assurance and Security (IAS 2008)*, (pp. 261–266), Naples, Italy: IEEE Computer Society.

Shaikh, R. A., Jameel, H., d'Auriol, B. J., Lee, S., Song, Y.-J., & Lee, H. (2009). Group-based Trust Management Scheme for Clustered Wireless Sensor Networks. *IEEE Transactions on Parallel and Distributed Systems, 20*(11), 1698–1712. doi:10.1109/TPDS.2008.258

Shaikh, R. A., Jameel, H., Lee, S., Rajput, S., & Song, Y. J. (2006a). Trust management problem in distributed wireless sensor networks. In *Proceedings of the 12th IEEE Int. Conf. on Embedded Real Time Computing Systems and its Applications*, (pp. 411–414). Sydney, Australia: IEEE Computer Society.

Shaikh, R. A., Lee, S., Khan, M. A. U., & Song, Y. J. (2006b). LSec: Lightweight security protocol for distributed wireless sensor network. In *11th IFIP Int. Conf. on*

Personal Wireless Comm., LNCS 4217, (pp. 367–377), Albacete, Spain: Springer-Verlag.

Shaikh, R. A., Lee, S., Song, Y. J., & Zhung, Y. (2006c). Securing distributed wireless sensor networks: Issues and guidelines. In *Proceedings of the IEEE Int. Conf. on Sensor Networks, Ubiquitous, and Trustworthy Computing- vol. 2 - Workshops*, pp. 226–231, Taiwan: IEEE Computer Society.

Shakkottai, S., Srikant, R., & Shroff, N. (2005). Unreliable sensor grids: Coverage, connectivity and diameter. *Ad Hoc Networks, 3*(6), 702–716. doi:10.1016/j.adhoc.2004.02.001

Shakya, M., Zhang, J., Zhang, P., & Lampe, M. (May, 2007). Design and optimization of wireless sensor network with mobile gateway. *21st International Conference on Advanced Information Networking and Applications Workshops*, (pp. 415-420). Niagara Falls, Canada.

Shamir, A. (1979). How to Share a Secret. *Communications of the ACM, 22*(11), 612–613. doi:10.1145/359168.359176

Shamir, A. (1984). Identity Based Cryptosystems and Signature Schemes. In . *Proceedings of CRYPTO, 84*, 47–53.

Shansi, R., Qun, L., Haining, W., Xin, C., & Xiaodong, Z. (2007). Design and Analysis of Sensing Scheduling Algorithms under Partial Coverage for Object Detection in Sensor Networks. *Transactions on Parallel and Distributed Systems, 18*(3), 334–350. doi:10.1109/TPDS.2007.41

Sharifi, M., Taleghan, M. E., & Taherkordi, A. (2006).). A Middleware Layer Mechanism for QoS Support in Wireless Sensor Networks. Paper presented at *the 4th IEEE International Conference on Networking* (ICN'06), Mauritius, April 23-29.

Shen, C., Srisathapornphat, C., & Jaikaeo, C. (2001). Sensor Information Networking Architecture and Applications. *IEEE Personal Communications, 8*(4), 52–59. doi:10.1109/98.944004

Shen, J., & Abraham, J. A. (1998). Native mode functional test generation for processors with applications to self test and design validation. In *Proceedings of the International Test Conference* (pp. 990-999).

Shen, X., Wang, Z., & Sun, Y. (2004). Wireless Sensor Networks for Industrial Applications. In *Proceedings of the 5Ih World Congress on Intelligent Control and Automation.* (pp 3636—3640) June 15-19, Hangzhou, P.R. China.

Shetty, S., Padala, P., & Frank, M. (2003). *A Survey of Market Based Approaches in Distributed Computing* (Technical Report TR03-13). Florida, USA: University of Florida Computer & Information Science & Engineering

Shields, C., & Levine, B. N. (2000). A protocol for anonymous communication over the internet. In *Proceedings of the 27th ACM conf. on Computer and communications security*, (pp. 33–42), Athens, Greece: ACM Press.

Shigang, C., & Zhan, Z. (2006). Localized algorithm for aggregate fairness in wireless sensor networks. Paper presented at the *Proceedings of the 12th annual international conference on Mobile computing and networking.*

Shigang, C., Shigang, C., Yuguang, F., & Ye, X. (2007). Lexicographic Maxmin Fairness for Data Collection in Wireless Sensor Networks. *Mobile Computing . IEEE Transactions, 6*(7), 762–776.

Shu, L. Wu. C., Zhang, Y., Chen, J., Wang, L., & Hauswirth, M. (2008). NetTopo: Beyond Simulator and Visualizor for Wireless Sensor Networks, In *Proceedings of the Second International Conference on Future Generation Communication and Networking* (FGCN 2008, pp. 17-20).

Simonis, I. (2004). Sensor Webs: A RoadMap. In *Proceedings. of the 1st Goettinger GI and Remote Sensing Days*, Institute for Geoinformatics, University of Muenster.

Simonis, I. (2007). *OGC Sensor Alert Service Implementation Specification*, Open Geospatial Consortium Inc. Retrieved November 25, 2008, from http://portal.opengeospatial.org/files/index.php?artifact_id=24780&version=1&format=pdf

Sohrabi, K., Gao, J., Ailawadhi, V., & Pottie, G. J. (2000). Protocols for self-organization of a wireless sensor network. *IEEE Personal Communications Magazine, 7*(5), 16–27. doi:10.1109/98.878532

Sohraby, K., Minoli, D., & Znati, T. (2007). *Wireless Sensor Networks - Technology, Protocols and Applications.* New York: Wiley-Interscience.

Somasundara, A. A., Ramamoorthy, A., & Srivastava, M. B. (December, 2004). Mobile Element Scheduling for Efficient Data Collection in Wireless Sensor Networks with Dynamic Deadlines. *25th IEEE International Real-Time Systems Symposium (RTSS)*, (pp. 296 - 305). Lisbon, Portugal.

Sridharan, A., & Krishnamachari, B. (2007). Maximizing Network Utilization with Max-Min Fairness in Wireless Sensor Networks. *Paper presented at the Modeling and Optimization in Mobile, Ad Hoc and Wireless Networks and Workshops, 2007. WiOpt 2007. 5th International Symposium.*

Srinivasan, A., & Wu, J. (2008). A Survey on Secure Localization in Wireless Sensor Networks. In Furht, B. (Ed.), *Encyclopedia of Wireless and Mobile Communications.* Boca Raton FL: CRC Press, Taylor and Francis Group.

Standard, I. E. E. E. 1609.2. (n.d.). IEEE Trial-Use Standard for Wireless Access in Vehicular Environments - *Security Services for Applications and Management Messages*, 2006.

Stankovic, J. A., Abdelzaher, T. F., Lu, C., Sha, L., & Hou, J. C. (2003). Real-time communication and coordination in embedded sensor networks. In *Proceedings of the IEEE, vol. 91, no. 7*, July 2003.

Stann, F., & Heidemann, J. (2003). RMST: reliable data transport in sensor networks. Paper presented at the Sensor Network Protocols and Applications, 2003. *Proceedings of the First IEEE. 2003 IEEE International Workshop.*

Stann, F., Heidemann, J., Shroff, R., & Murtaza, M. Z. (2006). RBP: Robust Broadcast Propagation in Wireless Networks. In *Proceedings of the ACM International Conference on Embedded Networked Sensor Systems* (pp. 85-98).

Stoianov, I., Nachman, L., & Madden, S. (2007). PIPENET: A Wireless Sensor Network for Pipeline Monitoring, In. *IPSN'07*, Cambridge, MA:ACM Press.

Stoleru, R., Stankovic, J., & Son, S. (2008). On Composability of Localization Protocols for Wireless Sensor Networks. *IEEE Network, 22*(4), 21–25. doi:10.1109/MNET.2008.4579767

Striki, M., & Baras, J. (2004, June). Towards Integrating Key Distribution with Entity Authentication for Efficient, Scalable and Secure Group Communication in MANETs. In *Proceedings of IEEE ICC'04, vol.7* pp. 4377-4381.

Stutzman, W. L., & Thiele, G. A. (1997). *Antenna Theory and Design (2nd ed.)*. New York: John Wiley & Sons.

Su, W. & Almaharmeh, B. (2008, April) QoS Integration of the Internet and Wireless Sensor Networks. *WSEAS Transactions on Computers. 4*, (7).

Su, W., & Lim, T. L. (2006). Cross-layer design and optimization for wireless sensor networks. In *Proceedings from SNPD-SAWN'06: The 7th ACIS Intl. Conference on Software Engineering, Artificial Intelligence, Networking, and Parallel/Distributed Computing*, Washington, USA.

Sugano, M., Kawazoe, T., Ohta, Y., & Murata, M. (2006). *Indoor Localization System Using RSSI Measurement of Wireless Sensor Network Based on ZigBee Standard*. The IASTED International Conference on Wireless Sensor Networks (WSN 2006), Banff, Canada.

Sukun, K., Fonseca, R., & Culler, D. (2004). Reliable transfer on wireless sensor networks. *Paper presented at the Sensor and Ad Hoc Communications and Networks, 2004. IEEE SECON 2004. 2004 1st Ann. IEEE Communications Society Conf.*

Suman, N., Jie, L., & Feng, Z. (2006). *Challenges in Building a Portal for Sensors World-Wide*. Paper presented at the First Workshop on World-Sensor-Web: Mobile Device Centric Sensory Networks and Applications, Boulder, CO.

Sumit, R., Ramakrishna, G., Ramesh, G., & Konstantinos, P. (2006). *Interference-aware fair rate control in wireless sensor networks* (Vol. 36, pp. 63-74): ACM.

Sun, Y. L., Zhu, H., & Liu, K. J. R. (2008). Defense of trust management vulnerabilities in distributed networks. *IEEE Communications Magazine, 46*(2), 112–119. doi:10.1109/MCOM.2008.4473092

SunSpotWorld. (2008) *SunSpotWorld - Home of Project Sun SPOT*. Retrieved June 6, 2008, from http://www.sunspotworld.com

Svaizer, P., Matassoni, M., & Omologo, M. (1997, April). Acoustic Source Location in a Three-Dimensional Space Using Crosspower Spectrum Phase. *IEEE international Conference on Acoustics, Speech, and Signal Processing: Vol. 1.* (pp. 231-234).

Szewczyk, R., Mainwaring, A. M., Polastre, J., Anderson, J., & Culler, D. E. (2004). *An analysis of a large scale habitat monitoring application.* Paper presented at the ACM Conference on Embedded Networked Sensor Systems.

Szewczyk, R., Mainwaring, A., Polastre, J., Anderson, J., & Culler, D. (2004). An Analysis of a Large Scale Habitat Monitoring Application. In *Proceedings from Sensys'04: The 2nd international conference on Embedded networked sensor systems*, Baltimore, MD.

Szudziejka, V., Kreylos, O., & Hamann, B. (2003). Visualization of environmental data generated by wireless sensor networks, In Copsey, D., (ed.), *Proceedings of the 2003 UC Davis Student Workshop on Computing* (TR CSE-2003-24, pp. 40-41).

Tacconi, D., Carreras, I., Miorandi, D., Chiti, F., Casile, A., & Fantacci, R. (March, 2007). *Mobile applications in disconnected WSN: System architecture, routing framework, applications* (Tech. Rep.). University of Florence, Florence.

Tai, S., Benkoczi, R. R., Hassanein, H., & Akl, S. G. (2007, June). QoS and data relaying for wireless sensor networks. *Journal of Parallel and Distributed Computing, 67*(6), 715–726. doi:10.1016/j.jpdc.2007.01.009

Tang, J., Xue, G., Chandler, C., & Zhang, W. (2006). Link scheduling with power control for throughput enhancement in multi-hop wireless networks. *IEEE Transactions on Vehicular Technology, 55*(3), 733–742. doi:10.1109/TVT.2006.873836

Tannenbaum, A. (2003). *Computer Networks PA.* Upper Saddle River, New Jersey: Prentice Hall.

Tannuri, E. A., & Morishita, E. A. (2006). Experimental and numerical evaluation of a typical dynamic positioning system. *Applied Ocean Research, 28,* 133–146. doi:10.1016/j.apor.2006.05.005

Tao, V., Liang, S., Croitoru, A., Haider, Z., & Wang, C. (2004). GeoSWIFT: Open Geospatial Sensing Services for Sensor Web. In: Stefanidis, A., Nittel, S. (eds), *GeoSensor Networks* (pp.267-274), Boca Raton, FL: CRC Press.

Texas Instruments Corporation. (n.d.). *MSP430 ultra-low power microcontrollers, MSP430x1xx – flash ROM no LCD – price list per 1000 units.* Retrieved February, 2007, from http://focus.ti.com/paramsearch/docs/parametricsearch.tsp?familyId=911&sectionId=95&tabId=1527&family=mcu

Texas Instruments. (2008) *CC1000 data sheet.* Retrieved December 11, 2008, from http://focus.ti.com/lit/ds/symlink/cc1000.pdf van Hoesel, L., & Havinga, P. (2004). A lightweight medium access protocol (LMAC) for wireless sensor networks. In *Proceedings from INSS'04: The First International Workshop on Networked Sensing Systems,* Tokyo, Japan.

Texas Instruments. (2008a) *CC2420 data sheet.,* Retrieved December 11, 2008, from http://focus.ti.com/lit/ds/symlink/cc2420.pdf

Tezcan, N., Wenye, W., & Mo-Yuen, C. (2005). A bidirectional reliable transport mechanism for wireless sensor networks. *Paper presented at the Military Communications Conference, 2005. MILCOM 2005. IEEE.*

Tham, C. K., & Buyya, R. (2005). SensorGrid: Integrating Sensor Networks and Grid Computing. *CSI communication* (Grid Computing).

Thatte, S. M., & Abraham, J. A. (1980). Test generation for microprocessors. *IEEE Transactions on Computers, C-29,* 429–441. doi:10.1109/TC.1980.1675602

The, A. RT-WiSe framework. (2008, May). *IPP Hurray.* Retrieved May, 2008 from, http://www.hurray.isep.ipp.pt/art-wise/

Theodorakopoulos, G., & Baras, J. S. (2006). On trust models and trust evaluation metrics for ad hoc networks. *IEEE Journal on Selected Areas in Communications, 24*(2), 318–328. doi:10.1109/JSAC.2005.861390

Tian, D., & Georganas, N. (2005). Connectivity maintenance and coverage preservation in wireless sensor networks. *Ad Hoc Networks, 3*(6), 744–761. doi:10.1016/j.adhoc.2004.03.001

Tilak, S., Abu-Ghazaleh, N. B., & Heinzelman, W. (2002). A taxonomy of wireless micro-sensor network models. *ACM SIGMOBILE Mobile Computing and Communications Review, 6*(2), 28–36. doi:10.1145/565702.565708

Tilak, S., Abu-Ghazaleh, N. B., & Heinzelman, W. (2002). Infrastructure tradeoffs for sensor networks. In *Proceedings of the ACM International Workshop on Wireless Sensor Networks and Applications Workshop, 2002.*

Tilak, S., Abu-Ghazaleh, N. B., & Heinzelman, W. (2005). *A Taxonomy of Wireless Micro-Sensor Network Models* (Tech. Rep.). New York: University Binghamton, Dept. of Computer Science, System Research Laboratory.

Tirkawi, F., & Fischer, S. (2009). Generality Challenges and Approaches in WSNs. *Journal of Communications* [IJCNS]. *Network and System Sciences, 1,* 58–63.

Toub, S. (n.d.). *Adaptive Huffman compression.* Retrieved July, 2002, from http://www.gotdotnet.com van de Goor, A. (1991). *Testing semiconductor memories: theory and practice.* New York: John Wiley & Sons.

Trumpler, E., & Han, R. (2006, May). *A systematic framework for evolving TinyOS.* Paper presented at the IEEE Workshop on Embedded Networked Sensors, Cambridge, MA.

Undercoffer, J., Avancha, S., Joshi, A., & Pinkston, J. (2002). Security for Sensor Networks. Paper presented at *2002 CADIP Research Symposium,* Baltimore, MD.

van de Goor, A. (1993). Using march tests to test SRAMs. *IEEE Design & Test of Computers, 10*(1), 8–14. doi:10.1109/54.199799

van de Goor, A., & Verhallen, T. (1992). Functional testing of current microprocessors (applied to the Intel i860). In *Proceedings of International Test Conference* (pp. 684-695).

Vaz, P. O., da Cunha, F. D., Almeida, J. M., Loureiro, A. A., & Mini, R. A. (2008). The problem of cooperation among different wireless sensor networks. In *Proceedings of the 11th international symposium on Modeling, analysis and simulation of wireless and mobile*. Vancouver, Canada (pp 86-91) ISBN:978-1-60558-235-1

Venugopal, S., Nadiminti, K., Gibbins, H., & Buyya, R. (2008). Designing a Resource Broker for Heterogeneous Grids. *Software, Practice & Experience, 38*(8), 793–825. doi:10.1002/spe.849

Vincze, Z., & Vida, R. (October, 2005). Multi-hop Wireless Sensor Networks with Mobile Sink. *ACM Conference on Emerging Network Experiment and Technology (CoNEXT '05)*. Toulouse, France.

Vitter, J. S. (1987). Design and analysis of dynamic huffman codes. *Journal of the ACM, 34*(4), 825–845. doi:10.1145/31846.42227

Vlajic, N., & Xia, D. (June 2006). Wireless Sensor Networks: To Cluster or Not To Cluster? *IEEE International Symposium on a World of Wireless and Multimedia Networks*. Niagara Falls, Buffalo-NY.

Völker, M., Katz, B., & Wagner, D. (2009). *On the complexity of scheduling with power control in geometric SINR*. Technical Report 15, ITI Wagner, Faculty of Informatics, Universität Karlsruhe.

Wadaa, A., Olariu, S., Wilson, L., Eltoweissy, M., & Jones, K. (2004). On providing anonymity in wireless sensor networks. In *Proceedings of the 10th Int. conf. on Parallel and Distributed Systems*, (pp. 411–418), California, USA: IEEE Computer Society.

Wallbaum, M. (2007). A priori error estimates for wireless local area network positioning systems. *Pervasive and Mobile Computing, 3*, 560–580. doi:10.1016/j.pmcj.2007.02.002

Wallner, D., Harder, E., & Agee, R. (1998). *Key management for multicast: issues and architectures*. RFC 2627.

Walsh, W. E., Wellman, M. P., Wurman, P. R., & MacKie-Mason, J. K. (1998). *Some Economics of Market-Based Distributed Scheduling*. Paper presented at the the 18th international Conference on Distributed Computing Systems.

Walters, J. P., Liang, Z., Shi, W., & Chaudhary, V. (2006). Wireless sensor network security: A survey. In Xiao, Y., (ed.), *Security in Distributed, Grid, and Pervasive Computing*, (pp. 367–410). CRC Press.

Wan, C., Eisenman, S. B., Campbell, A. T., & Crowcroft, J. (November, 2005). Siphon: overload traffic management using multi-radio virtual sinks in sensor networks. In *Proceedings of the 3rd international conference on Embedded networked sensor systems (SenSys '05)*, (pp. 116-129). San Diego, CA.

Wang, C., Sohraby, K., Li, B., Daneshmand, M., & Hu, Y. (2006) A Survey of Transport Protocols for Wireless Sensor Networks. *IEEE Network Magazine*. May/June 2006 (pp 34-40)

Wang, F., & Liu, J. (2007). *RBS: A Reliable Broadcast Service for Large-Scale Low Duty-Cycled Wireless Sensor Networks* (Tech. Rep.). Burnaby, British Columbia: Simon Fraser University, School of Computing Science.

Wang, G., Cao, G., Berman, P., La, P., & Thomas, F. (2007). Bidding Protocols for Deploying Mobile Sensors. *IEEE Transactions on Mobile Computing, 6*(5), 515–528. doi:10.1109/TMC.2007.1022

Wang, K., Chiasserini, C.-F., Rao, R. R., & Proakis, J. G. (2005). A joint solution to scheduling and power control for multicasting in wireless ad hoc networks. *EURASIP Journal on Applied Signal Processing*, (1): 144–152. doi:10.1155/ASP.2005.144

Wang, M. M., Cao, J. N., Li, J., & Das, S. K. (2008). Middleware for Wireless Sensor Networks: A Survey. *Journal of Computer Science And Technology, 23*(3), 3. doi:10.1007/s11390-008-9135-x

Wang, M. M., Cao, J. N., Li, J., & Das, S. K. (2008, May). Middleware for wireless sensor networks: A survey. *Journal of Computer Science and Technology, 23*(3), 305–326. doi:10.1007/s11390-008-9135-x

Wang, Q., Hossam, H., & Xu, K. (2006). A Practical Perspective on Wireless Sensor Networks. In Mahgoub, I. & Ilyas, M. (Eds.) *Smart Dust: Sensor Network Applications, Architecture, and Design.* CRC Press, Taylor & Francis Group.

Wang, X., Xing, G., Zhang, Y., Lu, C., Pless, R., & Gill, C. D. (2003). Integrated Coverage and Connectivity Configuration in Wireless Sensor Networks. *Proceedings of the First ACM Conference on Embedded Networked Sensor Systems* (pp. 28-39). Los Angeles, CA.

Wang, Y. C., Hu, C. C., & Tseng, Y. C. (2008). Efficient Placement and Dispatch of Sensors in a Wireless Sensor Network. *IEEE Transactions on Mobile Computing, 7*(2), 262–274. doi:10.1109/TMC.2007.70708

Wang, Y., & Li, X.-Y. (2003). Localized construction of bounded degree planar spanner for wireless ad hoc networks. In *Proceedings of the DIALM-POMC Joint Workshop on Foundations of Mobile Computing* (pp.59-68). San Diego, CA: ACM.

Ware, C. (2004). *Information Visualization Perception for Design.* San Francisco: Morgan Kaufmann Publishers.

Wattenhofer, R., Li, L., Bahl, P., & Wang, Y.-M. (2001). Distributed topology control for power efficient operation in multihop wireless ad hoc networks. In *20th IEEE International Conference on Computer Communications, Joint Conference of the IEEE Computer and Communications Societies (INFOCOM)* (pp. 1388-1397). Anchorage, AK: IEEE.

Wei, L., & Cassandras, C. G. (2005). A minimum-power wireless sensor network self-deployment scheme. *IEEE Wireless Communications and Networking Conference,* Volume 3, (pp. 1897–1902).

Weisstein, E. W. (2009a). *Reuleaux Triangle.* Retrieved (n.d.). From, http://mathworld.wolfram.com/Reuleaux-Triangle.html

Weisstein, E. W. (2009b). Reuleaux Tetrahedron. Retrieved (n.d.). From, http://mathworld.wolfram.com/ReuleauxTetrahedron.html

Welch, T. A. (1984). A technique for high-performance data compression. *IEEE Computer, 17*(6), 8–19.

Wenliang, D., Jing, D., Han, Y. S., & Shigang, C. (2004). A key management scheme for wireless sensor networks using deployment knowledge. *Twenty-third Annual Joint Conference of the IEEE Computer and Communications Societies Publication, Vol. 1.* (pp. 586-597).

Werner-Allen, G., Lorincz, K., Johnson, J., Lees, J., & Welsh, M. (2006). Fidelity and yield in a volcano monitoring sensor network. In *OSDI,* (pp. 381 - 396). Seattle, WA.

Wessner, C. W. (2006). *The Telecommunications Challenge: Changing Technologies and Evolving Policies,* Report of a Symposium. Washington DC: The National Academies Press.

Wikipedia, The free Encyclopedia (2009). *Renewable energy in Germany.* Retrieved January 27, 2009, from http://en.wikipedia.org/wiki/Renewable_energy_in_Germany

Wilson, P., Prashanth, D., & Aghajan, H. (2007). *Utilizing RFID signaling scheme for localization of stationary objects and speed estimation of mobile objects.* Paper presented at the IEEE International Conference on RFID.

Winkler, M., Tuchs, K. D., Hughes, K., & Barclay, G. (2008). Theoretical and practical aspects of military wireless sensor networks. *Journal of Telecommunications and Information Technology, 2,* 37–45.

Woo, A., & Culler, D. (2001, July). A Transmission Control Scheme for Media Access in Sensor Networks. In *Proceedings of the ACM/IEEE Conf. on Mobile Computing and Networks (MobiCOM 2001),* Rome.

Woo, A., Madden, S., & Govindan, R. (2004). Networking Support for Query Processing in Sensor Networks. *Communications of the ACM, 47*(6), 47–52. doi:10.1145/990680.990706

Wood, A. D., Fang, L., Stankovic, J. A., & He, T. (2006). SIGF: a family of configurable, secure routing protocols for wireless sensor networks. In *Proceedings of the 4th ACM workshop on Security of ad hoc and sensor networks,* (pp. 35–48). Alexandria, VA: ACM Press.

Wood, A. D., Stankovic, J. A., & Son, S. H. (2003). *JAM: A Jammed-Area Mapping Service for Sensor Networks.* Paper presented at the meetings of 24th IEEE International Real-Time System Symposium. California, USA

Wood, A. D., Stankovic, J. A., & Zhou, G. (2007). *DEEJAM: Defeating Energy-Efficient Jamming in IEEE 802.15.4-based Wireless Networks.* Paper presented at the 4th Annual IEEE Communications Society Conference on Sensor, Mesh and Ad Hoc Communications and Networks. San Diego, CA.

Wu, B., Cardei, M., & Wu, J. (2008). A Survey of Key Management in Mobile Ad Hoc Networks. In, ZhengJ., ZhangY., & Ma, M. (Eds.), *Hdbk. of Rsch on Wireless Security.* Hershey, PA: Idea Group Inc.

Wu, B., Chen, J., Wu, J., & Cardei, J. (2008). A Survey on Attacks and Countermeasures in Mobile Ad Hoc Networks, In Y. Xiao, X. Shen, & D. -Z. Du (Ed.), *Wireless/Mobile Network Security.* New York: Springer.

Wu, J., & Li, H. (1999). On calculating connected dominating set for efficient routing in ad hoc wireless networks, In *Proceedings of the Third International Workshop on Discrete Algorithms and Methods for Mobile Computing and Communications* (pp. 7–14). Seattle, WA: ACM.

Wu, Q. (1999). Performance of optimum transmitter power control in CDMA cellular mobile systems. *IEEE Transactions on Vehicular Technology, 48*(2), 571–575. doi:10.1109/25.752582

Xi, Y., Schwiebert, L., & Shi, W. (2006). Preserving source location privacy in monitoring-based wireless sensor networks. In *Proceedings of the Parallel and Distributed Processing Symposium (IPDPS 2006)*, Rhodes Island, Greece: IEEE Computer Society.

Xia, F. (2008). QoS Challenges and Opportunities in Wireless Sensor/Actuator Networks. *Sensors, 8*(2), 1099–1110. doi:10.3390/s8021099

Xia, F., & Zhao, W. H. (2007). Flexible Time-Triggered Sampling in Smart Sensor-Based Wireless Control Systems. *Sensors, 7*(11), 2548–2564. doi:10.3390/s7112548

Xia, F., Zhao, W., Sun, Y., & Tian, T. (2007, December). Fuzzy Logic Control Based QoS Management in Wireless Sensor/Actuator Networks. *Sensors, 7*(12), 3179–3191. doi:10.3390/s7123179

Xiao, B., Chen, H., & Zhou, S. (2008). Distributed Localization Using a Moving Beacon in Wireless Sensor Networks. *IEEE Transactions on Parallel and Distributed Systems, 19*(5), 587–600. doi:10.1109/TPDS.2007.70773

Xiao, Y., & Rayi, V. (2007). A survey of key management schemes in wireless sensor networks. *Computer Communications, 30*(11-12), 2314–2341. doi:10.1016/j.comcom.2007.04.009

Xing, G., Wang, X., Zhang, Y., Lu, C., Pless, R., & Gill, C. (2005). Integrated coverage and connectivity configuration for energy conservation in sensor networks. *ACM Transactions on Sensor Networks, 1*(1), 36–72. doi:10.1145/1077391.1077394

Xu, N., Rangwala, S., Chintalapudi, K., Ganesan, D., Broad, A., Govindan, R., et al. (Nov. 2004). A Wireless Sensor Network For Structural Monitoring. *ACM Conference on Embedded Networked Sensor Systems (SenSys 2004)*, (pp. 13-24). Baltimore, MD.

Xu, S., & Saadawi, T. (2001, June). Does the IEEE 802.11 MAC Protocol Work Well in Multihop Wireless Ad Hoc Networks? *IEEE Comm. Magazine*, June 2001.

Xu, W., Ma, K., Trappe, W., & Zhang, Y. (2006). Jamming Sensor Networks: Attacks and Defense Strategies. *IEEE Networks Special Issue on Sensor Networks, 20*, 41–47.

Xu, W., Trappe, W., & Zhang, Y. (2004). Channel Suffering and Spatial Retreats: Defenses against Wireless Denial of Service. *Proceedings of. 2004 ACM Workshop. Wireless Security* (pp. 80-89). ACM Press.

Xu, W., Trappe, W., & Zhang, Y. (2007) Channel surfing: defending wireless sensor networks from jamming and interference. *IPSN'07*, April 25-27 (pp 499-508) Cambridge MA.

Xu, W., Trappe, W., & Zhang, Y. (2008, April). Anti-jamming Timing Channels for Wireless Networks. *Proceedings of 2008 ACM Workshop. Wireless Security* (pp. 203-213). ACM Press.

Xu, W., Trappe, W., Zhang, Y., & Wood, T. (2005, May). *The feasibility of Launching and Detecting Jamming Attacks in Wireless Networks*, Paper presented at the 6th ACM International Symposium on Mobile Ad Hoc Networking and Computing.

Yamaji, M., Ishii, Y., Shimamura, T., & Yamamoto, S. (2008, June). *Wireless sensor network for industrial automation*. Paper presented at the International Conference on Networked Sensing Systems (INSS), Kanazawa, Japan.

Yan, T., He, T., & Stankovic, J. (2003). Differentiated Surveillance for Sensor Networks. In *Proceedings of the ACM International Conference on Embedded Networked Sensor Systems* (pp. 51-62).

Yang, G., Xiao, M., & Chen, C. (April, 2007). A Simple Energy-Balancing Method in RFID Sensor Networks. *2007 IEEE International Workshop on Anti-counterfeiting, Security, Identification*, (pp. 306-310).

Yang, S., Dai, F., Cardei, M., & Wu, J. (2006). On connected multiple point coverage in wireless sensor networks. *International Journal of Wireless Information Networks*, *13*(4), 289–301. doi:10.1007/s10776-006-0036-z

Yang, Y., Xia, P., Huang, L., & Zhou, Q. Xu, Y., & Li, X. (2006). SNAMP: A Multi-sniffer and Multi-view Visualization Platform for Wireless Sensor Networks, In *Proc. of the 2006 IST IEEE Conf. on Industrial Electronics and Applications* (pp. 1-4).

Yang, Z., Li, M., & Liu, Y. (2007). *Sea Depth Measurement with Restricted Floating Sensors*. Paper presented at the IEEE Real-Time Systems Symposium.

Yao, Z., Kim, D., & Doh, Y. (2006). PLUS: Parameterized and localized trust management scheme for sensor networks security. In *Proceedings of the 3rd IEEE Int. Conf. on Mobile Adhoc and Sensor Systems*, (pp. 437–446), Vancouver, Canada: IEEE Computer Society.

Yarvis, M., Kushalnagar, N., Singh, H., Rangarajan, A., Liu, Y., & Singh, S. (2005). Exploiting heterogeneity in sensor networks. In *Proceedings IEEE Infocom* (pp. 878-890). Miami, FL; IEEE.

Ye, F., Zhong, G., Cheng, J., Lu, S., & Zhang, L. (2003). PEAS: A Robust Energy Conserving Protocol for Long-Lived Sensor Networks. In *Proceedings ICDCS* (pp. 1-10). Providence, RI: IEEE.

Ye, W., Heidemann, J., & Estrin, D. (2002). An Energy-Efficient MAC protocol for Wireless Sensor Networks. In *Proceedings from Infocom'02: The 21st Conference on Computer Communications*, New York.

Yeh, J.-C., Wu, C.-F., Cheng, K.-L., Chou, Y.-F., Huang, C.-T., & Wu, C.-W. (2002). Flash memory built-in self-test using march-like algorithms. *IEEE International Workshop on Electronic Design, Test and Applications* (pp. 137-141).

Yener, B., Magdon-Ismail, M., & Sivrikaya, F. (2007). Joint problem of power optimal connectivity and coverage in wireless sensor networks. *Wireless Networks*, *13*(4), 537–550. doi:10.1007/s11276-006-5875-0

Yi, S., & Kravets, R. (2003, April). Moca: Mobile Certificate Authority for Wireless Ad Hoc Networks. In *Proceedings of Second Ann. PKI Research Workshop (PKI '03)*.

Yi, S., Naldurg, P., & Kravets, R. (2002). *Security Aware Ad hoc Routing for Wireless Networks*. (Tech. Rep. No. UIUCDCS-R-2002-2290). Illinois, USA: University of Illinois at Urbana-Champaign.

Yi, Z., & Krishnendu, C. (2004). Sensor Deployment and Target Localization in Distributed Sensor Networks. *ACM Transactions on Embedded Computing Systems*, *3*(1), 61–91. doi:10.1145/972627.972631

Yick, J., Mukherjee, B., & Ghosal, D. (2008). Wireless sensor network survey. *Journal of Computer Networks*, *52*, 2292–2330. doi:10.1016/j.comnet.2008.04.002

Yingshu, L., & Shan, G. (2008). Designing *k*-coverage schedules in wireless sensor networks. *Journal of Combinatorial Optimization*, *15*(2), 127–146. doi:10.1007/s10878-007-9072-6

Yogesh, S. zg, r, B. A., & Ian, F. A. (2003). ESRT: event-to-sink reliable transport in wireless sensor networks. Paper presented at the *Proceedings of the 4th ACM* *international symposium on Mobile ad hoc networking \& computing.*

Yoneki, E., & Bacon, J. (2005). *A survey of Wireless Sensor Network technologies: research trends and middleware's role (Technical Report UCAM-CL-TR646).* University of Cambridge, Cambridge, UK.

Yoon, H., Cheon, J. H., & Kim, Y. (2004). Batch verification with ID-based signatures. In *Proceedings of Information Security and Cryptology* (pp. 233-248).

Younis, M. F., Ghumman, K., & Eltoweissy, M. (2006). Location-Aware Combinatorial Key Management Scheme for Clustered Sensor Networks. In *IEEE Trans. Parallel and Distrib. Sys., vol.17,* 2006 (pp. 865-882).

Younis, M., & Akkaya, K. (2008). (in press). Strategies and Techniques for Node Placement in Wireless Sensor Networks: A Survey. *Ad Hoc Networks,* 6(4), 621–655. doi:10.1016/j.adhoc.2007.05.003

Younis, M., Akayya, K., Eltowiessy, M., & Wadaa, A. (2004, January). On Handling QoS Traffic in Wireless Sensor Networks. In *Proceedings of the International Conference HAWAII International Conference on System Sciences (HICSS-37) vol. 40, no. 8,* (pp. 102-116). Hawaii, USA.

Yu, Y., Govindan, R., & Estrin, D. (2001). *Geographical and Energy Aware Routing: a recursive data dissemination protocol for wireless sensor networks (Technical Report).* UCLA Computer Science Department, Los Angelas.

Yu, Y., Krishnamachari, B., & Prasanna, V. K. (2004). Energy-latency tradeoffs for data gathering in wireless sensor networks. In *Proceedings of IEEE Infocom 2004.*

Yu, Y., Krishnamachari, B., & Prasanna, V. K. (2004). Issues in Designing Middleware for Wireless Sensor Networks. *IEEE Network, 18,* 15–21. doi:10.1109/MNET.2004.1265829

Yu, Z., & Guan, Y. (2005, April). A key pre-distribution scheme using deployment knowledge for wireless sensor networks. In *IPSN '05: Proceedings of the 4th interna-tional symposium on Information Processing in Sensor Networks* (p. 261-268). Piscataway, NJ: IEEE Press.

Zander, J. (1992a). Performance of optimum transmitter power control in cellular radio systems. *IEEE Transactions on Vehicular Technology, 41*(1), 57–62. doi:10.1109/25.120145

Zander, J. (1992b). Distributed cochannel interference control in cellular radio systems. *IEEE Transactions on Vehicular Technology, 41*(3), 305–311. doi:10.1109/25.155977

Zapata, M. (200). *Secure Ad Hoc On-Demand Distance Vector (SAODV).* Retreived 2004, from, http://personals.ac.upc.edu/guerrero/papers/draft-guerrero-manet-saodv-01.txt

Zavlanos, M. M., & Pappas, G. J. (2008). Dynamic Assignment in Distributed Motion Planning With Local Coordination. *IEEE Transactions on Robotics, 24*(1), 232–242. doi:10.1109/TRO.2007.913992

Zhang, C., DeCleene, B., Kurose, J., & Towsley, D. (2002). Comparison of inter-area rekeying algorithms for secure wireless group communications. *Performance Evaluation, 49*(1-4), 1–20. doi:10.1016/S0166-5316(02)00120-7

Zhang, C., Lu, R., Ho, P., & Shen, X. (2008). An Efficient Identity-based Batch Verification Scheme for Vehicular Sensor Networks. In *Proceedings of IEEE INFOCOM.*

Zhang, F., & Kim, K. (2003). Efficient ID-based blind signature and proxy signature from bilinear pairings. In *. Proceedings of ACISP, LNCS, 2727,* 312–323.

Zhang, F., Safavi-Naini, R., & Susilo, W. (2003). Efficient verifiably encrypted signature and partially blind signature from bilinear pairings. In *. Proceedings of Indocrypt, LNCS, 2904,* 191–204.

Zhang, H., & Hou, J. (2005). Maintaining sensing coverage and connectivity in large sensor networks. *Ad Hoc & Sensor Wireless Networks, 1*(1-2), 89–124.

Zhang, H., & Hou, J. C. (2006). Is Deterministic Deployment Worse than Random Deployment for Wireless Sensor Networks? *IEEE International Conference on Computer Communications* (pp. 1-13). Barcelona, Spain.

Zhang, P., Sadler, C. M., Lyon, S. A., & Martonosi, M. (2004). *Hardware design experiences in ZebraNet.* Paper presented at the ACM Conference on Embedded Networked Sensor Systems.

Zhang, R. (2005). *Energy reduced software-based self-testing for wireless sensor network nodes.* Master of Engineering thesis, McGill University, Canada.

Zhang, R., Zilic, Z., & Radecka, K. (2006). Energy efficient software-based self-test for wireless sensor network nodes. In *Proceedings of the VLSI Test Symposium* (pp. 186-191).

Zhang, Y., Liu, W., & Lou, W. (2005, March). Anonymous Communications in Mobile Ad Hoc Networks. In *. Proceedings of IEEE INFOCOM*, *05*, 1940–1951.

Zhang, Y., Liu, W., Lou, W., & Fang, Y. (2006, Oct.-Dec.). Securing mobile ad hoc networks with certificateless public keys. In Proceedings of *IEEE Transactions on Dependable and Secure Computing* (Vol. 3, No. 4, pp 386-399).

Zhang, Y., Liu, W., Lou, W., Fang, Y., & Kwon, Y. (2005, May). AC-PKI: Anonymous and Certificateless Public-Key Infrastructure for Mobile Ad Hoc Networks. In *Proceedings of IEEE Int'l Conf. Comm* (pp. 3515-3519).

Zhao, F., & Guibas, L. J. (2004). *Wireless sensor networks: an information processing approach.* San Francisco: Morgan Kaufmann.

Zhao, J., & Govindan, R. (2003). Understanding packet delivery performance in dense wireless sensor networks. In *Proceedings ACM SenSys* (pp. 1-13). Los Angeles: ACM.

Zheng, T., Radhakrishnan, S., & Sarangan, V. (2005, April). *PMAC: an adaptive energy-efficient MAC protocol for wireless sensor networks.* Paper presented at the IEEE International Parallel and Distributed Processing Symposium (IPDPS), Denver, CO.

Zhipu, J., & Murray, R. M. (2007). Random consensus protocol in large-scale networks. *46th IEEE Conference on Decision and Control* (pp. 4227-4232).

Zhong, X., & Xu, C. Z. (2007). Energy-Aware Modeling and Scheduling for Dynamic Voltage Scaling with Statistical Real-Time Guarantee. *IEEE Transaction on Computers Journal*, *56*(3), 358–372. doi:10.1109/TC.2007.48

Zhong, X., & Xu, C. Z. (2007). Energy-Efficient Wireless Packet Scheduling with Quality of Service Control. *IEEE Transaction on Mobile Computing journal*, *6*(10), 1158-1170.

Zhou, G., He, T., Krishnamurthy, S., & Stankovic, J. (2004). Impact of Radio Irregularity on Wireless Sensor Networks. In *Proceedings MobiSys* (pp. 125-138). Boston: ACM.

Zhou, G., Stankovic, J., & Son, S. H. (2006). Crowded Spectrum in Wireless Sensor Networks. In *Proceedings from EmNetS'06: The third Workshop on Embedded Networked Sensors,* Cambridge, MA.

Zhou, L., & Haas, Z. (1999). Securing Ad Hoc Networks. *IEEE Network Magazine*, *13*(6), 24–30. doi:10.1109/65.806983

Zhou, P. (1995). Algorithms for Some problems in Geometric Covering. *Journal of Beijing Institute of Technology*, *15*(5), 21–25.

Zhou, Y., Fang, Y. G., & Zhang, Y. C. (2008). Securing Wireless Sensor Networks: A survey. *IEEE Communication Surveys & Tutorials*, *10*(3), 6–28. doi:10.1109/COMST.2008.4625802

Zhou, Y., Lyu, M. R., Liu, J., & Wang, H. (2005). *PORT: A Price-Oriented Reliable Transport protocol for wireless sensor networks*, Chicago, IL: IEEE.

Zhou, Z., Das, S., & Gupta, H. (2005). Fault tolerant connected sensor cover with variable sensing and transmission ranges. In *Proceedings IEEE SECON* (pp. 594-604). Santa Clara, CA: IEEE.

Zhu, B., Wan, Z., Kankanhalli, M. S., Bao, F., & Deng, R. H. (2004). Anonymous secure routing in mobile ad-hoc networks. In *Proceedings of the 29th IEEE Int. conf. on Local Computer Networks*, (pp. 102–108), Tampa, USA: IEEE Computer Society.

Zhu, S., Setia, S., & Jajodia, S. (2003). LEAP: efficient security mechanisms for large-scale distributed sensor networks. In *Proceedings of the 10th ACM Conf. on Computer and Comm. security*, (pp. 62–72). NY: ACM Press.

Zhu, S., Setia, S., & Jajodia, S. (2006). LEAP+: Efficient security mechanisms for large-scale distributed sensor networks. *ACM Trans. Sensor Networks*, *2*(4), 500–528. doi:10.1145/1218556.1218559

Zhu, S., Xu, S., Setia, S., & Jajodia, S. (2003). LHAP: A Lightweight Hop-by-Hop Authentication Protocol for Ad-Hoc Networks. In Proceedings of the *23rd International Conference on Distributed Computing Systems Workshops (ICDCSW'03)*.

ZigBee Standards Organization. (2008) *ZigBee Specifications*. Retreived march 20, 2008, from http://www.zigbee.org

Zorzi, M., & Rao, R. R. (2003). Geographic random forwarding(GeRaF) for ad hoc and sensor networks: energy and latency performance. *IEEE Transactions on Mobile Computing*, *2*(4), 349–365. doi:10.1109/TMC.2003.1255650

About the Contributors

Wenbin Jiang is an Associate Professor at the School of Computer Science and Technology of the Huazhong University of Science and Technology (HUST) in China. He received his Ph.D. in information and communication engineering from HUST in 2004. His research interests focus on Ubiquitous computing, P2P computing, Multimedia, computing system virtualization, etc. He has published about 30 papers. The projects that he has taken part in include National Science & Technology Pillar Program, 973 Plans, NFC project, etc. He is also a member of the creation group of the Natural Science Fund of Hubei. From 2005.12-2006.1, he has been to Aizu University for VLIW research. He has been the PC chairs, Publicity chairs, publication chairs, PC members of more than 40 international conferences including MUE'09, AINA'08, GPC'08, NPC'07, etc. He is a member of IEEE. Contact him at jwbhust@gmail.com.

Hai Jin is a Professor of Computer Science and Engineering at the Huazhong University of Science and Technology (HUST) in China. He is now the Dean of School of Computer Science and Technology at HUST. He received his Ph.D. in computer engineering from HUST in 1994. In 1996, he was awarded German Academic Exchange Service (DAAD) fellowship for visiting the Technical University of Chemnitz in Germany. He worked for the University of Hong Kong between 1998 and 2000 and participated in the HKU Cluster project. He worked as a visiting scholar at the University of Southern California between 1999 and 2000. He is the chief scientist of the largest grid computing project, ChinaGrid, in China. Dr. Jin is a senior member of IEEE and member of ACM. He is the member of Grid Forum Steering Group (GFSG). His research interests include computer architecture, cluster computing and grid computing, virtualization technology, peer-to-peer computing, network storage, network security. Contact him at hjin@hust.edu.cn.

* * *

Baha Alsaify received his Bachelor degree in Computer Engineering from the Jordan University of Science and Technology, Jordan in 2007. He moved after that to the US to pursue his higher education. Currently he is studying for his Master degree in Computer Engineering in the University of Arkansas, USA. His research interest is in Wireless Sensor Networks, especially in how to reduce energy consumption with in the Wireless Sensor Network environment by employing an appropriate scheduling algorithm to manage how and when the sensors in a wireless sensor environment goes to sleep, and for how long. He is a member of IEEE.

Habib M. Ammari is an Assistant Professor of Computer Science in the Department of Computer Science at Hofstra University and is the Founding Director of the Wireless Sensor and Mobile Ad-hoc Networks (WiSeMAN) Research Laboratory at Hofstra University. He received the Ph.D. degree in Computer Science and Engineering from The University of Texas at Arlington (UTA) in May 2008 and the Doctorat de Specialite and the Diploma of Engineering degrees in Computer Science from the Faculty of Sciences of Tunis, Tunisia, in 1996 and 1992, respectively. He was an Assistant Professor of Computer Science at Sup'Com Tunis from 1997-2005 and received tenure in 1999. His main research interests lie in the areas of wireless sensor and mobile ad hoc networking, and multihop mobile wireless Internet architectures and protocols. He received the US National Science Foundation (NSF) Research Grant Award and the Faculty Research and Development Grant Award from Hofstra College of Liberal Arts and Sciences, both in 2009. He published his first book "Challenges and Opportunities of Connected k-Covered Wireless Sensor Networks: From Sensor Deployment to Data Gathering" (Springer, 2009). He received the John Steven Schuchman Award for 2006-2007 Outstanding Research by a PhD Student and the Nortel Outstanding CSE Doctoral Dissertation Award, both from UTA in 2008 and 2009, respectively. He was a recipient of the TPC Best Paper Award from EWSN '08 and the Best Contribution Paper Award from IEEE PerCom '08—Google PhD forum. Also, he was an ACM Student Research Competition (ACM SRC) nominee at ACM MobiCom '05. He was selected for inclusion in the 2006 edition of Who's Who in America and the 2008-2009 Honors Edition of Madison Who's Who Among Executives and Professionals.

Junaid Ansari is currently a Ph.D. student and research assistant at the Department of Wireless Networks, RWTH Aachen University, Germany. He completed his M.Sc. in Communications Engineering from RWTH Aachen University, Germany in 2006 and his Bachelor's degree in Electrical Engineering from National University of Science and Technology, Pakistan in 2002. His current research interests include low-power design and energy efficient networking solutions for wireless sensor networks. He has been actively working and managing various large scale research projects related to wireless embedded networking funded by European Union and German government at the Department of Wireless Networks, RWTH Aachen University, Germany.

Rajkumar Buyya is an Associate Professor of Computer Science and Software Engineering; and Director of the Cloud Computing and Distributed Systems (CLOUDS) Laboratory at the University of Melbourne, Australia. He is also serving as the founding CEO of Manjrasoft Pty Ltd., a spin-off company of the University, commercialising its innovations in Grid and Cloud Computing. He has authored 300 publications and four text books. The books on emerging topics that Dr. Buyya edited include, High Performance Cluster Computing (Prentice Hall, USA, 1999), Content Delivery Networks (Springer, Germany, 2008) and Market-Oriented Grid and Utility Computing (Wiley, USA, 2009). Dr. Buyya has contributed to the creation of high-performance computing and communication system software for Indian PARAM supercomputers. He has pioneered Economic Paradigm for Service-Oriented Distributed Computing and developed key Grid and Cloud Computing technologies such as Gridbus and Aneka that power the emerging e-Science and e-Business applications. In this area, he has published hundreds of high quality and high impact research papers that are well referenced. The Journal of Information and Software Technology in Jan 2007 issue, based on an analysis of ISI citations, ranked Dr. Buyya's work (published in Software: Practice and Experience Journal in 2002) as one among the "Top 20 cited Software Engineering Articles in 1986-2005". He received the Chris Wallace Award for

Outstanding Research Contribution 2008 from the Computing Research and Education Association of Australasia. He is the recipient of "2009 IEEE Medal for Excellence in Scalable Computing". He has presented over 160 invited talks (keynotes, tutorials, and seminars) on his vision on IT Futures and advanced computing technologies at international conferences and institutions in Asia, Australia, Europe, North America, and South America. For further information on Dr. Buyya, please visit his Cyberhome: http://www.buyya.com.

Ruay-Shiung Chang received his B.S.E.E. degree from National Taiwan University in 1980 and his Ph.D. degree in Computer Science from National Tsing Hua University in 1988. He is now a professor in the Department of Computer Science and Information Engineering, National Dong Hwa University. His research interests include Internet, wireless networks, and grid computing. Dr. Chang is a member of ACM and IEICE, a senior member of IEEE, and founding member of ROC Institute of Information and Computing Machinery. Dr. Chang also served on the advisory council for the Public Interest Registry (www.pir.org) from 2004/5 to 2007/4.

Jianmin Chen received the B.S. degree (1988) in applied mathematics, M.S. degree (1992) in power mechanical engineering from Shanghai Jiaotong University, China. Also, she received the M.S. degree (1998) in computer science from Florida Atlantic University, USA. She is currently working toward the PhD degree in the Department of Computer Science and Engineering, Florida Atlantic University, under the supervision of Dr. Jie Wu. Her research interests include security, privacy and anonymity in Wireless Networks and Mobile Computing, Parallel and Distributed Systems.

Brian J. d'Auriol received the BSc(CS) and Ph.D. degrees from the University of New Brunswick in 1988 and 1995, respectively. He is now affiliated with the Dept. of Computer Engineering at Kyung Hee University, Korea. Previously, he had been a researcher at the Ohio Supercomputer Center, USA and assistant professor at The University of Texas at El Paso, The University of Akron, Wright State University and The University of Manitoba. He has organized and chaired the International Conference on Communications in Computing (CIC) 2000-2008 and the 11th Annual International Symposium on High Performance Computing Systems (HPCS'97) in 1997. His research includes information and data visualization with specialization in software, bioinformatics, and health care visualizations; optical bus parallel computing models, and recently, ubiquitous sensor networks. He is a member of the ACM and IEEE Computer Society.

Peng Deng is Research Fellow in the Cloud Computing and Distributed Systems (CLOUDS) Laboratory at the University of Melbourne, Australia.

Weiwei Fang received his B.S. Degree in the School of Computer and Information, Hefei University of Technology(HFUT) in Jun.2003. From Sep.2003 to Jun.2005, he was a M.S. student in the School of Computer Science and Engineering (SCSE), Beihang University(BUAA). In Sep.2005, he was enrolled as a Ph.D. student of BUAA without entrance exam for excellent work. From Sep. 2006 to Dec. 2006, he was a research assistant in Fraunhofer German-Sino Lab for Mobile Communications (MCI), TU Berlin.He is now working in the area of Wireless Sensor Network (WSN) with Prof. Depei Qian, and is a Ph.D. candidate in the Sino-German Joint Software Institute (JSI), BUAA.

Jaime Gómez (jgomez@tel.uva.es) was born in Aguilar de Bureba, Spain, in 1971. He received the M. S. Degree in Telecommunications Engineering in 2000, at University of Valladolid, Spain. He received the Ph.D. degree in 2005, at the same University. His PhD thesis was entitled "Guidance assistance, autonomous guidance and assisted distance guidance in Precision Agriculture". He is a lecturer of Department of Signal Theory, Communications and Telematic Engineering, at University of Valladolid, since 2001. His research interests include Communications, global and local positioning systems, GPS Applications in Agriculture, Machine Vision, Augmented Reality and sensorless control systems in DC motors and Brushless motors.

Ricardo H González G received an under-graduate degree in Computer Science from the Central University of Venezuela (UCV) in 1992, a Masters degree from Simon Bolivar University (USB) in 1998. He was a visiting researcher with University of the Balearic Islands (UIB), Spain, in 1998, and in Federal University of Minas Gerais (UFMG), Brasil, in 2007-2008. He is a Professor of Computer Science Department and a Ph.D. student at the Simon Bolivar University Caracas -Venezuela. His current research interests are in wireless Communications, Wireless Sensor Networks, Reliability and QoS.

Yuan He received his BE degree in Department of Computer Science and Technology from University of Science and Technology of China in 2003, and his ME degree in Institute of Software, Chinese Academy of Sciences, in 2006. He is now a PhD student in the Department of Computer Science and Engineering at Hong Kong University of Science and Technology, supervised by Dr. Yunhao Liu. His research interests include peer-to-peer computing, sensor networks, and pervasive computing. He is a student member of the IEEE and the IEEE Computer Society.

Tales Heimfarth received his PhD degree in Computer Science from the University of Paderborn, Germany, in 2007. His PhD dissertation was entitled "Biologically Inspired Methods for Organizing Distributed Services on Sensor Networks". He received a MSc degree in Computer Science in 2002, with the thesis "Real-time Communication Platform over an SCI Cluster", and a Bachelor degree in Computer Science in 2000, both from the Federal University of Rio Grande do Sul (UFRGS), Brazil. Since 2008 he holds a post-doc position at UFRGS, performing research in the area of sensor networks. He is author of several papers published in international conferences in the area of sensor networks, especially with focus on basic software and auto-organizing protocols. He participates of the ShoX project, which is an open-source, event-oriented simulator for ad-hoc networks, led by the University of Paderborn.

Juan Hernández-Serrano was born in Salamanca (Spain) in 1979. He received the M.S. degree in Telecommunications Engineering in 2002, and the Ph.D. degree in 2008, both from the Universitat Politècnica de Catalunya (UPC). In 2003 he joined the Information Security Group - ISG (http://isg.upc.es) within the Telematics Services Research Group - SERTEL (http://sertel.upc.es) at the Department of Telematics Engineering of the UPC (http://www-entel.upc.es). His research interests include wireless, sensor and cognitive radio security with special focus on group security in dynamic environments. He is currently working as assistant professor at the Escola Politècnica Superior de Castelldefels (EPSC) of the UPC.

Daniel Herrero de la Cal (Daniel.herrero@hc-technologies) was born in Aranda de Duero (Burgos), Spain, in 1978. He received his Industrial Engineering Bachelor Degree at University of Burgos,

Spain. He has worked in several companies such as Renault, Polymond or Enerman, and in 2006 he co-founded his own company called HC Technologies with Javier Herrero. He works as General Manager in HC Technologies. In his company, they develop wireless systems based on Zigbee, Bluetooth, Wifi, and control systems for remote monitoring (using GPRS and 3G technologies) and other different electronics systems.

Javier Herrero de la Cal (javier.herrero@hc-technologies.com) was born in Aranda de Duero, province of Burgos, in Spain in 1981. He received the M. S. Degree in Industrial Engineering in 2005 and the M. S. Degree in Automatic Control and Industrial Electronics Engineering in 2006, both at University of Valladolid, Spain. From 2006 to 2007 he worked as an Electrical Project Manager in Robert Bosch. He also co-founded his own company called HC Technologies where he works as Technical Manager. In his company, they develop wireless systems based on Zigbee, Bluetooth, Wifi, and control systems for remote monitoring (using GPRS and 3G technologies) and other different electronics systems.

Qiang-Sheng Hua is now a PhD candidate in the computer science department of The University of Hong Kong. His supervisor is Professor Francis C.M. Lau. He is expected to obtain his PhD degree in 2009.

Tomasz Kobialka is Research Fellow in the Cloud Computing and Distributed Systems (CLOUDS) Laboratory at the University of Melbourne, Australia. He completed BSc (Hons) from Australian National University.

Lars Kulik is a senior lecturer in the Department of Computer Science and Software Engineering at The University of Melbourne since 2004. He researches efficient algorithms for moving objects, information aggregation and dissemination algorithms for sensor networks, negotiation-based models for location privacy, spatial algorithms in pervasive computing environments and robust algorithms that cope with imperfection, especially in the context of mobile and location-aware computing. He was awarded his PhD at the University of Hamburg, Germany, in 2002.

Tony I. Larsson (M'90) became a Member (M) of IEEE in 1990. He received a MEng degree in 1974, a Tech.Lic. in 1986 in Computer Systems and a PhD degree in Computer Science in 1989, all at the Institute of Technology at Linköping University, Sweden. He worked for Ericsson AB from 1974 to 2002, in several different positions, such as engineer, manager, and expert (the latter in system design methods and architecture), in areas such as testing; computer aided design; radio network control, and distributed computer systems platforms for dependable telecommunication applications. He then worked for the Swedish defense material administration in the area of network based defense and is since 2003 Professor in Embedded Systems at Halmstad University.

Francis C.M. Lau received his PhD (1986) in computer science from the University of Waterloo, Canada. He is currently a professor of computer science at The University of Hong Kong where he has served since 1987. He is a visiting professor of the Institute of Theoretical Computer Science of Tsinghua University, China.

Heejo Lee is an associate professor at the Division of Computer and Communication Engineering, Korea University, Seoul, Korea. Before joining Korea University, he was at AhnLab, Inc. as a CTO from 2001 to 2003. From 2000 to 2001, he was a postdoc at the Department of Computer Sciences and the security center CERIAS, Purdue University. Dr. Lee received his BS, MS, PhD degree in Computer Science and Engineering from POSTECH, Pohang, Korea. Dr. Lee serves as an editor of Journal of Communications and Networks starting from 2007. He has been an advisory member of Korea Information Security Agency and Korea Supreme Prosecutor's Office. With the support of Korean government, he was working for constructing the Nat'l CERT in the Philippines (2006), the consultation of Cyber Security in Uzbekistan (2007). More information is available at http://ccs.korea.ac.kr.

Sungyoung Lee received his B.S. from Korea University, Seoul, Korea. He got his M.S. and PhD degrees in Computer Science from Illinois Institute of Technology (IIT), Chicago, Illinois, USA in 1987 and 1991 respectively. He has been a professor in the Dept. of Computer Engineering, Kyung Hee University, Korea since 1993. He is a founding director of the Ubiquitous Computing Laboratory, and has been affiliated with a director of Neo Medical ubiquitous-Life Care Information Technology Research Center, Kyung Hee University since 2006. Before joining Kyung Hee University, he was an assistant professor in the Dept. of Comp. Sci., Governors State University, Illinois, USA from 1992 to 1993. His current research focuses on Ubiquitous Computing and applications, Context-aware Middleware, Sensor Operating Systems, Real-Time Systems and Embedded Systems. He is a member of the ACM and IEEE.

Young-Koo Lee got his B.S., M.S. and PhD degrees in Computer Science from Korea Advanced Institute of Science and Technology, Korea. He is a professor in the Department of Computer Engineering at Kyung Hee University, Korea. His research interests include ubiquitous data management, data mining, and databases. He is a member of the IEEE, the IEEE Computer Society, and the ACM.

Chao Liu is currently a PHD candidate of the School of Computer Science and Technology at the Huazhong University of Science and Technology (HUST) in China. He obtained his Master's degree in computer engineering from China University of Geosciences (CUG) in 2006. Before joining HUST in Fall 2007, he served as a lecturer in Computer Science School of CUG. His current research interests includes distributed data processing, data integration and unstructured data management. The projects that he has taken part in include NFC project, National Science & Technology Pillar Program, etc. Contact him at cs.chaoliu@gmail.com.

Jiangchuan Liu received the B.Eng degree (cum laude) from Tsinghua University, Beijing, China, in 1999, and the PhD degree from The Hong Kong University of Science and Technology in 2003, both in computer science. He was a recipient of Microsoft Research Fellowship (2000), a recipient of Hong Kong Young Scientist Award (2003), and a co-inventor of one European patent and two US patents. He co-authored the Best Student Paper of IWQoS'08, and the 2009 Best Paper of IEEE Multimedia Communications Technical Committee. He is currently an Assistant Professor in the School of Computing Science, Simon Fraser University, British Columbia, Canada, and was an Assistant Professor in the Department of Computer Science and Engineering at The Chinese University of Hong Kong from 2003 to 2004. His research interests include multimedia systems and networks, wireless ad hoc and sensor networks, and peer-to-peer and overlay networks. He is an Associate Editor of IEEE Transactions on

Multimedia, and an editor of IEEE Communications Surveys and Tutorials. He is a Senior Member of IEEE and a member of Sigma Xi.

Yunhao Liu (SM '06/ACM '07) received his BS degree in Automation Department from Tsinghua University, China, in 1995, and an MA degree in Beijing Foreign Studies University, China, in 1997, and an MS and a Ph.D. degree in Computer Science and Engineering at Michigan State University in 2003 and 2004, respectively. Yunhao is now with the Department of Computer Science and Engineering at the Hong Kong University of Science and Technology. He is also an Adjunct Professor of Xi'an Jiaotong University, Jilin University, and Ocean University of China. His research interests include wireless sensor network, peer-to-peer computing, and pervasive computing. He is a senior member of the IEEE, and a member of the ACM. He received the Grand Prize of Hong Kong ICT Best Innovation and Research Award 2007.

Yunlu Liu received her B.S. & C.S. Degree in the School of Computer and Science, Shandong University in Jun 2003 & 2006 with the honor of Excellent Graduate Student. In Sep 2007, she was enrolled as a Ph.D. student in the School of Computer Science and Engineering, Beihang University under the supervision of Prof. Zhang Xiong. She will do research in Carnegie Mellon University under the joint supervision of Prof. Satyanarayanan from Sep 2009 to Dec 2010. Her resent research focuses on Wireless Sensor Networks. She has published 6 papers and applied 8 Chinese patents in this area.

Antonio Loureiro is a Professor of Computer Science at the Federal University of Minas Gerais (UFMG), Brazil. Professor Loureiro holds a PhD in Computer Science from the University of British Columbia, Canada, 1995. His main research areas are mobile, ubiquitous and autonomic computing, computer networks and distributed systems. In the last 10 years he has published over 100 papers in international conferences and journals. Professor Loureiro has also presented several tutorials in international conferences in the last five years. He was the TPC Chair for LANOMS 2001 (Latin American Network Operations and Management Symposium, sponsored by IEEE Communications Society) and for the 2005 ACM Workshop on Wireless Multimedia Networking and Performance Modeling

Petri Mähönen is currently a full professor and holds Ericsson Chair of Wireless Networks at the RWTH Aachen University in Germany. Before joining to RWTH Aachen in 2002, he was a research director and professor at the Centre for Wireless Communications and the University of Oulu, Finland. He has studied and worked in the United States, the United Kingdom and Finland. He has been a principal investigator in several international and national multi-million USD research projects. Dr. Mähönen has published ca. 200 papers in international journals and conferences and has been invited to deliver research talks at many universities, companies and conferences. He is a senior member of IEEE and ACM, and fellow of RAS. He is inventor or co-inventor for over 20 patents or patent applications. He is currently also a research area coordinator and one of the principal investigators for a newly formed Ultra High Speed Mobile Information and Communication (UMIC) research cluster at RWTH, which is one of the German national excellence clusters supported by the Federal Government of Germany established in 2006.

Bojan Mihajlović received the B.Eng. degree in Electrical Engineering from Ryerson University, Toronto, Canada, in 2004, and the M.Eng. degree from McGill University, Montreal, Canada, in 2008.

He is currently working toward the Ph.D. degree at McGill University, performing research in the area of computer architecture. He is the recipient of the Gerald G. Hatch Graduate Fellowship and a scholarship from the Natural Science and Engineering Research Council.

Raquel Mini holds a B.Sc., M.Sc. and Ph.D. in Computer Science from Federal University of Minas Gerais (UFMG), Brazil. Since 1999 she is an Associate Professor of Computer Science at PUC Minas, Brazil. In her Ph.D. thesis she studied the energy map construction in wireless sensor network. Since 2001 she has been studying energy aware protocols for these networks. In the last 3 years, she has published 11 papers about wireless sensor networks in international conferences and journals. Her main research areas are wireless sensor networks, distributed algorithms, and mobile computing.

Marimuthu Palaniswami received his ME from the Indian Institute of science, India, MEngSc from the University of Melbourne and Ph.D from the University of Newcastle, Australia before rejoining the University of Melbourne. He has been serving the University of Melbourne for over 21 years. He served as a co-director of an active research centre, the Centre of Expertise on Networked Decision & Sensor Systems (2002-2005). Presently, he is running the largest funded ARC Research Network on Intelligent Sensors, Sensor Networks and Information Processing (ISSNIP, http://www.issnip.unimelb.edu.au/) programme with $4.75 million ARC funding over 5 years and have structured it run as a network centre of excellence with complementary funding for fundamental research, test beds, international linkages and industry linkages.He has published more than 320 refereed papers and a huge proportion of them appeared in prestigious IEEE Journals and Conferences. He has won the University of Melbourne Knowledge Transfer Award in 2007 and 2008. He was given a Foreign Specialist Award by the Ministry of Education, Japan in recognition of his contributions to the field of Machine Learning. His academic excellence is recognised by several invited presentations of plenary/keynote lectures and panel member for several top International Conferences. He was the associate editor of Journals/transactions including IEEE Transaction on Neural Networks. His research interests include SVMs, Sensors and Sensor Networks, Machine Learning, Neural Network, Pattern Recognition, Signal Processing and Control. He is the co-director of an active research centre, Centre of Expertise on Networked Decision & Sensor Systems, attracting grants from industry and defence agencies in Australia and USA. He holds several large Australian Research Council Discovery and Linkage grants with a successful industry outreach programme. His research students have won five best student paper awards from conferences such as IEEE GLOBECOM and SAE; two of his project undergraduate students have won IEEE Region 10 best student paper awards, and three best undergraduate project awards including a state finalist for Siemens Prize have also been obtained. A commercialization grant to the tune of $180,000 has been awarded for a PhD project. He has successfully supervised 15 Ph.Ds and 11 M.Eng.Sc (Research). He is currently supervising 9 postgraduate students, 7 research fellows. He is also serving external boards as member and director.

Josep Pegueroles was born in Tortosa (Spain) in 1974. He received the M.S. degree in Telecommunications Engineering in 1999, and the Ph.D. degree in 2003, both from the Universitat Politècnica de Catalunya (UPC). In 1999 he joined the Information Security Workgroup - ISG (http://isg.upc.es) within the Telematics Services Research Group - SERTEL (http://sertel.upc.es) at the Department of Telematics Engineering - ENTEL (http://entel.upc.es) of the UPC (http://www.upc.es). Currently he works as assistant professor at the Telecommunications Engineering School in Barcelona - ETSETB (http://www.etsetb.upc.es). His research interests include security for multimedia networked services and secure group communications.

Edison Pignaton de Freitas has a position as Computer Engineer at the Brazilian Army. He is currently a PhD student at Halmstad University, Sweden, and Federal University of Rio Grande do Sul (UFRGS), Brazil, in the area of sensor networks, performing his research in the embedded systems groups of both universities, and has published several papers in this area. He got his Bachelor degree in Computer Engineering from the Military Institute of Engineering, Brazil, in 2003, and his MSc degree in Computer Science from UFRGS, in 2007. In 2001 and 2002 he participated in an interchange program for Engineering students between Brazil and France, studying at the *Génie Informatique Industriel* program of the Institut National des Sciences Appliquées, in Toulouse, and performing an internship at the Systems Department of the AIRBUS Central Entity, working in the A380 airplane project.

Katarzyna Radecka received the Ph.D. degree from McGill University, Montreal, Canada, in 2003. She received the MEng degree and BEng degree (honors) from the same university in 1997 and 1995. From 1996 to 1998, she was a member of the technical staff at Lucent Technologies in Allentown, Pennsylvania. Currently, she is an adjunct professor at McGill University in Montreal, Canada. Her research interests include verification and testing of digital hardware, as well as reversible and quantum computing. She coauthored a book on simulation-based verification in 2003.

Riaz Ahmed Shaikh received his B.S. degree in Computer Engineering from Sir Syed University of Engineering and Technology (SSUET), Karachi, Pakistan, in 2003, and his M.S. degree in Information Technology from National University of Sciences and Technology (NUST), Rawalpindi, Pakistan, in 2005. He has completed his Ph.D. from Comp. Eng., Dept. of Kyung Hee University, Suwon, South Korea. His research interests include privacy, security and trust management in wired and wireless networks. He is a member of the ACM, and ICST. More information is available at http://member.acm. org/~riaz289.

Haiying Shen received the BS degree in Computer Science and Engineering from Tongji University, China in 2000, and the MS and Ph.D. degrees in Computer Engineering from Wayne State University in 2004 and 2006, respectively. She is currently an Assistant Professor in the Department of Computer Science and Computer Engineering, and the Director of the Pervasive Communications Laboratory of the University of Arkansas. Her research interests include distributed and parallel computer systems and computer networks, with an emphasis on peer-to-peer and content delivery networks, wireless networks, resource management in cluster and grid computing, and data mining. Her research work has been published in top journals and conferences in these areas. She was the Program Co-Chair for a number of international conferences and member of the Program Committees of many leading conferences. She is a member of the IEEE and ACM.

Miguel Soriano was born in Barcelona (Spain) in 1967. He received the Telecommunication Engineering and the PhD. degree from the Universitat Politècnica de Catalunya (UPC). Now, he manages the Information Security Workgroup at the Department of Telematics Engineering of the UPC. Since 2007 he is professor at the UPC. His current research interests are related to information and network security including information hiding for copyright protection. He has participated in more than 30 R+D projects, both public (CICYT, DURSI, European Commission or CIRIT) and private funded, being the coordinator in 20 of them. He is co-author of 3 books, 2 patents, more than 20 ISI-JCR papers and more than 100 conference papers. He is Senior Member of IEEE since 2002. Moreover, he has been member of the program committee of many security conferences, and he is editor of the International Journal of Information Security.

George Spanogiannopoulos was born in Toronto, Canada and received his bachelors in Computer Science at York University in 2007. Currently, he is a graduate student and M.Sc candidate at York University's Department of Computer Science and Engineering researching various multipath routing techniques for streaming data in Wireless Sensor Networks which use ZigBee technology.

Dusan Stevanovic was born in Belgrade, Serbia, graduated with distinct honours from Department of Computer Science at University of Toronto in 2003. Previously, Dusan was employed for over 3 years in the telecom and financial industry as a software developer and consultant before joining the Department of Computer Science and Engineering at York University in 2007. He is currently finishing his Master thesis on Sink Mobility in ZigBee-based Wireless Sensor Networks and will be commencing Doctor of Philosophy degree at York University in the fall of 2009. He is the author of numerous papers published in prestigious networking conferences. His research interests are wireless sensor networks, ZigBee standard, network security and mobile networks.

Yan-Qiang Sun was born in 1985. He received the M. S. degree in Computer Science from National University of Defense Technology, and now is a Ph.D. candidate at this university. His research interests include the security and trustworthiness of mobile ad hoc networks and wireless sensor networks. (Email: yq_sun@nudt.edu.cn)

Roberto Vázquez Trueba (rvaztru@tel.uva.es) was born in Valladolid, Spain, in 1983. He received the M. S. Degree in Telecommunications Engineering in 2008, at University of Valladolid, Spain. In 2008 he worked for Junta de Castilla y León (Government of his region), in a social project that tried to approach Internet and new technologies to people that don't know them or don't have the possibility of use it regularly. Then, in summer of 2008 he started working in Altran Technologies (Spain) as a consultant and collaborated on projects with companies as Telefonica I+D, Vitaldent, Alcatel or Telefonica Mobile.

Juan Vera-del-Campo was born in Valencia (Spain) in 1980. He received a M.S. degree in Telecommunications Engineering from the Universidad Politécnica de Madrid in 2005. During that time, he was involved in developing advanced techniques for e-Learning. He is currently doing a Ph. D. at the Department of Telematics Engineering of the Universitat Politècnica de Catalunya, in the field of document location in peer-to-peer networks. He is also doing a B. degree in the business school of the Universitat Oberta de Catalunya. His research interests include peer-to-peer networks, ubiquitous computing, social networks and small devices, with a special focus on the trade-off between privacy and efficiency.

Natalija Vlajic (Ph.D. P.Eng) is an Assistant Professor at the Department of Computer Science and Engineering, York University, Toronto, Canada. Her main research interests are in the areas of wireless communication, sensor networks, computer networking, computer security, and neural networks. She is holds an NSERC UFA award, and has served as a technical program committee member and reviewer for numerous international conferences and journals.

Flávio Rech Wagner received a BSc degree in Electrical Engineering (1975) and an MSc degree in Computer Science (1977), both from the Federal University of Rio Grande do Sul (UFRGS), Brazil. He received a PhD degree in Computer Engineering from the University of Kaiserslautern, Germany, in 1983. In 1992 and 2002 he held post-doc positions at INPG (Institut National Polytechnique de Grenoble), France. He was invited professor at the University of Tübingen, Germany, in 1994. He is currently professor at UFRGS, position that he holds since 1977, and since 2006 he is the Director of the Institute of Informatics of UFRGS. He has been President of the Brazilian Computer Society, from 1999 to 2003. He is associate editor of the Design Automation for Embedded Systems journal and has been member of more than 30 program committees of international conferences, having published 3 books, 7 book chapters, and more than 120 papers in conference proceedings and journals.

Feng Wang received both his Bachelor's degree and Master's degree in Computer Science and Technology from Tsinghua University, Beijing, China, in 2002 and 2005, respectively. He is currently a Ph.D. student in School of Computing Science at Simon Fraser University, conducting researches on wireless sensor networks and peer-to-peer live streaming under the supervision of Prof. Jiangchuan Liu. His research interests include wireless sensor networks, peer-to-peer live streaming and distributed computing. He has been a Student Member of IEEE and IEEE Communications Society since 2007. In 2006 summer, he interned in Wireless and Networking Group at Microsoft Research Asia, where he conducted research on peer-to-peer live video streaming. He also conducted research in Department of Computing at Hong Kong Polytechnic University as a visiting PhD student in 2008 and 2009 springs, where his research was on data collections in wireless sensor networks. He was awarded Tsinghua University Scholarship for Excellent Student in 1998, 2000 and 2001. At Simon Fraser University, he was awarded the SFU-CS Graduate Entrance Scholarship in 2005 and the Graduate Fellowship in 2007, 2008 and 2009.

Shuo-Hung Wang received the Bachelor degree in information management from Chinese Culture University, Taiwan, in 1994, the M.S. degree in Department of Computer Science and Information Engineering from Dong-Hwa University, Taiwan, 2005. Now, he is a Ph.D. candidate in Dong-Hwa University for majoring information management. His research interests include distributed systems, wireless networks, and mobile computing with a focus on wireless sensor networks.

Xiao-Dong Wang was born in 1973.He received the Ph.D. degree in Computer Science from National University of Defense Technology in 2001. He is an associate professor and M.S. supervisor at National University of Defense Technology. His research interests include wireless networks and mobile computing.(Email: xdwang@nudt.edu.cn)

Jie Wu is a Professor at Department of Computer Science and Engineering, Florida Atlantic University. He has published over 200 papers in various journal and conference proceedings. His research interests are in the area of mobile computing, routing protocols, fault-tolerant computing, and interconnection networks. Dr. Wu is a Member of ACM and an IEEE fellow.

Zhang Xiong is a professor and Ph.D. supervisor in Computer Science and Engineering, Beihang University, and also the dean of China-France Engineering School, Beihang University. He is the member of the Computer Science and Technology Committee of Ministry of Education of the People's Republic of China, and the member of China Computer Federation Computer Architecture committee. He visited in the computer science department of Michigan State University in 1980's. He has published hundreds of papers and book chapters in the area of multi-media, artificial intelligence and wireless sensor networks and so on.

Chen Yu is an Associate Professor of College of Computer Science and Technology, Huazhong University of Science and Technology; Key Laboratory of Services Computing Technology and Systems, Ministry of Education; Key Laboratory of Cluster and Grid Computing, Hubei Province, China. He is also serving as a specific research fellow of College of Computer Science and Technology, Huazhong University of Science and Technology. He received his Bachelor and Master degree on Mathematics and Computer Science from Wuhan University, China, in 1998 and 2002, separately. And he received his Doctor degree on Information Science from TOHOKU University, Japan, in 2005. He served as a post doctoral fellow for JST (Japan Science and Technology Agency) and JSPS (Japan Society for the Promotion of Science) from 2005 to 2008. He has published more than twenty technical papers and received the best paper award in ICC 2005. His current research interests are Ad hoc and Wireless Sensor Networks, Pervasive Computing and Cloud Computing.

Xi Zhang is currently a Ph.D. student and research assistant at the Department of Wireless Networks, RWTH Aachen University, Germany. She completed her MEng. in Electrical and Electronic Engineering from Imperial College London, the U.K. in 2007 and spent the last year of her degree course as an exchange student at RWTH Aachen University. Her current research interests include cross-layer optimization of protocol stacks, mainly physical, medium access control and routing layer, in searching for low-power, real-time and intelligent solutions for wireless sensor networks. As a part of her research work she has participated actively in European research projects as well as research projects carried out in collaboration with industry.

Željko Žilić received the Ph.D. degree from the University of Toronto, Canada, in 1997. From 1997 to 1998, he was a Member of Technical Staff with the FPGA Division, Microelectronics Group, Lucent Technologies. He is currently an Associate Professor with McGill University, Montreal, Canada, where he has been granted the Chercheur Strategique Research Chair. He has published more than 100 research papers, coauthored two books, and is the holder of four patents. He is also on the Editorial Board of the two journals, and has served as a member of Technical Program Committees of several conferences. He was the recipient of *best paper* awards in 2001, 2005, and several *honorary mention* awards. For his undergraduate teaching, he was awarded with a Wighton Fellowship in 2006. As a Vice President of R&D, he helped start Monroi, Inc., which is a company that was given an Entrepreneur of the Year Award in 2004.

Index